Teacher, Student One-Stop Internet Resources

Log on to ips.msscience.com

ONLINE STUDY TOOLS
- Section Self-Check Quizzes
- Interactive Tutor
- Chapter Review Tests
- Standardized Test Practice
- Vocabulary PuzzleMaker

ONLINE RESEARCH
- WebQuest Projects
- Prescreened Web Links
- Career Links
- Internet Labs

INTERACTIVE ONLINE STUDENT EDITION
- Complete Interactive Student Edition available at mhln.com

FOR TEACHERS
- Teacher Bulletin Board
- Teaching Today—Professional Development

SAFETY SYMBOLS

SAFETY SYMBOLS	HAZARD	EXAMPLES	PRECAUTION	REMEDY
DISPOSAL	Special disposal procedures need to be followed.	certain chemicals, living organisms	Do not dispose of these materials in the sink or trash can.	Dispose of wastes as directed by your teacher.
BIOLOGICAL	Organisms or other biological materials that might be harmful to humans	bacteria, fungi, blood, unpreserved tissues, plant materials	Avoid skin contact with these materials. Wear mask or gloves.	Notify your teacher if you suspect contact with material. Wash hands thoroughly.
EXTREME TEMPERATURE	Objects that can burn skin by being too cold or too hot	boiling liquids, hot plates, dry ice, liquid nitrogen	Use proper protection when handling.	Go to your teacher for first aid.
SHARP OBJECT	Use of tools or glassware that can easily puncture or slice skin	razor blades, pins, scalpels, pointed tools, dissecting probes, broken glass	Practice common-sense behavior and follow guidelines for use of the tool.	Go to your teacher for first aid.
FUME	Possible danger to respiratory tract from fumes	ammonia, acetone, nail polish remover, heated sulfur, moth balls	Make sure there is good ventilation. Never smell fumes directly. Wear a mask.	Leave foul area and notify your teacher immediately.
ELECTRICAL	Possible danger from electrical shock or burn	improper grounding, liquid spills, short circuits, exposed wires	Double-check setup with teacher. Check condition of wires and apparatus.	Do not attempt to fix electrical problems. Notify your teacher immediately.
IRRITANT	Substances that can irritate the skin or mucous membranes of the respiratory tract	pollen, moth balls, steel wool, fiberglass, potassium permanganate	Wear dust mask and gloves. Practice extra care when handling these materials.	Go to your teacher for first aid.
CHEMICAL	Chemicals can react with and destroy tissue and other materials	bleaches such as hydrogen peroxide; acids such as sulfuric acid, hydrochloric acid; bases such as ammonia, sodium hydroxide	Wear goggles, gloves, and an apron.	Immediately flush the affected area with water and notify your teacher.
TOXIC	Substance may be poisonous if touched, inhaled, or swallowed.	mercury, many metal compounds, iodine, poinsettia plant parts	Follow your teacher's instructions.	Always wash hands thoroughly after use. Go to your teacher for first aid.
FLAMMABLE	Flammable chemicals may be ignited by open flame, spark, or exposed heat.	alcohol, kerosene, potassium permanganate	Avoid open flames and heat when using flammable chemicals.	Notify your teacher immediately. Use fire safety equipment if applicable.
OPEN FLAME	Open flame in use, may cause fire.	hair, clothing, paper, synthetic materials	Tie back hair and loose clothing. Follow teacher's instruction on lighting and extinguishing flames.	Notify your teacher immediately. Use fire safety equipment if applicable.

 Eye Safety Proper eye protection should be worn at all times by anyone performing or observing science activities.

 Clothing Protection This symbol appears when substances could stain or burn clothing.

 Animal Safety This symbol appears when safety of animals and students must be ensured.

 Handwashing After the lab, wash hands with soap and water before removing goggles.

Glencoe Science

Introduction to Physical Science

NATIONAL GEOGRAPHIC

McGraw-Hill Glencoe

New York, New York   Columbus, Ohio   Chicago, Illinois   Woodland Hills, California

Glencoe Science

Introduction to Physical Science

While gravity brings this kayaker to a landing, an opposing force, buoyancy, determines how deep in the landing pool the boat will submerge. To avoid underwater rocks, larger kayaks are chosen on creeks with steep drops in order to increase buoyancy.

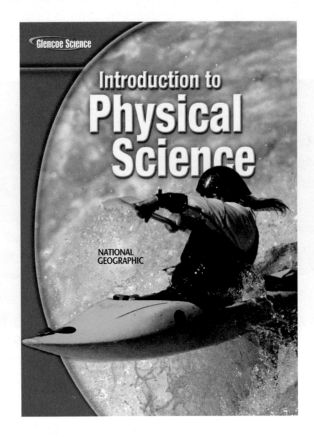

The McGraw·Hill Companies

Copyright © 2008 by The McGraw-Hill Companies, Inc. All rights reserved. Except as permitted under the United States Copyright Act, no part of this publication may be reproduced or distributed in any form or by any means, or stored in a database or retrieval system, without prior written permission of the publisher.

The National Geographic features were designed and developed by the National Geographic Society's Education Division. Copyright © National Geographic Society. The name "National Geographic Society" and the Yellow Border Rectangle are trademarks of the Society, and their use, without prior written permission, is strictly prohibited.

The "Science and Society" and the "Science and History" features that appear in this book were designed and developed by TIME School Publishing, a division of TIME Magazine. TIME and the red border are trademarks of Time Inc. All rights reserved.

Send all inquiries to:
Glencoe/McGraw-Hill
8787 Orion Place
Columbus, OH 43240-4027

ISBN: 978-0-07-877804-9
MHID: 0-07-877804-2

Printed in the United States of America.

5 6 7 8 9 10   RJE/LEH   12 11 10

Contents In Brief

unit 1 — The Nature of Science ... 2
- **Chapter 1** The Nature of Science ... 4
- **Chapter 2** Measurement ... 40

unit 2 — Matter ... 68
- **Chapter 3** Atoms, Elements, and the Periodic Table ... 70
- **Chapter 4** States of Matter ... 100
- **Chapter 5** Matter—Properties and Changes ... 132

unit 3 — Chemistry ... 158
- **Chapter 6** Atomic Structure and Chemical Bonds ... 160
- **Chapter 7** Chemical Reactions ... 188
- **Chapter 8** Substances, Mixtures, and Solubility ... 216
- **Chapter 9** Carbon Chemistry ... 248

unit 4 — Motion and Forces ... 278
- **Chapter 10** Motion and Momentum ... 280
- **Chapter 11** Force and Newton's Laws ... 308
- **Chapter 12** Forces and Fluids ... 338

unit 5 — Energy ... 370
- **Chapter 13** Energy and Energy Resources ... 372
- **Chapter 14** Work and Simple Machines ... 404
- **Chapter 15** Thermal Energy ... 432

unit 6 — Waves, Sound, and Light ... 458
- **Chapter 16** Waves ... 460
- **Chapter 17** Sound ... 488
- **Chapter 18** Electromagnetic Waves ... 518
- **Chapter 19** Light, Mirrors, and Lenses ... 548

unit 7 — Electricity and Magnetism ... 580
- **Chapter 20** Electricity ... 582
- **Chapter 21** Magnetism ... 612
- **Chapter 22** Electronics and Computers ... 640

Authors

Education Division
Washington, D.C.

Cathy Ezrailson
Science Department Head
Academy for Science and
Health Professions
Conroe, TX

Nicholas Hainen
Chemistry/Physics Teacher, Retired
Worthington City Schools
Worthington, OH

Patricia Horton
Mathematics and Science Teacher
Summit Intermediate School
Etiwanda, CA

Deborah Lillie
Math and Science Writer
Sudbury, MA

Thomas McCarthy, PhD
Science Department Chair
St. Edward's School
Vero Beach, FL

Eric Werwa, PhD
Department of Physics
and Astronomy
Otterbein College
Westerville, OH

Dinah Zike
Educational Consultant
Dinah-Might Activities, Inc.
San Antonio, TX

Margaret K. Zorn
Science Writer
Yorktown, VA

Series Consultants

CONTENT

Alton J. Banks, PhD
Director of the Faculty Center
for Teaching and Learning
North Carolina State University
Raleigh, NC

Jack Cooper
Ennis High School
Ennis, TX

Sandra K. Enger, PhD
Associate Director,
Associate Professor
UAH Institute for Science Education
Huntsville, AL

David G. Haase, PhD
North Carolina State University
Raleigh, NC

Michael A. Hoggarth, PhD
Department of Life and
Earth Sciences
Otterbein College
Westerville, OH

Jerome A. Jackson, PhD
Whitaker Eminent Scholar in Science
Program Director
Center for Science, Mathematics,
and Technology Education
Florida Gulf Coast University
Fort Meyers, FL

William C. Keel, PhD
Department of Physics
and Astronomy
University of Alabama
Tuscaloosa, AL

Linda McGaw
Science Program Coordinator
Advanced Placement Strategies, Inc.
Dallas, TX

Madelaine Meek
Physics Consultant Editor
Lebanon, OH

Robert Nierste
Science Department Head
Hendrick Middle School, Plano ISD
Plano, TX

Connie Rizzo, MD, PhD
Depatment of Science/Math
Marymount Manhattan College
New York, NY

Dominic Salinas, PhD
Middle School Science Supervisor
Caddo Parish Schools
Shreveport, LA

Cheryl Wistrom
St. Joseph's College
Rensselaer, IN

Carl Zorn, PhD
Staff Scientist
Jefferson Laboratory
Newport News, VA

MATH

Michael Hopper, DEng
Manager of Aircraft Certification
L-3 Communications
Greenville, TX

Teri Willard, EdD
Mathematics Curriculum Writer
Belgrade, MT

READING

Elizabeth Babich
Special Education Teacher
Mashpee Public Schools
Mashpee, MA

Barry Barto
Special Education Teacher
John F. Kennedy Elementary
Manistee, MI

Carol A. Senf, PhD
School of Literature,
Communication, and Culture
Georgia Institute of Technology
Atlanta, GA

Rachel Swaters-Kissinger
Science Teacher
John Boise Middle School
Warsaw, MO

SAFETY

Aileen Duc, PhD
Science 8 Teacher
Hendrick Middle School, Plano ISD
Plano, TX

Sandra West, PhD
Department of Biology
Texas State University-San Marcos
San Marcos, TX

ACTIVITY TESTERS

Nerma Coats Henderson
Pickerington Lakeview Jr. High School
Pickerington, OH

Mary Helen Mariscal-Cholka
William D. Slider Middle School
El Paso, TX

Science Kit and Boreal Laboratories
Tonawanda, NY

Reviewers

Deidre Adams
West Vigo Middle School
West Terre Haute, IN

Sharla Adams
IPC Teacher
Allen High School
Allen, TX

Maureen Barrett
Thomas E. Harrington Middle School
Mt. Laurel, NJ

John Barry
Seeger Jr.-Sr. High School
West Lebanon, IN

Desiree Bishop
Environmental Studies Center
Mobile County Public Schools
Mobile, AL

William Blair
Retired Teacher
J. Marshall Middle School
Billerica, MA

Tom Bright
Concord High School
Charlotte, NC

Nora M. Prestinari Burchett
Saint Luke School
McLean, VA

Lois Burdette
Green Bank Elementary-Middle School
Green Bank, WV

Marcia Chackan
Pine Crest School
Boca Raton, FL

Karen Curry
East Wake Middle School
Raleigh, NC

Joanne Davis
Murphy High School
Murphy, NC

Anthony J. DiSipio, Jr.
8th Grade Science
Octorana Middle School
Atglen, PA

Sueanne Esposito
Tipton High School
Tipton, IN

Sandra Everhart
Dauphin/Enterprise Jr. High Schools
Enterprise, AL

Mary Ferneau
Westview Middle School
Goose Creek, SC

Cory Fish
Burkholder Middle School
Henderson, NV

Linda V. Forsyth
Retired Teacher
Merrill Middle School
Denver, CO

George Gabb
Great Bridge Middle School
Chesapeake Public Schools
Chesapeake, VA

Annette D'Urso Garcia
Kearney Middle School
Commerce City, CO

Nerma Coats Henderson
Pickerington Lakeview Jr.
High School
Pickerington, OH

Lynne Huskey
Chase Middle School
Forest City, NC

Maria E. Kelly
Principal
Nativity School
Catholic Diocese of Arlington
Burke, VA

Michael Mansour
Board Member
National Middle Level Science
Teacher's Association
John Page Middle School
Madison Heights, MI

Mary Helen Mariscal-Cholka
William D. Slider Middle School
El Paso, TX

Michelle Mazeika-Simmons
Whiting Middle School
Whiting, IN

Sharon Mitchell
William D. Slider Middle School
El Paso, TX

Amy Morgan
Berry Middle School
Hoover, AL

Norma Neely, EdD
Associate Director for Regional
Projects
Texas Rural Systemic Initiative
Austin, TX

Annette Parrott
Lakeside High School
Atlanta, GA

Mark Sailer
Pioneer Jr.-Sr. High School
Royal Center, IN

Joanne Stickney
Monticello Middle School
Monticello, NY

Dee Stout
Penn State University
University Park, PA

Darcy Vetro-Ravndal
Hillsborough High School
Tampa, FL

Karen Watkins
Perry Meridian Middle School
Indianapolis, IN

Clabe Webb
Permian High School
Ector County ISD
Odessa, TX

Alison Welch
William D. Slider Middle School
El Paso, TX

Kim Wimpey
North Gwinnett High School
Suwanee, GA

Kate Ziegler
Durant Road Middle School
Raleigh, NC

Teacher Advisory Board

The Teacher Advisory Board gave the editorial staff and design team feedback on the content and design of the Student Edition. They provided valuable input in the development of the 2008 edition of *Glencoe Introduction to Physical Science.*

John Gonzales
Challenger Middle School
Tucson, AZ

Rachel Shively
Aptakisic Jr. High School
Buffalo Grove, IL

Roger Pratt
Manistique High School
Manistique, MI

Kirtina Hile
Northmor Jr. High/High School
Galion, OH

Marie Renner
Diley Middle School
Pickerington, OH

Nelson Farrier
Hamlin Middle School
Springfield, OR

Jeff Remington
Palmyra Middle School
Palmyra, PA

Erin Peters
Williamsburg Middle School
Arlington, VA

Rubidel Peoples
Meacham Middle School
Fort Worth, TX

Kristi Ramsey
Navasota Jr. High School
Navasota, TX

Student Advisory Board

The Student Advisory Board gave the editorial staff and design team feedback on the design of the Student Edition. We thank these students for their hard work and creative suggestions in making the 2008 edition of *Glencoe Introduction to Physical Science* student friendly.

Jack Andrews
Reynoldsburg Jr. High School
Reynoldsburg, OH

Peter Arnold
Hastings Middle School
Upper Arlington, OH

Emily Barbe
Perry Middle School
Worthington, OH

Kirsty Bateman
Hilliard Heritage Middle School
Hilliard, OH

Andre Brown
Spanish Emersion Academy
Columbus, OH

Chris Dundon
Heritage Middle School
Westerville, OH

Ryan Manafee
Monroe Middle School
Columbus, OH

Addison Owen
Davis Middle School
Dublin, OH

Teriana Patrick
Eastmoor Middle School
Columbus, OH

Ashley Ruz
Karrar Middle School
Dublin, OH

The Glencoe middle school science Student Advisory Board taking a timeout at COSI, a science museum in Columbus, Ohio.

HOW TO...
Use Your Science Book

Why do I need my science book?

Have you ever been in class and not understood all of what was presented? Or, you understood everything in class, but at home, got stuck on how to answer a question? Maybe you just wondered when you were ever going to use this stuff?

These next few pages are designed to help you understand everything your science book can be used for . . . besides a paperweight!

Before You Read

- **Chapter Opener** Science is occurring all around you, and the opening photo of each chapter will preview the science you will be learning about. The **Chapter Preview** will give you an idea of what you will be learning about, and you can try the **Launch Lab** to help get your brain headed in the right direction. The **Foldables** exercise is a fun way to keep you organized.

- **Section Opener** Chapters are divided into two to four sections. The **As You Read** in the margin of the first page of each section will let you know what is most important in the section. It is divided into four parts. **What You'll Learn** will tell you the major topics you will be covering. **Why It's Important** will remind you why you are studying this in the first place! The **Review Vocabulary** word is a word you already know, either from your science studies or your prior knowledge. The **New Vocabulary** words are words that you need to learn to understand this section. These words will be in **boldfaced** print and highlighted in the section. Make a note to yourself to recognize these words as you are reading the section.

As You Read

- **Headings** Each section has a title in large red letters, and is further divided into blue titles and small red titles at the beginnings of some paragraphs. To help you study, make an outline of the headings and subheadings.

- **Margins** In the margins of your text, you will find many helpful resources. The **Science Online** exercises and **Integrate** activities help you explore the topics you are studying. **MiniLabs** reinforce the science concepts you have learned.

- **Building Skills** You also will find an **Applying Math** or **Applying Science** activity in each chapter. This gives you extra practice using your new knowledge, and helps prepare you for standardized tests.

- **Student Resources** At the end of the book you will find **Student Resources** to help you throughout your studies. These include **Science, Technology,** and **Math Skill Handbooks,** an **English/Spanish Glossary,** and an **Index.** Also, use your **Foldables** as a resource. It will help you organize information, and review before a test.

- **In Class** Remember, you can always ask your teacher to explain anything you don't understand.

FOLDABLES Study Organizer

Science Vocabulary Make the following Foldable to help you understand the vocabulary terms in this chapter.

STEP 1 Fold a vertical sheet of notebook paper from side to side.

STEP 2 Cut along every third line of only the top layer to form tabs.

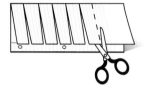

STEP 3 Label each tab with a vocabulary word from the chapter.

Build Vocabulary As you read the chapter, list the vocabulary words on the tabs. As you learn the definitions, write them under the tab for each vocabulary word.

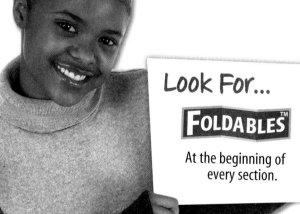

Look For... FOLDABLES At the beginning of every section.

In Lab

Working in the laboratory is one of the best ways to understand the concepts you are studying. Your book will be your guide through your laboratory experiences, and help you begin to think like a scientist. In it, you not only will find the steps necessary to follow the investigations, but you also will find helpful tips to make the most of your time.

- Each lab provides you with a **Real-World Question** to remind you that science is something you use every day, not just in class. This may lead to many more questions about how things happen in your world.

- Remember, experiments do not always produce the result you expect. Scientists have made many discoveries based on investigations with unexpected results. You can try the experiment again to make sure your results were accurate, or perhaps form a new hypothesis to test.

- Keeping a **Science Journal** is how scientists keep accurate records of observations and data. In your journal, you also can write any questions that may arise during your investigation. This is a great method of reminding yourself to find the answers later.

Look For...
- **Launch Labs** start every chapter.
- **MiniLabs** in the margin of each chapter.
- **Two Full-Period Labs** in every chapter.
- **EXTRA Try at Home Labs** at the end of your book.
- the **Web site** with laboratory demonstrations.

Before a Test

Admit it! You don't like to take tests! However, there *are* ways to review that make them less painful. Your book will help you be more successful taking tests if you use the resources provided to you.

- Review all of the **New Vocabulary** words and be sure you understand their definitions.
- Review the notes you've taken on your **Foldables,** in class, and in lab. Write down any question that you still need answered.
- Review the **Summaries** and **Self Check questions** at the end of each section.
- Study the concepts presented in the chapter by reading the **Study Guide** and answering the questions in the **Chapter Review.**

Look For...

- **Reading Checks** and **caption questions** throughout the text.
- The **Summaries** and **Self Check questions** at the end of each section.
- The **Study Guide** and **Review** at the end of each chapter.
- The **Standardized Test Practice** after each chapter.

Let's Get Started

To help you find the information you need quickly, use the Scavenger Hunt below to learn where things are located in Chapter 1.

1. What is the title of this chapter?

2. What will you learn in Section 1?

3. Sometimes you may ask, "Why am I learning this?" State a reason why the concepts from Section 2 are important.

4. What is the main topic presented in Section 2?

5. How many reading checks are in Section 1?

6. What is the Web address where you can find extra information?

7. What is the main heading above the sixth paragraph in Section 2?

8. There is an integration with another subject mentioned in one of the margins of the chapter. What subject is it?

9. List the new vocabulary words presented in Section 2.

10. List the safety symbols presented in the first Lab.

11. Where would you find a Self Check to be sure you understand the section?

12. Suppose you're doing the Self Check and you have a question about concept mapping. Where could you find help?

13. On what pages are the Chapter Study Guide and Chapter Review?

14. Look in the Table of Contents to find out on which page Section 2 of the chapter begins.

15. You complete the Chapter Review to study for your chapter test. Where could you find another quiz for more practice?

Contents

The Nature of Science—2

Chapter 1 The Nature of Science—4

- **Section 1** What is science? 6
- **Section 2** Science in Action 12
- **Section 3** Models in Science 21
- **Section 4** Evaluating Scientific Explanation 27
  - Lab What is the right answer? 31
  - Lab Identifying Parts of an Investigation 32

Chapter 2 Measurement—40

- **Section 1** Description and Measurement 42
- **Section 2** SI Units 50
  - Lab Scale Drawing 55
- **Section 3** Drawings, Tables, and Graphs 56
  - Lab: Design Your Own
    Pace Yourself 60

In each chapter, look for these opportunities for review and assessment:
- Reading Checks
- Caption Questions
- Section Review
- Chapter Study Guide
- Chapter Review
- Standardized Test Practice
- Online practice at ips.msscience.com

Get Ready to Read Strategies
- Preview 6A
- Identify the Main Idea 42A
- New Vocabulary 72A

Matter—68

Chapter 3 Atoms, Elements, and the Periodic Table—70

- **Section 1** Structure of Matter 72
- **Section 2** The Simplest Matter 80
  - Lab Elements and the Periodic Table 86
- **Section 3** Compounds and Mixtures 87
  - Lab Mystery Mixture 92

xiii

Contents

Chapter 4: States of Matter—100

- **Section 1** Matter 102
- **Section 2** Changes of State 107
  - Lab The Water Cycle 115
- **Section 3** Behavior of Fluids 116
  - Lab: Design Your Own
    Design Your Own Ship 124

Chapter 5: Matter–Properties and Changes—132

- **Section 1** Physical Properties 134
- **Section 2** Chemical Properties 139
- **Section 3** Physical and Chemical Changes 143
  - Lab Sunset in a Bag 149
  - Lab: Design Your Own
    Homemade pH Scale 150

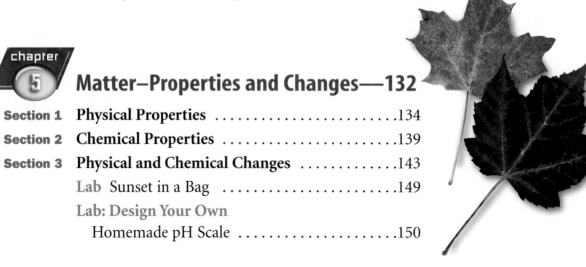

Contents

Unit 3 Chemistry—158

Chapter 6 Atomic Structure and Chemical Bonds—160

Section 1 Why do atoms combine? 162
Section 2 How Elements Bond 170
    Lab Ionic Compounds 179
    Lab: Model and Invent
        Atomic Structure 180

Chapter 7 Chemical Reactions—188

Section 1 Chemical Formulas and Equations 190
Section 2 Rates of Chemical Reactions 200
    Lab Physical or Chemical Change? 207
    Lab: Design Your Own
        Exothermic or Endothermic? 208

Chapter 8 Substances, Mixtures, and Solubility—216

Section 1 What is a solution? 218
Section 2 Solubility 224
    Lab Observing Gas Solubility 231
Section 3 Acidic and Basic Solutions 232
    Lab Testing pH Using Natural Indicators 240

Chapter 9 Carbon Chemistry—248

Section 1 Simple Organic Compounds 250
Section 2 Other Organic Compounds 257
    Lab Conversion of Alcohols 261
Section 3 Biological Compounds 262
    Lab Looking for Vitamin C 270

In each chapter, look for these opportunities for review and assessment:
- Reading Checks
- Caption Questions
- Section Review
- Chapter Study Guide
- Chapter Review
- Standardized Test Practice
- Online practice at **ips.msscience.com**

Get Ready to Read Strategies
- Monitor 102A
- Visualize 134A
- Questioning 162A
- Make Predictions 190A
- Identify Cause and Effect 218A
- Make Connections 250A

Contents

unit 4 Motion and Forces—278

chapter 10 Motion and Momentum—280

- **Section 1** What is motion? 282
- **Section 2** Acceleration 288
- **Section 3** Momentum 293
  - Lab Collisions 299
  - Lab: Design Your Own
    Car Safety Testing 300

chapter 11 Force and Newton's Laws—308

- **Section 1** Newton's First Law 310
- **Section 2** Newton's Second Law 316
- **Section 3** Newton's Third Law 323
  - Lab Balloon Races 329
  - Lab: Design Your Own
    Modeling Motion in Two Directions 330

In each chapter, look for these opportunities for review and assessment:
- Reading Checks
- Caption Questions
- Section Review
- Chapter Study Guide
- Chapter Review
- Standardized Test Practice
- Online practice at ips.msscience.com

chapter 12 Forces and Fluids—338

- **Section 1** Pressure 340
- **Section 2** Why do objects float? 348
  - Lab Measuring Buoyant Force 355
- **Section 3** Doing Work with Fluids 356
  - Lab Barometric Pressure and Weather 362

Contents

unit 5
Energy—370

chapter 13
Energy and Energy Resources—372

Section 1	**What is energy?** 374
Section 2	**Energy Transformations** 379
	Lab Hearing with Your Jaw 386
Section 3	**Sources of Energy** 387
	Lab: Use the Internet Energy to Power Your Life 396

chapter 14
Work and Simple Machines—404

Section 1	**Work and Power** 406
	Lab Building the Pyramids 411
Section 2	**Using Machines** 412
Section 3	**Simple Machines** 417
	Lab Pulley Power 424

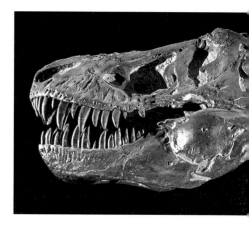

Get Ready to Read Strategies
- Summarize 282A
- Compare and Contrast 310A
- Make Inferences 340A
- Take Notes 374A
- Questions and Answers 406A

xvii

Contents

chapter 15

Thermal Energy—432

Section 1 Temperature and Thermal Energy 434
Section 2 Heat 438
    Lab Heating Up and Cooling Down 444
Section 3 Engines and Refrigerators 445
    Lab: Design Your Own
        Comparing Thermal Insulators 450

unit 6

Waves, Sound, and Light—458

chapter 16

Waves—460

Section 1 What are waves? 462
Section 2 Wave Properties 467
    Lab Waves on a Spring 472
Section 3 Wave Behavior 473
    Lab: Design Your Own
        Wave Speed 480

In each chapter, look for these opportunities for review and assessment:
- Reading Checks
- Caption Questions
- Section Review
- Chapter Study Guide
- Chapter Review
- Standardized Test Practice
- Online practice at ips.msscience.com

Contents

Chapter 17

Sound—488

Section 1 What is sound? 490
     Lab Observe and Measure Reflection
          of Sound 500
Section 2 Music 501
     Lab: Design Your Own
          Music 510

Get Ready to Read Strategies
- Identify the Main Idea 343A
- New Vocabulary 462A
- Monitor 490A
- Visualize 520A
- Questioning 550A

Chapter 18

Electromagnetic Waves—518

Section 1 The Nature of Electromagnetic
          Waves 520
Section 2 The Electromagnetic Spectrum 525
     Lab Prisms of Light 534
Section 3 Using Electromagnetic Waves 535
     Lab: Design Your Own
          Spectrum Inspection 540

Chapter 19

Light, Mirrors, and Lenses—548

Section 1 Properties of Light 550
Section 2 Reflection and Mirrors 555
     Lab Reflection from a Plane
          Mirror 561
Section 3 Refraction and Lenses 562
Section 4 Using Mirrors and Lenses 567
     Lab Image Formation by a
          Convex Lens 572

xix

Contents

unit 7 Electricity & Magnetism—580

In each chapter, look for these opportunities for review and assessment:
- Reading Checks
- Caption Questions
- Section Review
- Chapter Study Guide
- Chapter Review
- Standardized Test Practice
- Online practice at ips.msscience.com

chapter 20 Electricity—582

Section 1	Electric Charge 584
Section 2	Electric Current 591
Section 3	Electric Circuits 596
	Lab Current in a Parallel Circuit 603
	Lab A Model for Voltage and Current 604

chapter 21 Magnetism—612

Section 1	What is magnetism? 614
	Lab Make a Compass 620
Section 2	Electricity and Magnetism 621
	Lab How does an electric motor work? 632

Get Ready to Read Strategies
- Make Predictions 584A
- Identify Cause and Effect 614A
- Make Connections 642A

chapter 22 Electronics and Computers—640

Section 1	Electronics 642
	Lab Investigating Diodes 648
Section 2	Computers 649
	Lab: Use the Internet Does your computer have a virus? 660

xx

Contents

Student Resources—668

Science Skill Handbook—670
Scientific Methods 670
Safety Symbols 679
Safety in the Science Laboratory 680

Extra Try at Home Labs—682

Technology Skill Handbook—693
Computer Skills 693
Presentation Skills 696

Math Skill Handbook—697
Math Review 697
Science Applications 707

Reference Handbook—712
Periodic Table of the Elements 712

English/Spanish Glossary—714

Index—731

Credits—749

Cross-Curricular Readings

NATIONAL GEOGRAPHIC Unit Openers

Unit 1 How are Arms and Centimeters Connected?............. 2
Unit 2 How are Refrigerators and Frying Pans Connected?...... 68
Unit 3 How are Charcoal and Celebrations Connected?........ 158
Unit 4 How are City Streets and Zebra Mussels Connected?..... 278
Unit 5 How are Train Schedules and Oil Pumps Connected?...... 370
Unit 6 How are Radar and Popcorn Connected?............... 458
Unit 7 How are Cone-Bearing Trees and Static Electricity Connected?............... 582

NATIONAL GEOGRAPHIC VISUALIZING

1	The Modeling of King Tut.............................	24
2	Precision and Accuracy...............................	46
3	The Periodic Table...................................	82
4	States of Matter.....................................	110
6	Crystal Structure....................................	176
7	Chemical Reactions..................................	191
8	Acid Precipitation...................................	234
9	Organic Chemistry Nomenclature......................	255
10	Conservation of Momentum...........................	297
11	Newton's Laws in Sports.............................	325
12	Pressure at Varying Temperatures.....................	346
13	Energy Transformations..............................	382
14	Levers..	421
15	The Four-Stroke Cycle...............................	447
16	Interference..	478
17	The Doppler Effect..................................	497
18	The Universe.......................................	532
19	Reflections in Concave Mirrors.......................	559
20	Nerve Impulses.....................................	586
21	Voltmeters and Ammeters............................	623
22	A Hard Disk..	657

Cross-Curricular Readings

TIME Science and Society

9	From Plants to Medicine	272
11	Air Bag Safety	332
14	Bionic People	426
15	The Heat is On	452
17	It's a Wrap!	512
20	Fire in the Forest	606
22	E-Lectrifying E-Books	662

TIME Science and History

1	Women in Science	34
3	Ancient Views of Matter	94
5	Crumbling Monuments	152
7	Synthetic Diamonds	210
18	Hopping the Frequencies	542

Oops! Accidents in Science

4	The Incredible Stretching Goo	126
10	What Goes Around Comes Around	302
19	Eyeglasses: Inventor Unknown	574

Science and Language Arts

6	"Baring the Atom's Mother Heart"	182
12	"Hurricane"	364
21	"Aagjuuk and Sivulliit"	634

Science Stats

2	Biggest, Tallest, Loudest	62
8	Salty Solutions	242
13	Energy to Burn	398
16	Waves, Waves, and More Waves	482

Content Details

○ available as a video lab

Launch Lab

1. Observe How Gravity Accelerates Objects 5
2. Measuring Accurately 41
3. Observe Matter 71
4. Experiment with a Freezing Liquid 101
5. Classifying Different Types of Matter 133
6. Model the Energy of Electrons 161
7. Identify a Chemical Reaction 189
8. Particle Size and Dissolving Rates 217
9. Model Carbon's Bonding 249
10. Motion After a Collision 281
11. Forces and Motion 309
12. Forces Exerted by Air 339
13. Marbles and Energy 373
14. Compare Forces 405
15. Measuring Temperature 433
16. Waves and Energy 461
17. Making Human Sounds 489
18. Detecting Invisible Waves 519
19. Bending Light 549
20. Observing Electric Forces 583
21. Magnetic Forces 613
22. Electronic and Human Calculators 641

Mini Lab

1. Thinking Like a Scientist 23
2. Measuring Temperature 44
3. Investigating the Unseen 74
4. Observing Vaporization 112
5. Comparing Chemical Changes 145
6. Drawing Electron Dot Diagrams 168
7. Observing the Law of Conservation of Mass 194
8. Observing a Nail in a Carbonated Drink 233
9. Summing Up Protein 263
10. Modeling Acceleration 291
11. Measuring Force Pairs 327
12. Interpreting Footprints 342

xxiv

13	Building a Solar Collector	391
14	Observing Pulleys	422
15	Observing Convection	441
16	Comparing Sounds	465
17	Modeling a Stringed Instrument	504
18	Observing the Focusing of Infrared Rays	527
19	Forming an Image with a Lens	568
20	Identifying Simple Circuits	598
21	Observing Magnetic Fields	618

Mini LAB — Try at Home

1	Classifying Parts of a System	8
1	Forming a Hypothesis	14
2	Measuring Volume	52
3	Comparing Compounds	88
4	Predicting a Waterfall	119
5	Classifying Properties	136
6	Constructing a Model of Methane	173
7	Identifying Inhibitors	204
8	Observing Chemical Processes	228
9	Modeling Isomers	254
10	Measuring Average Speed	285
11	Observing Friction	314
12	Observing Bernoulli's Principle	359
13	Analyzing Energy Transformations	381
14	Work and Power	409
15	Comparing Rates of Melting	440
16	Observing How Light Refracts	474
17	Comparing and Contrasting Sounds	492
18	Observing Electric Fields	523
19	Observing Colors in the Dark	551
20	Investigating the Electric Force	592
21	Assembling an Electromagnet	622
22	Using Binary Numbers	650
22	Observing Memory	653

 available as a video lab

One-Page Labs

DVD	1 What is the right answer?...........................	31
	2 Scale Drawing.....................................	55
	3 Elements and the Periodic Table.....................	86
	4 The Water Cycle...................................	115
DVD	5 Sunset in a Bag...................................	149
	6 Ionic Compounds..................................	179
DVD	7 Physical or Chemical Change?.......................	207
	8 Observing Gas Solubility...........................	231
	9 Conversion of Alcohols.............................	261
DVD	10 Collisions...	299
DVD	11 Balloon Races.....................................	329
DVD	12 Measuring Buoyant Force...........................	355
	13 Hearing with Your Jaw.............................	386
DVD	14 Building the Pyramids	411
	15 Heating Up and Cooling Down	444
DVD	16 Waves on a Spring.................................	472
DVD	17 Observe and Measure Reflection of Sound	500
DVD	18 Prisms of Light....................................	534
	19 Reflection of a Plane Mirror	561
	20 Current in a Parallel Circuit	603
DVD	21 Make a Compass...................................	620
	22 Investigating Diodes...............................	648

xxvi

Two-Page Labs

- **1** Identifying Parts of an Investigation 32–33
- **3** Mystery Mixture 92–93
- **8** Testing pH Using Natural Indicators 240–241
- **9** Looking for Vitamin C 270–271
- **19** Image Formation by a Convex Lens 572–573
- **20** A Model for Voltage and Current 604–605
- **21** How does an electric motor work? 632–633

Design Your Own Labs

- **2** Pace Yourself 60–61
- **4** Design Your Own Ship 124–125
- **5** Homemade pH Scale 150–151
- **7** Exothermic or Endothermic? 208–209
- **10** Car Safety Testing 300–301
- **11** Modeling Motion in Two Directions 330–331
- **14** Pulley Power 424–425
- **15** Comparing Thermal Insulators 450–451
- **16** Wave Speed 480–481
- **17** Music 510–511
- **18** Spectrum Inspection 540–541

Model and Invent Labs

- **6** Atomic Structure 180–181

Use the Internet Labs

- **12** Barometric Pressure and Weather 362–363
- **13** Energy to Power Your Life 396–397
- **22** Does your computer have a virus? 660–661

xxvii

Activities

Applying Math

1	Seasonal Temperatures	17
2	Rounded Values	48
4	Calculating Density	121
5	Determining Density	135
7	Conserving Mass	196
10	Speed of a Swimmer	284
10	Acceleration of a Bus	290
10	Momentum of a Bicycle	294
11	Acceleration of a Car	319
14	Calculating Work	408
14	Calculating Power	409
14	Calculating Mechanical Advantage	413
14	Calculating Efficiency	415
15	Converting to Celsius	436
18	Wavelength of an FM Station	537
20	Voltage from a Wall Outlet	597
20	Electric Power Used by a Lightbulb	600

Applying Science

3	What's the best way to desalt ocean water?	89
4	How can ice save oranges?	111
6	How does the periodic table help you identify properties of elements?	167
8	How can you compare concentrations?	229
9	Which foods are the best for quick energy?	266
12	Layering Liquids	352
13	Is energy consumption outpacing production?	390
16	Can you create destructive interference?	477
17	How does Doppler radar work?	496
21	Finding the Magnetic Declination	617
22	How much information can be stored?	651

Activities

INTEGRATE

Astronomy: 51, 343, 496, 533, 536
Career: 13, 43, 78, 148, 165, 229, 264, 352, 448, 654
Chemistry: 593, 646
Earth Science: 91, 252, 388, 468
Environment: 221, 225, 656
Health: 9, 47, 137, 203, 470, 601, 630
History: 73, 104, 205, 317, 408, 529, 594, 629
Life Science: 90, 123, 140, 193, 236, 284, 311, 324, 361, 381, 383, 415, 419, 442, 443, 495, 530, 602, 618
Physics: 108, 171, 466, 557, 634
Social Studies: 294, 502

18, 22, 47, 58, 76, 81, 90, 105, 111, 113, 123, 138, 146, 164, 175, 195, 201, 219, 235, 238, 256, 267, 286, 296, 313, 324, 341, 358, 380, 390, 410, 413, 446, 471, 477, 495, 507, 521, 538, 558, 588, 601, 619, 627, 645, 652, 656, 658

Standardized Test Practice

38–39, 66–67, 98–99, 130–131, 156–157, 186–187, 214–215, 246–247, 276–277, 306–307, 336–337, 368–369, 402–403, 430–431, 456–457, 486–487, 516–517, 546–547, 578–579, 610–611, 638–639, 666–667

unit 1
The Nature of Science

How Are Arms & Centimeters Connected?

About 5,000 years ago, the Egyptians developed one of the earliest recorded units of measurement—the cubit, which was based on the length of the arm from elbow to fingertip. The Egyptian measurement system probably influenced later systems, many of which also were based on body parts such as arms and feet. Such systems, however, could be problematic, since arms and feet vary in length from one person to another. Moreover, each country had its own system, which made it hard for people from different countries to share information. The need for a precise, universal measurement system eventually led to the adoption of the meter as the basic international unit of length. A meter is defined as the distance that light travels in a vacuum in a certain fraction of a second—a distance that never varies. Meters are divided into smaller units called centimeters, which are seen on the rulers here.

unit ⚡ projects

Visit ips.msscience.com/unit_project to find project ideas and resources. Projects include:

- **History** Brainstorm characteristics of science fields, and then design a collage with science ideas for a ceiling tile or book cover.
- **Technology** Convert a family recipe from English measurement to SI units of measurement. Enjoy a classroom bake-off!
- **Model** Create a character in an SI world. Develop a picture storybook or comic book to demonstrate your knowledge of SI measurement.

WebQuest *The Nature of Science: Evaluating Bias in Advertisements* helps students to become informed about the techniques of advertising and evaluate bias in print media.

chapter 1

The Nature of Science

The BIG Idea
Science is an organized method of learning about the natural world.

SECTION 1
What is science?
Main Idea Science describes observations of the natural world and proposes explanations for those observations.

SECTION 2
Science in Action
Main Idea Scientific investigations follow a general pattern of observing, hypothesizing, investigating, analyzing, and concluding.

SECTION 3
Models in Science
Main Idea A model is a representation of an object or event that helps scientists understand the natural world.

SECTION 4
Evaluating Scientific Explanation
Main Idea How reliable an explanation is depends on the accuracy of the observations and conclusions supporting the explanation.

How is science a part of your life?
Scientists studying desert ecosystems in California wondered how such a dry environment could produce such beautiful, prolific flowers. Scientists began asking questions and performing investigations.

Science Journal Write down three examples of science in your everyday life.

Start-Up Activities

Observe How Gravity Accelerates Objects

Gravity is a familiar natural force that keeps you anchored on Earth, but how does it work? Scientists learn about gravity and other concepts by asking questions and making observations. By observing things in action scientists can study nature. Perform the lab below to see how gravity affects objects.

1. Collect three identical, unsharpened pencils.
2. Tape two of the pencils together.
3. Hold all the pencils at the same height as high as you can. Drop them together and observe what happens as they fall.
4. **Think Critically** Did the single pencil fall faster or slower than the pair? Predict in your Science Journal what would happen if you taped 30 pencils together and dropped them at the same time as you dropped a single pencil.

Preview this chapter's content and activities at
ips.msscience.com

FOLDABLES Study Organizer

Science Make the following Foldable to help identify what you already know, what you want to know, and what you learned about science.

STEP 1 Fold a vertical sheet of paper from side to side. Make the front edge about 1/2 inch shorter than the back edge.

STEP 2 Turn lengthwise and fold into thirds.

STEP 3 Unfold and cut only the top layer along both folds to make three tabs. Label each tab.

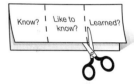

Identify Questions Before you read the chapter, write what you already know about science under the left tab of your Foldable, and write questions about what you'd like to know under the center tab. After you read the chapter, list what you learned under the right tab.

Get Ready to Read

Preview

① Learn It! If you know what to expect before reading, it will be easier to understand ideas and relationships presented in the text. Follow these steps to preview your reading assignments.

1. Look at the title and any illustrations that are included.
2. Read the headings, subheadings, and anything in bold letters.
3. Skim over the passage to see how it is organized. Is it divided into many parts?
4. Look at the graphics—pictures, maps, or diagrams. Read their titles, labels, and captions.
5. Set a purpose for your reading. Are you reading to learn something new? Are you reading to find specific information?

② Practice It! Take some time to preview this chapter. Skim all the main headings and subheadings. With a partner, discuss your answers to these questions.
- Which part of this chapter looks most interesting to you?
- Are there any words in the headings that are unfamiliar to you?
- Choose one of the lesson review questions to discuss with a partner.

③ Apply It! Now that you have skimmed the chapter, write a short paragraph describing one thing you want to learn from this chapter.

Target Your Reading

Use this to focus on the main ideas as you read the chapter.

As you preview the chapter, be sure to scan the illustrations, tables, and graphs. Skim the captions.

① **Before you read** the chapter, respond to the statements below on your worksheet or on a numbered sheet of paper.
- Write an **A** if you **agree** with the statement.
- Write a **D** if you **disagree** with the statement.

② **After you read** the chapter, look back to this page to see if you've changed your mind about any of the statements.
- If any of your answers changed, explain why.
- Change any false statements into true statements.
- Use your revised statements as a study guide.

Science Online
Print out a worksheet of this page at ips.msscience.com

Before You Read A or D		Statement	After You Read A or D
	1	A scientific theory is proposed before any investigations occur.	
	2	A scientific theory can eventually become a scientific law.	
	3	Technology is the practical use of science.	
	4	If a hypothesis is not supported by an investigation, the investigation is a waste of time.	
	5	Scientists can never know for sure whether an explanation is accurate, even after many investigations.	
	6	It is helpful to make changes to more than one variable during an experiment.	
	7	Models are only as accurate as the information used to create them.	
	8	Scientific data are reliable as long as they are observed at least one time.	
	9	Scientific conclusions are more reliable when other conclusions are ruled out.	

6 B

section 1
What is science?

as you read

What You'll Learn
- **Define** science and identify questions that science cannot answer.
- **Compare** and contrast theories and laws.
- **Identify** a system and its components.
- **Identify** the three main branches of science.

Why It's Important
Science can be used to learn more about the world you live in.

Review Vocabulary
theory: explanation of things or events that is based on knowledge gained from many observations and experiments

New Vocabulary
- science
- scientific theory
- scientific law
- system
- life science
- Earth science
- physical science
- technology

Learning About the World

When you think of a scientist, do you imagine a person in a laboratory surrounded by charts, graphs, glass bottles, and bubbling test tubes? It might surprise you to learn that anyone who tries to learn something about the natural world is a scientist. **Science** is a way of learning more about the natural world. Scientists want to know why, how, or when something occurred. This learning process usually begins by keeping your eyes open and asking questions about what you see.

Asking Questions Scientists ask many questions. How do things work? What do things look like? What are they made of? Why does something take place? Science can attempt to answer many questions about the natural world, but some questions cannot be answered by science. Look at the situations in **Figure 1**. Who should you vote for? What does this poem mean? Who is your best friend? Questions about art, politics, personal preference, or morality can't be answered by science. Science can't tell you what is right, wrong, good, or bad.

Figure 1 Questions about politics, literature, and art cannot be answered by science.

Figure 2 As new information becomes available, explanations can be modified or discarded and new explanations can be made.

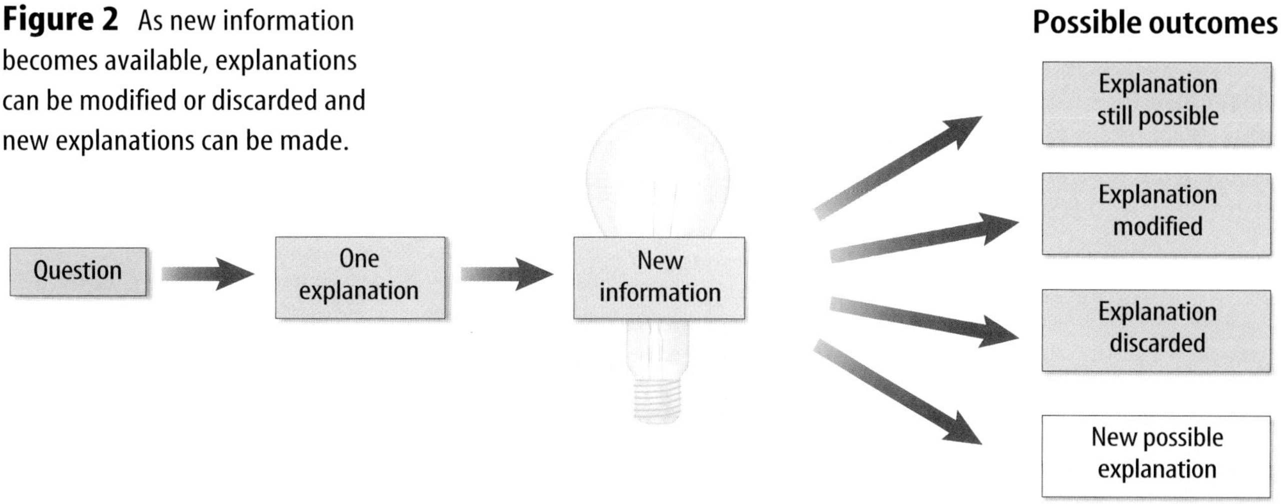

Question → One explanation → New information →

Possible outcomes
- Explanation still possible
- Explanation modified
- Explanation discarded
- New possible explanation

Possible Explanations If learning about your world begins with asking questions, can science provide answers to these questions? Science can answer a question only with the information available at the time. Any answer is uncertain because people will never know everything about the world around them. With new knowledge, they might realize that some of the old explanations no longer fit the new information. As shown in **Figure 2,** some observations might force scientists to look at old ideas and think of new explanations. Science can only provide possible explanations.

Reading Check *Why can't science answer questions with certainty?*

Scientific Theories An attempt to explain a pattern observed repeatedly in the natural world is called a **scientific theory.** Theories are not simply guesses or someone's opinions, nor are theories vague ideas. Theories in science must be supported by observations and results from many investigations. They are the best explanations that have been found so far. However, theories can change. As new data become available, scientists evaluate how the new data fit the theory. If enough new data do not support the theory, the theory can be changed to fit the new observations better.

Scientific Laws A rule that describes a pattern in nature is a **scientific law.** For an observation to become a scientific law, it must be observed repeatedly. The law then stands until someone makes observations that do not follow the law. A law helps you predict that an apple dropped from arm's length will always fall to Earth. The law, however, does not explain why gravity exists or how it works. A law, unlike a theory, does not attempt to explain why something happens. It simply describes a pattern.

Figure 3 Systems are a collection of structures, cycles, and processes.
Infer What systems can you identify in this classroom?

Classifying Parts of a System

Procedure
Think about how your school's cafeteria is run. Consider the physical structure of the cafeteria. How many people run it? Where does the food come from? How is it prepared? Where does it go? What other parts of the cafeteria system are necessary?

Analysis
Classify the parts of your school cafeteria's system as structures, cycles, or processes.

Try at Home

Systems in Science

Scientists can study many different things in nature. Some might study how the human body works or how planets move around the Sun. Others might study the energy carried in a lightning bolt. What do all of these things have in common? All of them are systems. A **system** is a collection of structures, cycles, and processes that relate to and interact with each other. The structures, cycles, and processes are the parts of a system, just like your stomach is one of the structures of your digestive system.

Reading Check *What is a system?*

Systems are not found just in science. Your school is a system with structures such as the school building, the tables and chairs, you, your teacher, the school bell, your pencil, and many other things. **Figure 3** shows some of these structures. Your school day also has cycles. Your daily class schedule and the calendar of holidays are examples of cycles. Many processes are at work during the school day. When you take a test, your teacher has a process. You might be asked to put your books and papers away and get out a pencil before the test is distributed. When the time is over, you are told to put your pencil down and pass your test to the front of the room.

Parts of a System Interact In a system, structures, cycles, and processes interact. Your daily schedule influences where you go and what time you go. The clock shows the teacher when the test is complete, and you couldn't complete the test without a pencil.

8 CHAPTER 1 The Nature of Science

Parts of a Whole All systems are made up of other systems. For example, you are part of your school. The human body is a system—within your body are other systems. Your school is part of a system—district, state, and national. You have your regional school district. Your district is part of a statewide school system. Scientists often break down problems by studying just one part of a system. A scientist might want to learn about how construction of buildings affects the ecosystem. Because an ecosystem has many parts, one scientist might study a particular animal, and another might study the effect of construction on plant life.

The Branches of Science

Science often is divided into three main categories, or branches—life science, Earth science, and physical science. Each branch asks questions about different kinds of systems.

Life Science The study of living systems and the ways in which they interact is called **life science.** Life scientists attempt to answer questions like "How do whales navigate the ocean?" and "How do vaccines prevent disease?" Life scientists can study living organisms, where they live, and how they interact. Dian Fossey, **Figure 4,** was a life scientist who studied gorillas, their habitat, and their behaviors.

People who work in the health field know a lot about the life sciences. Physicians, nurses, physical therapists, dietitians, medical researchers, and others focus on the systems of the human body. Some other examples of careers that use life science include biologists, zookeepers, botanists, farmers, and beekeepers.

Health Integration Systems The human body is composed of many different systems that all interact with one another to perform a function. The heart is like the control center. Even though not all systems report directly to the heart, they all interact with its function. If the heart is not working, the other systems fail as well. Research human body systems and explain how one system can affect another.

Figure 4 Over a span of 18 years, life scientist Dian Fossey spent much of her time observing mountain gorillas in Rwanda, Africa. She was able to interact with them as she learned about their behavior.

Figure 5 These volcanologists are studying the temperature of the lava flowing from a volcano.

Figure 6 Physical scientists study a wide range of subjects.

Earth Science The study of Earth systems and the systems in space is **Earth science.** It includes the study of nonliving things such as rocks, soil, clouds, rivers, oceans, planets, stars, meteors, and black holes. Earth science also covers the weather and climate systems that affect Earth. Earth scientists ask questions like "How can an earthquake be detected?" or "Is water found on other planets?" They make maps and investigate how geologic features formed on land and in the oceans. They also use their knowledge to search for fuels and minerals. Meteorologists study weather and climate. Geologists study rocks and geologic features. **Figure 5** shows a volcanologist—a person who studies volcanoes—measuring the temperature of lava.

Reading Check *What do Earth scientists study?*

Physical Science The study of matter and energy is **physical science.** Matter is anything that takes up space and has mass. The ability to cause change in matter is energy. Living and nonliving systems are made of matter. Examples include plants, animals, rocks, the atmosphere, and the water in oceans, lakes, and rivers. Physical science can be divided into two general fields—chemistry and physics. Chemistry is the study of matter and the interactions of matter. Physics is the study of energy and its ability to change matter. Figure 6 shows physical scientists at work.

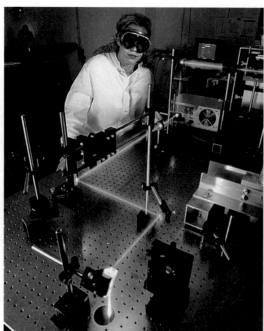

This chemist is studying the light emitted by certain compounds.

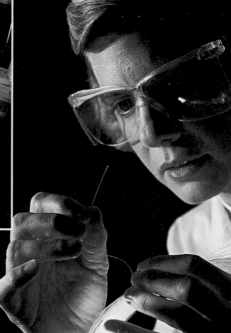

This physicist is studying light as it travels through optical fibers.

Careers Chemists ask questions such as "How can I make plastic stronger?" or "What can I do to make aspirin more effective?" Physicists might ask other types of questions, such as "How does light travel through glass fibers?" or "How can humans harness the energy of sunlight for their energy needs?"

Many careers are based on the physical sciences. Physicists and chemists are some obvious careers. Ultrasound and X-ray technicians working in the medical field study physical science because they study the energy in ultrasound or X rays and how it affects a living system.

Science and Technology Although learning the answers to scientific questions is important, these answers do not help people directly unless they can be applied in some way. **Technology** is the practical use of science, or applied science, as illustrated in **Figure 7**. Engineers apply science to develop technology. The study of how to use the energy of sunlight is science. Using this knowledge to create solar panels is technology. The study of the behavior of light as it travels through thin, glass, fiber-optic wires is science. The use of optical fibers to transmit information is technology. A scientist uses science to study how the skin of a shark repels water. The application of this knowledge to create a material that helps swimmers slip through the water faster is technology.

Figure 7 Solar-powered cars and the swimsuits worn in the Olympics are examples of technology—the application of science.

section 1 review

Summary

Learning About the World
- Scientists ask questions to learn how, why, or when something occurred.
- A theory is a possible explanation for observations that is supported by many investigations.
- A scientific law describes a pattern but does not explain why things happen.

Systems in Science
- A system is composed of structures, cycles, and processes that interact with each other.

The Branches of Science
- Science is divided into three branches—life science, Earth science, and physical science.
- Technology is the application of science in our everyday lives.

Self Check

1. **Compare and contrast** scientific theory and scientific law. Explain how a scientific theory can change.
2. **Explain** why science can answer some questions, but not others.
3. **Classify** the following statement as a theory or a law: Heating the air in a hot-air balloon causes the balloon to rise.
4. **Think Critically** Describe the importance of technology and how it relates to science.

Applying Skills

5. **Infer** Scientists ask questions and make observations. What types of questions and observations would you make if you were a scientist studying schools of fish in the ocean?

SECTION 1 What is science?

section 2
Science in Action

as you read

What You'll Learn
- **Identify** some skills scientists use.
- **Define** hypothesis.
- **Recognize** the difference between observation and inference.

Why It's Important
Science can be used to learn more about the world you live in.

Review Vocabulary
observation: a record or description of an occurrence or pattern in nature

New Vocabulary
- hypothesis
- infer
- controlled experiment
- variable
- independent variable
- dependent variable
- constant

Science Skills

You know that science involves asking questions, but how does asking questions lead to learning? Because no single way to gain knowledge exists, a scientist doesn't start with step one, then go to step two, and so on. Instead, scientists have a huge collection of skills from which to choose. Some of these skills include thinking, observing, predicting, investigating, researching, modeling, measuring, analyzing, and inferring. Science also can advance with luck and creativity.

Science Methods Investigations often follow a general pattern. As illustrated in **Figure 8,** most investigations begin by seeing something and then asking a question about what was observed. Scientists often perform research by talking with other scientists. They read books and scientific magazines to learn as much as they can about what is already known about their question. Usually, scientists state a possible explanation for their observation. To collect more information, scientists almost always make more observations. They might build a model of what they study or they might perform investigations. Often, they do both. How might you combine some of these skills in an investigation?

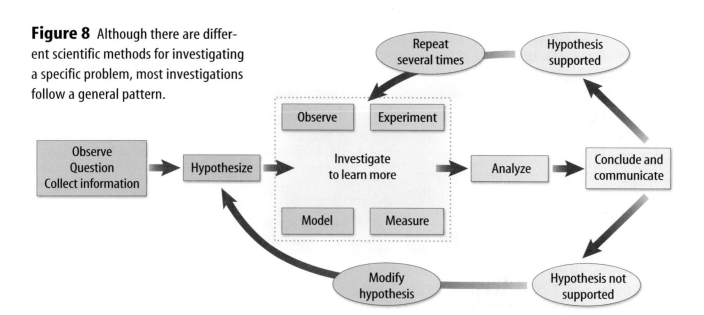

Figure 8 Although there are different scientific methods for investigating a specific problem, most investigations follow a general pattern.

12 CHAPTER 1 The Nature of Science

Figure 9 Investigations often begin by making observations and asking questions.

Questioning and Observing Ms. Clark placed a sealed shoe box on the table at the front of the laboratory. Everyone in the class noticed the box. Within seconds the questions flew. "What's in the box?" "Why is it there?"

Ms. Clark said she would like the class to see how they used some science skills without even realizing it.

"I think that she wants us to find out what's in it," Isabelle said to Marcus.

"Can we touch it?" asked Marcus.

"It's up to you," Ms. Clark said.

Marcus picked up the box and turned it over a few times.

"It's not heavy," Marcus observed. "Whatever is inside slides around." He handed the box to Isabelle.

Isabelle shook the box. The class heard the object strike the sides of the box. With every few shakes, the class heard a metallic sound. The box was passed around for each student to make observations and write them in his or her Science Journal. Some observations are shown in **Figure 9.**

Taking a Guess "I think it's a pair of scissors," said Marcus.

"Aren't scissors lighter than this?" asked Isabelle, while shaking the box. "I think it's a stapler."

"What makes you think so?" asked Ms. Clark.

"Well, staplers are small enough to fit inside a shoe box, and it seems to weigh about the same," said Isabelle.

"We can hear metal when we shake it," said Enrique.

"So, you are guessing that a stapler is in the box?"

"Yes," they agreed.

"You just stated a hypothesis," exclaimed Ms. Clark.

"A what?" asked Marcus.

Biologist Some naturalists study the living world, using mostly their observational skills. They observe animals and plants in their natural environment, taking care not to disturb the organisms they are studying. Make observations of organisms in a nearby park or backyard. Record your observations in your Science Journal.

Mini LAB

Forming a Hypothesis

Procedure
1. Fill a large **pot** with **water**. Place an **unopened can of diet soda** and an **unopened can of regular soda** into the pot of water and observe what each can does.
2. In your **Science Journal**, make a list of the possible explanations for your observation. Select the best explanation and write a hypothesis.
3. Read the nutritional facts on the back of each can and compare their ingredients.
4. Revise your hypothesis based on this new information.

Analysis
1. What did you observe when you placed the cans in the water?
2. How did the nutritional information on the cans change your hypothesis?
3. Infer why the two cans behaved differently in the water.

Try at Home

The Hypothesis "A **hypothesis** is a reasonable and educated possible answer based on what you know and what you observe."

"We know that a stapler is small, it can be heavy, and it is made of metal," said Isabelle.

"We observed that what is in the box is small, heavier than a pair of scissors, and made of metal," continued Marcus.

Analyzing Hypotheses "What other possible explanations fit with what you observed?" asked Ms. Clark.

"Well, it has to be a stapler," said Enrique.

"What if it isn't?" asked Ms. Clark. "Maybe you're overlooking explanations because your minds are made up. A good scientist keeps an open mind to every idea and explanation. What if you learn new information that doesn't fit with your original hypothesis? What new information could you gather to verify or disprove your hypothesis?"

"Do you mean a test or something?" asked Marcus.

"I know," said Enrique, "We could get an empty shoe box that is the same size as the mystery box and put a stapler in it. Then we could shake it and see whether it feels and sounds the same." Enrique's test is shown in **Figure 10**.

Making a Prediction "If your hypothesis is correct, what would you expect to happen?" asked Ms. Clark.

"Well, it would be about the same weight and it would slide around a little, just like the other box," said Enrique.

"It would have that same metallic sound when we shake it," said Marcus.

"So, you predict that the test box will feel and sound the same as your mystery box. Go ahead and try it," said Ms. Clark.

Figure 10 Comparing the known information with the unknown information can be valuable even though you cannot see what is inside the closed box.

14 CHAPTER 1 The Nature of Science

Testing the Hypothesis Ms. Clark gave the class an empty shoe box that appeared to be identical to the mystery box. Isabelle found a metal stapler. Enrique put the stapler in the box and taped the box closed. Marcus shook the box.

"The stapler does slide around but it feels just a little heavier than what's inside the mystery box," said Marcus. "What do you think?" he asked Isabelle as he handed her the box.

"It is heavier," said Isabelle "and as hard as I shake it, I can't get a metallic sound. What if we find the mass of both boxes? Then we'll know the exact mass difference between the two."

Using a balance, as shown in **Figure 11,** the class found that the test box had a mass of 410 g, and the mystery box had a mass of 270 g.

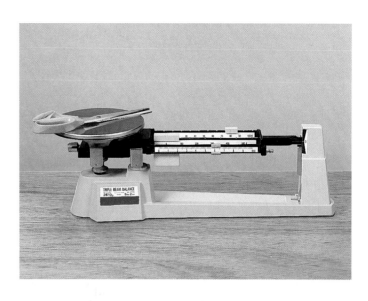

Figure 11 Laboratory balances are used to find the mass of objects.

Organizing Your Findings "Okay. Now you have some new information," said Ms. Clark. "But before you draw any conclusions, let's organize what we know. Then we'll have a summary of our observations and can refer back to them when we are drawing our conclusions."

"We could make a chart of our observations in our Science Journals," said Marcus.

"We could compare the observations of the mystery box with the observations of the test box," said Isabelle. The chart that the class made is shown in **Table 1.**

Table 1 Observation Chart		
Questions	**Mystery Box**	**Our Box**
Does it roll or slide?	It slides and appears to be flat.	It slides and appears to be flat.
Does it make any sounds?	It makes a metallic sound when it strikes the sides of the box.	The stapler makes a thudding sound when it strikes the sides of the box.
Is the mass evenly distributed in the box?	No. The object doesn't completely fill the box.	No. The mass of the stapler is unevenly distributed.
What is the mass of the box?	270 g	410 g

SECTION 2 Science in Action 15

Figure 12 Observations can be used to draw inferences.
Infer *Looking at both of these photos, what do you infer has taken place?*

Drawing Conclusions

"What have you learned from your investigation so far?" asked Ms. Clark.

"The first thing that we learned was that our hypothesis wasn't correct," answered Marcus.

"Would you say that your hypothesis was entirely wrong?" asked Ms. Clark.

"The boxes don't weigh the same, and the box with the stapler doesn't make the same sound as the mystery box. But there could be a difference in the kind of stapler in the box. It could be a different size or made of different materials."

"So you infer that the object in the mystery box is not exactly the same type of stapler, right?" asked Ms. Clark.

"What does *infer* mean?" asked Isabelle.

"To **infer** something means to draw a conclusion based on what you observe," answered Ms. Clark.

"So we inferred that the things in the boxes had to be different because our observations of the two boxes are different," said Marcus.

"I guess we're back to where we started," said Enrique. "We still don't know what's in the mystery box."

"Do you know more than you did before you started?" asked Ms. Clark.

"We eliminated one possibility," Isabelle added.

"Yes. We inferred that it's not a stapler, at least not like the one in the test box," said Marcus.

"So even if your observations don't support your hypothesis, you know more than you did when you started," said Ms. Clark.

Continuing to Learn "So when do we get to open the box and see what it is?" asked Marcus.

"Let me ask you this," said Ms. Clark. "Do you think scientists always get a chance to look inside to see if they are right?"

"If they are studying something too big or too small to see, I guess they can't," replied Isabelle. "What do they do in those cases?"

"As you learned, your first hypothesis might not be supported by your investigation. Instead of giving up, you continue to gather information by making more observations, making new hypotheses, and by investigating further. Some scientists have spent lifetimes researching their questions. Science takes patience and persistence," said Ms. Clark.

Communicating Your Findings It is not unusual for one scientist to continue the work of another or to try to duplicate the work of another scientist. It is important for scientists to communicate to others not only the results of the investigation, but also the methods by which the investigation was done. Scientists often publish reports in journals, books, and on the Internet to show other scientists the work that was completed. They also might attend meetings where they make speeches about their work.

Like the science-fair student in **Figure 13** demonstrates, an important part of doing science is the ability to communicate methods and results to others.

Reading Check *Why do scientists share information?*

Figure 13 Presentations are one way people in science communicate their findings.

Applying Math — Make a Data Table

SEASONAL TEMPERATURES Suppose you were given the average temperatures in a city for the four seasons over a three-year period: spring 1997 was 11°C; summer 1997 was 25°C; fall 1997 was 5°C; winter 1997 was −5°C; spring 1998 was 9°C; summer 1998 was 36°C; fall 1998 was 10°C; winter 1998 was −3°C; spring 1999 was 10°C; summer 1999 was 30°C; fall 1999 was 9°C; and winter 1999 was −2°C. How can you tell in which of the years each season had its coldest average?

Solution

1 *This is what you know:*   Temperatures were: 1997: 11°C, 25°C, 5°C, −5°C
                                                                                1998: 9°C, 36°C, 10°C, −3°C
                                                                                1999: 10°C, 30°C, 9°C, −2°C

2 *This is what you need to find out:*   Which of the years each season had its coldest temperature?

3 *This is the procedure you need to use:*
- Create a table with rows for seasons and columns for the years.
- Insert the values you were given.

4 *Check your answer:*   The four coldest seasons were spring 1998, summer 1997, fall 1997, and winter 1997.

Practice Problems

Use your table to find out which season had the greatest difference in temperatures over the three years from 1997 through 1999.

 For more practice, visit ips.msscience.com/math_practice

Topic: Scientific Method
Visit ips.msscience.com for Web links to information about the scientific method.

Activity Identify the three variables needed in every experiment and summarize the differences between them.

Experiments

Different types of questions call for different types of investigations. Ms. Clark's class made many observations about their mystery box and about their test box. They wanted to know what was inside. To answer their question, building a model—the test box—was an effective way to learn more about the mystery box. Some questions ask about the effects of one factor on another. One way to investigate these kinds of questions is by doing a controlled experiment. A **controlled experiment** involves changing one factor and observing its effect on another while keeping all other factors constant.

Variables and Constants Imagine a race in which the lengths of the lanes vary. Some lanes are 102 m long, some are 98 m long, and a few are 100 m long. When the first runner crosses the finish line, is he or she the fastest? Not necessarily. The lanes in the race have different lengths.

Variables are factors that can be changed in an experiment. Reliable experiments, like the race shown in **Figure 14,** attempt to change one variable and observe the effect of this change on another variable. The variable that is changed in an experiment is called the **independent variable.** The **dependent variable** changes as a result of a change in the independent variable. It usually is the dependent variable that is observed in an experiment. Scientists attempt to keep all other variables constant—or unchanged.

The variables that are not changed in an experiment are called **constants.** Examples of constants in the race include track material, wind speed, and distance. This way it is easier to determine exactly which variable is responsible for the runners' finish times. In this race, the runners' abilities were varied. The runners' finish times were observed.

Figure 14 The 400-m race is an example of a controlled experiment. The distance, track material, and wind speed are constants. The runners' abilities and their finish times are varied.

Figure 15 Safety is the most important aspect of any investigation.

Laboratory Safety

In your science class, you will perform many types of investigations. However, performing scientific investigations involves more than just following specific steps. You also must learn how to keep yourself and those around you safe by obeying the safety symbol warnings, shown in **Figure 16.**

In a Laboratory When scientists work in a laboratory, as shown in **Figure 15,** they take many safety precautions.

The most important safety advice in a science lab is to think before you act. Always check with your teacher several times in the planning stage of any investigation. Also make sure you know the location of safety equipment in the laboratory room and how to use this equipment, including the eyewashes, thermal mitts, and fire extinguisher.

Good safety habits include the following suggestions. Before conducting any investigation, find and follow all safety symbols listed in your investigation. You always should wear an apron and goggles to protect yourself from chemicals, flames, and pointed objects. Keep goggles on until activity, cleanup, and handwashing are complete. Always slant test tubes away from yourself and others when heating them. Never eat, drink, or apply makeup in the lab. Report all accidents and injuries to your teacher and always wash your hands after working with lab materials.

In the Field Investigations also take place outside the lab, in streams, farm fields, and other places. Scientists must follow safety regulations there, as well, such as wearing eye goggles and any other special safety equipment that is needed. Never reach into holes or under rocks. Always wash your hands after you've finished your field work.

Figure 16 Safety symbols are present on nearly every investigation you will do this year.
List *the safety symbols that should be on the lab the student is preparing to do in* **Figure 15.**

SECTION 2 Science in Action **19**

Figure 17 Accidents are not planned. Safety precautions must be followed to prevent injury.

Why have safety rules? Doing science in the class laboratory or in the field can be much more interesting than reading about it. However, safety rules must be strictly followed, so that the possibility of an accident greatly decreases. However, you can't predict when something will go wrong.

Think of a person taking a trip in a car. Most of the time when someone drives somewhere in a vehicle, an accident, like the one shown in **Figure 17,** does not occur. But to be safe, drivers and passengers always should wear safety belts. Likewise, you always should wear and use appropriate safety gear in the lab—whether you are conducting an investigation or just observing. The most important aspect of any investigation is to conduct it safely.

section 2 review

Summary

Science Skills
- The scientific method was developed to help scientists investigate their questions.
- Hypotheses are possible explanations for why something occurs.

Drawing Conclusions
- Scientists communicate with one another to share important information.

Experiments
- Controlled experiments test the effect of one factor on another.

Laboratory Safety
- Safety precautions must be followed when conducting any investigation.

Self Check

1. **Explain** the difference between an inference and an observation.
2. **Explain** the differences between independent and dependent variables.
3. **Think Critically** A classroom investigation lists bleach as an ingredient. Bleach can irritate your skin, damage your eyes, and stain your clothes. What safety symbols should be listed with this investigation? Explain.

Applying Skills

4. **Describe** the different types of safety equipment found in a scientific laboratory. From your list, which equipment should you use when working with a flammable liquid in the lab?

ips.msscience.com/self_check_quiz

section 3
Models in Science

Why are models necessary?

Just as you can take many different paths in an investigation, you can test a hypothesis in many different ways. Ms. Clark's class tested their hypothesis by building a model of the mystery box. A model is one way to test a hypothesis. In science, a **model** is any representation of an object or an event used as a tool for understanding the natural world.

Models can help you visualize, or picture in your mind, something that is difficult to see or understand. Ms. Clark's class made a model because they couldn't see the item inside the box. Models can be of things that are too small or too big to see. They also can be of things that can't be seen because they don't exist anymore or they haven't been created yet. Models also can show events that occur too slowly or too quickly to see. **Figure 18** shows different kinds of models.

as you read

What You'll Learn
- **Describe** various types of models.
- **Discuss** limitations of models.

Why It's Important
Models can be used to help understand difficult concepts.

Review Vocabulary
scientific method: processes scientists use to collect information and answer questions

New Vocabulary
- model

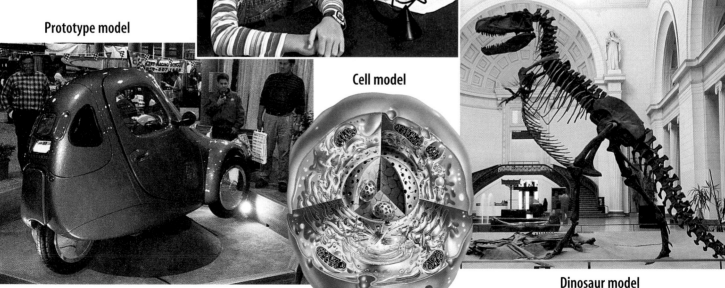

Figure 18 Models help scientists visualize and study complex things and things that can't be seen.

Prototype model

Solar system model

Cell model

Dinosaur model

SECTION 3 Models in Science **21**

Types of Models

Most models fall into three basic types—physical models, computer models, and idea models. Depending on the reason that a model is needed, scientists can choose to use one or more than one type of model.

Physical Models Models that you can see and touch are called physical models. Examples include things such as a tabletop solar system, a globe of Earth, a replica of the inside of a cell, or a gumdrop-toothpick model of a chemical compound. Models show how parts relate to one another. They also can be used to show how things appear when they change position or how they react when an outside force acts on them.

Computer Models Computer models are built using computer software. You can't touch them, but you can view them on a computer screen. Some computer models can model events that take a long time or take place too quickly to see. For example, a computer can model the movement of large plates in the Earth and might help predict earthquakes.

Computers also can model motions and positions of things that would take hours or days to calculate by hand or even using a calculator. They can also predict the effect of different systems or forces. **Figure 19** shows how computer models are used by scientists to help predict the weather based on the motion of air currents in the atmosphere.

Reading Check *What do computer models do?*

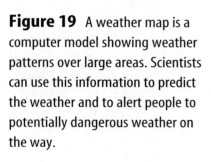

Science Online

Topic: Topographic Maps
Visit ips.msscience.com for Web links to information about topographic maps.

Activity List some of the different features found on topographic maps and explain their importance when reading and interpreting maps.

Figure 19 A weather map is a computer model showing weather patterns over large areas. Scientists can use this information to predict the weather and to alert people to potentially dangerous weather on the way.

Figure 20 Models can be created using various types of tools.

Idea Models Some models are ideas or concepts that describe how someone thinks about something in the natural world. Albert Einstein is famous for his theory of relativity, which involves the relationship between matter and energy. One of the most famous models Einstein used for this theory is the mathematical equation $E = mc^2$. This explains that mass, m, can be changed into energy, E. Einstein's idea models never could be built as physical models, because they are basically ideas.

Making Models

The process of making a model is something like a sketch artist at work, as shown in **Figure 20.** The sketch artist attempts to draw a picture from the description given by someone. The more detailed the description is, the better the picture will be. Like a scientist who studies data from many sources, the sketch artist can make a sketch based on more than one person's observation. The final sketch isn't a photograph, but if the information is accurate, the sketch should look realistic. Scientific models are made much the same way. The more information a scientist gathers, the more accurate the model will be. The process of constructing a model of King Tutankhamun, who lived more than 3,000 years ago, is shown in **Figure 21.**

Reading Check *How are sketches like specific models?*

Using Models

When you think of a model, you might think of a model airplane or a model of a building. Not all models are for scientific purposes. You use models, and you might not realize it. Drawings, maps, recipes, and globes are all examples of models.

Thinking Like a Scientist

Procedure
1. Pour 15 mL of **water** into a **test tube.**
2. Slowly pour 5 mL of **vegetable oil** into the test tube.
3. Add two drops of **food coloring** and observe the liquid for 5 min.

Analysis
1. Record your observations of the test tube's contents before and after the oil and the food coloring were added to it.
2. Infer a scientific explanation for your observations.

SECTION 3 Models in Science **23**

NATIONAL GEOGRAPHIC VISUALIZING THE MODELING OF KING TUT

Figure 21

More than 3,000 years ago, King Tutankhamun ruled over Egypt. His reign was a short one, and he died when he was just 18. In 1922, his mummified body was discovered, and in 1983 scientists recreated the face of this most famous of Egyptian kings. Some of the steps in building the model are shown here.

This is the most familiar image of the face of King Tut—the gold funerary mask that was found covering his skeletal face.

A First, a scientist used measurements and X rays to create a cast of the young king's skull. Depth markers (in red) were then glued onto the skull to indicate the likely thickness of muscle and other tissue.

B Clay was applied to fill in the area between the markers.

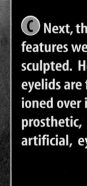

C Next, the features were sculpted. Here, eyelids are fashioned over inlaid prosthetic, or artificial, eyes.

D When this model of King Tut's face was completed, the long-dead ruler seemed to come to life.

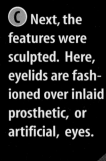

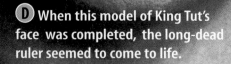

24 CHAPTER 1 The Nature of Science

Models Communicate Some models are used to communicate observations and ideas to other people. Often, it is easier to communicate ideas you have by making a model instead of writing your ideas in words. This way others can visualize them, too.

Models Test Predictions Some models are used to test predictions. Ms. Clark's class predicted that a box with a stapler in it would have characteristics similar to their mystery box. To test this prediction, the class made a model. Automobile and airplane engineers use wind tunnels to test predictions about how air will interact with their products.

Models Save Time, Money, and Lives Other models are used because working with and testing a model can be safer and less expensive than using the real thing. For example, the crash-test dummies shown in **Figure 22** are used in place of people when testing the effects of automobile crashes. To help train astronauts in the conditions they will encounter in space, NASA has built a special airplane. This airplane flies in an arc that creates the condition of freefall for 20 to 25 seconds. Making several trips in the airplane is easier, safer, and less expensive than making a trip into space.

Figure 22 Models are a safe and relatively inexpensive way to test ideas.

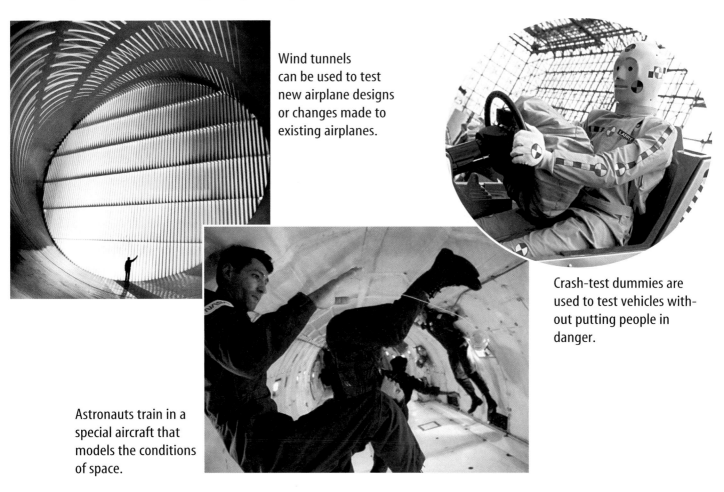

Wind tunnels can be used to test new airplane designs or changes made to existing airplanes.

Crash-test dummies are used to test vehicles without putting people in danger.

Astronauts train in a special aircraft that models the conditions of space.

Figure 23 The model of Earth's solar system changed as new information was gathered.

An early model of the solar system had Earth in the center with everything revolving around it.

Later on, a new model had the Sun in the center with everything revolving around it.

Limitations of Models

The solar system is too large to be viewed all at once, so models are made to understand it. Many years ago, scientists thought that Earth was the center of the universe and the sky was a blanket that covered the planet.

Later, through observation, it was discovered that the objects you see in the sky are the Sun, the Moon, stars, and other planets. This new model explained the solar system differently. Earth was still the center, but everything else orbited it as shown in **Figure 23.**

Models Change Still later, through more observation, it was discovered that the Sun is the center of the solar system. Earth, along with the other planets, orbits the Sun. In addition, it was discovered that other planets also have moons that orbit them. A new model was developed to show this.

Earlier models of the solar system were not meant to be misleading. Scientists made the best models they could with the information they had. More importantly, their models gave future scientists information to build upon. Models are not necessarily perfect, but they provide a visual tool to learn from.

section 3 review

Summary

Why are models necessary?
- Scientists develop models to help them visualize complex concepts.

Types of Models
- There are three types of models—physical models, computer models, and idea models.

Making Models
- The more information you have when creating a model, the more accurate the model will be.

Using Models
- Models are used to convey important information such as maps and schedules.

Limitations of Models
- Models can be changed over time as new information becomes available.

Self Check

1. **Infer** what types of models can be used to model weather. How are they used to predict weather patterns?
2. **Explain** how models are used in science.
3. **Describe** how consumer product testing services use models to ensure the safety of the final products produced.
4. **Describe** the advantages and limitations of the three types of models.
5. **Think Critically** Explain why some models are better than others for certain situations. Give one example.

Applying Math

6. **Use Proportions** On a map of a state, the scale shows that 1 cm is approximately 5 km. If the distance between two cities is 1.7 cm on the map, how many kilometers separate them?

26 CHAPTER 1 The Nature of Science

ips.msscience.com/self_check_quiz

section 4
Evaluating Scientific Explanation

Believe it or not?

Look at the photo in **Figure 24.** Do you believe what you see? Do you believe everything you read or hear? Think of something that someone told you that you didn't believe. Why didn't you believe it? Chances are you looked at the facts you were given and decided that there wasn't enough proof to make you believe it. What you did was evaluate, or judge the reliability of what you heard. When you hear a statement, you ask the question "How do you know?" If you decide that what you are told is reliable, then you believe it. If it seems unreliable, then you don't believe it.

Critical Thinking When you evaluate something, you use critical thinking. **Critical thinking** means combining what you already know with the new facts that you are given to decide if you should agree with something. You can evaluate an explanation by breaking it down into two parts. First you can look at and evaluate the observations. Based upon what you know, are the observations accurate? Then you can evaluate the inferences—or conclusions made about the observations. Do the conclusions made from the observations make sense?

as you read

What You'll Learn
- **Evaluate** scientific explanations.
- **Evaluate** promotional claims.

Why It's Important
Evaluating scientific claims can help you make better decisions.

Review Vocabulary
prediction: an educated guess as to what is going to happen based on observation

New Vocabulary
- critical thinking
- data

Figure 24 In science, observations and inferences are not always agreed upon by everyone.
Compare *Do you see the same things your classmates see in this photo?*

27

Table 2 Favorite Foods		
People's Preference	Tally	Frequency
Pepperoni pizza	ЖЖ ЖЖ ЖЖ ЖЖ ЖЖ ЖЖ ЖЖ II	37
Hamburgers with ketchup	ЖЖ ЖЖ ЖЖ ЖЖ ЖЖ III	28

Evaluating the Data

A scientific investigation always contains observations—often called **data.** Data are gathered during a scientific investigation and can be recorded in the form of descriptions, tables, graphs, or drawings. When evaluating a scientific claim, you might first look to see whether any data are given. You should be cautious about believing any claim that is not supported by data.

Are the data specific? The data given to back up a claim should be specific. That means they need to be exact. What if your friend tells you that many people like pizza more than they like hamburgers? What else do you need to know before you agree with your friend? You might want to hear about a specific number of people rather than unspecific words like *many* and *more.* You might want to know how many people like pizza more than hamburgers. How many people were asked about which kind of food they liked more? When you are given specific data, a statement is more reliable and you are more likely to believe it. An example of data in the form of a frequency table is shown in **Table 2.** A frequency table shows how many times types of data occur. Scientists must back up their scientific statements with specific data.

Figure 25 These scientists are writing down their observations during their investigation rather than waiting until they are back on land.
Draw Conclusions *Do you think this will increase or decrease the reliability of their data?*

Take Good Notes Scientists must take thorough notes at the time of an investigation, as the scientists shown in **Figure 25** are doing. Important details can be forgotten if you wait several hours or days before you write down your observations. It is also important for you to write down every observation, including ones that you don't expect. Often, great discoveries are made when something unexpected happens in an investigation.

Your Science Journal During this course, you will be keeping a science journal. You will write down what you do and see during your investigations. Your observations should be detailed enough that another person could read what you wrote and repeat the investigation exactly as you performed it. Instead of writing "the stuff changed color," you might say "the clear liquid turned to bright red when I added a drop of food coloring." Detailed observations written down during an investigation are more reliable than sketchy observations written from memory. Practice your observation skills by describing what you see in **Figure 26.**

Can the data be repeated? If your friend told you he could hit a baseball 100 m, but couldn't do it when you were around, you probably wouldn't believe him. Scientists also require repeatable evidence. When a scientist describes an investigation, as shown in **Figure 27,** other scientists should be able to do the investigation and get the same results. The results must be repeatable. When evaluating scientific data, look to see whether other scientists have repeated the data. If not, the data might not be reliable.

Evaluating the Conclusions

When you think about a conclusion that someone has made, you can ask yourself two questions. First, does the conclusion make sense? Second, are there any other possible explanations? Suppose you hear on the radio that your school will be running on a two-hour delay in the morning because of snow. You look outside. The roads are clear of snow. Does the conclusion that snow is the cause for the delay make sense? What else could cause the delay? Maybe it is too foggy or icy for the buses to run. Maybe there is a problem with the school building. The original conclusion is not reliable unless the other possible explanations are proven unlikely.

Figure 26 Detailed observations are important in order to get reliable data.
Observe *Use ten descriptive words to describe what you see happening in this photo.*

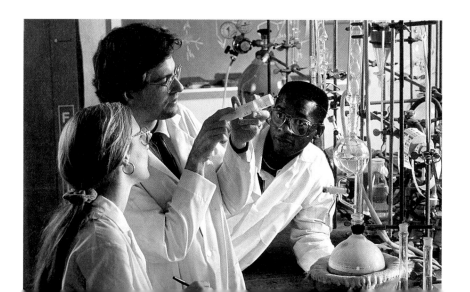

Figure 27 Working together is an important part of science. Several scientists must repeat an experiment and obtain the same results before data are considered reliable.

Evaluating Promotional Materials

Scientific processes are not used only in the laboratory. Suppose you saw an advertisement in the newspaper like the one in **Figure 28.** What would you think? First, you might ask, "Does this make sense?" It seems unbelievable. You would probably want to hear some of the scientific data supporting the claim before you would believe it. How was this claim tested? How is the amount of wrinkling in skin measured? You might also want to know if an independent laboratory repeated the results. An independent laboratory is one that is not related in any way to the company that is selling the product or service. It has nothing to gain from the sales of the product. Results from an independent laboratory usually are more reliable than results from a laboratory paid by the selling company. Advertising materials are designed to get you to buy a product or service. It is important that you carefully evaluate advertising claims and the data that support them before making a quick decision to spend your money.

Figure 28 All material should be read with an analytical mind. **Explain** *what this advertisement means.*

Section 4 Review

Summary

Believe it or not?
- By combining what you already know with new information as it becomes available, you can decide whether something is fact or fiction.
- Explanations should be evaluated by looking at both the observations and the conclusions the explanation is based on.

Evaluating the Data
- It is important to take thorough notes during any investigation.

Evaluating the Conclusions
- In order for a conclusion to be reliable, it must make sense.

Evaluating Promotional Materials
- Independent laboratories test products in order to provide more reliable results.

Self Check

1. **Describe** why it is important that scientific experiments be repeated.
2. **List** what types of scientific claims should be verified.
3. **Explain** how vague claims in advertising can be misleading.
4. **Think Critically** An advertisement on a food package claims it contains Glistain, a safe taste enhancer. Make a list of ten questions you would ask when evaluating this claim.

Applying Skills

5. **Classify** Watch three television commercials and read three magazine advertisements. Record the claims that each advertisement made. Classify each claim as being vague, misleading, reliable, and/or scientific.

ips.msscience.com/self_check_quiz

What is the right answer?

Scientists sometimes develop more than one explanation for observations. Can more than one explanation be correct? Do scientific explanations depend on judgment?

◉ Real-World Question

Can more than one explanation apply to the same observation?

Goals
- **Make a hypothesis** to explain an observation.
- **Construct** a model to support your hypothesis.
- **Refine** your model based on testing.

Materials
cardboard mailing tubes
*empty shoe boxes
length of rope
scissors
*Alternate materials

Safety Precautions

WARNING: *Be careful when punching holes with sharp tools.*

◉ Procedure

1. You will be shown a cardboard tube with four ropes coming out of it, one longer than the others. Your teacher will show you that when any of the three short ropes—A, C, or D—are pulled, the longer rope, B, gets shorter. Pulling on rope B returns the other ropes to their original lengths.
2. Make a hypothesis as to how the teacher's model works.
3. **Sketch** a model of a tube with ropes based on your hypothesis. Using a cardboard tube and two lengths of rope, build a model

according to your design. Test your model by pulling each of the ropes. If it does not perform as planned, modify your hypothesis and your model to make it work like your teacher's model.

◉ Conclude and Apply

1. **Compare** your model with those made by others in your class.
2. Can more than one design give the same result? Can more than one explanation apply to the same observation? Explain.
3. Without opening the tube, can you tell which model is exactly like your teacher's?

Communicating Your Data

Make a display of your working model. Include sketches of your designs. **For more help, refer to the** Science Skill Handbook.

LAB 31

LAB

Identifying Parts of an Investigation

Goals
- **Identify** parts of an experiment.
- **Identify** constants, variables, and controls in the experiment.
- **Graph** the results of the experiment and draw appropriate conclusions.

Materials
description of fertilizer experiment

Real-World Question

Science investigations contain many parts. How can you identify the various parts of an investigation? In addition to variables and constants, many experiments contain a control. A control is one test, or trial, where everything is held constant. A scientist compares the control trial to the other trials. What are the various parts of an experiment to test which fertilizer helps a plant grow best?

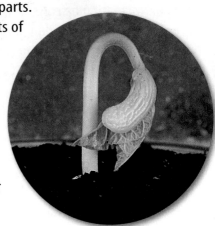

Procedure

1. **Read** the description of the fertilizer experiment.
2. **List** factors that remained constant in the experiment.
3. **Identify** any variables in the experiment.
4. **Identify** the control in the experiment.
5. **Identify** one possible hypothesis that the gardener could have tested in her investigation.
6. **Describe** how the gardener went about testing her hypothesis using different types of fertilizers.
7. **Graph** the data that the gardener collected in a line graph.

A gardener was interested in helping her plants grow faster. When she went to the nursery, she found three fertilizers available for her plants. One of those fertilizers, fertilizer A, was recommended to her. However, she decided to conduct a test to determine which of the three fertilizers, if any, helped her plants grow fastest. The gardener planted four seeds, each in a separate pot. She used the same type of pot and the same type of soil in each pot. She fertilized one seed

32 CHAPTER 1 The Nature of Science

with fertilizer A, one with fertilizer B, and one with fertilizer C. She did not fertilize the fourth seed. She placed the four pots near one another in her garden. She made sure to give each plant the same amount of water each day. She measured the height of the plants each week and recorded her data. After eight weeks of careful observation and record keeping, she had the following table of data.

Plant Height (cm)				
Week	Fertilizer A	Fertilizer B	Fertilizer C	No Fertilizer
1	0	0	0	0
2	2	4	1	1
3	5	8	5	4
4	9	13	8	7
5	14	18	12	10
6	20	24	15	13
7	27	31	19	16
8	35	39	22	20

Analyze Your Data

1. **Describe** the results indicated by your graph. What part of an investigation have you just done?

2. **Infer** Based on the results in the table and your graph, which fertilizer do you think the gardener should use if she wants her plants to grow the fastest? What part of an investigation have you just done?

3. **Define** Suppose the gardener told a friend who also grows these plants about her results. What is this an example of?

Conclude and Apply

1. **Interpret Data** Suppose fertilizer B is much more expensive than fertilizers A and C. Would this affect which fertilizer you think the gardener should buy? Why or why not?

2. **Explain** Does every researcher need the same hypothesis for an experiment? What is a second possible hypothesis for this experiment (different from the one you wrote in step 5 in the Procedure section)?

3. **Explain** if the gardener conducted an adequate test of her hypothesis.

Communicating Your Data

Compare your conclusions with those of other students in your class. **For more help, refer to the** Science Skill Handbook.

TIME SCIENCE AND HISTORY
SCIENCE CAN CHANGE THE COURSE OF HISTORY!

Women in Science

Is your family doctor a man or a woman? To your great-grandparents, such a question would likely have seemed odd. Why? Because 100 years ago, women weren't encouraged to study science as they are today. But that does not mean that there were no female scientists back in your great-grandparents' day. Many women managed to overcome great barriers and made discoveries that changed the world.

Maria Goeppert Mayer

"To my surprise, winning the prize wasn't half as exciting as doing the work itself. That was the fun—seeing it work out." Dr. Maria Goeppert Mayer won the Nobel Prize in Physics in 1963 for her work on the structure of an atom. Her model greatly increased human understanding of atoms, which make up all forms of matter.

Rita Levi-Montalcini

In 1986, Dr. Rita Levi-Montalcini was awarded the Nobel Prize in Medicine for her discovery of growth factors. Growth factors regulate the growth of cells and organs in the body. Because of her work, doctors are better able to understand why tumors form and wounds heal.

Rosalyn Sussman Yalow

"The world cannot afford the loss of the talents of half its people if we are to solve the many problems which beset us," Dr. Rosalyn Sussman Yalow said upon winning the Nobel Prize in Medicine in 1977 for discovering a way to measure tiny substances in the blood, such as hormones and drugs.

Her discovery made it possible for doctors to diagnose problems that they could not detect before.

Research Visit the link to the right to research some recent female Nobel prizewinners in physics, chemistry, and medicine. Write a short biography about their lives. How did their discoveries impact their scientific fields or people in general?

For more information, visit ips.msscience.com/time

chapter 1 Study Guide

Reviewing Main Ideas

Section 1 What is science?

1. Science is a way of learning more about the natural world. It can provide possible explanations for why and how things happen.
2. Systems are made up of structures, cycles, and processes that interact with one another.

Section 2 Science in Action

1. A hypothesis is a possible explanation based on what you know and what you observe.
2. It is important to always follow laboratory safety symbols and to wear and use appropriate gear during an experiment.

Section 3 Models in Science

1. Models are a graphic representation of an object or an event used to communicate ideas; test predictions; and save time, money, and lives.

Section 4 Evaluating Scientific Explanation

1. Reliable data are specific and repeatable by other scientists.
2. In order for a conclusion to be considered reliable, it must make sense and be the most likely explanation.

Visualizing Main Ideas

Copy and complete the following concept map.

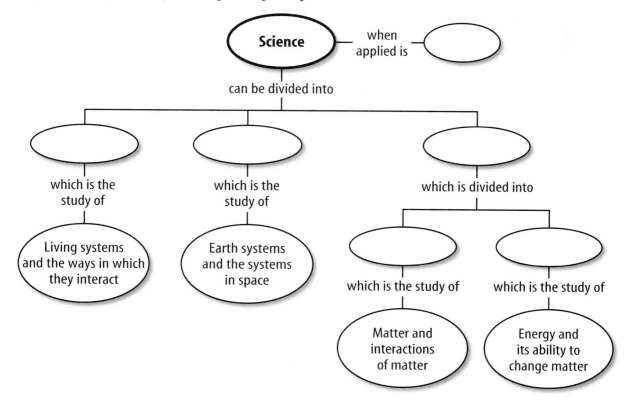

Chapter 1 Review

Using Vocabulary

constant p. 18
controlled experiment p. 18
critical thinking p. 27
data p. 28
dependent variable p. 18
Earth science p. 10
hypothesis p. 14
independent variable p. 18
infer p. 16
life science p. 9
model p. 21
physical science p. 10
science p. 6
scientific law p. 7
scientific theory p. 7
system p. 8
technology p. 11
variable p. 18

Explain the relationship between the words in the following sets.

1. hypothesis—scientific theory
2. constant—variable
3. science—technology
4. science—system
5. Earth science—physical science
6. critical thinking—infer
7. scientific law—observation
8. model—system
9. controlled experiment—variable
10. scientific theory—scientific law

Checking Concepts

Choose the word or phrase that best answers the question.

11. What does it mean to make an inference?
    A) make observations
    B) draw a conclusion
    C) replace
    D) test

12. Which of the following CANNOT protect you from splashing acid?
    A) goggles
    B) apron
    C) fire extinguisher
    D) gloves

13. If the results from your investigation do not support your hypothesis, what should you do?
    A) Should not do anything.
    B) Repeat the investigation until it agrees with the hypothesis.
    C) Modify your hypothesis.
    D) Change your data to fit your hypothesis.

14. Which of the following is NOT an example of a scientific hypothesis?
    A) Earthquakes happen because of stresses along continental plates.
    B) Some animals can detect ultrasound frequencies caused by earthquakes.
    C) Paintings are prettier than sculptures.
    D) Lava takes different forms depending on how it cools.

15. Using a computer to make a three-dimensional picture of a building is a type of which of the following?
    A) model
    B) hypothesis
    C) constant
    D) variable

16. Which of the following increases the reliability of a scientific explanation?
    A) vague statements
    B) notes taken after an investigation
    C) repeatable data
    D) several likely explanations

17. Which is an example of technology?
    A) a squirt bottle
    B) a poem
    C) a cat
    D) physical science

18. What explains something that takes place in the natural world?
    A) scientific law
    B) technology
    C) scientific theory
    D) experiments

19. An airplane model is an example of what type of model?
    A) physical
    B) computer
    C) idea
    D) mental

chapter 1 Review

Thinking Critically

20. **Draw Conclusions** When scientists study how well new medicines work, one group of patients receives the medicine while a second group does not. Why?

21. **Predict** How is using a rock hammer an example of technology?

22. **Compare and Contrast** How are scientific theories and scientific laws similar? How are they different?

Use the table below to answer question 23.

Hardness	
Object	Mohs Scale
copper	3.5
diamond	10
fingernail	2.5
glass	5.5
quartz	7
steel file	6.5

23. **Use Tables** Mohs hardness scale measures how easily an object can be scratched. The higher the number, the harder the material is. Use the table above to identify which material is the hardest and which is the softest.

24. **Make Operational Definitions** How does a scientific law differ from a state law? Give some examples of both types of laws.

25. **Infer** Why it is important to record and measure data accurately during an experiment?

26. **Predict** the quickest way to get to school in the morning. List some ways you could test your prediction.

Performance Activities

27. **Hypothesize** Using a basketball and a tennis ball, make a hypothesis about the number of times each ball will bounce when it hits the ground. Drop each ball from shoulder height five times, recording the number of bounces in a table. Which ball bounced more? Make a hypothesis to explain why.

28. **Observe** Pour some water in a small dish and sprinkle some pepper on top. Notice how the pepper floats on the water. Now add a few drops of liquid soap to the water. Write down your observations as you watch what happens to the pepper.

Applying Math

Use the illustration below to answer question 29.

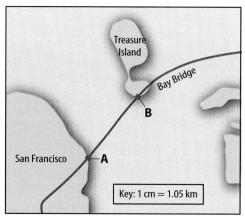

29. **Use Proportions** The map above shows the distance between two points. The scale shows that 1 cm is approximately 1.05 km. What is the approximate distance between Point A and Point B?

Chapter 1 Standardized Test Practice

Part 1 Multiple Choice

Record your answers on the answer sheet provided by your teacher or on a sheet of paper.

1. What is a rule describing a pattern in nature called?
   A. possible explanation
   B. scientific law
   C. scientific theory
   D. technology

Use the illustration below to answer questions 2–3.

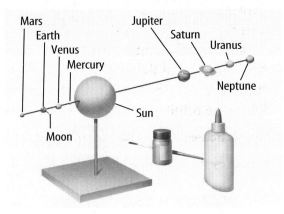

2. The model of the solar system best represents which kind of scientific model?
   A. idea
   B. computer
   C. physical
   D. realistic

3. All of the following are represented in the model EXCEPT which of the following?
   A. the sun
   B. the moon
   C. planets
   D. stars

4. Which of the following is not an example of a model?
   A. CD
   B. map
   C. recipe
   D. drawing

> **Test-Taking Tip**
>
> **Practice Practice** Remember that test taking skills can improve with practice. If possible, take at least one practice test and familiarize yourself with the test format and instructions.

5. Which of the following questions can science NOT answer?
   A. Why do the leaves on trees change colors in the fall?
   B. Why do bears hibernate in the winter?
   C. Where do waves in the ocean form?
   D. What is the most popular book?

6. What is it called when you combine what you already know with new facts?
   A. estimate
   B. hypothesis
   C. inference
   D. critical thinking

7. What are the variables that do not change in an experiment called?
   A. independent variables
   B. dependent variables
   C. constants
   D. inferences

8. An educated guess based on what you know and what you observe is called which of the following?
   A. prediction.
   B. hypothesis.
   C. conclusion.
   D. data.

Use the photo below to answer question 9.

9. What type of scientist could the person above be classified as?
   A. life scientist
   B. physical scientist
   C. Earth scientist
   D. medical doctor

Standardized Test Practice

Part 2 — Short Response/Grid In

Record your answers on the answer sheet provided by your teacher or on a sheet of paper.

Use the photo below to answer questions 10 and 11.

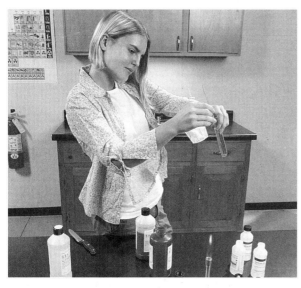

10. Look at the photo above and write down your immediate observations.

11. What safety precautions might this student want to take?

12. Explain why science can only provide possible explanations for occurrences in nature.

13. In class it is sometimes not common for students to share their answers on a test. Why is it important for scientists to share information?

14. Compare and contrast observation and inference.

15. Explain the relationship between science and technology.

16. What steps do scientists follow when investigating a problem?

17. List the three branches of science and give examples of questions that they ask.

18. What is the importance of scientific models?

Part 3 — Open Ended

Record your answers on a sheet of paper.

19. You want to know whether plants grow faster if there is music playing in their environment. How would you conduct this experiment? Be sure to identify the independent and dependent variables, and the constants.

20. Many outdoor clothing products are coated in a special waterproofing agent to protect the material from rain and snow. The manufacturers of the waterproofing agent hire independent field-testers to use their product in the field before marketing it to the public. Why would you want to know the results of the field-testers tests?

Use the illustrations below to answer questions 21–23.

21. What are the above drawings outlining?

22. Body systems interact with one another in order to function. What would happen if one system failed?

23. What is the importance of systems in science?

24. Make a frequency table from the following data. Make two observations about the data. 15 students prefer cold pizza for lunch; 10 students enjoy peanut butter with jelly; 3 students bring ham and cheese; and 5 students eat hot dogs and chips.

chapter 2

Measurement

The BIG Idea

Measurement is a description using numbers.

SECTION 1
Description and Measurement
Main Idea Measurements of a single item might vary depending on the accuracy of the measuring device or the precision of the measurements.

SECTION 2
SI Units
Main Idea The SI system of units provides a worldwide standard of measurement.

SECTION 3
Drawings, Tables, and Graphs
Main Idea Data can be displayed visually using drawings, tables, and graphs.

The Checkered Flag

Race car drivers win or lose by tenths of a second. The driver has to monitor his fuel usage, speed, and oil temperature in order to win the race. In this chapter, you will learn how scientists measure things like distance, time, volume, and temperature.

Science Journal As a member of the pit crew, how can you determine the miles per gallon the car uses? Write in your Science Journal how you would calculate this.

Start-Up Activities

Measuring Accurately

You make measurements every day. If you want to communicate those measurements to others, how can you be sure that they will understand exactly what you mean? Using vague words without units won't work. Do the lab below to see how confusion can result from using measurements that aren't standard.

1. As a class, choose six objects to measure in your classroom.
2. Measure each object using the width of your hand and write your measurements in your Science Journal.
3. Compare your measurements to those of your classmates.
4. **Think Critically** Describe in your Science Journal why it is better to switch from using hands to using units of measurement that are the same all the time.

 **Preview this chapter's content and activities at** ips.msscience.com

 **Measurement** Make the following Foldable to help you organize information about measurements.

STEP 1 Fold a sheet of paper in half two times lengthwise. Unfold.

STEP 2 Fold the paper widthwise in equal thirds and then in half.

STEP 3 Unfold, lay the paper lengthwise, and draw lines along the folds. Label your table as shown.

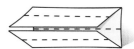

Estimates Before you read the chapter, select objects to measure and estimate their measurements. As you read the chapter, complete the table.

Get Ready to Read

Identify the Main Idea

① Learn It! Main ideas are the most important ideas in a paragraph, section, or chapter. Supporting details are facts or examples that explain the main idea. Understanding the main idea allows you to grasp the whole picture.

② Practice It! Read the following paragraph. Draw a graphic organizer like the one below to show the main idea and supporting details.

> What happens when you want to know the size of an object but you can't measure it? Perhaps it is too large to measure or you don't have a ruler handy. **Estimation** can help you make a rough measurement of an object. When you estimate, you can use your knowledge of the size of something familiar to estimate the size of a new object. Estimation is a skill based on previous experience and is useful when exact numbers are not required.
>
> —*from page 43*

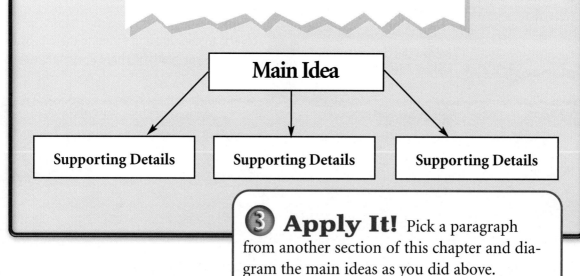

③ Apply It! Pick a paragraph from another section of this chapter and diagram the main ideas as you did above.

Target Your Reading

Reading Tip

The main idea is often the first sentence in a paragraph but not always.

Use this to focus on the main ideas as you read the chapter.

① **Before you read** the chapter, respond to the statements below on your worksheet or on a numbered sheet of paper.
- Write an **A** if you **agree** with the statement.
- Write a **D** if you **disagree** with the statement.

② **After you read** the chapter, look back to this page to see if you've changed your mind about any of the statements.
- If any of your answers changed, explain why.
- Change any false statements into true statements.
- Use your revised statements as a study guide.

Science Online
Print out a worksheet of this page at ips.msscience.com

Before You Read A or D		Statement	After You Read A or D
	1	Estimation is a guess made without any additional information.	
	2	A single measurement can't be precise.	
	3	A group of measurements that is precise must also be accurate.	
	4	Units of volume are SI base units.	
	5	The volume of a rock can be measured by multiplying length × width × height.	
	6	Mass and weight are the same thing.	
	7	A table and a graph can represent the same data in a different format.	
	8	A circle graph represents parts of a whole.	
	9	The x- and y-axes on a graph must always start at zero.	

42 B

section 1
Description and Measurement

as you read

What You'll Learn
- **Determine** how reasonable a measurement is by estimating.
- **Identify** and use the rules for rounding a number.
- **Distinguish** between precision and accuracy in measurements.

Why It's Important
Measurement helps you communicate information and ideas.

Review Vocabulary
description: an explanation of an observation

New Vocabulary
- measurement
- estimation
- precision
- accuracy

Measurement

How would you describe what you are wearing today? You might start with the colors of your outfit and perhaps you would even describe the style. Then you might mention sizes—size 7 shoes, size 14 shirt. Every day you are surrounded by numbers. **Measurement** is a way to describe the world with numbers. It answers questions such as how much, how long, or how far. Measurement can describe the amount of milk in a carton, the cost of a new compact disc, or the distance between your home and your school. It also can describe the volume of water in a swimming pool, the mass of an atom, or how fast a penguin's heart pumps blood.

The circular device in **Figure 1** is designed to measure the performance of an automobile in a crash test. Engineers use this information to design safer vehicles. In scientific endeavors, it is important that scientists rely on measurements instead of the opinions of individuals. You would not know how safe the automobile is if this researcher turned in a report that said, "Vehicle did fairly well in head-on collision when traveling at a moderate speed." What does "fairly well" mean? What is a "moderate speed"?

Figure 1 This device measures the range of motion of a seat-belted mannequin in a simulated accident.

Figure 2 Accurate measurement of distance and time is important for competitive sports like track and field.
Infer Why wouldn't a clock that measured in minutes be precise enough for this race?

Describing Events Measurement also can describe events such as the one shown in **Figure 2.** In the 1956 summer Olympics, sprinter Betty Cuthbert of Australia came in first in the women's 200-m dash. She ran the race in 23.4 s. In the 2000 summer Olympics, Marion Jones of the United States won the 100-m dash in a time of 10.75 s. In this example, measurements convey information about the year of the race, its length, the finishing order, and the time. Information about who competed and in what event are not measurements but help describe the event completely.

Estimation

What happens when you want to know the size of an object but you can't measure it? Perhaps it is too large to measure or you don't have a ruler handy. **Estimation** can help you make a rough measurement of an object. When you estimate, you can use your knowledge of the size of something familiar to estimate the size of a new object. Estimation is a skill based on previous experience and is useful when exact numbers are not required. Estimation is a valuable skill that improves with experience, practice, and understanding.

Reading Check *When should you not estimate a value?*

How practical is the skill of estimation? In many instances, estimation is used on a daily basis. A caterer prepares for each night's crowd based on an estimation of how many will order each entree. A chef makes her prize-winning chili. She doesn't measure the cumin; she adds "just that much." Firefighters estimate how much hose to pull off the truck when they arrive at a burning building.

Precision and Accuracy
A pharmacist has a very important job: making sure that patients receive the right medication at the correct dosage. Any error in dosage or type of pill could harm the patient. Explain how precision and accuracy play a role in the pharmacist's job. If a patient receives the wrong medication or an extra pill, how could that affect their health? Research some other careers that rely on precision and accuracy. How could errors in a measurement affect the professional's finished product?

SECTION 1 Description and Measurement

Figure 3 This student is about 1.5 m tall. **Estimate** *the height of the tree in the photo.*

Measuring Temperature

Procedure

1. Fill a **400-mL beaker** with **crushed ice**. Add enough **cold water** to fill the beaker.
2. Make three measurements of the temperature of the ice water using a **computer temperature probe.** Remove the computer probe and dry it with a **paper towel.** Record the measurement in your **Science Journal.** Allow the probe to warm to room temperature between each measurement.
3. Repeat step two using an **alcohol thermometer.**

Analysis

1. Average each set of measurements.
2. Which measuring device is more precise? Explain. Can you determine which is more accurate? How?

Using Estimation You can use comparisons to estimate measurements. For example, the tree in **Figure 3** is too tall to measure easily, but because you know the height of the student next to the tree, you can estimate the height of the tree. When you estimate, you often use the word *about*. For example, doorknobs are about 1 m above the floor, a sack of flour has a mass of about 2 kg, and you can walk about 5 km in an hour.

Estimation also is used to check that an answer is reasonable. Suppose you calculate your friend's running speed as 47 m/s. You are familiar with how long a second is and how long a meter is. Think about it. Can your friend really run a 50-m dash in 1 s? Estimation tells you that 47 m/s is unrealistically fast and you need to check your work.

Precision and Accuracy

One way to evaluate measurements is to determine whether they are precise. **Precision** is a description of how close measurements are to each other. Suppose you measure the distance between your home and your school five times with an odometer. Each time, you determine the distance to be 2.7 km. Suppose a friend repeated the measurements and measured 2.7 km on two days, 2.8 km on two days, and 2.6 km on the fifth day. Because your measurements were closer to each other than your friend's measurements, yours were more precise. The term *precision* also is used when discussing the number of decimal places a measuring device can measure. A clock with a second hand is considered more precise than one with only an hour hand.

44 CHAPTER 2 Measurement

Degrees of Precision The timing for Olympic events has become more precise over the years. Events that were measured in tenths of a second 100 years ago are measured to the hundredth of a second today. Today's measuring devices are more precise. **Figure 4** shows an example of measurements of time with varying degrees of precision.

Accuracy When you compare a measurement to the real, actual, or accepted value, you are describing **accuracy.** A watch with a second hand is more precise than one with only an hour hand, but if it is not properly set, the readings could be off by an hour or more. Therefore, the watch is not accurate. However, measurements of 1.03 m, 1.04 m, and 1.06 m compared to an actual value of 1.05 m are accurate, but not precise. **Figure 5** illustrates the difference between precision and accuracy.

Reading Check *What is the difference between precision and accuracy?*

Figure 4 Each of these clocks provides a different level of precision. **Infer** *which of the three you could use to be sure to make the 3:35 bus.*

For centuries, analog clocks—the kind with a face—were the standard.

Before the invention of clocks, as they are known today, a sundial was used. As the Sun passes through the sky, a shadow moves around the dial.

Digital clocks are now as common as analog ones.

SECTION 1 Description and Measurement **45**

NATIONAL GEOGRAPHIC VISUALIZING PRECISION AND ACCURACY

Figure 5

From golf to gymnastics, many sports require precision and accuracy. Archery—a sport that involves shooting arrows into a target—clearly shows the relationship between these two factors. An archer must be accurate enough to hit the bull's-eye and precise enough to do it repeatedly.

A The archer who shot these arrows is neither accurate nor precise—the arrows are scattered all around the target.

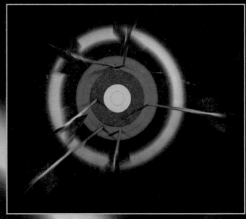

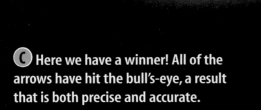

C Here we have a winner! All of the arrows have hit the bull's-eye, a result that is both precise and accurate.

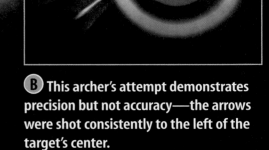

B This archer's attempt demonstrates precision but not accuracy—the arrows were shot consistently to the left of the target's center.

INTEGRATE Health

Precision and accuracy are important in many medical procedures. One of these procedures is the delivery of radiation in the treatment of cancerous tumors. Because radiation damages cells, it is important to limit the radiation to only the cancerous cells that are to be destroyed. A technique called Stereotactic Radiotherapy (SRT) allows doctors to be accurate and precise in delivering radiation to areas of the brain. The patient makes an impression of his or her teeth on a bite plate that is then attached to the radiation machine. This same bite plate is used for every treatment to position the patient precisely the same way each time. A CAT scan locates the tumor in relation to the bite plate, and the doctors can pinpoint with accuracy and precision where the radiation should go.

Rounding a Measurement Not all measurements have to be made with instruments that measure with great precision like the scale in **Figure 6.** Suppose you need to measure the length of the sidewalk outside your school. You could measure it to the nearest millimeter. However, you probably would need to know the length only to the nearest meter or tenth of a meter. So, if you found that the length was 135.841 m, you could round off that number to the nearest tenth of a meter and still be considered accurate. How would you round this number? To round a given value, follow these steps:

1. Look at the digit to the right of the place being rounded to.
   - If the digit to the right is 0, 1, 2, 3, or 4, the digit being rounded to remains the same.
   - If the digit to the right is 5, 6, 7, 8, or 9, the digit being rounded to increases by one.
2. The digits to the right of the digit being rounded to are deleted if they are also to the right of a decimal. If they are to the left of a decimal, they are changed to zeros.

Look back at the sidewalk example. If you want to round the sidewalk length of 135.841 to the tenths place, you look at the digit to the right of the 8. Because that digit is a 4, you keep the 8 and round it off to 135.8 m. If you want to round to the ones place, you look at the digit to the right of the 5. In this case you have an 8, so you round up, changing the 5 to a 6, and your answer is 136 m.

Science Online

Topic: Measurement
Visit ips.msscience.com for Web links to information about the importance of accuracy and precision in the medical field.

Activity Research a topic of interest on the Internet and present the topic and numeric data to your class. How might your classmates' understanding of the topic be affected if you presented crucial information inaccurately?

Figure 6 This laboratory scale measures to the nearest hundredth of a gram.

SECTION 1 Description and Measurement

Precision and Number of Digits When might you need to round a number? Suppose you want to divide a 2-L bottle of soft drink equally among seven people. When you divide 2 by 7, your calculator display reads as shown in **Figure 7.** Will you measure exactly 0.285 714 285 L for each person? No. All you need to know is that each person gets about 0.3 L of soft drink.

Using Precision and Significant Digits The number of digits that truly reflect the precision of a number are called the significant digits or significant figures. They are figured as follows:

- Digits other than zero are always significant.
- Final zeros after a decimal point (6.545 600 g) are significant.
- Zeros between any other digits (507.0301 g) are significant.
- Initial zeros (0.000 2030 g) are NOT significant.
- Zeros in a whole number (1650) may or may not be significant.
- A number obtained by counting instead of measuring, such as the number of people in a room or the number of meters in a kilometer, has infinite significant figures.

Applying Math — Rounding

ROUNDED VALUES The mass of one object is 6.941 g. The mass of a second object is 20.180 g. You need to know these values only to the nearest whole number to solve a problem. What are the rounded values?

Solution

1 *This is what you know:*
- mass of first object = 6.941 g
- mass of second object = 20.180 g

2 *This is what you need to find out:*
- the number to the right of the one's place
- first object: 9; second object: 1

3 *This is the procedure you need to use:*
digits 0, 1, 2, 3, 4 remain the same
for digits 5, 6, 7, 8, 9, round up

4 *Check your answer:*
- first object: 9 makes the 6 round up = 7
- second object: 1 makes the 0 remain the same = 20

Practice Problems

1. What are the rounded masses of the objects to the nearest tenth of a unit?
2. Round the following numbers: 25.643 to the ones place, 3.429 to the tenths place, 5.982 to the hundredths place, and 9.8210 to the tenths place.

For more practice, visit ips.msscience.com/math_practice

Following the Rules In the soft drink example you have an exact number, seven, for the number of people. This number has infinite significant digits. You also have the number two, for how many liters of soft drink you have. This has only one significant digit.

There are also rules to follow when deciding the number of significant digits in the answer to a calculation. They depend on what kind of calculation you are doing.

- For multiplication and division, you determine the number of significant digits in each number in your problem. The significant digits of your answer are determined by the number with fewer digits.

    $6.14 \times 5.6 = \boxed{34}.384$
    3 digits   2 digits   2 digits

- For addition and subtraction, you determine the place value of each number in your problem. The significant digits of the answer are determined by the number that is least precise.

    $\quad\;\; 6.14$ to the hundredths
    $\underline{+\;\; 5.6\;\;}$ to the tenths
    $\;\;\boxed{11.7}\,4$ to the tenths

In the soft drink example you are dividing and the number of significant digits is determined by the amount of soft drink, 2 L. There is one significant digit there; therefore, the amount of soft drink each person gets is rounded to 0.3 L.

Figure 7 Sometimes considering the size of each digit will help you realize they are unneeded. In this calculation, the seven ten-thousandths of a liter represents just a few drops of soft drink.

section 1 review

Summary

Measurement
- Measurement is used to answer questions such as how much, how long, or how far.

Estimation
- When making an estimate, rely on previous knowledge to make an educated guess about the size of an object.

Precision and Accuracy
- Precision is the ability to remain consistent. Accuracy compares a measurement to the real value of an object.
- Significant digits affect precision when calculating an answer and are determined by rules based on calculation.

Self Check

1. **Estimate** the distance between your desk and your teacher's desk. Explain the method you used.
2. **Infer** John's puppy has chewed on his ruler. Will John's measurements be accurate or precise? Why?
3. **Think Critically** Would the sum of 5.7 cm and 6.2 cm need to be rounded? Why or why not? Would the sum of 3.28 cm and 4.1 cm need to be rounded? Why or why not?

Applying Math

4. **Calculate** Perform the following calculations and express the answer using the correct number of significant digits: 42.35 + 214; 225/12. **For more help, refer to the Math Skill Handbook.**

section 2

SI Units

as you read

What You'll Learn
- Identify the purpose of SI.
- Identify the SI units of length, volume, mass, temperature, time, and rate.

Why It's Important
The SI system is used throughout the world, allowing you to measure quantities in the exact same way as other students around the world.

Review Vocabulary
variable: factors that can be changed in an experiment

New Vocabulary
- SI
- meter
- volume
- mass
- kilogram
- weight
- kelvin
- rate

The International System

Can you imagine how confusing it would be if people used different measuring systems? Sharing data would be complicated. To avoid confusion, scientists established the International System of Units, or **SI**. It was designed to provide a worldwide standard of physical measurement for science, industry, and commerce. SI base units are shown in **Table 1**. All other SI units can be created from these seven units.

✓ **Reading Check** *Why was SI established?*

The SI units are related by multiples of ten. Any SI unit can be converted to a smaller or larger SI unit by multiplying by a power of 10. For example, to rewrite a kilogram measurement in grams, you multiply by 1,000.

Ex. 5.67 kg × 1000 = 5670 grams

The new unit is renamed by changing the prefix, as shown in **Table 2**. For example, one millionth of a meter is one *micro*meter. One thousand grams is one *kilo*gram. **Table 3** shows some common objects and their measurements in SI units.

Table 1 SI Base Units		
Quantity	Unit	Symbol
length	meter	m
mass	kilogram	kg
temperature	kelvin	K
time	second	s
electric current	ampere	A
amount of substance	mole	mol
intensity of light	candela	cd

Table 2 SI Prefixes	
Prefix	Multiplier
giga-	1,000,000,000
mega-	1,000,000
kilo-	1,000
hecto-	100
deka-	10
[unit]	1
deci-	0.1
centi-	0.01
milli-	0.001
micro-	0.000 001
nano-	0.000 000 001

Length

Length is defined as the distance between two points. Lengths measured with different tools can describe a range of things from the distance from Earth to Mars to the thickness of a human hair. In your laboratory activities, you usually will measure length with a metric ruler or meterstick.

The **meter** (m) is the SI unit of length. One meter is about the length of a baseball bat. The size of a room or the dimensions of a building would be measured in meters. For example, the height of the Washington Monument in Washington, D.C. is 169 m.

Smaller objects can be measured in centimeters (cm) or millimeters (mm). The length of your textbook or pencil would be measured in centimeters. A twenty-dollar bill is 15.5 cm long. You would use millimeters to measure the width of the words on this page. To measure the length of small things such as blood cells, bacteria, or viruses, scientists use micrometers (millionths of a meter) and nanometers (billionths of a meter).

A Long Way Sometimes people need to measure long distances, such as the distance a migrating bird travels or the distance from Earth to the Moon. To measure such lengths, you use kilometers. Kilometers might be most familiar to you as the distance traveled in a car or the measure of a long-distance race, as shown in **Figure 8.** The course of a marathon is measured carefully so that the competitors run 42.2 km. When you drive from New York to Los Angeles, you cover 4,501 km.

Figure 8 These runners have just completed a 10-kilometer race—known as a 10K.
Estimate *how many kilometers is the distance between your home and your school.*

Measurement Accuracy
How important are accurate measurements? In 1999, the *Mars Climate Orbiter* disappeared as it was to begin orbiting Mars. NASA later discovered that a unit system error caused the flight path to be incorrect and the orbiter to be lost. Research the error and determine what systems of units were involved. How can using two systems of units cause errors?

Table 3 Common Objects in SI Measurements		
Object	Type of Measurement	Measurement
can of soft drink	volume	355 mL
bag of potatoes	mass	4.5 kg
fluorescent tube	length	1.2 m
refrigerator	temperature	276 K

Figure 9 A cubic meter equals the volume of a cube 1 m by 1 m by 1 m. **Infer** how many cubic centimeters are in a cubic meter.

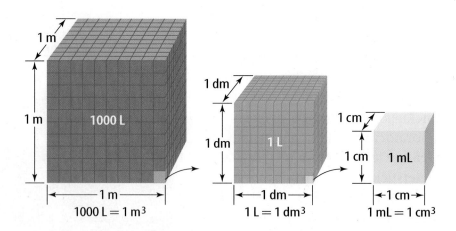

Volume

The amount of space an object occupies is its **volume.** Units of volume are created by multiplying units of length. Therefore, they are not base units and are not listed in **Table 1.** The cubic meter (m^3), shown in **Figure 9,** is the SI unit of volume. You can measure smaller volumes with the cubic centimeter (cm^3 or cc). To find the volume of a square or rectangular object, such as a brick or your textbook, measure its length, width, and height and multiply them together. What is the volume of a compact disc case?

You are probably familiar with a 2-L bottle. A liter is a measurement of liquid volume. A cube 10 cm by 10 cm by 10 cm holds 1 L (1,000 cm^3) of water. A cube 1 cm on each side holds 1 mL (1 cm^3) of water.

Volume by Immersion Not all objects have an even, regular shape. How can you find the volume of something irregular like a rock or a piece of metal?

Have you ever added ice cubes to a nearly full glass of water only to have the water overflow? Why did the water overflow? Did you suddenly have more water? The volume of water did not increase at all, but the water was displaced when the ice cubes were added. Each ice cube takes up space or has volume. The difference in the volume of water before and after the addition of the ice cubes equals the volume of the ice cubes that are under the surface of the water.

The ice cubes took up space and caused the total volume in the glass to increase. The volume of an irregular object can be measured the same way. Start with a known volume of water and drop in, or immerse, the object. The increase in the volume of water is equal to the volume of the object.

Measuring Volume

Procedure
1. Fill a plastic or glass **liquid measuring cup** until half full with **water.** Measure the volume.
2. Find an **object,** such as a rock, that will fit in your measuring cup.
3. Carefully lower the object into the water. If it floats, push it just under the surface with a **pencil.**
4. Record in your **Science Journal** the new volume of the water.

Analysis
1. How much space does the object occupy?
2. If 1 mL of water occupies exactly 1 cm^3 of space, what is the volume of the object in cm^3?

Figure 10 A triple beam balance compares an unknown mass to known masses.

Mass

The **mass** of an object measures the amount of matter in the object. The **kilogram** (kg) is the SI unit for mass. One liter of water has a mass of about 1 kg. Smaller masses are measured in grams (g). One gram is about the mass of a large paper clip.

You can determine mass with a triple-beam balance, shown in **Figure 10.** The balance compares an object to a known mass. It is balanced when the known standard mass of the slides on the balance is equal to the object on the pan.

Why use the word *mass* instead of *weight?* Weight and mass are not the same. Mass depends only on the amount of matter in an object. If you ride in an elevator in the morning and then ride in the space shuttle later that afternoon, your mass is the same. Mass does not change when only your location changes.

Weight Weight is a measurement of force. The SI unit for weight is the newton (N). Weight depends on gravity, which can change depending on where the object is located. A spring scale measures how a planet's gravitational force pulls on objects. Several spring scales are shown in **Figure 11.**

If you were to travel to other planets, your weight would change, even though you would still be the same size and have the same mass. This is because gravitational force is different on each planet. If you could take your bathroom scale, which uses a spring, to each of the planets in this solar system, you would find that you weigh much less on Mars and much more on Jupiter. A mass of 75 pounds, or 34 kg, on Earth is a weight of 332 N. On Mars, the same mass is 126 N, and on Jupiter it is 782 N.

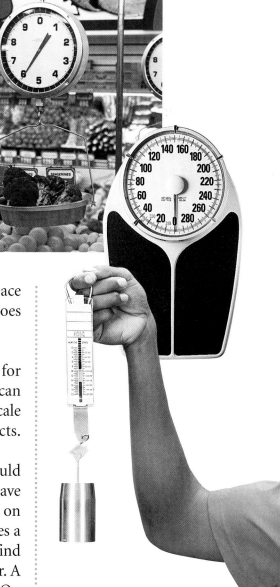

Figure 11 A spring scale measures an object's weight by how much it stretches a spring.

Reading Check *What does weight measure?*

SECTION 2 SI Units

Figure 12 The kelvin scale starts at 0 K. In theory, 0 K is the coldest temperature possible in nature.

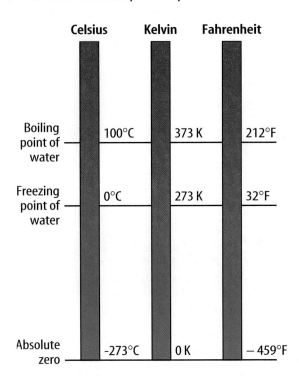

Temperature

The physical property of temperature is related to how hot or cold an object is. Temperature is a measure of the kinetic energy, or energy of motion, of the particles that make up matter.

Temperature is measured in SI with the **kelvin** (K) scale. The Fahrenheit and Celsius temperature scales are the two most common scales used on thermometers and in classroom laboratories. These two scales do not start at zero, as shown in **Figure 12**. A 1 K difference in temperature is the same as a 1°C difference in temperature.

Time and Rates

Time is the interval between two events. The SI unit of time is the second (s). Time also is measured in hours (h). Can you imagine hearing that a marathon was run in 7,620 s instead of 2 h and 7 min?

A **rate** is the amount of change of one measurement in a given amount of time. One rate you are familiar with is speed, which is the distance traveled in a given time. Speeds often are measured in kilometers per hour (km/h).

The unit that is changing does not necessarily have to be an SI unit. For example, you can measure the number of cars that pass through an intersection per hour in cars/h.

section 2 review

Summary

The International System
- The International System of Units, SI, was established to provide a standard of physical measurement and to reduce international confusion when comparing measurements.

Measurement
- Length is the distance between two points.
- Volume is the amount of space an object occupies.
- To calculate volume, multiply length by width by height.
- The amount of matter in an object is its mass.
- Weight is determined by gravitational pull.
- Celsius temperature scales are more common in laboratories than kelvin scales.

Self Check

1. **Describe** a situation in which different units of measure could cause confusion.
2. **Define** what type of quantity the cubic meter measures.
3. **Explain** how you would change a measurement in centimeters to kilometers.
4. **Identify** what SI unit replaces the pound. What does this measure?
5. **Think Critically** How would you find the mass of a metal cube?

Applying Math

6. **Measure** A block of wood is 0.2 m by 0.1 m by 0.5 m. Find its dimensions in centimeters. Then find its volume in cubic centimeters.

54 CHAPTER 2 Measurement

 ips.msscience.com/self_check_quiz

Scale Drawing

A scale drawing is used to represent something that is too large or too small to be drawn at its actual size. Blueprints for a house are a good example of a scale drawing.

Real-World Question

How can you represent your classroom accurately in a scale drawing?

Goals
- **Measure** using SI.
- **Make** a data table.
- **Calculate** new measurements.
- **Make** an accurate scale drawing.

Materials
1-cm graph paper
pencil
metric ruler
meterstick

Procedure

1. Use your meterstick to measure the length and width of your classroom. Note the locations and sizes of doors and windows.
2. **Record** the lengths of each item in a data table similar to the one below.
3. Use a scale of 2 cm = 1 m to calculate the lengths to be used in the drawing. Record them in your data table.
4. **Draw** the floor plan. Include the scale.

Room Dimensions

Part of Room	Distance in Room (m)	Distance on Drawing (cm)
Do not write in this book.		

Conclude and Apply

1. How did you calculate the lengths to be used on your drawing? Did you put a scale on your drawing?
2. **Infer** what your scale drawing would look like if you chose a different scale?
3. **Sketch** your room at home, estimating the distances. Compare this sketch to your scale drawing of the classroom. When would you use each type of illustration?
4. What measuring tool simplifies this task?

Communicating Your Data

Measure your room at home and compare it to the estimates on your sketch. Explain to someone at home what you did and how well you estimated the measurements. **For more help, refer to the** Science Skill Handbook.

LAB 55

section 3
Drawings, Tables, and Graphs

as you read

What You'll Learn
- **Describe** how to use pictures and tables to give information.
- **Identify** and use three types of graphs.
- **Distinguish** the correct use of each type of graph.

Why It's Important
Illustrations, tables, and graphs help you communicate data about the world around you in an organized and efficient way.

Review Vocabulary
model: a representation of an object or event used as a tool for understanding the natural world

New Vocabulary
- table
- graph
- line graph
- bar graph
- circle graph

Scientific Illustrations

Most science books include pictures. Photographs and drawings model and illustrate ideas and sometimes make new information clearer than written text can. For example, a drawing of an airplane engine shows how all the parts fit together much better than several pages of text could describe it.

Drawings A drawing is sometimes the best choice to show details. For example, a canyon cut through red rock reveals many rock layers. If the layers are all shades of red, a drawing can show exactly where the lines between the layers are. A drawing can emphasize only the things that are necessary to show.

A drawing also can show things you can't see. You can't see the entire solar system, but drawings show you what it looks like. Also, you can make quick sketches to help model problems. For example, you could draw the outline of two continents to show how they might have fit together at one time.

Drawings can show hidden things, as well. A drawing can show the details of the water cycle, as in **Figure 13.** Architects use drawings to show what the inside of a building will look like. Biologists use drawings to show where the nerves in your arm are found.

Figure 13 This drawing shows details of the water cycle that can't be seen in a photograph.

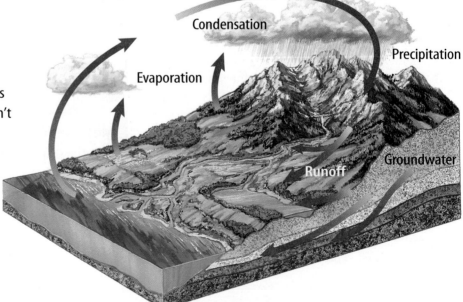

56 CHAPTER 2 Measurement

Photographs A still photograph shows an object exactly as it is at a single moment in time. Movies show how an object moves and can be slowed down or sped up to show interesting features. In your schoolwork, you might use photographs in a report. For example, you could show the different types of trees in your neighborhood for a report on ecology.

Tables and Graphs

Everyone who deals with numbers and compares measurements needs an organized way to collect and display data. A **table** displays information in rows and columns so that it is easier to read and understand, as seen in **Table 4.** The data in the table could be presented in a paragraph, but it would be harder to pick out the facts or make comparisons.

A **graph** is used to collect, organize, and summarize data in a visual way. The relationships between the data often are seen more clearly when shown in a graph. Three common types of graphs are line, bar, and circle graphs.

Line Graph A **line graph** shows the relationship between two variables. A variable is something that can change, or vary, such as the temperature of a liquid or the number of people in a race. Both variables in a line graph must be numbers. An example of a line graph is shown in **Figure 14.** One variable is shown on the horizontal axis, or *x*-axis, of the graph. The other variable is placed along the vertical axis, or *y*-axis. A line on the graph shows the relationship between the two variables.

Table 4 Endangered Animal Species in the United States

Year	Number of Endangered Animal Species
1984	192
1986	213
1988	245
1990	263
1992	284
1994	321
1996	324
1998	357
2000	379
2002	389

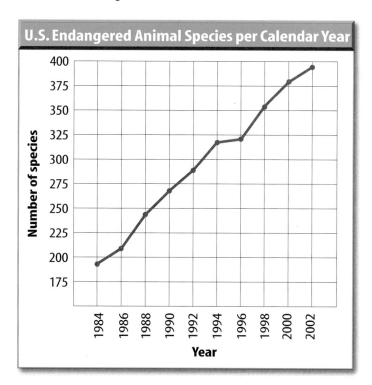

Figure 14 To find the number of endangered animal species in 1992, find that year on the *x*-axis and see what number corresponds to it on the *y*-axis.
Interpret Data *How many species were endangered in 1998?*

SECTION 3 Drawings, Tables, and Graphs **57**

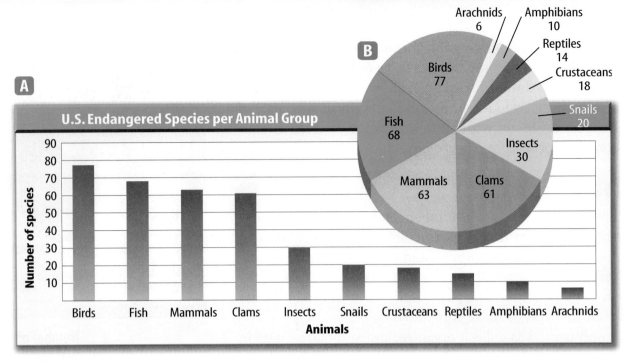

Figure 15 **A** Bar graphs allow you to picture the results easily. **B** On this circle graph, you can see what part of the whole each animal represents.
Infer *Which category of animals has the most endangered species?*

Bar Graph A **bar graph** uses rectangular blocks, or bars, of varying sizes to show the relationships among variables. One variable is divided into parts. It can be numbers, such as the time of day, or a category, such as an animal. The second variable must be a number. The bars show the size of the second variable. For example, if you made a bar graph of the endangered species data from **Figure 14,** the bar for 1990 would represent 263 species. An example of a bar graph is shown in **Figure 15A.**

Circle Graph Suppose you want to show the relationship among the types of endangered species. A **circle graph** shows the parts of a whole. Circle graphs are sometimes called pie graphs. Each piece of pie visually represents a fraction of the total. Looking at the circle graph in **Figure 15B,** you see quickly which animals have the highest number of endangered species by comparing the sizes of the pieces of pie.

A circle has a total of 360°. To make a circle graph, you need to determine what fraction of 360 each part should be. First, determine the total of the parts. In **Figure 15B,** the total of the parts, or endangered species, is 367. One fraction of the total, *Mammals,* is 63 of 367 species. What fraction of 360 is this? To determine this, set up a ratio and solve for *x:*

$$\frac{63}{367} = \frac{x}{360°} \qquad x = 61.8°$$

Mammals will have an angle of 61.8° in the graph. The other angles in the circle are determined the same way.

Reading Check *What is another name for a circle graph?*

Topic: scientific Data
Visit ips.msscience.com for Web links to information about scientific illustrations, tables, and graphs and their importance in the scientific community.

Activity Create a table or graph using data collected from a classroom observation.

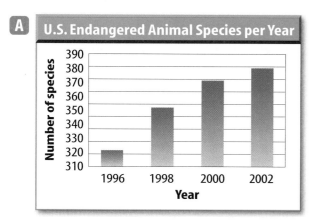
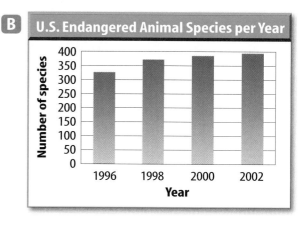

Reading Graphs When you are using or making graphs to display data, be careful—the scale of a graph can be misleading. The way the scale on a graph is marked can create the wrong impression, as seen in **Figure 16A.** Until you see that the *y*-axis doesn't start at zero, it appears that the number of endangered species has quadrupled in just six years.

This is called a broken scale and is used to highlight small but significant changes, just as an inset on a map draws attention to a small area of a larger map. **Figure 16B** shows the same data on a graph that does not have a broken scale. The number of species has only increased 20 percent from 1996 to 2002. Both graphs have correct data, but must be read carefully. Always analyze the measurements and graphs that you come across. If there is a surprising result, look closer at the scale.

Figure 16 Careful reading of graphs is important. A This graph does not start at zero, which makes it appear that the number of species has more than quadrupled from 1996–2002. B The actual increase is about 20 percent, as you can see from this full graph. The broken scale must be noted in order to interpret the results correctly.

section 3 review

Summary

Scientific Illustrations
- Drawings and illustrations can help people visualize complex concepts.
- A drawing can include details you see and those that are hidden.
- Photographs are an exact representation of an object at a single moment in time.

Tables and Graphs
- Tables display information while graphs are used to summarize data.
- A line graph shows the relationship between two variables, a bar graph shows the relationship among variables, and a circle graph shows the parts of a whole.
- It is important to pay close attention to the scale on graphs in order to analyze the information.

Self Check

1. **Explain** how to use **Figure 16** to find the number of endangered species in 1998.
2. **Infer** what type of graph you would use to display data gathered in a survey about students' after-school activities.
3. **Think Critically** Why is it important to be careful when making or using graphs?
4. **Describe** a time when an illustration would be helpful in everyday activities.
5. **Identify** when you would use a broken scale.

Applying Skills

6. **Use a Spreadsheet** Make a spreadsheet to display how the total mass of a 500-kg elevator changes as 50-kg passengers are added one at a time.

LAB Design Your Own

PACE YOURSELF

Goals
- **Design** an experiment that allows you to measure speed for each member of your group accurately.
- **Display** data in a table and a graph.

Possible Materials
meterstick
stopwatch
*watch with a second hand
*Alternate materials

Safety Precautions
Work in an area where it is safe to run. Participate only if you are physically able to exercise safely. As you design your plan, make a list of all the specific safety and health precautions you will take as you perform the investigation. Get your teacher's approval of the list before you begin.

Real-World Question
Track meets and other competitions require participants to walk, run, or wheel a distance that has been precisely measured. Officials make sure all participants begin at the same time, and each person's time is stopped at the finish line. If you are practicing for a local marathon or 10K, you need to know your speed or pace in order to compare it with those of other participants. How can your performance be measured accurately? How will you measure the speed of each person in your group? How will you display these data?

Form a Hypothesis
Think about the information you have learned about precision, measurement, and graphing. In your group, make a hypothesis about a technique that will provide you with the most precise measurement of each person's pace.

Using Scientific Methods

▶ Test Your Hypothesis

Make a Plan

1. As a group, decide what materials you will need.
2. How far will you travel? How will you measure that distance? How precise can you be?
3. How will you measure time? How precise can you be?
4. List the steps and materials you will use to test your hypothesis. Be specific. Will you try any part of your test more than once?
5. Before you begin, create a data table. Your group must decide on its design. Be sure to leave enough room to record the results for each person's time. If more than one trial is to be run for each measurement, include room for the additional data.

Follow Your Plan

1. Make sure that your teacher approves your plan before you start.
2. Carry out the experiment as planned and approved.
3. Be sure to record your data in the data table as you proceed with the measurements.

▶ Analyze Your Data

1. **Graph** your data. What type of graph would be best?
2. Are your data table and graph easy to understand? Explain.
3. How do you know that your measurements are precise?
4. Do any of your data appear to be out of line with the rest?

▶ Conclude and Apply

1. **Explain** how it is possible for different members of a group to find different times while measuring the same event.
2. **Infer** what tools would help you collect more precise data.
3. What other data displays could you use? What are the advantages and disadvantages of each?

Communicating Your Data

Make a larger version of your graph to display in your classroom with the graphs of other groups. **For more help, refer to the** Science Skill Handbook.

LAB 61

SCIENCE Stats

Biggest, Tallest, Loudest

Did you know...

... The world's most massive flower belongs to a species called *Rafflesia* (ruh FLEE zhee uh) and has a mass of up to 11 kg. The diameter, or the distance across the flower's petals, can measure up to 1 m.

... The world's tallest building is the Petronus Towers in Kuala Lumpur, Malaysia. It is 452 m tall. The tallest building in the United States is Chicago's Sears Tower, shown here, which measures 442 m.

Applying Math How many of the largest rafflesia petals would you have to place side by side to equal the height of the Sears Tower?

...One of the loudest explosions on Earth was the 1883 eruption of Krakatau (krah kuh TAHEW), an Indonesian volcano. It was heard from more than 3,500 km away.

Write About It

Visit ips.msscience.com/science_stats to find facts that describe some of the shortest, smallest, or fastest things on Earth. Create a class bulletin board with the facts you and your classmates find.

Chapter 2 Study Guide

Reviewing Main Ideas

Section 1 — Description and Measurement

1. Length, volume, mass, temperature, and rates are used to describe objects and events.
2. Estimation is used to make an educated guess at a measurement.
3. Accuracy describes how close a measurement is to the true value. Precision describes how close measurements are to each other.

Section 2 — SI Units

1. The international system of measurement is called SI. It is used throughout the world for communicating data.
2. The SI unit of length is the meter. Volume—the amount of space an object occupies—can be measured in cubic meters. The mass of an object is measured in kilograms.

Section 3 — Drawings, Tables, and Graphs

1. Tables, photographs, drawings, and graphs are tools used to collect, organize, summarize, and display data in a way that is easy to use and understand.
2. Line graphs show the relationship between two variables that are numbers on an *x*-axis and a *y*-axis. Bar graphs divide a variable into parts to show a relationship. Circle graphs show the parts of a whole like pieces of a pie.

Visualizing Main Ideas

Copy and complete the following concept map.

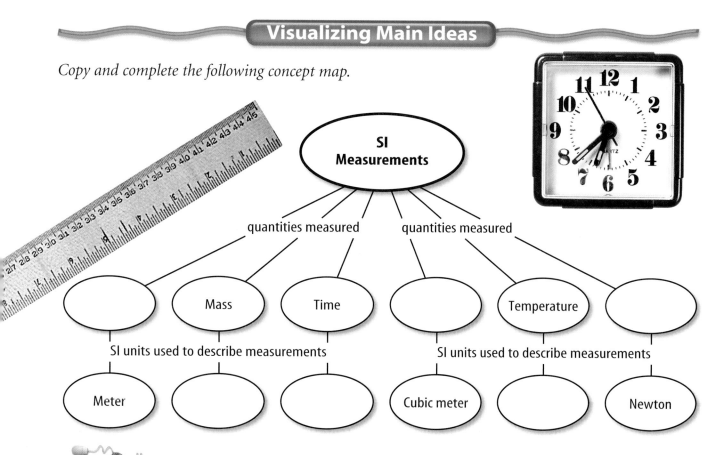

chapter 2 Review

Using Vocabulary

accuracy p. 45
bar graph p. 58
circle graph p. 58
estimation p. 43
graph p. 57
kelvin p. 54
kilogram p. 53
line graph p. 57
mass p. 53
measurement p. 42
meter p. 51
precision p. 44
rate p. 54
SI p. 50
table p. 57
volume p. 52
weight p. 53

Each phrase below describes a vocabulary word. Write the word that matches the phrase describing it.

1. the SI unit for length
2. a description with numbers
3. a method of making a rough measurement
4. the amount of matter in an object
5. a graph that shows parts of a whole
6. a description of how close measurements are to each other
7. the SI unit for temperature
8. an international system of units
9. the amount of space an object occupies

Checking Concepts

Choose the word or phrase that best answers the question.

10. The measurement 25.81 g is precise to the nearest
    A) gram.
    B) kilogram.
    C) tenth of a gram.
    D) hundredth of a gram.

11. What is the SI unit of mass?
    A) kilometer   C) liter
    B) meter   D) kilogram

12. What would you use to measure length?
    A) graduated cylinder
    B) triple beam balance
    C) meterstick
    D) spring scale

13. The cubic meter is the SI unit of what?
    A) volume   C) mass
    B) weight   D) distance

14. Which term describes how close measurements are to each other?
    A) significant digits
    B) estimation
    C) accuracy
    D) precision

15. Which of the following is a temperature scale?
    A) volume   C) Celsius
    B) mass   D) Mercury

16. Which of the following is used to organize data?
    A) table   C) precision
    B) rate   D) meterstick

17. To show the number of wins for each football team in your district, which of the following would you use?
    A) photograph   C) bar graph
    B) line graph   D) SI

18. To show 25 percent on a circle graph, the section must measure what angle?
    A) 25°   C) 180°
    B) 90°   D) 360°

64 CHAPTER REVIEW ips.msscience.com/vocabulary_puzzlemaker

chapter 2 Review

Thinking Critically

19. **Infer** How would you estimate the volume your backpack could hold?

20. **Explain** Why do scientists in the United States use SI rather than the English system (feet, pounds, pints, etc.) of measurement?

21. **List** the following in order from smallest to largest: 1 m, 1 mm, 10 km, 100 mm.

22. **Describe** Give an example of an instance when you would use a line graph. Could you use a bar graph for the same purpose?

23. **Compare and contrast** volume, length, and mass. How are they similar? Different? Give several examples of units that are used to measure each quantity. Which units are SI?

24. **Infer** Computer graphics artists can specify the color of a point on a monitor by using characters for the intensities of three colors of light. Why was this method of describing color invented?

Use the photo below to answer question 25.

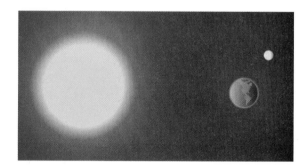

25. **Interpreting Scientific Illustrations** What does the figure show? How has this drawing been simplified?

Performance Activities

26. **Newspaper Search** Look through a week's worth of newspapers and evaluate any graphs or tables that you find.

Applying Math

Use the table below to answer question 27.

Areas of Bodies of Water	
Body of Water	Area (km^2)
Currituck Sound (North Carolina)	301
Pocomoke Sound (Maryland/Virginia)	286
Chincoteague Bay (Maryland/Virginia)	272
Core Sound (North Carolina)	229

27. **Make and Use Graphs** The table shows the area of several bodies of water. Make a bar graph of the data.

Use the illustration below to answer question 28.

28. **Travel Distances** The map above shows the driving distance from New York City to Denver, Colorado in kilometers. Convert the distance to meters and then find out how many meters are in a mile and convert the distance to miles.

29. **Round Digits** Round the following numbers to the correct number of significant digits.

$$42.86 \text{ kg} \times 38.703 \text{ kg}$$
$$10 \text{ g} \times 25.05 \text{ g}$$
$$5.8972 \text{ nm} \times 34.15731 \text{ nm}$$

Chapter 2 Standardized Test Practice

Part 1 Multiple Choice

Record your answers on the answer sheet provided by your teacher or on a sheet of paper.

1. Which best describes measurements that are accurate?
   A. They are very close to an accepted value.
   B. They are based on an estimate.
   C. They are very close to each other.
   D. They are not based on numbers.

2. The mass of a sample of calcium chloride is 33.755 grams. Round to the nearest hundredth of a gram.
   A. 33.8 g
   B. 34 g
   C. 33.76 g
   D. 33.75 g

Use the illustration below to answer questions 3 and 4.

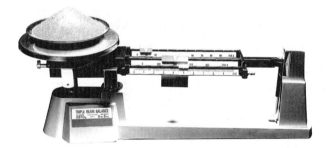

3. Which quantity is measured using this tool?
   A. weight
   B. mass
   C. volume
   D. length

4. Which measurement does this balance show?
   A. about 315 g
   B. about 326 g
   C. about 325 g
   D. about 215 g

Test-Taking Tip

Take Your Time Read carefully and make notes of the units used in any measurement.

Use the graph below to answer questions 5 and 6.

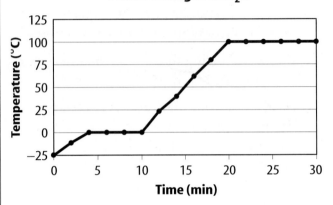

5. The graph shows data from an experiment in which ice was heated until it melted, then became steam. In what phase is the H_2O at 16 minutes?
   A. solid
   B. liquid
   C. gas
   D. plasma

6. Which statement describes the trend evident in the data?
   A. Temperature continually increased as time increased.
   B. Temperature did not change as time increased.
   C. Temperature increases were divided by plateaus as time increased.
   D. Temperature continually decreased as time increased.

7. Which of these represents 1/1000th of a meter?
   A. mm
   B. km
   C. ms
   D. dm

8. Which is NOT a unit of volume?
   A. milliliter
   B. cubic centimeter
   C. deciliter
   D. kelvin

Standardized Test Practice

Part 2 Short Response/Grid In

Record your answers on the answer sheet provided by your teacher or on a sheet of paper.

9. 35.77 g of Solid A are mixed with 95.3 g of Solid B. Write the mass of the mixture with the correct number of significant digits. Explain your answer.

10. Arrange these measuring tools in order from least to most precise: stopwatch measuring to 1/100ths of a second, atomic clock, sundial, wall clock with 2 hands.

Use the illustration below to answer questions 11 and 12.

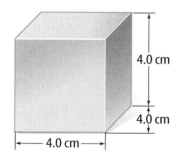

11. Define the term volume. Calculate the volume of the cube shown above. Give your answer in cm^3 and mL.

12. Describe how you would find the volume of the cube using the immersion method.

13. Create a table which shows the differences between mass and weight.

14. How do graphs make it easier to analyze data?

15. Why are drawings an effective way to communicate information?

16. Explain the difference between precision and accuracy.

Part 3 Open Ended

Record your answers on a sheet of paper.

17. A recipe calls for 1 cup sugar, 1 cup flour, and 1 cup milk. Define the term precision, and explain how to measure these ingredients precisely using kitchen tools.

18. While shopping, you find a rug for your room. Without a ruler, you must estimate the rug's size to determine if it will fit in your room. How will you proceed?

19. Identify the most appropriate SI length unit to measure the following: your height, the distance between two cities, the width of a computer screen, the radius of a coin, the length of a muscle cell. How are units converted in the SI system?

Use the figure below to answer questions 20 and 21.

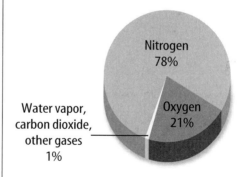

20. Calculate the angle of each section in this circle graph.

21. Create a bar graph using this data.

22. Measurement is a part of everyday life. Describe measurements someone might make as part of his or her normal activities.

23. You must decide what items to pack for a hiking trip. Space is limited, and you must carry all items during hikes. What measurements are important in your preparation?

unit 2 Matter

How Are Refrigerators & Frying Pans Connected?

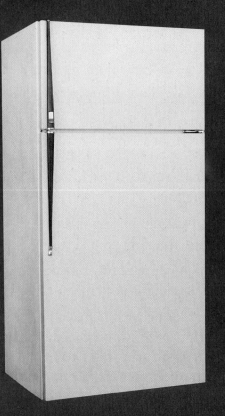

In the late 1930s, scientists were experimenting with a gas that they hoped would work as a new coolant in refrigerators. They filled several metal canisters with the gas and stored the canisters on dry ice. Later, when they opened the canisters, they were surprised to find that the gas had disappeared and that the inside of each canister was coated with a slick, powdery, white solid. The gas had undergone a chemical change. That is, the chemical bonds in its molecules had broken and new bonds had formed, turning one kind of matter into a completely different kind of matter. Strangely, the mysterious white powder proved to be just about the slipperiest substance that anyone had ever encountered. Years later, a creative Frenchman obtained some of the slippery stuff and tried applying it to his fishing tackle to keep the lines from tangling. His wife noticed what he was doing and suggested putting the substance on the inside of a frying pan to keep food from sticking. He did, and nonstick cookware was born!

unit projects

Visit ips.msscience.com/unit_project to find project ideas and resources. Projects include:
- **History** Research the French chemist Antoine-Laurent Lavoisier. Design a time line with 20 of his contributions to chemistry.
- **Technology** Design a classroom periodic table wall mural. Use the information as a learning tool and review game.
- **Model** Demonstrate your knowledge of the characteristics of physical and chemical change by preparing a simple snack to share.

 Investigating the Sun is an exploration into the composition of our nearest star, the energy it produces, and the possibilities of harnessing that energy for everyday use.

chapter 3

Atoms, Elements, and the Periodic Table

The BIG Idea

All types of matter—elements, compounds, and mixtures—are made of atoms.

SECTION 1
Structure of Matter
Main Idea Atoms contain protons and neutrons in a tiny nucleus and electrons in a cloud around the nucleus.

SECTION 2
The Simplest Matter
Main Idea An element is made of atoms that have the same number of protons.

SECTION 3
Compounds and Mixtures
Main Idea Compounds contain different types of atoms bonded together. Mixtures contain different substances mixed together.

What a fun ride!

The pilot gives the propane burner a long, loud blast that will heat the air in the balloon, making you soar higher. During your ride, you might start thinking about what is keeping you airborne. In this chapter, as you learn about atoms and elements, you will be able to understand more about matter.

Science Journal Make a list of three questions that you think of when you look at the picture.

Start-Up Activities

Observe Matter

You've just finished playing basketball. You're hot and thirsty. You reach for your bottle of water and take a drink. Releasing your grip, you notice that the bottle is nearly empty. Is the bottle really almost empty? According to the dictionary, *empty* means "containing nothing." When you have finished all the water in the bottle, will it be empty or full?

1. Wad up a dry paper towel or tissue and tape it to the inside of a plastic cup as shown.
2. Fill a bowl or sink with water. Turn the cup upside down and slowly push the cup straight down into the water as far as you can.
3. Slowly raise the cup straight up and out of the water. Remove the paper towel or tissue paper and examine it.
4. **Think Critically** In your Science Journal, describe the lab and its results. Explain what you think happened. Was anything in the cup besides the paper? If so, what was it?

Atoms, Elements, and the Periodic Table Make the following Foldable to help you identify the main ideas about atoms, elements, compounds, and mixtures.

STEP 1 Draw a mark at the midpoint of a sheet of paper along the side edge. Then **fold** the top and bottom edges in to touch the midpoint.

STEP 2 **Fold** in half from side to side.

STEP 3 Open and cut along the inside fold lines to form four tabs.

STEP 4 Label each tab as shown.

Read and Write As you read the chapter, list several everyday examples of atoms, elements, compounds, and mixtures on the back of the appropriate tab.

 Preview this chapter's content and activities at ips.msscience.com

71

Get Ready to Read

New Vocabulary

① Learn It!
What should you do if you find a word you don't know or understand? Here are some suggested strategies:

1. Use context clues (from the sentence or the paragraph) to help you define it.
2. Look for prefixes, suffixes, or root words that you already know.
3. Write it down and ask for help with the meaning.
4. Guess at its meaning.
5. Look it up in the glossary or a dictionary.

② Practice It!
Look at the word *charge* in the following passage. See how context clues can help you understand its meaning.

Context Clue
Like charges repel each other and opposite charges attract each other.

Context Clue
…cathode rays were made up of negatively charged particles.

Context Clue
Negatively charged particles are called electrons.

> Thomson knew that like charges repel each other and opposite charges attract each other. When he saw that the [cathode] rays traveled toward a positively charged plate, he concluded that the cathode rays were made up of negatively charged particles. These invisible, negatively charged particles are called **electrons**.
>
> —*from page 76*

③ Apply It!
Make a vocabulary bookmark with a strip of paper. As you read, keep track of words you do not know or want to learn more about.

Target Your Reading

Reading Tip

Read a paragraph containing a vocabulary term from beginning to end. Then, go back to determine the meaning of the term.

Use this to focus on the main ideas as you read the chapter.

① **Before you read** the chapter, respond to the statements below on your worksheet or on a numbered sheet of paper.
- Write an **A** if you **agree** with the statement.
- Write a **D** if you **disagree** with the statement.

② **After you read** the chapter, look back to this page to see if you've changed your mind about any of the statements.
- If any of your answers changed, explain why.
- Change any false statements into true statements.
- Use your revised statements as a study guide.

Science Online
Print out a worksheet of this page at ips.msscience.com

Before You Read A or D		Statement	After You Read A or D
	1	Matter cannot be created or destroyed.	
	2	The model of the atom has remained mostly unchanged since the idea of atoms was first proposed.	
	3	Atoms contain mostly empty space.	
	4	All atoms contain at least one neutron.	
	5	Two atoms of the same element might contain different numbers of neutrons.	
	6	If you are given the element name, you can determine the mass number of an atom.	
	7	A mixture is a type of substance.	
	8	Substances that contain the same elements will have the same chemical and physical properties.	
	9	Both compounds and mixtures contain more than one type of element.	

section 1
Structure of Matter

as you read

What You'll Learn
- **Describe** characteristics of matter.
- **Identify** what makes up matter.
- **Identify** the parts of an atom.
- **Compare** the models that are used for atoms.

Why It's Important
Matter makes up almost everything we see—and much of what we can't see.

Review Vocabulary
density: the mass of an object divided by its volume

New Vocabulary
- matter
- atom
- law of conservation of matter
- electron
- nucleus
- proton
- neutron

What is matter?

Is a glass with some water in it half empty or half full? Actually, neither is correct. The glass is completely full—half full of water and half full of air. What is air? Air is a mixture of several gases, including nitrogen and oxygen, which are kinds of matter. **Matter** is anything that has mass and takes up space. So, even though you can't see it or hold it in your hand, air is matter. What about all the things you can see, taste, smell, and touch? Most are made of matter, too. Look at the things pictured in **Figure 1** and determine which of them are matter.

What isn't matter?

You can see the words on this page because of the light from the Sun or from a fixture in the room. Does light have mass or take up space? What about the warmth from the Sun or the heat from the heater in your classroom? Light and heat do not take up space, and they have no mass. Therefore, they are not forms of matter. Emotions, thoughts, and ideas are not matter either. Does this information change your mind about the items in **Figure 1**?

Reading Check *Why is air matter, but light is not?*

Figure 1 A rainbow is formed when light filters through the raindrops, a plant grows from a seed in the ground, and a statue is sculpted from bronze.
Identify *which are matter.*

72 CHAPTER 3 Atoms, Elements, and the Periodic Table

Figure 2 Early Beliefs About the Composition of Matter

Many Indian Philosophers (1,000 B.C.)	Kashyapa, an Indian Philosopher (1,000 B.C.)	Many Greek Philosophers (500–300 B.C.)	Democritus (380 B.C.)	Aristotle (330 B.C.)	Chinese Philosophers (300 B.C.)
• Ether—an invisible substance that filled the heavens • Earth • Water • Air • Fire	• Five elements broken down into smaller units called parmanu • Parmanu of earth elements are heavier than air elements	• Earth • Water • Air • Fire	• Tiny individual particles he called *atomos* • Empty space through which atoms move • Each substance composed of one type of *atomos*	• Empty space could not exist • Earth • Water • Air • Fire	• Metal • Earth • Water • Air • Fire

What makes up matter?

Suppose you cut a chunk of wood into smaller and smaller pieces. Do the pieces seem to be made of the same matter as the large chunk you started with? If you could cut a small enough piece, would it still have the same properties as the first chunk? Would you reach a point where the last cut resulted in a piece that no longer resembled the first chunk? Is there a limit to how small a piece can be? For centuries, people have asked questions like these and wondered what matter is made of.

An Early Idea Democritus, who lived from about 460 B.C. to 370 B.C., was a Greek philosopher who thought the universe was made of empty space and tiny bits of stuff. He believed that the bits of stuff were so small they could no longer be divided into smaller pieces. He called these tiny pieces atoms. The term *atom* comes from a Greek word that means "cannot be divided." Today an **atom** is defined as a small particle that makes up most types of matter. **Figure 2** shows the difference between Democritus's ideas and those of other early scientists and philosophers. Democritus thought that different types of atoms existed for every type of matter and that the atom's identity explained the characteristics of each type of matter. Democritus's ideas about atoms were a first step toward understanding matter. However, his ideas were not accepted for over 2,000 years. It wasn't until the early 1800s that scientists built upon the concept of atoms to form the current atomic theory of matter.

INTEGRATE History

Atomism Historians note that Leucippus developed the idea of the atom around 440 B.C. He and his student, Democritus, refined the idea of the atom years later. Their concept of the atom was based on five major points: (1) all matter is made of atoms, (2) there are empty spaces between atoms, (3) atoms are complete solids, (4) atoms do not have internal structure, and (5) atoms are different in size, shape, and weight.

Figure 3 When wood burns, matter is not lost. The total mass of the wood and the oxygen it combines with during a fire equals the total mass of the ash, water vapor, carbon dioxide, and other gases produced.
Infer *When you burn wood in a fireplace, what is the source of oxygen?*

wood + oxygen = ash + gases + water vapor

Investigating the Unseen

Procedure
1. Your teacher will give you a **sealed shoe box** that contains **one or more items**.
2. Try to find out how many and what kinds of items are inside the box. You cannot look inside the box. The only observations you can make are by handling the box.

Analysis
1. How many items do you infer are in the box? Sketch the apparent shapes of the items and identify them if you can.
2. Compare your procedure with how scientists perform experiments and make models to find out more about the atom.

Lavoisier's Contribution Lavoisier (la VWAH see ay), a French chemist who lived about 2,000 years after Democritus, also was curious about matter—especially when it changed form. Before Lavoisier, people thought matter could appear and disappear because of the changes they saw as matter burned or rusted. You might have thought that matter can disappear if you've ever watched wood burn in a fireplace or at a bonfire. Lavoisier showed that wood and the oxygen it combines with during burning have the same mass as the ash, water, carbon dioxide, and other gases that are produced, as shown in **Figure 3**. In a similar way, an iron bar, oxygen, and water have the same mass as the rust that forms when they interact. From Lavoisier's work came the **law of conservation of matter,** which states that matter is not created or destroyed—it only changes form.

Models of the Atom

Models are often used for things that are too small or too large to be observed or that are too difficult to be understood easily. One way to make a model is to make a smaller version of something large. If you wanted to design a new sailboat, would you build a full-sized boat and hope it would float? It would be more efficient, less expensive, and safer to build and test a smaller version first. Then, if it didn't float, you could change your design and build another model. You could keep trying until the model worked.

In the case of atoms, scientists use large models to explain something that is too small to be looked at. These models of the atom were used to explain data or facts that were gathered experimentally. As a result, these models are also theories.

Dalton's Atomic Model In the early 1800s, an English schoolteacher and chemist named John Dalton studied the experiments of Lavoisier and others. Dalton thought he could design an atomic model that explained the results of those experiments. Dalton's atomic model was a set of ideas—not a physical object. Dalton believed that matter was made of atoms that were too small to be seen by the human eye. He also thought that each type of matter was made of only one kind of atom. For example, gold atoms make up a gold nugget and give a gold ring its shiny appearance. Likewise, iron atoms make up an iron bar and give it unique properties, and so on. Because predictions using Dalton's model were supported by data, the model became known as the atomic theory of matter.

Sizes of Atoms Atoms are so small it would take about 1 million of them lined up in a row to equal the thickness of a human hair. For another example of how small atoms are, look at **Figure 4.** Imagine you are holding an orange in your hand. If you wanted to be able to see the individual atoms on the orange's surface, the size of the orange would have to be increased to the size of Earth. Then, imagine the Earth-sized orange covered with billions and billions of marbles. Each marble would represent one of the atoms on the skin of the orange. No matter what kind of model you use to picture it, the result is the same—an atom is an extremely small particle of matter.

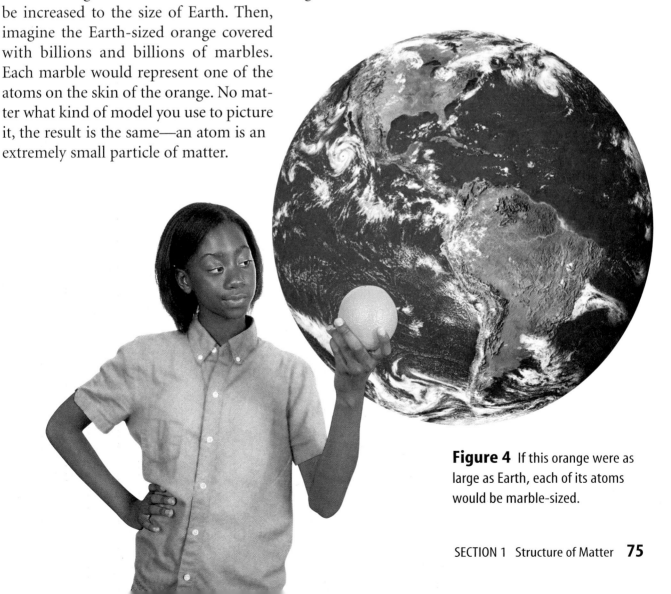

Figure 4 If this orange were as large as Earth, each of its atoms would be marble-sized.

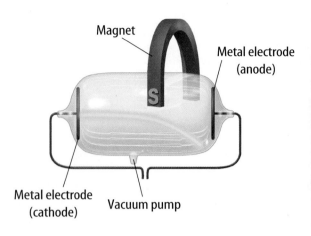

Figure 5 In Thomson's experiment, the magnet caused the cathode rays inside the tube to bend. **Describe** what you think would happen to the cathode rays if the magnet were removed.

Topic: Subatomic Particles
Visit ips.msscience.com for Web links to information about particles that make up atoms.

Activity Can any of the particles be divided further? Display your data in a table.

Discovering the Electron One of the many pioneers in the development of today's atomic model was J.J. Thomson, an English scientist. He conducted experiments using a cathode ray tube, which is a glass tube sealed at both ends out of which most of the air has been pumped. Thomson's tube had a metal plate at each end. The plates were connected to a high-voltage electrical source that gave one of the plates—the anode—a positive charge and the other plate—the cathode—a negative charge. During his experiments, Thomson observed rays that traveled from the cathode to the anode. These cathode rays were bent by a magnet, as seen in **Figure 5,** showing that they were made up of particles that had mass and charge. Thomson knew that like charges repel each other and opposite charges attract each other. When he saw that the rays traveled toward a positively charged plate, he concluded that the cathode rays were made up of negatively charged particles. These invisible, negatively charged particles are called **electrons**.

Reading Check *Why were the cathode rays in Thomson's cathode ray tube bent by a magnet?*

Try to imagine Thomson's excitement at this discovery. He had shown that atoms are not too tiny to divide after all. Rather, they are made up of even smaller subatomic particles. Other scientists soon built upon Thomson's results and found that the electron had a small mass. In fact, an electron is 1/1,837 the mass of the lightest atom, the hydrogen atom. In 1906, Thomson received the Nobel Prize in Physics for his work on the discovery of the electron.

Matter that has an equal amount of positive and negative charge is said to be neutral—it has no net charge. Because most matter is neutral, Thomson pictured the atom as a ball of positive charge with electrons embedded in it. It was later determined that neutral atoms contained an equal number of positive and negative charges.

Thomson's Model Thomson's model, shown in **Figure 6,** can be compared to chocolate chips spread throughout a ball of cookie dough. However, the model did not provide all the answers to the questions that puzzled scientists about atoms.

Rutherford—The Nucleus Scientists still had questions about how the atom was arranged and about the presence of positively charged particles. In about 1910, a team of scientists led by Ernest Rutherford worked on these questions. In their experiment, they bombarded an extremely thin piece of gold foil with alpha particles. Alpha particles are tiny, high-energy, positively charged particles that he predicted would pass through the foil. Most of the particles passed straight through the foil as if it were not there at all. However, other particles changed direction, and some even bounced back. Rutherford thought the result was so remarkable that he later said, "It was almost as incredible as if you had fired a 15-inch shell at a piece of tissue paper, and it came back and hit you."

Positive Center Rutherford concluded that because so many of the alpha particles passed straight through the gold foil, the atoms must be made of mostly empty space. However, because some of the positively charged alpha particles bounced off something, the gold atoms must contain some positively charged object concentrated in the midst of this empty space. Rutherford called the positively charged, central part of the atom the **nucleus** (NEW klee us). He named the positively charged particles in the nucleus **protons**. He also suggested that electrons were scattered in the mostly empty space around the nucleus, as shown in **Figure 7.**

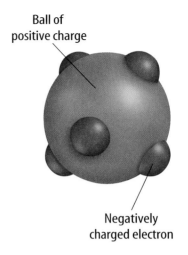

Figure 6 Thomson's model shows the atom as electrons embedded in a ball of positive charge.
Explain *how Thomson knew atoms contained positive and negative charges.*

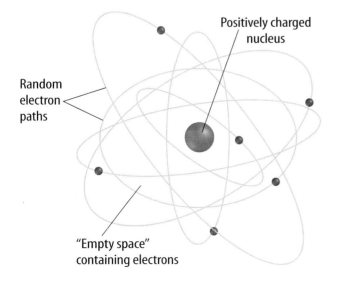

Figure 7 Rutherford concluded that the atom must be mostly empty space in which electrons travel in random paths around the nucleus. He also thought the nucleus of the atom must be small and positively charged.
Identify *where most of the mass of an atom is concentrated.*

SECTION 1 Structure of Matter **77**

Discovering the Neutron Rutherford had been puzzled by one observation from his experiments with nuclei. After the collisions, the nuclei seemed to be heavier. Where did this extra mass come from? James Chadwick, a student of Rutherford's, answered this question. The alpha particles themselves were not heavier. The atoms that had been bombarded had given off new particles. Chadwick experimented with these new particles and found that, unlike electrons, the paths of these particles were not affected by an electric field. To explain his observations, he said that these particles came from the nucleus and had no charge. Chadwick called these uncharged particles **neutrons** (NEW trahnz). His proton-neutron model of the atomic nucleus is still accepted today.

Improving the Atomic Model

Early in the twentieth century, a scientist named Niels Bohr found evidence that electrons in atoms are arranged according to energy levels. The lowest energy level is closest to the nucleus and can hold only two electrons. Higher energy levels are farther from the nucleus and can contain more electrons. To explain these energy levels, some scientists thought that the electrons might orbit an atom's nucleus in paths that are specific distances from the nucleus, as shown in **Figure 8.** This is similar to how the planets orbit the Sun.

The Modern Atomic Model As a result of continuing research, scientists now realize that because electrons have characteristics that are similar to waves and particles, their energy levels are not defined, planet-like orbits around the nucleus. Rather, it seems most likely that electrons move in what is called the atom's electron cloud, as shown in **Figure 9.**

Physicists and Chemists Physicists generally study the physical atom. The physical atom includes the inner components of an atom such as protons and neutrons, the forces that hold or change their positions in space and the bulk properties of elements such as melting point. Chemists, on the other hand, study the chemical atom. The chemical atom refers to the manner in which different elements relate to each other and the new substances formed by their union.

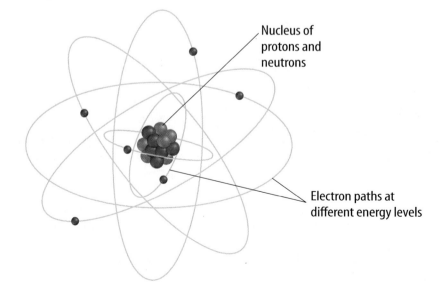

Figure 8 This simplified Bohr model shows a nucleus of protons and neutrons and electron paths based on energy levels.

The Electron Cloud The electron cloud is a spherical cloud of varying density surrounding the nucleus. The varying density shows where an electron is more or less likely to be. Atoms with electrons in higher energy levels have electron clouds of different shapes that also show where those electrons are likely to be. Generally, the electron cloud has a radius 10,000 times that of the nucleus.

Further Research By the 1930s, it was recognized that matter was made up of atoms, which were, in turn, made up of protons, neutrons, and electrons. But scientists, called physicists, continued to study the basic parts of this atom. Today, they have succeeded in breaking down protons and neutrons into even smaller particles called quarks. These particles can combine to make other kinds of tiny particles, too. The six types of quarks are *up, down, strange, charmed, top,* and *bottom*. Quarks have fractional electric charges of +2/3 or −1/3, unlike the +1 charge of a proton or the −1 charge of an electron. Research will continue as new discoveries are made about the structure of matter.

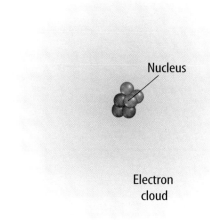

Figure 9 This model of the atom shows the electrons moving around the nucleus in a region called an electron cloud. The dark cloud of color represents the area where the electron is more likely to be found.
Infer *What does the intensity of color near the nucleus suggest?*

section 1 review

Summary

What is matter?
- Matter is anything that has mass and takes up space.
- Matter is composed of atoms.

Models of the Atom
- Democritus introduced the idea of an atom. Lavoisier showed matter is neither created nor destroyed, just changed.
- Dalton's ideas led to the atomic theory of matter.
- Thomson discovered the electron.
- Rutherford discovered protons exist in the nucleus.
- Chadwick discovered the neutron.

Improving the Atomic Model
- Niels Bohr suggested electrons move in energy levels.
- More recent physicists introduced the idea of the electron cloud and were able to break down protons and neutrons into smaller particles called quarks.

Self Check

1. **List** five examples of matter and five examples that are not matter. Explain your answers.
2. **Describe** and name the parts of the atom.
3. **Explain** why the word *atom* was an appropriate term for Democritus's idea.
4. **Think Critically** When neutrons were discovered, were these neutrons created in the experiment? How does Lavoisier's work help answer this question?
5. **Explain** the law of conservation of matter using your own examples.
6. **Think Critically** How is the electron cloud model different from Bohr's atomic model?

Applying Skills

7. **Classify** each scientist and his contribution according to the type of discovery each person made. Explain why you grouped certain scientists together.
8. **Evaluate Others' Data and Conclusions** Analyze, review, and critique the strengths and weaknesses of Thomson's "cookie dough" theory using the results of Rutherford's gold foil experiment.

section 2
The Simplest Matter

as you read

What You'll Learn
- **Describe** the relationship between elements and the periodic table.
- **Explain** the meaning of atomic mass and atomic number.
- **Identify** what makes an isotope.
- **Contrast** metals, metalloids, and nonmetals.

Why It's Important
Everything on Earth is made of the elements that are listed on the periodic table.

Review Vocabulary
mass: a measure of the amount of matter an object has

New Vocabulary
- element
- atomic number
- isotope
- mass number
- atomic mass
- metal
- nonmetal
- metalloid

The Elements

Have you watched television today? TV sets are common, yet each one is a complex system. The outer case is made mostly of plastic, and the screen is made of glass. Many of the parts that conduct electricity are metals or combinations of metals. Other parts in the interior of the set contain materials that barely conduct electricity. All of the different materials have one thing in common: they are made up of even simpler materials. In fact, if you had the proper equipment, you could separate the plastics, glass, and metals into these simpler materials.

One Kind of Atom Eventually, though, you would separate the materials into groups of atoms. At that point, you would have a collection of elements. An **element** is matter made of only one kind of atom. At least 110 elements are known and at least 90 of them occur naturally on Earth. These elements make up gases in the air, minerals in rocks, and liquids such as water. Examples of naturally occurring elements include the oxygen and nitrogen in the air you breathe and the metals gold, silver, aluminum, and iron. The other elements are known as synthetic elements. These elements have been made in nuclear reactions by scientists with machines called particle accelerators, like the one shown in **Figure 10.** Some synthetic elements have important uses in medical testing and are found in smoke detectors and heart pacemaker batteries.

Figure 10 The Tevatron has a circumference of 6.3 km—a distance that allows particles to accelerate to high speeds. These high-speed collisions can create synthetic elements.

Figure 11 When you look for information in the library, a system of organization called the Dewey Decimal Classification System helps you find a book quickly and efficiently.

Dewey Decimal Classification System	
000	Computers, information and general reference
100	Philosophy and psychology
200	Religion
300	Social sciences
400	Languages
500	Science
600	Technology
700	Arts and recreation
800	Literature
900	History and geography

The Periodic Table

Suppose you go to a library, like the one shown in **Figure 11,** to look up information for a school assignment. How would you find the information? You could look randomly on shelves as you walk up and down rows of books, but the chances of finding your book would be slim. To avoid such haphazard searching, some libraries use the Dewey Decimal Classification System to categorize and organize their volumes and to help you find books quickly and efficiently.

Charting the Elements Chemists have created a chart called the periodic table of the elements to help them organize and display the elements. **Figure 12** shows how scientists changed their model of the periodic table over time.

On the inside back cover of this book, you will find a modern version of the periodic table. Each element is represented by a chemical symbol that contains one to three letters. The symbols are a form of chemical shorthand that chemists use to save time and space—on the periodic table as well as in written formulas. The symbols are an important part of an international system that is understood by scientists everywhere.

The elements are organized on the periodic table by their properties. There are rows and columns that represent relationships between the elements. The rows in the table are called periods. The elements in a row have the same number of energy levels. The columns are called groups. The elements in each group have similar properties related to their structure. They also tend to form similar bonds.

Topic: New Elements
Visit ips.msscience.com for Web links to information about new elements.

Activity Research physical properties of two synthetic elements.

SECTION 2 The Simplest Matter **81**

NATIONAL GEOGRAPHIC VISUALIZING THE PERIODIC TABLE

Figure 12

The familiar periodic table that adorns many science classrooms is based on a number of earlier efforts to identify and classify the elements. In the 1790s, one of the first lists of elements and their compounds was compiled by French chemist Antoine-Laurent Lavoisier, who is shown in the background picture with his wife and assistant, Marie Anne. Three other tables are shown here.

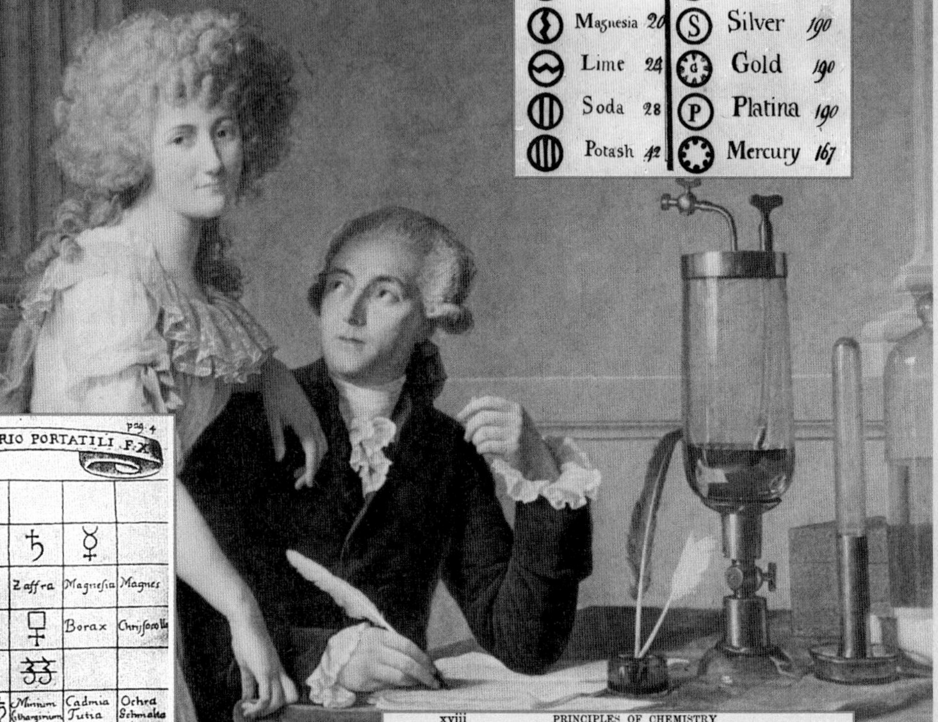

John Dalton (Britain, 1803) used symbols to represent elements. His table also assigned masses to each element.

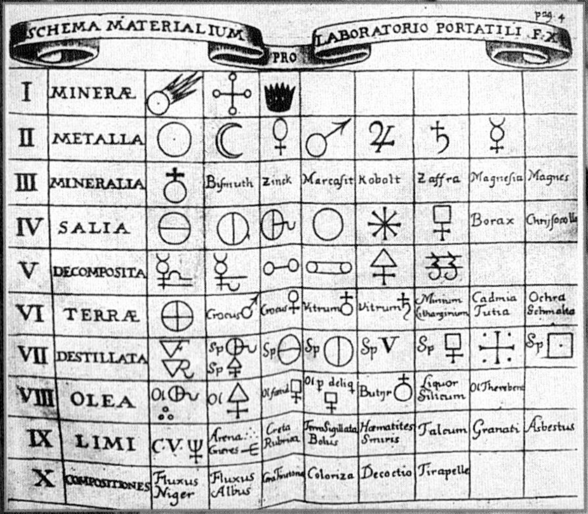

An early alchemist put together this table of elements and compounds. Some of the symbols have their origin in astrology.

Dmitri Mendeleev (Russia, 1869) arranged the 63 elements known to exist at that time into groups based on their chemical properties and atomic weights. He left gaps for elements he predicted were yet to be discovered.

82 CHAPTER 3 Atoms, Elements, and the Periodic Table

Identifying Characteristics

Each element is different and has unique properties. These differences can be described in part by looking at the relationships between the atomic particles in each element. The periodic table contains numbers that describe these relationships.

Number of Protons and Neutrons Look up the element chlorine on the periodic table found on the inside back cover of your book. Cl is the symbol for chlorine, as shown in **Figure 13,** but what are the two numbers? The top number is the element's **atomic number.** It tells you the number of protons in the nucleus of each atom of that element. Every atom of chlorine, for example, has 17 protons in its nucleus.

Reading Check *What are the atomic numbers for Cs, Ne, Pb, and U?*

Isotopes Although the number of protons changes from element to element, every atom of the same element has the same number of protons. However, the number of neutrons can vary even for one element. For example, some chlorine atoms have 18 neutrons in their nucleus while others have 20. These two types of chlorine atoms are chlorine-35 and chlorine-37. They are called **isotopes** (I suh tohps), which are atoms of the same element that have different numbers of neutrons.

You can tell someone exactly which isotope you are referring to by using its mass number. An atom's **mass number** is the number of protons plus the number of neutrons it contains. The numbers 35 and 37, which were used to refer to chlorine, are mass numbers. Hydrogen has three isotopes with mass numbers of 1, 2, and 3. They are shown in **Figure 14.** Each hydrogen atom always has one proton, but in each isotope the number of neutrons is different.

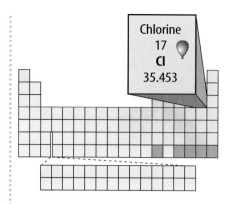

Figure 13 The periodic table block for chlorine shows its symbol, atomic number, and atomic mass.
Determine *if chlorine atoms are more or less massive than carbon atoms.*

Figure 14 Three isotopes of hydrogen are known to exist. They have zero, one, and two neutrons in addition to their one proton. Protium, with only the one proton, is the most abundant isotope.

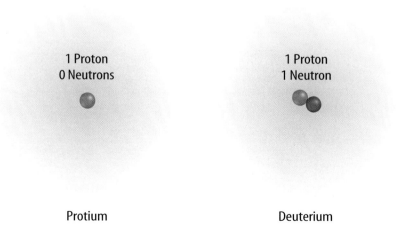

SECTION 2 The Simplest Matter **83**

Circle Graph Showing Abundance of Chlorine Isotopes

Average atomic mass = 35.45 u

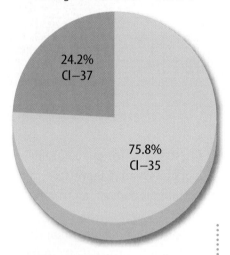

Figure 15 If you have 1,000 atoms of chlorine, about 758 will be chlorine-35 and have a mass of 34.97 u each. About 242 will be chlorine-37 and have a mass of 36.97 u each. The total mass of the 1,000 atoms is 35,454 u, so the average mass of one chlorine atom is about 35.45 u.

Atomic Mass The **atomic mass** is the weighted average mass of the isotopes of an element. The atomic mass is the number found below the element symbol in **Figure 13**. The unit that scientists use for atomic mass is called the atomic mass unit, which is given the symbol u. It is defined as 1/12 the mass of a carbon-12 atom.

The calculation of atomic mass takes into account the different isotopes of the element. Chlorine's atomic mass of 35.45 u could be confusing because there aren't any chlorine atoms that have that exact mass. About 76 percent of chlorine atoms are chlorine-35 and about 24 percent are chlorine-37, as shown in **Figure 15**. The weighted average mass of all chlorine atoms is 35.45 u.

Classification of Elements

Elements fall into three general categories—metals, metalloids (ME tuh loydz), and nonmetals. The elements in each category have similar properties.

Metals generally have a shiny or metallic luster and are good conductors of heat and electricity. All metals, except mercury, are solids at room temperature. Metals are malleable (MAL yuh bul), which means they can be bent and pounded into various shapes. The beautiful form of the shell-shaped basin in **Figure 16** is a result of this characteristic. Metals are also ductile, which means they can be drawn into wires without breaking. If you look at the periodic table, you can see that most of the elements are metals.

Figure 16 The artisan is chasing, or chiseling, the malleable metal into the desired form.

Other Elements Nonmetals are elements that are usually dull in appearance. Most are poor conductors of heat and electricity. Many are gases at room temperature, and bromine is a liquid. The solid nonmetals are generally brittle, meaning they cannot change shape easily without breaking. The nonmetals are essential to the chemicals of life. More than 97 percent of your body is made up of various nonmetals, as shown in **Figure 17.** You can see that, except for hydrogen, the nonmetals are found on the right side of the periodic table.

Metalloids are elements that have characteristics of metals and nonmetals. On the periodic table, metalloids are found between the metals and nonmetals. All metalloids are solids at room temperature. Some metalloids are shiny and many are conductors, but they are not as good at conducting heat and electricity as metals are. Some metalloids, such as silicon, are used to make the electronic circuits in computers, televisions, and other electronic devices.

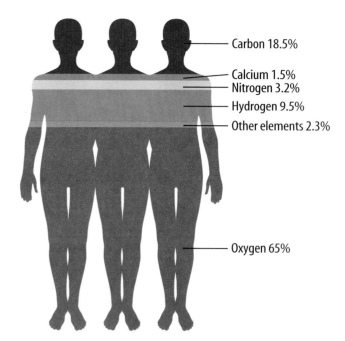

Figure 17 You are made up of mostly nonmetals.

Reading Check *What is a metalloid?*

section 2 review

Summary

The Elements
- An element is matter made of only one type of atom.
- Some elements occur naturally on Earth. Synthetic elements are made in nuclear reactions in particle accelerators.
- Elements are divided into three categories based on certain properties.

The Periodic Table
- The periodic table arranges and displays all known elements in an orderly way.
- Each element has a chemical symbol.

Identifying Characteristics
- Each element has a unique number of protons, called the atomic mass number.
- Isotopes of an element are important when determining the atomic mass of an element.

Self Check

1. **Explain** some of the uses of metals based on their properties.
2. **Describe** the difference between atomic number and atomic mass.
3. **Define** the term *isotope.* Explain how two isotopes of an element are different.
4. **Identify** the isotopes of hydrogen.
5. **Think Critically** Describe how to find the atomic number for the element oxygen. Explain what this information tells you about oxygen.

Applying Math

6. **Simple Equation** An atom of niobium has a mass number of 93. How many neutrons are in the nucleus of this atom? An atom of phosphorus has 15 protons and 15 neutrons in the nucleus. What is the mass number of this isotope?

Elements and the Periodic Table

The periodic table organizes the elements, but what do they look like? What are they used for? In this lab, you'll examine some elements and share your findings with your classmates.

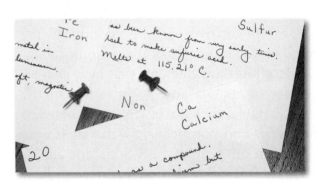

Real-World Question

What are some of the characteristics and purposes of the chemical elements?

Goals
- **Classify** the chemical elements.
- **Organize** the elements into the groups and periods of the periodic table.

Materials
colored markers
large index cards
Merck Index
encyclopedia
*other reference materials
large bulletin board
8½-in × 14-in paper
thumbtacks
*pushpins

*Alternate materials

Safety Precaution

WARNING: Use care when handling sharp objects.

Procedure

1. Select the assigned number of elements from the list provided by your teacher.
2. **Design** an index card for each of your selected elements. On each card, mark the element's atomic number in the upper left-hand corner and write its symbol and name in the upper right-hand corner.
3. **Research** each of the elements and write several sentences on the card about its appearance, its other properties, and its uses.
4. **Classify** each of your elements as a metal, a metalloid, or a nonmetal based upon its properties.
5. **Write** the appropriate classification on each of your cards using the colored marker chosen by your teacher.
6. Work with your classmates to make a large periodic table. Use thumbtacks to attach your cards to a bulletin board in their proper positions on the periodic table.
7. **Draw** your own periodic table. Place the elements' symbols and atomic numbers in the proper locations on your table.

Conclude and Apply

1. **Interpret** the class data and classify the elements into the categories metal, metalloid, and nonmetal. Highlight each category in a different color on your periodic table.
2. **Predict** the properties of a yet-undiscovered element located directly under francium on the periodic table.

86 CHAPTER 3 Atoms, Elements, and the Periodic Table

section 3

Compounds and Mixtures

Substances

Scientists classify matter in several ways that depend on what it is made of and how it behaves. For example, matter that has the same composition and properties throughout is called a **substance**. Elements, such as a bar of gold or a sheet of aluminum, are substances. When different elements combine, other substances are formed.

Compounds What do you call the colorless liquid that flows from the kitchen faucet? You probably call it water, but maybe you've seen it written H_2O. The elements hydrogen and oxygen exist as separate, colorless gases. However, these two elements can combine, as shown in **Figure 18,** to form the compound water, which is different from the elements that make it up. A **compound** is a substance whose smallest unit is made up of atoms of more than one element bonded together.

Compounds often have properties that are different from the elements that make them up. Water is distinctly different from the elements that make it up. It is also different from another compound made from the same elements. Have you ever used hydrogen peroxide (H_2O_2) to disinfect a cut? This compound is a different combination of hydrogen and oxygen and has different properties from those of water.

Water is a nonirritating liquid that is used for bathing, drinking, cooking, and much more. In contrast, hydrogen peroxide carries warnings on its labels such as *Keep Hydrogen Peroxide Out of the Eyes.* Although it is useful in solutions for cleaning contact lenses, it is not safe for your eyes as it comes from the bottle.

as you read

What You'll Learn
- **Identify** the characteristics of a compound.
- **Compare and contrast** different types of mixtures.

Why It's Important
The food you eat, the materials you use, and all matter can be classified by compounds or mixtures.

Review Vocabulary
formula: shows which elements and how many atoms of each make up a compound

New Vocabulary
- substance
- mixture
- compound

Figure 18 A space shuttle is powered by the reaction between liquid hydrogen and liquid oxygen. The reaction produces a large amount of energy and the compound water.
Explain *why a car that burns hydrogen rather than gasoline would be friendly to the environment.*

SECTION 3 Compounds and Mixtures **87**

Figure 19 The elements hydrogen and oxygen can form two compounds—water and hydrogen peroxide. Note the differences in their structure.

Mini LAB

Comparing Compounds

Procedure

1. Collect the following substances—**granular sugar, rubbing alcohol,** and **salad oil.**
2. Observe the color, appearance, and state of each substance. Note the thickness or texture of each substance.
3. Stir a spoonful of each substance into separate **glasses** of **hot tap water** and observe.

Analysis

1. Compare the different properties of the substances.
2. The formulas of the three substances are made of only carbon, hydrogen, and oxygen. Infer how they can have different properties.

Try at Home

Compounds Have Formulas What's the difference between water and hydrogen peroxide? H_2O is the chemical formula for water, and H_2O_2 is the formula for hydrogen peroxide. The formula tells you which elements make up a compound as well as how many atoms of each element are present. Look at **Figure 19.** The subscript number written below and to the right of each element's symbol tells you how many atoms of that element exist in one unit of that compound. For example, hydrogen peroxide has two atoms of hydrogen and two atoms of oxygen. Water is made up of two atoms of hydrogen and one atom of oxygen.

Carbon dioxide, CO_2, is another common compound. Carbon dioxide is made up of one atom of carbon and two atoms of oxygen. Carbon and oxygen also can form the compound carbon monoxide, CO, which is a gas that is poisonous to all warm-blooded animals. As you can see, no subscript is used when only one atom of an element is present. A given compound always is made of the same elements in the same proportion. For example, water always has two hydrogen atoms for every oxygen atom, no matter what the source of the water is. No matter what quantity of the compound you have, the formula of the compound always remains the same. If you have 12 atoms of hydrogen and six atoms of oxygen, the compound is still written H_2O, but you have six molecules of H_2O (6 H_2O), not $H_{12}O_6$. The formula of a compound communicates its identity and makeup to any scientist in the world.

Reading Check *Propane has three carbon and eight hydrogen atoms. What is its chemical formula?*

88 CHAPTER 3 Atoms, Elements, and the Periodic Table

Mixtures

When two or more substances (elements or compounds) come together but don't combine to make a new substance, a **mixture** results. Unlike compounds, the proportions of the substances in a mixture can be changed without changing the identity of the mixture. For example, if you put some sand into a bucket of water, you have a mixture of sand and water. If you add more sand or more water, it's still a mixture of sand and water. Its identity has not changed. Air is another mixture. Air is a mixture of nitrogen, oxygen, and other gases, which can vary at different times and places. Whatever the proportion of gases, it is still air. Even your blood is a mixture that can be separated, as shown in **Figure 20,** by a machine called a centrifuge.

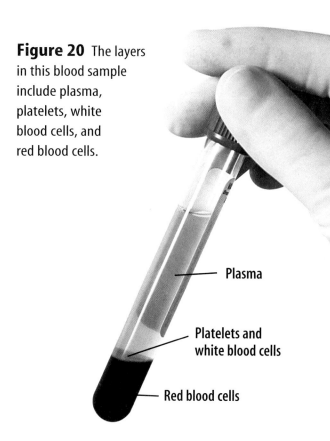

Figure 20 The layers in this blood sample include plasma, platelets, white blood cells, and red blood cells.

 *How do the proportions of a mixture relate to its identity?*

Applying Science

What's the best way to desalt ocean water?

You can't drink ocean water because it contains salt and other suspended materials. Or can you? In many areas of the world where drinking water is in short supply, methods for getting the salt out of salt water are being used to meet the demand for fresh water. Use your problem-solving skills to find the best method to use in a particular area.

Methods for Desalting Ocean Water			
Process	Amount of Water a Unit Can Desalt in a Day (m³)	Special Needs	Number of People Needed to Operate
Distillation	1,000 to 200,000	lots of energy to boil the water	many
Electrodialysis	10 to 4,000	stable source of electricity	1 to 2 persons

Identifying the Problem

The table above compares desalting methods. In distillation, the ocean water is heated. Pure water boils off and is collected, and the salt is left behind. Electrodialysis uses an electric current to pull salt particles out of water.

Solving the Problem

1. What method(s) might you use to desalt the water for a large population where energy is plentiful?
2. What method(s) would you choose to use in a single home?

Figure 21 Mixtures are part of your everyday life.

Topic: Mixtures
Visit ips.msscience.com for Web links to information about separating mixtures.

Activity Describe the difference between mixtures and compounds.

 Your blood is a mixture made up of elements and compounds. It contains white blood cells, red blood cells, water, and a number of dissolved substances. The different parts of blood can be separated and used by doctors in different ways. The proportions of the substances in your blood change daily, but the mixture does not change its identity.

Separating Mixtures Sometimes you can use a liquid to separate a mixture of solids. For example, if you add water to a mixture of sugar and sand, only the sugar dissolves in the water. The sand then can be separated from the sugar and water by pouring the mixture through a filter. Heating the remaining solution will separate the water from the sugar.

At other times, separating a mixture of solids of different sizes might be as easy as pouring them through successively smaller sieves or filters. A mixture of marbles, pebbles, and sand could be separated in this way.

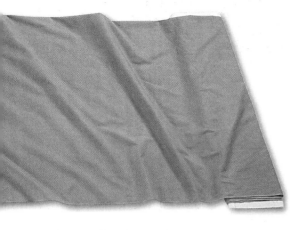

Homogeneous or Heterogeneous Mixtures, such as the ones shown in **Figure 21,** can be classified as homogeneous or heterogeneous. *Homogeneous* means "the same throughout." You can't see the different parts in this type of mixture. In fact, you might not always know that homogeneous mixtures are mixtures because you can't tell by looking. Which mixtures in **Figure 21** are homogeneous? No matter how closely you look, you can't see the individual parts that make up air or the parts of the mixture called brass in the lamp shown. Homogeneous mixtures can be solids, liquids, or gases.

A heterogeneous mixture has larger parts that are different from each other. You can see the different parts of a heterogeneous mixture, such as sand and water. How many heterogeneous mixtures are in **Figure 21?** A pepperoni and mushroom pizza is a tasty kind of heterogeneous mixture. Other examples of this kind of mixture include tacos, vegetable soup, a toy box full of toys, or a toolbox full of nuts and bolts.

Rocks and Minerals
Scientists called geologists study rocks and minerals. A mineral is composed of a pure substance. Rocks are mixtures and can be described as being homogeneous or heterogeneous. Research to learn more about rocks and minerals and note some examples of homogeneous and heterogeneous rocks in your Science Journal.

section 3 review

Summary

Substances
- A substance can be either an element or a compound.
- A compound contains more than one kind of element bonded together.
- A chemical formula shows which elements and how many atoms of each make up a compound.

Mixtures
- A mixture contains substances that are not chemically bonded together.
- There are many ways to separate mixtures, based on their physical properties.
- Homogeneous mixtures are those that are the same throughout. These types of mixtures can be solids, liquids, or gases.
- Heterogeneous mixtures have larger parts that are different from each other.

Self Check

1. **List** three examples of compounds and three examples of mixtures. Explain your choices.
2. **Determine** A container contains a mixture of sand, salt, and pebbles. How can each substance be separated from the others?
3. **Think Critically** Explain whether your breakfast was a compound, a homogeneous mixture, or a heterogeneous mixture.

Applying Skills

4. **Compare and contrast** compounds and mixtures based on what you have learned from this section.
5. **Use a Database** Use a computerized card catalog or database to find information about one element from the periodic table. Include information about the properties and uses of the mixtures and/or compounds in which the element is frequently found.

Mystery Mixture

Real-World Question

You will encounter many compounds that look alike. For example, a laboratory stockroom is filled with white powders. It is important to know what each is. In a kitchen, cornstarch, baking powder, and powdered sugar are compounds that look alike. To avoid mistaking one for another, you can learn how to identify them. Different compounds can be identified by using chemical tests. For example, some compounds react with certain liquids to produce gases. Other combinations produce distinctive colors. Some compounds have high melting points. Others have low melting points. How can the compounds in an unknown mixture be identified by experimentation?

Goals
- **Test** for the presence of certain compounds.
- **Decide** which of these compounds are present in an unknown mixture.

Materials
test tubes (4)
cornstarch
powdered sugar
baking soda
mystery mixture
small scoops (3)
dropper bottles (2)
iodine solution
white vinegar
hot plate
250-mL beaker
water (125 mL)
test-tube holder
small pie pan

Safety Precautions

WARNING: *Use caution when handling hot objects. Substances could stain or burn clothing. Be sure to point the test tube away from your face and your classmates while heating.*

Using Scientific Methods

Procedure

1. Copy the data table into your Science Journal. Record your results carefully for each of the following steps.
2. Place a small scoopful of cornstarch on the pie pan. Do the same for the sugar and baking soda making separate piles. Add a drop of vinegar to each. Wash and dry the pan after you record your observations.
3. Again, place a small scoopful of cornstarch, sugar, and baking soda on the pie pan. Add a drop of iodine solution to each one. Wash and dry the pan after you record your observations.
4. Again place a small scoopful of each compound in a separate test tube. Hold the test tube with the test-tube holder and with an oven mitt. Gently heat the test tube in a beaker of boiling water on a hot plate.
5. Follow steps 2 through 4 to test your mystery mixture for each compound.

Identifying Presence of Compounds

Substance to Be Tested	Fizzes with Vinegar	Turns Blue with Iodine	Melts When Heated
Cornstarch			
Sugar		Do not write in this book.	
Baking soda			
Mystery mix			

Analyze Your Data

Identify from your data table which compound(s) you have.

Conclude and Apply

1. **Describe** how you decided which substances were in your unknown mixture.
2. **Explain** how you would be able to tell if all three compounds were not in your mystery substance.
3. **Draw a Conclusion** What would you conclude if you tested baking powder from your kitchen and found that it fizzed with vinegar, turned blue with iodine, and did not melt when heated?

Communicating Your Data

Make a different data table to display your results in a new way. **For more help, refer to the** Science Skill Handbook.

TIME SCIENCE AND HISTORY
SCIENCE CAN CHANGE THE COURSE OF HISTORY!

Ancient Views of Matter

Two cultures observed the world around them differently

The world's earliest scientists were people who were curious about the world around them and who tried to develop explanations for the things they observed. This type of observation and inquiry flourished in ancient cultures such as those found in India and China. Read on to see how the ancient Indians and Chinese defined matter.

Indian Ideas

To Indians living about 3,000 years ago, the world was made up of five elements: fire, air, earth, water, and ether, which they thought of as an unseen substance that filled the heavens. Building upon this concept, the early Indian philosopher Kashyapa (kah SHI ah pah) proposed that the five elements could be broken down into smaller units called parmanu (par MAH new). Parmanu were similar to atoms in that they were too small to be seen but still retained the properties of the original element. Kashyapa also believed that each type of parmanu had unique physical and chemical properties.

Parmanu of earth elements, for instance, were heavier than parmanu of air elements. The different properties of the parmanu determined the characteristics of a substance. Kashyapa's ideas about matter are similar to those of the Greek philosopher Democritus, who lived centuries after Kashyapa.

Chinese Ideas

The ancient Chinese also broke matter down into five elements: fire, wood, metal, earth, and water. Unlike the early Indians, however, the Chinese believed that the elements constantly changed form. For example, wood can be burned and thus changes to fire. Fire eventually dies down and becomes ashes, or earth. Earth gives forth metals from the ground. Dew or water collects on these metals, and the water then nurtures plants that grow into trees, or wood.

This cycle of constant change was explained in the fourth century B.C. by the philosopher Tsou Yen. Yen, who is known as the founder of Chinese scientific thought, wrote that all changes that took place in nature were linked to changes in the five elements.

Research Write a brief paragraph that compares and contrasts the ancient Indian and Chinese views of matter. How are they different? Similar? Which is closer to the modern view of matter? Explain.

For more information, visit ips.msscience.com/time

chapter 3 Study Guide

Reviewing Main Ideas

Section 1 Structure of Matter

1. Matter is anything that occupies space and has mass.
2. Matter is made up of atoms.
3. Atoms are made of smaller parts called protons, neutrons, and electrons.
4. Many models of atoms have been created as scientists try to discover and define the atom's internal structure. Today's model has a central nucleus with the protons and neutrons, and an electron cloud surrounding it.

Section 2 The Simplest Matter

1. Elements are the building blocks of matter.

2. An element's atomic number tells how many protons its atoms contain, and its atomic mass tells the average mass of its atoms.
3. Isotopes are two or more atoms of the same element that have different numbers of neutrons.

Section 3 Compounds and Mixtures

1. Compounds are substances that are produced when elements combine. Compounds contain specific proportions of the elements that make them up.
2. Mixtures are combinations of compounds and elements that have not formed new substances. Their proportions can change.

Visualizing Main Ideas

Copy and complete the following concept map.

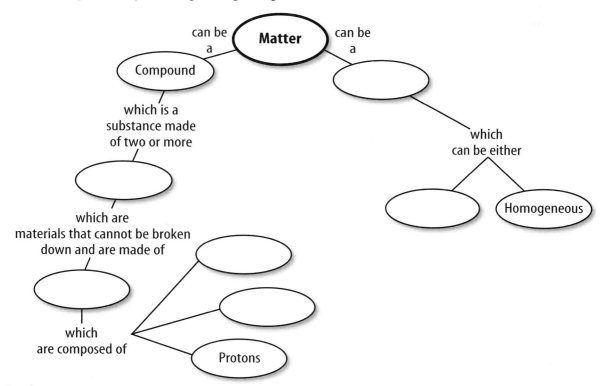

chapter 3 Review

Using Vocabulary

atom p. 73	matter p. 72
atomic mass p. 84	metal p. 84
atomic number p. 83	metalloid p. 85
compound p. 87	mixture p. 89
electron p. 76	neutron p. 78
element p. 80	nonmetal p. 85
isotope p. 83	nucleus p. 77
law of conservation of matter p. 74	proton p. 77
	substance p. 87
mass number p. 83	

Fill in the blanks with the correct vocabulary word or words.

1. The _____ is the particle in the nucleus of the atom that carries a positive charge and is counted to identify the atomic number.

2. The new substance formed when elements combine chemically is a(n) _____.

3. Anything that has mass and takes up space is _____.

4. The particles in the atom that account for most of the mass of the atom are protons and _____.

5. Elements that are shiny, malleable, ductile, good conductors of heat and electricity, and make up most of the periodic table are _____.

Checking Concepts

Choose the word or phrase that best answers the question.

6. What is a solution an example of?
   A) element
   B) heterogeneous mixture
   C) compound
   D) homogeneous mixture

7. The nucleus of one atom contains 12 protons and 12 neutrons, while the nucleus of another atom contains 12 protons and 16 neutrons. What are the atoms?
   A) chromium atoms
   B) two different elements
   C) two isotopes of an element
   D) negatively charged

8. What is a compound?
   A) a mixture of chemicals and elements
   B) a combination of two or more elements
   C) anything that has mass and occupies space
   D) the building block of matter

9. What does the atom consist of?
   A) electrons, protons, and alpha particles
   B) neutrons and protons
   C) electrons, protons, and neutrons
   D) elements, protons, and electrons

10. In an atom, where is an electron located?
    A) in the nucleus with the proton
    B) on the periodic table of the elements
    C) with the neutron
    D) in a cloudlike formation surrounding the nucleus

11. How is matter defined?
    A) the negative charge in an atom
    B) anything that has mass and occupies space
    C) the mass of the nucleus
    D) sound, light, and energy

12. What are two atoms that have the same number of protons called?
    A) metals
    B) nonmetals
    C) isotopes
    D) metalloids

13. Which is a heterogeneous mixture?
    A) air   C) a salad
    B) brass   D) apple juice

96 CHAPTER REVIEW   ips.msscience.com/vocabulary_puzzlemaker

Chapter 3 Review

Use the illustration below to answer questions 14 and 15.

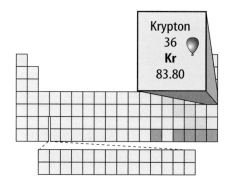

14. Using the figure above, krypton has
    A) an atomic number of 84.
    B) an atomic number of 36.
    C) an atomic mass of 36.
    D) an atomic mass of 72.

15. From the figure, the element krypton is
    A) a solid.   C) a mixture.
    B) a liquid.   D) a gas.

Thinking Critically

16. **Analyze Information** A chemical formula is written to indicate the makeup of a compound. What is the ratio of sulfur atoms to oxygen atoms in SO_2?

17. **Determine** which element contains seven protons.

18. **Describe** Using the periodic table, what are the atomic numbers for carbon, sodium, and nickel?

19. **Explain** how cobalt-60 and cobalt-59 can be the same element but have different mass numbers.

20. **Analyze Information** What did Rutherford's gold foil experiment tell scientists about atomic structure?

21. **Predict** Suppose Rutherford had bombarded aluminum foil with alpha particles instead of the gold foil he used in his experiment. What observations do you predict Rutherford would have made? Explain your prediction.

22. **Draw Conclusions** You are shown a liquid that looks the same throughout. You're told that it contains more than one type of element and that the proportion of each varies throughout the liquid. Is this an element, a compound, or a mixture?

Use the illustrations below to answer question 23.

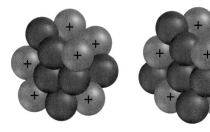

23. **Interpret Scientific Illustrations** Look at the two carbon atoms above. Explain whether or not the atoms are isotopes.

24. **Explain** how the atomic mass of an element is determined.

Performance Activities

25. **Newspaper Article** As a newspaper reporter in the year 1896, you have heard about the discovery of the electron. Research and write an article about the scientist and the discovery.

Applying Math

26. **Atomic Mass** Krypton has six naturally occurring isotopes with atomic masses of 78, 80, 82, 83, 84, and 86. Make a table of the number of protons, electrons, and neutrons in each isotope.

27. **Atomic Ratio** A researcher is analyzing two different compounds, sulfuric acid (H_2SO_4) and hydrogen peroxide (H_2O_2). What is the ratio of hydrogen to oxygen in sulfuric acid? What is the ratio of hydrogen to oxygen in hydrogen peroxide?

chapter 3 Standardized Test Practice

Part 1 Multiple Choice

Record your answers on the answer sheet provided by your teacher or on a sheet of paper.

1. Which of the scientists below introduced the idea that matter is made up of tiny, individual bits called atoms?
   A. Arrhenius
   B. Avogadro
   C. Chadwick
   D. Democritus

Use the illustration below to answer questions 2 and 3.

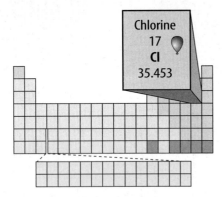

2. The periodic table block shown above lists properties of the element chlorine. What does the number 35.453 mean?
   A. the number of neutrons and in every chlorine atom
   B. the number of neutrons and protons in every chlorine atom
   C. the average number of neutrons in a chlorine atom
   D. the average number of neutrons and protons in a chlorine atom

3. According to the periodic table block, how many electrons does an uncharged atom of chlorine have?
   A. 17
   B. 18
   C. 35
   D. 36

Test-Taking Tip

Full Understanding Read each question carefully for full understanding.

4. Which of the following scientists envisioned the atom as a ball of positive charge with electrons embedded in it, much like chocolate chips spread through cookie dough?
   A. Crookes
   B. Dalton
   C. Thomson
   D. Rutherford

Use the illustration below to answer questions 5 and 6.

| 1 Proton | 1 Proton | 1 Proton |
| 0 Neutrons | 1 Neutron | 2 Neutrons |

5. Which of the following correctly identifies the three atoms shown in the illustration above?
   A. hydrogen, lithium, sodium
   B. hydrogen, helium, lithium
   C. hydrogen, helium, helium
   D. hydrogen, hydrogen, hydrogen

6. What is the mass number for each of the atoms shown in the illustration?
   A. 0, 1, 2
   B. 1, 1, 1
   C. 1, 2, 2
   D. 1, 2, 3

7. Which of the following are found close to the right side of the periodic table?
   A. metals
   B. lanthanides
   C. nonmetals
   D. metalloids

8. Which of the following is a characteristic that is typical of a solid, nonmetal element?
   A. shiny
   B. brittle
   C. good heat conductor
   D. good electrical conductor

Standardized Test Practice

Part 2 Short Response/Grid In

Record your answers on the answer sheet provided by your teacher or on a sheet of paper.

9. Are electrons more likely to be in an energy level close to the nucleus or far away from the nucleus? Why?

10. How many naturally-occurring elements are listed on the periodic table?

11. Is the human body made of mostly metal, nonmetals, or metalloids?

12. A molecule of hydrogen peroxide is composed of two atoms of hydrogen and two atoms of oxygen. What is the formula for six molecules of hydrogen peroxide?

13. What is the modern-day name for cathode rays?

Use the illustration below to answer questions 14 and 15.

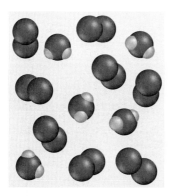

14. The illustration above shows atoms of an element and molecules of a compound that are combined without making a new compound. What term describes a combination such as this?

15. If the illustration showed only the element or only the compound, what term would describe it?

Part 3 Open Ended

Record your answers on a sheet of paper.

16. Describe Dalton's ideas about the composition of matter, including the relationship between atoms and elements.

Use the illustration below to answer questions 17 and 18.

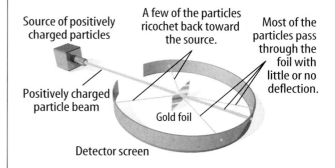

17. The illustration above shows Rutherford's gold foil experiment. Describe the setup shown. What result did Rutherford expect from his experiment?

18. What is the significance of the particles that reflected back from the gold foil? How did Rutherford explain his results?

19. Describe three possible methods for separating mixtures. Give an example for each method.

20. What are the rows and columns on the periodic table called? How are elements in the rows similar, and how are elements in the columns similar?

21. Describe how Thomson was able to show that cathode rays were streams of particles, not light.

22. Describe how the mass numbers, or atomic masses, listed on the periodic table for the elements are calculated.

chapter 4

States of Matter

The BIG Idea

The particles in solids, liquids, and gases are always in motion.

SECTION 1
Matter
Main Idea The state of matter depends on the motion of the particles and on the attractions between them.

SECTION 2
Changes of State
Main Idea When matter changes state, its thermal energy changes.

SECTION 3
Behavior of Fluids
Main Idea The particles in a fluid, a liquid, or a gas exert a force on everything they touch.

Ahhh!

A long, hot soak on a snowy day! This Asian monkey called a macaque is experiencing the effects of heat—the transfer of thermal energy from a warmer object to a colder object. In this chapter, you will learn about heat and the three common states of matter on Earth.

Science Journal Write why you think there is snow on the ground but the water is not frozen.

Start-Up Activities

Experiment with a Freezing Liquid

Have you ever thought about how and why you might be able to ice-skate on a pond in the winter but swim in the same pond in the summer? Many substances change form as temperature changes.

1. Make a table to record temperature and appearance. Obtain a test tube containing an unknown liquid from your teacher. Place the test tube in a rack.
2. Insert a thermometer into the liquid. **WARNING:** *Do not allow the thermometer to touch the bottom of the test tube.* Starting immediately, observe and record the substance's temperature and appearance every 30 s.
3. Continue making measurements and observations until you're told to stop.
4. **Think Critically** In your Science Journal, describe your investigation and observations. Did anything unusual happen while you were observing? If so, what?

 **Preview this chapter's content and activities at** ips.msscience.com

FOLDABLES Study Organizer

Changing States of Matter Make the following Foldable to help you study the changes in water.

STEP 1 Fold a vertical sheet of paper from left to right two times. Unfold.

STEP 2 Fold the paper in half from top to bottom two times.

STEP 3 Unfold and draw lines along the folds.

STEP 4 Label the top row and first column as shown below.

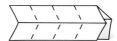

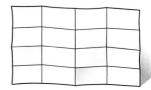

Read and Write As you read the chapter, define the states of matter as listed on your Foldable in the *Define States* column. Write what happens when heat is added to or lost from the three states of matter.

101

Get Ready to Read

Monitor

① Learn It! An important strategy to help you improve your reading is monitoring, or finding your reading strengths and weaknesses. As you read, monitor yourself to make sure the text makes sense. Discover different monitoring techniques you can use at different times, depending on the type of test and situation.

② Practice It! The paragraph below appears in Section 1. Read the passage and answer the questions that follow. Discuss your answers with other students to see how they monitor their reading.

> All matter is made up of tiny particles, such as atoms, molecules, or ions. Each particle attracts other particles. In other words, each particle pulls other particles toward itself. These particles also are constantly moving. The motion of the particles and the strength of attraction between the particles determine a material's state of matter.
>
> — *from page 102*

- What questions do you still have after reading?
- Do you understand all of the words in the passage?
- Did you have to stop reading often? Is the reading level appropriate for you?

③ Apply It! Identify one paragraph that is difficult to understand. Discuss it with a partner to improve your understanding.

Target Your Reading

Reading Tip
Monitor your reading by slowing down or speeding up depending on your understanding of the text.

Use this to focus on the main ideas as you read the chapter.

① **Before you read** the chapter, respond to the statements below on your worksheet or on a numbered sheet of paper.
- Write an **A** if you **agree** with the statement.
- Write a **D** if you **disagree** with the statement.

② **After you read** the chapter, look back to this page to see if you've changed your mind about any of the statements.
- If any of your answers changed, explain why.
- Change any false statements into true statements.
- Use your revised statements as a study guide.

Science Online
Print out a worksheet of this page at ips.msscience.com

Before You Read A or D		Statement	After You Read A or D
	1	Particles in solids vibrate in place.	
	2	A water spider can walk on water because of uneven forces acting on the surface water molecules.	
	3	Particles in a gas are far apart with empty space between them.	
	4	A large glass of warm water has the same amount of thermal energy as a smaller glass of water at the same temperature.	
	5	Boiling and evaporation are two types of vaporization.	
	6	While a substance is boiling, its temperature increases.	
	7	Pressure is, in part, related to the area over which a force is distributed.	
	8	At sea level, the air exerts a pressure of about 101,000 N per square meter.	
	9	An object will float in a fluid that is denser than itself.	

102 B

section 1
Matter

as you read

What You'll Learn
- **Recognize** that matter is made of particles in constant motion.
- **Relate** the three states of matter to the arrangement of particles within them.

Why It's Important
Everything you can see, taste, and touch is matter.

Review Vocabulary
atom: a small particle that makes up most types of matter

New Vocabulary
- matter
- solid
- liquid
- viscosity
- surface tension
- gas

What is matter?

Take a look at the beautiful scene in **Figure 1.** What do you see? Perhaps you notice the water and ice. Maybe you are struck by the Sun in the background. All of these images show examples of matter. **Matter** is anything that takes up space and has mass. Matter doesn't have to be visible—even air is matter.

States of Matter All matter is made up of tiny particles, such as atoms, molecules, or ions. Each particle attracts other particles. In other words, each particle pulls other particles toward itself. These particles also are constantly moving. The motion of the particles and the strength of attraction between the particles determine a material's state of matter.

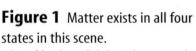 *What determines a material's state of matter?*

There are three familiar states of matter—solid, liquid, and gas. A fourth state of matter known as plasma occurs at extremely high temperatures. Plasma is found in stars, lightning, and neon lights. Although plasma is common in the universe, it is not common on Earth. For that reason, this chapter will focus only on the three states of matter that are common on Earth.

Figure 1 Matter exists in all four states in this scene.
Identify *the solid, liquid, gas, and plasma in this photograph.*

Solids

What makes a substance a solid? Think about some familiar solids. Chairs, floors, rocks, and ice cubes are a few examples of matter in the solid state. What properties do all solids share? A **solid** is matter with a definite shape and volume. For example, when you pick up a rock from the ground and place it in a bucket, it doesn't change shape or size. A solid does not take the shape of a container in which it is placed. This is because the particles of a solid are packed closely together, as shown in **Figure 2**.

Particles in Motion The particles that make up all types of matter are in constant motion. Does this mean that the particles in a solid are moving too? Although you can't see them, a solid's particles are vibrating in place. The particles do not have enough energy to move out of their fixed positions.

Reading Check *What motion do solid particles have?*

Figure 2 The particles in a solid vibrate in place while maintaining a constant shape and volume.

Crystalline Solids In some solids, the particles are arranged in a repeating, three-dimensional pattern called a crystal. These solids are called crystalline solids. In **Figure 3** you can see the arrangement of particles in a crystal of sodium chloride, which is table salt. The particles in the crystal are arranged in the shape of a cube. Diamond, another crystalline solid, is made entirely of carbon atoms that form crystals that look more like pyramids. Sugar, sand, and snow are other crystalline solids.

Figure 3 The particles in a crystal of sodium chloride (NaCl) are arranged in an orderly pattern.

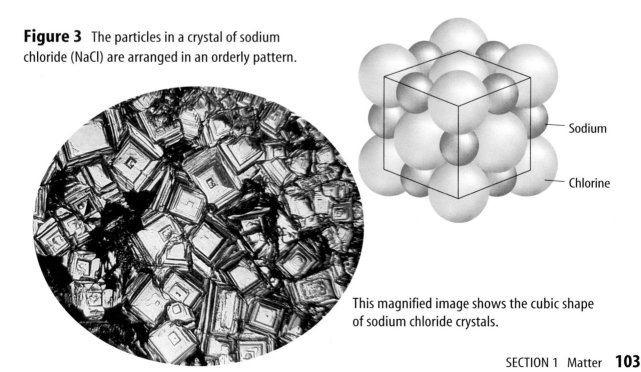

This magnified image shows the cubic shape of sodium chloride crystals.

Amorphous Solids Some solids come together without forming crystal structures. These solids often consist of large particles that are not arranged in a repeating pattern. Instead, the particles are found in a random arrangement. These solids are called amorphous (uh MOR fuhs) solids. Rubber, plastic, and glass are examples of amorphous solids.

Fresh Water Early settlers have always decided to build their homes near water. The rivers provided ways for people to travel, drinking water for themselves and their animals, and irrigation for farming. Over time, small communities became larger communities with industry building along the same water.

✓ Reading Check *How is a crystalline solid different from an amorphous solid?*

Liquids

From the orange juice you drink with breakfast to the water you use to brush your teeth at night, matter in the liquid state is familiar to you. How would you describe the characteristics of a liquid? Is it hard like a solid? Does it keep its shape? A **liquid** is matter that has a definite volume but no definite shape. When you pour a liquid from one container to another, the liquid takes the shape of the container. The volume of a liquid, however, is the same no matter what the shape of the container. If you pour 50 mL of juice from a carton into a pitcher, the pitcher will contain 50 mL of juice. If you then pour that same juice into a glass, its shape will change again but its volume will not.

Free to Move The reason that a liquid can have different shapes is because the particles in a liquid move more freely, as shown in **Figure 4,** than the particles in a solid. The particles in a liquid have enough energy to move out of their fixed positions but not enough energy to move far apart.

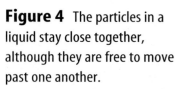

Figure 4 The particles in a liquid stay close together, although they are free to move past one another.

Liquid

104 CHAPTER 4 States of Matter

Viscosity Do all liquids flow the way water flows? You know that honey flows more slowly than water and you've probably heard the phrase "slow as molasses." Some liquids flow more easily than others. A liquid's resistance to flow is known as the liquid's **viscosity.** Honey has a high viscosity. Water has a lower viscosity. The slower a liquid flows, the higher its viscosity is. The viscosity results from the strength of the attraction between the particles of the liquid. For many liquids, viscosity increases as the liquid becomes colder.

Surface Tension If you're careful, you can float a needle on the surface of water. This is because attractive forces cause the particles on the surface of a liquid to pull themselves together and resist being pushed apart. You can see in **Figure 5** that particles beneath the surface of a liquid are pulled in all directions. Particles at the surface of a liquid are pulled toward the center of the liquid and sideways along the surface. No liquid particles are located above to pull on them. The uneven forces acting on the particles on the surface of a liquid are called **surface tension.** Surface tension causes the liquid to act as if a thin film were stretched across its surface. As a result you can float a needle on the surface of water. For the same reason, the water spider can move around on the surface of a pond or lake. When a liquid is present in small amounts, surface tension causes the liquid to form small droplets.

Science Online

Topic: Plasma
Visit ips.msscience.com for Web links to information about the states of matter.

Activity List four ways that plasma differs from the other three states of matter

Figure 5 Surface tension exists because the particles at the surface experience different forces than those at the center of the liquid.

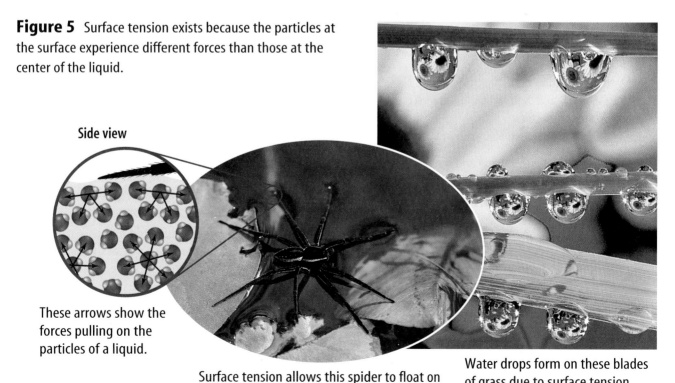

Side view

These arrows show the forces pulling on the particles of a liquid.

Surface tension allows this spider to float on water as if the water had a thin film.

Water drops form on these blades of grass due to surface tension.

Gases

Unlike solids and liquids, most gases are invisible. The air you breathe is a mixture of gases. The gas in the air bags in **Figure 6** and the helium in some balloons are examples of gases. **Gas** is matter that does not have a definite shape or volume. The particles in gas are much farther apart than those in a liquid or solid. Gas particles move at high speeds in all directions. They will spread out evenly, as far apart as possible. If you poured a small volume of a liquid into a container, the liquid would stay in the bottom of the container. However, if you poured the same volume of a gas into a container, the gas would fill the container completely. A gas can expand or be compressed. Decreasing the volume of the container squeezes the gas particles closer together.

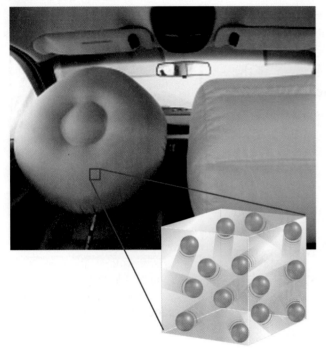

Figure 6 The particles in gas move at high speeds in all directions. The gas inside these air bags spreads out to fill the entire volume of the bag.

Vapor Matter that exists in the gas state but is generally a liquid or solid at room temperature is called vapor. Water, for example, is a liquid at room temperature. Thus, water vapor is the term for the gas state of water.

section 1 review

Summary

What is matter?
- Matter is anything that takes up space and has mass. Solid, liquid, and gas are the three common states of matter.

Solids
- Solids have a definite volume and shape.
- Solids with particles arranged in order are called crystalline solids. The particles in amorphous solids are not in any order.

Liquids
- Liquids have definite volume but no defined shape.
- Viscosity is a measure of how easily liquids flow.

Gases
- Gases have no definite volume or shape.
- Vapor refers to gaseous substances that are normally liquids or solids at room temperature.

Self Check

1. **Define** the two properties of matter that determine its state.
2. **Describe** the movement of particles within solids, liquids, and gases.
3. **Name** the property that liquids and solids share. What property do liquids and gases share?
4. **Infer** A scientist places 25 mL of a yellow substance into a 50-mL container. The substance quickly fills the entire container. Is it a solid, liquid, or gas?
5. **Think Critically** The particles in liquid A have a stronger attraction to each other than the particles in liquid B. If both liquids are at the same temperature, which liquid has a higher viscosity? Explain.

Applying Skills

6. **Concept Map** Draw a Venn diagram in your Science Journal and fill in the characteristics of the states of matter.

ips.msscience.com/self_check_quiz

section 2

Changes of State

Thermal Energy and Heat

Shards of ice fly from the sculptor's chisel. As the crowd looks on, a swan slowly emerges from a massive block of ice. As the day wears on, however, drops of water begin to fall from the sculpture. Drip by drip, the sculpture is transformed into a puddle of liquid water. What makes matter change from one state to another? To answer this question, you need to think about the particles that make up matter.

Energy Simply stated, energy is the ability to do work or cause change. The energy of motion is called kinetic energy. Particles within matter are in constant motion. The amount of motion of these particles depends on the kinetic energy they possess. Particles with more kinetic energy move faster and farther apart. Particles with less energy move more slowly and stay closer together.

The total kinetic and potential energy of all the particles in a sample of matter is called **thermal energy.** Thermal energy, an extensive property, depends on the number of particles in a substance as well as the amount of energy each particle has. If either the number of particles or the amount of energy in each particle changes, the thermal energy of the sample changes. With identically sized samples, the warmer substance has the greater thermal energy. In **Figure 7,** the particles of hot water from the hot spring have more thermal energy than the particles of snow on the surrounding ground.

as you read

What **You'll Learn**
- **Define and compare** thermal energy and temperature.
- **Relate** changes in thermal energy to changes of state.
- **Explore** energy and temperature changes on a graph.

Why **It's Important**
Matter changes state as it heats up or cools down.

Review Vocabulary
energy: the ability to do work or cause change

New Vocabulary
- thermal energy
- temperature
- heat
- melting
- freezing
- vaporization
- condensation

Figure 7 These girls are enjoying the water from the hot spring. **Infer** why the girls appear to be comfortable in the hot spring while there is snow on the ground.

SECTION 2 Changes of State **107**

Figure 8 The particles in hot tea move faster than those in iced tea. The temperature of hot tea is higher than the temperature of iced tea.
Identify *which tea has the higher kinetic energy.*

Temperature Not all of the particles in a sample of matter have the same amount of energy. Some have more energy than others. The average kinetic energy of the individual particles is the **temperature,** an intensive property, of the substance. You can find an average by adding up a group of numbers and dividing the total by the number of items in the group. For example, the average of the numbers 2, 4, 8, and 10 is (2 + 4 + 8 + 10) ÷ 4 = 6. Temperature is different from thermal energy because thermal energy is a total and temperature is an average.

You know that the iced tea is colder than the hot tea, as shown in **Figure 8.** Stated differently, the temperature of iced tea is lower than the temperature of hot tea. You also could say that the average kinetic energy of the particles in the iced tea is less than the average kinetic energy of the particles in the hot tea.

Heat When a warm object is brought near a cooler object, thermal energy will be transferred from the warmer object to the cooler one. The movement of thermal energy from a substance at a higher temperature to one at a lower temperature is called **heat.** When a substance is heated, it gains thermal energy. Therefore, its particles move faster and its temperature rises. When a substance is cooled, it loses thermal energy, which causes its particles to move more slowly and its temperature to drop.

Types of Energy Thermal energy is one of several different forms of energy. Other forms include the chemical energy in chemical compounds, the electrical energy used in appliances, the electromagnetic energy of light, and the nuclear energy stored in the nucleus of an atom. Make a list of examples of energy that you are familiar with.

Reading Check *How is heat related to temperature?*

Specific Heat

As you study more science, you will discover that water has many unique properties. One of those is the amount of heat required to increase the temperature of water as compared to most other substances. The specific heat of a substance is the amount of heat required to raise the temperature of 1 g of a substance 1°C.

Substances that have a low specific heat, such as most metals and the sand in **Figure 9,** heat up and cool down quickly because they require only small amounts of heat to cause their temperatures to rise. A substance with a high specific heat, such as the water in **Figure 9,** heats up and cools down slowly because a much larger quantity of heat is required to cause its temperature to rise or fall by the same amount.

Changes Between the Solid and Liquid States

Matter can change from one state to another when thermal energy is absorbed or released. This change is known as change of state. The graph in **Figure 11** shows the changes in temperature as thermal energy is gradually added to a container of ice.

Melting As the ice in **Figure 11** is heated, it absorbs thermal energy and its temperature rises. At some point, the temperature stops rising and the ice begins to change into liquid water. The change from the solid state to the liquid state is called **melting.** The temperature at which a substance changes from a solid to a liquid is called the melting point. The melting point of water is 0°C.

Amorphous solids, such as rubber and glass, don't melt in the same way as crystalline solids. Because they don't have crystal structures to break down, these solids get softer and softer as they are heated, as you can see in **Figure 10.**

Figure 9 The specific heat of water is greater than that of sand. The energy provided by the Sun raises the temperature of the sand much faster than the water.

Figure 10 Rather than melting into a liquid, glass gradually softens. Glass blowers use this characteristic to shape glass into beautiful vases while it is hot.

SECTION 2 Changes of State 109

NATIONAL GEOGRAPHIC VISUALIZING STATES OF MATTER

Figure 11

Like most substances, water can exist in three distinct states—solid, liquid, or gas. At certain temperatures, water changes from one state to another. This diagram shows what changes occur as water is heated or cooled.

MELTING When ice melts, its temperature remains constant until all the ice turns to water. Continued heating of liquid water causes the molecules to vibrate even faster, steadily raising the temperature.

FREEZING When liquid water freezes, it releases thermal energy and turns into the solid state, ice.

VAPORIZATION When water reaches its boiling point of 100°C, water molecules are moving so fast that they break free of the attractions that hold them together in the liquid state. The result is vaporization—the liquid becomes a gas. The temperature of boiling water remains constant until all of the liquid turns to steam.

CONDENSATION When steam is cooled, it releases thermal energy and turns into its liquid state. This process is called condensation.

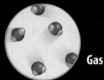

Solid state: ice

Liquid state: water

Gaseous state: steam

Freezing The process of melting a crystalline solid can be reversed if the liquid is cooled. The change from the liquid state to the solid state is called **freezing.** As the liquid cools, it loses thermal energy. As a result, its particles slow down and come closer together. Attractive forces begin to trap particles, and the crystals of a solid begin to form. As you can see in **Figure 11,** freezing and melting are opposite processes.

The temperature at which a substance changes from the liquid state to the solid state is called the freezing point. The freezing point of the liquid state of a substance is the same temperature as the melting point of the solid state. For example, solid water melts at 0°C and liquid water freezes at 0°C.

During freezing, the temperature of a substance remains constant while the particles in the liquid form a crystalline solid. Because particles in a liquid have more energy than particles in a solid, energy is released during freezing. This energy is released into the surroundings. After all of the liquid has become a solid, the temperature begins to decrease again.

Topic: Freezing Point Study
Visit ips.msscience.com for Web links to information about freezing.

Activity Make a list of several substances and the temperatures at which they freeze. Find out how the freezing point affects how the substance is used.

Applying Science

How can ice save oranges?

During the spring, Florida citrus farmers carefully watch the fruit when temperatures drop close to freezing. When the temperatures fall below 0°C, the liquid in the cells of oranges can freeze and expand. This causes the cells to break, making the oranges mushy and the crop useless for sale. To prevent this, farmers spray the oranges with water just before the temperature reaches 0°C. How does spraying oranges with water protect them?

Identifying the Problem

Using the diagram in **Figure 11,** consider what is happening to the water at 0°C. Two things occur. What are they?

Solving the Problem
1. What change of state and what energy changes occur when water freezes?
2. How does the formation of ice on the orange help the orange?

SECTION 2 Changes of State **111**

Mini LAB

Observing Vaporization

Procedure
1. Use a **dropper** to place one drop of **rubbing alcohol** on the back of your hand.
2. Describe how your hand feels during the next 2 min.
3. Wash your hands.

Analysis
1. What changes in the appearance of the rubbing alcohol did you notice?
2. What sensation did you feel during the 2 min? How can you explain this sensation?
3. Infer how sweating cools the body.

Changes Between the Liquid and Gas States

After an early morning rain, you and your friends enjoy stomping through the puddles left behind. But later that afternoon when you head out to run through the puddles once more, the puddles are gone. The liquid water in the puddles changed into a gas. Matter changes between the liquid and gas states through vaporization and condensation.

Vaporization As liquid water is heated, its temperature rises until it reaches 100°C. At this point, liquid water changes into water vapor. The change from a liquid to a gas is known as **vaporization** (vay puh ruh ZAY shun). You can see in **Figure 11** that the temperature of the substance does not change during vaporization. However, the substance absorbs thermal energy. The additional energy causes the particles to move faster until they have enough energy to escape the liquid as gas particles.

Two forms of vaporization exist. Vaporization that takes place below the surface of a liquid is called boiling. When a liquid boils, bubbles form within the liquid and rise to the surface, as shown in **Figure 12**. The temperature at which a liquid boils is called the boiling point. The boiling point of water is 100°C.

Vaporization that takes place at the surface of a liquid is called evaporation. Evaporation, which occurs at temperatures below the boiling point, explains how puddles dry up. Imagine that you could watch individual water molecules in a puddle. You would notice that the molecules move at different speeds. Although the temperature of the water is constant, remember that temperature is a measure of the average kinetic energy of the molecules. Some of the fastest-moving molecules overcome the attractive forces of other molecules and escape from the surface of the water.

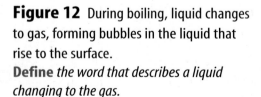

Figure 12 During boiling, liquid changes to gas, forming bubbles in the liquid that rise to the surface.
Define *the word that describes a liquid changing to the gas.*

112 CHAPTER 4 States of Matter

Figure 13 The drops of water on these glasses and pitcher of lemonade were formed when water vapor in the air lost enough energy to return to the liquid state. This process is called condensation.

Location of Molecules It takes more than speed for water molecules to escape the liquid state. During evaporation, these faster molecules also must be near the surface, heading in the right direction, and they must avoid hitting other water molecules as they leave. With the faster particles evaporating from the surface of a liquid, the particles that remain are the slower, cooler ones. Evaporation cools the liquid and anything near the liquid. You experience this cooling effect when perspiration evaporates from your skin.

Condensation Pour a nice, cold glass of lemonade and place it on the table for a half hour on a warm day. When you come back to take a drink, the outside of the glass will be covered by drops of water, as shown in **Figure 13.** What happened? As a gas cools, its particles slow down. When particles move slowly enough for their attractions to bring them together, droplets of liquid form. This process, which is the opposite of vaporization, is called **condensation.** As a gas condenses to a liquid, it releases the thermal energy it absorbed to become a gas. During this process, the temperature of the substance does not change. The decrease in energy changes the arrangement of particles. After the change of state is complete, the temperature continues to drop, as you saw in **Figure 11.**

Topic: Condensation
Visit ips.msscience.com for Web links to information about how condensation is involved in weather.

Activity Find out how condensation is affected by the temperature as well as the amount of water in the air.

 *What energy change occurs during condensation?*

Condensation formed the droplets of water on the outside of your glass of lemonade. In the same way, water vapor in the atmosphere condenses to form the liquid water droplets in clouds. When the droplets become large enough, they can fall to the ground as rain.

SECTION 2 Changes of State

Figure 14 The solid carbon dioxide (dry ice) at the bottom of this beaker of water is changing directly into gaseous carbon dioxide. This process is called sublimation.

Changes Between the Solid and Gas States

Some substances can change from the solid state to the gas state without ever becoming a liquid. During this process, known as sublimation, the surface particles of the solid gain enough energy to become a gas. One example of a substance that undergoes sublimation is dry ice. Dry ice is the solid form of carbon dioxide. It often is used to keep materials cold and dry. At room temperature and pressure, carbon dioxide does not exist as a liquid. Therefore, as dry ice absorbs thermal energy from the objects around it, it changes directly into a gas. When dry ice becomes a gas, it absorbs thermal energy from water vapor in the air. As a result, the water vapor cools and condenses into liquid water droplets, forming the fog you see in **Figure 14.**

---

section 2 review

Summary

Thermal Energy and Heat
- Thermal energy depends on the amount of the substance and the kinetic energy of particles in the substance.
- Heat is the movement of thermal energy from a warmer substance to a cooler one.

Specific Heat
- Specific heat is a measure of the amount of energy required to raise 1 g of a substance 1°C.

Changes Between Solid and Liquid States
- During all changes of state, the temperature of a substance stays the same.

Changes Between Liquid and Gas States
- Vaporization is the change from the liquid state to a gaseous state.
- Condensation is the change from the gaseous state to the liquid state.

Changes Between Solid and Gas States
- Sublimation is the process of a substance going from the solid state to the gas state without ever being in the liquid state.

Self Check

1. **Describe** how thermal energy and temperature are similar. How are they different?
2. **Explain** how a change in thermal energy causes matter to change from one state to another. Give two examples.
3. **List** the three changes of state during which energy is absorbed.
4. **Describe** the two types of vaporization.
5. **Think Critically** How can the temperature of a substance remain the same even if the substance is absorbing thermal energy?
6. **Write** a paragraph in your Science Journal that explains why you can step out of the shower into a warm bathroom and begin to shiver.

Applying Math

7. **Make and Use Graphs** Use the data you collected in the Launch Lab to plot a temperature-time graph. At what temperature does the graph level off? What was the liquid doing during this time period?
8. **Use Numbers** If sample A requires 10 calories to raise the temperature of a 1-g sample 1°C, how many calories does it take to raise a 5-g sample 10°C?

ips.msscience.com/self_check_quiz

The Water Cycle

Water is all around us and you've used water in all three of its common states. This lab will give you the opportunity to observe the three states of matter and to discover for yourself if ice really melts at 0°C and if water boils at 100°C.

Characteristics of Water Sample

Time (min)	Temperature (°C)	Physical State
	Do not write in this book.	

Real-World Question

How does the temperature of water change as it is heated from a solid to a gas?

Goals
- **Measure** the temperature of water as it heats.
- **Observe** what happens as the water changes from one state to another.
- **Graph** the temperature and time data.

Materials
hot plate
ice cubes (100 mL)
Celsius thermometer
*electronic temperature probe
wall clock
*watch with second hand
stirring rod
250-mL beaker
*Alternate materials

Safety Precautions

Procedure

1. Make a data table similar to the table shown.
2. Put 150 mL of water and 100 mL of ice into the beaker and place the beaker on the hot plate. Do not touch the hot plate.
3. Put the thermometer into the ice/water mixture. Do not stir with the thermometer or allow it to rest on the bottom of the beaker. After 30 s, read and record the temperature in your data table.
4. Plug in the hot plate and turn the temperature knob to the medium setting.
5. Every 30 s, read and record the temperature and physical state of the water until it begins to boil. Use the stirring rod to stir the contents of the beaker before making each temperature measurement. Stop recording. Allow the water to cool.

Analyze Your Data

Use your data to make a graph plotting time on the x-axis and temperature on the y-axis. Draw a smooth curve through the data points.

Conclude and Apply

1. **Describe** how the temperature of the ice/water mixture changed as you heated the beaker.
2. **Describe** the shape of the graph during any changes of state.

Communicating Your Data

Add labels to your graph. Use the detailed graph to explain to your class how water changes state. **For more help, refer to the Science Skill Handbook.**

LAB **115**

section 3
Behavior of Fluids

as you read

What You'll Learn
- **Explain** why some things float but others sink.
- **Describe** how pressure is transmitted through fluids.

Why It's Important
Pressure enables you to squeeze toothpaste from a tube, and buoyant force helps you float in water.

Review Vocabulary
force: a push or pull

New Vocabulary
- pressure
- buoyant force
- Archimedes' principle
- density
- Pascal's principle

Pressure

It's a beautiful summer day when you and your friends go outside to play volleyball, much like the kids in **Figure 15**. There's only one problem—the ball is flat. You pump air into the ball until it is firm. The firmness of the ball is the result of the motion of the air particles in the ball. As the air particles in the ball move, they collide with one another and with the inside walls of the ball. As each particle collides with the inside walls, it exerts a force, pushing the surface of the ball outward. A force is a push or a pull. The forces of all the individual particles add together to make up the pressure of the air.

Pressure is equal to the force exerted on a surface divided by the total area over which the force is exerted.

$$\text{pressure} = \frac{\text{force}}{\text{area}}$$

When force is measured in newtons (N) and area is measured in square meters (m^2), pressure is measured in newtons per square meter (N/m^2). This unit of pressure is called a pascal (Pa). A more useful unit when discussing atmospheric pressure is the kilopascal (kPa), which is 1,000 pascals.

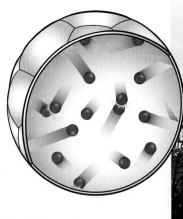

Figure 15 Without the pressure of air inside this volleyball, the ball would be flat.

116 CHAPTER 4 States of Matter

Figure 16 The force of the dancer's weight on pointed toes results in a higher pressure than the same force on flat feet. **Explain** why the pressure is higher.

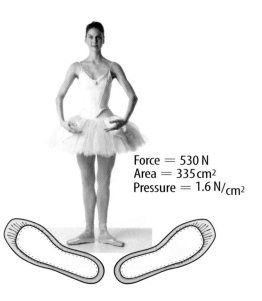

Force = 530 N
Area = 335 cm²
Pressure = 1.6 N/cm²

Force = 530 N
Area = 37 cm²
Pressure = 14 N/cm²

Force and Area You can see from the equation on the opposite page that pressure depends on the quantity of force exerted and the area over which the force is exerted. As the force increases over a given area, pressure increases. If the force decreases, the pressure will decrease. However, if the area changes, the same amount of force can result in different pressure. **Figure 16** shows that if the force of the ballerina's weight is exerted over a smaller area, the pressure increases. If that same force is exerted over a larger area, the pressure will decrease.

Reading Check *What variables does pressure depend on?*

Atmospheric Pressure You can't see it and you usually can't feel it, but the air around you presses on you with tremendous force. The pressure of air also is known as atmospheric pressure because air makes up the atmosphere around Earth. Atmospheric pressure is 101.3 kPa at sea level. This means that air exerts a force of about 101,000 N on every square meter it touches. This is approximately equal to the weight of a large truck.

It might be difficult to think of air as having pressure when you don't notice it. However, you often take advantage of air pressure without even realizing it. Air pressure, for example, enables you to drink from a straw. When you first suck on a straw, you remove the air from it. As you can see in **Figure 17,** air pressure pushes down on the liquid in your glass then forces liquid up into the straw. If you tried to drink through a straw inserted into a sealed, airtight container, you would not have any success because the air would not be able to push down on the surface of the drink.

Figure 17 The downward pressure of air pushes the juice up into the straw.

Air pressure

SECTION 3 Behavior of Fluids **117**

Balanced Pressure If air is so forceful, why don't you feel it? The reason is that the pressure exerted outward by the fluids in your body balances the pressure exerted by the atmosphere on the surface of your body. Look at **Figure 18.** The atmosphere exerts a pressure on all surfaces of the dancer's body. She is not crushed by this pressure because the fluids in her body exert a pressure that balances atmospheric pressure.

Variations in Atmospheric Pressure Atmospheric pressure changes with altitude. Altitude is the height above sea level. As altitude increases atmospheric pressure decreases. This is because fewer air particles are found in a given volume. Fewer particles have fewer collisions, and therefore exert less pressure. This idea was tested in the seventeenth century by a French physician named Blaise Pascal. He designed an experiment in which he filled a balloon only partially with air. He then had the balloon carried to the top of a mountain. **Figure 19** shows that as Pascal predicted, the balloon expanded while being carried up the mountain. Although the amount of air inside the balloon stayed the same, the air pressure pushing in on it from the outside decreased. Consequently, the particles of air inside the balloon were able to spread out further.

Figure 18 Atmospheric pressure exerts a force on all surfaces of this dancer's body.
Explain *why she can't feel this pressure.*

Figure 19 Notice how the balloon expands as it is carried up the mountain. The reason is that atmospheric pressure decreases with altitude. With less pressure pushing in on the balloon, the gas particles within the balloon are free to expand.

118 CHAPTER 4 States of Matter

Air Travel If you travel to higher altitudes, perhaps flying in an airplane or driving up a mountain, you might feel a popping sensation in your ears. As the air pressure drops, the air pressure in your ears becomes greater than the air pressure outside your body. The release of some of the air trapped inside your ears is heard as a pop. Airplanes are pressurized so that the air pressure within the cabin does not change dramatically throughout the course of a flight.

Changes in Gas Pressure

In the same way that atmospheric pressure can vary as conditions change, the pressure of gases in confined containers also can change. The pressure of a gas in a closed container changes with volume and temperature.

Pressure and Volume If you squeeze a portion of a filled balloon, the remaining portion of the balloon becomes more firm. By squeezing it, you decrease the volume of the balloon, forcing the same number of gas particles into a smaller space. As a result, the particles collide with the walls more often, thereby producing greater pressure. This is true as long as the temperature of the gas remains the same. You can see the change in the motion of the particles in **Figure 20.** What will happen if the volume of a gas increases? If you make a container larger without changing its temperature, the gas particles will collide less often and thereby produce a lower pressure.

Predicting a Waterfall
Procedure
1. Fill a **plastic cup** to the brim with **water.**
2. Cover the top of the cup with an **index card.**
3. Predict what will happen if you turn the cup upside down.
4. While holding the index card in place, turn the cup upside down over a sink. Then let go of the card.

Analysis
1. What happened to the water when you turned the cup?
2. How can you explain your observation in terms of the concept of fluid pressure?

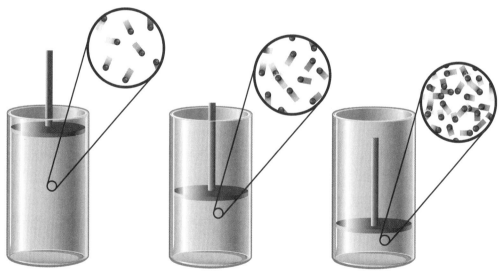

Figure 20 As volume decreases, pressure increases.

As the piston is moved down, the gas particles have less space and collide more often. The pressure increases.

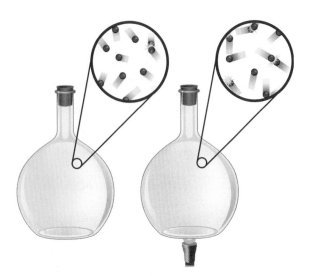

Pressure and Temperature When the volume of a confined gas remains the same, the pressure can change as the temperature of the gas changes. You have learned that temperature rises as the kinetic energy of the particles in a substance increases. The greater the kinetic energy is, the faster the particles move. The faster the speed of the particles is, the more they collide and the greater the pressure is. If the temperature of a confined gas increases, the pressure of the gas will increase, as shown in **Figure 21.**

 *Why would a sealed container of air be crushed after being frozen?*

Figure 21 Even though the volume of this container does not change, the pressure increases as the substance is heated.
Describe *what will happen if the substance is heated too much.*

Float or Sink

You may have noticed that you feel lighter in water than you do when you climb out of it. While you are under water, you experience water pressure pushing on you in all directions. Just as air pressure increases as you walk down a mountain, water pressure increases as you swim deeper in water. Water pressure increases with depth. As a result, the pressure pushing up on the bottom of an object is greater than the pressure pushing down on it because the bottom of the object is deeper than the top.

The difference in pressure results in an upward force on an object immersed in a fluid, as shown in **Figure 22.** This force is known as the **buoyant force.** If the buoyant force is equal to the weight of an object, the object will float. If the buoyant force is less than the weight of an object, the object will sink.

Figure 22 The pressure pushing up on an immersed object is greater than the pressure pushing down on it. This difference results in the buoyant force.

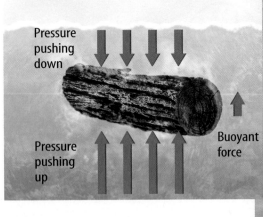

Weight is a force in the downward direction. The buoyant force is in the upward direction. An object will float if the upward force is equal to the downward force.

120 CHAPTER 4 States of Matter

Archimedes' Principle What determines the buoyant force? According to **Archimedes'** (ar kuh MEE deez) **principle,** the buoyant force on an object is equal to the weight of the fluid displaced by the object. In other words, if you place an object in a beaker that already is filled to the brim with water, some water will spill out of the beaker, as in **Figure 23.** If you weigh the spilled water, you will find the buoyant force on the object.

Density Understanding density can help you predict whether an object will float or sink. **Density** is mass divided by volume.

$$\text{density} = \frac{\text{mass}}{\text{volume}}$$

An object will float in a fluid that is more dense than itself and sink in a fluid that is less dense than itself. If an object has the same density, the object will neither sink nor float but instead stay at the same level in the fluid.

Figure 23 When the golf ball was dropped in the large beaker, it displaced some of the water, which was collected and placed into the smaller beaker. **Communicate** what you know about the weight and the volume of the displaced water.

Applying Math — Find an Unknown

CALCULATING DENSITY You are given a sample of a solid that has a mass of 10.0 g and a volume of 4.60 cm³. Will it float in liquid water, which has a density of 1.00 g/cm³?

Solution

1 *This is what you know:*
- mass = 10.0 g
- volume = 4.60 cm³
- density of water = 1.00 g/cm³

2 *This is what you need to find:* the density of the sample

3 *This is the procedure you need to use:*
- density = mass/volume
- density = 10.0 g/4.60 cm³ = 2.17 g/cm³
- The density of the sample is greater than the density of water. The sample will sink.

4 *Check your answer:*
- Find the mass of your sample by multiplying the density and the volume.

Practice Problems

1. A 7.40-cm³ sample of mercury has a mass of 102 g. Will it float in water?
2. A 5.0-cm³ sample of aluminum has a mass of 13.5 g. Will it float in water?

For more practice, visit
ips.msscience.com/
math_practice

SECTION 3 Behavior of Fluids

Figure 24 A hydraulic lift utilizes Pascal's principle to help lift this car and this dentist's chair.

Pascal's Principle

What happens if you squeeze a plastic container filled with water? If the container is closed, the water has nowhere to go. As a result, the pressure in the water increases by the same amount everywhere in the container—not just where you squeeze or near the top of the container. When a force is applied to a confined fluid, an increase in pressure is transmitted equally to all parts of the fluid. This relationship is known as **Pascal's principle.**

Hydraulic Systems You witness Pascal's principle when a car is lifted up to have its oil changed or if you are in a dentist's chair as it is raised or lowered, as shown in **Figure 24.** These devices, known as hydraulic (hi DRAW lihk) systems, use Pascal's principle to increase force. Look at the tube in **Figure 25.** The force applied to the piston on the left increases the pressure within the fluid. That increase in pressure is transmitted to the piston on the right. Recall that pressure is equal to force divided by area. You can solve for force by multiplying pressure by area.

$$\text{pressure} = \frac{\text{force}}{\text{area}} \quad \text{or} \quad \text{force} = \text{pressure} \times \text{area}$$

If the two pistons on the tube have the same area, the force will be the same on both pistons. If, however, the piston on the right has a greater surface area than the piston on the left, the resulting force will be greater. The same pressure multiplied by a larger area equals a greater force. Hydraulic systems enable people to lift heavy objects using relatively small forces.

Figure 25 By increasing the area of the piston on the right side of the tube, you can increase the force exerted on the piston. In this way a small force pushing down on the left piston can result in a large force pushing up on the right piston. The force can be great enough to lift a car.

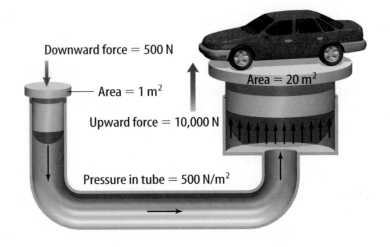

122 CHAPTER 4 States of Matter

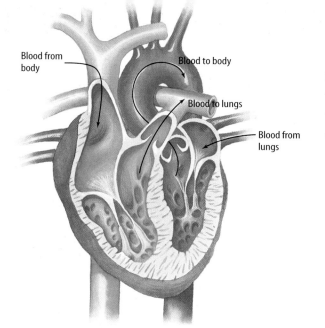

Figure 26 The heart is responsible for moving blood throughout the body. Two force pumps work together to move blood to and from the lungs and to the rest of the body.

Force Pumps If an otherwise closed container has a hole in it, any fluid in the container will be pushed out the opening when you squeeze it. This arrangement, known as a force pump, makes it possible for you to squeeze toothpaste out of a tube or mustard from a plastic container.

 Your heart has two force pumps. One pump pushes blood to the lungs, where it picks up oxygen. The other force pump pushes the oxygen-rich blood to the rest of your body. These pumps are shown in **Figure 26**.

Topic: Blood Pressure
Visit ips.msscience.com for Web links to information about blood pressure. Find out what the term means, how it changes throughout the human body, and why it is unhealthy to have high blood pressure.

Activity Write a paragraph in your Science Journal that explains why high blood pressure is dangerous.

section 3 review

Summary

Pressure
- Pressure depends on force and area.
- The air around you exerts a pressure.
- The pressure inside your body matches the pressure exerted by air.

Changes in Gas Pressure
- The pressure exerted by a gas depends on its volume and its temperature.

Float or Sink
- Whether an object floats or sinks depends on its density relative to the density of the fluid it's in.

Pascal's Principle
- This principle relates pressure and area to force.

Self Check

1. **Describe** what happens to pressure as the force exerted on a given area increases.
2. **Describe** how atmospheric pressure changes as altitude increases.
3. **State** Pascal's principle in your own words.
4. **Infer** An object floats in a fluid. What can you say about the buoyant force on the object?
5. **Think Critically** All the air is removed from a sealed metal can. After the air has been removed, the can looks as if it were crushed. Why?

Applying Math

6. **Simple Equations** What pressure is created when 5.0 N of force are applied to an area of 2.0 m^2? How does the pressure change if the force is increased to 10.0 N? What about if instead the area is decreased to 1.0 m^2?

LAB Design Your Own

Design Your Own Ship

Goals
- **Design** an experiment that uses Archimedes' principle to determine the size of ship needed to carry a given amount of cargo in such a way that the top of the ship is even with the surface of the water.

Possible Materials
balance
small plastic cups (2)
graduated cylinder
metric ruler
scissors
marbles (cupful)
sink
*basin, pan, or bucket
*Alternate materials

Safety Precautions

Real-World Question
It is amazing to watch ships that are taller than buildings float easily on water. Passengers and cargo are carried on these ships in addition to the tremendous weight of the ship itself. How can you determine the size of a ship needed to keep a certain mass of cargo afloat?

Form a Hypothesis
Think about Archimedes' principle and how it relates to buoyant force. Form a hypothesis to explain how the volume of water displaced by a ship relates to the mass of cargo the ship can carry.

Cargo ship

Test Your Hypothesis

Make a Plan

1. Obtain a set of marbles or other items from your teacher. This is the cargo that your ship must carry. Think about the type of ship

you will design. Consider the types of materials you will use. Decide how your group is going to test your hypothesis.

2. **List** the steps you need to follow to test your hypothesis. Include in your plan how you will measure the mass of your ship and cargo, calculate the volume of water your ship must displace in order to float with its cargo, and measure the volume and mass of the displaced water. Also, explain how you will design your ship so that it will float with the top of the ship even with the surface of the water. Make the ship.

3. **Prepare** a data table in your Science Journal to use as your group collects data. Think about what data you need to collect.

Follow Your Plan

1. Make sure your teacher approves your plan before you start.

2. Perform your experiment as planned. Be sure to follow all proper safety procedures. In particular, clean up any spilled water immediately.

3. Record your observations carefully and complete the data table in your Science Journal.

Analyze Your Data

1. **Write** your calculations showing how you determined the volume of displaced water needed to make your ship and cargo float.

2. Did your ship float at the water's surface, sink, or float above the water's surface? Draw a diagram of your ship in the water.

3. **Explain** how your experimental results agreed or failed to agree with your hypothesis.

Conclude and Apply

1. If your ship sank, how would you change your experiment or calculations to correct the problem? What changes would you make if your ship floated too high in the water?

2. What does the density of a ship's cargo have to do with the volume of cargo the ship can carry? What about the density of the water?

Communicating Your Data

Compare your results with other students' data. Prepare a combined data table or summary showing how the calculations affect the success of the ship. **For more help, refer to the** Science Skill Handbook.

Oops! Accidents in SCIENCE

SOMETIMES GREAT DISCOVERIES HAPPEN BY ACCIDENT!

The Incredible Stretching Goo

A serious search turns up a toy

During World War II, when natural resources were scarce and needed for the war effort, the U.S. government asked an engineer to come up with an inexpensive alternative to synthetic rubber. While researching the problem and looking for solutions, the engineer dropped boric acid into silicone oil. The result of these two substances mixing together was—a goo!

Because of its molecular structure, the goo could bounce and stretch in all directions. The engineer also discovered the goo could break into pieces. When strong pressure is applied to the substance, it reacts like a solid and breaks apart. Even though the combination was versatile—and quite amusing, the U.S. government decided the new substance wasn't a good substitute for synthetic rubber.

A few years later, the recipe for the stretch material fell into the hands of a businessperson, who saw the goo's potential—as a toy. The toymaker paid $147 for rights to the boric acid and silicone oil mixture. And in 1949 it was sold at toy stores for the first time. The material was packaged in a plastic egg and it took the U.S. by storm. Today, the acid and oil mixture comes in a multitude of colors and almost every child has played with it at some time.

The substance can be used for more than child's play. Its sticky consistency makes it good for cleaning computer keyboards and removing small specks of lint from fabrics.

People use it to make impressions of newspaper print or comics. Athletes strengthen their grips by grasping it over and over. Astronauts use it to anchor tools on spacecraft in zero gravity. All in all, a most *eggs-cellent* idea!

Research As a group, examine a sample of the colorful, sticky, stretch toy made of boric acid and silicone oil. Then brainstorm some practical—and impractical—uses for the substance.

Science online
For more information, visit ips.msscience.com/oops

chapter 4 Study Guide

Reviewing Main Ideas

Section 1 Matter

1. All matter is composed of tiny particles that are in constant motion.
2. In the solid state, the attractive force between particles holds them in place to vibrate.
3. Particles in the liquid state have defined volumes and are free to move about within the liquid.

Section 2 Changes of State

1. Thermal energy is the total energy of the particles in a sample of matter. Temperature is the average kinetic energy of the particles in a sample.
2. An object gains thermal energy when it changes from a solid to a liquid, or when it changes from a liquid to a gas.
3. An object loses thermal energy when it changes from a gas to a liquid, or when it changes from a liquid to a solid.

Section 3 Behavior of Fluids

1. Pressure is force divided by area.
2. Fluids exert a buoyant force in the upward direction on objects immersed in them.
3. An object will float in a fluid that is more dense than itself.
4. Pascal's principle states that pressure applied to a liquid is transmitted evenly throughout the liquid.

Visualizing Main Ideas

Copy and complete the following concept map on matter.

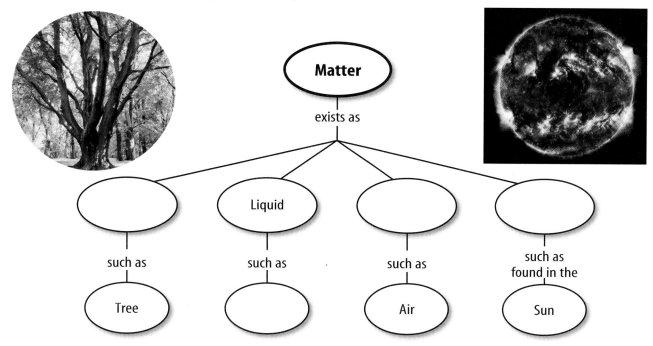

Chapter 4 Review

Using Vocabulary

Archimedes' principle p. 121
buoyant force p. 120
condensation p. 113
density p. 121
freezing p. 111
gas p. 106
heat p. 108
liquid p. 104
matter p. 102
melting p. 109
Pascal's principle p. 122
pressure p. 116
solid p. 103
surface tension p. 105
temperature p. 108
thermal energy p. 107
vaporization p. 112
viscosity p. 105

Fill in the blanks with the correct vocabulary word.

1. A(n) _____ can change shape and volume.
2. A(n) _____ has a different shape but the same volume in any container.
3. _____ is thermal energy moving from one substance to another.
4. _____ is a measure of the average kinetic energy of the particles of a substance.
5. A substance changes from a gas to a liquid during the process of _____.
6. A liquid becomes a gas during _____.
7. _____ is mass divided by volume.
8. _____ is force divided by area.
9. _____ explains what happens when force is applied to a confined fluid.

Checking Concepts

Choose the word or phrase that best answers the question.

10. Which of these is a crystalline solid?
    A) glass
    B) sugar
    C) rubber
    D) plastic

11. Which description best describes a solid?
    A) It has a definite shape and volume.
    B) It has a definite shape but not a definite volume.
    C) It adjusts to the shape of its container.
    D) It can flow.

12. What property enables you to float a needle on water?
    A) viscosity
    B) temperature
    C) surface tension
    D) crystal structure

13. What happens to an object as its kinetic energy increases?
    A) It holds more tightly to nearby objects.
    B) Its mass increases.
    C) Its particles move more slowly.
    D) Its particles move faster.

14. During which process do particles of matter release energy?
    A) melting
    B) freezing
    C) sublimation
    D) boiling

15. How does water vapor in air form clouds?
    A) melting
    B) evaporation
    C) condensation
    D) sublimation

16. Which is a unit of pressure?
    A) N
    B) kg
    C) g/cm^3
    D) N/m^2

17. Which change results in an increase in gas pressure in a balloon?
    A) decrease in temperature
    B) decrease in volume
    C) increase in volume
    D) increase in altitude

18. In which case will an object float on a fluid?
    A) Buoyant force is greater than weight.
    B) Buoyant force is less than weight.
    C) Buoyant force equals weight.
    D) Buoyant force equals zero.

chapter 4 Review

Use the photo below to answer question 19.

19. In the photo above, the water in the small beaker was displaced when the golf ball was added to the large beaker. What principle does this show?
    A) Pascal's principle
    B) the principle of surface tension
    C) Archimedes' principle
    D) the principle of viscosity

20. Which is equal to the buoyant force on an object?
    A) volume of the object
    B) weight of the displaced fluid
    C) weight of object
    D) volume of fluid

Thinking Critically

21. **Explain** why steam causes more severe burns than boiling water.

22. **Explain** why a bathroom mirror becomes fogged while you take a shower.

23. **Form Operational Definitions** Write operational definitions that explain the properties of and differences among solids, liquids, and gases.

24. **Determine** A king's crown has a volume of 110 cm³ and a mass of 1,800 g. The density of gold is 19.3 g/cm³. Is the crown pure gold?

25. **Infer** Why do some balloons pop when they are left in sunlight for too long?

Performance Activities

26. **Storyboard** Create a visual-aid storyboard to show ice changing to steam. There should be a minimum of five frames.

Applying Math

Use the graph below to answer question 27.

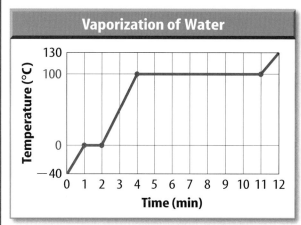

27. **Explain** how this graph would change if a greater volume of water were heated. How would it stay the same?

Use the table below to answer question 28.

Water Pressure			
Depth (m)	Pressure (atm)	Depth (m)	Pressure (atm)
0	1.0	100	11.0
25	3.5	125	13.5
50	6.0	150	16.0
75	8.5	175	18.5

28. **Make and Use Graphs** In July of 2001, Yasemin Dalkilic of Turkey dove to a depth of 105 m without any scuba equipment. Make a depth-pressure graph for the data above. Based on your graph, how does water pressure vary with depth? Note: The pressure at sea level, 101.3 kPa, is called one atmosphere (atm).

Chapter 4 Standardized Test Practice

Part 1 Multiple Choice

Record your answers on the answer sheet provided by your teacher or on a sheet of paper.

1. In which state of matter do particles stay close together, yet are able to move past one another?
   A. solid   C. liquid
   B. gas   D. plasma

Use the illustration below to answer questions 2 and 3.

2. Which statement is true about the volume of the water displaced when the golf ball was dropped into the large beaker?
   A. It is equal to the volume of the golf ball.
   B. It is greater than the volume of the golf ball.
   C. It is less than the volume of the golf ball.
   D. It is twice the volume of a golf ball.

3. What do you know about the buoyant force on the golf ball?
   A. It is equal to the density of the water displaced.
   B. It is equal to the volume of the water displaced.
   C. It is less than the weight of the water displaced.
   D. It is equal to the weight of the water displaced.

4. What is the process called when a gas cools to form a liquid?
   A. condensation   C. boiling
   B. sublimation   D. freezing

5. Which of the following is an amorphous solid?
   A. diamond   C. glass
   B. sugar   D. sand

6. Which description best describes a liquid?
   A. It has a definite shape and volume.
   B. It has a definite volume but not a definite shape.
   C. It expands to fill the shape and volume of its container.
   D. It cannot flow.

7. During which processes do particles of matter absorb energy?
   A. freezing and boiling
   B. condensation and melting
   C. melting and vaporization
   D. sublimation and freezing

Use the illustration below to answer questions 8 and 9.

8. What happens as the piston moves down?
   A. The volume of the gas increases.
   B. The volume of the gas decreases.
   C. The gas particles collide less often.
   D. The pressure of the gas decreases.

9. What relationship between the volume and pressure of a gas does this illustrate?
   A. As volume decreases, pressure decreases.
   B. As volume decreases, pressure increases.
   C. As volume decreases, pressure remains the same.
   D. As the volume increases, pressure remains the same.

Standardized Test Practice

Part 2 | Short Response/Grid In

Record your answers on the answer sheet provided by your teacher or on a sheet of paper.

10. A balloon filled with helium bursts in a closed room. What space will the helium occupy?

Use the illustration below to answer questions 11 and 12.

11. If the force exerted by the dancer is 510 N, what is the pressure she exerts if the area is 335 cm^2 on the left and 37 cm^2 on the right?

12. Compare the pressure the dancer would exert on the floor if she were wearing large clown shoes to the photo on the left.

13. If a balloon is blown up and tied closed, air is held inside it. What will happen to the balloon if it is then pushed into hot water or held over a heater? Why does this happen?

14. What is the relationship of heat and thermal energy?

15. Why are some insects able to move around on the surface of a lake or pond?

16. How does the weight of a floating object compare with the buoyant force acting on the object?

17. What is the mass of an object that has a density of 0.23 g/cm^3 and whose volume is 52 cm^3?

Part 3 | Open Ended

Record your answer on a sheet of paper.

18. Compare and contrast evaporation and boiling.

Use the illustration below to answer questions 19 and 20.

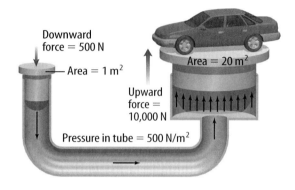

19. Name and explain the principle that is used in lifting the car.

20. Explain what would happen if you doubled the area of the piston on the right side of the hydraulic system.

21. Explain why a woman might put dents in a wood floor when walking across it in high-heeled shoes, but not when wearing flat sandals.

22. Explain why the tires on a car might become flattened on the bottom after sitting outside in very cold weather.

23. Compare the arrangement and movement of the particles in a solid, a liquid, and a gas.

24. Explain why the water in a lake is much cooler than the sand on the beach around it on a sunny summer day.

Test-Taking Tip

Show Your Work For open-ended questions, show all of your work and any calculations on your answer sheet.

Hint: In question 20, the pressure in the tube does not change.

chapter 5

Matter— Properties and Changes

The BIG Idea
Properties and changes of matter can be classified as either chemical or physical.

SECTION 1
Physical Properties
Main Idea Physical properties of a substance can be observed without changing the identity of the substance.

SECTION 2
Chemical Properties
Main Idea A chemical property is a substance's ability to change into another substance.

SECTION 3
Physical and Chemical Changes
Main Idea Unlike a physical change, a chemical change involves changing one substance into a different substance.

Why do icebergs float?
This iceberg once was part of an Antarctic ice shelf. Density is a physical property. Ice floats in water because it is less dense than water. It underwent a physical change when it became an iceberg. In this chapter you'll learn about other physical and chemical properties and the changes associated with them.

Science Journal What happens to a swimming pool when the correct chemicals are not added to the water?

Start-Up Activities

Classifying Different Types of Matter

Using your senses to observe characteristics of matter will help you classify, or categorize, it. This will help you understand what the types of matter are and can help you identify unknown types of matter. In this lab, you will observe and compare the characteristics of two items that you might be familiar with.

1. Obtain a table-tennis ball and a golf ball from your teacher.
2. How are the two balls similar?
3. Which ball is heavier?
4. Compare the surfaces of the table-tennis ball and the golf ball. How are their surfaces different?
5. Place each ball in water and observe.
6. **Think Critically** Create a classification system to classify different kinds of balls. Which characteristics might you use? Describe your classification system in your Science Journal.

 Preview this chapter's content and activities at ips.msscience.com

 **Properties and Changes of Matter** Make the following Foldable to help you organize types of properties and changes into groups based on their common features.

STEP 1 Fold a sheet of paper in half lengthwise. Make the back edge about 1.25 cm longer than the front edge.

STEP 2 Fold in half, then fold in half again to make three folds.

STEP 3 Unfold and cut only the top layer along the three folds to make four tabs.

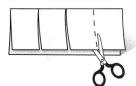

STEP 4 Label the tabs as shown.

Find Main Ideas As you read the chapter, list examples of each type of property and each type of change under the appropriate tabs.

Get Ready to Read

Visualize

① Learn It! Visualize by forming mental images of the text as you read. Imagine how the text descriptions look, sound, feel, smell, or taste. Look for any pictures or diagrams on the page that may help you add to your understanding.

② Practice It! Read the following paragraph. As you read, use the underlined details to form a picture in your mind.

> Unprotected cars driven on salted roads and steel structures like the one shown in **Figure 14** can begin to rust after only a few winters. A shiny copper penny becomes dull and dark. An apple left out too long begins to turn brown. What do all these changes have in common? Each of these changes is a chemical change.
>
> — from page 145

Based on the description above, try to visualize chemical changes. Now look at the photo on page 145.
- How closely does it match your mental picture?
- Reread the passage and look at the picture again. Did your ideas change?
- Compare your image with what others in your class visualized.

③ Apply It! Read the chapter and list three subjects you were able to visualize. Make a rough sketch showing what you visualized.

Target Your Reading

Reading Tip

Forming your own mental images will help you remember what you read.

Use this to focus on the main ideas as you read the chapter.

① **Before you read** the chapter, respond to the statements below on your worksheet or on a numbered sheet of paper.
- Write an **A** if you **agree** with the statement.
- Write a **D** if you **disagree** with the statement.

② **After you read** the chapter, look back to this page to see if you've changed your mind about any of the statements.
- If any of your answers changed, explain why.
- Change any false statements into true statements.
- Use your revised statements as a study guide.

Science Online

Print out a worksheet of this page at ips.msscience.com

Before You Read A or D		Statement	After You Read A or D
	1	A large ice cube is denser than a small ice cube.	
	2	All acids are too dangerous to touch.	
	3	Liquid water has different chemical properties than solid water.	
	4	Soap is an example of a base.	
	5	A chemical property of a substance describes its ability to change to a different substance.	
	6	Salts are the result of the reaction between an acid and a base.	
	7	A chemical change occurs when water boils.	
	8	Chemical changes occur in your body.	
	9	A bubbling liquid is a sure sign of a chemical change.	

section 1

Physical Properties

as you read

What You'll Learn
- **Describe** the common physical properties of matter.
- **Explain** how to find the density of a substance.
- **Compare and contrast** the properties of acids and bases.

Why It's Important
When you learn about physical properties, you can better describe the world around you.

Review Vocabulary
matter: anything that has mass and takes up space

New Vocabulary
- physical property
- density
- state of matter
- size-dependent property
- size-independent property

Physical Properties

Have you ever been asked by a teacher to describe something that you saw on a field trip? How would you describe the elephant in the exhibit shown in **Figure 1?** What features can you use in your description—color, shape, size, and texture? These features are all properties, or characteristics, of the elephant. Scientists use the term *physical property* to describe a characteristic of matter that you can detect with your senses. A **physical property** is any characteristic of matter that can be observed without changing the identity of the material. All matter, such as the elephant, has physical properties.

Common Physical Properties You probably are familiar with some physical properties, such as color, shape, smell, and taste. You might not be as familiar with others, such as mass, volume, and density. Mass (m) is the amount of matter in an object. A golf ball has more mass than a table-tennis ball. Volume (V) is the amount of space that matter takes up. A swimming pool holds a larger volume of water than a paper cup does. **Density** (D) is the amount of mass in a given volume. A golf ball is more dense than a table-tennis ball. Density is determined by finding the mass of a sample of matter and dividing this mass by the volume of the sample.

Formula for Density
Density = mass/volume   or $D = \dfrac{m}{V}$

Figure 1 This large gray African elephant is displayed on the main floor of the National Museum of Natural History in Washington, D.C.

Density A table-tennis ball and a golf ball are about the same volume. When you decided which had a higher density, you compared their masses. Because they are about the same volume, the one with more mass had the higher density. Suppose you were asked if all the bowling balls in **Figure 2** were identical. They appear to be the same size, shape, and color, but do they all have the same mass? If you could pick up these bowling balls, you would discover that their masses differ. You also might notice that the heavier balls strike the pins harder. Although the volumes of the balls are nearly identical, the densities of the bowling balls are different because their masses are different.

Identifying Unknown Substances In some cases, density also can be used to identify unknown compounds and elements. The element silver, for example, has a density of 10.5 g/cm^3 at 20°C. Suppose you want to know whether or not a ring is pure silver. You can find the ring's density by dividing the mass of the ring by its volume. If the density of the ring is determined to be 11.3 g/cm^3, then the ring is not pure silver.

Figure 2 These bowling balls look the same but have different densities. **Identify** *the types of matter you think you would see, hear, taste, touch, and smell at a bowling alley.*

Applying Math — Solve a One-Step Equation

DETERMINING DENSITY An antique dealer decided to use density to help determine the material used to make a statue. The volume of the statue is 1,000 cm^3 and the mass is 8,470 g. What is its density?

Solution

1 *This is what you know:*
- density = mass/volume = m/V
- m = 8,470 g, V = 1,000 cm^3

2 *This is what you need to find out:* Find the density (D)

3 *This is the procedure you need to use:*
- $D = m/V$
- $D = m/V$ = 8,470 g/1,000 cm^3
  = 8.470 g/cm^3

4 *Check your answer:* Substitute the density and one of the knowns back into the main equation. Did you calculate the other known?

Practice Problems

1. If a candlestick has a mass of 8.5 g and a volume of 0.96 cm^3, what is its density?

2. If the density of a plastic ball is 5.4 g/cm^3 and the volume is 7.5 cm^3, what is the mass of the plastic ball?

For more practice, visit ips.msscience.com/math_practice

Figure 3 All three states of water are present here—solid, liquid, and gas—but you can only see the solid and liquid states. The water vapor in the air is not visible.

State of Matter State of matter is another physical property. The **state of matter** tells you whether a sample of matter is a solid, a liquid, or a gas. This property depends on the temperature and pressure of the matter. The ice in **Figure 3** is water in the solid state. Water in the liquid state can be seen in the ocean and in the clouds. Gaseous water cannot be seen but exists as vapor in the air. In each case, each molecule of water is the same—two hydrogen atoms and one oxygen atom. But water appears to be different because it exists in different states, as shown in **Figure 3**.

Size-Dependent and Size-Independent Properties Some physical properties change when the size of an object changes. These properties are called **size-dependent properties.** For example, a wooden block might have a volume of 30 cm³ and a mass of 20 g. A larger block might have a volume of 60 cm³ and a mass of 40 g. The volume and mass of the block change when the size of the block changes. However, the density of both blocks is 0.67 g/cm³. Some physical properties do not change when an object changes size. Density is an example of a **size-independent property.** Other examples of size-dependent and size-independent properties are shown in **Table 1.**

Classifying Properties
Procedure
1. Obtain three different-sized **blocks** of the same type of wood.
2. Write all your observations of each block in your **Science Journal** as you make your measurements.
3. Measure the length, width, height, and mass of each block. Calculate the volume and density of each block.

Analysis
1. Which properties were size-dependent?
2. Which properties were size-independent?

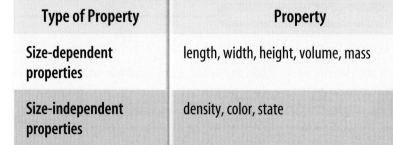

Table 1 Physical Properties	
Type of Property	Property
Size-dependent properties	length, width, height, volume, mass
Size-independent properties	density, color, state

Physical Properties of Acids and Bases

One way to describe matter is to classify it as either an acid or a base. The concentration of an acid or base can be determined by finding the pH of the sample. The pH scale has a range of 0 to 14. Acids have a pH below 7. Bases have a pH above 7. A sample with a pH of exactly 7 is neutral—neither acidic nor basic. Pure water is a substance with a pH of exactly 7.

Properties of Acids What do you think of when you hear the word *acid*? Do you picture a dangerous chemical that can burn your skin, make holes in your clothes, and even destroy metal? Some acids, such as concentrated hydrochloric acid, are like that. But some acids are edible. One example is shown in **Figure 4.** Carbonated soft drinks contain acids. Every time you eat a citrus fruit such as an orange or a grapefruit, you eat citric and ascorbic (uh SOR bihk) acids. What properties do these and other acids have in common?

Imagine the sharp smell of a freshly sliced lemon. That scent comes from the citric acid in the fruit. Take a big bite out of the fruit shown in **Figure 5** and you would immediately notice a sour taste. If you then rubbed your molars back and forth, your teeth would squeak. All of these physical properties are common in acids.

Reading Check *What are two uses of an acid?*

Figure 4 When you sip a carbonated soft drink, you drink carbonic and phosphoric (faws FOR ihk) acids.
Identify *an area of your body where acids are found.*

Aging Vitamin C and alpha-hydroxy acids are found in fruits and are the active ingredient in some anti-aging skin creams. It is believed that these ingredients slow down the aging process. Researchers examine safety issues regarding these products as well as their components.

Figure 5 All citrus fruits contain citric and ascorbic acids, which is why these fruits taste sour.

SECTION 1 Physical Properties

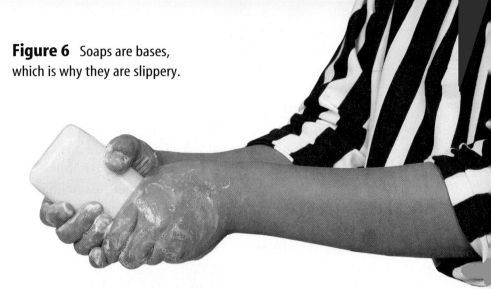

Figure 6 Soaps are bases, which is why they are slippery.

Topic: Acids and Bases
Visit ips.msscience.com for Web links to information about acid and base reactions.

Activity List some common and industrial uses for acids and bases.

Physical Properties of Bases Bases have physical properties that are different from acids. A familiar example of a base is ammonia (uh MOH nyuh), often used for household cleaning. If you got a household cleaner that contained ammonia on your fingers and then rubbed your fingers together, they would feel slippery. Another familiar base is soap, shown in **Figure 6,** which also has a slippery feel. You shouldn't taste soap, but if you accidentally did, you'd notice a bitter taste. A bitter taste and a slippery feel are physical properties of bases.

Reading Check What are two examples of products that contain bases?

It is important to note that you should never taste, touch, or smell anything in a lab unless your teacher tells you to do so.

section 1 review

Summary

Physical Properties
- Characteristics that can be observed without changing the identity of the object.
- Color, shape, smell, taste, mass, volume, and density are all physical properties.
- Mass and volume are size-dependent properties of an object.

Density
- density = mass/volume

Properties of Acids and Bases
- Acids smell sharp, taste sour, and have a pH below 7.
- Bases are slippery, taste bitter, and have a pH above 7.

Self Check

1. **Describe** the physical properties of a baseball.
2. **Explain** why density is a size-independent property. How does it differ from a size-dependent property?
3. **Describe** Give an example of an acid and a base. How do they differ from a neutral substance?
4. **Think Critically** How could you identify a pure metal if you have a balance, a graduated cylinder, and a table of densities for metals? (1 mL = 1 cm^3)

Applying Math

5. **Solve One-Step Equations** What is the density of a substance with a mass of 65.7 g and a volume of 3.40 cm^3?

section 2
Chemical Properties

A Complete Description

You've observed that the density of a table-tennis ball is less than the density of a golf ball. You also have noticed the state of water in an ice cube and in a lake. You've noticed the taste of acid in a lemon and the slippery feel of a base such as soap. However, a description of something using only physical properties is not complete. What type of property describes how matter behaves?

Common Chemical Properties If you strike a match on a hard, rough surface, the match probably will start to burn. Phosphorus (FAWS for us) compounds on the match head and the wood in the match combine with oxygen to form new materials. Why does that happen? The phosphorus compounds and the wood have the ability to burn. The ability to burn is a chemical property. A **chemical property** is a characteristic of matter that allows it to change to a different type of matter.

Reading Check *What is a chemical property?*

You see an example of a chemical property when you leave a half-eaten apple on your desk, and the exposed part turns brown. The property you observe is the ability to react with oxygen. Two other chemical properties are shown in **Figure 7**.

as you read

What You'll Learn
- **Describe** chemical properties of matter.
- **Explain** the chemical properties of acids and bases.
- **Explain** how a salt is formed.

Why It's Important
Chemical properties can help you predict how matter will change.

Review Vocabulary
solubility: the amount of a substance that will dissolve in a given amount of another substance

New Vocabulary
- chemical property
- reactivity
- salts

Figure 7 The chemical properties of a material often require a warning about its careful use. Gas pumps warn customers not to get near them with anything that might start the gasoline burning. Workers who use toxic chemicals have to wear protective clothing.

SECTION 2 Chemical Properties **139**

Figure 8 Gold and iron have different chemical properties that make them suitable for uses in a wide variety of jewelry and tools.

Choosing Materials Look at **Figure 8.** Would you rather wear a bracelet made of gold or one made of iron? Why? Iron is less attractive and less valuable than gold. It also has an important chemical property that makes it unsuitable for jewelry. Think about what happens to iron when it is left out in moist air. Iron rusts easily because of its high reactivity (ree ak TIH vuh tee) with oxygen and moisture in the air. **Reactivity** is how easily one thing reacts with something else. The low reactivity of silver and gold, in addition to their desirable physical properties, makes those metals good choices for jewelry.

 *Why is a fiberglass boat hull better than one made of a metal?*

 **Chemical Properties and Pools** The "chlorine" added to swimming pools is actually a compound called hypochlorous acid, which forms when chlorine reacts with water. This acid kills bacteria, insects, algae, and plants. The person in **Figure 9** is testing the pool water to see whether it has the correct amount of chlorine.

Any time you have standing water, mosquitoes and other insects can lay eggs in it. Various plants and algae can turn a sparkling blue pool into a slimy green mess. Bacteria are another problem. When you go swimming, you bring along millions of uninvited guests—the normal bacteria that live on your skin. The chlorine compounds kill the bacteria—as well as insects, algae, and plants that might be in the pool.

Hypochlorous acid can cause problems as well. It combines with nitrogen in the pool to form chloramines. Have your eyes ever burned after swimming in a pool? Chloramines can irritate the skin and eyes of swimmers.

Figure 9 Pool water must be tested to keep the water safe for swimmers.
Determine *How do physical and chemical properties differ?*

Chemical Properties of Acids and Bases You have learned that acids and bases have physical properties that make acids taste sour and bases taste bitter and feel slippery. The chemical properties of acids and bases are what make them both useful but sometimes harmful. Several acids and bases are shown in **Table 2**.

Acids Many acids react with, or corrode, certain metals. Have you ever used aluminum foil to cover leftover spaghetti or tomato sauce? **Figure 10** shows what you might see the next day. You might see small holes in the foil where it has come into contact with the tomatoes in the sauce. The acids in tomato sauce, oranges, carbonated soft drinks, and other foods are edible. However, many acids can damage plant and animal tissue. Small amounts of nitric (NI trihk) acid and sulfuric (sul FYOOR ihk) acid are found in rain. This rain, called acid rain, harms plant and animal life in areas where acid rain falls. Sulfuric acid that has no water mixed with it is useful in many industries because it removes water from certain materials. However, that same property causes burns on skin that touches sulfuric acid.

Figure 10 Aluminum reacts easily with acids, which is why acidic food, such as tomatoes, should not be cooked or stored in aluminum.

Table 2 Common Acids and Bases		
Name of Acid	**Formula**	**Where It's Found**
Acetic acid	CH_3COOH	Vinegar
Acetylsalicylic acid	$C_9H_8O_4$	Aspirin
Ascorbic acid (vitamin C)	$C_6H_8O_6$	Citrus fruits, tomatoes
Carbonic acid	H_2CO_3	Carbonated drinks
Hyrdrochloric acid	HCl	Gastric juice in stomach
Name of Base		
Aluminum hydroxide	$Al(OH)_3$	Deodorant, antacid
Calcium hydroxide	$Ca(OH)_2$	Leather tanning, manufacture of mortar and plaster
Magnesium hydroxide	$Mg(OH)_2$	Laxative, antacid
Sodium hydroxide	NaOH	Drain cleaner, soap making
Ammonia	NH_3	Household cleaners, fertilizer, production of rayon and nylon

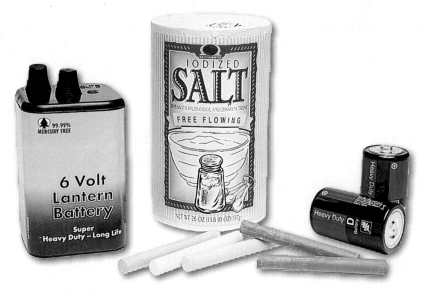

Figure 11 These everyday items contain salts.

Bases A concentrated base is as dangerous as a concentrated acid. A base, such as sodium hydroxide (hi DRAHK side) can damage living tissue. It is not uncommon for someone who smells strong ammonia to get a bloody nose or to get a burn if a strong base is touched. Ammonia feels slippery to the touch because the base reacts with the proteins in the tissues on your fingertips, which results in damaged tissue.

Salts What happens in reactions between acids and bases? Acids and bases often are studied together because they react with each other to form water and other useful compounds called salts. **Salts** are compounds made of a metal and nonmetal that are formed when acids and bases react. Look at **Figure 11.** That white solid in your salt shaker—table salt—is the most common salt. Table salt, sodium chloride, can be formed by the reaction between the base sodium hydroxide and hydrochloric acid. Other useful salts are calcium carbonate, which is chalk, and ammonium chloride, which is used in some types of batteries.

section 2 review

Summary

Chemical Properties
- These properties have characteristics that cannot be observed without altering the identity of the substance.

Chemical Properties of Acids and Bases
- Strong acids and bases can be equally dangerous.
- Strong acids react with and corrode metals.
- Ammonia and sodium hydroxide are examples of bases.

Salts
- A salt is composed of a metal and a nonmetal.
- An acid and a base combine to form a salt and water.

Self Check

1. **Compare and Contrast** How do chemical and physical properties differ?
2. **Describe** three chemical properties of an acid.
3. **Identify** two different salts and their uses.
4. **Think Critically** Think about safety precautions you take around your home. Which ones are based on physical properties and which ones are based on chemical properties? Explain.

Applying Skills

5. **Classify** each of the following properties as being physical or chemical: iron rusting, gasoline burning, solid sulfur shattering, and lye feeling slippery.

142 CHAPTER 5 Matter—Properties and Changes

ips.mssscience.com/self_check_quiz

section 3
Physical and Chemical Changes

Physical Change

The crowd gathers at a safe distance and the cameras from the news media are rolling. A sense of excitement, fear, and anticipation fills the air. The demolition experts are making their final inspections. Then, in just a few seconds, the old stadium becomes a pile of rubble. The appearance of the stadium changed.

What is physical change? Most matter can undergo physical change. A **physical change** is any change in size, shape, form, or state where the identity of the matter stays the same. Only the physical properties change. The stadium in **Figure 12** underwent a physical change from its original form to a pile of steel and concrete. The materials are the same; they just look different.

Reading Check *What is a physical change?*

as you read

What **You'll Learn**
- **Identify** physical and chemical changes.
- **Exemplify** how physical and chemical changes affect the world you live in.

Why **It's Important**
Chemical changes are all around us, from the leaves changing color in the fall to the baking of bread.

Review Vocabulary
weathering: the action of the elements in altering the color, texture, composition or form of exposed objects

New Vocabulary
- physical change
- chemical change

Figure 12 This stadium underwent a physical change—its form changed.

Examples of Physical Changes How can you recognize a physical change? Just look to see whether or not the matter has changed size, shape, form, or state. If you cut a watermelon into chunks, the watermelon has changed size and shape. That's a physical change. If you pop one of those chunks into your mouth and bite it, you have changed the watermelon's size and shape again.

Change of State Matter can undergo a physical change in another way, too. It can change from one state to another. Suppose it's a hot day. You and your friends decide to make snow cones. A snow cone is a mixture of water, sugar, food coloring, and flavoring. The water in the snow cone is solid, but in the hot sunshine, it begins to warm. When the temperature of the water reaches its melting point, the solid water begins to melt. The chemical composition of the water—two hydrogens and one oxygen—does not change. However, its form changes. This is an example of a physical change. The solid water becomes a liquid and drips onto the sidewalk. As the drops of liquid sit in the sunshine, the water changes state again, evaporating to become a gas. Water also can change from a solid to liquid by melting. Other examples of change of state are shown in **Figure 13.**

Figure 13 The four most common changes of state are shown here.
Explain *if physical changes can be reversed.*

A solid will melt, becoming a liquid.

As it cools, this liquid metal will become solid steel.

Water vapor in the air changes to liquid water when dew forms.

Liquid water in perspiration changes to a gas when it evaporates from your skin.

Figure 14 Chemical changes occur all around you.

This unprotected car fender was exposed to salt and water which caused it to rust.

This bridge support will have to be repaired or replaced because of the rust damage.

Apples and pennies darken due to chemical changes.

Chemical Changes

Unprotected cars driven on salted roads and steel structures like the one shown in **Figure 14** can begin to rust after only a few winters. A shiny copper penny becomes dull and dark. An apple left out too long begins to turn brown. What do all these changes have in common? Each of these changes is a chemical change. A **chemical change** occurs when one type of matter changes into a different type of matter with different properties.

 *What happens during a chemical change?*

Examples of Chemical Change Chemical changes are going on around you—and inside you—every day. Plants use photosynthesis to produce food—the product of chemical changes. When you eat fruits and vegetables produced by photosynthesis, these products must be chemically changed again so that the cells in your body can use them as food. There are many chemical changes occurring outside of your body, too. Silver tarnishing, copper forming a green coating, iron rusting, and petroleum products combusting are all examples of chemical changes that are occurring around you. Although these reactions may be occurring at different rates and producing different products, they are still examples of chemical changes.

Mini LAB

Comparing Chemical Changes

Procedure

1. Separate a piece of **fine steel wool** into two halves.
2. Dip one half in **tap water** and the other half in the same amount of **salt water.**
3. Place both pieces of steel wool on a **paper plate** and label them. Observe every day for five days.

Analysis
1. What happened to the steel wool that was dipped in the salt water?
2. What might be a common problem with machinery that is operated near an ocean?

SECTION 3 Physical and Chemical Changes

Figure 15 Chemical changes are common when food, such as cake, is cooked. **Determine** How is a chemical change different from a physical change?

New Materials Are Formed Ice melts, paper is cut, metal is hammered into sheets, and clay is molded into a vase. Seeing signs of these physical changes is easy—something changes shape, size, form, or state.

The only sure way to know whether a chemical change has occurred is if a new type of matter is formed that is chemically different from the starting matter. A chemical change cannot be reversed easily. For example, when wood burns, you see it change to ash and gases that have properties that are different from the wood and oxygen that burned. You can't put the ash and gases back together to make wood. When the cake shown in **Figure 15** is baked, changes occur that make the cake batter become solid. The chemical change that occurs when baking powder mixes with water results in bubbles that make the cake rise. Raw egg in the batter undergoes changes that make the egg solid. These changes cannot be reversed.

Topic: Chemical Changes
Visit ips.msscience.com for Web links to information about physical and chemical changes.

Activity Describe the physical and chemical changes that are involved in making and baking a yeast bread.

 How can you be sure that a chemical change has occurred?

Signs of Chemical Change In these examples, you know that a chemical change occurred because you can see that a new substance forms. It's not always easy to tell when new substances are formed. What are other signs of chemical change?

One sign of a chemical change is the release or absorption of energy in the form of light, heat, or sound. Release of energy is obvious when something burns—light and heat are given off. Sometimes an energy change is so small or slow that it is difficult to notice, like when something rusts. Another sign that indicates a chemical change is the formation of a gas or a solid that is not the result of a change of state.

146 CHAPTER 5 Matter—Properties and Changes

Chemical and Physical Changes in Nature

Often, a color is evidence of a chemical change, an example of which is shown in **Figure 16.** Year round, leaves contain yellow, red, and orange pigments that are masked, or hidden, by large amounts of green chlorophyll. In autumn, changes in temperature and rainfall amounts cause trees to stop producing chlorophyll. When chlorophyll production stops, the masked pigments become visible.

Physical Weathering Some physical changes occur quickly. Others take place over a long time. Physical weathering is a physical change that is responsible for much of the shape of Earth's surface. Examples are shown in **Figure 17.** Examples also can be found in your own school yard. All of the soil that you see comes from physical weathering. Wind and water erode rocks, breaking them into small bits. Water fills cracks in rocks. When it freezes and thaws several times, the rock splits into smaller pieces. No matter how small the pieces of rock are, they are made up of the same things that made up the bigger pieces of rock. The rock simply has undergone a physical change. Gravity, plants, animals, and the movement of land during earthquakes also help cause physical changes on Earth.

Figure 16 Chemical changes that occur in the fall bring about the color changes in these leaves.

Figure 17 You can see dramatic examples of physical weathering caused by water and wind on rocky coastlines.

SECTION 3 Physical and Chemical Changes **147**

Figure 18 Over many years, acidic rainwater slowly reacts with layers of limestone rock. It forms caves and collects minerals that it later deposits as cave formations.

Geologists Carlsbad Caverns in New Mexico contain cave formations similar to the ones shown here. Geologists study the history of the Earth as recorded in rocks and often investigate deep within Earth's caves. Stalagmites are cave formations that form on the floor of the cave and grow upward. Inside Carlsbad Caverns you will find a stalagmite called the Giant Dome that is 19 m tall. Research and find out more information about geologists and this huge cave.

Chemical Weathering Cave formations like the one in **Figure 18** form by chemical weathering. As drops of water drip through the rocks above this cavern room, minerals become dissolved in the water. These icicle shapes, or stalactites, are formed when the water evaporates leaving the mineral deposits. There are instances of unnatural chemical weathering. The acid in acid rain can chemically weather marble buildings and statues, and other outdoor objects.

section 3 review

Summary

Physical Changes
- Physical changes are changes in the size, shape, form, or state of an object where its identity remains the same.

Chemical Changes
- A chemical change occurs when one type of matter changes into another type of matter with different properties.
- Energy, in the form of light, heat, or sound can be released or absorbed.
- A gas or a solid, not resulting from a change of state, can be formed.
- Physical and chemical weathering also occur in nature.

Self Check

1. **List** five physical changes that you can observe in your home.
2. **Describe** how physical changes can alter Earth's surface.
3. **Explain** what happens when carbon burns. List the signs that a chemical change has occurred.
4. **Think Critically** Which of the following involves a chemical change: combining an acid and a base, dew forming, or souring milk.

Applying Skills

5. **Draw Conclusions** A log is reduced to a small pile of ash when it burns. Explain the difference in mass between the log and the ash.

148 CHAPTER 5 Matter—Properties and Changes

 ips.msscience.com/self_check_quiz

Sunset in a Bag

Real-World Question

How do you know when a chemical change occurs? You'll see some signs of chemical change in this lab.

Goals
- **Observe** a chemical change.
- **Identify** some signs of chemical change.

Materials
baking soda
calcium chloride
phenol red solution
warm water
teaspoons (2)
resealable plastic bag
graduated cylinder

Safety Precautions

Procedure

1. Add 20 mL of warm water to the plastic bag. Add a teaspoon of calcium chloride to the water, seal the bag, and slosh the contents to mix the solution. Record your observations.

2. Add 5 mL of phenol red solution to the same bag. Seal the bag, slosh the contents, and record your observations.

3. Open the bag and quickly add a teaspoon of baking soda. Seal the bag and slosh the contents to mix the ingredients together. Observe what happens.

Conclude and Apply

1. **Identify** in which step a physical change occurred. In which step did a chemical change occur? How do you know?

2. **Predict** if a change in energy always indicates a chemical change. Why or why not?

Communicating Your Data

Compare your conclusions with those of other students in your class. **For more help, refer to the** Science Skill Handbook.

LAB **149**

LAB Design Your Own

Homemade pH Scale

Goals
- **Design** an experiment that allows you to test solutions to find the pH of each.
- **Classify** a solution as an acid or a base according to its pH.

Possible Materials
vial of pH paper
1–14 pH color chart
distilled water
fruit juices
vinegar
salt
sugar
soft drinks
household cleaners
soaps and detergents
antacids

Safety Precautions

WARNING: Never eat, taste, smell, or touch any chemical during a lab.

Real-World Question

The more concentrated an acid or base is, the more likely it is to be harmful to living organisms. A pH scale is used to measure the concentration of acids and bases. A solution with a pH below 7 is acidic, a pH of 7 is neutral, and a pH above 7 is basic. In this lab, you will measure the pH of some things using treated paper. When it is dipped into a solution, this paper changes color. Check the color against the chart below to find the pH of the solution. How acidic or basic are some common household items?

Form a Hypothesis

Form a hypothesis to explain which kinds of solutions you are testing are acids and which kinds are bases. Copy and complete the table below.

pH of Solutions		
Solution To Be Tested	pH	Acid, Base, or Neutral
Do not write in this book.		

pH	Color	pH	Color
1		8	
2		9	
3		10	
4		11	
5		12	
6		13	
7		14	

Using Scientific Methods

▶ Test Your Hypothesis

Make a Plan

1. As a group, decide which materials you will test. If a material is not a liquid, dissolve it in water so you can test the solution.
2. List the steps and materials that you need to test your hypothesis. Be specific. What parts of the experiment will you repeat, if any?
3. Before you begin, copy a data table like the one shown into your Science Journal. Be sure to leave room to record results for each solution tested. If there is to be more than one trial for each solution, include room for the additional trials.
4. Reread the entire experiment to make sure that all the steps are in logical order.

Follow Your Plan

1. Make sure your teacher approves your plan and data table. Be sure that you have included any suggested changes.
2. Carry out the experiment as planned and approved. Wash your hands when you are done.
3. **Record** the pH value of each solution in the data table as you complete each test. Determine whether each solution is acidic, basic, or neutral.

▶ Analyze Your Data

1. **Infer** Were any materials neither acids nor bases? How do you know?
2. **Interpret Data** Using your data table, conclude which types of materials are usually acidic and which are usually basic.
3. **Draw Conclusions** At what pH do you think acids become too dangerous to touch? Bases? Explain your answers.
4. **Analyze Results** What is the pH range of the foods that you tested?

▶ Conclude and Apply

Determine Perhaps you have been told that you can use vinegar to dissolve hard-water deposits because vinegar is an acid. If you run out of vinegar, which of the items you tested could you most likely use instead of vinegar for this purpose?

Compare your findings with those of other student groups. Discuss why any differences in the data might have occurred.

LAB **151**

TIME SCIENCE AND HISTORY

SCIENCE CAN CHANGE THE COURSE OF HISTORY!

CRUMBLING MONUMENTS

Acid rain is eroding some of the world's most famous monuments

The Taj Mahal in India, the Acropolis in Greece, and the Colosseum in Italy, have stood for centuries. They've survived wars, souvenir-hunters, and natural weathering from wind and rain. But now, something far worse threatens their existence—acid rain. Over the last few decades, this form of pollution has eaten away at some of history's greatest monuments.

Acid rain leads to health and environmental risks. It also harms human-made structures. Most of these structures are made of sandstone, limestone, and marble. Acid rain causes the calcium in these stones to form calcium sulfate, or gypsum. Gypsum's powdery little blotches are sometimes called "marble cancer." When it rains, the gypsum washes away, along with some of the surface of the monument. In many cases, acidic soot falls into the cracks of monuments. When rainwater seeps into the cracks, acidic water is formed, which further damages the structure.

In London, acid rain has forced workers to repair and replace so much of Westminster Abbey that the structure is becoming a mere copy of the original. Because of pollution, many corroding statues displayed outdoors have been brought inside museums.

Throughout the world, acid rain has weathered many structures more in the last 20 years than in the prior 2,000 years. This is one reason some steps have been taken in Europe and the United States to reduce emissions from the burning of fossil fuels.

Identify Which monuments and buildings represent the United States? Brainstorm a list with your class. Then choose a monument, and using your school's media center, learn more about it. Is acid rain affecting it in any way?

For more information, visit ips.msscience.com/time

chapter 5 Study Guide

Reviewing Main Ideas

Section 1 Physical Properties

1. A physical property can be observed without changing the makeup of the material.
2. Acids and bases have physical properties. Acids have a sharp smell and a sour taste. Bases have a bitter taste and feel slippery.
3. Mass, volume, state of matter, and density are examples of physical properties.

Section 2 Chemical Properties

1. A chemical property is a characteristic of matter that allows it to change to a different type of matter.
2. Acids and bases are in many household products.
3. Acids and bases react with each other to produce water and a salt.

Section 3 Physical and Chemical Changes

1. A physical change is a change in the size, shape, form, or state of matter. The chemical makeup of the matter stays the same.
2. Water undergoes a change of state when it changes from a solid to a liquid or a liquid to a gas.
3. In chemical changes, new matter is changed to a different type of matter.
4. Evidence that a chemical change might have occurred includes a color or energy change or the formation of a gas or solid.

Visualizing Main Ideas

Copy and complete the following concept map about matter.

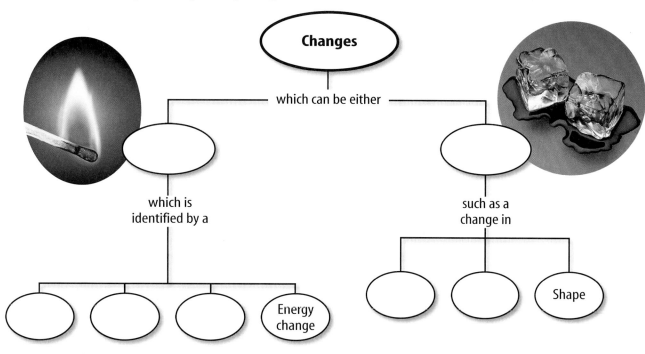

chapter 5 Review

Using Vocabulary

chemical change p. 145
chemical property p. 139
density p. 134
physical change p. 143
physical property p. 134
reactivity p. 140
salt p. 142
size-dependent property p. 136
size-independent property p. 136
state of matter p. 136

Answer the following questions using complete sentences.

1. Mass divided by volume is the formula for which physical property?
2. Which type of properties include color, shape, size, and state?
3. Snow melting in sunshine is an example of which type of change?
4. Acid rain damaging marble statues is an example of which type of change?
5. Iron rusts in moist air. Which chemical property is this?

Checking Concepts

Choose the word or phrase that best answers the question.

6. Which of the following is a chemical property of a substance?
   A) density   C) mass
   B) white powder   D) reacts with HCl

7. Which item below is a sign of a chemical change?
   A) change of water vapor to liquid
   B) release of energy
   C) change from a liquid to a solid
   D) change in shape

8. Which type of change listed below results in new compounds being formed?
   A) chemical   C) seasonal
   B) physical   D) state

9. Salts are formed when which of the following react?
   A) solids and gases
   B) acids and bases
   C) bases and gases
   D) acids and solids

10. Which of the following physical properties does a base have?
    A) cold to touch
    B) gives off gas
    C) slippery and bitter taste
    D) sharp smell and sour taste

11. Which of the following changes when water evaporates?
    A) the physical properties of the water
    B) the chemical properties of the water
    C) the color of the water
    D) the mass of the water

Use the illustrations below to answer question 12.

12. Which figure above clearly identifies a chemical change?
    A) A
    B) B
    C) C
    D) D

154 CHAPTER REVIEW

Thinking Critically

13. **Explain** Think about what you know about density. Could a bag of feathers have more mass than the same size bag of rocks? Explain.

14. **Classify** each of the following as either a physical property or a chemical property.
    a. Sulfur shatters when hit.
    b. Copper statues turn green.
    c. Baking soda is a white powder.
    d. Newspaper turns brown when it is exposed to air and light.

Use the photo below to answer questions 15 and 16.

15. **Identify** The antacid dissolves in water. Is this a physical or a chemical property of the antacid? Explain.

16. **Draw Conclusions** Think about what you know about density. Is the antacid tablet more or less dense than water. Explain.

17. **Determine** A jeweler bends gold into a beautiful ring. What type of change is this? Explain.

18. **Compare and Contrast** Relate such human characteristics as hair and eye color, height, and weight to physical properties of matter. Relate human behavior to chemical properties. Think about how you observe these properties.

19. **Identify** Sugar dissolves in water. Is this a physical property or a chemical property of sugar?

20. **Evaluate** When butane burns, it combines with oxygen in the air to form carbon dioxide and water. Which two elements must be present in butane?

21. **Identify** each of the following as a physical change or a chemical change.
    a. Metal is drawn out into a wire.
    b. Sulfur in eggs tarnishes silver.
    c. Baking powder bubbles when water is added to it.

Performance Activities

22. **Display** Create a display that demonstrates the characteristics of a chemical change. Be sure your display shows release of energy, change of color, and the formation of a solid.

Applying Math

Use the table below to answer questions 23–25.

Using the Density Formula

Sample	Mass (g)	Volume (cm^3)	Density (g/cm^3)
A	5.4	3.8	
B	6.8		0.65
C		8.6	2.18

23. **Determine Density** Knowing the mass and volume of Sample A, calculate its density.

24. **Determine** the mass of Sample C from its density and volume.

25. **Determine the volume** of Sample B from its mass and density.

Chapter 5 Standardized Test Practice

Part 1 | Multiple Choice

Record your answers on the answer sheet provided by your teacher or on a sheet of paper.

1. Which of the following properties is a size-independent property?
   A. state
   B. length
   C. mass
   D. volume

Use the table below to answer questions 2, 3 and 4.

Density of Some Pure Metals	
Metal	Density (g/cm^3)
Copper	8.96
Iron	7.87
Lead	11.3
Magnesium	1.74
Silver	10.5
Zinc	7.14

2. According to the table above, what is the mass of a 6.37-cm^3 sample of iron?
   A. 0.809 g
   B. 1.24 g
   C. 7.87 g
   D. 50.1 g

3. According to the table above, what is the volume of a 25.1 g piece of silver?
   A. 8.25 cm^3
   B. 2.39 cm^3
   C. 5.73 cm^3
   D. 3.46 cm^3

4. During an experiment, you measure the mass of an unknown substance as 28.4 g and its volume as 2.5 cm^3. Use the table above to determine the most likely identity of the unknown substance.
   A. zinc
   B. magnesium
   C. lead
   D. iron

5. Which of the following terms describes a substance with a pH of 7?
   A. acidic
   B. neutral
   C. basic
   D. bitter

Use the photograph below to answer questions 6 and 7.

6. Which of the following statements best explains why the nails in the photograph above rusted?
   A. The nails drew rust from the air.
   B. Iron in the nails changed into other elements.
   C. The rust formed when iron mixed with oxygen and moisture in the air.
   D. A temperature change drew rust out of the nails.

7. Which of the following best describes rust?
   A. reversible physical change
   B. irreversible physical change
   C. reversible chemical change
   D. irreversible chemical change

8. Which of the following describes a chemical change?
   A. water freezing into ice
   B. a match burning
   C. dew forming on a leaf
   D. magnetization of an iron nail

9. The density of a 30-g sample of gold is 19.3 g/cm^3. What is the density of a 90-g sample?
   A. 6.43 g
   B. 19.3 g
   C. 57.9 g
   D. 79.3 g

Standardized Test Practice

Part 2 Short Response/Grid In

Record your answers on the answer sheet provided by your teacher or on a sheet of paper.

10. Tell whether each of the following words describes an acid or a base: bitter, slippery, sour.

11. What is the density of a 25.3-g sample of quartz that has a volume of 9.55 cm^3?

12. Are all acids and bases dangerous? Explain why or why not, and give examples to support your answer.

13. Describe a chemical and a physical change that occurs in food as you chew it.

14. The density of stainless steel is 8.02 g/cm^3. What is the volume of a 9.25 g piece of stainless steel?

Use the photograph below to answer questions 15 and 16.

15. The photograph above shows a campfire with water heating in a cooking pot. Does the water experience a chemical or physical change as its temperature rises? Explain.

16. Are the logs on the campfire experiencing a chemical or physical change? How can you tell?

Part 3 Open Ended

Record your answers on a sheet of paper.

Use the photograph below to answer questions 17 and 18.

17. What would happen if you left the glass of cold water shown in the photograph above in sunlight for several hours? Describe how some physical properties of the water would change.

18. What properties of the water would not change? Explain why the density of the water would or would not change.

19. Suppose you have three different-sized balls, each having a different mass. Can you tell which has the greatest density simply by feeling which is heavier? Explain why or why not.

20. Describe the pH scale. What is its use?

21. What are some things you might observe during an experiment that would indicate a chemical change may be occurring? Explain which of these may, instead, be the result of a physical change.

Test-Taking Tip

Answer Every Part Make sure each part of the question is answered when listing discussion points. For example, if the question asks you to compare and contrast, make sure you list both similarities and differences.

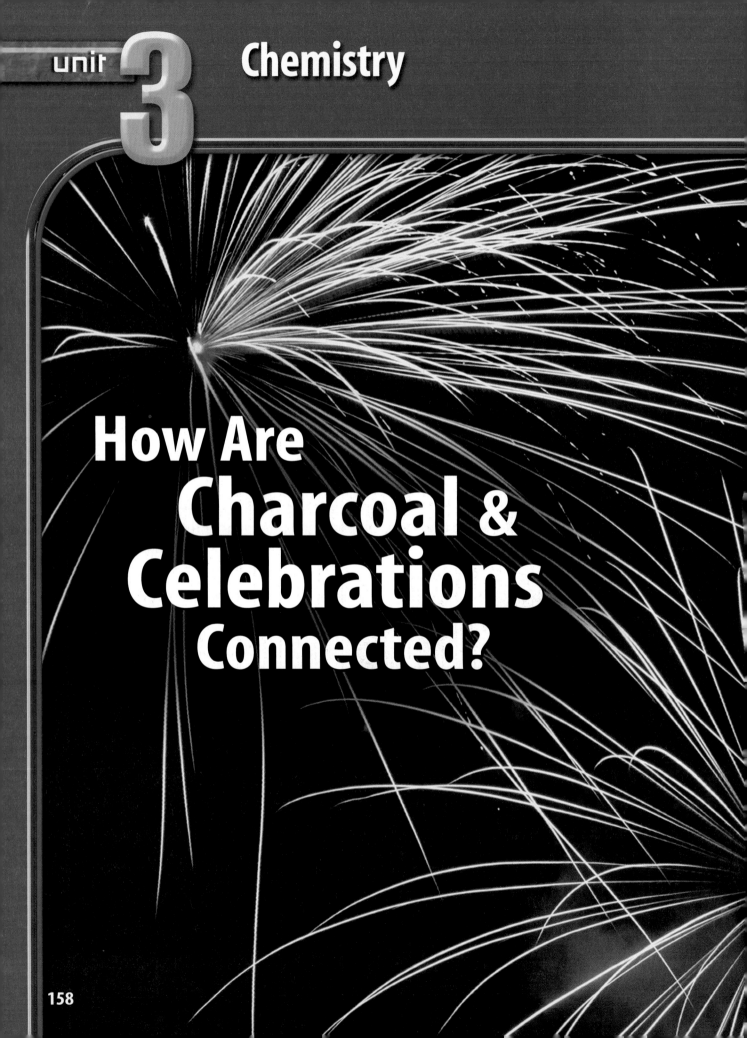

unit 3 Chemistry

How Are Charcoal & Celebrations Connected?

NATIONAL GEOGRAPHIC

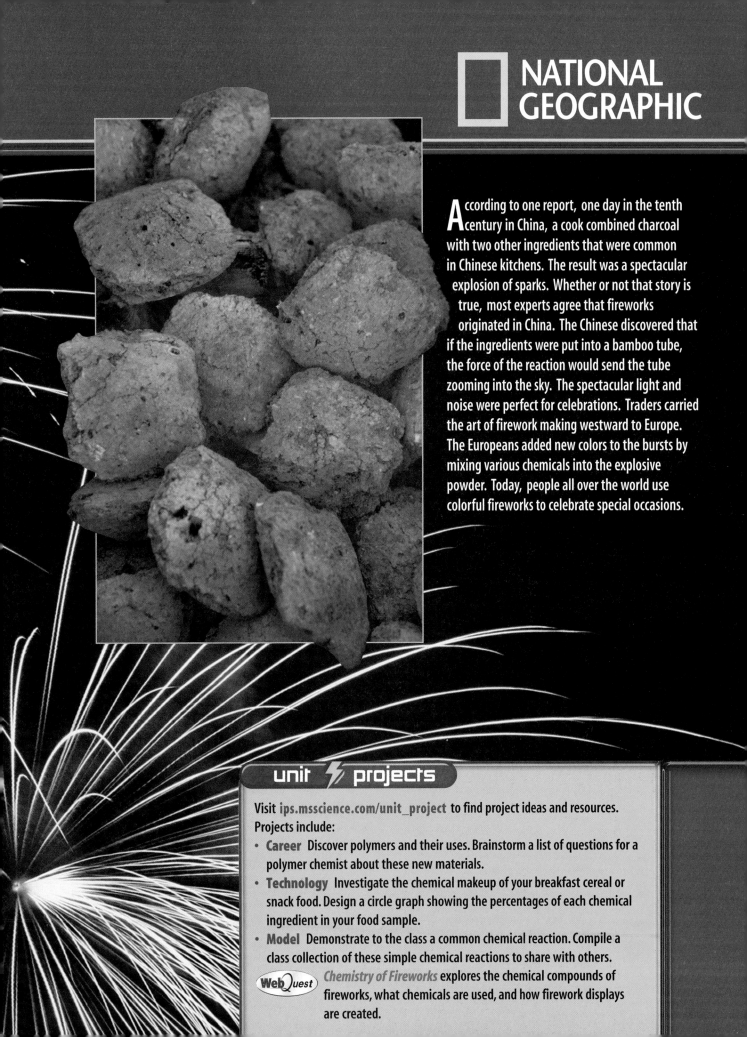

According to one report, one day in the tenth century in China, a cook combined charcoal with two other ingredients that were common in Chinese kitchens. The result was a spectacular explosion of sparks. Whether or not that story is true, most experts agree that fireworks originated in China. The Chinese discovered that if the ingredients were put into a bamboo tube, the force of the reaction would send the tube zooming into the sky. The spectacular light and noise were perfect for celebrations. Traders carried the art of firework making westward to Europe. The Europeans added new colors to the bursts by mixing various chemicals into the explosive powder. Today, people all over the world use colorful fireworks to celebrate special occasions.

unit projects

Visit **ips.msscience.com/unit_project** to find project ideas and resources. Projects include:

- **Career** Discover polymers and their uses. Brainstorm a list of questions for a polymer chemist about these new materials.
- **Technology** Investigate the chemical makeup of your breakfast cereal or snack food. Design a circle graph showing the percentages of each chemical ingredient in your food sample.
- **Model** Demonstrate to the class a common chemical reaction. Compile a class collection of these simple chemical reactions to share with others.

WebQuest *Chemistry of Fireworks* explores the chemical compounds of fireworks, what chemicals are used, and how firework displays are created.

chapter 6

Atomic Structure and Chemical Bonds

The BIG Idea
An atom's structure affects how it bonds to other atoms.

SECTION 1
Why do atoms combine?
Main Idea When atoms combine, they become more stable.

SECTION 2
How Elements Bond
Main Idea Elements bond by transferring electrons or by sharing electrons.

The Noble Family

Blimps, city lights, and billboards, all have something in common—they use gases that are members of the same element family. In this chapter, you'll learn about the unique properties of element families. You'll also learn how electrons can be lost, gained, and shared by atoms to form chemical bonds.

Science Journal Write a sentence comparing household glue to chemical bonds.

Start-Up Activities

Model the Energy of Electrons

It's time to clean out your room—again. Where do all these things come from? Some are made of cloth and some of wood. The books are made of paper and an endless array of things are made of plastic. Fewer than 100 different kinds of naturally occurring elements are found on Earth. They combine to make all these different substances. What makes elements form chemical bonds with other elements?

1. Pick up a paper clip with a magnet. Touch that paper clip to another paper clip and pick it up.
2. Continue picking up paper clips this way until you have a strand of them and no more will attach.
3. Then, gently pull off the paper clips one by one.
4. **Think Critically** In your Science Journal, discuss which paper clip was easiest to remove and which was hardest. Was the clip that was easiest to remove closer to or farther from the magnet?

Chemical Bonds Make the following Foldable to help you classify information by diagramming ideas about chemical bonds.

STEP 1 Fold a vertical sheet of paper in half from top to bottom.

STEP 2 Fold in half from side to side with the fold at the top.

STEP 3 Unfold the paper once. Cut only the fold of the top flap to make two tabs.

STEP 4 Turn the paper vertically and label the tabs as shown.

Summarize As you read the chapter, identify the main ideas of bonding under the appropriate tabs. After you have read the chapter, explain the difference between polar covalent bonds and covalent bonds on the inside portion of your Foldable.

Preview this chapter's content and activities at
ips.msscience.com

161

Get Ready to Read

Questioning

① Learn It! Asking questions helps you to understand what you read. As you read, think about the questions you'd like answered. Often you can find the answer in the next paragraph or section. Learn to ask good questions by asking *who, what, when, where, why,* and *how*.

② Practice It! Read the following passage from Section 2.

> In medieval times, alchemists (AL kuh mists) were the first to explore the world of chemistry. Although many of them believed in magic and mystical transformations, alchemists did learn much about the properties of some elements. They even used symbols to represent them in chemical processes.
>
> —*from page 177*

Here are some questions you might ask about this paragraph:

- Who were alchemists?
- What was their contribution to chemistry?
- What symbols did alchemists use to represent these elements?
- How do modern element symbols compare to the symbols the alchemists used?

③ Apply It! As you read the chapter, look for answers to section headings that are in the form of questions.

Target Your Reading

Reading Tip

Test yourself. Create questions and then read to find answers to your own questions.

Use this to focus on the main ideas as you read the chapter.

① Before you read the chapter, respond to the statements below on your worksheet or on a numbered sheet of paper.
- Write an **A** if you **agree** with the statement.
- Write a **D** if you **disagree** with the statement.

② After you read the chapter, look back to this page to see if you've changed your mind about any of the statements.
- If any of your answers changed, explain why.
- Change any false statements into true statements.
- Use your revised statements as a study guide.

Before You Read A or D		Statement	After You Read A or D
	1	All matter, including solids like wood and steel, contains mostly empty space.	
	2	Scientists are able to pinpoint the exact location of an electron in an atom.	
	3	Electrons orbit a nucleus just like planets orbit the Sun.	
	4	The number of electrons in a neutral atom is the same as the atom's atomic number.	
	5	Noble gases react easily with other elements.	
	6	All elements transfer the same number of electrons when bonding with other elements.	
	7	Electrons in metals can move freely among all the ions in the metal.	
	8	Some atoms bond by sharing electrons between the two atoms.	
	9	Water molecules have two opposite ends, like poles in a magnet.	

Science Online
Print out a worksheet of this page at ips.msscience.com

section 1

Why do atoms combine?

as you read

What You'll Learn
- **Identify** how electrons are arranged in an atom.
- **Compare** the relative amounts of energy of electrons in an atom.
- **Compare** how the arrangement of electrons in an atom is related to its place in the periodic table.

Why It's Important
Chemical reactions take place all around you.

Review Vocabulary
atom: the smallest part of an element that keeps all the properties of that element

New Vocabulary
- electron cloud
- energy level
- electron dot diagram
- chemical bond

Figure 1 You can compare and contrast electrons with planets.

Atomic Structure

When you look at your desk, you probably see it as something solid. You might be surprised to learn that all matter, even solids like wood and metal contain mostly empty space. How can this be? The answer is that although there might be little or no space between atoms, a lot of empty space lies within each atom.

At the center of every atom is a nucleus containing protons and neutrons. This nucleus represents most of the atom's mass. The rest of the atom is empty except for the atom's electrons, which are extremely small compared with the nucleus. Although the exact location of any one electron cannot be determined, the atom's electrons travel in an area of space around the nucleus called the **electron cloud.**

To visualize an atom, picture the nucleus as the size of a penny. In this case, electrons would be smaller than grains of dust and the electron cloud would extend outward as far as 20 football fields.

Electrons You might think that electrons resemble planets circling the Sun, but they are very different, as you can see in **Figure 1.** First, planets have no charges, but the nucleus of an atom has a positive charge and electrons have negative charges.

Second, planets travel in predictable orbits—you can calculate exactly where one will be at any time. This is not true for electrons. Although electrons do travel in predictable areas, it is impossible to calculate the exact position of any one electron. Instead scientists use a mathematical model that predicts where an electron is most likely to be.

Planets travel in well-defined paths.

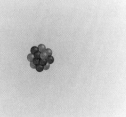

Electrons travel around the nucleus. However, their paths are not well-defined.

162 CHAPTER 6 Atomic Structure and Chemical Bonds

Element Structure Each element has a unique atomic structure consisting of a specific number of protons, neutrons, and electrons. The number of protons and electrons is always the same for a neutral atom of a given element. **Figure 2** shows a two-dimensional model of the electron structure of a lithium atom, which has three protons and four neutrons in its nucleus, and three electrons moving around its nucleus.

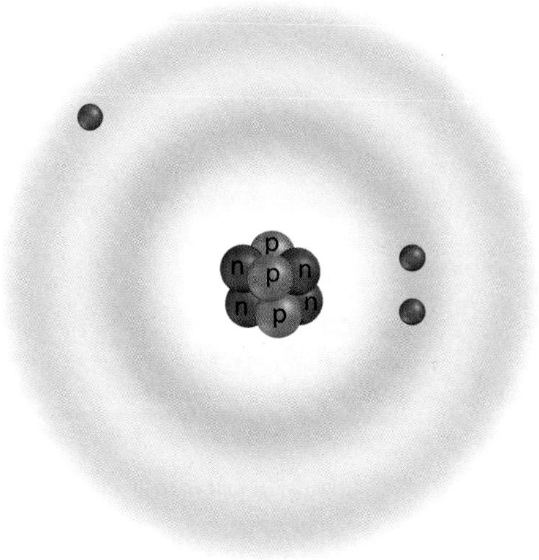

Electron Arrangement

The number and arrangement of electrons in the electron cloud of an atom are responsible for many of the physical and chemical properties of that element.

Electron Energy Although all the electrons in an atom are somewhere in the electron cloud, some electrons are closer to the nucleus than others. The different areas for an electron in an atom are called **energy levels. Figure 3** shows a model of what these energy levels might look like. Each level represents a different amount of energy.

Number of Electrons Each energy level can hold a maximum number of electrons. The farther an energy level is from the nucleus, the more electrons it can hold. The first energy level, energy level 1, can hold one or two electrons, the second, energy level 2, can hold up to eight, the third can hold up to 18, and the fourth energy level can hold a maximum of 32 electrons.

Figure 2 This neutral lithium atom has three positively charged protons, three negatively charged electrons, and four neutral neutrons.

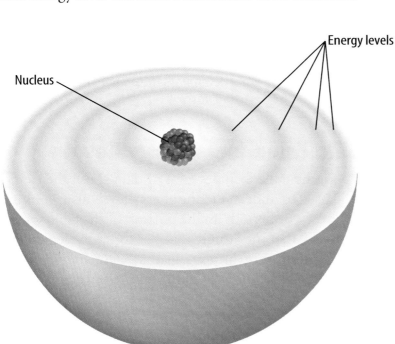

Figure 3 Electrons travel in three dimensions around the nucleus of an atom. The dark bands in this diagram show the energy levels where electrons are most likely to be found.
Identify *the energy level that can hold the most electrons.*

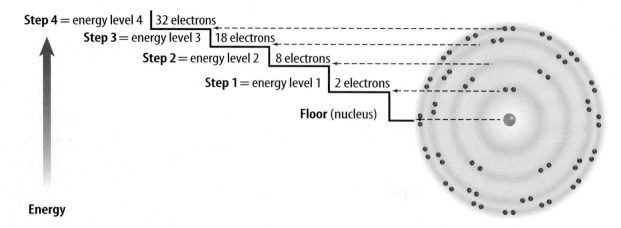

Figure 4 The farther an energy level is from the nucleus, the more electrons it can hold.
Identify the energy level with the least energy and the energy level with the most energy.

Energy Steps The stairway, shown in **Figure 4,** is a model that shows the maximum number of electrons each energy level can hold in the electron cloud. Think of the nucleus as being at floor level. Electrons within an atom have different amounts of energy, represented by energy levels. These energy levels are represented by the stairsteps in **Figure 4.** Electrons in the level closest to the nucleus have the lowest amount of energy and are said to be in energy level one. Electrons farthest from the nucleus have the highest amount of energy and are the easiest to remove. To determine the maximum number of electrons that can occupy an energy level, use the formula, $2n^2$, where n equals the number of the energy level.

Recall the Launch Lab at the beginning of the chapter. It took more energy to remove the paper clip that was closest to the magnet than it took to remove the one that was farthest away. That's because the closer a paper clip was to the magnet, the stronger the magnet's attractive force was on the clip. Similarly, the closer a negatively charged electron is to the positively charged nucleus, the more strongly it is attracted to the nucleus. Therefore, removing electrons that are close to the nucleus takes more energy than removing those that are farther away from the nucleus.

 What determines the amount of energy an electron has?

Science Online
Topic: Electrons
Visit ips.msscience.com for Web links to information about electrons and their history.

Activity Research why scientists cannot locate the exact positions of an electron.

Periodic Table and Energy Levels

The periodic table includes a lot of data about the elements and can be used to understand the energy levels also. Look at the horizontal rows, or periods, in the portion of the table shown in **Figure 5.** Recall that the atomic number for each element is the same as the number of protons in that element and that the number of protons equals the number of electrons because an atom is electrically neutral. Therefore, you can determine the number of electrons in an atom by looking at the atomic number written above each element symbol.

CHAPTER 6 Atomic Structure and Chemical Bonds

Electron Configurations

If you look at the periodic table shown in **Figure 5,** you can see that the elements are arranged in a specific order. The number of electrons in a neutral atom of the element increases by one from left to right across a period. For example, the first period consists of hydrogen with one electron and helium with two electrons in energy level one. Recall from **Figure 4** that energy level one can hold up to two electrons. Therefore, helium's outer energy level is complete. Atoms with a complete outer energy level are stable. Therefore, helium is stable.

Reading Check *What term is given to the rows of the periodic table?*

The second period begins with lithium, which has three electrons—two in energy level one and one in energy level two. Lithium has one electron in its outer energy level. To the right of lithium is beryllium with two outer-level electrons, boron with three, and so on until you reach neon with eight.

Look again at **Figure 4.** You'll see that energy level two can hold up to eight electrons. Not only does neon have a complete outer energy level, but also this configuration of exactly eight electrons in an outer energy level is stable. Therefore, neon is stable. The third period elements fill their outer energy levels in the same manner, ending with argon. Although energy level three can hold up to 18 electrons, argon has eight electrons in its outer energy level—a stable configuration. Each period in the periodic table ends with a stable element.

Nobel Prize Winner Ahmed H. Zewail is a professor of chemistry and physics and the director of the Laboratory for Molecular Sciences at the California Institute of Technology. He was awarded the 1999 Nobel Prize in Chemistry for his research. Zewail and his research team use lasers to record the making and breaking of chemical bonds.

Figure 5 This portion of the periodic table shows the electron configurations of some elements. Count the electrons in each element and notice how the number increases across a period.

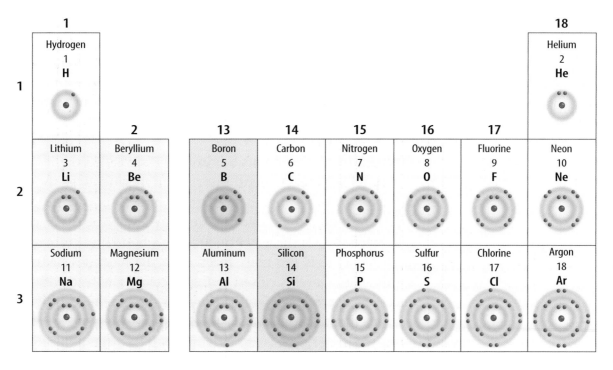

SECTION 1 Why do atoms combine? **165**

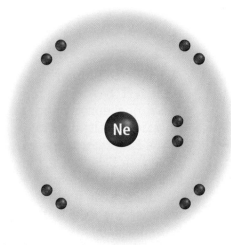

Figure 6 The noble gases are stable elements because their outer energy levels are complete or have a stable configuration of eight electrons like neon shown here.

Figure 7 The halogen element fluorine has seven electrons in its outer energy level.
Determine *how many electrons the halogen family member bromine has in its outer energy level.*

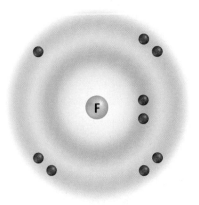

Element Families

Elements can be divided into groups, or families. Each column of the periodic table in **Figure 5** contains one element family. Hydrogen is usually considered separately, so the first element family begins with lithium and sodium in the first column. The second family starts with beryllium and magnesium in the second column, and so on. Just as human family members often have similar looks and traits, members of element families have similar chemical properties because they have the same number of electrons in their outer energy levels.

It was the repeating pattern of properties that gave Russian chemist Dmitri Mendeleev the idea for his first periodic table in 1869. While listening to his family play music, he noticed how the melody repeated with increasing complexity. He saw a similar repeating pattern in the elements and immediately wrote down a version of the periodic table that looks much as it does today.

Noble Gases Look at the structure of neon in **Figure 6**. Neon and the elements below it in Group 18 have eight electrons in their outer energy levels. Their energy levels are stable, so they do not combine easily with other elements. Helium, with two electrons in its lone energy level, is also stable. At one time these elements were thought to be completely unreactive, and therefore became known as the inert gases. When chemists learned that some of these gases can react, their name was changed to noble gases. They are still the most stable element group.

This stability makes possible one widespread use of the noble gases—to protect filaments in lightbulbs. Another use of noble gases is to produce colored light in signs. If an electric current is passed through them they emit light of various colors—orange-red from neon, lavender from argon, and yellowish-white from helium.

Halogens The elements in Group 17 are called the halogens. A model of the element fluorine in period 2 is shown in **Figure 7.** Like all members of this family, fluorine needs one electron to obtain a stable outer energy level. The easier it is for a halogen to gain this electron to form a bond, the more reactive it is. Fluorine is the most reactive of the halogens because its outer energy level is closest to the nucleus. The reactivity of the halogens decreases down the group as the outer energy levels of each element's atoms get farther from the nucleus. Therefore, bromine in period 4 is less reactive than fluorine in period 2.

Alkali Metals Look at the element family in Group 1 on the periodic table at the back of this book, called the alkali metals. The first members of this family, lithium and sodium, have one electron in their outer energy levels. You can see in **Figure 8** that potassium also has one electron in its outer level. Therefore, you can predict that the next family member, rubidium, does also. These electron arrangements are what determines how these metals react.

Reading Check *How many electrons do the alkali metals have in their outer energy levels?*

The alkali metals form compounds that are similar to each other. Alkali metals each have one outer energy level electron. It is this electron that is removed when alkali metals react. The easier it is to remove an electron, the more reactive the atom is. Unlike halogens, the reactivities of alkali metals increase down the group; that is, elements in the higher numbered periods are more reactive than elements in the lower numbered periods. This is because their outer energy levels are farther from the nucleus. Less energy is needed to remove an electron from an energy level that is farther from the nucleus than to remove one from an energy level that is closer to the nucleus. For this reason, cesium in period 6 loses an electron more readily and is more reactive than sodium in period 3.

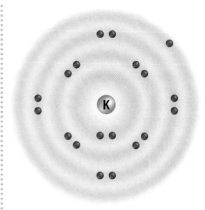

Figure 8 Potassium, like lithium and sodium, has only one electron in its outer level.

Applying Science

How does the periodic table help you identify properties of elements?

The periodic table displays information about the atomic structure of the elements. This information includes the properties, such as the energy level, of the elements. Can you identify an element if you are given information about its energy level? Use your ability to interpret the periodic table to find out.

Identifying the Problem

Recall that elements in a group in the periodic table contain the same number of electrons in their outer levels. The number of electrons increases by one from left to right across a period. Refer to **Figure 5.** Can you identify an unknown element or the group a known element belongs to?

Solving the Problem

1. An unknown element in Group 2 has a total number of 12 electrons and two electrons in its outer level. What is it?
2. Name the element that has eight electrons, six of which are in its outer level.
3. Silicon has a total of 14 electrons, four electrons in its outer level, and three energy levels. What group does silicon belong to?
4. Three elements have the same number of electrons in their outer energy levels. One is oxygen. Using the periodic table, what might the other two be?

SECTION 1 Why do atoms combine?

Mini LAB

Drawing Electron Dot Diagrams

Procedure

1. Draw a periodic table that includes the first 18 elements—the elements from hydrogen through argon. Make each block a 3-cm square.
2. Fill in each block with the electron dot diagram of the element.

Analysis

1. What do you observe about the electron dot diagram of the elements in the same group?
2. Describe any changes you observe in the electron dot diagrams across a period.

Electron Dot Diagrams

You have read that the number of electrons in the outer energy level of an atom determines many of the chemical properties of the atom. Because these electrons are so important in determining the chemical properties of atoms, it can be helpful to make a model of an atom that shows only the outer electrons. A model like this can be used to show what happens to these electrons during reactions.

Drawing pictures of the energy levels and electrons in them takes time, especially when a large number of electrons are present. If you want to see how atoms of one element will react, it is handy to have an easier way to represent the atoms and the electrons in their outer energy levels. You can do this with electron dot diagrams. An **electron dot diagram** is the symbol for the element surrounded by as many dots as there are electrons in its outer energy level. Only the outer energy level electrons are shown because these are what determine how an element can react.

How to Write Them How do you know how many dots to make? For Groups 1 and 2, and 13–18, you can use the periodic table or the portion of it shown in **Figure 5**. Group 1 has one outer electron. Group 2 has two. Group 13 has three, Group 14, four, and so on to Group 18. All members of Group 18 have stable outer energy levels. From neon down, they have eight electrons. Helium has only two electrons, because that is all that its single energy level can hold.

The dots are written in pairs on four sides of the element symbol. Start by writing one dot on the top of the element symbol, then work your way around, adding dots to the right, bottom, and left. Add a fifth dot to the top to make a pair. Continue in this manner until you reach eight dots to complete the level.

The process can be demonstrated by writing the electron dot diagram for the element nitrogen. First, write N—the element symbol for nitrogen. Then, find nitrogen in the periodic table and see what group it is in. It's in Group 15, so it has five electrons in its outer energy level. The completed electron dot diagram for nitrogen can be seen in **Figure 9**.

The electron dot diagram for iodine can be drawn the same way. The completed diagram is shown on the right in **Figure 9**.

Figure 9 Electron dot diagrams show only the electrons in the outer energy level.
Explain why only the outer energy level electrons are shown.

Nitrogen contains five electrons in its outer energy level.

Iodine contains seven electrons in its outer energy level.

Figure 10 Some models are made by gluing pieces together. The glue that holds elements together in a chemical compound is the chemical bond.

Using Dot Diagrams Now that you know how to write electron dot diagrams for elements, you can use them to show how atoms bond with each other. A **chemical bond** is the force that holds two atoms together. Chemical bonds unite atoms in a compound much as glue unites the pieces of the model in **Figure 10**. Atoms bond with other atoms in such a way that each atom becomes more stable. That is, their outer energy levels will resemble those of the noble gases.

Reading Check *What is a chemical bond?*

section 1 review

Summary

Atom Structure
- At the center of the atom is the nucleus.
- Electrons exist in an area called the electron cloud.
- Electrons have a negative charge.

Electron Arrangement
- The different regions for an electron in an atom are called energy levels.
- Each energy level can hold a maximum number of electrons.

The Periodic Table
- The number of electrons is equal to the atomic number.
- The number of electrons in a neutral atom increases by one from left to right across a period.

Self Check

1. **Determine** how many electrons nitrogen has in its outer energy level. How many does bromine have?
2. **Solve** for the number of electrons that oxygen has in its first energy level. Second energy level?
3. **Identify** which electrons in oxygen have more energy, those in the first energy level or those in the second.
4. **Think Critically** Atoms in a group of elements increase in size as you move down the columns in the periodic table. Explain why this is so.

Applying Math

5. **Solve One-Step Equations** You can calculate the maximum number of electrons each energy level can hold using the formula $2n^2$. Calculate the number of electrons in the first five energy levels where n equals the number of energy levels.

ips.msscience.com/self_check_quiz

section 2
How Elements Bond

as you read

What You'll Learn
- **Compare and contrast** ionic and covalent bonds.
- **Distinguish** between compounds and molecules.
- **Identify** the difference between polar and nonpolar covalent bonds.
- **Interpret** chemical shorthand.

Why It's Important
Chemical bonds join the atoms in the materials you use every day.

Review Vocabulary
electron: a negatively charged particle that exists in an electron cloud around an atom's nucleus

New Vocabulary
- ion
- ionic bond
- compound
- metallic bond
- covalent bond
- molecule
- polar bond
- chemical formula

Ionic Bonds—Loss and Gain

When you put together the pieces of a jigsaw puzzle, they stay together only as long as you wish. When you pick up the completed puzzle, it falls apart. When elements are joined by chemical bonds, they do not readily fall apart. What would happen if suddenly the salt you were shaking on your fries separated into sodium and chlorine? Atoms form bonds with other atoms using the electrons in their outer energy levels. They have four ways to do this—by losing electrons, by gaining electrons, by pooling electrons, or by sharing electrons with another element.

Sodium is a soft, silvery metal as shown in **Figure 11.** It can react violently when added to water or to chlorine. What makes sodium so reactive? If you look at a diagram of its energy levels below, you will see that sodium has only one electron in its outer level. Removing this electron empties this level and leaves the completed level below. By removing one electron, sodium's electron configuration becomes the same as that of the stable noble gas neon.

Chlorine forms bonds in a way that is the opposite of sodium—it gains an electron. When chlorine accepts an electron, its electron configuration becomes the same as that of the noble gas argon.

Figure 11 Sodium and chlorine react, forming white crystalline sodium chloride.

Sodium

Sodium is a silvery metal that can be cut with a knife. Chlorine is a greenish, poisonous gas.

Chlorine

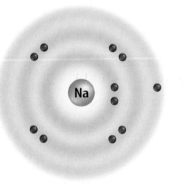
Sodium atom

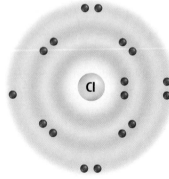
Chlorine atom

If chlorine receives an electron from sodium, both atoms will become stable and a bond will form.

170 CHAPTER 6 Atomic Structure and Chemical Bonds

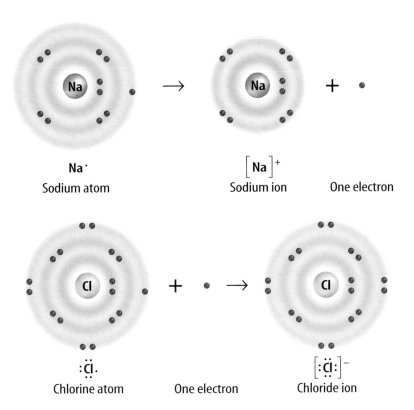

Na·
Sodium atom

[Na]⁺
Sodium ion

One electron

:Cl̈·
Chlorine atom

One electron

[:C̈l:]⁻
Chloride ion

Figure 12 Ions form when elements lose or gain electrons. When sodium comes into contact with chlorine, an electron is transferred from the sodium atom to the chlorine atom. Na becomes a Na⁺ ion. Cl becomes a Cl⁻ ion.

Ions When ions dissolve in water, they separate. Because of their positive and negative charges, the ions can conduct an electric current. If wires are placed in such a solution and the ends of the wires are connected to a battery, the positive ions move toward the negative terminal and the negative ions move toward the positive terminal. This flow of ions completes the circuit.

Ions—A Question of Balance As you just learned, a sodium atom loses an electron and becomes more stable. But something else happens also. By losing an electron, the balance of electric charges changes. Sodium becomes a positively charged ion because there is now one fewer electron than there are protons in the nucleus. In contrast, chlorine becomes an ion by gaining an electron. It becomes negatively charged because there is one more electron than there are protons in the nucleus.

An atom that is no longer neutral because it has lost or gained an electron is called an **ion** (I ahn). A sodium ion is represented by the symbol Na⁺ and a chloride ion is represented by the symbol Cl⁻. **Figure 12** shows how each atom becomes an ion.

Bond Formation The positive sodium ion and the negative chloride ion are strongly attracted to each other. This attraction, which holds the ions close together, is a type of chemical bond called an **ionic bond**. In **Figure 13**, sodium and chloride ions form an ionic bond. The compound sodium chloride, or table salt, is formed. A **compound** is a pure substance containing two or more elements that are chemically bonded.

Na° + ·C̈l: → [Na]⁺[:C̈l:]⁻

Figure 13 An ionic bond forms between atoms of opposite charges. **Describe** *how an atom becomes positive or negative.*

SECTION 2 How Elements Bond **171**

Figure 14 Magnesium has two electrons in its outer energy level.

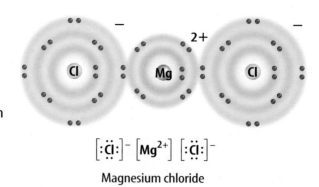

Magnesium chloride

If one electron is lost to each of two chlorine atoms, magnesium chloride forms.

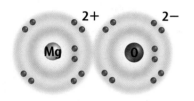

Magnesium oxide

If both electrons are lost to one oxygen atom, magnesium oxide forms.

Determine *the electron arrangement for magnesium sulfide and calcium oxide.*

More Gains and Losses You have seen what happens when elements gain or lose one electron, but can elements lose or gain more than one electron? The element magnesium, Mg, in Group 2 has two electrons in its outer energy level. Magnesium can lose these two electrons and achieve a completed energy level. These two electrons can be gained by two chlorine atoms. As shown in **Figure 14,** a single magnesium ion represented by the symbol Mg^{2+} and two chloride ions are generated. The two negatively charged chloride ions are attracted to the positively charged magnesium ion forming ionic bonds. As a result of these bonds, the compound magnesium chloride ($MgCl_2$) is produced.

Some atoms, such as oxygen, need to gain two electrons to achieve stability. The two electrons released by one magnesium atom could be gained by a single atom of oxygen. When this happens, magnesium oxide (MgO) is formed, as shown in **Figure 14.** Oxygen can form similar compounds with any positive ion from Group 2.

Metallic Bonding—Pooling

You have just seen how metal atoms form ionic bonds with atoms of nonmetals. Metals can form bonds with other metal atoms, but in a different way. In a metal, the electrons in the outer energy levels of the atoms are not held tightly to individual atoms. Instead, they move freely among all the ions in the metal, forming a shared pool of electrons, as shown in **Figure 15. Metallic bonds** form when metal atoms share their pooled electrons. This bonding affects the properties of metals. For example, when a metal is hammered into sheets or drawn into a wire, it does not break. Instead, layers of atoms slide over one another. The pooled electrons tend to hold the atoms together. Metallic bonding also is the reason that metals conduct electricity well. The outer electrons in metal atoms readily move from one atom to the next to transmit current.

Figure 15 In metallic bonding, the outer electrons of the silver atoms are not attached to any one silver atom. This allows them to move and conduct electricity.

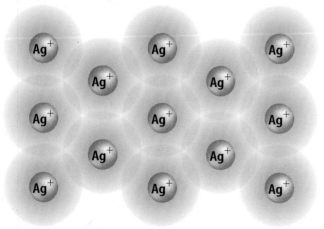

Covalent Bonds—Sharing

Some atoms are unlikely to lose or gain electrons because the number of electrons in their outer levels makes this difficult. For example, carbon has six protons and six electrons. Four of the six electrons are in its outer energy level. To obtain a more stable structure, carbon would either have to gain or lose four electrons. This is difficult because gaining and losing so many electrons takes so much energy. The alternative is sharing electrons.

The Covalent Bond Atoms of many elements become more stable by sharing electrons. The chemical bond that forms between nonmetal atoms when they share electrons is called a **covalent** (koh VAY luhnt) **bond.** Shared electrons are attracted to the nuclei of both atoms. They move back and forth between the outer energy levels of each atom in the covalent bond. So, each atom has a stable outer energy level some of the time. Covalently bonded compounds are called molecular compounds.

Reading Check *How do atoms form covalent bonds?*

The atoms in a covalent bond form a neutral particle, which contains the same numbers of positive and negative charges. The neutral particle formed when atoms share electrons is called a **molecule** (MAH lih kyewl). A molecule is the basic unit of a molecular compound. You can see how molecules form by sharing electrons in **Figure 16.** Notice that no ions are involved because no electrons are gained or lost. Crystalline solids, such as sodium chloride, are not referred to as molecules, because their basic units are ions, not molecules.

Mini LAB

Constructing a Model of Methane

Procedure
1. Using **circles of colored paper** to represent protons, neutrons, and electrons, build paper models of one carbon atom and four hydrogen atoms.
2. Use your models of atoms to construct a molecule of methane by forming covalent bonds. The methane molecule has four hydrogen atoms chemically bonded to one carbon atom.

Analysis
1. In the methane molecule, do the carbon and hydrogen atoms have the same arrangement of electrons as two noble gas elements? Explain your answer.
2. Does the methane molecule have a charge?

Try at Home

Figure 16 Covalent bonding is another way that atoms become more stable. Sharing electrons allows each atom to have a stable outer energy level. These atoms form a single covalent bond.

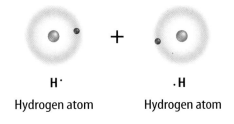
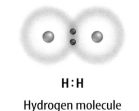
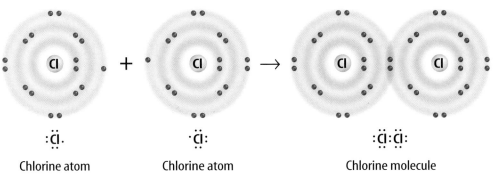

Figure 17 An atom can also form a covalent bond by sharing two or three electrons.

$$\cdot \overset{\cdot}{\underset{}{C}} \cdot \;+\; \cdot \overset{\cdot\cdot}{\underset{\cdot\cdot}{O}} \colon \;+\; \cdot \overset{\cdot\cdot}{\underset{\cdot\cdot}{O}} \colon \;\rightarrow\; \colon \overset{\cdot\cdot}{\underset{\cdot\cdot}{O}} \colon\colon C \colon\colon \overset{\cdot\cdot}{\underset{\cdot\cdot}{O}} \colon$$

Carbon atom   Oxygen atoms   Carbon dioxide molecule

In carbon dioxide, carbon shares two electrons with each of two oxygen atoms forming two double bonds. Each oxygen atom shares two electrons with the carbon atom.

Nitrogen atoms   Nitrogen molecule

Each nitrogen atom shares three electrons in forming a triple bond.

Double and Triple Bonds Sometimes an atom shares more than one electron with another atom. In the molecule carbon dioxide, shown in **Figure 17,** each of the oxygen atoms shares two electrons with the carbon atom. The carbon atom shares two of its electrons with each oxygen atom. When two pairs of electrons are involved in a covalent bond, the bond is called a double bond. **Figure 17** also shows the sharing of three pairs of electrons between two nitrogen atoms in the nitrogen molecule. When three pairs of electrons are shared by two atoms, the bond is called a triple bond.

✓ **Reading Check** *How many pairs of electrons are shared in a double bond?*

Polar and Nonpolar Molecules

You have seen how atoms can share electrons and that they become more stable by doing so, but do they always share electrons equally? The answer is no. Some atoms have a greater attraction for electrons than others do. Chlorine, for example, attracts electrons more strongly than hydrogen does. When a covalent bond forms between hydrogen and chlorine, the shared pair of electrons tends to spend more time near the chlorine atom than the hydrogen atom.

This unequal sharing makes one side of the bond more negative than the other, like poles on a battery. This is shown in **Figure 18.** Such bonds are called polar bonds. A **polar bond** is a bond in which electrons are shared unevenly. The bonds between the oxygen atom and hydrogen atoms in the water molecule are another example of polar bonds.

Figure 18 Hydrogen chloride is a polar covalent molecule.

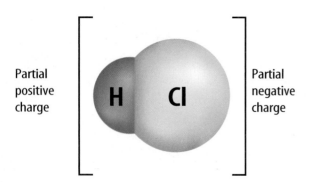

Partial positive charge   Partial negative charge

174 CHAPTER 6 Atomic Structure and Chemical Bonds

The Polar Water Molecule Water molecules form when hydrogen and oxygen share electrons. **Figure 19** shows how this sharing is unequal. The oxygen atom has a greater share of the electrons in each bond—the oxygen end of a water molecule has a slight negative charge and the hydrogen end has a slight positive charge. Because of this, water is said to be polar—having two opposite ends or poles like a magnet.

When they are exposed to a negative charge, the water molecules line up like magnets with their positive ends facing the negative charge. You can see how they are drawn to the negative charge on the balloon in **Figure 19.** Water molecules also are attracted to each other. This attraction between water molecules accounts for many of the physical properties of water.

Molecules that do not have these uneven charges are called nonpolar molecules. Because each element differs slightly in its ability to attract electrons, the only completely nonpolar bonds are bonds between atoms of the same element. One example of a nonpolar bond is the triple bond in the nitrogen molecule.

Like ionic compounds, some molecular compounds can form crystals, in which the basic unit is a molecule. Often you can see the pattern of the units in the shape of ionic and molecular crystals, as shown in **Figure 20.**

Topic: Polar Molecules
Visit ips.msscience.com for Web links to information about soaps and detergents.

Activity Oil and water are not soluble in one another. However, if you add a few grams of a liquid dish detergent, the oil will become soluble in the water. Instead of two layers, there will be only one. Explain why soap can help the oil become soluble in water.

Figure 19 Two hydrogen atoms share electrons with one oxygen atom, but the sharing is unequal. The electrons are more likely to be closer to the oxygen than the hydrogens. The space-saving model shows how the charges are separated or polarized.
Define the term polar.

The positive ends of the water molecules are attracted to the negatively charged balloon, causing the stream of water to bend.

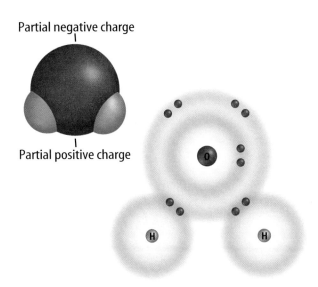

SECTION 2 How Elements Bond **175**

NATIONAL GEOGRAPHIC VISUALIZING CRYSTAL STRUCTURE

Figure 20

Many solids exist as crystals. Whether tiny grains of table salt or big, chunky blocks of quartz you might find rock hunting, a crystal's shape is often a reflection of the arrangement of its particles. Knowing a solid's crystal structure helps researchers understand its physical properties. Some crystals with cubic and hexagonal shapes are shown here.

HEXAGONAL Quartz crystals, above, are six sided, just as a snowflake, above right, has six points. This is because the molecules that make up both quartz and snowflakes arrange themselves into hexagonal patterns.

CUBIC Salt, left, and fluorite, above, form cube-shaped crystals. This shape is a reflection of the cube-shaped arrangement of the ions in the crystal.

Chemical Shorthand

In medieval times, alchemists (AL kuh mists) were the first to explore the world of chemistry. Although many of them believed in magic and mystical transformations, alchemists did learn much about the properties of some elements. They even used symbols to represent them in chemical processes, some of which are shown in **Figure 21.**

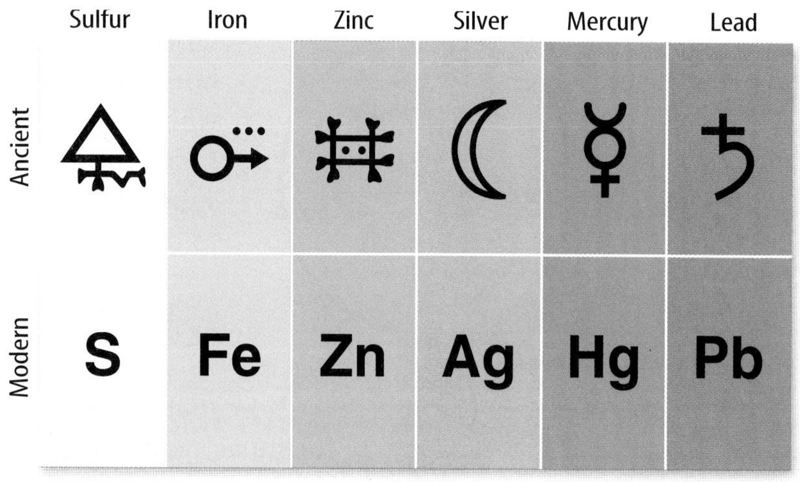

Figure 21 Alchemists used elaborate symbols to describe elements and processes. Modern chemical symbols are letters that can be understood all over the world.

Symbols for Atoms Modern chemists use symbols to represent elements, too. These symbols can be understood by chemists everywhere. Each element is represented by a one letter-, two letter-, or three-letter symbol. Many symbols are the first letters of the element's name, such as H for hydrogen and C for carbon. Others are the first letters of the element's name in another language, such as K for potassium, which stands for kalium, the Latin word for potassium.

Symbols for Compounds Compounds can be described using element symbols and numbers. For example, **Figure 22** shows how two hydrogen atoms join together in a covalent bond. The resulting hydrogen molecule is represented by the symbol H_2. The small 2 after the H in the formula is called a subscript. *Sub* means "below" and *script* means "write," so a subscript is a number that is written a little below a line of text. The subscript 2 means that two atoms of hydrogen are in the molecule.

Figure 22 Chemical formulas show you the kind and number of atoms in a molecule. **Describe** *the term* subscript.

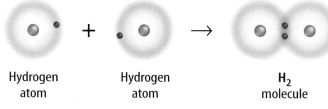

Hydrogen atom + Hydrogen atom → H_2 molecule

The subscript 2 after the H indicates that the hydrogen molecule contains two atoms of hydrogen.

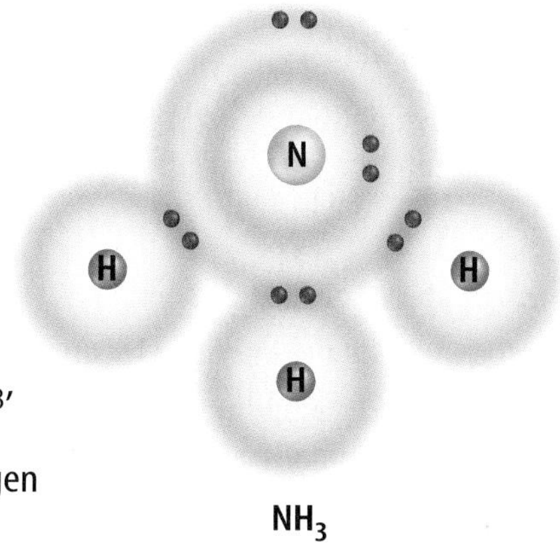

The formula for ammonia, NH_3, tells you that the ratio is one nitrogen atom to three hydrogen atoms.

NH_3

SECTION 2 How Elements Bond 177

Chemical Formulas A **chemical formula** is a combination of chemical symbols and numbers that shows which elements are present in a compound and how many atoms of each element are present. When no subscript is shown, the number of atoms is understood to be one.

Reading Check *What is a chemical formula and what does it tell you about a compound?*

Now that you know a few of the rules for writing chemical formulas, you can look back at other chemical compounds shown earlier in this chapter and begin to predict their chemical formulas. A water molecule contains one oxygen atom and two hydrogen atoms, so its formula is H_2O. Ammonia, shown in **Figure 22,** is a covalent compound that contains one nitrogen atom and three hydrogen atoms. Its chemical formula is NH_3.

The black tarnish that forms on silver, shown in **Figure 23,** is a compound made up of the elements silver and sulfur in the proportion of two atoms of silver to one atom of sulfur. If alchemists knew the composition of silver tarnish, how might they have written a formula for the compound? The modern formula for silver tarnish is Ag_2S. The formula tells you that it is a compound that contains two silver atoms and one sulfur atom.

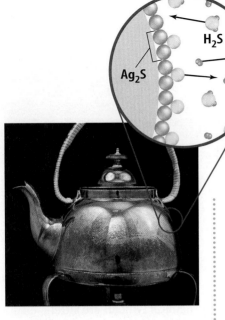

Figure 23 Silver tarnish is the compound silver sulfide, Ag_2S. The formula shows that two silver atoms are combined with one sulfur atom.

section 2 review

Summary

Four Types of Bonds
- Ionic bond is the attraction that holds ions close together.
- Metallic bonds form when metal atoms pool their electrons.
- Covalent bonds form when atoms share electrons.
- A polar covalent bond is a bond in which electrons are shared unevenly.

Chemical Shorthand
- Compounds can be described by using element symbols and numbers.
- A chemical formula is a combination of element symbols and numbers.

Self Check

1. **Determine** Use the periodic table to decide whether lithium forms a positive or negative ion. Does fluorine form a positive or negative ion? Write the formula for the compound formed from these two elements.
2. **Compare and contrast** polar and nonpolar bonds.
3. **Explain** how a chemical formula indicates the ratio of elements in a compound.
4. **Think Critically** Silicon has four electrons in its outer energy level. What type of bond is silicon most likely to form with other elements? Explain.

Applying Skills

5. **Predict** what type of bonds that will form between the following pairs of atoms: carbon and oxygen, potassium and bromine, fluorine and fluorine.

Ionic Compounds

Metals in Groups 1 and 2 often lose electrons and form positive ions. Nonmetals in Groups 16 and 17 often gain electrons and become negative ions. How can compounds form between these four groups of elements?

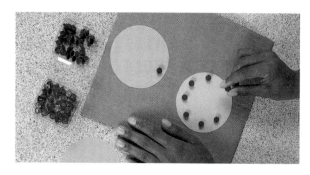

Real-World Question
How do different atoms combine with each other to form compounds?

Goals
- **Construct** models of electron gain and loss.
- **Determine** formulas for the ions and compounds that form when electrons are gained or lost.

Materials
paper (8 different colors)
tacks (2 different colors)
corrugated cardboard
scissors

Safety Precautions

Procedure

1. Cut colored-paper disks 7 cm in diameter to represent the elements Li, S, Mg, O, Ca, Cl, Na, and I. Label each disk with one symbol.
2. Lay circles representing the atoms Li and S side by side on cardboard.
3. Choose colored thumbtacks to represent the outer electrons of each atom. Place the tacks evenly around the disks to represent the outer electron levels of the elements.
4. Move electrons from the metal atom to the nonmetal atom so that both elements achieve noble gas arrangements of eight outer electrons. If needed, cut additional paper disks to add more atoms of one element.
5. Write the formula for each ion and the compound formed when you shift electrons.
6. Repeat steps 2 through 6 to combine Mg and O, Ca and Cl, and Na and I.

Conclude and Apply

1. **Draw** electron dot diagrams for all of the ions produced.
2. **Identify** the noble gas elements having the same electron arrangements as the ions you made in this lab.
3. **Analyze Results** Why did you have to use more than one atom in some cases? Why couldn't you take more electrons from one metal atom or add extra ones to a nonmetal atom?

Communicating Your Data

Compare your compounds and dot diagrams with those of other students in your class. **For more help, refer to the Science Skill Handbook.**

LAB Model and Invent

Atomic Structure

Goals
- **Design** a model of a chosen element.
- **Observe** the models made by others in the class and identify the elements they represent.

Possible Materials
magnetic board
rubber magnetic strips
candy-coated chocolates
scissors
paper
marker
coins

Safety Precautions

WARNING: Never eat any food in the laboratory. Wash hands thoroughly.

Real-World Question

As more information has become known about the structure of the atom, scientists have developed new models. Making your own model and studying the models of others will help you learn how protons, neutrons, and electrons are arranged in an atom. Can an element be identified based on a model that shows the arrangement of the protons, neutrons, and electrons of an atom? How will your group construct a model of an element that others will be able to identify?

Make A Model

1. Choose an element from periods 2 or 3 of the periodic table. How can you determine the number of protons, neutrons, and electrons in an atom given the atom's mass number?

2. How can you show the difference between protons and neutrons? What materials will you use to represent the electrons of the atom? How will you represent the nucleus?

180 CHAPTER 6

3. How will you model the arrangement of electrons in the atom? Will the atom have a charge? Is it possible to identify an atom by the number of protons it has?

4. Make sure your teacher approves your plan before you proceed.

Test Your Model

1. **Construct** your model. Then record your observations in your Science Journal and include a sketch.

2. **Construct** another model of a different element.

3. **Observe** the models made by your classmates. Identify the elements they represent.

Analyze Your Data

1. **State** what elements you identified using your classmates' models.

2. **Identify** which particles always are present in equal numbers in a neutral atom.

3. **Predict** what would happen to the charge of an atom if one of the electrons were removed.

4. **Describe** what happens to the charge of an atom if two electrons are added. What happens to the charge of an atom if one proton and one electron are removed?

5. **Compare and contrast** your model with the electron cloud model of the atom. How is your model similar? How is it different?

Conclude and Apply

1. **Define** the minimum amount of information that you need to know in order to identify an atom of an element.

2. **Explain** If you made models of the isotopes boron-10 and boron-11, how would these models be different?

Communicating Your Data

Compare your models with those of other students. Discuss any differences you find among the models.

Science and Language Arts

"Baring the Atom's Mother Heart"
from Selu: Seeking the Corn-Mother's Wisdom
by Marilou Awiakta

Author Marilou Awiakta was raised near Oak Ridge National Laboratory, a nuclear research laboratory in Tennessee where her father worked. She is of Cherokee and Irish descent. This essay resulted from conversations the author had with writer Alice Walker. It details the author's concern with nuclear technology.

"What is the atom, Mother? Will it hurt us?"

I was nine years old. It was December 1945. Four months earlier, in the heat of an August morning—Hiroshima. Destruction. Death. Power beyond belief, released from something invisible[1]. Without knowing its name, I'd already felt the atoms' power in another form…

"What is the atom, Mother? Will it hurt us?"

"It can be used to hurt everybody, Marilou. It killed thousands[2] of people in Hiroshima and Nagasaki. But the atom itself…? It's invisible, the smallest bit of matter. And it's in everything. Your hand, my dress, the milk you're drinking—…

…Mother already had taught me that beyond surface differences, everything is [connected]. It seemed natural for the atom to be part of this connection. At school, when I was introduced to Einstein's theory of relativity—that energy and matter are one—I accepted the concept easily.

1 can't see
2 10,500

Understanding Literature

Refrain Refrains are emotionally charged words or phrases that are repeated throughout a literary work and can serve a number of purposes. In this work, the refrain is when the author asks, "What is the atom, Mother? Will it hurt us?" Do you think the refrain helps the reader understand the importance of the atom?

Respond to the Reading

1. How did the author's mother explain the atom to her?
2. Is this a positive or negative explanation of the atom?
3. **Linking Science and Writing** Write a short poem about some element you learned about in this chapter.

Nuclear fission, or splitting atoms, is the breakdown of an atom's nucleus. It occurs when a particle, such as a neutron, strikes the nucleus of a uranium atom, splitting the nucleus into two fragments, called fission fragments, and releasing two or three neutrons. These released neutrons ultimately cause a chain reaction by splitting more nuclei and releasing more neutrons. When it is uncontrolled, this chain reaction results in a devastating explosion.

chapter 6 Study Guide

Reviewing Main Ideas

Section 1 Why do atoms combine?

1. The electrons in the electron cloud of an atom are arranged in energy levels.
2. Each energy level can hold a specific number of electrons.
3. The periodic table supplies a great deal of information about the elements.
4. The number of electrons in an atom increases across each period of the periodic table.
5. The noble gas elements are stable because their outer energy levels are stable.
6. Electron dot diagrams show the electrons in the outer energy level of an atom.

Section 2 How Elements Bond

1. An atom can become stable by gaining, losing, or sharing electrons so that its outer energy level is full.
2. Ionic bonds form when a metal atom loses one or more electrons and a nonmetal atom gains one or more electrons.
3. Covalent bonds are created when two or more nonmetal atoms share electrons.
4. The unequal sharing of electrons results in a polar covalent bond.
5. A chemical formula indicates the kind and number of atoms in a compound.

Visualizing Main Ideas

Copy and complete the following concept map on types of bonds.

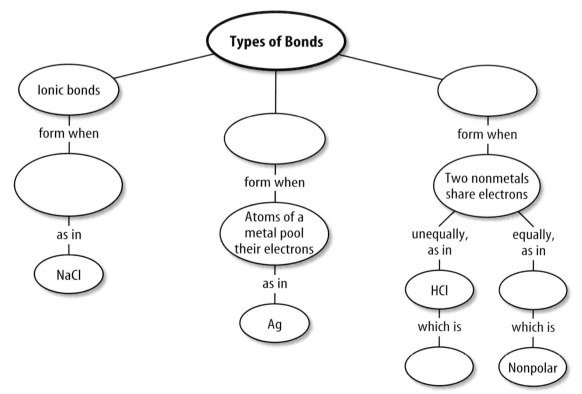

chapter 6 Review

Using Vocabulary

chemical bond p. 169	energy level p. 163
chemical formula p. 178	ion p. 171
compound p. 171	ionic bond p. 171
covalent bond p. 173	metallic bond p. 172
electron cloud p. 162	molecule p. 173
electron dot diagram p. 168	polar bond p. 174

Distinguish between the terms in each of the following pairs.

1. ion—molecule
2. molecule—compound
3. electron dot diagram—ion
4. chemical formula—molecule
5. ionic bond—covalent bond
6. electron cloud—electron dot diagram
7. covalent bond—polar bond
8. compound—formula
9. metallic bond—ionic bond

Checking Concepts

Choose the word or phrase that best answers the question.

10. Which of the following is a covalently bonded molecule?
    A) Cl_2 C) Ne
    B) air   D) salt

11. What is the number of the group in which the elements have a stable outer energy level?
    A) 1   C) 16
    B) 13   D) 18

12. Which term describes the units that make up substances formed by ionic bonding?
    A) ions   C) acids
    B) molecules D) atoms

13. Which of the following describes what is represented by the symbol Cl^-?
    A) an ionic compound
    B) a polar molecule
    C) a negative ion
    D) a positive ion

14. What happens to electrons in the formation of a polar covalent bond?
    A) They are lost.
    B) They are gained.
    C) They are shared equally.
    D) They are shared unequally.

15. Which of the following compounds is unlikely to contain ionic bonds?
    A) NaF   C) LiCl
    B) CO   D) $MgBr_2$

16. Which term describes the units that make up compounds with covalent bonds?
    A) ions   C) salts
    B) molecules D) acids

17. In the chemical formula CO_2, the subscript 2 shows which of the following?
    A) There are two oxygen ions.
    B) There are two oxygen atoms.
    C) There are two CO_2 molecules.
    D) There are two CO_2 compounds.

Use the figure below to answer question 18.

18. Which is NOT true about the molecule H_2O?
    A) It contains two hydrogen atoms.
    B) It contains one oxygen atom.
    C) It is a polar covalent compound.
    D) It is an ionic compound.

chapter 6 Review

Thinking Critically

19. **Explain** why Groups 1 and 2 form many compounds with Groups 16 and 17.

Use the illustration below to answer questions 20 and 21.

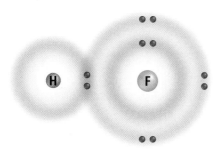

20. **Explain** what type of bond is shown here.

21. **Predict** In the HF molecule above, predict if the electrons are shared equally or unequally between the two atoms. Where do the electrons spend more of their time?

22. **Analyze** When salt dissolves in water, the sodium and chloride ions separate. Explain why this might occur.

23. **Interpret Data** Both cesium, in period 6, and lithium, in period 2, are in the alkali metals family. Cesium is more reactive. Explain this using the energy step diagram in **Figure 4**.

24. **Explain** Use the fact that water is a polar molecule to explain why water has a much higher boiling point than other molecules of its size.

25. **Predict** If equal masses of CuCl and $CuCl_2$ decompose into their components—copper and chlorine—predict which compound will yield more copper. Explain.

26. **Concept Map** Draw a concept map starting with the term *Chemical Bond* and use all the vocabulary words.

27. **Recognize Cause and Effect** A helium atom has only two electrons. Why does helium behave as a noble gas?

28. **Draw a Conclusion** A sample of an element can be drawn easily into wire and conducts electricity well. What kind of bonds can you conclude are present?

Performance Activities

29. **Display** Make a display featuring one of the element families described in this chapter. Include electronic structures, electron dot diagrams, and some compounds they form.

Applying Math

Use the table below to answer question 30.

Formulas of Compounds		
Compound	Number of Metal Atoms	Number of Nonmetal Atoms
Cu_2O		
Al_2S_3	Do not write in this book.	
NaF		
$PbCl_4$		

30. **Make and Use Tables** Fill in the second column of the table with the number of metal atoms in one unit of the compound. Fill in the third column with the number of atoms of the nonmetal in one unit.

31. **Molecules** What are the percentages of each atom for this molecule, K_2CO_3?

32. **Ionic Compounds** Lithium, as a positive ion, is written as Li^{1+}. Nitrogen, as a negative ion, is written as N^{3-}. In order for the molecule to be neutral, the plus and minus charges have to equal zero. How many lithium atoms are needed to make the charges equal to zero?

33. **Energy Levels** Calculate the maximum number of electrons in energy level 6.

ips.msscience.com/chapter_review

Chapter 6 Standardized Test Practice

Part 1 Multiple Choice

Record your answers on the answer sheet provided by your teacher or on a sheet of paper.

1. Sodium combines with fluorine to produce sodium fluoride (NaF), an active ingredient in toothpaste. In this form, sodium has the electron configuration of which other element?
   A. neon
   B. magnesium
   C. lithium
   D. chlorine

Use the illustration below to answer questions 2 and 3.

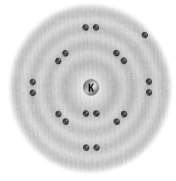

2. The illustration above shows the electron configuration for potassium. How many electrons does potassium need to gain or lose to become stable?
   A. gain 1
   B. gain 2
   C. lose 1
   D. lose 2

3. Potassium belongs to the Group 1 family of elements on the periodic table. What is the name of this group?
   A. halogens
   B. alkali metals
   C. noble gases
   D. alkaline metals

4. What type of bond connects the atoms in a molecule of nitrogen gas (N_2)?
   A. ionic
   B. single
   C. double
   D. triple

Use the illustration below to answer questions 5 and 6.

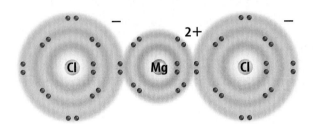

Magnesium chloride

5. The illustration above shows the electron distribution for magnesium chloride. Which of the following is the correct way to write the formula for magnesium chloride?
   A. Mg_2Cl
   B. $MgCl_2$
   C. $MgCl$
   D. Mg_2Cl_2

6. Which of the following terms best describes the type of bonding in magnesium chloride?
   A. ionic
   B. pooling
   C. metallic
   D. covalent

7. What is the maximum number of electrons in the third energy level?
   A. 8
   B. 16
   C. 18
   D. 24

Standardized Test Practice

Part 2 — Short Response/Grid In

Record your answers on the answer sheet provided by your teacher or on a sheet of paper.

8. What is an electron cloud?

9. Explain what is wrong with the following statement: All covalent bonds between atoms are polar to some degree because each element differs slightly in its ability to attract electrons. Give an example to support your answer.

Use the illustration below to answer questions 10 and 11.

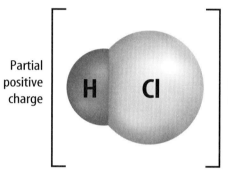

10. The illustration above shows how hydrogen and chlorine combine to form a polar molecule. Explain why the bond is polar.

11. What is the electron dot diagram for the molecule in the illustration?

12. What is the name of the family of elements in Group 17 of the periodic table?

13. Name two ways that electrons around a nucleus are different from planets circling the Sun.

14. Which family of elements used to be known as inert gases? Why was the name changed?

Test-Taking Tip

Take Your Time Stay focused during the test and don't rush, even if you notice that other students are finishing the test early.

Part 3 — Open Ended

Record your answers on a sheet of paper.

15. Scientific experiments frequently require an oxygen-free environment. Such experiments often are performed in containers flooded with argon gas. Describe the arrangement of electrons in an argon atom. Why is argon often a good choice for these experiments?

16. Which group of elements is called the halogen elements? Describe their electron configurations and discuss their reactivity. Name two elements that belong to this group.

17. What is an ionic bond? Describe how sodium chloride forms an ionic bond.

18. Explain metallic bonding. What are some ways this affects the properties of metals?

19. Explain why polar molecules exist, but polar ionic compounds do not exist.

Use the illustration below to answer questions 20 and 21.

20. Explain what is happening in the photograph above. What would happen if the balloon briefly touched the water?

21. Draw a model showing the electron distribution for a water molecule. Explain how the position of the electrons causes the effect shown in your illustration.

chapter 7

Chemical Reactions

The BIG Idea

In chemical reactions, atoms in reactants are rearranged to form products with different chemical properties.

SECTION 1
Chemical Formulas and Equations
Main Idea Atoms are not created or destroyed in chemical reactions—they are just rearranged.

SECTION 2
Rates of Chemical Reactions
Main Idea Reaction rates are affected by several things, including temperature, concentration, surface area, inhibitors, and catalysts.

What chemical reactions happen at chemical plants?

Chemical plants like this one provide the starting materials for thousands of chemical reactions. The compact discs you listen to, personal items, such as shampoo and body lotion, and medicines all have their beginnings in a chemical plant.

Science Journal What additional types of products do you think are manufactured in a chemical plant?

Start-Up Activities

Identify a Chemical Reaction

You can see substances changing every day. Fuels burn, giving energy to cars and trucks. Green plants convert carbon dioxide and water into oxygen and sugar. Cooking an egg or baking bread causes changes too. These changes are called chemical reactions. In this lab you will observe a common chemical change.

WARNING: *Do not touch the test tube. It will be hot. Use extreme caution around an open flame. Point test tubes away from you and others.*

1. Place 3 g of sugar into a large test tube.
2. Carefully light a laboratory burner.
3. Using a test-tube holder, hold the bottom of the test tube just above the flame for 45 s or until something happens with the sugar.
4. Observe any change that occurs.
5. **Think Critically** Describe in your Science Journal the changes that took place in the test tube. What do you think happened to the sugar? Was the substance that remained in the test tube after heating the same as the substance you started with?

Foldables Study Organizer

Chemical Reaction Make the following Foldable to help you understand chemical reactions.

STEP 1 Fold a vertical sheet of notebook paper in half lengthwise.

STEP 2 Cut along every third line of only the top layer to form tabs.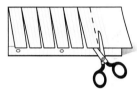

STEP 3 Label each tab.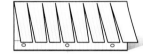

Research Information Before you read the chapter, write several questions you have about chemical reactions on the front of the tabs. As you read, add more questions. Under the tabs of your Foldable, write answers to the questions you recorded on the tabs.

 Preview this chapter's content and activities at ips.msscience.com

Get Ready to Read

Make Predictions

① Learn It! A prediction is an educated guess based on what you already know. One way to predict while reading is to guess what you believe the author will tell you next. As you are reading, each new topic should make sense because it is related to the previous paragraph or passage.

② Practice It! Read the excerpt below from Section 1. Based on what you have read, make predictions about what you will read in the rest of the lesson. After you read Section 1, go back to your predictions to see if they were correct.

Predict what type of properties chemical changes affect.

Do you think melting is a chemical or physical change?

Predict what might happen to the atoms in water when it undergoes a chemical change.

> Matter can undergo two kinds of changes—physical and chemical. Physical changes in a substance affect only the physical properties, such as size and shape, or whether it is a solid, liquid, or gas. For example, when water freezes, its physical state changes from liquid to solid, but it is still water.
>
> —*from page 190*

③ Apply It! Before you read, skim the questions in the Chapter Review. Choose three questions and predict the answers.

Target Your Reading

Reading Tip

As you read, check the predictions you made to see if they were correct.

Use this to focus on the main ideas as you read the chapter.

1. **Before you read** the chapter, respond to the statements below on your worksheet or on a numbered sheet of paper.
   - Write an **A** if you **agree** with the statement.
   - Write a **D** if you **disagree** with the statement.

2. **After you read** the chapter, look back to this page to see if you've changed your mind about any of the statements.
   - If any of your answers changed, explain why.
   - Change any false statements into true statements.
   - Use your revised statements as a study guide.

Science Online
Print out a worksheet of this page at ips.msscience.com

Before You Read A or D		Statement	After You Read A or D
	1	Burning is an example of a chemical change.	
	2	A chemical equation only tells the names of reactants and products.	
	3	When a substance burns, atoms disappear and new atoms are created.	
	4	When balancing a chemical equation, it's okay to change the subscripts of a chemical formula.	
	5	Some reactions release energy and some adsorb energy.	
	6	During chemical reactions, bonds in the reactants break and new bonds form.	
	7	Reactions that release energy do not need any energy to start the reaction.	
	8	Increasing temperature will speed up most chemical reactions.	

190 B

section 1
Chemical Formulas and Equations

as you read

What You'll Learn
- **Determine** whether or not a chemical reaction is occurring.
- **Determine** how to read and understand a balanced chemical equation.
- **Examine** some reactions that release energy and others that absorb energy.
- **Explain** the law of conservation of mass.

Why It's Important
Chemical reactions warm your home, cook your meals, digest your food, and power cars and trucks.

Review Vocabulary
atom: the smallest piece of matter that still retains the property of the element

New Vocabulary
- chemical reaction
- reactant
- product
- chemical equation
- endothermic reaction
- exothermic reaction

Physical or Chemical Change?

You can smell a rotten egg and see the smoke from a campfire. Signs like these tell you that a chemical reaction is taking place. Other evidence might be less obvious, but clues are always present to announce that a reaction is under way.

Matter can undergo two kinds of changes—physical and chemical. Physical changes in a substance affect only physical properties, such as its size and shape, or whether it is a solid, liquid, or gas. For example, when water freezes, its physical state changes from liquid to solid, but it's still water.

In contrast, chemical changes produce new substances that have properties different from those of the original substances. The rust on a bike's handlebars, for example, has properties different from those of the metal around it. Another example is the combination of two liquids that produce a precipitate, which is a solid, and a liquid. The reaction of silver nitrate and sodium chloride forms solid silver chloride and liquid sodium nitrate. A process that produces chemical change is a **chemical reaction.**

To compare physical and chemical changes, look at the newspaper shown in **Figure 1.** If you fold it, you change its size and shape, but it is still newspaper. Folding is a physical change. If you use it to start a fire, it will burn. Burning is a chemical change because new substances result. How can you recognize a chemical change? **Figure 2** shows what to look for.

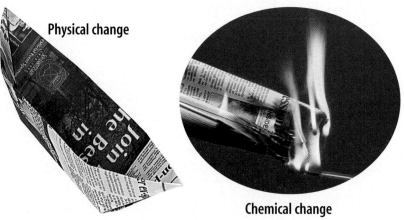

Physical change

Chemical change

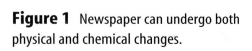

Figure 1 Newspaper can undergo both physical and chemical changes.

190 CHAPTER 7 Chemical Reactions

NATIONAL GEOGRAPHIC VISUALIZING CHEMICAL REACTIONS

Figure 2

Chemical reactions take place when chemicals combine to form new substances. Your senses—sight, taste, hearing, smell, and touch—can help you detect chemical reactions in your environment.

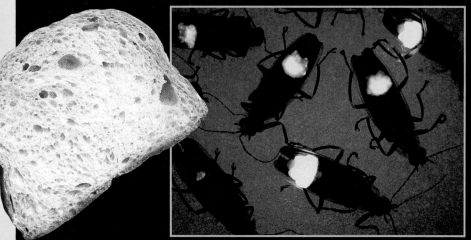

▲ **SIGHT** When you spot a firefly's bright glow, you are seeing a chemical reaction in progress—two chemicals are combining in the firefly's abdomen and releasing light in the process. The holes in a slice of bread are visible clues that sugar molecules were broken down by yeast cells in a chemical reaction that produces carbon dioxide gas. The gas caused the bread dough to rise.

▼ **TASTE** A boy grimaces after sipping milk that has gone sour due to a chemical reaction.

▲ **SMELL AND TOUCH** Billowing clouds of acrid smoke and waves of intense heat indicate that chemical reactions are taking place in this burning forest.

▲ **HEARING** A Russian cosmonaut hoists a flare into the air after landing in the ocean during a training exercise. The hissing sound of the burning flare is the result of a chemical reaction.

SECTION 1 Chemical Formulas and Equations **191**

Chemical Equations

To describe a chemical reaction, you must know which substances react and which substances are formed in the reaction. The substances that react are called the reactants (ree AK tunts). **Reactants** are the substances that exist before the reaction begins. The substances that form as a result of the reaction are called the **products.**

When you mix baking soda and vinegar, a vigorous chemical reaction occurs. The mixture bubbles and foams up inside the container, as you can see in **Figure 3.**

Baking soda and vinegar are the common names for the reactants in this reaction, but they also have chemical names. Baking soda is the compound sodium hydrogen carbonate (often called sodium bicarbonate), and vinegar is a solution of acetic (uh SEE tihk) acid in water. What are the products? You saw bubbles form when the reaction occurred, but is that enough of a description?

Describing What Happens Bubbles tell you that a gas has been produced, but they don't tell you what kind of gas. Are bubbles of gas the only product, or do some atoms from the vinegar and baking soda form something else? What goes on in the chemical reaction can be more than what you see with your eyes. Chemists try to find out which reactants are used and which products are formed in a chemical reaction. Then, they can write it in a shorthand form called a chemical equation. A **chemical equation** tells chemists at a glance the reactants, products, and proportions of each substance present. Some equations also tell the physical state of each substance.

Reading Check *What does a chemical equation tell chemists?*

Figure 3 The bubbles tell you that a chemical reaction has taken place.
Predict *how you might find out whether a new substance has formed.*

Table 1 Reactions Around the Home

Reactants		Products
Baking soda + Vinegar	→	Gas + White solid
Charcoal + Oxygen	→	Ash + Gas + Heat
Iron + Oxygen + Water	→	Rust
Silver + Hydrogen sulfide	→	Black tarnish + Gas
Gas (kitchen range) + Oxygen	→	Gas + Heat
Sliced apple + Oxygen	→	Apple turns brown

Using Words One way you can describe a chemical reaction is with an equation that uses words to name the reactants and products. The reactants are listed on the left side of an arrow, separated from each other by plus signs. The products are placed on the right side of the arrow, also separated by plus signs. The arrow between the reactants and products represents the changes that occur during the chemical reaction. When reading the equation, the arrow is read as *produces*.

You can begin to think of processes as chemical reactions even if you do not know the names of all the substances involved. **Table 1** can help you begin to think like a chemist. It shows the word equations for chemical reactions you might see around your home. See how many other reactions you can find. Look for the signs you have learned that indicate a reaction might be taking place. Then, try to write them in the form shown in the table.

Using Chemical Names Many chemicals used around the home have common names. For example, acetic acid dissolved in water is called vinegar. Some chemicals, such as baking soda, have two common names—it also is known as sodium bicarbonate. However, chemical names are usually used in word equations instead of common names. In the baking soda and vinegar reaction, you already know the chemical names of the reactants—sodium hydrogen carbonate and acetic acid. The names of the products are sodium acetate, water, and carbon dioxide. The word equation for the reaction is as follows.

Acetic acid + Sodium hydrogen carbonate →
    Sodium acetate + Water + Carbon dioxide

Autumn Leaves A color change can indicate a chemical reaction. When leaves change colors in autumn, the reaction may not be what you expect. The bright yellow and orange are always in the leaves, but masked by green chlorophyll. When the growth season ends, more chlorophyll is broken down than produced. The orange and yellow colors become visible.

Mini LAB

Observing the Law of Conservation of Mass

Procedure

1. Place a piece of **steel wool** into a **medium test tube**. Seal the end of the test tube with a **balloon**.
2. Find the mass.
3. Using a test-tube holder, heat the bottom of the tube for two minutes in a **hot water bath** provided by your teacher. Allow the tube to cool completely.
4. Find the mass again.

Analysis

1. What did you observe that showed a chemical reaction took place?
2. Compare the mass before and after the reaction.
3. Why was it important for the test tube to be sealed?

Figure 4 The law of conservation of mass states that the number and kind of atoms must be equal for products and reactants.

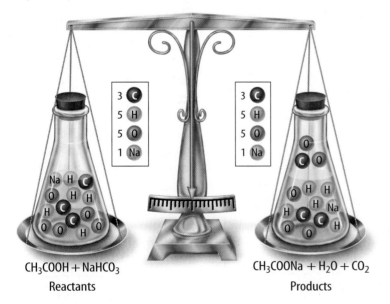

Using Formulas The word equation for the reaction of baking soda and vinegar is long. That's why chemists use chemical formulas to represent the chemical names of substances in the equation. You can convert a word equation into a chemical equation by substituting chemical formulas for the chemical names. For example, the chemical equation for the reaction between baking soda and vinegar can be written as follows:

$$CH_3COOH + NaHCO_3 \rightarrow CH_3COONa + H_2O + CO_2$$

Acetic acid (vinegar)   Sodium hydrogen carbonate (baking soda)   Sodium acetate   Water   Carbon dioxide

Subscripts When you look at chemical formulas, notice the small numbers written to the right of the atoms. These numbers, called subscripts, tell you the number of atoms of each element in that compound. For example, the subscript 2 in CO_2 means that each molecule of carbon dioxide has two oxygen atoms. If an atom has no subscript, it means that only one atom of that element is in the compound, so carbon dioxide has only one carbon atom.

Conservation of Mass

What happens to the atoms in the reactants when they are converted into products? According to the law of conservation of mass, the mass of the products must be the same as the mass of the reactants in that chemical reaction. This principle was first stated by the French chemist Antoine Lavoisier (1743–1794), who is considered the first modern chemist. Lavoisier used logic and scientific methods to study chemical reactions. He proved by his experiments that nothing is lost or created in chemical reactions.

He showed that chemical reactions are much like mathematical equations. In math equations, the right and left sides of the equation are numerically equal. Chemical equations are similar, but it is the number and kind of atoms that are equal on the two sides. Every atom that appears on the reactant side of the equation also appears on the product side, as shown in **Figure 4**. Atoms are never lost or created in a chemical reaction; however, they do change partners.

194 CHAPTER 7 Chemical Reactions

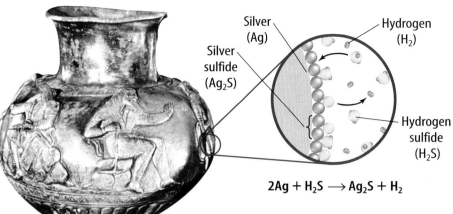

Figure 5 Keeping silver bright takes frequent polishing, especially in homes heated by gas. Sulfur compounds found in small concentrations in natural gas react with silver, forming black silver sulfide, Ag_2S.

$$2Ag + H_2S \rightarrow Ag_2S + H_2$$

Balancing Chemical Equations

When you write the chemical equation for a reaction, you must observe the law of conservation of mass. Look back at **Figure 4.** It shows that when you count the number of carbon, hydrogen, oxygen, and sodium atoms on each side of the arrow in the equation, you find equal numbers of each kind of atom. This means the equation is balanced and the law of conservation of mass is observed.

Not all chemical equations are balanced so easily. For example, silver tarnishes, as in **Figure 5,** when it reacts with sulfur compounds in the air, such as hydrogen sulfide. The following unbalanced equation shows what happens when silver tarnishes.

$$\underset{\text{Silver}}{Ag} + \underset{\text{Hydrogen sulfide}}{H_2S} \rightarrow \underset{\text{Silver sulfide}}{Ag_2S} + \underset{\text{Hydrogen}}{H_2}$$

Count the Atoms Count the number of atoms of each type in the reactants and in the products. The same numbers of hydrogen and sulfur atoms are on each side, but one silver atom is on the reactant side and two silver atoms are on the product side. This cannot be true. A chemical reaction cannot create a silver atom, so this equation does not represent the reaction correctly. Place a 2 in front of the reactant Ag and check to see if the equation is balanced. Recount the number of atoms of each type.

$$2Ag + H_2S \rightarrow Ag_2S + H_2$$

The equation is now balanced. There are an equal number of silver atoms in the reactants and the products. When balancing chemical equations, numbers are placed before the formulas as you did for Ag. These are called coefficients. However, never change the subscripts written to the right of the atoms in a formula. Changing these numbers changes the identity of the compound.

Topic: Chemical Equations
Visit ips.msscience.com for Web links to information about chemical equations and balancing them.

Activity Find a chemical reaction that takes place around your home or school. Write a chemical equation describing it.

SECTION 1 Chemical Formulas and Equations

Energy in Chemical Reactions

Often, energy is released or absorbed during a chemical reaction. The energy for the welding torch in **Figure 6** is released when hydrogen and oxygen combine to form water.

$$2H_2 + O_2 \rightarrow 2H_2O + \text{energy}$$

Energy Released Where does this energy come from? To answer this question, think about the chemical bonds that break and form when atoms gain, lose, or share electrons. When such a reaction takes place, bonds break in the reactants and new bonds form in the products. In reactions that release energy, the products are more stable, and their bonds have less energy than those of the reactants. The extra energy is released in various forms—light, sound, and thermal energy.

Applying Math — Balancing Equations

CONSERVING MASS Methane and oxygen react to form carbon dioxide and water. You can see how mass is conserved by balancing the equation: $CH_4 + O_2 \rightarrow CO_2 + H_2O$.

Solution

1 *This is what you know:*

The number of atoms of C, H, and O in reactants and products.

2 *This is what you need to do:*

Make sure that the reactants and products have equal numbers of atoms of each element. Start with the reactant having the greatest number of different elements.

Reactants	Products	Action
$CH_4 + O_2$ have 4 H atoms	$CO_2 + H_2O$ have 2 H atoms	Need 2 more H atoms in Products. Multiply H_2O by 2 to give 4 H atoms
$CH_4 + O_2$ have 2 O atoms	$CO_2 + 2H_2O$ have 4 O atoms	Need 2 more O atoms in Reactants. Multiply O_2 by 2 to give 4 O atoms

The balanced equation is $CH_4 + 2O_2 \rightarrow CO_2 + 2H_2O$.

3 *Check your answer:*

Count the carbons, hydrogens, and oxygens on each side.

Practice Problems

1. Balance the equation $Fe_2O_3 + CO \rightarrow Fe_3O_4 + CO_2$.
2. Balance the equation $Al + I_2 \rightarrow AlI_3$.

For more practice, visit
ips.msscience.com/
math_practice

Figure 6 This welding torch burns hydrogen and oxygen to produce temperatures above 3,000°C. It can even be used underwater.
Identify *the products of this chemical reaction.*

Energy Absorbed What happens when the reverse situation occurs? In reactions that absorb energy, the reactants are more stable, and their bonds have less energy than those of the products.

$$2H_2O \text{ (Water)} + \text{energy} \rightarrow 2H_2 \text{ (Hydrogen)} + O_2 \text{ (Oxygen)}$$

In this reaction the extra energy needed to form the products can be supplied in the form of electricity, as shown in **Figure 7.**

Reactions can release or absorb several kinds of energy, including electricity, light, sound, and thermal energy. When thermal energy is gained or lost in reactions, special terms are used. **Endothermic** (en doh THUR mihk) **reactions** absorb thermal energy. **Exothermic** (ek soh THUR mihk) **reactions** release thermal energy. The root word *therm* refers to heat, as it does in thermos bottles and thermometers.

Energy Released Several types of reactions that release thermal energy. Burning is an exothermic chemical reaction in which a substance combines with oxygen to produce thermal energy along with light, carbon dioxide, and water.

Reading Check *What type of chemical reaction is burning?*

Rapid Release Sometimes energy is released rapidly. For example, charcoal lighter fluid combines with oxygen in the air and produces enough thermal energy to ignite a charcoal fire within a few minutes.

Figure 7 Electrical energy is needed to break water into its components. This is the reverse of the reaction that takes place in the welding torch shown in **Figure 6.**

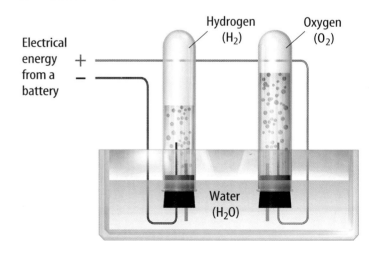

SECTION 1 Chemical Formulas and Equations **197**

Figure 8 Two exothermic reactions are shown. The charcoal fire to cook the food was started when lighter fluid combined rapidly with oxygen in air. The iron in the wheelbarrow combined slowly with oxygen in the air to form rust.

Fast reaction

Slow reaction

Slow Release Other materials also combine with oxygen but release thermal energy so slowly that you cannot see or feel it happen. This is the case when iron combines with oxygen in the air to form rust. The slow release of thermal energy from a reaction also is used in heat packs that can keep your hands warm for several hours. Fast and slow energy release are compared in **Figure 8.**

Energy Absorbed Some chemical reactions and physical processes need to have thermal energy added before they can proceed. An example of an endothermic physical process that absorbs thermal energy is the cold pack shown in **Figure 9.**

The heavy plastic cold pack holds ammonium nitrate and water. The two substances are separated by a plastic divider. When you squeeze the bag, you break the divider so that the ammonium nitrate dissolves in the water. The dissolving process absorbs thermal energy, which must come from the surrounding environment—the surrounding air or your skin after you place the pack on the injury.

Figure 9 The thermal energy needed to dissolve the ammonium nitrate in this cold pack comes from the surrounding environment.

198 CHAPTER 7 Chemical Reactions

Energy in the Equation The word *energy* often is written in equations as either a reactant or a product. Energy written as a reactant helps you think of energy as a necessary ingredient for the reaction to take place. For example, electrical energy is needed to break up water into hydrogen and oxygen. It is important to know that energy must be added to make this reaction occur.

Similarly, in the equation for an exothermic reaction, the word *energy* often is written along with the products. This tells you that energy is released. You include energy when writing the reaction that takes place between oxygen and methane in natural gas when you cook on a gas range, as shown in **Figure 10.** This thermal energy cooks your food.

$$CH_4 + 2O_2 \rightarrow CO_2 + 2H_2O + \text{energy}$$
Methane Oxygen   Carbon   Water
                    dioxide

Although it is not necessary, writing the word *energy* can draw attention to an important aspect of the equation.

Figure 10 Energy from a chemical reaction is used to cook. **Determine** if energy is used as a reactant or a product in this reaction.

section 1 review

Summary

Physical or Chemical Change?
- Matter can undergo physical and chemical changes.
- A chemical reaction produces chemical changes.

Chemical Equations
- A chemical equation describes a chemical reaction.
- Chemical formulas represent chemical names for substances.
- A balanced chemical equation has the same number of atoms of each kind on both sides of the equation.

Energy in Chemical Reactions
- Endothermic reactions absorb thermal energy.
- Exothermic reactions release thermal energy.

Self Check

1. **Determine** if each of these equations is balanced. Why or why not?
   a. $Ca + Cl_2 \rightarrow CaCl_2$
   b. $Zn + Ag_2S \rightarrow ZnS + Ag$
2. **Describe** what evidence might tell you that a chemical reaction has occurred.
3. **Think Critically** After a fire, the ashes have less mass and take up less space than the trees and vegetation before the fire. How can this be explained in terms of the law of conservation of mass?

Applying Math

4. **Calculate** The equation for the decomposition of silver oxide is $2Ag_2O \rightarrow 4Ag + O_2$. Set up a ratio to calculate the number of oxygen molecules released when 1 g of silver oxide is broken down. There are 2.6×10^{21} molecules in 1 g of silver oxide.

section 2

Rates of Chemical Reactions

as you read

What You'll Learn
- **Determine** how to describe and measure the speed of a chemical reaction.
- **Identify** how chemical reactions can be sped up or slowed down.

Why It's Important
Speeding up useful reactions and slowing down destructive ones can be helpful.

Review Vocabulary
state of matter: physical property that is dependent on temperature and pressure and occurs in four forms—solid, liquid, gas, or plasma

New Vocabulary
- activation energy
- rate of reaction
- concentration
- inhibitor
- catalyst
- enzyme

How Fast?

Fireworks explode in rapid succession on a summer night. Old copper pennies darken slowly while they lie forgotten in a drawer. Cooking an egg for two minutes instead of five minutes makes a difference in the firmness of the yolk. The amount of time you leave coloring solution on your hair must be timed accurately to give the color you want. Chemical reactions are common in your life. However, notice from these examples that time has something to do with many of them. As you can see in **Figure 11,** not all chemical reactions take place at the same rate.

Some reactions, such as fireworks or lighting a campfire, need help to get going. You may also notice that others seem to start on their own. In this section, you will also learn about factors that make reactions speed up or slow down once they get going.

Figure 11 Reaction speeds vary greatly. Fireworks are over in a few seconds. However, the copper coating on pennies darkens slowly as it reacts with substances it touches.

Activation Energy—Starting a Reaction

Before a reaction can start, molecules of the reactants have to bump into each other, or collide. This makes sense because to form new chemical bonds, atoms have to be close together. But, not just any collision will do. The collision must be strong enough. This means the reactants must smash into each other with a certain amount of energy. Anything less, and the reaction will not occur. Why is this true?

To form new bonds in the product, old bonds must break in the reactants, and breaking bonds takes energy. To start any chemical reaction, a minimum amount of energy is needed. This energy is called the **activation energy** of the reaction.

Reading Check *What term describes the minimum amount of energy needed to start a reaction?*

What about reactions that release energy? Is there an activation energy for these reactions too? Yes, even though they release energy later, these reactions also need enough energy to start.

One example of a reaction that needs energy to start is the burning of gasoline. You have probably seen movies in which a car plunges over a cliff, lands on the rocks below, and suddenly bursts into flames. But if some gasoline is spilled accidentally while filling a gas tank, it probably will evaporate harmlessly in a short time.

Why doesn't this spilled gasoline explode as it does in the movies? The reason is that gasoline needs energy to start burning. That is why there are signs at filling stations warning you not to smoke. Other signs advise you to turn off the ignition, not to use mobile phones, and not to reenter the car until fueling is complete.

This is similar to the lighting of the Olympic Cauldron, as shown in **Figure 12**. Cauldrons designed for each Olympics contain highly flammable materials that cannot be extinguished by high winds or rain. However, they do not ignite until the opening ceremonies when a runner lights the cauldron using a flame that was kindled in Olympia, Greece, the site of the original Olympic Games.

Topic: Olympic Torch
Visit ips.msscience.com for Web links to information about the Olympic Torch.

Activity With each new Olympics, the host city devises a new Olympic Torch. Research the process that goes into developing the torch and the fuel it uses.

Figure 12 Most fuels need energy to ignite. The Olympic Torch, held by Cathy Freeman in the 2000 Olympics, provided the activation energy required to light the fuel in the cauldron.

Figure 13 The diminishing amount of wax in this candle as it burns indicates the rate of the reaction.

Reaction Rate

Many physical processes are measured in terms of a rate. A rate tells you how much something changes over a given period of time. For example, the rate or speed at which you run or ride your bike is the distance you move divided by the time it took you to move that distance. You may jog at a rate of 8 km/h.

Chemical reactions have rates, too. The **rate of reaction** tells how fast a reaction occurs after it has started. To find the rate of a reaction, you can measure either how quickly one of the reactants is consumed or how quickly one of the products is created, as in **Figure 13**. Both measurements tell how the amount of a substance changes per unit of time.

✓ **Reading Check** *What can you measure to determine the rate of a reaction?*

Reaction rate is important in industry because the faster the product can be made, the less it usually costs. However, sometimes fast rates of reaction are undesirable such as the rates of reactions that cause fruit to ripen. The slower the reaction rate, the longer the food will stay edible. What conditions control the reaction rate, and how can the rate be changed?

Temperature Changes Rate You can slow the ripening of some fruits by putting them in a refrigerator, as in **Figure 14**. Ripening is caused by a series of chemical reactions. Lowering the temperature of the fruit slows the rates of these reactions.

Figure 14 Tomatoes are often picked green and then held in refrigerated storage until they can be delivered to grocery stores.

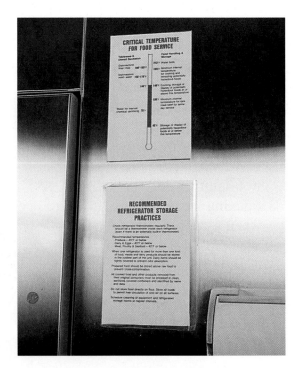

202 CHAPTER 7 Chemical Reactions

INTEGRATE Health

Meat and fish decompose faster at higher temperatures, producing toxins that can make you sick. Keeping these foods chilled slows the decomposition process. Bacteria grow faster at higher temperatures, too, so they reach dangerous levels sooner. Eggs may contain such bacteria, but the heat required to cook eggs also kills bacteria, so hard-cooked eggs are safer to eat than soft-cooked or raw eggs.

Temperature Affects Rate Most chemical reactions speed up when temperature increases. This is because atoms and molecules are always in motion, and they move faster at higher temperatures, as shown in **Figure 15.** Faster molecules collide with each other more often and with greater energy than slower molecules do, so collisions are more likely to provide enough energy to break the old bonds. This is the activation energy.

The high temperature inside an oven speeds up the chemical reactions that turn a liquid cake batter into a more solid, spongy cake. This works the other way, too. Lowering the temperature slows down most reactions. If you set the oven temperature too low, your cake will not bake properly.

Concentration Affects Rate The closer reactant atoms and molecules are to each other, the greater the chance of collisions between them and the faster the reaction rate. It's like the situation shown in **Figure 16.** When you try to walk through a crowded train station, you're more likely to bump into other people than if the station were not so crowded. The amount of substance present in a certain volume is called the **concentration** of that substance. If you increase the concentration, you increase the number of particles of a substance per unit of volume.

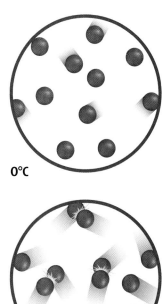
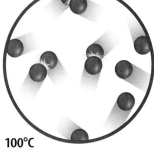

Figure 15 Molecules collide more frequently at higher temperatures than at lower temperatures. This means they are more likely to react.

Collisions are more frequent in a concentrated solution.

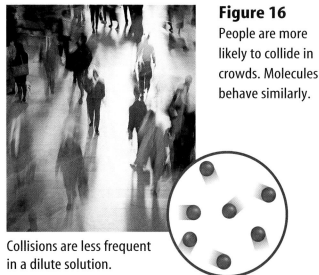

Collisions are less frequent in a dilute solution.

Figure 16 People are more likely to collide in crowds. Molecules behave similarly.

Figure 17 Iron atoms trapped inside this steel beam cannot react with oxygen quickly. More iron atoms are exposed to oxygen molecules in this steel wool, so the reaction speeds up.

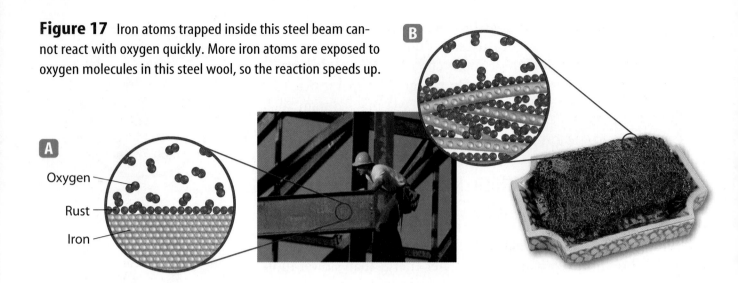

Identifying Inhibitors

Procedure

1. Look at the ingredients listed on **packages of cereals** and **crackers** in your kitchen.
2. Note the preservatives listed. These are chemical inhibitors.
3. Compare the date on the box with the approximate date the box was purchased to estimate shelf life.

Analysis

1. What is the average shelf life of these products?
2. Why is increased shelf life of such products important?

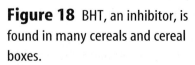

Surface Area Affects Rate The exposed surface area of reactant particles also affects how fast the reaction can occur. You can quickly start a campfire with small twigs, but starting a fire with only large logs would probably not work.

Only the atoms or molecules in the outer layer of the reactant material can touch the other reactants and react. **Figure 17A** shows that when particles are large, most of the iron atoms are stuck inside and can't react. In **Figure 17B,** more of the reactant atoms are exposed to the oxygen and can react.

Slowing Down Reactions

Sometimes reactions occur too quickly. For example, food and medications can undergo chemical reactions that cause them to spoil or lose their effectiveness too rapidly. Luckily, these reactions can be slowed down.

A substance that slows down a chemical reaction is called an **inhibitor.** An inhibitor makes the formation of a certain amount of product take longer. Some inhibitors completely stop reactions. Many cereals and cereal boxes contain the compound butylated hydroxytoluene, or BHT. The BHT slows the spoiling of the cereal and increases its shelf life.

Figure 18 BHT, an inhibitor, is found in many cereals and cereal boxes.

Speeding Up Reactions

Is it possible to speed up a chemical reaction? Yes, you can add a catalyst (KAT uh lihst). A **catalyst** is a substance that speeds up a chemical reaction. Catalysts do not appear in chemical equations, because they are not changed permanently or used up. A reaction using a catalyst will not produce more product than a reaction without a catalyst, but it will produce the same amount of product faster.

Reading Check *What does a catalyst do in a chemical reaction?*

How does a catalyst work? Many catalysts speed up reaction rates by providing a surface for the reaction to take place. Sometimes the reacting molecules are held in a particular position that favors reaction. Other catalysts reduce the activation energy needed to start the reaction.

Catalytic Converters Catalysts are used in the exhaust systems of cars and trucks to aid fuel combustion. The exhaust passes through the catalyst, often in the form of beads coated with metals such as platinum or rhodium. Catalysts speed the reactions that change incompletely burned substances that are harmful, such as carbon monoxide, into less harmful substances, such as carbon dioxide. Similarly, hydrocarbons are changed into carbon dioxide and water. The result of these reactions is cleaner air. These reactions are shown in **Figure 19**.

Breathe Easy The Clean Air Act of 1970 required the reduction of 90 percent of automobile tailpipe emissions. The reduction of emissions included the amount of hydrocarbons and carbon monoxide released. Automakers needed to develop technology to meet this new standard. After much hard work, the result of this legislation was the introduction of the catalytic converter in 1975.

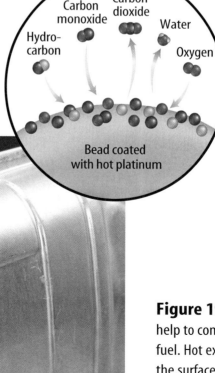

Figure 19 Catalytic converters help to complete combustion of fuel. Hot exhaust gases pass over the surfaces of metal-coated beads. On the surface of the beads, carbon monoxide and hydrocarbons are converted to CO_2 and H_2O.

Enzymes Are Specialists Some of the most effective catalysts are at work in thousands of reactions that take place in your body. These catalysts, called **enzymes,** are large protein molecules that speed up reactions needed for your cells to work properly. They help your body convert food to fuel, build bone and muscle tissue, convert extra energy to fat, and even produce other enzymes.

These are complex reactions. Without enzymes, they would occur at rates that are too slow to be useful or they would not occur at all. Enzymes make it possible for your body to function. Like other catalysts, enzymes function by positioning the reacting molecules so that their structures fit together properly. Enzymes are a kind of chemical specialist—enzymes exist to carry out each type of reaction in your body.

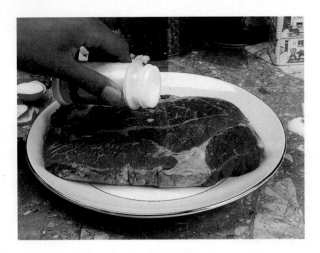

Figure 20 The enzymes in meat tenderizer break down protein in meat, making it more tender.

Other Uses Enzymes work outside your body, too. One class of enzymes, called proteases (PROH tee ay ses), specializes in protein reactions. They work within cells to break down large, complex molecules called proteins. The meat tenderizer shown in **Figure 20** contains proteases that break down protein in meat, making it more tender. Contact lens cleaning solutions also contain proteases that break down proteins from your eyes that can collect on your lenses and cloud your view.

section 2 review

Summary

Chemical Reactions
- To form new bonds in the product, old bonds must break in the reactants. This takes energy.
- Activation energy is the minimum quantity of energy needed to start a reaction.

Reaction Rate
- The rate of reaction tells you how fast a reaction occurs.
- Temperature, concentration, and surface area affect the rate of reaction.

Inhibitors and Catalysts
- Inhibitors slow down reactions. Catalysts speed up reactions.
- Enzymes are catalysts that speed up or slow down reactions for your cells.

Self Check

1. **Describe** how you can measure reaction rates.
2. **Explain** in the general reaction A + B + energy → C, how the following will affect the reaction rate.
   a. increasing the temperature
   b. decreasing the reactant concentration
3. **Describe** how catalysts work to speed up chemical reactions.
4. **Think Critically** Explain why a jar of spaghetti sauce can be stored for weeks on the shelf in the market but must be placed in the refrigerator after it is opened.

Applying Math

5. **Solve One-Step Equations** A chemical reaction is proceeding at a rate of 2 g of product every 45 s. How long will it take to obtain 50 g of product?

Physical or Chemical Change?

Real-World Question

Matter can undergo two kinds of changes—physical and chemical. A physical change affects only physical properties. When a chemical change takes place, a new product is produced. How can a scientist tell if a chemical change took place?

Goals
- **Determine** if a physical or chemical change took place.

Materials
500-mL Erlenmeyer flask
100-mL graduated cylinder
one-hole stopper with 15-cm length of glass tube inserted
1,000-mL beaker
45-cm length of rubber (or plastic) tubing
stopwatch or clock with second hand
weighing dish   balance
baking soda   vinegar

Safety Precautions

WARNING: *Vinegar (acetic acid) may cause skin and eye irritation.*

Procedure

1. Add about 300 mL of water to the 500-mL Erlenmeyer flask.
2. Weigh 5 g of baking soda. Carefully pour the baking soda into the flask. Swirl the flask until the solution is clear.
3. Insert the rubber stopper with the glass tubing into the flask.
4. Add about 600 mL of water to the 1,000-mL beaker.
5. Attach one end of the rubber tubing to the top of the glass tubing. Place the other end of the rubber tubing in the beaker. Be sure the rubber tubing remains under the water.
6. Remove the stopper from the flask. Carefully add 80 mL of vinegar to the flask. Replace the stopper.
7. Count the number of bubbles coming into the beaker for 20 s. Repeat this two more times.
8. Record your data in your Science Journal.

Conclude and Apply

1. **Describe** what you observed in the flask after the acid was added to the baking soda solution.
2. **Classify** Was this a physical or chemical change? How do you know?
3. **Analyze Results** Was this process endothermic or exothermic?
4. **Calculate** the average reaction rate based on the number of bubbles per second.

Compare your results with those of other students in your class.

Design Your Own

Exothermic or Endothermic?

Goals
- **Design** an experiment to test whether a reaction is exothermic or endothermic.
- **Measure** the temperature change caused by a chemical reaction.

Possible Materials
test tubes (8)
test-tube rack
3% hydrogen peroxide solution
raw liver
raw potato
thermometer
stopwatch
clock with second hand
25-mL graduated cylinder

Safety Precautions

WARNING: *Hydrogen peroxide can irritate skin and eyes and damage clothing. Be careful when handling glass thermometers. Test tubes containing hydrogen peroxide should be placed and kept in racks. Dispose of materials as directed by your teacher. Wash your hands when you complete this lab.*

● *Real-World Question*

Energy is always a part of a chemical reaction. Some reactions need a constant supply of energy to proceed. Other reactions release energy into the environment. What evidence can you find to show that a reaction between hydrogen peroxide and liver or potato is exothermic or endothermic?

● *Form a Hypothesis*

Make a hypothesis that describes how you can use the reactions between hydrogen peroxide and liver or potato to determine whether a reaction is exothermic or endothermic.

● *Test Your Hypothesis*

Make a Plan

1. As a group, look at the list of materials. Decide which procedure you will use to test your hypothesis, and which measurements you will make.
2. **Decide** how you will detect the heat released to the environment during the reaction. Determine how many measurements you will need to make during a reaction.
3. You will get more accurate data if you repeat each experiment several times. Each repeated experiment is called a trial. Use the average of all the trials as your data for supporting your hypothesis.
4. **Decide** what the variables are and what your control will be.
5. **Copy** the data table in your Science Journal before you begin to carry out your experiment.

Follow Your Plan

1. Make sure your teacher approves your plan before you start.
2. Carry out your plan.
3. **Record** your measurements immediately in your data table.
4. **Calculate** the averages of your trial results and record them in your Science Journal.

208 CHAPTER 7 Chemical Reactions

Using Scientific Methods

▶ Analyze Your Data

1. Can you infer that a chemical reaction took place? What evidence did you observe to support this?
2. **Identify** what the variables were in this experiment.
3. **Identify** the control.

Temperature After Adding Liver/Potato				
Trial	Temperature After Adding Liver (°C)		Temperature After Adding Potato (°C)	
	Starting	After ____ min	Starting	After ____ min
1				
2	Do not write in this book.			
3				
4				

▶ Conclude and Apply

1. Do your observations allow you to distinguish between an exothermic reaction and an endothermic reaction? Use your data to explain your answer.
2. Where do you think that the energy involved in this experiment came from? Explain your answer.

Communicating Your Data

Compare the results obtained by your group with those obtained by other groups. Are there differences? **Explain** how these might have occurred.

TIME SCIENCE AND HISTORY

SCIENCE CAN CHANGE THE COURSE OF HISTORY!

Synthetic Diamonds

Natural Diamond

Almost the Real Thing

Synthetic Diamond

Diamonds are the most dazzling, most dramatic, most valuable natural objects on Earth. Strangely, these beautiful objects are made of carbon, the same material graphite—the stuff found in pencils—is made of. So why is a diamond hard and clear and graphite soft and black? A diamond's hardness is a result of how strongly its atoms are linked. What makes a diamond transparent is the way its crystals are arranged. The carbon in a diamond is almost completely pure, with trace amounts of boron and nitrogen in it. These elements account for the many shades of color found in diamonds.

A diamond is the hardest naturally occurring substance on Earth. It's so hard, only a diamond can scratch another diamond. Diamonds are impervious to heat and household chemicals. Their crystal structure allows them to be split (or crushed) along particular lines.

Diamonds are made when carbon is squeezed at high pressures and temperatures in Earth's upper mantle, about 150 km beneath the surface. At that depth, the temperature is about 1,400°C, and the pressure is about 55,000 atmospheres greater than the pressure at sea level.

As early as the 1850s, scientists tried to convert graphite into diamonds. It wasn't until 1954 that researchers produced the first synthetic diamonds by compressing carbon under extremely high pressure and heat. Scientists converted graphite powder into tiny diamond crystals using pressure of more than 68,000 atm, and a temperature of about 1,700°C for about 16 hours.

Synthetic diamonds are human-made, but they're not fake. They have all the properties of natural diamonds, from hardness to excellent heat conductivity. Experts claim to be able to detect synthetics because they contain tiny amounts of metal (used in their manufacturing process) and have a different luminescence than natural diamonds. In fact, most synthetics are made for industrial use. One major reason is that making small synthetic diamonds is cheaper than finding small natural ones. The other reason is that synthetics can be made to a required size and shape. Still, if new techniques bring down the cost of producing large, gem-quality synthetic diamonds, they may one day compete with natural diamonds as jewelry.

Research Investigate the history of diamonds—natural and synthetic. Explain the differences between them and their uses. Share your findings with the class.

For more information, visit ips.msscience.com/time

chapter 7 Study Guide

Reviewing Main Ideas

Section 1 — Formulas and Chemical Equations

1. Chemical reactions often cause observable changes, such as a change in color or odor, a release or absorption of heat or light, or a release of gas.

2. A chemical equation is a shorthand method of writing what happens in a chemical reaction. Chemical equations use symbols to represent the reactants and products of a reaction, and sometimes show whether energy is produced or absorbed.

3. The law of conservation of mass requires a balanced chemical reaction that contains the same number of atoms of each element in the products as in the reactants. This is true in every balanced chemical equation.

Section 2 — Rates of Chemical Reactions

1. The rate of reaction is a measure of how quickly a reaction occurs.

2. All reactions have an activation energy—a certain minimum amount of energy required to start the reaction.

3. The rate of a chemical reaction can be influenced by the temperature, the concentration of the reactants, and the exposed surface area of the reactant particles.

4. Catalysts can speed up a reaction without being used up. Inhibitors slow down the rate of reaction.

5. Enzymes are protein molecules that act as catalysts in your body's cells.

Visualizing Main Ideas

Copy and complete the following concept map on chemical reactions.

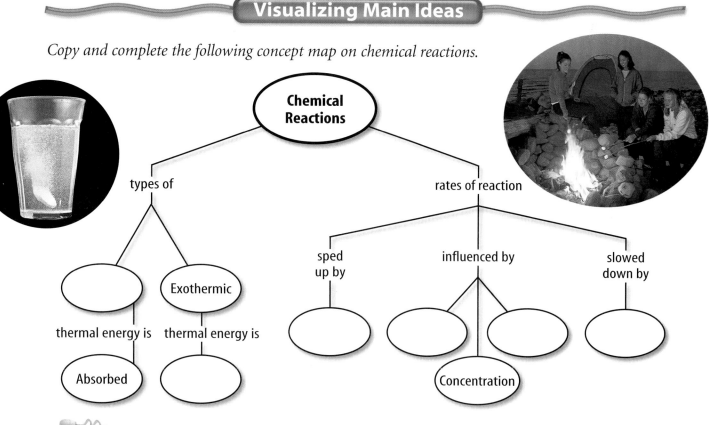

Chapter 7 Review

Using Vocabulary

activation energy p. 201
catalyst p. 205
chemical equation p. 192
chemical reaction p. 190
concentration p. 203
endothermic reaction p. 197
enzyme p. 206
exothermic reaction p. 197
inhibitor p. 204
product p. 192
rate of reaction p. 202
reactant p. 192

Explain the differences between the vocabulary terms in each of the following sets.

1. exothermic reaction—endothermic reaction
2. activation energy—rate of reaction
3. reactant—product
4. catalyst—inhibitor
5. concentration—rate of reaction
6. chemical equation—reactant
7. inhibitor—product
8. catalyst—chemical equation
9. rate of reaction—enzyme

Checking Concepts

Choose the word or phrase that best answers the question.

10. Which statement about the law of conservation of mass is NOT true?
    A) The mass of reactants must equal the mass of products.
    B) All the atoms on the reactant side of an equation are also on the product side.
    C) The reaction creates new types of atoms.
    D) Atoms are not lost, but are rearranged.

11. To slow down a chemical reaction, what should you add?
    A) catalyst   C) inhibitor
    B) reactant   D) enzyme

12. Which of these is a chemical change?
    A) Paper is shredded.
    B) Liquid wax turns solid.
    C) A raw egg is broken.
    D) Soap scum forms.

13. Which of these reactions releases thermal energy?
    A) unbalanced   C) exothermic
    B) balanced   D) endothermic

14. A balanced chemical equation must have the same number of which of these on both sides of the equation?
    A) atoms   C) molecules
    B) reactants   D) compounds

15. What does NOT affect reaction rate?
    A) balancing   C) surface area
    B) temperature   D) concentration

16. Which is NOT a balanced equation?
    A) $CuCl_2 + H_2S \rightarrow CuS + 2HCl$
    B) $AgNO_3 + NaI \rightarrow AgI + NaNO_3$
    C) $2C_2H_6 + 7O_2 \rightarrow 4CO_2 + 6H_2O$
    D) $MgO + Fe \rightarrow Fe_2O_3 + Mg$

17. Which is NOT evidence that a chemical reaction has occurred?
    A) Milk tastes sour.
    B) Steam condenses on a cold window.
    C) A strong odor comes from a broken egg.
    D) A slice of raw potato darkens.

18. Which of the following would decrease the rate of a chemical reaction?
    A) increase the temperature
    B) reduce the concentration of a reactant
    C) increase the concentration of a reactant
    D) add a catalyst

19. Which of these describes a catalyst?
    A) It is a reactant.
    B) It speeds up a reaction.
    C) It appears in the chemical equation.
    D) It can be used in place of an inhibitor.

chapter 7 Review

Thinking Critically

20. **Cause and Effect** Pickled cucumbers remain edible much longer than fresh cucumbers do. Explain.

21. **Analyze** A beaker of water in sunlight becomes warm. Has a chemical reaction occurred? Explain.

22. **Distinguish** if $2Ag + S$ is the same as Ag_2S. Explain.

23. **Infer** Apple slices can be kept from browning by brushing them with lemon juice. Infer what role lemon juice plays in this case.

24. **Draw a Conclusion** Chili can be made using ground meat or chunks of meat. Which would you choose, if you were in a hurry? Explain.

Use the graph below to answer question 25.

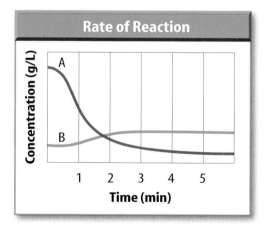

Rate of Reaction

25. **Interpret Scientific Illustrations** The two curves on the graph represent the concentrations of compounds A (blue) and B (red) during a chemical reaction.
    a. Which compound is a reactant?
    b. Which compound is a product?
    c. During which time period is the concentration of the reactant changing most rapidly?

26. **Form a Hypothesis** You are cleaning out a cabinet beneath the kitchen sink and find an unused steel wool scrub pad that has rusted completely. Will the remains of this pad weigh more or less than when it was new? Explain.

Performance Activities

27. **Poster** Make a list of the preservatives in the food you eat in one day. Present your findings to your class in a poster.

Applying Math

Use the graph below to answer question 28.

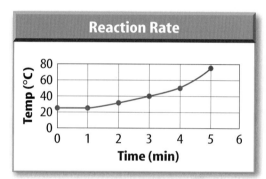

Reaction Rate

28. **Reaction Rates** In the reaction graph above, how long does it take the reaction to reach 50°C?

29. **Chemical Equation** In the following chemical equation, $3Na + AlCl_3 \rightarrow 3NaCl + Al$, how many aluminum molecules will be produced if you have 30 molecules of sodium?

30. **Catalysis** A zinc catalyst is used to reduce the reaction time by 30%. If the normal time for the reaction to finish is 3 h, how long will it take with the catalyst?

31. **Molecules** Silver has 6.023×10^{23} molecules per 107.9 g. How many molecules are there if you have
    a. 53.95 g?
    b. 323.7 g?
    c. 10.79 g?

chapter 7 Standardized Test Practice

Part 1 | Multiple Choice

Record your answers on the answer sheet provided by your teacher or on a sheet of paper.

Use the photo below to answer questions 1 and 2.

1. The photograph shows the reaction of copper (Cu) with silver nitrate (AgNO$_3$) to produce copper nitrate (Cu(NO$_3$)$_2$) and silver (Ag). The chemical equation that describes this reaction is the following:

   $$2AgNO_3 + Cu \rightarrow Cu(NO_3)_2 + 2Ag$$

   What term describes what is happening in the reaction?
   A. catalyst
   B. chemical change
   C. inhibitor
   D. physical change

2. Which of the following terms describes the copper on the left side of the equation?
   A. reactant   C. enzyme
   B. catalyst   D. product

3. Which term best describes a chemical reaction that absorbs thermal energy?
   A. catalytic   C. endothermic
   B. exothermic   D. acidic

4. What should be balanced in a chemical equation?
   A. compounds   C. molecules
   B. atoms   D. molecules and atoms

Test-Taking Tip

Read All Questions Never skip a question. If you are unsure of an answer, mark your best guess on another sheet of paper and mark the question in your test booklet to remind you to come back to it at the end of the test.

Use the photo below to answer questions 5 and 6.

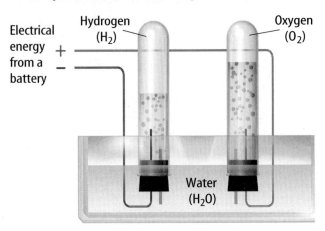

5. The photograph above shows a demonstration of electrolysis, in which water is broken down into hydrogen and oxygen. Which of the following is the best way to write the chemical equation for this process?
   A. $H_2O + energy \rightarrow H_2 + O_2$
   B. $H_2O + energy \rightarrow 2H_2 + O_2$
   C. $2H_2O + energy \rightarrow 2H_2 + O_2$
   D. $2H_2O + energy \rightarrow 2H_2 + 2O_2$

6. For each atom of hydrogen that is present before the reaction begins, how many atoms of hydrogen are present after the reaction?
   A. 1   C. 4
   B. 2   D. 8

7. What is the purpose of an inhibitor in a chemical reaction?
   A. decrease the shelf life of food
   B. increase the surface area
   C. decrease the speed of a chemical reaction
   D. increase the speed of a chemical reaction

214 STANDARDIZED TEST PRACTICE

Part 2 Short Response/Grid In

Record your answers on the answer sheet provided by your teacher or on a sheet of paper.

8. If the volume of a substance changes but no other properties change, is this a physical or a chemical change? Explain.

Use the equation below to answer question 9.

$$CaCl_2 + 2AgNO_3 \rightarrow 2\boxed{} + Ca(NO_3)_2$$

9. When solutions of calcium chloride ($CaCl_2$) and silver nitrate ($AgNO_3$) are mixed, calcium nitrate ($Ca(NO_3)_2$) and a white precipitate, or residue, form. Determine the chemical formula of the precipitate.

Use the illustration below to answer questions 10 and 11.

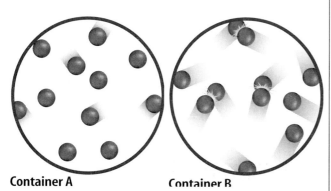

Container A   Container B

10. The figure above demonstrates the movement of atoms at temperatures of 0°C and 100°C. What would happen to the movement of the atoms if the temperature dropped far below 0°C?

11. Describe how the difference in the movement of the atoms at two different temperatures affects the rate of most chemical reactions.

12. Is activation energy needed for reactions that release energy? Explain.

Part 3 Open Ended

Record your answers on a sheet of paper.

Use the illustration below to answer questions 13 and 14.

13. The photograph above shows a forest fire that began when lightning struck a tree. Describe the chemical reaction that occurs when trees burn. Is the reaction endothermic or exothermic? What does this mean? Why does this cause a forest fire to spread?

14. The burning of logs in a forest fire is a chemical reaction. What prevents this chemical reaction from occurring when there is no lightning to start a fire?

15. Explain how the surface area of a material can affect the rate at which the material reacts with other substances. Give an example to support your answer.

16. One of the chemical reactions that occurs in the formation of glass is the combining of calcium carbonate ($CaCO_3$) and silica (SiO_2) to form calcium silicate ($CaSiO_3$) and carbon dioxide (CO_2):

$$CaCO_3 + SiO_2 \rightarrow CaSiO_3 + CO_2$$

Describe this reaction using the names of the chemicals. Discuss which bonds are broken and how atoms are rearranged to form new bonds.

chapter 8

Substances, Mixtures, and Solubility

The BIG Idea

Matter can be classified as a substance (element or compound) or a mixture (homogeneous or heterogeneous).

SECTION 1
What is a solution?
Main Idea Solutions are homogeneous mixtures that can be solids, liquids, or gases.

SECTION 2
Solubility
Main Idea Solubility refers to the amount of a solute that can dissolve in a solvent at a given temperature and pressure.

SECTION 3
Acidic and Basic Solutions
Main Idea When dissolved in water, acids produce hydronium (H_3O^+) ions, and bases produce hydroxide (OH^-) ions.

Big-Band Mixtures

It's a parade and the band plays. Just as the mixing of notes produces music, the mixing of substances produces many of the things around you. From the brass in tubas to the lemonade you drink, you live in a world of mixtures. In this chapter, you'll learn why some substances form mixtures and others do not.

Science Journal Find and name four items around you that are mixtures.

Start-Up Activities

Particle Size and Dissolving Rates

Why do drink mixes come in powder form? What would happen if you dropped a big chunk of drink mix into the water? Would it dissolve quickly? Powdered drink mix dissolves faster in water than chunks do because it is divided into smaller particles, exposing more of the mix to the water. See for yourself how particle size affects the rate at which a substance dissolves.

1. Pour 400 mL of water into each of two 600-mL beakers.
2. Carefully grind a bouillon cube into powder using a mortar and pestle.
3. Place the bouillon powder into one beaker and drop a whole bouillon cube into the second beaker.
4. Stir the water in each beaker for 10 s and observe.
5. **Think Critically** Write a paragraph in your Science Journal comparing the color of the two liquids and the amount of undissolved bouillon at the bottom of each beaker. How does the particle size affect the rate at which a substance dissolves?

Solutions Make the following Foldable to help classify solutions based on their common features.

STEP 1 Fold a vertical sheet of paper from side to side. Make the front edge about 1.25 cm shorter than the back edge.

STEP 2 Turn lengthwise and fold into thirds.

STEP 3 Unfold and cut only the top layer along both folds to make three tabs.

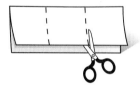

STEP 4 Label each tab as shown.

Find Main Ideas As you read the chapter, classify solutions based on their states and list them under the appropriate tabs. On your Foldable, circle the solutions that are acids and underline the solutions that are bases.

Preview this chapter's content and activities at
ips.msscience.com

217

Get Ready to Read

Identify Cause and Effect

1 Learn It! A *cause* is the reason something happens. The result of what happens is called an effect. Learning to identify causes and effects helps you understand why things happen. By using graphic organizers, you can sort and analyze causes and effects as you read.

2 Practice It! Read the following paragraph. Then use the graphic organizer below to show what might happen when a solution is cooled.

> Under certain conditions, a solute can come back out of solution and form a solid. This process is called crystallization. Sometimes this occurs when the solution is cooled or when some of the solvent evaporates. Crystallization is the result of a physical change.
>
> — from page 220

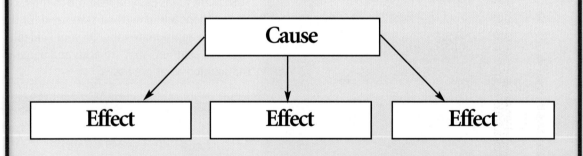

3 Apply It! As you read the chapter, be aware of causes and effects of dissolving. Find at least one cause and its effect.

Target Your Reading

Reading Tip

Graphic organizers such as the Cause-Effect organizer help you organize what you are reading so you can remember it later.

Use this to focus on the main ideas as you read the chapter.

① **Before you read** the chapter, respond to the statements below on your worksheet or on a numbered sheet of paper.
- Write an **A** if you **agree** with the statement.
- Write a **D** if you **disagree** with the statement.

② **After you read** the chapter, look back to this page to see if you've changed your mind about any of the statements.
- If any of your answers changed, explain why.
- Change any false statements into true statements.
- Use your revised statements as a study guide.

Science Online
Print out a worksheet of this page at ips.msscience.com

Before You Read A or D		Statement	After You Read A or D
	1	Only a chemical process can change one substance into one or more new substances.	
	2	Fruit drink is an example of a substance.	
	3	Brass, a type of metal, is an example of a solution.	
	4	The solubility of a solute in a solvent varies with temperature.	
	5	A solute that dissolves quickly is more soluble than one that dissolves slowly.	
	6	You can increase the solubility of a solute by stirring it in the solvent.	
	7	Concentration is a measure of how much solute is dissolved.	
	8	A strong acid is one that is concentrated.	
	9	The more hydrogen atoms an acid contains, the stronger it is.	

218 B

section 1
What is a solution?

as you read

What You'll Learn
- **Distinguish** between substances and mixtures.
- **Describe** two different types of mixtures.
- **Explain** how solutions form.
- **Describe** different types of solutions.

Why It's Important
The air you breathe, the water you drink, and even parts of your body are all solutions.

Review Vocabulary
proton: positively charged particle located in the nucleus of an atom

New Vocabulary
- substance
- heterogeneous mixture
- homogeneous mixture
- solution
- solute
- solvent
- precipitate

Substances

Water, salt water, and pulpy orange juice have some obvious differences. These differences can be explained by chemistry. Think about pure water. No matter what you do to it physically—freeze it, boil it, stir it, or strain it—it still is water. On the other hand, if you boil salt water, the water turns to gas and leaves the salt behind. If you strain pulpy orange juice, it loses its pulp. How does chemistry explain these differences? The answer has to do with the chemical compositions of the materials.

Atoms and Elements Recall that atoms are the basic building blocks of matter. Each atom has unique chemical and physical properties which are determined by the number of protons it has. For example, all atoms that have eight protons are oxygen atoms. A **substance** is matter that has the same fixed composition and properties. It can't be broken down into simpler parts by ordinary physical processes, such as boiling, grinding, or filtering. Only a chemical process can change a substance into one or more new substances. **Table 1** lists some examples of physical and chemical processes. An element is an example of a pure substance; it cannot be broken down into simpler substances. The number of protons in an element, like oxygen, are fixed—it cannot change unless the element changes.

Compounds Water is another example of a substance. It is always water even when you boil it or freeze it. Water, however, is not an element. It is an example of a compound which is made of two or more elements that are chemically combined. Compounds also have fixed compositions. The ratio of the atoms in a compound is always the same. For example, when two hydrogen atoms combine with one oxygen atom, water is formed. All water—whether it's in the form of ice, liquid, or steam—has the same ratio of hydrogen atoms to oxygen atoms.

Table 1 Examples of Physical and Chemical Processes	
Physical Processes	**Chemical Processes**
Boiling	Burning
Changing pressure	Reacting with other chemicals
Cooling	Reacting with light
Sorting	

218 CHAPTER 8 Substances, Mixtures, and Solubility

Figure 1 Mixtures can be separated by physical processes.
Explain why the iron-sand mixture and the pulpy lemonade are not pure substances.

Separation by magnetism

Separation by straining

Mixtures

Imagine drinking a glass of salt water. You would know right away that you weren't drinking pure water. Like salt water, many things are not pure substances. Salt water is a mixture of salt and water. Mixtures are combinations of substances that are not bonded together and can be separated by physical processes. For example, you can boil salt water to separate the salt from the water. If you had a mixture of iron filings and sand, you could separate the iron filings from the sand with a magnet. **Figure 1** shows some mixtures being separated.

Unlike compounds, mixtures do not always contain the same proportions of the substances that they are composed of. Lemonade is a mixture that can be strong tasting or weak tasting, depending on the amounts of water and lemon juice that are added. It also can be sweet or sour, depending on how much sugar is added. But whether it is strong, weak, sweet, or sour, it is still lemonade.

Heterogeneous Mixtures It is easy to tell that some things are mixtures just by looking at them. A watermelon is a mixture of fruit and seeds. The seeds are not evenly spaced through the whole melon—one bite you take might not have any seeds in it and another bite might have several seeds. A type of mixture where the substances are not mixed evenly is called a **heterogeneous** (he tuh ruh JEE nee us) **mixture.** The different areas of a heterogeneous mixture have different compositions. The substances in a heterogeneous mixture are usually easy to tell apart, like the seeds from the fruit of a watermelon. Other examples of heterogeneous mixtures include a bowl of cold cereal with milk and the mixture of pens, pencils, and books in your backpack.

Topic: Desalination
Visit ips.msscience.com for Web links to information about how salt is removed from salt water to provide drinking water.

Activity Compare and contrast the two most common methods used for desalination.

SECTION 1 What is a solution? **219**

Homogeneous Mixtures Your shampoo contains many ingredients, but you can't see them when you look at the shampoo. It is the same color and texture throughout. Shampoo is an example of a homogeneous (hoh muh JEE nee us) mixture. A **homogeneous mixture** contains two or more substances that are evenly mixed on a molecular level but still are not bonded together. Another name for a homogeneous mixture is a **solution.** The sugar and water in the frozen pops shown in **Figure 2,** are a solution—the sugar is evenly distributed in the water, and you can't see the sugar.

Figure 2 Molecules of sugar and water are evenly mixed in frozen pops.

Reading Check *What is another name for a homogeneous mixture?*

How Solutions Form

How do you make sugar water for a hummingbird feeder? You might add sugar to water and heat the mixture until the sugar disappears. The sugar molecules would spread out until they were evenly spaced throughout the water, forming a solution. This is called dissolving. The substance that dissolves—or seems to disappear—is called the **solute.** The substance that dissolves the solute is called the **solvent.** In the hummingbird feeder solution, the solute is the sugar and the solvent is water. The substance that is present in the greatest quantity is the solvent.

Figure 3 Minerals and soap react to form soap scum, which comes out of the water solution and coats the tiles of a shower.

Forming Solids from Solutions Under certain conditions, a solute can come back out of its solution and form a solid. This process is called crystallization. Sometimes this occurs when the solution is cooled or when some of the solvent evaporates. Crystallization is the result of a physical change. When some solutions are mixed, a chemical reaction occurs, forming a solid. This solid is called a **precipitate** (prih SIH puh tayt). A precipitate is the result of a chemical change. Precipitates probably have formed in your sink or shower because of chemical reactions. Minerals that are dissolved in tap water react chemically with soap. The product of this reaction leaves the water as a precipitate called soap scum, shown in **Figure 3.**

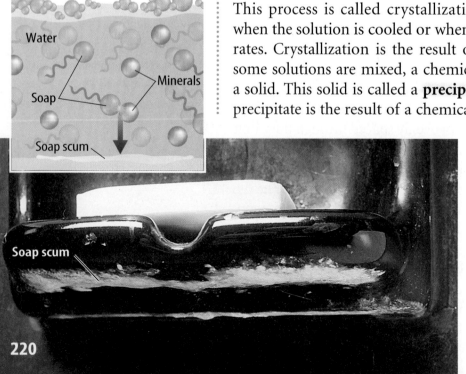

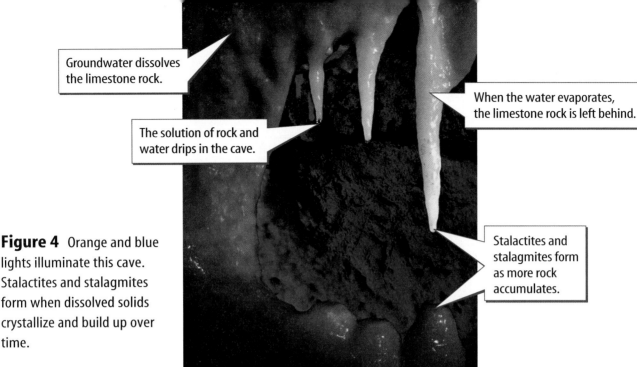

Figure 4 Orange and blue lights illuminate this cave. Stalactites and stalagmites form when dissolved solids crystallize and build up over time.

- Groundwater dissolves the limestone rock.
- The solution of rock and water drips in the cave.
- When the water evaporates, the limestone rock is left behind.
- Stalactites and stalagmites form as more rock accumulates.

Stalactites and stalagmites in caves are formed from solutions, as shown in **Figure 4.** First, minerals dissolve in water as it flows through rocks at the top of the cave. This solution of water and dissolved minerals drips from the ceiling of the cave. When drops of the solution evaporate from the roof of the cave, the minerals are left behind. They create the hanging rock formations called stalactites. When drops of the solution fall onto the floor of the cave and evaporate, they form stalagmites. Very often, a stalactite develops downward while a stalagmite develops upward until the two meet. One continuous column of minerals is formed. This process will be discussed later.

Types of Solutions

So far, you've learned about types of solutions in which a solid solute dissolves in a liquid solvent. But solutions can be made up of different combinations of solids, liquids, and gases, as shown in **Table 2.**

Table 2 Examples of Common Solutions

	Solvent/State	Solute/State	State of Solution
Earth's atmosphere	nitrogen/gas	oxygen/gas carbon dioxide/gas argon/gas	gas
Ocean water	water/liquid	salt/solid oxygen/gas carbon dioxide/gas	liquid
Carbonated beverage	water/liquid	carbon dioxide/gas	liquid
Brass	copper/solid	zinc/solid	solid

SECTION 1 What is a solution? **221**

Figure 5 Acetic acid (a liquid), carbon dioxide (a gas), and drink-mix crystals (a solid) can be dissolved in water (a liquid). **Determine** whether one liquid solution could contain all three different kinds of solute.

Liquid Solutions

You're probably most familiar with liquid solutions like the ones shown in **Figure 5,** in which the solvent is a liquid. The solute can be another liquid, a solid, or even a gas. You've already learned about liquid-solid solutions such as sugar water and salt water. When discussing solutions, the state of the solvent usually determines the state of the solution.

Liquid-Gas Solutions Carbonated beverages are liquid-gas solutions—carbon dioxide is the gaseous solute, and water is the liquid solvent. The carbon dioxide gas gives the beverage its fizz and some of its tartness. The beverage also might contain other solutes, such as the compounds that give it its flavor and color.

Reading Check *What are the solutes in a carbonated beverage?*

Liquid-Liquid Solutions In a liquid-liquid solution, both the solvent and the solute are liquids. Vinegar, which you might use to make salad dressing, is a liquid-liquid solution made of 95 percent water (the solvent) and 5 percent acetic acid (the solute).

Gaseous Solutions

In gaseous solutions, a smaller amount of one gas is dissolved in a larger amount of another gas. This is called a gas-gas solution because both the solvent and solute are gases. The air you breathe is a gaseous solution. Nitrogen makes up about 78 percent of dry air and is the solvent. The other gases are the solutes.

CHAPTER 8 Substances, Mixtures, and Solubility

Figure 6 Metal alloys can contain either metal or nonmetal solutes dissolved in a metal solvent.

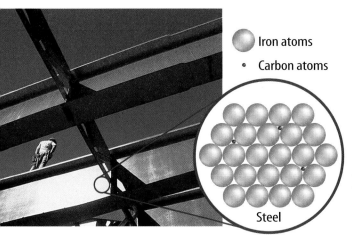

Steel is a solid solution of the metal iron and the nonmetal carbon.

Brass is a solid solution made of copper and zinc.

Solid Solutions In solid solutions, the solvent is a solid. The solute can be a solid, liquid, or gas. The most common solid solutions are solid-solid solutions—ones in which the solvent and the solute are solids. A solid-solid solution made from two or more metals is called an alloy. It's also possible to include elements that are not metals in alloys. For example, steel is an alloy that has carbon dissolved in iron. The carbon makes steel much stronger and yet more flexible than iron. Two alloys are shown in **Figure 6**.

section 1 review

Summary

Substances
- Elements are substances that cannot be broken down into simpler substances.
- A compound is made up of two or more elements bonded together.

Mixtures and Solutions
- Mixtures are either heterogeneous or homogeneous.
- Solutions have two parts—solute and solvent.
- Crystallization and precipitation are two ways that solids are formed from solutions.

Types of Solutions
- The solutes and solvents can be solids, liquids, or gases.

Self Check

1. **Compare and contrast** substances and mixtures. Give two examples of each.
2. **Describe** how heterogeneous and homogeneous mixtures differ.
3. **Explain** how a solution forms.
4. **Identify** the common name for a solid-solid solution of metals.
5. **Think Critically** The tops of carbonated-beverage cans usually are made with a different aluminum alloy than the pull tabs are made with. Explain.

Applying Skills

6. **Compare and contrast** the following solutions: a helium-neon laser, bronze (a copper-tin alloy), cloudy ice cubes, and ginger ale.

ips.msscience.com/self_check_quiz

section 2
Solubility

as you read

What You'll Learn
- **Explain** why water is a good general solvent.
- **Describe** how the structure of a compound affects which solvents it dissolves in.
- **Identify** factors that affect how much of a substance will dissolve in a solvent.
- **Describe** how temperature affects reaction rate.
- **Explain** how solute particles affect physical properties of water.

Why It's Important
How you wash your hands, clothes, and dishes depends on which substances can dissolve in other substances.

Review Vocabulary
polar bond: a bond resulting from the unequal sharing of electrons

New Vocabulary
- aqueous
- solubility
- saturated
- concentration

Water—The Universal Solvent

In many solutions, including fruit juice and vinegar, water is the solvent. A solution in which water is the solvent is called an **aqueous** (A kwee us) solution. Because water can dissolve so many different solutes, chemists often call it the universal solvent. To understand why water is such a great solvent, you must first know a few things about atoms and bonding.

Molecular Compounds When certain atoms form compounds, they share electrons. Sharing electrons is called covalent bonding. Compounds that contain covalent bonds are called molecular compounds, or molecules.

If a molecule has an even distribution of electrons, like the one in **Figure 7,** it is called nonpolar. The atoms in some molecules do not have an even distribution of electrons. For example, in a water molecule, two hydrogen atoms share electrons with a single oxygen atom. However, as **Figure 7** shows, the electrons spend more time around the oxygen atom than they spend around the hydrogen atoms. As a result, the oxygen portion of the water molecule has a partial negative charge and the hydrogen portions have a partial positive charge. The overall charge of the water molecule is neutral. Such a molecule is said to be polar, and the bonds between its atoms are called polar covalent bonds.

Figure 7 Some atoms share electrons to form covalent bonds.

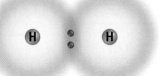

Two atoms of hydrogen share their electrons equally. Such a molecule is nonpolar.

The electrons spend more time around the oxygen atom than the hydrogen atoms. Such a molecule is polar.

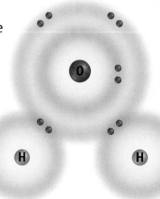

(Partial negative charge)

(Partial positive charge)

Ionic Bonds Some atoms do not share electrons when they join with other atoms to form compounds. Instead, these atoms lose or gain electrons. When they do, the number of protons and electrons within an atom are no longer equal, and the atom becomes positively or negatively charged. Atoms with a charge are called ions. Bonds between ions that are formed by the transfer of electrons are called ionic bonds, and the compound that is formed is called an ionic compound. Table salt is an ionic compound that is made of sodium ions and chloride ions. Each sodium atom loses one electron to a chlorine atom and becomes a positively charged sodium ion. Each chlorine atom gains one electron from a sodium atom, becoming a negatively charged chloride ion.

Reading Check *How does an ionic compound differ from a molecular compound?*

How Water Dissolves Ionic Compounds Now think about the properties of water and the properties of ionic compounds as you visualize how an ionic compound dissolves in water. Because water molecules are polar, they attract positive and negative ions. The more positive part of a water molecule—where the hydrogen atoms are—is attracted to negatively charged ions. The more negative part of a water molecule—where the oxygen atom is—attracts positive ions. When an ionic compound is mixed with water, the different ions of the compound are pulled apart by the water molecules. **Figure 8** shows how sodium chloride dissolves in water.

Solutions Seawater is a solution that contains nearly every element found on Earth. Most elements are present in tiny quantities. Sodium and chloride ions are the most common ions in seawater. Several gases, including oxygen, nitrogen, and carbon dioxide, also are dissolved in seawater.

Figure 8 Water dissolves table salt because its partial charges are attracted to the charged ions in the salt.

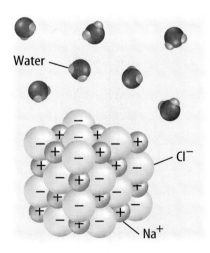

The partially negative oxygen in the water molecule is attracted to a positive sodium ion.

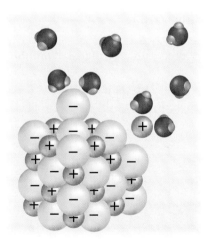

The partially positive hydrogen atoms in another water molecule are attracted to a negative chloride ion.

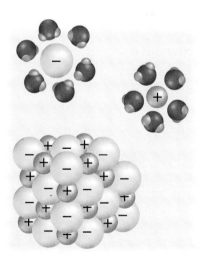

The sodium and chloride ions are pulled apart from each other, and more water molecules are attracted to them.

SECTION 2 Solubility **225**

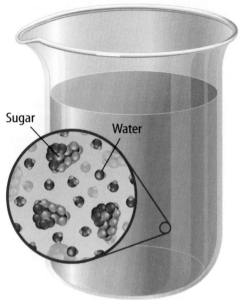

Figure 9 Sugar molecules that are dissolved in water spread out until they are spaced evenly in the water.

How Water Dissolves Molecular Compounds

Can water also dissolve molecular compounds that are not made of ions? Water does dissolve molecular compounds, such as sugar, although it doesn't break each sugar molecule apart. Water simply moves between different molecules of sugar, separating them. Like water, a sugar molecule is polar. Polar water molecules are attracted to the positive and negative portions of the polar sugar molecules. When the sugar molecules are separated by the water and spread throughout it, as **Figure 9** shows, they have dissolved.

What will dissolve?

When you stir a spoonful of sugar into iced tea, all of the sugar dissolves but none of the metal in the spoon does. Why does sugar dissolve in water, but metal does not? A substance that dissolves in another is said to be soluble in that substance. You would say that the sugar is soluble in water but the metal of the spoon is insoluble in water, because it does not dissolve readily.

Like Dissolves Like When trying to predict which solvents can dissolve which solutes, chemists use the rule of "like dissolves like." This means that polar solvents dissolve polar solutes and nonpolar solvents dissolve nonpolar solutes. In the case of sugar and water, both are made up of polar molecules, so sugar is soluble in water. In the case of salt and water, the sodium and chloride ion pair is like the water molecule because it has a positive charge at one end and a negative charge at the other end.

Reading Check *What does "like dissolves like" mean?*

On the other hand, if a solvent and a solute are not similar, the solute won't dissolve. For example, oil and water do not mix. Oil molecules are nonpolar, so polar water molecules are not attracted to them. If you pour vegetable oil into a glass of water, the oil and the water separate into layers instead of forming a solution, as shown in **Figure 10.** You've probably noticed the same thing about the oil-and-water mixtures that make up some salad dressings. The oil stays on the top. Oils generally dissolve better in solvents that have nonpolar molecules.

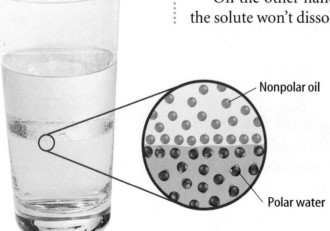

Figure 10 Water and oil do not mix because water molecules are polar and oil molecules are nonpolar.

How much will dissolve?

Even though sugar is soluble in water, if you tried to dissolve 1 kg of sugar into one small glass of water, not all of the sugar would dissolve. **Solubility** (sahl yuh BIH luh tee) is a measurement that describes how much solute dissolves in a given amount of solvent. The solubility of a material has been described as the amount of the material that can dissolve in 100 g of solvent at a given temperature. Some solutes are highly soluble, meaning that a large amount of solute can be dissolved in 100 g of solvent. For example, 63 g of potassium chromate can be dissolved in 100 g of water at 25°C. On the other hand, some solutes are not very soluble. For example, only 0.00025 g of barium sulfate will dissolve in 100 g of water at 25°C. When a substance has an extremely low solubility, like barium sulfate does in water, it usually is considered insoluble.

Reading Check *What is an example of a substance that is considered to be insoluble in water?*

Solubility in Liquid-Solid Solutions Did you notice that the temperature was included in the explanation about the amount of solute that dissolves in a quantity of solvent? The solubility of many solutes changes if you change the temperature of the solvent. For example, if you heat water, not only does the sugar dissolve at a faster rate, but more sugar can dissolve in it. However, some solutes, like sodium chloride and calcium carbonate, do not become more soluble when the temperature of water increases. The graph in **Figure 11** shows how the temperature of the solvent affects the solubility of some solutes.

Solubility in Liquid-Gas Solutions Unlike liquid-solid solutions, an increase in temperature decreases the solubility of a gas in a liquid-gas solution. You might notice this if you have ever opened a warm carbonated beverage and it bubbled up out of control while a chilled one barely fizzed. Carbon dioxide is less soluble in a warm solution. What keeps the carbon dioxide from bubbling out when it is sitting at room temperature on a supermarket shelf? When a bottle is filled, extra carbon dioxide gas is squeezed into the space above the liquid, increasing the pressure in the bottle. This increased pressure increases the solubility of gas and forces most of it into the solution. When you open the cap, the pressure is released and the solubility of the carbon dioxide decreases.

Reading Check *Why does a bottle of carbonated beverage go "flat" after it has been opened for a few days?*

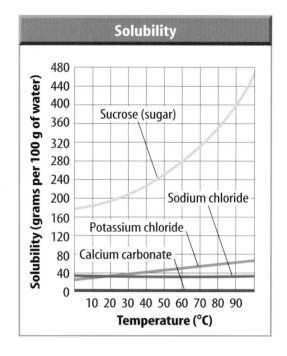

Figure 11 The solubility of some solutes changes as the temperature of the solvent increases.
Use a Graph *According to the graph, is it likely that warm ocean water contains any more sodium chloride than cold ocean water does?*

Mini LAB

Observing Chemical Processes

Procedure

1. Pour **two small glasses of milk.**
2. Place one glass of milk in the **refrigerator.** Leave the second glass on the counter.
3. Allow the milk to sit overnight. **WARNING:** *Do not drink the milk that sat out overnight.*
4. On the following day, smell both glasses of milk. Record your observations.

Analysis

1. Compare and contrast the smell of the refrigerated milk to the non-refrigerated milk.
2. Explain why refrigeration is needed.

Try at Home

Figure 12 The Dead Sea has an extremely high concentration of dissolved minerals. When the water evaporates, the minerals are left behind and form pillars.

Saturated Solutions If you add calcium carbonate to 100 g of water at 25°C, only 0.0014 g of it will dissolve. Additional calcium carbonate will not dissolve. Such a solution—one that contains all of the solute that it can hold under the given conditions—is called a **saturated** solution. **Figure 12** shows a saturated solution. If a solution is a liquid-solid solution, the extra solute that is added will settle to the bottom of the container. It's possible to make solutions that have less solute than they would need to become saturated. Such solutions are unsaturated. An example of an unsaturated solution is one containing 50 g of sugar in 100 g of water at 25°C. That's much less than the 204 g of sugar the solution would need to be saturated.

A hot solvent usually can hold more solute than a cool solvent can. When a saturated solution cools, some of the solute usually falls out of the solution. But if a saturated solution is cooled slowly, sometimes the excess solute remains dissolved for a period of time. Such a solution is said to be supersaturated, because it contains more than the normal amount of solute.

Rate of Dissolving

Solubility does not tell you how fast a solute will dissolve—it tells you only how much of a solute will dissolve at a given temperature. Some solutes dissolve quickly, but others take a long time to dissolve. A solute dissolves faster when the solution is stirred or shaken or when the temperature of the solution is increased. These methods increase the rate at which the surfaces of the solute come into contact with the solvent. Increasing the area of contact between the solute and the solvent can also increase the rate of dissolving. This can be done by breaking up the solute into smaller pieces, which increases the surface area of the solute that is exposed to the solvent.

Molecules are always moving and colliding. The collisions must take place for chemical processes to occur. The chemical processes take place at a given rate of reaction. Temperature has a large effect on that rate. The higher the temperature, the more collisions occur and the higher the rate of reaction. The opposite is also true. The lower the temperature, the less collisions occur and the lower the rate of reaction. Refrigerators are an example of slowing the reaction rate—and therefore the chemical process—down to prevent food spoilage.

Concentration

What makes strong lemonade strong and weak lemonade weak? The difference between the two drinks is the amount of water in each one compared to the amount of lemon. The lemon is present in different concentrations in the solution. The **concentration** of a solution tells you how much solute is present compared to the amount of solvent. You can give a simple description of a solution's concentration by calling it either concentrated or dilute. These terms are used when comparing the concentrations of two solutions with the same type of solute and solvent. A concentrated solution has more solute per given amount of solvent than a dilute solution.

Measuring Concentration Can you imagine a doctor ordering a dilute intravenous, or IV, solution for a patient? Because dilute is not an exact measurement, the IV could be made with a variety of amounts of medicine. The doctor would need to specify the exact concentration of the IV solution to make sure that the patient is treated correctly.

Pharmacist Doctors rely on pharmacists to formulate IV solutions. Pharmacists begin with a concentrated form of the drug, which is supplied by pharmaceutical companies. This is the solute of the IV solution. The pharmacist adds the correct amount of solvent to a small amount of the solute to achieve the concentration requested by the doctor. There may be more than one solute per IV solution in varying concentrations.

Applying Science

How can you compare concentrations?

A solute is a substance that can be dissolved in another substance called a solvent. Solutions vary in concentration, or strength, depending on the amount of solute and solvent being used. Fruit drinks are examples of such a solution. Stronger fruit drinks appear darker in color and are the result of more drink mix being dissolved in a given amount of water. What would happen if more water were added to the solution?

Glucose Solutions (g/100 mL)		
Solute Glucose (g)	Solvent Water (mL)	Solution Concentration of Glucose (%)
2	100	2
4	100	4
10	100	10
20	100	20

Identifying the Problem

The table on the right lists different concentration levels of glucose solutions, a type of carbohydrate your body uses as a source of energy. The glucose is measured in grams, and the water is measured in milliliters.

Solving the Problem

A physician writes a prescription for a patient to receive 1,000 mL of a 20 percent solution of glucose. How many grams of glucose must the pharmacist add to 1,000 mL of water to prepare this 20 percent concentration level?

Figure 13 Concentrations can be stated in percentages.
Identify *the percentage of this fruit drink that is water, assuming there are no other dissolved substances.*

One way of giving the exact concentration is to state the percentage of the volume of the solution that is made up of solute. Labels on fruit drinks show their concentration like the one in **Figure 13**. When a fruit drink contains 15 percent fruit juice, the remaining 85 percent of the drink is water and other substances such as sweeteners and flavorings. This drink is more concentrated than another brand that contains 10 percent fruit juice, but it's more dilute than pure juice, which is 100 percent juice. Another way to describe the concentration of a solution is to give the percentage of the total mass that is made up of solute.

Effects of Solute Particles All solute particles affect the physical properties of the solvent, such as its boiling point and freezing point. The effect that a solute has on the freezing or boiling point of a solvent depends on the number of solute particles.

When a solvent such as water begins to freeze, its molecules arrange themselves in a particular pattern. Adding a solute such as sodium chloride to this solvent changes the way the molecules arrange themselves. To overcome this interference of the solute, a lower temperature is needed to freeze the solvent.

When a solvent such as water begins to boil, the solvent molecules are gaining enough energy to move from the liquid state to the gaseous state. When a solute such as sodium chloride is added to the solvent, the solute particles interfere with the evaporation of the solvent particles. More energy is needed for the solvent particles to escape from the liquid, and the boiling point of the solution will be higher.

section 2 review

Summary

The Universal Solvent
- Water is known as the universal solvent.
- A molecule that has an even distribution of electrons is a nonpolar molecule.
- A molecule that has an uneven distribution of electrons is a polar molecule.
- A compound that loses or gains electrons is an ionic compound.

Dissolving a Substance
- Chemists use the rule "like dissolves like."

Concentration
- Concentration is the quantity of solute present compared to the amount of solvent.

Self Check

1. **Identify** the property of water that makes it the universal solvent.
2. **Describe** the two methods to increase the rate at which a substance dissolves.
3. **Infer** why it is important to add sodium chloride to water when making homemade ice cream.
4. **Think Critically** Why can the fluids used to dry-clean clothing remove grease even when water cannot?

Applying Skills

5. **Recognize Cause and Effect** Why is it more important in terms of reaction rate to take groceries straight home from the store when it is 25°C than when it is 2°C?

Observing Gas Solubility

On a hot day, a carbonated beverage will cool you off. If you leave the beverage uncovered at room temperature, it quickly loses its fizz. However, if you cap the beverage and place it in the refrigerator, it will still have its fizz hours later. In this lab you will explore why this happens.

Real-World Question

What effect does temperature have on the fizz, or carbon dioxide, in your carbonated beverage?

Goals
- **Observe** the effect that temperature has on solubility of a gas in a liquid.
- **Compare** the amount of carbon dioxide released at room temperature and in hot tap water.

Materials
carbonated beverages in plastic bottles, thoroughly chilled (2)
balloons (2)   *ruler
tape   container
fabric tape measure   hot tap water
*string   *Alternative materials

Safety Precautions

WARNING: *DO NOT point the bottles at anyone at any time during the lab.*

Procedure

1. Carefully remove the caps from the thoroughly chilled plastic bottles one at a time. Create as little agitation as possible.
2. Quickly cover the opening of each bottle with an uninflated balloon.
3. Use tape to secure and tightly seal the balloons to the top of the bottles.
4. Gently agitate one bottle from side to side for two minutes. Measure the circumference of the balloon.

WARNING: *Contents under pressure can cause serious accidents. Be sure to wear safety goggles, and DO NOT point the bottles at anyone.*

5. Gently agitate the second bottle in the same manner as in step 4. Then, place the bottle in a container of hot tap water for ten minutes. Measure the circumference of the balloon.

Conclude and Apply

1. **Contrast** the relative amounts of carbon dioxide gas released from the cold and the warm carbonated beverages.
2. **Infer** Why does the warmed carbonated beverage release a different amount of carbon dioxide than the chilled one?

Communicating Your Data

Compare the circumferences of your balloons with those of members of your class. **For more help, refer to the** Science Skill Handbook.

section 3
Acidic and Basic Solutions

as you read

What You'll Learn
- **Compare** acids and bases and their properties.
- **Describe** practical uses of acids and bases.
- **Explain** how pH is used to describe the strength of an acid or base.
- **Describe** how acids and bases react when they are brought together.

Why It's Important
Many common products, such as batteries and bleach, work because of acids or bases.

Review Vocabulary
physical property: any characteristic of a material that can be seen or measured without changing the material

New Vocabulary
- acid
- hydronium ion
- base
- pH
- indicator
- neutralization

Acids

What makes orange juice, vinegar, dill pickles, and grapefruit tangy? Acids cause the sour taste of these and other foods. **Acids** are substances that release positively charged hydrogen ions, H^+, in water. When an acid mixes with water, the acid dissolves, releasing a hydrogen ion. The hydrogen ion then combines with a water molecule to form a hydronium ion, as shown in **Figure 14**. **Hydronium ions** are positively charged and have the formula H_3O^+.

Properties of Acidic Solutions Sour taste is one of the properties of acidic solutions. The taste allows you to detect the presence of acids in your food. However, even though you can identify acidic solutions by their sour taste, you should never taste anything in the laboratory, and you should never use taste to test for the presence of acids in an unknown substance. Many acids can cause serious burns to body tissues.

Another property of acidic solutions is that they can conduct electricity. The hydronium ions in an acidic solution can carry the electric charges in a current. This is why some batteries contain an acid. Acidic solutions also are corrosive, which means they can break down certain substances. Many acids can corrode fabric, skin, and paper. The solutions of some acids also react strongly with certain metals. The acid-metal reaction forms metallic compounds and hydrogen gas, leaving holes in the metal in the process.

$$H^+ + H_2O \rightarrow H_3O^+$$

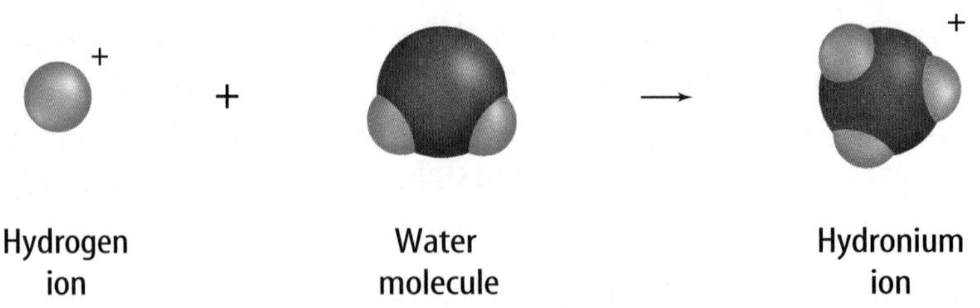

Hydrogen ion + Water molecule → Hydronium ion

Figure 14 One hydrogen ion can combine with one water molecule to form one positively charged hydronium ion.
Identify what kinds of substances are sources of hydrogen ions.

232 **CHAPTER 8** Substances, Mixtures, and Solubility

Figure 15 Each of these products contains an acid or is made with the help of an acid.
Describe how your life would be different if acids were not available to make these products.

Uses of Acids You're probably familiar with many acids. Vinegar, which is used in salad dressing, contains acetic acid. Lemons, limes, and oranges have a sour taste because they contain citric acid. Your body needs ascorbic acid, which is vitamin C. Ants that sting inject formic acid into their victims.

Figure 15 shows other products that are made with acids. Sulfuric acid is used in the production of fertilizers, steel, paints, and plastics. Acids often are used in batteries because their solutions conduct electricity. For this reason, it sometimes is referred to as battery acid. Hydrochloric acid, which is known commercially as muriatic acid, is used in a process called pickling. Pickling is a process that removes impurities from the surfaces of metals. Hydrochloric acid also can be used to clean mortar from brick walls. Nitric acid is used in the production of fertilizers, dyes, and plastics.

Acid in the Environment Carbonic acid plays a key role in the formation of caves and of stalactites and stalagmites. Carbonic acid is formed when carbon dioxide in soil is dissolved in water. When this acidic solution comes in contact with calcium carbonate—or limestone rock—it can dissolve it, eventually carving out a cave in the rock. A similar process occurs when acid rain falls on statues and eats away at the stone, as shown in **Figure 16.** When this acidic solution drips from the ceiling of the cave, water evaporates and carbon dioxide becomes less soluble, forcing it out of solution. The solution becomes less acidic and the limestone becomes less soluble, causing it to come out of solution. These solids form stalactites and stalagmites.

Observing a Nail in a Carbonated Drink

Procedure
1. Observe the initial appearance of an **iron nail.**
2. Pour enough **carbonated soft drink** into a **cup or beaker** to cover the nail.
3. Drop the nail into the soft drink and observe what happens.
4. Leave the nail in the soft drink overnight and observe it again the next day.

Analysis
1. Describe what happened when you first dropped the nail into the soft drink and the appearance of the nail the following day.
2. Based upon the fact that the soft drink was carbonated, explain why you think the drink reacted with the nail as you observed.

SECTION 3 Acidic and Basic Solutions

NATIONAL GEOGRAPHIC VISUALIZING ACID PRECIPITATION

Figure 16

When fossil fuels such as coal and oil are burned, a variety of chemical compounds are produced and released into the air. In the atmosphere, some of these compounds form acids that mix with water vapor and fall back to Earth as acid precipitation—rain, sleet, snow, or fog. The effects of acid precipitation on the environment can be devastating. Winds carry these acids hundreds of miles from their source, damaging forests, corroding statues, and endangering human health.

B Sulfur dioxide and nitrogen oxides react with water vapor in the air to form highly acidic solutions of nitric acid (HNO_3) and sulfuric acid (H_2SO_4). These solutions eventually return to Earth as acid precipitation.

C Some acid rain in the United States has a pH as low as 2.3—close to the acidity of stomach acid.

A Power plants and cars burn fossil fuels to generate energy for human use. In the process, sulfur dioxide (SO_2) and nitrogen oxides are released into the atmosphere.

Bases

People often use ammonia solutions to clean windows and floors. These solutions have different properties from those of acidic solutions. Ammonia is called a base. **Bases** are substances that can accept hydrogen ions. When bases dissolve in water, some hydrogen atoms from the water molecules are attracted to the base. A hydrogen atom in the water molecule leaves behind the other hydrogen atom and oxygen atom. This pair of atoms is a negatively charged ion called a hydroxide ion. A hydroxide ion has the formula OH^-. Most bases contain a hydroxide ion, which is released when the base dissolves in water. For example, sodium hydroxide is a base with the formula NaOH. When NaOH dissolves in water, a sodium ion and the hydroxide ion separate.

Topic: Calcium Hydroxide
Visit ips.msscience.com for Web links to information about the uses for calcium hydroxide.

Activity Describe the chemical reaction that converts limestone (calcium carbonate) to calcium hydroxide.

Properties of Basic Solutions Most soaps are bases, so if you think about how soap feels, you can figure out some of the properties of basic solutions. Basic solutions feel slippery. Acids in water solution taste sour, but bases taste bitter—as you know if you have ever accidentally gotten soap in your mouth.

Like acids, bases are corrosive. Bases can cause burns and damage tissue. You should never touch or taste a substance to find out whether it is a base. Basic solutions contain ions and can conduct electricity. Basic solutions are not as reactive with metals as acidic solutions are.

Uses of Bases Many uses for bases are shown in **Figure 17**. Bases give soaps, ammonia, and many other cleaning products some of their useful properties. The hydroxide ions produced by bases can interact strongly with certain substances, such as dirt and grease.

Chalk and oven cleaner are examples of familiar products that contain bases. Your blood is a basic solution. Calcium hydroxide, often called lime, is used to mark the lines on athletic fields. It also can be used to treat lawns and gardens that have acidic soil. Sodium hydroxide, known as lye, is a strong base that can cause burns and other health problems. Lye is used to make soap, clean ovens, and unclog drains.

Figure 17 Many products, including soaps, cleaners, and plaster contain bases or are made with the help of bases.

pH Levels Most life-forms can't exist at extremely low pH levels. However, some bacteria thrive in acidic environments. Acidophils are bacteria that exist at low pH levels. These bacteria have been found in the Hot Springs of Yellowstone National Park in areas with pH levels ranging from 1 to 3.

What is pH?

You've probably heard of pH-balanced shampoo or deodorant, and you might have seen someone test the pH of the water in a swimming pool. **pH** is a measure of how acidic or basic a solution is. The pH scale ranges from 0 to 14. Acidic solutions have pH values below 7. A solution with a pH of 0 is very acidic. Hydrochloric acid can have a pH of 0. A solution with a pH of 7 is neutral, meaning it is neither acidic nor basic. Pure water is neutral. Basic solutions have pH values above 7. A solution with a pH of 14 is very basic. Sodium hydroxide can have a pH of 14. **Figure 18** shows where various common substances fall on the pH scale.

The pH of a solution is related directly to its concentrations of hydronium ions (H_3O^+) and hydroxide ions (OH^-). Acidic solutions have more hydronium ions than hydroxide ions. Neutral solutions have equal numbers of the two ions. Basic solutions have more hydroxide ions than hydronium ions.

 *In a neutral solution, how do the numbers of hydronium ions and hydroxide ions compare?*

pH Scale The pH scale is not a simple linear scale like mass or volume. For example, if one book has a mass of 2 kg and a second book has a mass of 1 kg, the mass of the first book is twice that of the second. However, a change of 1 pH unit represents a tenfold change in the acidity of the solution. For example, if one solution has a pH of 1 and a second solution has a pH of 2, the first solution is not twice as acidic as the second—it is ten times more acidic. To determine the difference in pH strength, use the following calculation: 10^n, where n = the difference between pHs. For example: pH3 − pH1 = 2, 10^2 = 100 times more acidic.

Figure 18 The pH scale classifies a solution as acidic, basic, or neutral.

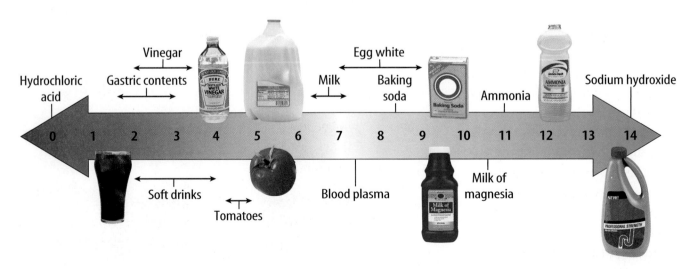

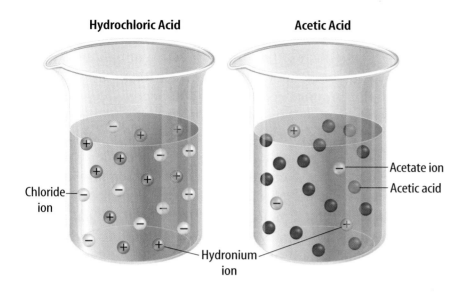

Figure 19 Hydrochloric acid separates into ions more readily than acetic acid does when it dissolves in water. Therefore, hydrochloric acid exists in water as separated ions. Acetic acid exists in water almost entirely as molecules.

Strengths of Acids and Bases You've learned that acids give foods a sour taste but also can cause burns and damage tissue. The difference between food acids and the acids that can burn you is that they have different strengths. The acids in food are fairly weak acids, while the dangerous acids are strong acids. The strength of an acid is related to how easily the acid separates into ions, or how easily a hydrogen ion is released, when the acid dissolves in water. Look at **Figure 19.** In the same concentration, a strong acid—like hydrochloric acid—forms more hydronium ions in solution than a weak acid does—like acetic acid. More hydronium ions means the strong-acid solution has a lower pH than the weak-acid solution. Similarly, the strength of a base is related to how easily the base separates into ions, or how easily a hydroxide ion is released, when the base dissolves in water. The relative strengths of some common acids and bases are shown in **Table 3.**

Reading Check *What determines the strength of an acid or a base?*

An acid containing more hydrogen atoms, such as carbonic acid, H_2CO_3, is not necessarily stronger than an acid containing fewer hydrogen atoms, such as nitric acid, HNO_3. An acid's strength is related to how easily a hydrogen ion separates—not to how many hydrogen atoms it has. For this reason, nitric acid is stronger than carbonic acid.

Table 3 Strengths of Some Acids and Bases

	Acid	Base
Strong	hydrochloric (HCl) sulfuric (H_2SO_4) nitric (HNO_3)	sodium hydroxide (NaOH) potassium hydroxide (KOH)
Weak	acetic (CH_3COOH) carbonic (H_2CO_3) ascorbic ($H_2C_6H_6O_6$)	ammonia (NH_3) aluminum hydroxide ($Al(OH)_3$) iron (III) hydroxide ($Fe(OH)_3$)

Topic: Indicators
Visit ips.msscience.com for Web links to information about the types of pH indicators.

Activity Describe how plants can act as indicators in acidic and basic solutions.

Indicators

What is a safe way to find out how acidic or basic a solution is? **Indicators** are compounds that react with acidic and basic solutions and produce certain colors, depending on the solution's pH.

Because they are different colors at different pHs, indicators can help you determine the pH of a solution. Some indicators, such as litmus, are soaked into paper strips. When litmus paper is placed in an acidic solution, it turns red. When placed in a basic solution, litmus paper turns blue. Some indicators can change through a wide range of colors, with each different color appearing at a different pH value.

Neutralization

Perhaps you've heard someone complain about heartburn or an upset stomach after eating spicy food. To feel better, the person might have taken an antacid. Think about the word *antacid* for a minute. How do antacids work?

Heartburn or stomach discomfort is caused by excess hydrochloric acid in the stomach. Hydrochloric acid helps break down the food you eat, but too much of it can irritate your stomach or digestive tract. An antacid product, often made from the base magnesium hydroxide, $Mg(OH)_2$, neutralizes the excess acid. **Neutralization** (new truh luh ZAY shun) is the reaction of an acid with a base. It is called this because the properties of both the acid and base are diminished, or neutralized. In most cases, the reaction produces a water and a salt. **Figure 20** illustrates the relative amounts of hydronium and hydroxide ions between pH 0 and pH 14.

Reading Check *What are the products of neutralization?*

Figure 20 The pH of a solution is more acidic when greater amounts of hydronium ions are present.
Define *what makes a pH 7 solution neutral.*

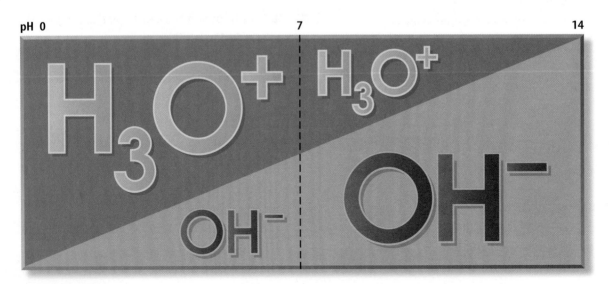

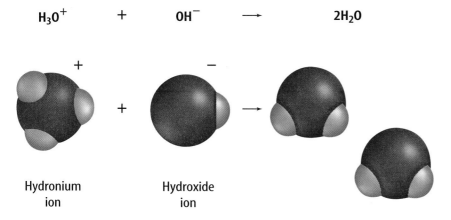

Figure 21 When acidic and basic solutions react, hydronium and hydroxide ions react to form water.
Determine why the pH of the solution changes.

How does neutralization occur? Recall that every water molecule contains two hydrogen atoms and one oxygen atom. As **Figure 21** shows, when one hydronium ion reacts with one hydroxide ion, the product is two water molecules. This reaction occurs during acid-base neutralization. Equal numbers of hydronium ions from the acidic solution and hydroxide ions from the basic solution react to produce water. Pure water has a pH of 7, which means that it's neutral.

 *What happens to acids and bases during neutralization?*

section 3 review

Summary

Acids and Bases
- Acids are substances that release positively charged hydrogen ions in water.
- Substances that accept hydrogen ions in water are bases.
- Acidic and basic solutions can conduct electricity.

pH
- pH measures how acidic or basic a solution is.
- The scale ranges from 0 to 14.

Neutralization
- Neutralization is the interaction between an acid and a base to form water and a salt.

Self Check

1. **Identify** what ions are produced by acids in water and bases in water. Give two properties each of acids and bases.
2. **Name** three acids and three bases and list an industrial or household use of each.
3. **Explain** how the concentration of hydronium ions and hydroxide ions are related to pH.
4. **Think Critically** In what ways might a company that uses a strong acid handle an acid spill on the factory floor?

Applying Math

5. **Solve One-Step Equations** How much more acidic is a solution with a pH of 2 than one with a pH of 6? How much more basic is a solution with a pH of 13 than one with a pH of 10?

Testing pH Using Natural Indicators

Goals
- **Determine** the relative acidity or basicity of several common solutions.
- **Compare** the strengths of several common acids and bases.

Materials
small test tubes (9)
test-tube rack
concentrated red cabbage juice in a dropper bottle
labeled bottles containing:
   household ammonia, baking soda solution, soap solution, 0.1M hydrochloric acid solution, white vinegar, colorless carbonated soft drink, borax soap solution, distilled water
grease pencil
droppers (9)

Safety Precautions

WARNING: *Many acids and bases are poisonous, can damage your eyes, and can burn your skin. Wear goggles and gloves AT ALL TIMES. Tell your teacher immediately if a substance spills. Wash your hands after you finish but before removing your goggles.*

Real-World Question
You have learned that certain substances, called indicators, change color when the pH of a solution changes. The juice from red cabbage is a natural indicator. How do the pH values of various solutions compare to each other? How can you use red cabbage juice to determine the relative pH of several solutions?

Procedure

1. **Design** a data table to record the names of the solutions to be tested, the colors caused by the added cabbage juice indicator, and the relative strengths of the solutions.
2. Mark each test tube with the identity of the acid or base solution it will contain.
3. Half-fill each test tube with the solution to be tested.
   WARNING: *If you spill any liquids on your skin, rinse the area immediately with water. Alert your teacher if any liquid spills in the work area or on your skin.*

240 CHAPTER 8 Substances, Mixtures, and Solubility

Using Scientific Methods

4. Add ten drops of the cabbage juice indicator to each of the solutions to be tested. Gently agitate or wiggle each test tube to mix the cabbage juice with the solution.
5. **Observe** and record the color of each solution in your data table.

Determining pH Values

Cabbage Juice Color	Relative Strength of Acid or Base
	strong acid
	medium acid
	weak acid
	neutral
	weak base
	medium base
	strong base

▶ Analyze Your Data

1. **Compare** your observations with the table above. Record in your data table the relative acid or base strength of each solution you tested.
2. **List** the solutions by pH value starting with the most acidic and finishing with the most basic.

▶ Conclude and Apply

1. **Classify** which solutions were acidic and which were basic.
2. **Identify** which solution was the weakest acid. The strongest base? The closest to neutral?
3. **Predict** what ion might be involved in the cleaning process based upon your data for the ammonia, soap, and borax soap solutions.

▶ Form a Hypothesis

Form a hypothesis that explains why the borax soap solution was less basic than an ammonia solution of approximately the same concentration.

Communicating Your Data

Use your data to create labels for the solutions you tested. Include the relative strength of each solution and any other safety information you think is important on each label. **For more help, refer to the Science Skill Handbook.**

SCIENCE Stats

Salty Solutions

Did you know...

...Seawater is certainly a salty solution. Ninety-nine percent of all salt ions in the sea are sodium, chlorine, sulfate, magnesium, calcium, and potassium. The major gases in the sea are nitrogen, oxygen, carbon dioxide, argon, neon, and helium.

...Tears and saliva have a lot in common. Both are salty solutions that protect you from harmful bacteria, keep tissues moist, and help spread nutrients. Bland-tasting saliva, however, is 99 percent water. The remaining one percent is a combination of many ions, including sodium and several proteins.

...The largest salt lake in the United States is the Great Salt Lake. It covers more than 4,000 km^2 in Utah and is up to 13.4 m deep. The Great Salt Lake and the Salt Lake Desert were once part of the enormous, prehistoric Lake Bonneville, which was 305 m deep at some points.

Applying Math At its largest, Lake Bonneville covered about 32,000 km^2. What percentage of that area does the Great Salt Lake now cover?

...Salt can reduce pain. Gargled salt water is a disinfectant; it fights the bacteria that cause some sore throats.

Graph It

Visit **ips.msscience.com/science_stats** to research and learn about other elements in seawater. Create a graph that shows the amounts of the ten most common elements in 1 L of seawater.

Chapter 8 Study Guide

Reviewing Main Ideas

Section 1 What is a solution?

1. Elements and compounds are pure substances, because their compositions are fixed. Mixtures are not pure substances.
2. Heterogeneous mixtures are not mixed evenly. Homogeneous mixtures, also called solutions, are mixed evenly on a molecular level.
3. Solutes and solvents can be gases, liquids, or solids, combined in many different ways.

Section 2 Solubility

1. Because water molecules are polar, they can dissolve many different solutes. Like dissolves like.
2. Temperature and pressure can affect solubility.

3. Solutions can be unsaturated, saturated, or supersaturated, depending on how much solute is dissolved compared to the solubility of the solute in the solvent.
4. The concentration of a solution is the amount of solute in a particular volume of solvent.

Section 3 Acidic and Basic Solutions

1. Acids release H+ ions and produce hydronium ions when they are dissolved in water. Bases accept H+ ions and produce hydroxide ions when dissolved in water.
2. pH expresses the concentrations of hydronium ions and hydroxide ions in aqueous solutions.
3. In a neutralization reaction, an acid reacts with a base to form water and a salt.

Visualizing Main Ideas

Copy and complete the concept map on the classification of matter.

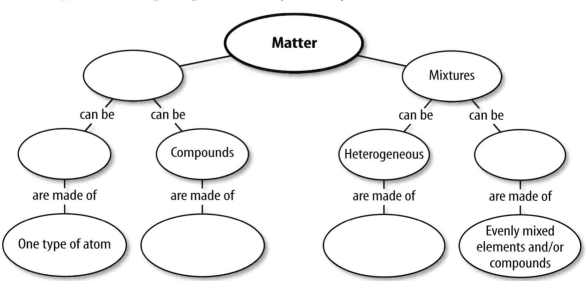

chapter 8 Review

Using Vocabulary

acid p. 232	neutralization p. 238
aqueous p. 224	pH p. 236
base p. 235	precipitate p. 220
concentration p. 229	saturated p. 228
heterogeneous	solubility p. 227
mixture p. 219	solute p. 220
homogeneous	solution p. 220
mixture p. 220	solvent p. 220
hydronium ion p. 232	substance p. 218
indicator p. 238	

Fill in the blanks with the correct vocabulary word.

1. A base has a(n) _____ value above 7.
2. A measure of how much solute is in a solution is its _____.
3. The amount of a solute that can dissolve in 100 g of solvent is its _____.
4. The _____ is the substance that is dissolved to form a solution.
5. The reaction between an acidic and basic solution is called _____.
6. A(n) _____ has a fixed composition.

Checking Concepts

Choose the word or phrase that best answers the question.

7. Which of the following is a solution?
   A) pure water
   B) an oatmeal-raisin cookie
   C) copper
   D) vinegar

8. What type of compounds will not dissolve in water?
   A) polar   C) nonpolar
   B) ionic   D) charged

9. What type of molecule is water?
   A) polar   C) nonpolar
   B) ionic   D) precipitate

10. When chlorine compounds are dissolved in pool water, what is the water?
    A) the alloy
    B) the solvent
    C) the solution
    D) the solute

11. A solid might become less soluble in a liquid when you decrease what?
    A) particle size   C) temperature
    B) pressure   D) container size

12. Which acid is used in the industrial process known as pickling?
    A) hydrochloric   C) sulfuric
    B) carbonic   D) nitric

13. A solution is prepared by adding 100 g of solid sodium hydroxide, NaOH, to 1,000 mL of water. What is the solid NaOH called?
    A) solution   C) solvent
    B) solute   D) mixture

14. Given equal concentrations, which of the following will produce the most hydronium ions in an aqueous solution?
    A) a strong base   C) a strong acid
    B) a weak base   D) a weak acid

15. Bile, an acidic body fluid used in digestion, has a high concentration of hydronium ions. Predict its pH.
    A) 11   C) less than 7
    B) 7   D) greater than 7

16. When you swallow an antacid, what happens to your stomach acid?
    A) It is more acidic.
    B) It is concentrated.
    C) It is diluted.
    D) It is neutralized.

chapter 8 Review

Thinking Critically

17. **Infer** why deposits form in the steam vents of irons in some parts of the country.

18. **Explain** if it is possible to have a dilute solution of a strong acid.

19. **Draw Conclusions** Antifreeze is added to water in a car's radiator to prevent freezing in cold months. It also prevents overheating or boiling. Explain how antifreeze does both.

Use the illustration below to answer question 20.

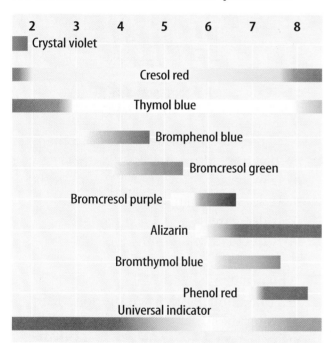

20. **Interpret** Chemists use a variety of indicators. Using the correct indicator is important. The color change must occur at the proper pH or the results could be misleading. Looking at the indicator chart, what indicators could be used to produce a color change at both pH 2 and pH 8?

21. **Explain** Water molecules can break apart to form H^+ ions and OH^- ions. Water is known as an amphoteric substance, which is something that can act as an acid or a base. Explain how this can be so.

22. **Describe** how a liquid-solid solution forms. How is this different from a liquid-gas solution? How are these two types of solutions different from a liquid-liquid solution? Give an example of each with your description.

23. **Compare and contrast** examples of heterogeneous and homogeneous mixtures from your daily life.

24. **Form a Hypotheses** A warm carbonated beverage seems to fizz more than a cold one when it is opened. Explain this based on the solubility of carbon dioxide in water.

Performance Activities

25. **Poem** Write a poem that explains the difference between a substance and a mixture.

Applying Math

Use the graph below to answer question 26.

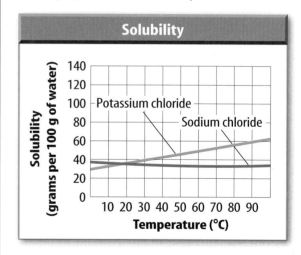

26. **Solubility** Using the solubility graph above, estimate the solubilities of potassium chloride and sodium chloride in grams per 100 g of water at 80°C.

27. **Juice Concentration** You made a one-liter (1,000 mL) container of juice. How much concentrate, in mL, did you add to make a concentration of 18 percent?

Chapter 8 Standardized Test Practice

Part 1 | Multiple Choice

Record your answers on the answer sheet provided by your teacher or on a sheet of paper.

Use the illustration below to answer questions 1 and 2.

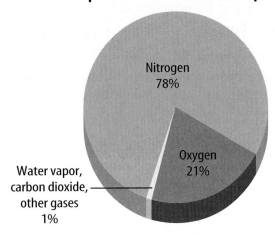

Composition of Earth's Atmosphere
- Nitrogen 78%
- Oxygen 21%
- Water vapor, carbon dioxide, other gases 1%

1. Which term best describes Earth's atmosphere?
   - A. saturated
   - B. solution
   - C. precipitate
   - D. indicator

2. Which of these is the solvent in Earth's atmosphere?
   - A. nitrogen
   - B. oxygen
   - C. water vapor
   - D. carbon dioxide

3. What characteristic do aqueous solutions share?
   - A. They contain more than three solutes.
   - B. No solids or gases are present as solutes in them.
   - C. All are extremely concentrated.
   - D. Water is the solvent in them.

Test-Taking Tip

Start the Day Right The morning of the test, eat a healthy breakfast with a balanced amount of protein and carbohydrates.

Use the illustration below to answer questions 4 and 5.

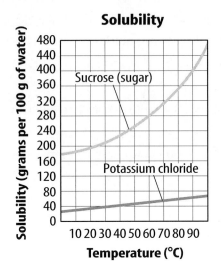

Solubility graph showing Sucrose (sugar) and Potassium chloride, Solubility (grams per 100 g of water) vs Temperature (°C)

4. How does the solubility of sucrose change as the temperature increases?
   - A. It increases.
   - B. It does not change.
   - C. It decreases.
   - D. It fluctuates randomly.

5. Which statement is TRUE?
   - A. Potassium chloride is more soluble in water than sucrose.
   - B. As water temperature increases, the solubility of potassium chloride decreases.
   - C. Sucrose is more soluble in water than potassium chloride.
   - D. Water temperature has no effect on the solubility of these two chemicals.

6. Which of these is a property of acidic solutions?
   - A. They taste sour.
   - B. They feel slippery.
   - C. They are in many cleaning products.
   - D. They taste bitter.

Standardized Test Practice

Part 2 Short Response/Grid In

Record your answers on the answer sheet provided by your teacher or on a sheet of paper.

7. Identify elements present in the alloy steel. Compare the flexibility and strength of steel and iron.

Use the illustration below to answer questions 8 and 9.

8. How can you tell that the matter in this bowl is a mixture?

9. What kind of mixture is this? Define this type of mixture, and give three additional examples.

10. Explain why a solute broken into small pieces will dissolve more quickly than the same type and amount of solute in large chunks.

11. Compare the concentration of two solutions: Solution A is composed of 5 grams of sodium chloride dissolved in 100 grams of water. Solution B is composed of 27 grams of sodium chloride dissolved in 100 grams of water.

12. Give the pH of the solutions vinegar, blood plasma, and ammonia. Compare the acidities of soft drinks, tomatoes, and milk.

13. Describe how litmus paper is used to determine the pH of a solution.

Part 3 Open Ended

Record your answers on a sheet of paper.

14. Compare and contrast crystallization and a precipitation reaction.

15. Why is a carbonated beverage defined as a liquid-gas solution? In an open container, the ratio of liquid solvent to gas solute changes over time. Explain.

Use the illustration below to answer questions 16 and 17.

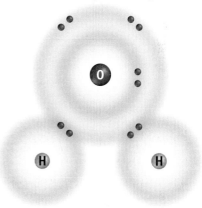

(Partial negative charge)

(Partial positive charge)

16. The diagram shows a water molecule. Use the distribution of electrons to describe this molecule's polarity.

17. Explain how the polarity of water molecules makes water effective in dissolving ionic compounds.

18. Marble statues and building facades in many of the world's cities weather more quickly today than when first constructed. Explain how the pH of water plays a role in this process.

19. Acetic acid, CH_3COOH, has more hydrogen atoms than the same concentration of hydrochloric acid, HCl. Hydrogen ions separate more easily from hydrochloric than acetic acid. Which acid is strongest? Why?

chapter 9

Carbon Chemistry

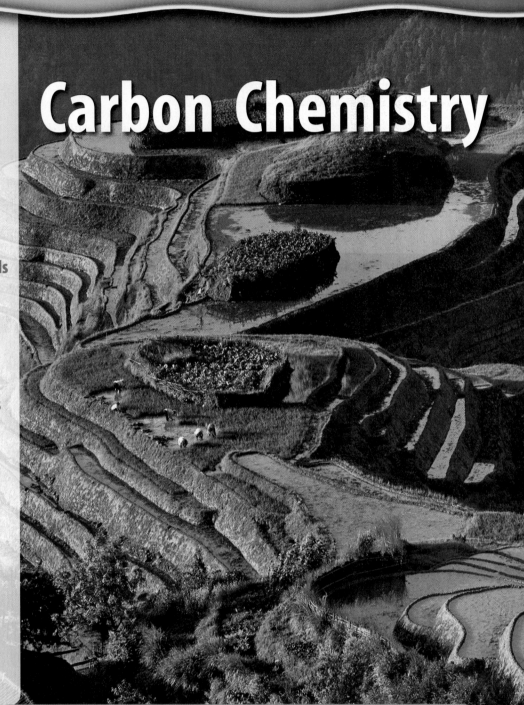

The BIG Idea
All living things are made of carbon-containing compounds.

SECTION 1
Simple Organic Compounds
Main Idea Carbon can form covalent bonds with four other atoms including other carbon atoms.

SECTION 2
Other Organic Compounds
Main Idea Substituted hydrocarbons are formed when hydrogen atoms are replaced by groups of atoms.

SECTION 3
Biological Compounds
Main Idea Proteins, carbohydrates, and lipids are biological compounds formed from chains or rings of carbon atoms.

Fields of ...Methane?
These fields produce an important food—rice. But, the flooded soil is also ideal for the production of methane—the simplest organic compound. The methane is produced when plants decay under water without oxygen.

Science Journal Find and name four items around your classroom that are made from carbon compounds.

Start-Up Activities

Model Carbon's Bonding

Many of the compounds that compose your body and other living things are carbon compounds. This lab demonstrates some of the atomic combinations possible with one carbon and four other atoms. The ball represents a carbon atom. The toothpicks represent chemical bonds.

WARNING: *Do not eat any foods from this lab. Wash your hands before and after this lab.*

1. Insert four toothpicks into a small clay or plastic foam ball so they are evenly spaced around the sphere.

2. Make models of as many molecules as possible by adding raisins, grapes, and gumdrops to the ends of the toothpicks. Use raisins to represent hydrogen atoms, grapes to represent chlorine atoms, and gumdrops to represent fluorine atoms.

3. **Think Critically** Draw each model and write the formula for it in your Science Journal. What can you infer about the number of compounds a carbon atom can form?

Preview this chapter's content and activities at
ips.msscience.com

Hydrocarbons Make the following Foldable to help you learn the definitions of vocabulary words. This will help you understand the chapter content.

STEP 1 Fold a sheet of paper in half lengthwise. Make the back edge about 1.25 cm longer than the front edge.

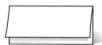

STEP 2 Fold in half, then fold in half again to make three folds.

STEP 3 Unfold and cut only the top layer along the three folds to make four tabs.

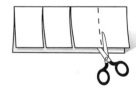

STEP 4 Label the tabs as shown.

Find Main Ideas As you read the chapter, find the definitions for each vocabulary word and write them under the appropriate tabs. Add additional words and definitions to help you understand your reading. List examples of each type of hydrocarbon under the appropriate tab.

Get Ready to Read

Make Connections

① Learn It! Make connections between what you read and what you already know. Connections can be based on personal experiences (text-to-self), what you have read before (text-to-text), or events in other places (text-to-world).

As you read, ask connecting questions. Are you reminded of a personal experience? Have you read about the topic before? Did you think of a person, a place, or an event in another part of the world?

② Practice It! Read the excerpt below and make connections to your own knowledge and experience.

Text-to-self: Have you ever snapped together blocks in different arrangements?

Text-to-self: What is an organic molecule?

Text-to-self: Have you ever used a product that contained butane or isobutane?

> Suppose you had ten blocks that could be snapped together in different arrangements. Each arrangement of the same ten blocks is different. The atoms in an organic molecule also can have different arrangements but still have the same molecular formula. Compounds that have the same molecular formula but different arrangements, or structures, are called **isomers** (I suh murz). Two isomers, butane and isobutane, are shown in Figure 7.
>
> — *from page 254*

③ Apply It! As you read this chapter, choose five words or phrases that make a connection to something you already know.

Reading Tip

Make connections with things that you use or see every day.

Target Your Reading

Use this to focus on the main ideas as you read the chapter.

① **Before you read** the chapter, respond to the statements below on your worksheet or on a numbered sheet of paper.
- Write an **A** if you **agree** with the statement.
- Write a **D** if you **disagree** with the statement.

② **After you read** the chapter, look back to this page to see if you've changed your mind about any of the statements.
- If any of your answers changed, explain why.
- Change any false statements into true statements.
- Use your revised statements as a study guide.

Before You Read A or D		Statement	After You Read A or D
	1	Only living things can make carbon compounds.	
	2	Hydrogen atoms often bond with carbon to form compounds.	
	3	Simple sugars are the building blocks of proteins.	
	4	Carbon atoms can form single, double, and triple covalent bonds.	
	5	The suffix in the name of an organic compound indicates the kind of bonds joining the carbon atoms.	
	6	Unsaturated fats contain only single covalent bonds.	
	7	Sugars, starches, and cellulose are carbohydrates.	
	8	Alcohols contain the hydroxyl group.	
	9	Carboxylic acids and amino acids contain nitrogen.	

Science Online
Print out a worksheet of this page at ips.msscience.com

250 B

section 1

Simple Organic Compounds

as you read

What You'll Learn
- **Explain** why carbon is able to form many compounds.
- **Describe** how saturated and unsaturated hydrocarbons differ.
- **Identify** isomers of organic compounds.

Why It's Important
Plants, animals, and many of the things that are part of your life are made of organic compounds.

Review Vocabulary
chemical bond: force that holds two atoms together

New Vocabulary
- organic compound
- hydrocarbon
- saturated hydrocarbon
- unsaturated hydrocarbon
- isomer

Organic Compounds

Earth's crust contains less than one percent carbon, yet all living things on Earth are made of carbon-containing compounds. Carbon's ability to bond easily and form compounds is the basis of life on Earth. A carbon atom has four electrons in its outer energy level, so it can form four covalent bonds with as many as four other atoms. When carbon atoms form four covalent bonds, they obtain the stability of a noble gas with eight electrons in their outer energy level. One of carbon's most frequent partners in forming covalent bonds is hydrogen.

Substances can be classified into two groups—those derived from living things and those derived from nonliving things, as shown in **Figure 1.** Most of the substances associated with living things contain carbon and hydrogen. These substances were called organic compounds, which means "derived from a living organism." However, in 1828, scientists discovered that living organisms are not necessary to form organic compounds. Despite this, scientists still use the term **organic compound** for most compounds that contain carbon.

Reading Check *What is the origin of the term organic compound?*

Figure 1 Most substances can be classified as living or nonliving things.

Living things and products made from living things such as this wicker chair contain carbon.

Most of the things in this photo are nonliving and are composed of elements other than carbon.

Hydrocarbons

Many compounds are made of only carbon and hydrogen. A compound that contains only carbon and hydrogen atoms is called a **hydrocarbon.** The simplest hydrocarbon is methane, the primary component of natural gas. If you have a gas stove or gas furnace in your home, methane usually is the fuel that is burned in these appliances. Methane consists of a single carbon atom covalently bonded to four hydrogen atoms. The formula for methane is CH_4. **Figure 2** shows a model of the methane molecule and its structural formula. In a structural formula, the line between one atom and another atom represents a pair of electrons shared between the two atoms. This pair forms a single bond. Methane contains four single bonds.

Now, visualize the removal of one of the hydrogen atoms from a methane molecule, as in **Figure 3A.** A fragment of the molecule called a methyl group, $-CH_3$, would remain. The methyl group then can form a single bond with another methyl group. If two methyl groups bond with each other, the result is the two-carbon hydrocarbon ethane, C_2H_6, which is shown with its structural formula in **Figure 3B.**

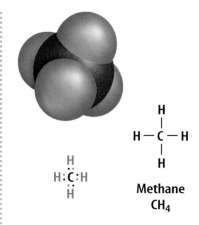

Figure 2 Methane is the simplest hydrocarbon molecule. **Explain** why this is true.

Figure 3 Here's a way to visualize how larger hydrocarbons are built up.

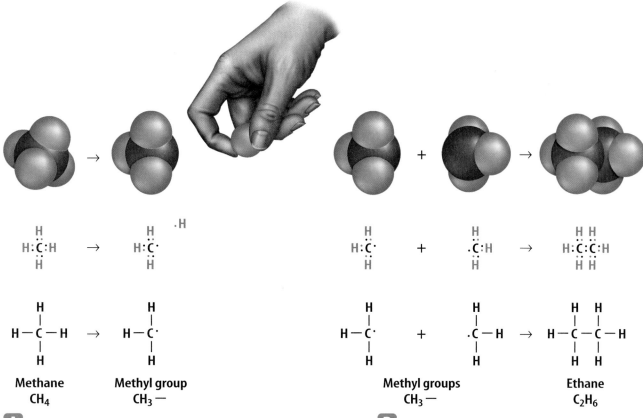

A A hydrogen is removed from a methane molecule, forming a methyl group.

B Each carbon atom in ethane has four bonds after the two methyl groups join.

SECTION 1 Simple Organic Compounds **251**

Figure 4 Propane and butane are two useful fuels. **Explain** why they are called "saturated."

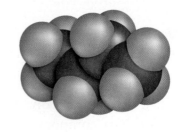

Propane
C$_3$H$_8$

When propane burns, it releases energy as the chemical bonds are broken. Propane often is used to fuel camp stoves and outdoor grills.

Butane
C$_4$H$_{10}$

Butane also releases energy when it burns. Butane is the fuel that is used in disposable lighters.

Hydrocarbons Petroleum is a mixture of hydrocarbons that formed from plants and animals that lived in seas and lakes hundreds of millions of years ago. With the right temperature and pressure, this plant and animal matter, buried deep under Earth's surface, decomposed to form petroleum. Why is petroleum a nonrenewable resource?

Saturated Hydrocarbons Methane and ethane are members of a series of molecules in which carbon and hydrogen atoms are joined by single covalent bonds. When all the bonds in a hydrocarbon are single bonds, the molecule is called a **saturated hydrocarbon.** It is called *saturated* because no additional hydrogen atoms can be added to the molecule. The carbon atoms are saturated with hydrogen atoms. The formation of larger hydrocarbons occurs in a way similar to the formation of ethane. A hydrogen atom is removed from ethane and replaced by a —CH$_3$ group. Propane, with three carbon atoms, is the third member of the series of saturated hydrocarbons. Butane has four carbon atoms. Both of these hydrocarbons are shown in **Figure 4.** The names and the chemical formulas of a few of the smaller saturated hydrocarbons are listed in **Table 1.** Saturated hydrocarbons are named with an *-ane* ending. Another name for these hydrocarbons is alkanes.

Reading Check *What is a saturated hydrocarbon?*

These short hydrocarbon chains have low boiling points, so they evaporate and burn easily. That makes methane a good fuel for your stove or furnace. Propane is used in gas grills, lanterns, and to heat air in hot-air balloons. Butane often is used as a fuel for camp stoves and lighters. Longer hydrocarbons are found in oils and waxes. Carbon can form long chains that contain hundreds or even thousands of carbon atoms. These extremely long chains make up many of the plastics that you use.

Unsaturated Hydrocarbons Carbon also forms hydrocarbons with double and triple bonds. In a double bond, two pairs of electrons are shared between two atoms, and in a triple bond, three pairs of electrons are shared. Hydrocarbons with double or triple bonds are called **unsaturated hydrocarbons.** This is because the carbon atoms are not saturated with hydrogen atoms.

Ethene, the simplest unsaturated hydrocarbon, has two carbon atoms joined by a double bond. Propene is an unsaturated hydrocarbon with three carbons. Some unsaturated hydrocarbons have more than one double bond. Butadiene (byew tuh DI een) has four carbon atoms and two double bonds. The structures of ethene, propene, and butadiene are shown in **Figure 5.**

Unsaturated compounds with at least one double bond are named with an -*ene* ending. Notice that the names of the compounds below have an -*ene* ending. These compounds are called alkenes.

Table 1 The Structures of Hydrocarbons

Name	Structural Formula	Chemical Formula
Methane	H–C–H with H above and below	CH_4
Ethane	H–C–C–H with H's above and below	C_2H_6
Propane	H–C–C–C–H with H's above and below	C_3H_8
Butane	H–C–C–C–C–H with H's above and below	C_4H_{10}

Reading Check *What type of bonds are found in unsaturated hydrocarbons?*

Figure 5 You'll find unsaturated hydrocarbons in many of the products you use every day.

Ethene

Ethene helps ripen fruits and vegetables. It's also used to make milk and soft-drink bottles.

Propene

This detergent bottle contains the tough plastic polypropylene, which is made from propene.

Butadiene

Butadiene made it possible to replace natural rubber with synthetic rubber.

SECTION 1 Simple Organic Compounds

Figure 6 In the welder's torch, ethyne, also called acetylene, is combined with oxygen to form a mixture that burns, releasing intense light and heat. The two carbon atoms in ethyne are joined by a triple bond.

H—C≡C—H

Ethyne or Acetylene
C_2H_2

Mini LAB

Modeling Isomers

Procedure

WARNING: *Do not eat any foods in this lab.*

1. Construct a model of pentane, C_5H_{12}. Use **toothpicks** for covalent bonds and small balls of different **colored clay or gumdrops** for carbon and hydrogen atoms.
2. Using the same materials, build a molecule with a different arrangement of the atoms. Are there any other possibilities?
3. Make a model of hexane, C_6H_{14}.
4. Arrange the atoms of hexane in different ways.

Analysis
1. How many isomers of pentane did you build? How many isomers of hexane?
2. Do you think there are more isomers of heptane, C_7H_{16}, than hexane? Explain.

Try at Home

Triple Bonds Unsaturated hydrocarbons also can have triple bonds, as in the structure of ethyne (EH thine) shown in **Figure 6**. Ethyne, commonly called acetylene (uh SE tuh leen), is a gas used for welding because it produces high heat as it burns. Welding torches mix acetylene and oxygen before burning. These unsaturated compounds are called alkynes.

Hydrocarbon Isomers Suppose you had ten blocks that could be snapped together in different arrangements. Each arrangement of the same ten blocks is different. The atoms in an organic molecule also can have different arrangements but still have the same molecular formula. Compounds that have the same molecular formula but different arrangements, or structures, are called **isomers** (I suh murz). Two isomers, butane and isobutane, are shown in **Figure 7**. They have different chemical and physical properties because of their different structures. As the size of a hydrocarbon molecule increases, the number of possible isomers also increases.

By now, you might be confused about how organic compounds are named. **Figure 8** explains the system that is used to name simple organic compounds.

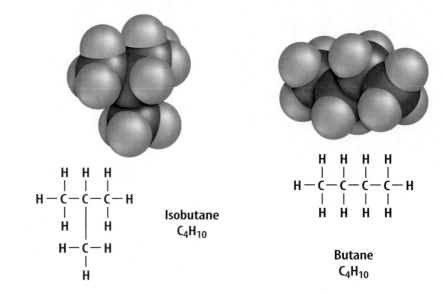

Figure 7 Butane and isobutane both have four carbons and ten hydrogens but their structures and properties are different.

NATIONAL GEOGRAPHIC VISUALIZING ORGANIC CHEMISTRY NOMENCLATURE

Figure 8

More than one million organic compounds have been discovered and created, and thousands of new ones are synthesized in laboratories every year. To keep track of all these carbon-containing molecules, the International Union of Pure and Applied Chemistry, or IUPAC, devised a special naming system (a nomenclature) for organic compounds. As shown here, different parts of an organic compound's name—its root, suffix, or prefix—give information about its size and structure.

Carbon atoms	Name	Molecular formula
1	Methane	CH_4
2	Ethane	CH_3CH_3
3	Propane	$CH_3CH_2CH_3$
4	Butane	$CH_3CH_2CH_2CH_3$
5	Pentane	$CH_3CH_2CH_2CH_2CH_3$
6	Hexane	$CH_3CH_2CH_2CH_2CH_2CH_3$
7	Heptane	$CH_3CH_2CH_2CH_2CH_2CH_2CH_3$
8	Octane	$CH_3CH_2CH_2CH_2CH_2CH_2CH_2CH_3$
9	Nonane	$CH_3CH_2CH_2CH_2CH_2CH_2CH_2CH_2CH_3$
10	Decane	$CH_3CH_2CH_2CH_2CH_2CH_2CH_2CH_2CH_2CH_3$

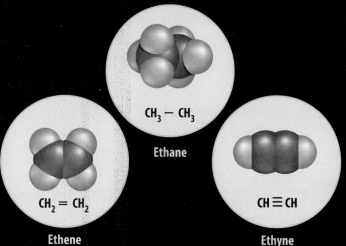

Ethene — $CH_2 = CH_2$
Ethane — $CH_3 - CH_3$
Ethyne — $CH \equiv CH$

ROOT WORDS The key to every name given to a compound in organic chemistry is its root word. This word tells how many carbon atoms are found in the longest continuous carbon chain in the compound. Except for compounds with one to four carbon atoms, the root word is based on Greek numbers.

SUFFIXES The suffix of the name for an organic compound indicates the kind of covalent bonds joining the compound's carbon atoms. If the atoms are joined by single covalent bonds, the compound's name will end in *-ane*. If there is a double covalent bond in the carbon chain, the compound's name ends in *-ene*. Similarly, if there is a triple bond in the chain, the compound's name will end in *-yne*.

PREFIXES The prefix of the name for an organic compound describes how the carbon atoms in the compound are arranged. Organic molecules that have names with the prefix *cyclo-* contain a ring of carbon atoms. For example, cyclopentane contains five carbon atoms all joined by single bonds in a ring.

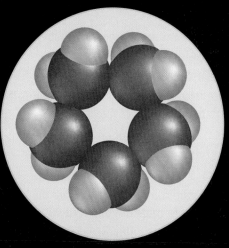

Cyclopentane

SECTION 1 Simple Organic Compounds

Figure 9 Visualize a hydrogen atom removed from a carbon atom on both ends of a hexane chain. The two end carbons form a bond with each other. **Describe** *how the chemical formula changes.*

Hexane
C_6H_{14}

Cyclohexane
C_6H_{12}

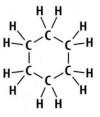

Topic: Cyclohexane
Visit ips.msscience.com for Web links to information about the manufacturing and the uses of cyclohexane.

Activity List the physical properties of cyclohexane. How does it compare with the physical properties of hexane?

Hydrocarbons in Rings You might be thinking that all hydrocarbons are chains of carbon atoms with two ends. Some molecules contain rings. You can see the structures of two different molecules in **Figure 9.** The carbon atoms of hexane bond together to form a closed ring containing six carbons. Each carbon atom still has four bonds. The prefix *cyclo-* in their names tells you that the molecules are cyclic, or ring shaped.

Ring structures are not uncommon in chemical compounds. Many natural substances such as sucrose, glucose, and fructose are ring structures. Ring structures can contain one or more double bonds.

Reading Check *What does the prefix* cyclo- *tell you about a molecule?*

section 1 review

Summary

Organic Compounds
- All living things contain carbon.
- Carbon atoms form covalent bonds.

Hydrocarbons
- Hydrocarbons are compounds that contain only hydrogen and carbon.
- The simplest hydrocarbon is methane.
- Saturated hydrocarbons are compounds that contain only single covalent bonds.
- Unsaturated hydrocarbons are compounds that form double and triple bonds.

Isomers
- Isomers are compounds that have the same molecular formula but different structures.
- Isomers also have different chemical and physical properties.

Self Check

1. **Describe** a carbon atom.
2. **Identify** Give one example of each of the following: a compound with a single bond, a compound with a double bond, and a compound with a triple bond. Write the chemical formula and draw the structure for each.
3. **Draw** all the possible isomers for heptane, C_7H_{16}.
4. **Think Critically** Are propane and cylcopropane isomers? Draw their structures. Use the structures and formulas to explain your answer.

Applying Math

5. **Make and Use Graphs** From **Table 1,** plot the number of carbon atoms on the *x*-axis and the number of hydrogen atoms on the *y*-axis. Predict the formula for the saturated hydrocarbon that has 11 carbon atoms.

section 2
Other Organic Compounds

Substituted Hydrocarbons

Suppose you pack an apple in your lunch every day. One day, you have no apples, so you substitute a pear. When you eat your lunch, you'll notice a difference in the taste and texture of your fruit. Chemists make substitutions, too. They change hydrocarbons to make compounds called substituted hydrocarbons. To make a substituted hydrocarbon, one or more hydrogen atoms are replaced by atoms such as halogens or by groups of atoms. Such changes result in compounds with chemical properties different from the original hydrocarbon. When one or more chlorine atoms are added to methane in place of hydrogens, new compounds are formed. **Figure 10** shows the four possible compounds formed by substituting chlorine atoms for hydrogen atoms in methane.

as you read

What You'll Learn
- **Describe** how new compounds are formed by substituting hydrogens in hydrocarbons.
- **Identify** the classes of compounds that result from substitution.

Why It's Important

Many natural and manufactured organic compounds are formed by replacing hydrogen with other atoms.

Review Vocabulary
chemical formula: chemical shorthand that uses symbols to tell what elements are in a compound and their ratios

New Vocabulary
- hydroxyl group
- amino group
- carboxyl group
- amino acid

Chloromethane contains a single chlorine atom. CH_3Cl

Trichloromethane, or chloroform, has three chlorine atoms. It is used in the production of fluoropolymers—one of the raw materials used to make nonstick coating. $CHCl_3$

Dichloromethane contains two chlorine atoms. This is used in some paint and varnish removers. CH_2Cl_2

Carbon tetrachloride is a fully substituted methane molecule with four chlorines. CCl_4

Figure 10 Chlorine can replace hydrogen atoms in methane.

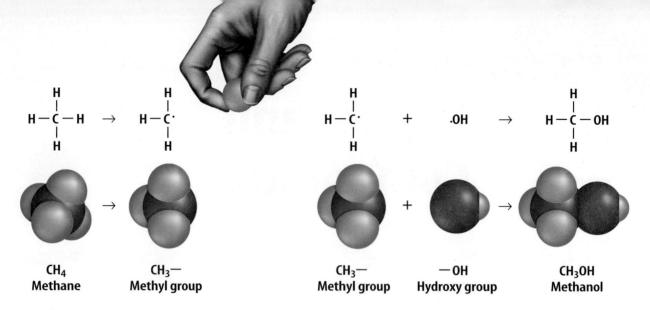

Figure 11 After the methane molecule loses one of its hydrogen atoms, it has an extra electron to share, as does the hydroxyl group. **Identify** the type of bond formed.

Alcohols Groups of atoms also can be added to hydrocarbons to make different compounds. The **hydroxyl** (hi DROK sul) **group** is made up of an oxygen atom and a hydrogen atom joined by a covalent bond. A hydroxyl group is represented by the formula $-OH$. An alcohol is formed when a hydroxyl group replaces a hydrogen atom in a hydrocarbon. **Figure 11** shows the formation of the alcohol methanol. A hydrogen atom in the methane molecule is replaced by a hydroxyl group.

Reading Check *What does the formula* $-OH$ *represent?*

Larger alcohol molecules are formed by adding more carbon atoms to the chain. Ethanol is an alcohol produced naturally when sugar in corn, grains, and fruit ferments. It is a combination of ethane, which contains two carbon atoms, and an $-OH$ group. Its formula is C_2H_5OH. Isopropyl alcohol forms when the hydroxyl group is substituted for a hydrogen atom on the middle carbon of propane rather than one of the end carbons. **Table 2** lists three alcohols with their structures and uses. You've probably used isopropyl alcohol to disinfect injuries. Did you know that ethanol can be added to gasoline and used as a fuel for your car?

Table 2 Common Alcohols			
Uses	Methanol	Ethanol	Isopropyl Alcohol
Fuel	yes	yes	no
Cleaner	yes	yes	yes
Disinfectant	no	yes	yes
Manufacturing chemicals	yes	yes	yes

Carboxylic Acids Have you ever tasted vinegar? Vinegar is a solution of acetic acid and water. You can think of acetic acid as the hydrocarbon methane with a carboxyl (car BOK sul) group substituted for a hydrogen. A **carboxyl group** consists of a carbon atom that has a double bond with one oxygen atom and a single bond with a hydroxyl group. Its formula is $-COOH$. When a carboxyl group is substituted in a hydrocarbon, the substance formed is called a carboxylic acid. The simplest carboxylic acid is formic acid. Formic acid consists of a single hydrogen atom and a carboxyl group. You can see the structures of formic acid and acetic acid in **Figure 12.**

You probably can guess that many other carboxylic acids are formed from longer hydrocarbons. Many carboxylic acids occur in foods. Citric acid is found in citrus fruits such as oranges and grapefruit. Lactic acid is present in sour milk. Acetic acid dissolved in water—vinegar—often is used in salad dressings.

Figure 12 *Crematogaster* ants make the simplest carboxylic acid, formic acid. Notice the structure of the $-COOH$ group.
Describe how the structures of formic acid and acetic acid differ.

Amines Substituted hydrocarbons, called amines, formed when an amino (uh ME noh) group replaces a hydrogen atom. An **amino group** is a nitrogen atom joined by covalent bonds to two hydrogen atoms. It has the formula $-NH_2$. Methylamine, shown in **Figure 13,** is formed when one of the hydrogens in methane is replaced with an amino group. A more complex amine is the novocaine dentists once used to numb your mouth during dental work. Amino groups are important because they are a part of many biological compounds that are essential for life. When an amino group bonds with one additional hydrogen atom, the result is ammonia, NH_3.

Figure 13 Complex amines account for the strong smells of cheeses such as these.

Amino Acids You have seen that a carbon group can be substituted onto one end of a chain to make a new molecule. It's also possible to substitute groups on both ends of a chain and even to replace hydrogen atoms bonded to carbon atoms in the middle of a chain. When both an amino group ($-NH_2$) and a carboxyl acid group ($-COOH$) replace hydrogens on the same carbon atom in a molecule, a type of compound known as an **amino acid** is formed. Amino acids are essential for human life.

Methylamine
CH_3NH_2

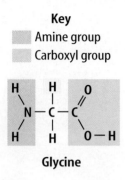
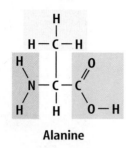

Glycine   Alanine

Figure 14 The 20 amino acids needed for protein synthesis each contain a central carbon atom bonded to an amine group, a hydrogen atom, and a carboxyl group. The fourth bond, shown in yellow, is different for each amino acid.

The Building Blocks of Protein

Amino acids are the building blocks of proteins, which are an important class of biological molecules needed by living cells. Twenty different amino acids bond in different combinations to form the variety of proteins that are needed in the human body. Glycine and alanine are shown in **Figure 14**.

Glycine is the simplest amino acid. It is a methane molecule in which one hydrogen atom has been replaced by an amine group and another has been replaced by a carboxyl group. The other 19 amino acids are formed by replacing the yellow highlighted hydrogen atom with different groups. For example, in alanine, one hydrogen atom is replaced by a methyl ($-CH_3$) group.

Reading Check *What are the building blocks of protein?*

Some amino acids, such as glycine and alanine, are manufactured within the human body. They are called nonessential amino acids. This means that it is not essential to consume these types of amino acids. More than half of the twenty amino acids are considered nonessential. The essential amino acids, those that must be consumed, are obtained by eating protein-rich foods. These foods include meat, eggs, and milk.

section 2 review

Summary

Substituted Hydrocarbons

- A substituted hydrocarbon has one or more hydrogen atoms replaced.
- The chemical properties of the substituted hydrocarbon are different from the original hydrocarbon.

Types of Substitutions

- Alcohols are made when a hydroxyl group is substituted for a hydrogen atom.
- Carboxylic acids are formed when the carboxyl group is substituted for a hydrogen atom.
- When an amino group is substituted for hydrogen, an amine is formed.
- Amino acids have both an amino group and a carboxyl group.
- Twenty amino acids are building blocks of proteins needed in the human body.

Self Check

1. **Draw** Tetrafluoroethylene is a substituted hydrocarbon in which all four of the hydrogen atoms are replaced by fluorine. Draw the structural formula for this molecule.
2. **Describe** how the 20 amino acids differ from each other.
3. **Identify** Starting with a hexane molecule, C_6H_{14}, draw and label each new molecule when adding an alcohol group, a carboxylic group, and an amino group.
4. **Think Critically** The formula for one compound that produces the odor in skunk spray is $CH_3CH_2CH_2CH_2SH$. Draw and examine the structural formula. Does it fit the definition of a substituted hydrocarbon? Explain.

Applying Skills

5. **Define** Compounds in which hydrogen atoms have been replaced by chlorine and fluorine atoms are known as chlorofluorocarbons (CFCs). Draw the structures of the four CFCs using CH_4 as the starting point.

Conversion of Alcohols

Have you ever wondered how chemists change one substance into another? In this lab, you will change an alcohol into an acid.

Alcohol Conversion	
Procedure Step	Observations
Step 2	
Step 3	Do not write in this book.
Step 4	
Step 5	

Real-World Question

What changes occur when ethanol is exposed to conditions like those produced by exposure to air and bacteria?

Goals
- **Observe** a chemical change in an alcohol.
- **Infer** the product of the chemical change.

Materials
test tube and stopper
test-tube rack
pH test paper
10-mL graduated cylinders (2)
dropper
0.01M potassium permanganate solution (1 mL)
6.0M sodium hydroxide solution (1 mL)
ethanol (3 drops)

Safety Precautions

WARNING: *Handle these chemicals with care. Immediately flush any spills with water and call your teacher. Keep your hands away from your face.*

Procedure

1. Measure 1 mL of potassium permanganate solution and pour it into a test tube. Carefully measure 1 mL of sodium hydroxide solution and add it to the test tube.
2. With your teacher's help, dip a piece of pH paper into the mixture in the test tube. Record the result in your Science Journal.
3. Add three drops of ethanol to the test tube. Put a stopper on the test tube and gently shake it for one minute. Record any changes.
4. Place the test tube in a test-tube rack. Observe and record any changes you notice during the next five minutes.
5. Test the sample with pH paper again. Record what you observe.
6. Your teacher will dispose of the solutions.

Conclude and Apply

1. **Analyze Results** Did a chemical reaction take place? What leads you to infer this?
2. **Predict** Alcohols can undergo a chemical reaction to form carboxylic acids in the presence of potassium permanganate. If the alcohol used is ethanol, what would you predict to be the chemical formula of the acid produced?

Communicating Your Data

Compare your conclusions with other students in your class. **For more help, refer to the** Science Skill Handbook.

LAB **261**

section 3
Biological Compounds

as you read

What You'll Learn
- **Describe** how large organic molecules are made.
- **Explain** the roles of organic molecules in the body.
- **Explain** why eating a balanced diet is important for good health.

Why It's Important
Polymers are organic molecules that are important to your body processes and everyday living.

Review Vocabulary
chemical reaction: process that produces chemical change, resulting in new substances that have properties different from those of the original substances

New Vocabulary
- polymer
- monomer
- polymerization
- protein
- carbohydrate
- sugars
- starches
- lipids
- cholesterol

What's a polymer?

Now that you know about some simple organic molecules, you can begin to learn about more complex molecules. One type of complex molecule is called a polymer (PAH luh mur). A **polymer** is a molecule made up of many small organic molecules linked together with covalent bonds to form a long chain. The small, organic molecules that link together to form polymers are called **monomers.** Polymers can be produced by living organisms or can be made in a laboratory. Polymers produced by living organisms are called natural polymers. Polymers made in a laboratory are called synthetic polymers.

Reading Check *What is a polymer, and how does it resemble a chain?*

To picture what polymers are, it is helpful to start with small synthetic polymers. You use such polymers every day. Plastics, synthetic fabrics, and nonstick surfaces on cookware are polymers. The unsaturated hydrocarbon ethylene, C_2H_4, is the monomer of a common polymer used often in plastic bags. The monomers are bonded together in a chemical reaction called **polymerization** (puh lih muh ruh ZAY shun). As you can see in **Figure 15,** the double bond breaks in each ethylene molecule. The two carbon atoms then form new bonds with carbon atoms in other ethylene molecules. This process is repeated many times and results in a much larger molecule called polyethylene. A polyethylene molecule can contain 10,000 ethylene units.

Figure 15 Small molecules called monomers link into long chains to form polymers.

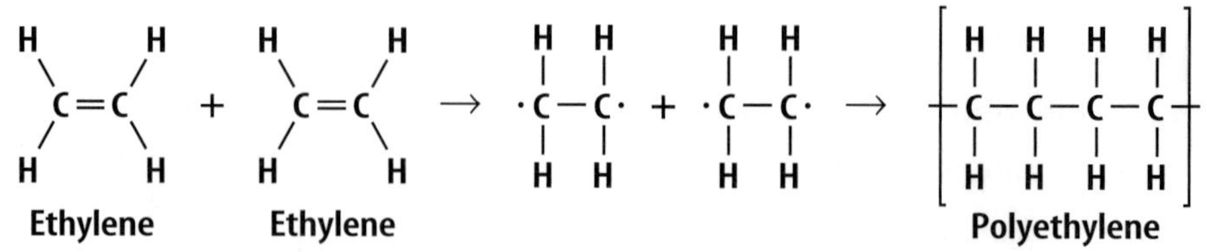

The carbon atoms that were joined by the double bond each have an electron to share with another carbon in another molecule of ethylene. The process goes on until a long molecule is formed.

Glycine + Alanine → (dipeptide) + H₂O

Proteins are Polymers

You've probably heard about proteins when you've been urged to eat healthful foods. A **protein** is a polymer that consists of a chain of individual amino acids linked together. Your body cannot function properly without them. Proteins in the form of enzymes serve as catalysts and speed up chemical reactions in cells. Some proteins make up the structural materials in ligaments, tendons, muscles, cartilage, hair, and fingernails. Hemoglobin, which carries oxygen through the blood, is a protein polymer, and all body cells contain proteins.

The various functions in your body are performed by different proteins. Your body makes many of these proteins by assembling 20 amino acids in different ways. Nine of the amino acids that are needed to make proteins cannot be produced by your body. These amino acids, which are called essential amino acids, must come from the food you eat. That's why you need to eat a diet containing protein-rich foods, like those in **Table 3**.

The process by which your body converts amino acids to proteins is shown in **Figure 16**. In this reaction, the amino group of the amino acid alanine forms a bond with the carboxyl group of the amino acid glycine, and a molecule of water is released. Each end of this new molecule can form similar bonds with another amino acid. The process continues in this way until the amino acid chain, or protein, is complete.

Reading Check *How is an amino acid converted to protein?*

Figure 16 Both ends of an amino acid can link with other amino acids.
Identify *the molecule that is released in the process.*

Mini LAB

Summing Up Protein

Procedure
1. Make a list of the foods you ate during the last 24 h.
2. Use the data your teacher gives you to find the total number of grams of protein in your diet for the day. Multiply the grams of protein in one serving of food by the number of units of food you ate. The recommended daily allowance (RDA) of protein for girls 11 to 14 years old is 46 g per day. For boys 11 to 14 years old, the RDA is 45 g per day.

Analysis
1. Was your total greater or less than the RDA?
2. Which of the foods you ate supplied the largest amount of protein? What percent of the total grams of protein did that food supply?

Table 3 Protein Content (Approximate)	
Foods	Protein Content (g)
Chicken breast (113 g)	28
Eggs (2)	12
Whole milk (240 mL)	8
Peanut butter (30 g)	8
Kidney beans (127 g)	8

Figure 17 These foods contain a high concentration of carbohydrates.

Nutrition Science A dietician studies the science of nutrition. Their main focus is to help people maintain good health and prevent and control disease through the foods they eat. Dieticians assess the individual's nutritional needs and build a dietary program specifically to meet these needs. Dieticians work in a variety of fields from schools and company cafeterias to hospitals and nursing homes.

Carbohydrates

The day before a race, athletes often eat large amounts of foods containing carbohydrates like the ones in **Figure 17.** What's in pasta and other foods like bread and fruit that gives the body a lot of energy? These foods contain sugars and starches, which are members of the family of organic compounds called carbohydrates. A **carbohydrate** is an organic compound that contains only carbon, hydrogen, and oxygen, usually in a ratio of two hydrogen atoms to one oxygen atom. In the body, carbohydrates are broken down into simple sugars that the body can use for energy. The different types of carbohydrates are divided into groups—sugars, starches, and cellulose.

Table 4 below gives some approximate carbohydrate content for some of the common foods.

Table 4 Carbohydrates in Foods (Approximate)	
Foods	Carbohydrate Content (g)
Apple (1)	21
White rice ($\frac{1}{2}$ cup)	17
Baked potato ($\frac{1}{2}$ cup)	15
Wheat bread (1 slice)	13
Milk (240 mL)	12

Figure 18 Glucose and fructose are simple six-carbon carbohydrates found in many fresh and packaged foods. Glucose and fructose are isomers.
Explain why they are isomers.

Sugars If you like chocolate-chip cookies or ice cream, then you're familiar with sugars. They are the substances that make fresh fruits and candy sweet. Simple **sugars** are carbohydrates containing five, six, or seven carbon atoms arranged in a ring. The structures of glucose and fructose, two common simple sugars, are shown in **Figure 18.** Glucose forms a six-carbon ring. It is found in many naturally sweet foods, such as grapes and bananas. Fructose is the sweet substance found in ripe fruit and honey. It often is found in corn syrup and added to many foods as a sweetener. The sugar you probably have in your sugar bowl or use in baking a cake is sucrose. Sucrose, shown in **Figure 19,** is a combination of the two simple sugars glucose and fructose. In the body, sucrose cannot move through cell membranes. It must be broken down into glucose and fructose to enter cells. Inside the cells, these simple sugars are broken down further, releasing energy for cell functions.

Starches Starches are large carbohydrates that exist naturally in grains such as rice, wheat, corn, potatoes, lima beans, and peas. **Starches** are polymers of glucose monomers in which hundreds or even thousands of glucose molecules are joined together. Because each sugar molecule releases energy when it is broken down, starches are sources of large amounts of energy.

Figure 19 Sucrose is a molecule of glucose combined with a molecule of fructose.
Identify What small molecule must be added to sucrose when it separates to form the two six-carbon sugars?

SECTION 3 Biological Compounds **265**

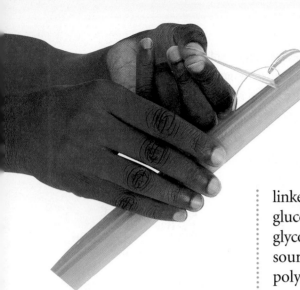

Figure 20 Your body cannot chemically break down the long cellulose fibers in celery, but it needs fiber to function properly.

Other Glucose Polymers Two other important polymers that are made up of glucose molecules are cellulose and glycogen. Cellulose makes up the long, stiff fibers found in the walls of plant cells, like the strands that pull off the celery stalk in **Figure 20**. It is a polymer that consists of long chains of glucose units linked together. Glycogen is a polymer that also contains chains of glucose units, but the chains are highly branched. Animals make glycogen and store it mainly in their muscles and liver as a ready source of glucose. Although starch, cellulose, and glycogen are polymers of glucose, humans can't use cellulose as a source of energy. The human digestive system can't convert cellulose into sugars. Grazing animals, such as cows, have special digestive systems that allow them to break down cellulose into sugars.

Reading Check How do the location and structure of glycogen and cellulose differ?

Applying Science

Which foods are best for quick energy?

Foods high in carbohydrates are sources of energy.

Identifying the Problem

The chart shows some foods and their carbohydrate count. Look at the differences in how much energy they might provide, given their carbohydrate count.

Solving the Problem

1. Create a high-energy meal with the most carbohydrates. Include one choice from each category. Create another meal that contains a maximum of 60 g of carbohydrates.
2. Meat and many vegetables have only trace amounts of carbohydrates. How many grams of carbohydrates would a meal of turkey, stuffing, lettuce salad, and lemonade contain?

Carbohydrate Counts for Common Foods

Main Dish		Side Dish		Drink	
two slices wheat bread	26 g	fudge brownie	25 g	orange juice	27 g
macaroni and cheese	29 g	apple	21 g	cola	38 g
two pancakes	28 g	baked beans	27 g	sweetened iced tea	22 g
chicken and noodles	39 g	blueberry muffin	27 g	lemon-lime soda	38 g
hamburger with bun	34 g	cooked carrots	8 g	hot cocoa	25 g
hot oatmeal	25 g	banana	28 g	apple juice	29 g
plain bagel	38 g	baked potato	34 g	lemonade	28 g
bran flakes with raisins	47 g	stuffing	22 g	whole milk	12 g
lasagna	50 g	brown rice	22 g	chocolate milk	26 g
spaghetti with marinara	50 g	corn on the cob	14 g	sports drink	24 g

Lipids

A **lipid** is an organic compound that contains the same elements as carbohydrates—carbon, hydrogen, and oxygen—but in different proportions. They are the reaction products of glycerol, which has three –OH groups and three long-chain carboxylic acids, as pictured in **Figure 21.** Lipids are in many of the foods you eat such as the ones shown in **Figure 22.** Lipids are commonly called fats and oils, but they also are found in greases and waxes such as beeswax. Wax is a lipid, but it is harder than fat because of its chemical composition. Bees secrete wax from a gland in the abdomen to form beeswax, which is part of the honeycomb.

Lipids Store Energy Lipids store energy in their bonds, just as carbohydrates do, but they are a more concentrated source of energy than carbohydrates. If you eat more food than your body needs to supply you with the energy for usual activities, the excess energy from the food is stored by producing lipids.

How can energy be stored in a molecule? The chemical reaction that produces lipids is endothermic. An endothermic reaction is one in which energy is absorbed. This means that energy is stored in the chemical bonds of lipids. When your body needs energy, the bonds are broken and energy is released. This process protects your body when you need extra energy or when you might not be able to eat. If you regularly eat more food than you need, large amounts of lipids will be produced and stored as fat on your body.

Reading Check *What is a lipid and how does your body use lipids to store energy?*

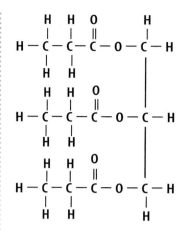

Figure 21 Lipids consist of two parts—glycerol and three molecules of carboxylic acid.
Identify *which portion is from glycerol and which portion is from carboxylic acid.*

Topic: Lipids
Visit ips.msscience.com for Web links to information about your body's requirement for lipids.

Activity Lipids (fats) are important to your body's dietary needs. Make a table of saturated and unsaturated lipids that you consume. Try to list a total of ten.

Figure 22 Many of the foods that you eat contain fats and oils, which are lipids.

SECTION 3 Biological Compounds **267**

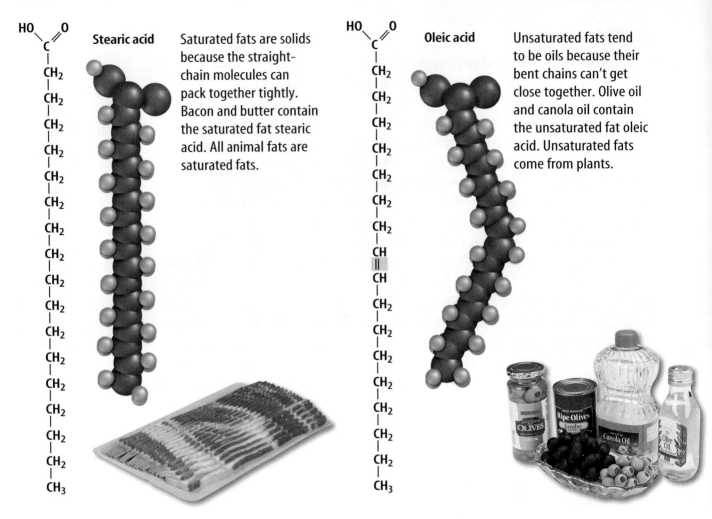

Figure 23 Whether a lipid is a liquid or a solid depends on the type of bonds it has.

Saturated and Unsaturated Lipids Not all lipids are the same. Recall the difference between saturated and unsaturated hydrocarbons. Unsaturated molecules have one or more double or triple bonds between carbon atoms. Lipid molecules also can be saturated or unsaturated. As you can see in **Figure 23,** when a lipid is saturated, the acid chains are straight because all the bonds are single bonds. They are able to pack together closely. A compact arrangement of the molecules is typical of a solid such as margarine or shortening. These solid lipids consist mainly of saturated fats.

When a lipid is unsaturated, as in **Figure 23,** the molecule bends wherever there is a double bond. This prevents the chains from packing close together, so these lipids tend to be liquid oils such as olive or corn oil.

Doctors have observed that people who eat a diet high in saturated fats have an increased risk of developing cardiovascular problems such as heart disease. The effect of saturated fat seems to be increased blood cholesterol, which may be involved in the formation of deposits on artery walls. Fortunately, many foods containing unsaturated fats are available. Making wise choices in the foods that you eat can help keep your body healthy.

Cholesterol

Cholesterol is a complex lipid that is present in foods that come from animals, such as meat, butter, eggs, and cheese. However, cholesterol is not a fat. Even if you don't eat foods containing cholesterol, your body makes its own supply. Your body needs cholesterol for building cell membranes. Cholesterol is not found in plants, so oils derived from plants are free of cholesterol. However, the body can convert fats in these oils to cholesterol.

Deposits of cholesterol, called plaque, can build up on the inside walls of arteries. This condition, known as atherosclerosis, is shown in **Figure 24.** When arteries become clogged, the flow of blood is restricted, which results in high blood pressure. This, in turn, can lead to heart disease. Although the cause of plaque build up on the inside walls of arteries is unknown, limiting the amount of saturated fat and cholesterol might help to lower cholesterol levels in the blood and might help reduce the risk of heart problems.

Heart disease is a major health concern in the United States. As a result, many people are on low-cholesterol diets. What types of foods should people choose to lower their cholesterol level?

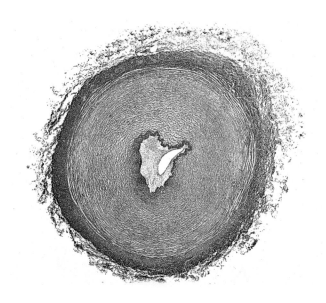

Figure 24 This view of an artery shows atherosclerosis, a dangerous condition in which arteries in the body become clogged. Deposits build up on the inside walls of the artery, leaving less room for blood to flow.

 *What is atherosclerosis and why is this condition dangerous?*

section 3 review

Summary

Polymers and Proteins

- A polymer is a molecule made up of small, repeating units.
- The small molecules that link together to form a polymer are called monomers.
- A protein is a polymer that consists of individual amino acids linked together.

Carbohydrates, Lipids, and Cholesterol

- Carbohydrates and lipids are organic compounds that contain carbon, hydrogen, and oxygen.
- Lipids store energy in their bonds.
- Unsaturated lipids have one or more double or triple bonds between carbon atoms.
- Cholesterol is a complex lipid.

Self Check

1. **Define** the process by which proteins are made. What other product is formed along with a protein molecule?
2. **Explain** how carbohydrates, proteins, and lipids are important to body functions.
3. **Analyze** how cellulose, starch, and glycogen are different.
4. **Describe** how your body obtains and uses cholesterol.
5. **Think Critically** Explain why even people who eat a healthful diet might gain weight if they don't get enough exercise.

Applying Skills

6. **Draw a Conclusion** Polyunsaturated fats are recommended for a healthful diet. Using what you know about lipids, what might *polyunsaturated* mean?

LAB

Looking for Vitamin C

Goals
- **Prepare** an indicator solution.
- **Verify** a positive test by using a known material.
- **Apply** the test to various foods.
- **Infer** some foods your diet should contain.

Possible Materials
starch solution
iodine solution
vitamin-C solution
water
droppers (10)
15-mL test tubes (10)
test-tube rack
250-mL beaker
stirrer
10-mL graduated cylinder
mortar and pestle
liquid foods such as milk, orange juice, vinegar, and cola
solid foods such as tomatoes, onions, citrus fruits, potatoes, bread, salt, and sugar

Safety Precautions

WARNING: *Do not taste any materials used in the lab. Use care when mashing food samples.*

Real-World Question
Vitamin C is essential to humans for good health and the prevention of disease. Your body cannot produce this necessary organic molecule, so you must consume it in your food. How do you know which foods are good sources of vitamin C? Reactions that cause color changes are useful as chemical tests. This activity uses the disappearance of the dark-blue color of a solution of starch and iodine to show the presence of vitamin C. Can you test foods for vitamin C? Could the starch-iodine solution be used to show the presence of vitamin C in food?

Procedure
1. Collect all the materials and equipment you will need.
2. Obtain 10 mL of the starch solution from your teacher.
3. Add the starch solution to 200 mL of water in a 250-mL beaker. Stir.
4. Add four drops of iodine solution to the beaker to make a dark-blue indicator solution. Stir in the drops.

270 CHAPTER 9 Carbon Chemistry

Using Scientific Methods

5. Obtain your teacher's approval of your indicator solution before proceeding.
6. **Measure** and place 5 mL of the indicator solution in a clean test tube.
7. Obtain 5 mL of vitamin-C solution from your teacher.
8. Using a clean dropper, add one drop of the vitamin-C solution to the test tube. Stir. Continue adding drops and stirring until you notice a color change. Place a piece of white paper behind the test tube to show the color clearly. Record the number of drops added and any observations.

9. Using a clean test tube and dropper for each test, repeat steps 6 through 8, replacing the vitamin-C solution with other liquids and solids. Add drops of liquid foods or juices until a color change is noted or until you have added about 40 drops of the liquid. Mash solid foods such as onion and potato. Add about 1 g of the food directly to the test tube and stir. Test at least four liquids and four solids.
10. Record the amount of each food added and observations in a table.

Analyze Your Data

1. **Infer** What indicates a positive test for vitamin C? How do you know?
2. **Describe** a negative test for vitamin C.
3. **Observe** Which foods tested positive for vitamin C? Which foods, if any, tested negative for vitamin C?

Conclude and Apply

1. **Explain** which foods you might include in your diet to make sure you get vitamin C every day.
2. **Determine** if a vitamin-C tablet could take the place of these foods. Explain.

Compare your results with other class members. Were your results consistent? Make a record of the foods you eat for two days. Does your diet contain the minimum RDA of vitamin C?

TIME SCIENCE AND Society
SCIENCE ISSUES THAT AFFECT YOU!

From Plants to Medicine

Wild plants help save lives

Look carefully at those plants growing in your backyard or neighborhood. With help from scientists, they could save a life. Many of the medicines that doctors prescribe were first developed from plants. For example, aspirin was extracted from the bark of a willow tree. A cancer medication was extracted from the bark of the Pacific yew tree. Aspirin and the cancer medication are now made synthetically—their carbon structures are duplicated in the lab and factory.

Throughout history, and in all parts of the world, traditonal healers have used different parts of plants and flowers to help treat people. Certain kinds of plants have been mashed up and applied to the body to heal burns and sores, or have been swallowed or chewed to help people with illnesses.

Modern researchers are studying the medicinal value of plants and then figuring out the plants' properties and makeup. This is giving scientists important information as they turn to more and more plants to help make medicine in the lab. Studying these plants—and how people in different cultures use them—is the work of scientists called ethnobotanists (eth noh BAH tuhn ihsts). Ethnobotanist Memory Elvin-Lewis notes that plants help treat illnesses.

She visits healers who show her the plants that they find most useful. "Plants are superior chemists producing substances with sophisticated molecular structures that protect the plant from injury and disease," writes Professor Michele L. Trankina. It's these substances in plants that are used as sources of medicines. And it's these substances that are giving researchers and chemists leads to making similar substances in the lab. That can only mean good news—and better health—for people!

Promising cancer medications are made from the bark of the Pacific yew tree.

Memory Elvin-Lewis has spent part of her career studying herbal medicines.

Investigate Research the work of people like Carole L. Cramer. She's modifying common farm plants so that they produce human antibodies used to treat human illnesses. Use the link on the right or your school's media center to get started in your search.

For more information, visit
ips.msscience.com/time

chapter 9 Study Guide

Reviewing Main Ideas

Section 1 Simple Organic Compounds

1. Hydrocarbons are compounds containing only carbon and hydrogen.

2. If a hydrocarbon has only single bonds, it is called a saturated hydrocarbon.

3. Unsaturated hydrocarbons have one or more double or triple bonds in their structure.

Section 2 Other Organic Compounds

1. Hydrogens can be substituted with other atoms or with groups of atoms.

2. An amino acid contains an amino group and a carboxyl group substituted on the same carbon atom.

3. An alcohol is formed when a hydroxyl group is substituted for a hydrogen atom in a hydrocarbon.

4. A carboxylic acid is made when a carboxyl group is substituted and an amine is formed when an amino group ($-NH_2$) is substituted.

Section 3 Biological Compounds

1. Many biological compounds are large molecules called polymers.

2. Proteins serve a variety of functions, including catalyzing many cell reactions.

3. Carbohydrates and lipids are energy sources and the means of storing energy.

Visualizing Main Ideas

Copy and complete the following concept map on simple organic compounds.

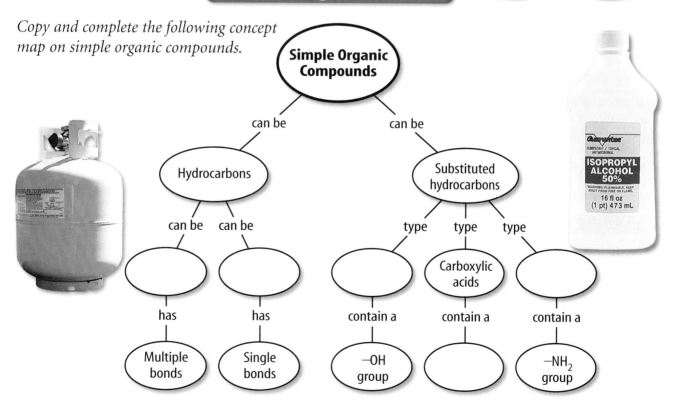

Chapter 9 Review

Using Vocabulary

amino acid p. 259
amino group p. 259
carbohydrate p. 264
carboxyl group p. 259
cholesterol p. 269
hydrocarbon p. 251
hydroxyl group p. 258
isomer p. 254
lipid p. 267
monomer p. 262
organic compound p. 250
polymer p. 262
polymerization p. 262
protein p. 263
saturated hydrocarbon p. 252
starches p. 265
sugars p. 265
unsaturated hydrocarbon p. 253

Answer the following questions using complete sentences.

1. Explain the difference between an amino group and an amino acid.
2. How does a hydroxyl group differ from a carboxyl group?
3. Explain why eating carbohydrates would be beneficial to an athlete before a race.
4. What is the connection between a polymer and a protein?
5. What do carbohydrates and lipids have in common?
6. Explain the difference between a saturated and an unsaturated compound.

Checking Concepts

Choose the word or phrase that best answers the question.

7. A certain carbohydrate molecule has ten oxygen atoms. How many hydrogen atoms does it contain?
    A) five
    B) 20
    C) ten
    D) 16

8. Which is NOT a group that can be substituted in a hydrocarbon?
    A) amino
    B) carboxyl
    C) hydroxyl
    D) lipid

9. Which chemical formula represents an alcohol?
    A) CH_3COOH
    B) CH_3NH_2
    C) CH_3OH
    D) CH_4

10. Which substance can build up in arteries and lead to heart disease?
    A) cholesterol
    B) fructose
    C) glucose
    D) starch

11. What is an organic molecule that contains a triple bond called?
    A) polymer
    B) saturated hydrocarbon
    C) isomer
    D) unsaturated hydrocarbon

12. What is the name of the substituted hydrocarbon with the chemical formula CH_2F_2?
    A) methane
    B) fluoromethane
    C) difluoromethane
    D) trifluoromethane

13. Which chemical formula below represents an amino acid?
    A) CH_3COOH
    B) CH_3NH_2
    C) NH_2CH_2COOH
    D) CH_4

14. Proteins are biological polymers made up of what type of monomers?
    A) alcohols
    B) amino acids
    C) ethene molecules
    D) propene molecules

15. Excess energy is stored in your body as which of the following?
    A) proteins
    B) isomers
    C) lipids
    D) saturated hydrocarbons

16. Which is a ring-shaped molecule?
    A) acetone
    B) ethylene
    C) cyclopentane
    D) dichloroethane

chapter 9 Review

Thinking Critically

Use the figures below to answer question 17.

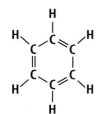

17. **Compare and Contrast** Benzene and cyclohexane are both ring molecules. Discuss the similarities and differences of the two molecules.

18. **Explain** Ethanol is used as a fuel for cars. Explain how energy is obtained from ethanol to fuel a car.

19. **Analyze** Candle wax is one of the longer hydrocarbons. Explain why heat and light are produced in a burning candle.

20. **Infer** In the polymerization of amino acids to make proteins, water molecules are produced as part of the reaction. However, in the polymerization of ethylene, no water is produced. Explain.

21. **Recognize Cause and Effect** Marathon runners go through a process known as hitting the wall. They have used up all their readily available glucose and start using stored lipids as fuel. What is the advantage of eating lots of carbohydrates the day before a race?

22. **Hypothesize** PKU is a genetic disorder that can lead to brain damage. People with this disorder cannot process one of the amino acids. Luckily, damage can be prevented by a proper diet. How is this possible?

23. **Explain** Medicines previously obtained from plants are now manufactured. Can these two medicines be the same?

Performance Activities

24. **Scientific Drawing** Research an amino acid that was not mentioned in the chapter. Draw its structural formula and highlight the portion that substitutes for a hydrogen atom.

Applying Math

Use the graph below to answer question 25.

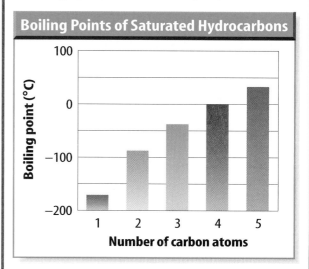

25. **Hydrocarbons** Using the graph above, explain how the boiling point varies with the number of carbon atoms. What do you predict would be the approximate boiling point of hexane?

Use the figure below to answer question 26.

Sucrose

26. **Simple Sugar** What are the percentages of carbon, oxygen, and hydrogen in a sucrose molecule?

27. **Polyethylene** If one polyethylene molecule contains 10,000 ethylene units, how many can be made from 3 million ethylene units?

Chapter 9 Standardized Test Practice

Part 1 Multiple Choice

Record your answers on the answer sheet provided by your teacher or on a sheet of paper.

Use the table below to answer questions 1 and 2.

Double Bonded Hydrocarbons			
Name	Formula	Name	Formula
Ethene	C_2H_4	Pentene	C_5H_{10}
Propene	C_3H_6	Octene	?
Butene	C_4H_8	Decene	$C_{10}H_{20}$

1. What is the general formula for this family?
   A. $C_{2n}H_n$
   B. C_nH_{2n+2}
   C. C_nH_{2n}
   D. C_nH_{2n-2}

2. What is the formula of octene?
   A. C_6H_{12}
   B. C_8H_{16}
   C. C_6H_{10}
   D. C_8H_{18}

3. Based on its root name and suffix, what is the structural formula of propyne?
   A. $H-C\equiv C-CH_3$
   B. $CH_3-CH_2-CH_3$
   C. $H_2C=CHCH_3$
   D. $HC\equiv CH$

4. As five amino acids polymerize to form a protein, how many water molecules split off?
   A. 6
   B. 5
   C. 4
   D. none

5. As a NH_2 group replaces a hydrogen in a hydrocarbon, which type of compound is formed?
   A. carboxylic acid
   B. amino acid
   C. alcohol
   D. amine

6. One of the freons used in refrigerators is dichloro-difluoromethane. How many H atoms are in this molecule?
   A. 4
   B. 2
   C. 1
   D. none

Use the structures below to answer questions 7–9.

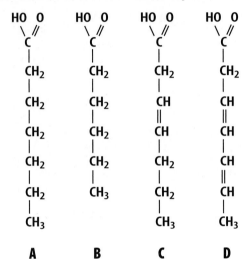

7. Which is saturated and has the fewest number of carbon atoms?
   A. A
   B. B
   C. C
   D. D

8. Which is a polyunsaturated acid?
   A. A
   B. B
   C. C
   D. D

9. These are all considered to be carboxylic acids because they contain which of the following?
   A. a $-COOH$ group
   B. a $-CH_3$ group
   C. a double bond
   D. C, H, and O atoms

Test-Taking Tip

Figures and Illustrations Be sure you understand all symbols in a figure or illustration before attempting to answer any questions about them.

Question 7 Even though the hydrogen molecules are shown on the same side of the carboxylic acid molecules, the structural formula places one hydrogen on either side of the carbon.

Standardized Test Practice

Part 2 Short Response/Grid In

Record answers on the answer sheet provided by your teacher or on a sheet of paper.

Use the illustration below to answer questions 10 and 11.

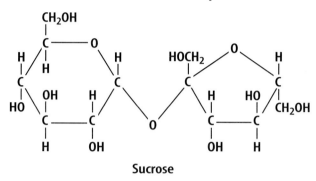

Sucrose

10. If its formula is given as $C_xH_yO_z$ what are the values of x, y, and z?

11. Sucrose is a carbohydrate. What ratio between atoms denotes a carbohydrate?

12. What is the structural formula of the propyl group?

13. What is the molecular formula of propylamine?

Use the following explanation to answer questions 14 and 15.

A carbon atom has a mass of 12 units and hydrogen 1. A molecule of methane has a molecular mass of $(12 \times 1) + (1 \times 4) = 16$ units, and is $\frac{4}{16} = 25\%$ hydrogen by mass.

Methane

Ethane

14. What is molecular mass of C_2H_6?

15. What is the percent carbon by mass for methane and ethane?

Part 3 Open Ended

Record your answers on a sheet of paper.

Use the following explanation to answer questions 16–18.

Glycol is a chief component of antifreeze. It IUPAC name is 1, 2-ethanediol.

16. The root "eth" indicates how many carbon atoms?

17. The "ane" suffix indicates which bond between carbons: single, double, or triple?

18. What functional group does "ol" indicate?

Use the table below to answer questions 19 and 20.

Hydrocarbon Isomers	
Formula	Number of Isomers
C_2H_6	1
C_4H_{10}	2
C_6H_{14}	5
C_8H_{18}	18
$C_{10}H_{22}$	75

19. Sketch a graph of this data. Is it linear?

20. Predict how many isomers of C_5H_{12} might exist. Draw them.

21. Draw a reasonable structural formula for carbon dioxide. Recall how many bonds each carbon atom can form.

unit 4
Motion and Forces

How Are
City Streets &
Zebra Mussels
Connected?

NATIONAL GEOGRAPHIC

As long as people and cargo have traveled the open seas, ships have taken on extra deadweight, or ballast. Ballast helps adjust the ships' depth in the water and counteracts uneven cargo loads. For centuries, ballast was made up of solid materials—usually rocks, bricks, or sand. These materials had to be unloaded by hand when heavier cargo was taken on board. Many port cities used the discarded ballast stones to pave their dirt roads. The new streets were called "cobblestone." By the mid-1800s, shipbuilding and pump technology had improved, so that taking on and flushing out large quantities of water was relatively easy. Water began to replace rocks and sand as ballast. This new form of deadweight often contained living creatures. When the tanks were flushed at the ships' destinations, those sea creatures would be expelled as well. One such creature was the zebra mussel, which is believed to have been introduced to North America in the 1980s as ballast on a cargo ship. Since then, the population of zebra mussels in the Great Lakes has increased rapidly. Some native species in these lakes are now being threatened by the invading zebra mussel population.

unit projects

Visit **ips.msscience.com/unit_project** to find project ideas and resources.
Projects include:

- **History** Write a 60-second Moment in History on the life and scientific contributions of Sir Isaac Newton.
- **Technology** Dissect gears in clocks and explore how clocks work. Design a flow chart of the system where every minute counts.
- **Model** Design a tower system that will keep a ball moving down the track using limited supplies. This time, slower is better!

WebQuest *Roller Coaster Physics* is an investigation of acceleration, laws of motion, gravity, and coaster design. Create your own roller coaster.

chapter 10

Motion and Momentum

The BIG Idea

The motion of an object can be described by its velocity.

SECTION 1
What is motion?
Main Idea Motion is a change in position.

SECTION 2
Acceleration
Main Idea Acceleration occurs when an object speeds up, slows down, or changes direction.

SECTION 3
Momentum
Main Idea In a collision, momentum can be transferred from one object to another.

Springing into Action

The hunt might just be over for this pouncing leopard. A leopard can run as fast as 60 km/h over short distances and can jump as high as 3 m. However, a leopard must be more than fast and strong. To catch its prey, it must also be able to change its speed and direction quickly.

Science Journal Describe how your motion changed as you moved from your school's entrance to your classroom.

280

Start-Up Activities

Motion After a Collision

How is it possible for a 70-kg football player to knock down a 110-kg football player? The smaller player usually must be running faster. Mass makes a difference when two objects collide, but the speed of the objects also matters. Explore the behavior of colliding objects during this lab.

1. Space yourself about 2 m away from a partner. Slowly roll a baseball on the floor toward your partner, and have your partner roll a baseball quickly into your ball.
2. Have your partner slowly roll a baseball as you quickly roll a tennis ball into the baseball.
3. Roll two tennis balls toward each other at the same speed.
4. **Think Critically** In your Science Journal, describe how the motion of the balls changed after the collisions, including the effects of speed and type of ball.

 **Preview this chapter's content and activities at** ips.msscience.com

 **Motion and Momentum**
Make the following Foldable to help you understand the vocabulary terms in this chapter.

STEP 1 Fold a vertical sheet of notebook paper from side to side.

STEP 2 Cut along every third line of only the top layer to form tabs.

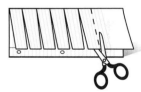

STEP 3 Label each tab.

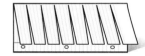

Build Vocabulary As you read the chapter, list the vocabulary words about motion and momentum on the tabs. As you learn the definitions, write them under the tab for each vocabulary word.

Get Ready to Read

Summarize

① Learn It! Summarizing helps you organize information, focus on main ideas, and reduce the amount of information to remember. To summarize, restate the important facts in a short sentence or paragraph. Be brief and do not include too many details.

② Practice It! Read the text on page 295 labeled *Conservation of Momentum*. Then read the summary below and look at the important facts from that passage.

Important Facts

- A collision doesn't change the total momentum of a group of objects.
- Friction is an outside force that causes the total momentum to decrease.
- Objects can bounce off each other when they collide.
- Objects can stick together when they collide.

Summary

The total momentum of a group of objects remains constant unless an outside force acts on the objects.

③ Apply It! Practice summarizing as you read this chapter. Stop after each section and write a brief summary.

Target Your Reading

Reading Tip
Reread your summary to make sure you didn't change the author's original meaning or ideas.

Use this to focus on the main ideas as you read the chapter.

① **Before you read** the chapter, respond to the statements below on your worksheet or on a numbered sheet of paper.
- Write an **A** if you **agree** with the statement.
- Write a **D** if you **disagree** with the statement.

② **After you read** the chapter, look back to this page to see if you've changed your mind about any of the statements.
- If any of your answers changed, explain why.
- Change any false statements into true statements.
- Use your revised statements as a study guide.

Before You Read A or D		Statement	After You Read A or D
	1	Distance traveled and displacement are always equal.	
	2	When an object changes direction, it is accelerating.	
	3	A horizontal line on a distance-time graph means the speed is zero.	
	4	If two objects are moving at the same speed, the heavier object is harder to stop.	
	5	The instantaneous speed of an object is always equal to its average speed.	
	6	Momentum equals mass divided by velocity.	
	7	Speed always is measured in kilometers per hour.	
	8	An object's momentum increases if its speed increases.	
	9	If a car is accelerating, its speed must be increasing.	
	10	Speed and velocity are the same thing.	

Science Online
Print out a worksheet of this page at ips.msscience.com

282 B

section 1

What is motion?

as you read

What You'll Learn
- **Define** distance, speed, and velocity.
- **Graph** motion.

Why It's Important
The different motions of objects you see every day can be described in the same way.

Review Vocabulary
meter: SI unit of distance, abbreviated m; equal to 39.37 in

New Vocabulary
- speed
- average speed
- instantaneous speed
- velocity

Matter and Motion

All matter in the universe is constantly in motion, from the revolution of Earth around the Sun to electrons moving around the nucleus of an atom. Leaves rustle in the wind. Lava flows from a volcano. Bees move from flower to flower as they gather pollen. Blood circulates through your body. These are all examples of matter in motion. How can the motion of these different objects be described?

Changing Position

To describe an object in motion, you must first recognize that the object is in motion. Something is in motion if it is changing position. It could be a fast-moving airplane, a leaf swirling in the wind, or water trickling from a hose. Even your school, attached to Earth, is moving through space. When an object moves from one location to another, it is changing position. The runners shown in **Figure 1** sprint from the start line to the finish line. Their positions change, so they are in motion.

Figure 1 These sprinters are in motion because their positions change.

282 **CHAPTER 10** Motion and Momentum

Relative Motion Determining whether something changes position requires a point of reference. An object changes position if it moves relative to a reference point. To visualize this, picture yourself competing in a 100-m dash. You begin just behind the start line. When you pass the finish line, you are 100 m from the start line. If the start line is your reference point, then your position has changed by 100 m relative to the start line, and motion has occurred. Look at **Figure 2.** How can you determine that the dog has been in motion?

Reading Check *How do you know if an object has changed position?*

Distance and Displacement Suppose you are to meet your friends at the park in five minutes. Can you get there on time by walking, or should you ride your bike? To help you decide, you need to know the distance you will travel to get to the park. This distance is the length of the route you will travel from your house to the park.

Suppose the distance you traveled from your house to the park was 200 m. When you get to the park, how would you describe your location? You could say that your location was 200 m from your house. However, your final position depends on both the distance you travel and the direction. Did you go 200 m east or west? To describe your final position exactly, you also would have to tell the direction from your starting point. To do this, you would specify your displacement. Displacement includes the distance between the starting and ending points and the direction in which you travel. **Figure 3** shows the difference between distance and displacement.

Figure 2 Motion occurs when something changes position relative to a reference point.
Explain *how the position of the dog changed.*

Figure 3 Distance is how far you have walked. Displacement is the direction and difference in position between your starting and ending points.

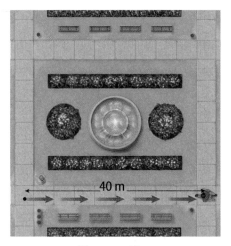

Distance: 40 m
Displacement: 40 m east

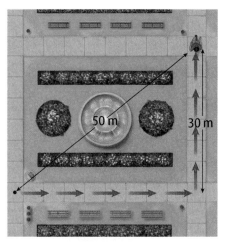

Distance: 70 m
Displacement: 50 m northeast

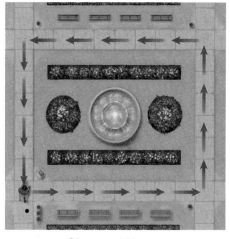

Distance: 140 m
Displacement: 0 m

SECTION 1 What is motion? **283**

Animal Speeds Different animals can move at different top speeds. What are some of the fastest animals? Research the characteristics that help animals run, swim, or fly at high speed.

Speed

To describe motion, you usually want to describe how fast something is moving. The faster something is moving, the greater the distance it can travel in a unit of time, such as one second or one hour. **Speed** is the distance an object travels in a unit of time. For example, an object with a speed of 5 m/s can travel 5 m in 1 s. Speed can be calculated from this equation:

Speed Equation

$$\text{speed (in meters/second)} = \frac{\text{distance (in meters)}}{\text{time (in seconds)}}$$

$$s = \frac{d}{t}$$

The unit for speed is the unit of distance divided by the unit of time. In SI units, speed is measured in units of m/s—meters per second. However, speed can be calculated using other units such as kilometers for distance and hours for time.

Applying Math — Solve a Simple Equation

SPEED OF A SWIMMER Calculate the speed of a swimmer who swims 100 m in 56 s.

Solution

1. *This is what you know:*
   - distance: $d = 100$ m
   - time: $t = 56$ s

2. *This is what you need to know:*
   - speed: $s = ?$ m/s

3. *This is the procedure you need to use:*
   Substitute the known values for distance and time into the speed equation and calculate the speed:
   $$s = \frac{d}{t} = \frac{100 \text{ m}}{56 \text{ s}} = \frac{100}{56} \frac{\text{m}}{\text{s}} = 1.8 \text{ m/s}$$

4. *Check your answer:*
   Multiply your answer by the time. You should get the distance that was given.

Practice Problems

1. A runner completes a 400-m race in 43.9 s. In a 100-m race, he finishes in 10.4 s. In which race was his speed faster?

2. A passenger train travels from Boston to New York, a distance of 350 km, in 3.5 h. What is the train's speed?

For more practice, visit ips.msscience.com/math_practice

Average Speed A car traveling in city traffic might have to speed up and slow down many times. How could you describe the speed of an object whose speed is changing? One way is to determine the object's average speed between where it starts and stops. The speed equation on the previous page can be used to calculate the average speed. **Average speed** is found by dividing the total distance traveled by the total time taken.

Reading Check *How is average speed calculated?*

Instantaneous Speed An object in motion can change speeds many times as it speeds up or slows down. The speed of an object at one instant of time is the object's **instantaneous speed**. To understand the difference between average and instantaneous speeds, think about walking to the library. If it takes you 0.5 h to walk 2 km to the library, your average speed would be as follows:

$$s = \frac{d}{t}$$
$$= \frac{2 \text{ km}}{0.5 \text{ h}} = 4 \text{ km/h}$$

However, you might not have been moving at the same speed throughout the trip. At a crosswalk, your instantaneous speed might have been 0 km/h. If you raced across the street, your speed might have been 7 km/h. If you were able to walk at a steady rate of 4 km/h during the entire trip, you would have moved at a constant speed. Average speed, instantaneous speed, and constant speed are illustrated in **Figure 4**.

Measuring Average Speed

Procedure
1. Choose two points, such as two doorways, and mark each with a small piece of **masking tape**.
2. Measure the distance between the two points.
3. Use a **watch, clock,** or **timer** that indicates seconds to time yourself walking from one mark to the other.
4. Time yourself walking slowly, walking safely and quickly, and walking with a varying speed; for example, slow/fast/slow.

Analysis
1. Calculate your average speed in each case.
2. Predict how long it would take you to walk 100 m slowly, at your normal speed, and quickly.

Try at Home

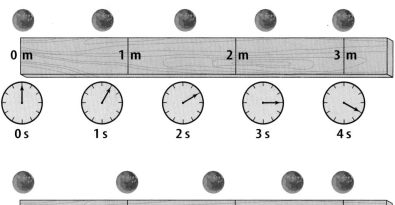

Figure 4 The average speed of each ball is the same from 0 s to 4 s.

The top ball is moving at a constant speed. In each second, the ball moves the same distance.

The bottom ball has a varying speed. Its instantaneous speed is fastest between 0 s and 1 s, slower between 2 s and 3 s, and slowest between 3 s and 4 s.

SECTION 1 What is motion? **285**

Graphing Motion

You can represent the motion of an object with a distance-time graph. For this type of graph, time is plotted on the horizontal axis and distance is plotted on the vertical axis. **Figure 5** shows the motion of two students who walked across a classroom plotted on a distance-time graph.

Distance-Time Graphs and Speed A distance-time graph can be used to compare the speeds of objects. Look at the graph shown in **Figure 5.** According to the graph, after 1 s student A traveled 1 m. Her average speed during the first second is as follows:

$$\text{speed} = \frac{\text{distance}}{\text{time}} = \frac{1 \text{ m}}{1 \text{ s}} = 1 \text{ m/s}$$

Student B, however, traveled only 0.5 m in the first second. His average speed is

$$\text{speed} = \frac{\text{distance}}{\text{time}} = \frac{0.5 \text{ m}}{1 \text{ s}} = 0.5 \text{ m/s}$$

So student A traveled faster than student B. Now compare the steepness of the lines on the graph in **Figure 5.** The line representing the motion of student A is steeper than the line for student B. A steeper line on the distance-time graph represents a greater speed. A horizontal line on the distance-time graph means that no change in position occurs. In that case, the speed of the object is zero.

Topic: Land Speed Record
Visit ips.msscience.com for Web links to information about how the land speed record has changed over the past century.

Activity Make a graph showing the increase in the land speed over time.

Figure 5 The motion of two students walking across a classroom is plotted on this distance-time graph.
Use the graph to determine which student had the faster average speed.

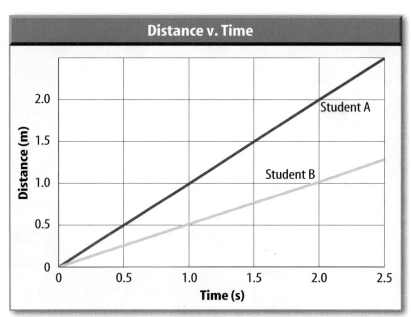

286 CHAPTER 10 Motion and Momentum

Velocity

The motion of an object also depends on the direction in which the object is moving. The direction of an object's motion can be described with its velocity. The **velocity** of an object is the speed of the object and the direction of its motion. For example, if a car is moving west with a speed of 80 km/h, the car's velocity is 80 km/h west. The velocity of an object is sometimes represented by an arrow. The arrow points in the direction in which the object is moving. **Figure 6,** uses arrows to show the velocities of two hikers.

The velocity of an object can change if the object's speed changes, its direction of motion changes, or they both change. For example, suppose a car is traveling at a speed of 40 km/h north and then turns left at an intersection and continues on with a speed of 40 km/h. The speed of the car is constant at 40 km/h, but the velocity changes from 40 km/h north to 40 km/h west. Why can you say the velocity of a car changes as it comes to a stop at an intersection?

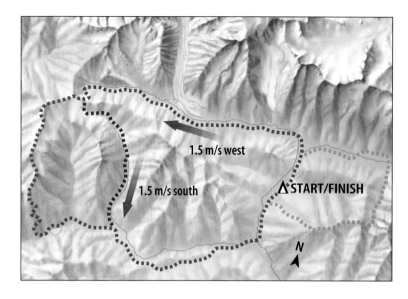

Figure 6 The arrows show the velocities of two hikers. Although the hikers have the same speed, they have different velocities because they are moving in different directions.

section 1 review

Summary

Changing Position
- An object is in motion if it changes position relative to a reference point.
- Motion can be described by distance, speed, displacement, and velocity, where displacement and velocity also include direction.

Speed and Velocity
- The speed of an object can be calculated by dividing the distance traveled by the time needed to travel the distance.
- For an object traveling at constant speed, its average speed is the same as its instantaneous speed.
- The velocity of an object is the speed of the object and its direction of motion.

Graphing Motion
- A line on a distance-time graph becomes steeper as an object's speed increases.

Self Check

1. **Identify** the two pieces of information you need to know the velocity of an object.
2. **Make and Use Graphs** You walk forward at 1.5 m/s for 8 s. Your friend decides to walk faster and starts out at 2.0 m/s for the first 4 s. Then she slows down and walks forward at 1.0 m/s for the next 4 s. Make a distance-time graph of your motion and your friend's motion. Who walked farther?
3. **Think Critically** A bee flies 25 m north of the hive, then 10 m east, 5 m west, and 10 m south. How far north and east of the hive is it now? Explain how you calculated your answer.

Applying Math

4. **Calculate** the average velocity of a dancer who moves 5 m toward the left of the stage over the course of 15 s.
5. **Calculate Travel Time** An airplane flew a distance of 650 km at an average speed of 300 km/h. How much time did the flight take?

section 2
Acceleration

as you read

What You'll Learn
- **Define** acceleration.
- **Predict** what effect acceleration will have on motion.

Why It's Important
Whenever the motion of an object changes, it is accelerating.

Review Vocabulary
kilogram: SI unit of mass, abbreviated kg; equal to approximately 2.2 lbs

New Vocabulary
- acceleration

Acceleration and Motion

When you watch the first few seconds of a liftoff, a rocket barely seems to move. With each passing second, however, you can see it move faster until it reaches an enormous speed. How could you describe the change in the rocket's motion? When an object changes its motion, it is accelerating. **Acceleration** is the change in velocity divided by the time it takes for the change to occur.

Like velocity, acceleration has a direction. If an object speeds up, the acceleration is in the direction that the object is moving. If an object slows down, the acceleration is opposite to the direction that the object is moving. What if the direction of the acceleration is at an angle to the direction of motion? Then the direction of motion will turn toward the direction of the acceleration.

Speeding Up You get on a bicycle and begin to pedal. The bike moves slowly at first, but speeds up as you keep pedaling. Recall that the velocity of an object is the speed of an object and its direction of motion. Acceleration occurs whenever the velocity of an object changes. Because the bike's speed is increasing, the velocity of the bike is changing. As a result, the bike is accelerating.

For example, the toy car in **Figure 7** is accelerating because it is speeding up. The speed of the car is 10 cm/s after 1s, 20 cm/s after 2s, and 30 cm/s after 3s. Here the direction of the car's acceleration is in the same direction as the car's velocity—to the right.

Figure 7 The toy car is accelerating because its speed is increasing.

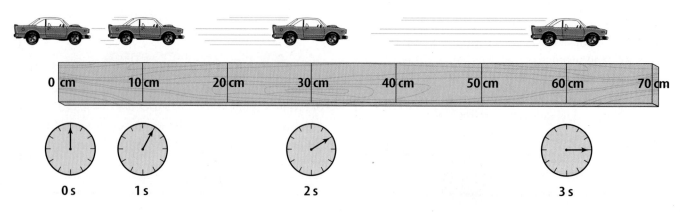

288 CHAPTER 10 Motion and Momentum

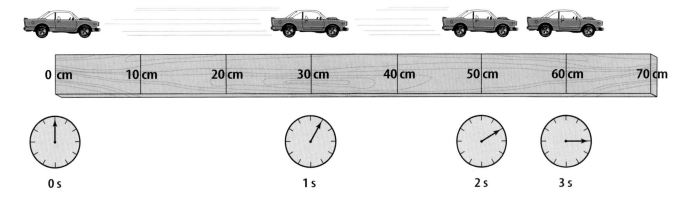

Slowing Down Now suppose you are biking at a speed of 4 m/s and you apply the brakes. This causes you to slow down. When you slow down, your velocity changes because your speed decreases. This means that acceleration occurs when an object slows down, as well as when it speeds up. The car in **Figure 8** is slowing down. During each time interval, the car travels a smaller distance, so its speed is decreasing.

In both of these examples, speed is changing, so acceleration is occurring. Because speed is decreasing in the second example, the direction of the acceleration is opposite to the direction of motion. Any time an object slows down, its acceleration is in the direction opposite to the direction of its motion.

Changing Direction The velocity of an object also changes if the direction of motion changes. Then the object doesn't move in a straight line, but instead moves in a curved path. The object is accelerating because its velocity is changing. In this case the direction of acceleration is at an angle to the direction of motion.

Again imagine yourself riding a bicycle. When you turn the handlebars, the bike turns. Because the direction of the bike's motion has changed, the bike has accelerated. The acceleration is in the direction that the bicycle turned.

Figure 9 shows another example of an object that is accelerating. The ball starts moving upward, but its direction of motion changes as its path turns downward. Here the acceleration is downward. The longer the ball accelerates, the more its path turns toward the direction of acceleration.

Reading Check What are three ways to accelerate?

Figure 9 The ball starts out by moving forward and upward, but the acceleration is downward, so the ball's path turns in that direction.

Figure 8 The car is moving to the right but accelerating to the left. In each time interval, it covers less distance and moves more slowly.
Determine how the car's velocity is changing.

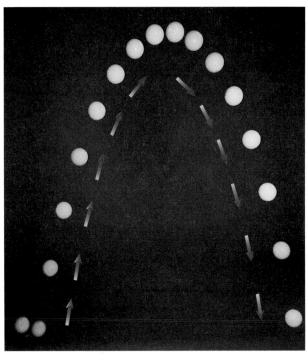

SECTION 2 Acceleration **289**

Calculating Acceleration

If an object is moving in only one direction, its acceleration can be calculated using this equation.

Acceleration Equation

acceleration (in m/s^2) = $\dfrac{(\text{final speed (in m/s)} - \text{initial speed (in m/s)})}{\text{time (in s)}}$

$$a = \frac{(s_f - s_i)}{t}$$

In this equation, time is the length of time over which the motion changes. In SI units, acceleration has units of meters per second squared (m/s^2).

Applying Math — Solve a Simple Equation

ACCELERATION OF A BUS Calculate the acceleration of a bus whose speed changes from 6 m/s to 12 m/s over a period of 3 s.

Solution

1 *This is what you know:*
- initial speed: s_i = 6 m/s
- final speed: s_f = 12 m/s
- time: t = 3 s

2 *This is what you need to know:* acceleration: a = ? m/s^2

3 *This is the procedure you need to use:* Substitute the known values of initial speed, final speed and time in the acceleration equation and calculate the acceleration:

$$a = \frac{(s_f - s_i)}{t} = \frac{(12 \text{ m/s} - 6 \text{ m/s})}{3 \text{ s}} = 6 \frac{\text{m}}{\text{s}} \times \frac{1}{3 \text{ s}} = 2 \text{ m/s}^2$$

4 *Check your answer:* Multiply the calculated acceleration by the known time. Then add the known initial speed. You should get the final speed that was given.

Practice Problems

1. Find the acceleration of a train whose speed increases from 7 m/s to 17 m/s in 120 s.
2. A bicycle accelerates from rest to 6 m/s in 2 s. What is the bicycle's acceleration?

For more practice, visit ips.msscience.com/math_practice

Figure 10 When skidding to a stop, you are slowing down. This means you have a negative acceleration.

Positive and Negative Acceleration An object is accelerating when it speeds up, and the acceleration is in the same direction as the motion. An object also is accelerating when it slows down, but the acceleration is in the direction opposite to the motion, such as the bicycle in **Figure 10.** How else is acceleration different when an object is speeding up and slowing down?

Suppose you were riding your bicycle in a straight line and increased your speed from 4 m/s to 6 m/s in 5 s. You could calculate your acceleration from the equation on the previous page.

$$a = \frac{(s_f - s_i)}{t}$$
$$= \frac{(6 \text{ m/s} - 4 \text{ m/s})}{5 \text{ s}} = \frac{+2 \text{ m/s}}{5 \text{ s}}$$
$$= +0.4 \text{ m/s}^2$$

When you speed up, your final speed always will be greater than your initial speed. So subtracting your initial speed from your final speed gives a positive number. As a result, your acceleration is positive when you are speeding up.

Suppose you slow down from a speed of 4 m/s to 2 m/s in 5 s. Now the final speed is less than the initial speed. You could calculate your acceleration as follows:

$$a = \frac{(s_f - s_i)}{t}$$
$$= \frac{(2 \text{ m/s} - 4 \text{ m/s})}{5 \text{ s}} = \frac{-2 \text{ m/s}}{5 \text{ s}}$$
$$= -0.4 \text{ m/s}^2$$

Because your final speed is less than your initial speed, your acceleration is negative when you slow down.

Modeling Acceleration

Procedure
1. Use **masking tape** to lay a course on the floor. Mark a starting point and place marks along a straight path at 10 cm, 40 cm, 90 cm, 160 cm, and 250 cm from the start.
2. Clap a steady beat. On the first beat, the person walking the course should be at the starting point. On the second beat, the walker should be on the first mark, and so on.

Analysis
1. Describe what happens to your speed as you move along the course. Infer what would happen if the course were extended farther.
2. Repeat step 2, starting at the other end. Are you still accelerating? Explain.

SECTION 2 Acceleration **291**

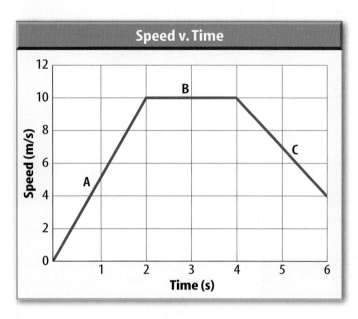

Figure 11 A speed-time graph can be used to find acceleration. When the line rises, the object is speeding up. When the line falls, the object is slowing down.
Infer *what acceleration a horizontal line represents.*

Graphing Accelerated Motion The motion of an object that is moving in a single direction can be shown with a graph. For this type of graph, speed is plotted on the vertical axis and time on the horizontal axis. Take a look at **Figure 11**. On section A of the graph, the speed increases from 0 m/s to 10 m/s during the first 2 s, so the acceleration is +5 m/s². The line in section A slopes upward to the right. An object that is speeding up will have a line on a speed-time graph that slopes upward.

Now look at section C. Between 4 s and 6 s, the object slows down from 10 m/s to 4 m/s. The acceleration is −3 m/s². On the speed-time graph, the line in section C is sloping downward to the right. An object that is slowing down will have a line on a speed-time graph that slopes downward.

On section B, where the line is horizontal, the change in speed is zero. So a horizontal line on the speed-time graph represents an acceleration of zero or constant speed.

section 2 review

Summary

Acceleration and Motion

- Acceleration is the change in velocity divided by the time it takes to make the change. Acceleration has direction.
- Acceleration occurs whenever an object speeds up, slows down, or changes direction.

Calculating Acceleration

- For motion in a single direction, acceleration can be calculated from this equation:

$$a = \frac{s_f - s_i}{t}$$

- If an object is speeding up, its acceleration is positive; if an object is slowing down, its acceleration is negative.
- On a speed-time graph, a line sloping upward represents increasing speed, a line sloping downward represents decreasing speed, and a horizontal line represents zero acceleration or constant speed.

Self Check

1. **Compare and contrast** speed, velocity, and acceleration.
2. **Infer** the motion of a car whose speed-time graph shows a horizontal line, followed by a straight line that slopes downward to the bottom of the graph.
3. **Think Critically** You start to roll backward down a hill on your bike, so you use the brakes to stop your motion. In what direction did you accelerate?

Applying Math

4. **Calculate** the acceleration of a runner who accelerates from 0 m/s to 3 m/s in 12 s.
5. **Calculate Speed** An object falls with an acceleration of 9.8 m/s². What is its speed after 2 s?
6. **Make and Use a Graph** A sprinter had the following speeds at different times during a race: 0 m/s at 0 s, 4 m/s at 2 s, 7 m/s at 4 s, 10 m/s at 6 s, 12 m/s at 8 s, and 10 m/s at 10 s. Plot these data on a speed-time graph. During what time intervals is the acceleration positive? Negative? Is the acceleration ever zero?

section 3

Momentum

Collisions

A collision occurs when a moving object collides with other objects. What happens when a cue ball collides with another ball in a game of pool? The answer is that the velocities of the two balls change. The collision can change the speed of each ball, the direction of motion of each ball, or both. When a collision occurs, changes in motion of the colliding objects depend on their masses and their velocities before the collision.

Mass and Inertia

The mass of an object affects how easy it is to change its motion. **Mass** is the amount of matter in an obejct. Imagine a person rushing toward you. To stop the person, you would have to push on him or her. However, you would have to push harder to stop an adult than to stop a child. The child would be easier to stop because it has less mass than the adult. The more mass an object has, the harder it is to change its motion. In **Figure 12,** the tennis ball has more mass than the table-tennis ball. A big racquet rather than a small paddle is used to change its motion. The tendency of an object to resist a change in its motion is called **inertia**. The amount of resistance to a change in motion increases as an object's mass increases.

Reading Check *What is inertia?*

as you read

What You'll Learn
- **Define** momentum.
- **Explain** why momentum might not be conserved after a collision.
- **Predict** motion using the law of conservation of momentum.

Why It's Important
Objects in motion have momentum. The motion of objects after they collide depends on their momentum.

Review Vocabulary
triple-beam balance: scientific instrument used to measure mass precisely by comparing the mass of a sample to known masses

New Vocabulary
- mass
- inertia
- momentum
- law of conservation of momentum

Figure 12 A tennis ball has more mass than a table-tennis ball. The tennis ball must be hit harder than the table-tennis ball to change its velocity by the same amount.

Forensics and Momentum Forensic investigations of accidents and crimes often involve determining the momentum of an object. For example, the law of conservation of momentum sometimes is used to reconstruct the motion of vehicles involved in a collision. Research other ways momentum is used in forensic investigations.

Momentum

You know that the faster a bicycle moves, the harder it is to stop. Just as increasing the mass of an object makes it harder to stop, so does increasing the speed or velocity of the object. The **momentum** of an object is a measure of how hard it is to stop the object, and it depends on the object's mass and velocity. Momentum is usually symbolized by p.

Momentum Equation

momentum (in kg · m/s) = **mass** (in kg) × **velocity** (in m/s)

$$p = mv$$

Mass is measured in kilograms and velocity has units of meters per second, so momentum has units of kilograms multiplied by meters per second (kg · m/s). Also, because velocity includes a direction, momentum has a direction that is the same as the direction of the velocity.

 Explain how an object's momentum changes as its velocity changes.

Applying Math — Solve a Simple Equation

MOMENTUM OF A BICYCLE Calculate the momentum of a 14-kg bicycle traveling north at 2 m/s.

Solution

1 *This is what you know:*
- mass: $m = 14$ kg
- velocity: $v = 2$ m/s north

2 *This is what you need to find:*
momentum: $p = ?$ kg · m/s

3 *This is the procedure you need to use:*
Substitute the known values of mass and velocity into the momentum equation and calculate the momentum:

$p = mv = (14$ kg$)(2$ m/s north$) = 28$ kg · m/s north

4 *Check your answer:*
Divide the calculated momentum by the mass of the bicycle. You should get the velocity that was given.

Practice Problems

1. A 10,000-kg train is traveling east at 15 m/s. Calculate the momentum of the train.
2. What is the momentum of a car with a mass of 900 kg traveling north at 27 m/s?

 For more practice, visit ips.msscience.com/math_practice

Conservation of Momentum

If you've ever played billiards, you know that when the cue ball hits another ball, the motions of both balls change. The cue ball slows down and may change direction, so its momentum decreases. Meanwhile, the other ball starts moving, so its momentum increases.

In any collision, momentum is transferred from one object to another. Think about the collision between two billiard balls. If the momentum lost by one ball equals the momentum gained by the other ball, then the total amount of momentum doesn't change. When the total momentum of a group of objects doesn't change, momentum is conserved.

The Law of Conservation of Momentum According to the **law of conservation of momentum,** the total momentum of a group of objects remains constant unless outside forces act on the group. The moving cue ball and the other billiard balls in **Figure 13** are a group of objects. The law of conservation of momentum means that collisions between these objects don't change the total momentum of all the objects in the group.

Only an outside force, such as friction between the billiard balls and the table, can change the total momentum of the group of objects. Friction will cause the billiard balls to slow down as they roll on the table and the total momentum will decrease.

Types of Collisions Objects can collide with each other in different ways. **Figure 14** shows two examples. Sometimes objects will bounce off each other like the bowling ball and bowling pins. In other collisions, objects will collide and stick to each other, as when one football player tackles another.

Figure 13 When the cue ball hits the other billiard balls, it slows down because it transfers some of its momentum to the other billiard balls.
Predict *what would happen to the speed of the cue ball if all of its momentum were transferred to the other billiard balls.*

Figure 14 When objects collide, they can bounce off each other, or they can stick to each other.

When the bowling ball hits the pins, the ball and the pins bounce off each other. The momentum of the ball and the pins changes during the collision.

When one player tackles the other, they stick together. The momentum of each player changes during the collision.

Before the student on skates and the backpack collide, she is not moving.

After the collision, the student and the backpack move together at a slower speed than the backpack had before the collision.

Figure 15 Momentum is transferred from the backpack to the student.

Using Momentum Conservation The law of momentum conservation can be used to predict the velocity of objects after they collide. To use the law of conservation of momentum, assume that the total momentum of the colliding objects doesn't change.

For example, imagine being on skates when someone throws a backpack to you, as in **Figure 15.** The law of conservation of momentum can be used to find your velocity after you catch the backpack. Suppose a 2-kg backpack initially has a velocity of 5 m/s east. Your mass is 48 kg, and initially you're at rest. Then the total initial momentum is:

total momentum = momentum of backpack + your momentum
= 2 kg × 5 m/s east + 48 kg × 0 m/s
= 10 kg · m/s east

After the collision, the total momentum remains the same, and only one object is moving. Its mass is the sum of your mass and the mass of the backpack. You can use the equation for momentum to find the final velocity

total momentum = (mass of backpack + your mass) × velocity
10 kg · m/s east = (2 kg + 48 kg) × velocity
10 kg · m/s east = (50 kg) × velocity
0.2 m/s east = velocity

This is your velocity right after you catch the backpack. The velocity of you and the backpack together is much smaller than the initial velocity of the backpack. **Figure 16** shows the results of collisions between two objects that don't stick to each other.

Topic: Collisions
Visit ips.msscience.com for Web links to information about collisions between objects with different masses.

Activity Draw diagrams showing the results of collisions between a bowling ball and a tennis ball if they are moving in the same direction and if they are in opposite directions.

NATIONAL GEOGRAPHIC VISUALIZING CONSERVATION OF MOMENTUM

Figure 16

The law of conservation of momentum can be used to predict the results of collisions between different objects, whether they are subatomic particles smashing into each other at enormous speeds, or the collisions of marbles, as shown on this page. What happens when one marble hits another marble initially at rest? The results of the collisions depend on the masses of the marbles.

A Here, a less massive marble strikes a more massive marble that is at rest. After the collision, the smaller marble bounces off in the opposite direction. The larger marble moves in the same direction that the small marble was initially moving.

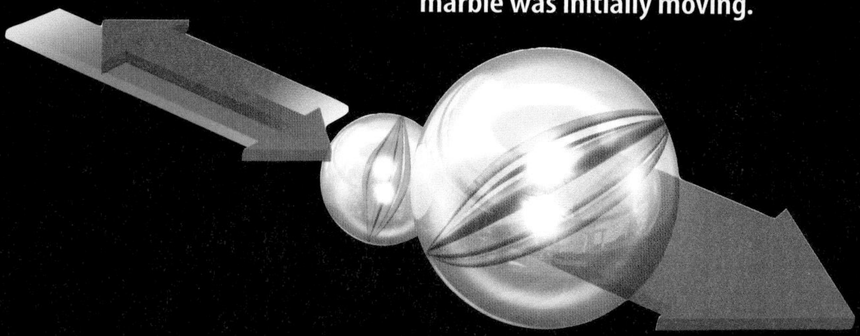

B Here, the large marble strikes the small marble that is at rest. After the collision, both marbles move in the same direction. The less massive marble always moves faster than the more massive one.

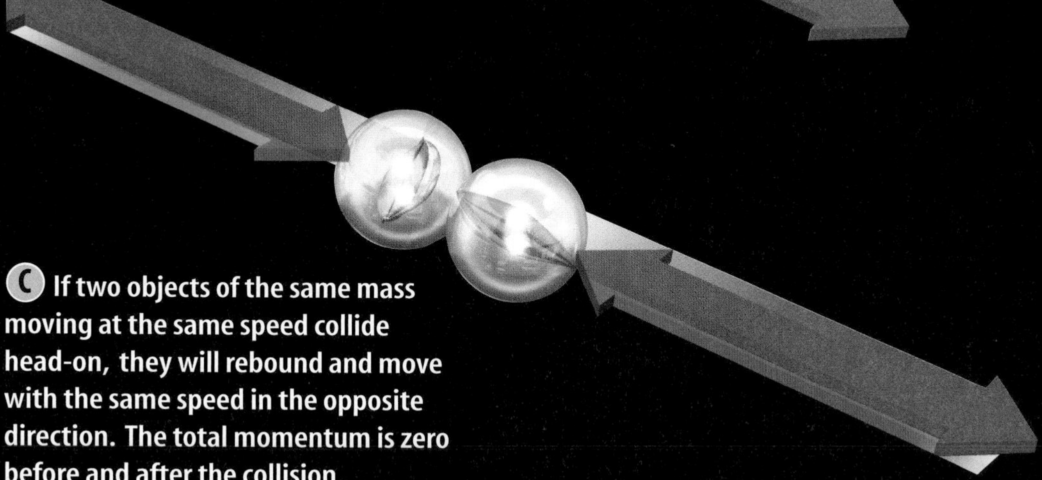

C If two objects of the same mass moving at the same speed collide head-on, they will rebound and move with the same speed in the opposite direction. The total momentum is zero before and after the collision.

SECTION 3 Momentum

Figure 17 When bumper cars collide, they bounce off each other, and momentum is transferred.

Colliding and Bouncing Off In some collisions, the objects involved, like the bumper cars in **Figure 17**, bounce off each other. The law of conservation of momentum can be used to determine how these objects move after they collide.

For example, suppose two identical objects moving with the same speed collide head on and bounce off. Before the collision, the momentum of each object is the same, but in opposite directions. So the total momentum before the collision is zero. If momentum is conserved, the total momentum after the collision must be zero also. This means that the two objects must move in opposite directions with the same speed after the collision. Then the total momentum once again is zero.

section 3 review

Summary

Inertia and Momentum

- Inertia is the tendency of an object to resist a change in motion.
- The momentum of an object in motion is related to how hard it is to stop the object, and can be calculated from the following equation:
  $$p = mv$$
- The momentum of an object is in the same direction as its velocity.

The Law of Conservation of Momentum

- According to the law of conservation of momentum, the total momentum of a group of objects remains constant unless outside forces act on the group.
- When objects collide, they can bounce off each other or stick together.

Self Check

1. **Explain** how momentum is transferred when a golfer hits a ball with a golf club.
2. **Determine** if the momentum of an object moving in a circular path at constant speed is constant.
3. **Explain** why the momentum of a billiard ball rolling on a billiard table changes.
4. **Think Critically** Two identical balls move directly toward each other with equal speeds. How will the balls move if they collide and stick together?

Applying Math

5. **Momentum** What is the momentum of a 0.1-kg mass moving with a velocity of 5 m/s west?
6. **Momentum Conservation** A 1-kg ball moving at 3 m/s east strikes a 2-kg ball and stops. If the 2-kg ball was initially at rest, find its velocity after the collision.

Science online ips.msscience.com/self_check_quiz

Collisions

A collision occurs when a baseball bat hits a baseball or a tennis racket hits a tennis ball. What would happen if you hit a baseball with a table-tennis paddle or a table-tennis ball with a baseball bat? How do the masses of colliding objects change the results of collisions?

Real-World Question

How does changing the size and number of objects in a collision affect the collision?

Goals
- **Compare and contrast** different collisions.
- **Determine** how the velocities after a collision depend on the masses of the colliding objects.

Materials
small marbles (5)   metersticks (2)
large marbles (2)   tape

Safety Precautions

Procedure

1. Tape the metersticks next to each other, slightly farther apart than the width of the large marbles. This limits the motion of the marbles to nearly a straight line.

2. Place a small target marble in the center of the track formed by the metersticks. Place another small marble at one end of the track. Flick the small marble toward the target marble. Describe the collision.

3. Repeat step 2, replacing the two small marbles with the two large marbles.

4. Repeat step 2, replacing the small shooter marble with a large marble.

5. Repeat step 2, replacing the small target marble with a large marble.

6. Repeat step 2, replacing the small target marble with four small marbles that are touching.

7. Place two small marbles at opposite ends of the metersticks. Shoot the marbles toward each other and describe the collision.

8. Place two large marbles at opposite ends of the metersticks. Shoot the marbles toward each other and describe the collision.

9. Place a small marble and a large marble at opposite ends of the metersticks. Shoot the marbles toward each other and describe the collision.

Conclude and Apply

1. **Describe** In which collisions did the shooter marble change direction? How did the mass of the target marble compare with the mass of the shooter marble in these collisions?

2. **Describe** how the velocity of the shooter marble changed when the target marble had the same mass and was at rest.

Communicating Your Data

Make a chart showing your results. You might want to make before-and-after sketches, with short arrows to show slow movement and long arrows to show fast movement.

Design Your Own

Car Safety Testing

Goals
- **Construct** a fast car.
- **Design** a safe car that will protect a plastic egg from the effects of inertia when the car crashes.

Possible Materials
insulated foam meat trays or fast food trays
insulated foam cups
straws, narrow and wide
straight pins
tape
plastic eggs

Safety Precautions

WARNING: *Protect your eyes from possible flying objects.*

● Real-World Question

Imagine that you are a car designer. How can you create an attractive, fast car that is safe? When a car crashes, the passengers have inertia that can keep them moving. How can you protect the passengers from stops caused by sudden, head-on impacts?

● Form a Hypothesis

Develop a hypothesis about how to design a car to deliver a plastic egg quickly and safely through a race course and a crash at the end.

● Test Your Hypothesis

Make a Plan

1. Be sure your group has agreed on the hypothesis statement.
2. **Sketch** the design for your car. List the materials you will need. Remember that to make the car move smoothly, narrow straws will have to fit into the wider straws.

300 CHAPTER 10 Motion and Momentum

Using Scientific Methods

3. As a group, make a detailed list of the steps you will take to test your hypothesis.
4. Gather the materials you will need to carry out your experiment.

Follow Your Plan

1. Make sure your teacher approves your plan before you start. Include any changes suggested by your teacher in your plans.
2. Carry out the experiment as planned.
3. **Record** any observations that you made while doing your experiment. Include suggestions for improving your design.

◯ Analyze Your Data

1. **Compare** your car design to the designs of the other groups. What made the fastest car fast? What slowed the slowest car?
2. **Compare** your car's safety features to those of the other cars. What protected the eggs the best? How could you improve the unsuccessful designs?
3. **Predict** What effect would decreasing the speed of your car have on the safety of the egg?

◯ Conclude and Apply

1. **Summarize** How did the best designs protect the egg?
2. **Apply** If you were designing cars, what could you do to better protect passengers from sudden stops?

Communicating Your Data

Write a descriptive paragraph about ways a car could be designed to protect its passengers effectively. Include a sketch of your ideas.

Oops! Accidents in SCIENCE

SOMETIMES GREAT DISCOVERIES HAPPEN BY ACCIDENT!

What Goes Around Comes Around
The Story of Boomerangs

Imagine a group gathered on a flat, yellow plain on the Australian Outback. One youth steps forward and, with the flick of an arm, sends a long, flat, angled stick soaring and spinning into the sky. The stick's path curves until it returns right back into the thrower's hand. Thrower after thrower steps forward, and the contest goes on all afternoon.

This contest involved throwing boomerangs—elegantly curved sticks. Because of how boomerangs are shaped, they always return to the thrower's hand

This amazing design is over 15,000 years old. Scientists believe that boomerangs developed from simple clubs thrown to stun and kill animals for food. Differently shaped clubs flew in different ways. As the shape of the club was refined, people probably started throwing them for fun too. In fact, today, using boomerangs for fun is still a popular sport, as world-class throwers compete in contests of strength and skill.

Boomerangs come in several forms, but all of them have several things in common. First a boomerang is shaped like an airplane's wing: flat on one side and curved on the other. Second, boomerangs are angled, which makes them spin as they fly. These two features determine the aerodynamics that give the boomerang its unique flight path.

From its beginning as a hunting tool to its use in today's World Boomerang Championships, the boomerang has remained a source of fascination for thousands of years.

Design Boomerangs are made from various materials. Research to find instructions for making boomerangs. After you and your friends build some boomerangs, have a competition of your own.

For more information, visit ips.msscience.com/oops

Chapter 10 Study Guide

Reviewing Main Ideas

Section 1 — What is motion?

1. The position of an object depends on the reference point that is chosen.
2. An object is in motion if the position of the object is changing.
3. The speed of an object equals the distance traveled divided by the time:
$$s = \frac{d}{t}$$
4. The velocity of an object includes the speed and the direction of motion.
5. The motion of an object can be represented on a speed-time graph.

Section 2 — Acceleration

1. Acceleration is a measure of how quickly velocity changes. It includes a direction.
2. An object is accelerating when it speeds up, slows down, or turns.
3. When an object moves in a straight line, its acceleration can be calculated by
$$a = \frac{(s_f - s_i)}{t}$$

Section 3 — Momentum

1. Momentum equals the mass of an object times its velocity:
$$p = mv$$
2. Momentum is transferred from one object to another in a collision.
3. According to the law of conservation of momentum, the total amount of momentum of a group of objects doesn't change unless outside forces act on the objects.

Visualizing Main Ideas

Copy and complete the following table on motion.

Describing Motion

Quantity	Definition	Direction
Distance	length of path traveled	no
Displacement	direction and change in position	
Speed		no
Velocity	rate of change in position and direction	
Acceleration		
Momentum		yes

chapter 10 Review

Using Vocabulary

acceleration p. 288
average speed p. 285
inertia p. 293
instantaneous speed p. 285
law of conservation
    of momentum p. 295
mass p. 293
momentum p. 294
speed p. 284
velocity p. 287

Explain the relationship between each pair of terms.

1. speed—velocity
2. velocity—acceleration
3. velocity—momentum
4. momentum—law of conservation of momentum
5. mass—momentum
6. mass—inertia
7. momentum—inertia
8. average speed—instantaneous speed

Checking Concepts

Choose the word or phrase that best answers the question.

9. What measures the quantity of matter?
   A) speed
   B) weight
   C) acceleration
   D) mass

10. Which of the following objects is NOT accelerating?
    A) a jogger moving at a constant speed
    B) a car that is slowing down
    C) Earth orbiting the Sun
    D) a car that is speeding up

11. Which of the following equals speed?
    A) acceleration/time
    B) (change in velocity)/time
    C) distance/time
    D) displacement/time

12. Which of these is an acceleration?
    A) 5 m east
    B) 15 m/s east
    C) 52 m/s² east
    D) 32 s² east

13. Resistance to a change in motion increases when which of these increases?
    A) velocity
    B) speed
    C) instantaneous speed
    D) mass

14. What is 18 cm/h north an example of?
    A) speed
    B) velocity
    C) acceleration
    D) momentum

15. Which is true when the velocity and the acceleration of an object are in the same direction?
    A) The object's speed is constant.
    B) The object changes direction.
    C) The object speeds up.
    D) The object slows down.

16. Which of the following equals the change in velocity divided by the time?
    A) speed
    B) displacement
    C) momentum
    D) acceleration

17. You travel to a city 200 km away in 2.5 hours. What is your average speed in km/h?
    A) 180 km/h
    B) 12.5 km/h
    C) 80 km/h
    D) 500 km/h

18. A cue ball hits another billiard ball and slows down. Why does the speed of the cue ball decrease?
    A) The cue ball's momentum is positive.
    B) The cue ball's momentum is negative.
    C) Momentum is transferred to the cue ball.
    D) Momentum is transferred from the cue ball.

chapter 10 Review

Thinking Critically

19. **Explain** You run 100 m in 25 s. If you later run the same distance in less time, explain if your average speed increase or decrease.

Use the graph below to answer questions 20 and 21.

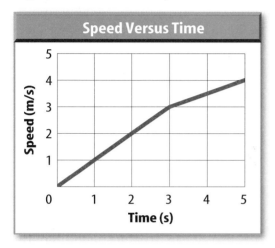

20. **Compare** For the motion of the object plotted on the speed-time graph above, how does the acceleration between 0 s and 3 s compare to the acceleration between 3 s and 5 s?

21. **Calculate** the acceleration of the object over the time interval from 0 s to 3 s.

22. **Infer** The molecules in a gas are often modeled as small balls. If the molecules all have the same mass, infer what happens if two molecules traveling at the same speed collide head on.

23. **Calculate** What is your displacement if you walk 100 m north, 20 m east, 30 m south, 50 m west, and then 70 m south?

24. **Infer** You are standing on ice skates and throw a basketball forward. Infer how your velocity after you throw the basketball compares with the velocity of the basketball.

25. **Determine** You throw a ball upward and then it falls back down. How does the velocity of the ball change as it rises and falls?

Use the graph below to answer question 26.

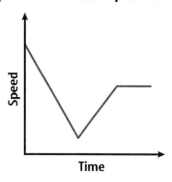

26. **Make and Use Graphs** The motion of a car is plotted on the speed-time graph above. Over which section of the graph is the acceleration of the car zero?

Performance Activities

27. **Demonstrate** Design a racetrack and make rules that specify the types of motion allowed. Demonstrate how to measure distance, measure time, and calculate speed accurately.

Applying Math

28. **Velocity of a Ball** Calculate the velocity of a 2-kg ball that has a momentum of 10 kg · m/s west.

29. **Distance Traveled** A car travels for a half hour at a speed of 40 km/h. How far does the car travel?

Use the graph below to answer question 30.

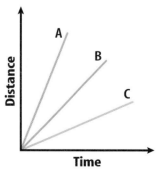

30. **Speed** From the graph determine which object is moving the fastest and which is moving the slowest.

Chapter 10 Standardized Test Practice

Part 1 Multiple Choice

Record your answers on the answer sheet provided by your teacher or on a sheet of paper.

1. What is the distance traveled divided by the time taken to travel that distance?
   A. acceleration   C. speed
   B. velocity   D. inertia

2. Sound travels at a speed of 330 m/s. How long does it take for the sound of thunder to travel 1,485 m?
   A. 45 s   C. 4,900 s
   B. 4.5 s   D. 0.22 s

Use the figure below to answer questions 3 and 4.

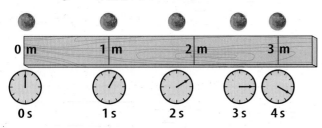

3. During which time period is the ball's average speed the fastest?
   A. between 0 and 1 s
   B. between 1 and 2 s
   C. between 2 and 3 s
   D. between 3 and 4 s

4. What is the average speed of the ball?
   A. 0.75 m/s   C. 10 m/s
   B. 1 m/s   D. 1.3 m/s

5. A car accelerates from 15 m/s to 30 m/s in 3.0 s. What is the car's acceleration?
   A. 10 m/s^2 C. 15 m/s^2
   B. 25 m/s^2 D. 5.0 m/s^2

6. Which of the following can occur when an object is accelerating?
   A. It speeds up.   C. It changes direction.
   B. It slows down.   D. all of the above

7. What is the momentum of a 21-kg bicycle traveling west at 3.0 m/s?
   A. 7 kg · m/s west   C. 18 kg · m/s west
   B. 63 kg · m/s west D. 24 kg · m/s west

Use the figure below to answer questions 8–10.

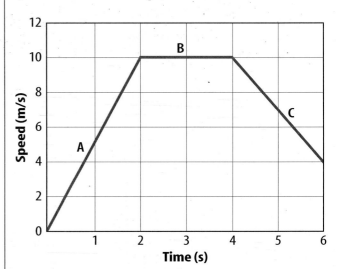

8. What is the acceleration between 0 and 2 s?
   A. 10 m/s^2 C. 0 m/s^2
   B. 5 m/s^2 D. −5 m/s^2

9. During what time period does the object have a constant speed?
   A. between 1 and 2 s
   B. between 2 and 3 s
   C. between 4 and 5 s
   D. between 5 and 6 s

10. What is the acceleration between 4 and 6 s?
    A. 10 m/s^2 C. 6 m/s^2
    B. 4 m/s^2 D. −3 m/s^2

11. An acorn falls from the top of an oak and accelerates at 9.8 m/s^2. It hits the ground in 1.5 s. What is the speed of the acorn when it hits the ground?
    A. 9.8 m/s   C. 15 m/s
    B. 20 m/s   D. 30 m/s

Standardized Test Practice

Part 2 — Short Response/Grid In

Record your answers on the answer sheet provided by your teacher or on a sheet of paper.

12. Do two objects that have the same mass always have the same momentum? Why or why not?

13. What is the momentum of a 57 kg cheetah running north at 27 m/s?

14. A sports car and a moving van are traveling at a speed of 30 km/h. Which vehicle will be easier to stop? Why?

Use the figure below to answer questions 15 and 16.

15. What happens to the momentum of the bowling ball when it hits the pins?

16. What happens to the speed of the ball and the speed of the pins?

17. What is the speed of a race horse that runs 1,500 m in 125 s?

18. A car travels for 5.5 h at an average speed of 75 km/h. How far did it travel?

19. If the speedometer on a car indicates a constant speed, can you be certain the car is not accelerating? Explain.

20. A girl walks 2 km north, then 2 km east, then 2 km south, then 2 km west. What distance does she travel? What is her displacement?

Part 3 — Open Ended

Record your answers on a sheet of paper.

Use the figure below to answer questions 21 and 22.

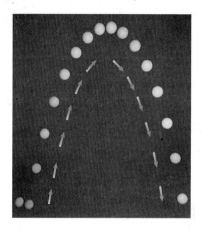

21. Describe the motion of the ball in terms of its speed, velocity, and acceleration.

22. During which part of its path does the ball have positive acceleration? During which part of its path does it have negative acceleration? Explain.

23. Describe what will happen when a baseball moving to the left strikes a bowling ball that is at rest.

24. A girl leaves school at 3:00 and starts walking home. Her house is 2 km from school. She gets home at 3:30. What was her average speed? Do you know her instantaneous speed at 3:15? Why or why not?

25. Why is it dangerous to try to cross a railroad track when a very slow-moving train is approaching?

Test-Taking Tip

Look for Missing Information Questions sometimes will ask about missing information. Notice what is missing as well as what is given.

chapter 11

The BIG Idea
An object's motion changes if the forces acting on the object are unbalanced.

SECTION 1
Newton's First Law
Main Idea If the net force on an object is zero, the motion of the object does not change.

SECTION 2
Newton's Second Law
Main Idea An object's acceleration equals the net force divided by the mass.

SECTION 3
Newton's Third Law
Main Idea Forces act in equal but opposite pairs.

Force and Newton's Laws

Moving at a Crawl
This enormous vehicle is a crawler that moves a space shuttle to the launch pad. The crawler and space shuttle together have a mass of about 7,700,000 kg. To move the crawler at a speed of about 1.5 km/h requires a force of about 10,000,000 N. This force is exerted by 16 electric motors in the crawler.

Science Journal Describe three examples of pushing or pulling an object. How did the object move?

Start-Up Activities

Forces and Motion

Imagine being on a bobsled team speeding down an icy run. Forces are exerted on the sled by the ice, the sled's brakes and steering mechanism, and gravity. Newton's laws predict how these forces cause the bobsled to turn, speed up, or slow down. Newton's Laws tell how forces cause the motion of any object to change.

1. Lean two metersticks parallel, less than a marble width apart on three books as shown on the left. This is your ramp.
2. Tap a marble so it rolls up the ramp. Measure how far up the ramp it travels before rolling back.
3. Repeat step 2 using two books, one book, and zero books. The same person should tap with the same force each time.
4. **Think Critically** Make a table to record the motion of the marble for each ramp height. What would happen if the ramp were perfectly smooth and level?

Newton's Laws Make the following Foldable to help you organize your thoughts about Newton's laws.

STEP 1 **Fold** a sheet of paper in half lengthwise. Make the back edge about 5 cm longer than the front edge.

STEP 2 **Turn** the paper so the fold is on the bottom. Then **fold** it into thirds.

STEP 3 **Unfold and cut** only the top layer along both folds to make three tabs.

STEP 4 **Label** the foldable as shown.

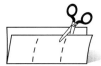

Make a Concept Map As you read the chapter, record what you learn about each of Newton's laws in your concept map.

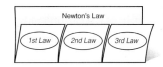

Preview this chapter's content and activities at
ips.msscience.com

Get Ready to Read

Compare and Contrast

① Learn It! Good readers compare and contrast information as they read. This means they look for similarities and differences to help them to remember important ideas. Look for signal words in the text to let you know when the author is comparing or contrasting.

Compare and Contrast Signal Words	
Compare	**Contrast**
as	but
like	or
likewise	unlike
similarly	however
at the same time	although
in a similar way	on the other hand

② Practice It! Read the excerpt below and notice how the author uses contrast signal words to describe the differences between weight and mass.

> When you stand on a bathroom scale, you are measuring the pull of the Earth's gravity — a force. **However,** mass is the amount of matter in an object, and doesn't depend on location. Weight will vary with location, **but** mass will remain constant.
>
> —*from page 318*

③ Apply It! Compare and contrast sliding friction on page 314 and air resistance on page 321.

Target Your Reading

Reading Tip

As you read, use other skills, such as summarizing and connecting, to help you understand comparisons and contrasts.

Use this to focus on the main ideas as you read the chapter.

1 Before you read the chapter, respond to the statements below on your worksheet or on a numbered sheet of paper.
- Write an **A** if you **agree** with the statement.
- Write a **D** if you **disagree** with the statement.

2 After you read the chapter, look back to this page to see if you've changed your mind about any of the statements.
- If any of your answers changed, explain why.
- Change any false statements into true statements.
- Use your revised statements as a study guide.

Before You Read A or D		Statement	After You Read A or D
	1	If an object is moving, unbalanced forces are acting on the object.	
	2	When you jump up into the air, the ground exerts a force on you.	
	3	A force is a push or a pull.	
	4	Gravity does not pull on astronauts while in orbit around Earth.	
	5	Objects must be touching each other to apply forces on one another.	
	6	An object traveling in a circle at a constant speed is not accelerating.	
	7	Action and reaction force pairs cancel each other because they are equal in size but opposite in direction.	
	8	Gravity pulls on all objects that have mass.	
	9	An object at rest can have forces acting on it.	

Science Online
Print out a worksheet of this page at ips.msscience.com

310 B

section 1
Newton's First Law

as you read

What You'll Learn
- **Distinguish** between balanced and net forces.
- **Describe** Newton's first law of motion.
- **Explain** how friction affects motion.

Why It's Important
Forces can cause the motion of objects to change.

Review Vocabulary
velocity: the speed and direction of a moving object

New Vocabulary
- force
- net force
- balanced forces
- unbalanced forces
- Newton's first law of motion
- friction

Force

A soccer ball sits on the ground, motionless, until you kick it. Your science book sits on the table until you pick it up. If you hold your book above the ground, then let it go, gravity pulls it to the floor. In every one of these cases, the motion of the ball or book was changed by something pushing or pulling on it. An object will speed up, slow down, or turn only if something is pushing or pulling on it.

A **force** is a push or a pull. Examples of forces are shown in **Figure 1.** Think about throwing a ball. Your hand exerts a force on the ball, and the ball accelerates forward until it leaves your hand. After the ball leaves your hand, the force of gravity causes its path to curve downward. When the ball hits the ground, the ground exerts a force, stopping the ball.

A force can be exerted in different ways. For instance, a paper clip can be moved by the force a magnet exerts, the pull of Earth's gravity, or the force you exert when you pick it up. These are all examples of forces acting on the paper clip.

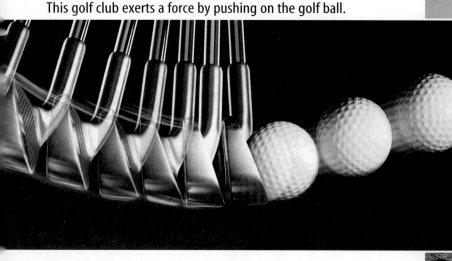

The magnet on the crane pulls the pieces of scrap metal upward.

Figure 1 A force is a push or a pull.

This golf club exerts a force by pushing on the golf ball.

310 CHAPTER 11 Force and Newton's Laws

This door is not moving because the forces exerted on it are equal and in opposite directions.

The door is closing because the force pushing the door closed is greater than the force pushing it open.

Figure 2 When the forces on an object are balanced, no change in motion occurs. A change in motion occurs only when the forces acting on an object are unbalanced.

Combining Forces More than one force can act on an object at the same time. If you hold a paper clip near a magnet, you, the magnet, and gravity all exert forces on the paper clip. The combination of all the forces acting on an object is the **net force.** When more than one force acts on an object, the net force determines how the motion of an object changes. If the motion of an object changes, its velocity changes. A change in velocity means the object is accelerating.

How do forces combine to form the net force? If the forces are in the same direction, they add together to form the net force. If two forces are in opposite directions, then the net force is the difference between the two forces, and it is in the direction of the larger force.

Balanced and Unbalanced Forces A force can act on an object without causing it to accelerate if other forces cancel the push or pull of the force. Look at **Figure 2.** If you and your friend push on a door with the same force in opposite directions, the door does not move. Because you both exert forces of the same size in opposite directions on the door, the two forces cancel each other. Two or more forces exerted on an object are **balanced forces** if their effects cancel each other and they do not change the object's velocity. If the forces on an object are balanced, the net force is zero. If the net force is not zero, the forces are **unbalanced forces.** Then the effects of the forces don't cancel, and the object's velocity changes.

Biomechanics Whether you run, jump, or sit, forces are being exerted on different parts of your body. Biomechanics is the study of how the body exerts forces and how it is affected by forces acting on it. Research how biomechanics has been used to reduce job-related injuries. Write a paragraph on what you've learned in your Science Journal.

SECTION 1 Newton's First Law

Newton's First Law of Motion

If you stand on a skateboard and someone gives you a push, then you and your skateboard will start moving. You will begin to move when the force was applied. An object at rest—like you on your skateboard—remains at rest unless an unbalanced force acts on it and causes it to move.

Because a force had to be applied to make you move when you and your skateboard were at rest, you might think that a force has to be applied continually to keep an object moving. Surprisingly, this is not the case. An object can be moving even if the net force acting on it is zero.

The Italian scientist Galileo Galilei, who lived from 1564 to 1642, was one of the first to understand that a force doesn't need to be constantly applied to an object to keep it moving.

Galileo's ideas helped Isaac Newton to better understand the nature of motion. Newton, who lived from 1642 to 1727, explained the motion of objects in three rules called Newton's laws of motion.

Newton's first law of motion describes how an object moves when the net force acting on it is zero. According to **Newton's first law of motion,** if the net force acting on an object is zero, the object remains at rest, or if the object is already moving, continues to move in a straight line with constant speed.

Figure 3 When two objects in contact try to slide past each other, friction keeps them from moving or slows them down.

Without friction, the rock climber would slide down the rock.

Friction

Galileo realized the motion of an object doesn't change until an unbalanced force acts on it. Every day you see moving objects come to a stop. The force that brings nearly everything to a stop is **friction,** which is the force that acts to resist sliding between two touching surfaces, as shown in **Figure 3.** Friction is why you never see objects moving with constant velocity unless a net force is applied. Friction is the force that eventually brings your skateboard to a stop unless you keep pushing on it. Friction also acts on objects that are sliding or moving through substances such as air or water.

Friction slows down this sliding baseball player.

Friction Opposes Sliding Although several different forms of friction exist, they all have one thing in common. If two objects are in contact, frictional forces always try to prevent one object from sliding on the other object. If you rub your hand against a tabletop, you can feel the friction push against the motion of your hand. If you rub the other way, you can feel the direction of friction change so it is again acting against your hand's motion.

 *What do the different forms of friction have in common?*

Topic: Galileo and Newton
Visit ips.msscience.com for Web links to information about the lives of Galileo and Newton.

Activity Make a time line showing important events in the lives of either Galileo or Newton.

Older Ideas About Motion It took a long time to understand motion. One reason was that people did not understand the behavior of friction and that friction was a force. Because moving objects eventually come to a stop, people thought the natural state of an object was to be at rest. For an object to be in motion, something always had to be pushing or pulling it to keep the object moving. As soon as the force stopped, the object would stop moving.

Galileo understood that an object in constant motion is as natural as an object at rest. It was usually friction that made moving objects slow down and eventually come to a stop. To keep an object moving, a force had to be applied to overcome the effects of friction. If friction could be removed, an object in motion would continue to move in a straight line with constant speed. **Figure 4** shows motion where there is almost no friction.

Figure 4 In an air hockey game, the puck floats on a layer of air, so that friction is almost eliminated. As a result, the puck moves in a straight line with nearly constant speed after it's been hit.
Infer *how the puck would move if there was no layer of air.*

SECTION 1 Newton's First Law **313**

Mini LAB

Observing Friction

Procedure
1. Lay a **bar of soap**, a **flat eraser**, and a **key** side by side on one end of a **hard-sided notebook**.
2. At a constant rate, slowly lift the end of notebook with objects on it. Note the order in which the objects start sliding.

Analysis
1. For which object was static friction the greatest? For which object was it the smallest? Explain, based on your observations.
2. Which object slid the fastest? Which slid the slowest? Explain why there is a difference in speed.
3. How could you increase and decrease the amount of friction between two materials?

Try at Home

Static Friction If you've ever tried pushing something heavy, like a refrigerator, you might have discovered that nothing happened at first. Then as you push harder and harder, the object suddenly will start to move. When you first start to push, friction between the heavy refrigerator and the floor opposes the force you are exerting and the net force is zero. The type of friction that prevents an object from moving when a force is applied is called static friction.

Static friction is caused by the attraction between the atoms on the two surfaces that are in contact. This causes the surfaces to stick or weld together where they are in contact. Usually, as the surface gets rougher and the object gets heavier, the force of static friction will be larger. To move the object, you have to exert a force large enough to break the bonds holding two surfaces together.

Sliding Friction While static friction acts on an object at rest, sliding friction slows down an object that slides. When you push a box over the floor, sliding friction acts in the direction opposite to the motion of the box. If you stop pushing, sliding friction causes the box to stop. Sliding friction is due to the microscopic roughness of two openers, as shown in **Figure 5**. The surfaces tend to stick together where they touch. The bonds between the surfaces are broken and form again as the surfaces slide past each other. This causes sliding friction. **Figure 6** shows that sliding friction is produced when the brake pad in a bicycle's brakes rub against the wheel.

Reading Check What is the difference between static friction and sliding friction?

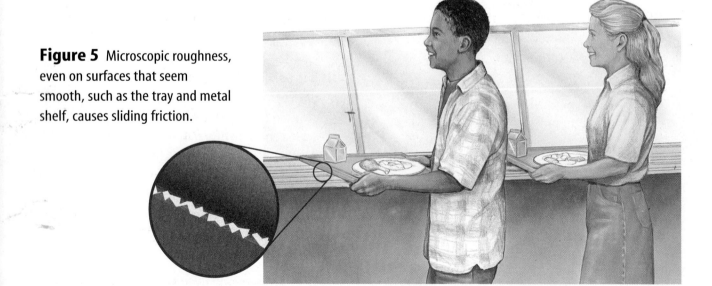

Figure 5 Microscopic roughness, even on surfaces that seem smooth, such as the tray and metal shelf, causes sliding friction.

Figure 6 Sliding friction and rolling friction can act on a bicycle.

Sliding friction between the brake pads and the wheel causes the wheel to slow down.

Rolling friction acts between the ground and the bicycle tires when the wheels are rolling.

Rolling Friction If you're coasting on a bicycle or on a skateboard, you slow down and eventually stop because of another type of frictional force. Rolling friction occurs when an object rolls across a surface. It is rolling friction between the bicycle tires and the ground, as in **Figure 6,** that slows a moving bicycle.

The size of the rolling friction force due usually is much less than the force of sliding friction between the same surfaces. This is why it takes less force to pull a box on a wagon or cart with wheels, than to drag the box along the ground. Rolling friction between the wheels and the ground is less than the sliding friction between the box and the ground.

section 1 review

Summary

Force
- A force is a push or a pull.
- The net force on an object is the combination of all the forces acting on the object.
- The forces acting on an object can be balanced or unbalanced. If the forces are balanced, the net force is zero.

Newton's First Law of Motion
- If the net force on an object at rest is zero, the object remains at rest, or if the object is moving, it continues moving in a straight line with constant speed.

Friction
- Friction is the force that acts to resist sliding between two surfaces that are touching.
- Three types of friction are static friction, sliding friction, and rolling friction.

Self Check

1. **Explain** whether a force is acting on a car that is moving at 20 km/h and turns to the left.
2. **Describe** the factors that cause static friction between two surfaces to increase.
3. **Discuss** why friction made it difficult to discover Newton's first law of motion.
4. **Discuss** whether an object can be moving if the net force acting on the object is zero.
5. **Think Critically** For the following actions, explain whether the forces involved are balanced or unbalanced.
   a. You push a box until it moves.
   b. You push a box but it doesn't move.
   c. You stop pushing a box and it slows down.

Applying Skills

6. **Compare and contrast** static, sliding, and rolling friction.

ips.msscience.com/self_check_quiz

section 2
Newton's Second Law

as you read

What You'll Learn
- **Explain** Newton's second law of motion.
- **Explain** why the direction of force is important.

Why It's Important
Newton's second law of motion explains how changes in motion can be calculated.

Review Vocabulary
acceleration: the change in velocity divided by the time over which the change occurred

New Vocabulary
- Newton's second law of motion
- weight
- center of mass

Figure 7 The force needed to change the motion of an object depends on its mass.
Predict which grocery cart would be easier to stop.

Force and Acceleration

When you go shopping in a grocery store and push a cart, you exert a force to make the cart move. If you want to slow down or change the direction of the cart, a force is required to do this, as well. Would it be easier for you to stop a full or empty grocery cart suddenly, as in **Figure 7**? When the motion of an object changes, the object is accelerating. Acceleration occurs any time an object speeds up, slows down, or changes its direction of motion.

Newton's second law of motion connects the net force on an object, its mass, and its acceleration. According to **Newton's second law of motion**, the acceleration of an object equals the net force divided by the mass and is in the direction of the net force. The acceleration can be calculated using this equation:

Newton's Second Law Equation

$$\text{acceleration (in meters/second}^2\text{)} = \frac{\text{net force (in newtons)}}{\text{mass (in kilograms)}}$$

$$a = \frac{F_{net}}{m}$$

In this equation, a is the acceleration, m is the mass, and F_{net} is the net force. If both sides of the above equation are multiplied by the mass, the equation can be written this way:

$$F_{net} = ma$$

Reading Check *What is Newton's second law?*

Units of Force Force is measured in newtons, abbreviated N. Because the SI unit for mass is the kilogram (kg) and acceleration has units of meters per second squared (m/s²), 1 N also is equal to 1 kg·m/s². In other words, to calculate a force in newtons from the equation shown on the prior page, the mass must be given in kg and the acceleration in m/s².

Gravity

One force that you are familiar with is gravity. Whether you're coasting down a hill on a bike or a skateboard or jumping into a pool, gravity is pulling you downward. Gravity also is the force that causes Earth to orbit the Sun and the Moon to orbit Earth.

What is gravity? The force of gravity exists between any two objects that have mass. Gravity always is attractive and pulls objects toward each other. A gravitational attraction exists between you and every object in the universe that has mass. However, the force of gravity depends on the mass of the objects and the distance between them. The gravitational force becomes weaker the farther apart the objects are and also decreases as the masses of the objects involved decrease.

For example, there is a gravitational force between you and the Sun and between you and Earth. The Sun is much more massive than Earth, but is so far away that the gravitational force between you and the Sun is too weak to notice. Only Earth is close enough and massive enough to exert a noticeable gravitational force on you. The force of gravity between you and Earth is about 1,650 times greater than between you and the Sun.

Newton and Gravity
Isaac Newton was the first to realize that gravity—the force that made objects fall to Earth—was also the force that caused the Moon to orbit Earth and the planets to orbit the Sun. In 1687, Newton published a book that included the law of universal gravitation. This law showed how to calculate the gravitational force between any two objects. Using the law of universal gravitation, astronomers were able to explain the motions of the planets in the solar system, as well as the motions of distant stars and galaxies.

Weight When you stand on a bathroom scale, what are you measuring? The **weight** of an object is the size of the gravitational force exerted on an object. Your weight on Earth is the gravitational force between you and Earth. On Earth, weight is calculated from this equation:

$$W = m\,(9.8 \text{ m/s}^2)$$

In this equation, W is the weight in N, and m is the mass in kg. Your weight would change if you were standing on a planet other than Earth, as shown in **Table 1.** Your weight on a different planet would be the gravitational force between you and the planet.

Table 1	Weight of 60-kg Person on Different Planets	
Place	Weight in Newtons if Your Mass were 60 kg	Percent of Your Weight on Earth
Mars	221	37.7
Earth	588	100.0
Jupiter	1,390	236.4
Pluto	35	5.9

Figure 8 The sled speeds up when the net force on the sled is in the same direction as the sled's velocity.

Figure 9 The sled slows down when the net force on the sled is in the direction opposite to the sled's velocity.

Weight and Mass Weight and mass are different. Weight is a force, just like the push of your hand is a force, and is measured in newtons. When you stand on a bathroom scale, you are measuring the pull of Earth's gravity—a force. However, mass is the amount of matter in an object, and doesn't depend on location. Weight will vary with location, but mass will remain constant. A book with a mass of 1 kg has a mass of 1 kg on Earth or on Mars. However, the weight of the book would be different on Earth and Mars. The two planets would exert a different gravitational force on the book.

Using Newton's Second Law

The second law of motion tells how to calculate the acceleration of an object if the object's mass and the forces acting on it are known. Recall that the acceleration equals the change in velocity divided by the change in time. If an object's acceleration is known, the change in its velocity can be determined

Speeding Up When does an unbalanced force cause an object to speed up? If an object is moving, a net force applied in the same direction as the object is moving causes the object to speed up. For exmple, in **Figure 8** the applied force is in the same direction as the sled's velocity. This makes the sled speed up and its velocity increase.

The net force on a ball falling to the ground is downward. This force is in the same direction that the ball is moving. Because the net force on the ball is in the same direction as the ball's velocity, the ball speeds up as it falls.

Slowing Down If the net force on an object is in the direction opposite to the object's velocity, the object slows down. In **Figure 9,** the force of sliding friction becomes larger when the boy puts his feet in the snow. The net force on the sled is the combination of gravity and sliding friction. When the sliding friction force becomes large enough, the net force is opposite to the sled's velocity. This causes the sled to slow down.

Calculating Acceleration Newton's second law of motion can be used to calculate acceleration. For example, suppose you pull a 10-kg box so that the net force on the box is 5 N. The acceleration can be found as follows:

$$a = \frac{F_{net}}{m} = \frac{5 \text{ N}}{10 \text{ kg}} = 0.5 \text{ m/s}^2$$

The box keeps accelerating as long as you keep pulling on it. The acceleration does not depend on how fast the box is moving. It depends only on the net force and the mass of the box.

Applying Math — Solve a Simple Equation

ACCELERATION OF A CAR A net force of 4,500 N acts on a car with a mass of 1,500 kg. What is the acceleration of the car?

Solution

1 *This is what you know:*
- net force: $F_{net} = 4{,}500$ N
- mass: $m = 1{,}500$ kg

2 *This is what you need to find:*
acceleration: $a = ?$ m/s^2

3 *This is the procedure you need to use:*
Substitute the known values for net force and mass into the equation for Newton's second law of motion to calculate the acceleration:

$$a = \frac{F_{net}}{m} = \frac{4{,}500 \text{ N}}{1{,}500 \text{ kg}} = 3.0 \frac{\text{N}}{\text{kg}} = 3.0 \text{ m/s}^2$$

4 *Check your answer:*
Multiply your answer by the mass, 1,500 kg. The result should be the given net force, 4,500 N.

Practice Problems

1. A book with a mass of 2.0 kg is pushed along a table. If the net force on the book is 1.0 N, what is the book's acceleration?

2. A baseball has a mass of 0.15 kg. What is the net force on the ball if its acceleration is 40.0 m/s^2?

For more practice visit ips.msscience.com/math_practice

Figure 10 The force due to gravity on the ball is at an angle to the ball's velocity. This causes the ball to move in a curved path. **Infer** *how the ball would move if it were thrown horizontally.*

Turning Sometimes the net force on an object and the object's velocity, are neither in the same direction nor in the opposite direction. Then the object will move in a curved path, instead of a straight line.

When you shoot a basketball, it doesn't continue to move in a straight line after it leaves your hand. Instead the ball starts to curve downward, as shown in **Figure 10.** Gravity pulls the ball downward, so the net force on the ball is at an angle to the ball's velocity. This causes the ball to move in a curved path.

Circular Motion

A rider on a merry-go-round ride moves in a circle. This type of motion is called circular motion. If you are in circular motion, your direction of motion is constantly changing. This means you are constantly accelerating. According to Newton's second law of motion, if you are constantly accelerating, there must be a non-zero net force acting on you the entire time.

To cause an object to move in circular motion with constant speed, the net force on the object must be at right angles to the velocity. When an object moves in circular motion, the net force on the object is called centripetal force. The direction of the centripetal force is toward the center of the object's circular path.

Satellite Motion Objects that orbit Earth are satellites of Earth. Some satellites go around Earth in nearly circular orbits. The centripetal force is due to gravity between the satellite and Earth. Gravity causes the net force on the satellite to always point toward Earth, which is center of the satellite's orbit. Why doesn't a satellite fall to Earth like a baseball does? Actually, a satellite is falling to Earth just like a baseball.

Suppose Earth were perfectly smooth and you throw a baseball horizontally. Gravity pulls the baseball downward so it travels in a curved path. If the baseball is thrown faster, its path is less curved, and it travels farther before it hits the ground. If the baseball were traveling fast enough, as it fell, its curved path would follow the curve of Earth's surface as shown in **Figure 11.** Then the baseball would never hit the ground. Instead, it would continue to fall around Earth.

Satellites in orbit are being pulled toward Earth just as baseballs are. The difference is that satellites are moving so fast horizontally that Earth's surface curves downward at the same rate that the satellites are falling downward. The speed at which a object must move to go into orbit near Earth's surface is about 8 km/s, or about 29,000 km/h.

Figure 11 The faster a ball is thrown, the farther it travels before gravity pulls it to Earth. If the ball is traveling fast enough, Earth's surface curves away from it as fast as it falls downward. Then the ball never hits the ground.

Air Resistance

Whether you are walking, running, or biking, air is pushing against you. This push is air resistance. Air resistance is a form of friction that acts to slow down any object moving in the air. Air resistance is a force that gets larger as an object moves faster. Air resistance also depends on the shape of an object. A piece of paper crumpled into a ball falls faster than a flat piece of paper falls.

When an object falls it speeds up as gravity pulls it downward. At the same time, the force of air resistance pushing up on the object is increasing as the object moves faster. Finally, the upward air resistance force becomes large enough to equal the downward force of gravity.

When the air resistance force equals the weight, the net force on the object is zero. By Newton's second law, the object's acceleration then is zero, and its speed no longer increases. When air resistance balances the force of gravity, the object falls at a constant speed called the terminal velocity.

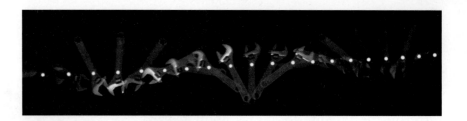

Figure 12 The wrench is spinning as it slides across the table. The center of mass of the wrench, shown by the dots, moves as if the net force is acting at that point.

Center of Mass

When you throw a stick, the motion of the stick might seem to be complicated. However, there is one point on the stick, called the center of mass, that moves in a smooth path. The **center of mass** is the point in an object that moves as if all the object's mass were concentrated at that point. For a symmetrical object, such as a ball, the center of mass is at the object's center. However, for any object the center of mass moves as if the net force is being applied there.

Figure 12 shows how the center of mass of a wrench moves as it slides across a table. The net force on the wrench is the force of friction between the wrench and the table. This causes the center of mass to move in a straight line with decreasing speed.

section 2 review

Summary

Force and Acceleration
- According to Newton's second law, the net force on an object, its mass, and its acceleration are related by
  $$F_{net} = ma$$

Gravity
- The force of gravity between any two objects is always attractive and depends on the masses of the objects and the distance between them.

Using Newton's Second Law
- A moving object speeds up if the net force is in the direction of the motion.
- A moving object slows down if the net force is in the direction opposite to the motion.
- A moving object turns if the net force is at an angle to the direction of motion.

Circular Motion
- In circular motion with constant speed, the net force is called the centripetal force and points toward the center of the circular path.

Self Check

1. **Make a diagram** showing the forces acting on a coasting bike rider traveling at 25 km/h on a flat roadway.
2. **Analyze** how your weight would change with time if you were on a space ship traveling away from Earth toward the Moon.
3. **Explain** how the force of air resistance depends on an object's speed.
4. **Infer** the direction of the net force acting on a car as it slows down and turns right.
5. **Think Critically** Three students are pushing on a box. Under what conditions will the motion of the box change?

Applying Math

6. **Calculate Net Force** A car has a mass of 1,500 kg. If the car has an acceleration of 2.0 m/s^2, what is the net force acting on the car?
7. **Calculate Mass** During a softball game, a softball is struck by a bat and has an acceleration of 1,500 m/s^2. If the net force exerted on the softball by the bat is 300 N, what is the softball's mass?

 ips.msscience.com/self_check_quiz

section 3
Newton's Third Law

Action and Reaction

Newton's first two laws of motion explain how the motion of a single object changes. If the forces acting on the object are balanced, the object will remain at rest or stay in motion with constant velocity. If the forces are unbalanced, the object will accelerate in the direction of the net force. Newton's second law tells how to calculate the acceleration, or change in motion, of an object if the net force acting on it is known.

Newton's third law describes something else that happens when one object exerts a force on another object. Suppose you push on a wall. It may surprise you to learn that if you push on a wall, the wall also pushes on you. According to **Newton's third law of motion**, forces always act in equal but opposite pairs. Another way of saying this is for every action, there is an equal but opposite reaction. This means that when you push on a wall, the wall pushes back on you with a force equal in strength to the force you exerted. When one object exerts a force on another object, the second object exerts the same size force on the first object, as shown in **Figure 13.**

as you read

What You'll Learn
- Identify the relationship between the forces that objects exert on each other.

Why It's Important
Newton's third law can explain how birds fly and rockets move.

Review Vocabulary
force: a push or a pull

New Vocabulary
- Newton's third law of motion

Figure 13 The car jack is pushing up on the car with the same amount of force with which the car is pushing down on the jack.
Identify *the other force acting on the car.*

Figure 14 In this collision, the first car exerts a force on the second. The second exerts the same force in the opposite direction on the first car.
Explain whether both cars will have the same acceleration.

Topic: How Birds Fly
Visit ips.msscience.com for Web links to information about how birds and other animals fly.

Activity Make a diagram showing the forces acting on a bird as it flies.

Figure 15 When the child pushes against the wall, the wall pushes against the child.

Action and Reaction Forces Don't Cancel The forces exerted by two objects on each other are often called an action-reaction force pair. Either force can be considered the action force or the reaction force. You might think that because action-reaction forces are equal and opposite that they cancel. However, action and reaction force pairs don't cancel because they act on different objects. Forces can cancel only if they act on the same object.

For example, imagine you're driving a bumper car and are about to bump a friend in another car, as shown in **Figure 14**. When the two cars collide, your car pushes on the other car. By Newton's third law, that car pushes on your car with the same force, but in the opposite direction. This force causes you to slow down. One force of the action-reaction force pair is exerted on your friend's car, and the other force of the force pair is exerted on your car. Another example of an action-reaction pair is shown in **Figure 15**.

You constantly use action-reaction force pairs as you move about. When you jump, you push down on the ground. The ground then pushes up on you. It is this upward force that pushes you into the air. **Figure 16** shows some examples of how Newton's laws of motion are demonstrated in sporting events.

INTEGRATE Life Science Birds and other flying creatures also use Newton's third law. When a bird flies, its wings push in a downward and a backward direction. This pushes air downward and backward. By Newton's third law, the air pushes back on the bird in the opposite directions—upward and forward. This force keeps a bird in the air and propels it forward.

324   CHAPTER 11 Force and Newton's Laws

NATIONAL GEOGRAPHIC VISUALIZING NEWTON'S LAWS IN SPORTS

Figure 16

Although it is not obvious, Newton's laws of motion are demonstrated in sports activities all the time. According to the first law, if an object is in motion, it moves in a straight line with constant speed unless a net force acts on it. If an object is at rest, it stays at rest unless a net force acts on it. The second law states that a net force acting on an object causes the object to accelerate in the direction of the force. The third law can be understood this way—for every action force, there is an equal and opposite reaction force.

▲ **NEWTON'S FIRST LAW** According to Newton's first law, the diver does not move in a straight line with constant speed because of the force of gravity.

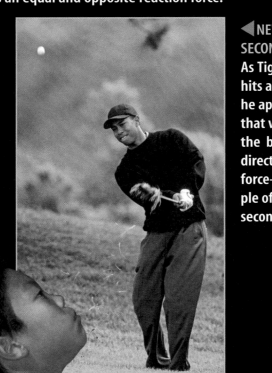

◀ **NEWTON'S SECOND LAW** As Tiger Woods hits a golf ball, he applies a force that will drive the ball in the direction of that force—an example of Newton's second law.

▶ **NEWTON'S THIRD LAW** Newton's third law applies even when objects do not move. Here a gymnast pushes downward on the bars. The bars push back on the gymnast with an equal force.

SECTION 3 Newton's Third Law

Figure 17 The force of the ground on your foot is equal and opposite to the force of your foot on the ground. If you push back harder, the ground pushes forward harder.
Determine *In what direction does the ground push on you if you are standing still?*

Change in Motion Depends on Mass Often you might not notice the effects of the forces in an action-reaction pair. This can hapen if one of the objects involved is much more massive than the other. Then the massice object might seem to remain motionless when one of the action-reaction forces acts on it. For example, when you walk forward, as in **Figure 17,** you push backward on the ground. Earth then pushes you forward with the same size force. Because Earth's mass is so large, the force you exert causes Earth to have an extremely small acceleration. This acceleration is so small that Earth's change in motion is undetectable as you walk.

A Rocket Launch The launching of a space shuttle is a spectacular example of Newton's third law. Three rocket engines supply the force, called thrust, that lifts the rocket. When the rocket fuel is ignited, a hot gas is produced. The gas molecules collide with the inside of the engine, as shown in **Figure 18.** As these collisions occur, the engine pushes the hot gases downward. According to Newton's third law of motion, the hot gases push upward on the engine. The upward force exerted by the gases on the rocket propels the rocket upward.

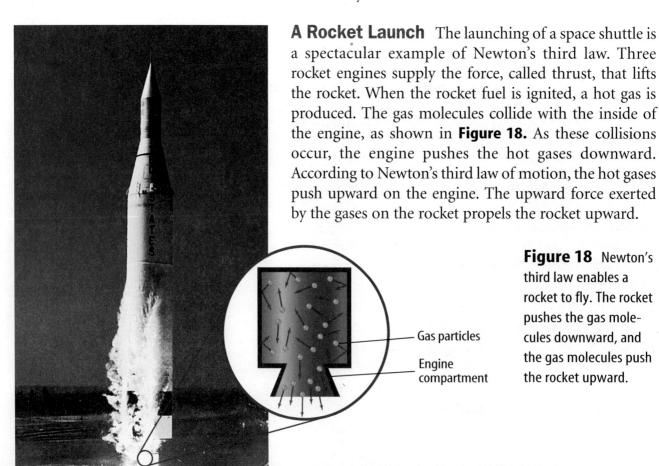

Figure 18 Newton's third law enables a rocket to fly. The rocket pushes the gas molecules downward, and the gas molecules push the rocket upward.

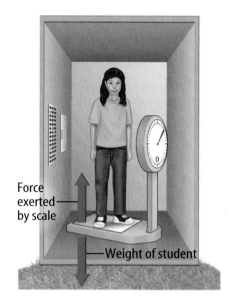
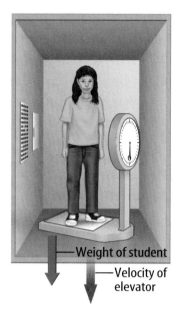

Figure 19 Whether you are standing on Earth or falling, the force of Earth's gravity on you doesn't change. However, your weight measured by a scale would change.

Weightlessness

You might have seen pictures of astronauts floating inside a space shuttle as it orbits Earth. The astronauts are said to be weightless, as if Earth's gravity were no longer pulling on them. Yet the force of gravity on the shuttle is almost 90 percent as large as at Earth's surface. Newton's laws of motion can explain why the astronauts float as if there were no forces acting on them.

Measuring Weight Think about how you measure your weight. When you stand on a scale, your weight pushes down on the scale. This causes the scale pointer to point to your weight. At the same time, by Newton's third law the scale pushes up on you with a force equal to your weight, as shown in **Figure 19.** This force balances the downward pull of gravity on you.

Free Fall and Weightlessness Now suppose you were standing on a scale in an elevator that is falling, as shown in **Figure 19.** A falling object is in free fall when the only force acting on the object is gravity. Inside the free-falling elevator, you and the scale are both in free fall. Because the only force acting on you is gravity, the scale no longer is pushing up on you. According to Newton's third law, you no longer push down on the scale. So the scale pointer stays at zero and you seem to be weightless. Weightlessness is the condition that occurs in free fall when the weight of an object seems to be zero.

However, you are not really weightless in free fall because Earth is still pulling down on you. With nothing to push up on you, such as your chair, you would have no sensation of weight.

Measuring Force Pairs

Procedure
1. Work in pairs. Each person needs a **spring scale.**
2. Hook the two scales together. Each person should pull back on a scale. Record the two readings. Pull harder and record the two readings.
3. Continue to pull on both scales, but let the scales move toward one person. Do the readings change?
4. Try to pull in such a way that the two scales have different readings.

Analysis
1. What can you conclude about the pair of forces in each situation?
2. Explain how this experiment demonstrates Newton's third law.

SECTION 3 Newton's Third Law **327**

Figure 20 These oranges seem to be floating because they are falling around Earth at the same speed as the space shuttle and the astronauts. As a result, they aren't moving relative to the astronauts in the cabin.

Weightlessness in Orbit To understand how objects move in the orbiting space shuttle, imagine you were holding a ball in the free-falling elevator. If you let the ball go, the position of the ball relative to you and the elevator wouldn't change, because you, the ball, and the elevator are moving at the same speed.

However, suppose you give the ball a gentle push downward. While you are pushing the ball, this downward force adds to the downward force of gravity. According to Newton's second law, the acceleration of the ball increases. So while you are pushing, the acceleration of the ball is greater than the acceleration of both you and the elevator. This causes the ball to speed up relative to you and the elevator. After it speeds up, it continues moving faster than you and the elevator, and it drifts downward until it hits the elevator floor.

When the space shuttle orbits Earth, the shuttle and all the objects in it are in free fall. They are falling in a curved path around Earth, instead of falling straight downward. As a result, objects in the shuttle appear to be weightless, as shown in **Figure 20**. A small push causes an object to drift away, just as a small downward push on the ball in the free-falling elevator caused it to drift to the floor.

section 3 review

Summary

Action and Reaction

- According to Newton's third law, when one object exerts a force on another object, the second object exerts the same size force on the first object.
- Either force in an action-reaction force pair can be the action force or the reaction force.
- Action and reaction force pairs don't cancel because they are exerted on different objects.
- When action and reaction forces are exerted by two objects, the accelerations of the objects depend on the masses of the objects.

Weightlessness

- A falling object is in free fall if the only force acting on it is gravity.
- Weightlessness occurs in free fall when the weight of an object seems to be zero.
- Objects orbiting Earth appear to be weightless because they are in free fall in a curved path around Earth.

Self Check

1. **Evaluate** the force a skateboard exerts on you if your mass is 60 kg and you push on the skateboard with a force of 60 N.
2. **Explain** why you move forward and a boat moves backward when you jump from a boat to a pier.
3. **Describe** the action and reaction forces when a hammer hits a nail.
4. **Infer** You and a child are on skates and you give each other a push. If the mass of the child is half your mass, who has the greater acceleration? By what factor?
5. **Think Critically** Suppose you are walking in an airliner in flight. Use Newton's third law to describe the effect of your walk on the motion on the airliner.

Applying Math

6. **Calculate Acceleration** A person standing in a canoe exerts a force of 700 N to throw an anchor over the side. Find the acceleration of the canoe if the total mass of the canoe and the person is 100 kg.

BALLOON RACES

▶ Real-World Question

The motion of a rocket lifting off a launch pad is determined by Newton's laws of motion. Here you will make a balloon rocket that is powered by escaping air. How do Newton's laws of motion explain the motion of balloon rockets?

Goals
- **Measure** the speed of a balloon rocket.
- **Describe** how Newton's laws explain a rocket's motion.

Materials
balloons
drinking straws
string
tape
meterstick
stopwatch
*clock
*Alternate materials

Safety Precautions

▶ Procedure

1. Make a rocket path by threading a string through a drinking straw. Run the string across the classroom and fasten at both ends.
2. Blow up a balloon and hold it tightly at the end to prevent air from escaping. Tape the balloon to the straw on the string.
3. Release the balloon so it moves along the string. Measure the distance the balloon travels and the time it takes.
4. Repeat steps 2 and 3 with different balloons.

▶ Conclude and Apply

1. **Compare and contrast** the distances traveled. Which rocket went the greatest distance?
2. **Calculate** the average speed for each rocket. Compare and contrast them. Which rocket has the greatest average speed?
3. **Infer** which aspects of these rockets made them travel far or fast.
4. **Draw** a diagram showing all the forces acting on a balloon rocket.
5. Use Newton's laws of motion to explain the motion of a balloon rocket from launch until it comes to a stop.

𝒞ommunicating Your Data

Discuss with classmates which balloon rocket traveled the farthest. Why? **For more help, refer to the** Science Skill Handbook.

LAB 329

LAB Design Your Own

MODELING MOTION IN TWO DIRECTIONS

Goals
- **Move** the skid across the ground using two forces.
- **Measure** how fast the skid can be moved.
- **Determine** how smoothly the direction can be changed.

Possible Materials
masking tape
stopwatch
*watch or clock with a second hand
meterstick
*metric tape measure
spring scales marked in newtons (2)
plastic lid
golf ball
*tennis ball
*Alternate materials

Safety Precautions

Real-World Question

When you move a computer mouse across a mouse pad, how does the rolling ball tell the computer cursor to move in the direction that you push the mouse? Inside the housing for the mouse's ball are two or more rollers that the ball rubs against as you move the mouse. They measure up-and-down and back-and-forth motions. The motion of the cursor on the screen is based on the movement of the up-and-down rollers and the back-and-forth rollers. Can any object be moved along a path by a series of motions in only two directions?

Form a Hypothesis

How can you combine forces to move in a straight line, along a diagonal, or around corners? Place a golf ball on something that will slide, such as a plastic lid. The plastic lid is called a skid. Lay out a course to follow on the floor. Write a plan for moving your golf ball along the path without having the golf ball roll away.

Test Your Hypothesis

Make a Plan

1. Lay out a course that involves two directions, such as always moving forward or left.
2. Attach two spring scales to the skid. One always will pull straight forward. One always will pull to one side. You cannot turn the skid. If one scale is pulling toward the door of your classroom, it always must pull in that direction. (It can pull with zero force if needed, but it can't push.)
3. How will you handle movements along diagonals and turns?
4. How will you measure speed?

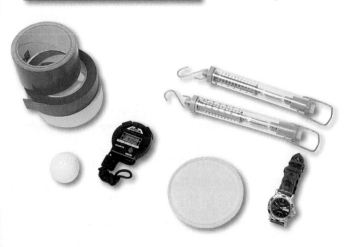

330 CHAPTER 11 Force and Newton's Laws

Using Scientific Methods

5. **Experiment** with your skid. How hard do you have to pull to counteract sliding friction at a given speed? How fast can you accelerate? Can you stop suddenly without spilling the golf ball, or do you need to slow down?

6. **Write** a plan for moving your golf ball along the course by pulling only forward or to one side. Be sure you understand your plan and have considered all the details.

Follow Your Plan

1. Make sure your teacher approves your plan before you start.
2. Move your golf ball along the path.
3. Modify your plan, if needed.
4. **Organize** your data so they can be used to run your course and write them in your Science Journal.
5. **Test** your results with a new route.

Analyze Your Data

1. What was the difference between the two routes? How did this affect the forces you needed to use on the golf ball?
2. How did you separate and control variables in this experiment?
3. Was your hypothesis supported? Explain.

Conclude and Apply

1. What happens when you combine two forces at right angles?
2. If you could pull on all four sides (front, back, left, right) of your skid, could you move anywhere along the floor? Make a hypothesis to explain your answer.

Compare your conclusions with those of other students in your class. **For more help, refer to** the Science Skill Handbook.

LAB 331

TIME SCIENCE AND Society
SCIENCE ISSUES THAT AFFECT YOU!

Air Bag Safety

After complaints and injuries, air bags in cars are helping all passengers

The car in front of yours stops suddenly. You hear the crunch of car against car and feel your seat belt grab you. Your mom is covered with, not blood, thank goodness, but with a big white cloth. Your seat belts and air bags worked perfectly.

Popcorn in the Dash

Air bags have saved more than a thousand lives since 1992. They are like having a giant popcorn kernel in the dashboard that pops and becomes many times its original size. But unlike popcorn, an air bag is triggered by impact, not temperature. In a crash, a chemical reaction produces a gas that expands in a split second, inflating a balloonlike bag to cushion the driver and possibly the front-seat passenger. The bag deflates quickly so it doesn't trap people in the car.

Newton and the Air Bag

When you're traveling in a car, you move with it at whatever speed it is going. According to Newton's first law, you are the object in motion, and you will continue in motion unless acted upon by a force, such as a car crash.

Unfortunately, a crash stops the car, but it doesn't stop you, at least, not right away. You continue moving forward if your car doesn't have air bags or if you haven't buckled your seat belt. You stop when you strike the inside of the car. You hit the dashboard or steering wheel while traveling at the speed of the car. When an air bag inflates, you come to a stop more slowly, which reduces the force that is exerted on you.

A test measures the speed at which an air bag deploys.

Measure Hold a paper plate 26 cm in front of you. Use a ruler to measure the distance. That's the distance drivers should have between the chest and the steering wheel to make air bags safe. Inform adult drivers in your family about this safety distance.

For more information, visit ips.msscience.com/time

chapter 11 Study Guide

Reviewing Main Ideas

Section 1 · Newton's First Law

1. A force is a push or a pull.
2. Newton's first law states that objects in motion tend to stay in motion and objects at rest tend to stay at rest unless acted upon by a nonzero net force.
3. Friction is a force that resists motion between surfaces that are touching each other.

Section 2 · Newton's Second Law

1. Newton's second law states that an object acted upon by a net force will accelerate in the direction of this force.
2. The acceleration due to a net force is given by the equation $a = F_{net}/m$.

3. The force of gravity between two objects depends on their masses and the distance between them.
4. In circular motion, a force pointing toward the center of the circle acts on an object.

Section 3 · Newton's Third Law

1. According to Newton's third law, the forces two objects exert on each other are always equal but in opposite directions.
2. Action and reaction forces don't cancel because they act on different objects.
3. Objects in orbit appear to be weightless because they are in free fall around Earth.

Visualizing Main Ideas

Copy and complete the following concept map on Newton's laws of motion.

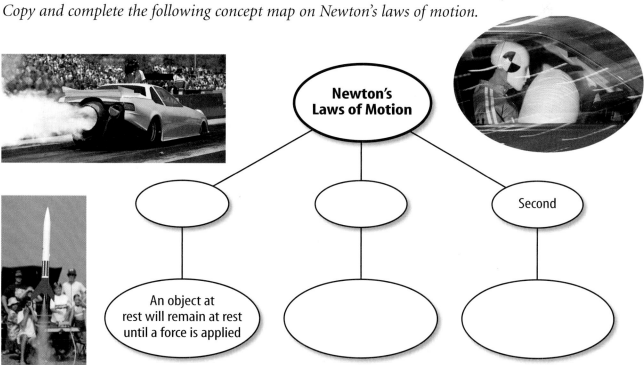

chapter 11 Review

Using Vocabulary

balanced forces p. 311
center of mass p. 322
force p. 310
friction p. 312
net force p. 311
Newton's first law of motion p. 312
Newton's second law of motion p. 316
Newton's third law of motion p. 323
unbalanced forces p. 311
weight p. 317

Explain the differences between the terms in the following sets.

1. force—inertia—weight
2. Newton's first law of motion—Newton's third law of motion
3. friction—force
4. net force—balanced forces
5. weight—weightlessness
6. balanced forces—unbalanced forces
7. friction—weight
8. Newton's first law of motion—Newton's second law of motion
9. friction—unbalanced force
10. net force—Newton's third law of motion

Checking Concepts

Choose the word or phrase that best answers the question.

11. Which of the following changes when an unbalanced force acts on an object?
    A) mass   C) inertia
    B) motion   D) weight

12. Which of the following is the force that slows a book sliding on a table?
    A) gravity
    B) static friction
    C) sliding friction
    D) inertia

Use the illustration below to answer question 13.

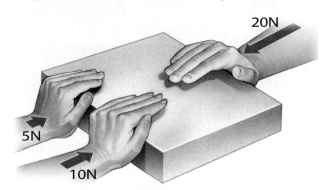

13. Two students are pushing on the left side of a box and one student is pushing on the right. The diagram above shows the forces they exert. Which way will the box move?
    A) up   C) down
    B) left   D) right

14. What combination of units is equivalent to the newton?
    A) m/s^2 C) $kg \cdot m/s^2$
    B) $kg \cdot m/s$ D) kg/m

15. Which of the following is a push or a pull?
    A) force   C) acceleration
    B) momentum   D) inertia

16. An object is accelerated by a net force in which direction?
    A) at an angle to the force
    B) in the direction of the force
    C) in the direction opposite to the force
    D) Any of these is possible.

17. You are riding on a bike. In which of the following situations are the forces acting on the bike balanced?
    A) You pedal to speed up.
    B) You turn at constant speed.
    C) You coast to slow down.
    D) You pedal at constant speed.

18. Which of the following has no direction?
    A) force   C) weight
    B) acceleration D) mass

chapter 11 Review

Thinking Critically

19. **Explain** why the speed of a sled increases as it moves down a snow-covered hill, even though no one is pushing on the sled.

20. **Explain** A baseball is pitched east at a speed of 40 km/h. The batter hits it west at a speed of 40 km/h. Did the ball accelerate?

21. **Form a Hypothesis** Frequently, the pair of forces acting between two objects are not noticed because one of the objects is Earth. Explain why the force acting on Earth isn't noticed.

22. **Identify** A car is parked on a hill. The driver starts the car, accelerates until the car is driving at constant speed, drives at constant speed, and then brakes to put the brake pads in contact with the spinning wheels. Explain how static friction, sliding friction, rolling friction, and air resistance are acting on the car.

23. **Draw Conclusions** You hit a hockey puck and it slides across the ice at nearly a constant speed. Is a force keeping it in motion? Explain.

24. **Infer** Newton's third law describes the forces between two colliding objects. Use this connection to explain the forces acting when you kick a soccer ball.

25. **Recognize Cause and Effect** Use Newton's third law to explain how a rocket accelerates upon takeoff.

26. **Predict** Two balls of the same size and shape are dropped from a helicopter. One ball has twice the mass of the other ball. On which ball will the force of air resistance be greater when terminal velocity is reached?

Use the figure below to answer question 27.

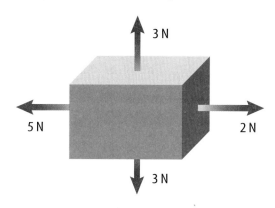

27. **Interpreting Scientific Illustrations** Is the force on the box balanced? Explain.

Performance Activities

28. **Oral Presentation** Research one of Newton's laws of motion and compose an oral presentation. Provide examples of the law. You might want to use a visual aid.

29. **Writing in Science** Create an experiment that deals with Newton's laws of motion. Document it using the following subject heads: *Title of Experiment, Partners' Names, Hypothesis, Materials, Procedures, Data, Results,* and *Conclusion.*

Applying Math

30. **Acceleration** If you exert a net force of 8 N on a 2-kg object, what will its acceleration be?

31. **Force** You push against a wall with a force of 5 N. What is the force the wall exerts on your hands?

32. **Net Force** A 0.4-kg object accelerates at 2 m/s^2. Find the net force.

33. **Friction** A 2-kg book is pushed along a table with a force of 4 N. Find the frictional force on the book if the book's acceleration is 1.5 m/s^2.

ips.msscience.com/chapter_review

Chapter 11 Standardized Test Practice

Part 1 Multiple Choice

Record your answers on the answer sheet provided by your teacher or on a sheet of paper.

1. Which of the following descriptions of gravitational force is *not* true?
   A. It depends on the mass of objects.
   B. It is a repulsive force.
   C. It depends on the distance between objects.
   D. It exists between all objects.

Use the table below to answer questions 2 and 3.

Mass of Common Objects	
Object	Mass (g)
Cup	380
Book	1,100
Can	240
Ruler	25
Stapler	620

2. Which object would have an acceleration of 0.89 m/s² if you pushed on it with a force of 0.55 N?
   A. book   C. ruler
   B. can   D. stapler

3. Which object would have the greatest acceleration if you pushed on it with a force of 8.2 N?
   A. can   C. ruler
   B. stapler D. book

Test-Taking Tip

Check Symbols Be sure you understand all symbols on a table or graph before answering any questions about the table or graph.

Question 3 The mass of the objects are given in grams, but the force is given in newtons which is a kg·m/s². The mass must be converted from grams to kilograms.

4. What is the weight of a book that has a mass of 0.35 kg?
   A. 0.036 N   C. 28 N
   B. 3.4 N   D. 34 N

5. If you swing an object on the end of a string around in a circle, the string pulls on the object to keep it moving in a circle. What is the name of this force?
   A. inertial   C. resistance
   B. centripetal   D. gravitational

6. What is the acceleration of a 1.4-kg object if the gravitational force pulls downward on it, but air resistance pushes upward on it with a force of 2.5 N?
   A. 11.6 m/s², downward
   B. 11.6 m/s², upward
   C. 8.0 m/s², downward
   D. 8.0 m/s², upward

Use the figure below to answer questions 7 and 8.

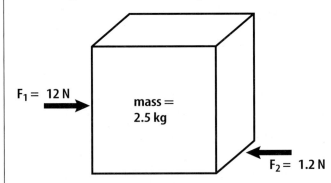

7. The figure above shows the horizontal forces that act on a box that is pushed from the left with a force of 12 N. What force is resisting the horizontal motion in this illustration?
   A. friction   C. inertia
   B. gravity   D. momentum

8. What is the acceleration of the box?
   A. 27 m/s²   C. 4.3 m/s²
   B. 4.8 m/s²   D. 0.48 m/s²

Standardized Test Practice

Part 2 — Short Response/Grid In

Record your answers on the answer sheet provided by your teacher or on a sheet of paper.

9. A skater is coasting along the ice without exerting any apparent force. Which law of motion explains the skater's ability to continue moving?

10. After a soccer ball is kicked into the air, what force or forces are acting on it?

11. What is the force on an 8.55-kg object that accelerates at 5.34 m/s²?

Use the figure below to answer questions 12 and 13.

12. Two bumper cars collide and then move away from each other. How do the forces the bumper cars exert on each other compare?

13. After the collision, determine whether both bumper cars will have the same acceleration.

14. Does acceleration depend on the speed of an object? Explain.

15. An object acted on by a force of 2.8 N has an acceleration of 3.6 m/s². What is the mass of the object?

16. What is the acceleration a 1.4-kg object falling through the air if the force of air resistance on the object is 2.5 N?

17. Name three ways you could accelerate if you were riding a bicycle.

Part 3 — Open Ended

Record your answers on a sheet of paper.

18. When astronauts orbit Earth, they float inside the spaceship because of weightlessness. Explain this effect.

19. Describe how satellites are able to remain in orbit around Earth.

Use the figure below to answer questions 20 and 21.

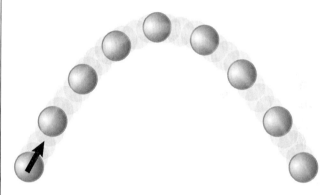

20. The figure above shows the path a ball thrown into the air follows. What causes the ball to move along a curved path?

21. What effect would throwing the ball harder have on the ball's path? Explain.

22. How does Newton's second law determine the motion of a book as you push it across a desktop?

23. A heavy box sits on a sidewalk. If you push against the box, the box moves in the direction of the force. If the box is replaced with a ball of the same mass, and you push with the same force against the ball, will it have the same acceleration as the box? Explain.

24. According to Newton's third law of motion, a rock sitting on the ground pushes against the ground, and the ground pushes back against the rock with an equal force. Explain why this force doesn't cause the rock to accelerate upward from the ground according to Newton's second law.

chapter 12

Forces and Fluids

The BIG Idea
A fluid exerts forces on an object in the fluid.

SECTION 1
Pressure
Main Idea The pressure exerted by a fluid increases as depth increases.

SECTION 2
Why do objects float?
Main Idea A fluid exerts an upward buoyant force on any object placed in the fluid.

SECTION 3
Doing Work with Fluids
Main Idea Fluids can be made to exert forces that do useful work.

A Very Fluid Situation
This swimmer seems to be defying gravity as he skims over the water. Even though the swimmer floats, a rock with the same mass would sink to the bottom of the pool. What forces cause something to float? Why does a swimmer float, while a rock sinks?

Science Journal Compare and contrast five objects that float with five objects that sink.

Start-Up Activities

Forces Exerted by Air

When you are lying down, something is pushing down on you with a force equal to the weight of several small cars. What is the substance that is applying all this pressure on you? It's air, a fluid that exerts forces on everything it is in contact with.

1. Suck water into a straw. Try to keep the straw completely filled with water.
2. Quickly cap the top of the straw with your finger and observe what happens to the water.
3. Release your finger from the top of the straw for an instant and replace it as quickly as possible. Observe what happens to the water.
4. Release your finger from the top of the straw and observe.
5. **Think Critically** Write a paragraph describing your observations of the water in the straw. When were the forces acting on the water in the straw balanced and when were they unbalanced?

Fluids Make the following Foldable to compare and contrast the characteristics of two types of fluids—liquids and gases.

STEP 1 Fold one sheet of paper lengthwise.

STEP 2 Fold into thirds.

STEP 3 Unfold and draw overlapping ovals. Cut the top sheet along the folds.

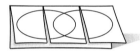

STEP 4 Label the ovals as shown.

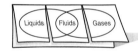

Construct a Venn Diagram As you read the chapter, list the characteristics of liquids under the left tab, those characteristics of gases under the right tab, and those characteristics common to both under the middle tab.

Preview this chapter's content and activities at
ips.msscience.com

Get Ready to Read

Make Inferences

① Learn It! When you make inferences, you draw conclusions that are not directly stated in the text. This means you "read between the lines." You interpret clues and draw upon prior knowledge. Authors rely on a reader's ability to infer because all the details are not always given.

② Practice It! Read the excerpt below and pay attention to highlighted words as you make inferences. Use this Think-Through chart to help you make inferences.

The buoyant force pushes an object in a fluid upward, but **gravity pulls the object downward**. If the **weight of the object** is greater than the buoyant force, the **net force** on the object is downward and it sinks.

— *from page 349*

Text	Question	Inferences
Gravity pulls the object downward	What is gravity?	A downward force on the object?
Weight of the object	Is weight a force?	Weight is downward pull due to gravity?
Net force	What is the net force?	Combination of all forces on an object?

③ Apply It! As you read this chapter, practice your skill at making inferences by making connections and asking questions.

Target Your Reading

Reading Tip

Sometimes you make inferences by using other reading skills, such as questioning and predicting.

Use this to focus on the main ideas as you read the chapter.

1. **Before you read** the chapter, respond to the statements below on your worksheet or on a numbered sheet of paper.
   - Write an **A** if you **agree** with the statement.
   - Write a **D** if you **disagree** with the statement.

2. **After you read** the chapter, look back to this page to see if you've changed your mind about any of the statements.
   - If any of your answers changed, explain why.
   - Change any false statements into true statements.
   - Use your revised statements as a study guide.

Science Online
Print out a worksheet of this page at ips.msscience.com

Before You Read A or D		Statement	After You Read A or D
	1	The pressure due to a force depends on the area over which a force is exerted.	
	2	Fluids are substances that can flow.	
	3	The pressure exerted by water depends on the shape of its container.	
	4	The pressure in a fluid decreases with depth.	
	5	A rock sinks in water because the rock's weight is less than the upward force exerted by the water.	
	6	A metal boat floats in water because the density of the boat is less than the density of water.	
	7	Squeezing on a closed plastic water bottle causes the pressure in the water to decrease.	
	8	The pressure exerted by fluid decreases as the speed of the fluid increases.	
	9	The lift on a wing depends on the size of the wing.	

340 B

section 1
Pressure

as you read

What You'll Learn
- **Define and calculate** pressure.
- **Model** how pressure varies in a fluid.

Why It's Important
Some of the processes that help keep you alive, such as inhaling and exhaling, depend on differences in pressure.

Review Vocabulary
weight: on Earth, the gravitational force between an object and Earth

New Vocabulary
- pressure
- fluid

What is pressure?

What happens when you walk in deep, soft snow or dry sand? Your feet sink into the snow or sand and walking can be difficult. If you rode a bicycle with narrow tires over these surfaces, the tires would sink even deeper than your feet.

How deep you sink depends on your weight as well as the area over which you make contact with the sand or snow. Like the person in **Figure 1,** when you stand on two feet, you make contact with the sand over the area covered by your feet. However, if you were to stand on a large piece of wood, your weight would be distributed over the area covered by the wood.

In both cases, your weight exerted a downward force on the sand. What changed was the area of contact between you and the sand. By changing the area of contact, you changed the pressure you exerted on the sand due to your weight. **Pressure** is the force per unit area that is applied on the surface of an object. When you stood on the board, the area of contact increased, so that the same force was applied over a larger area. As a result, the pressure that was exerted on the sand decreased and you didn't sink as deep.

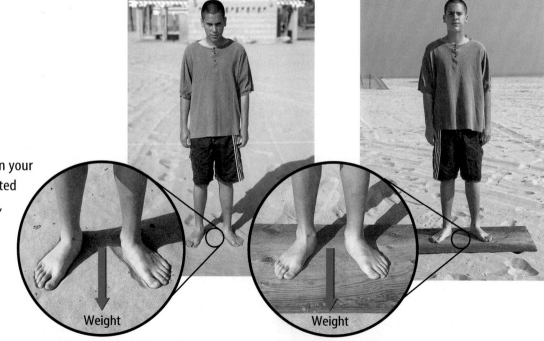

Figure 1 When your weight is distributed over a larger area, the pressure you exert on the sand decreases.

340 CHAPTER 12 Forces and Fluids

Calculating Pressure What would happen to the pressure exerted by your feet if your weight increased? You might expect that you would sink deeper in the sand, so the pressure also would increase. Pressure increases if the force applied increases, and decreases if the area of contact increases. Pressure can be calculated from this formula.

Pressure Equation

$$\text{Pressure (in pascals)} = \frac{\text{force (in newtons)}}{\text{area (in meters squared)}}$$

$$P = \frac{F}{A}$$

The unit of pressure in the SI system is the pascal, abbreviated Pa. One pascal is equal to a force of 1 N applied over an area of 1 m², or 1 Pa = 1 N/m². The weight of a dollar bill resting completely flat on a table exerts a pressure of about 1 Pa on the table. Because 1 Pa is a small unit of pressure, pressure sometimes is expressed in units of kPa, which is 1,000 Pa.

Topic: Snowshoes
Visit ips.msscience.com for Web links to information about the history and use of snowshoes. These devices have been used for centuries in cold, snowy climates.

Activity Use simple materials, such as pipe cleaners, string, or paper, to make a model of a snowshoe.

Applying Math — Solve One-Step Equations

CALCULATING PRESSURE A water glass sitting on a table weighs 4 N. The bottom of the water glass has a surface area of 0.003 m². Calculate the pressure the water glass exerts on the table.

Solution

1. *This is what you know:*
   - force: $F = 4$ N
   - area: $A = 0.003$ m²

2. *This is what you need to find out:*
   - pressure: $P = ?$ Pa

3. *This is the procedure you need to use:*
   Substitute the known values for force and area into the pressure equation and calculate the pressure:
   $$P = \frac{F}{A} = \frac{(4 \text{ N})}{(0.003 \text{ m}^2)}$$
   $$= 1{,}333 \text{ N/m}^2 = 1{,}333 \text{ Pa}$$

4. *Check your answer:*
   Multiply pressure by the given area. You should get the force that was given.

Practice Problems

1. A student weighs 600 N. The student's shoes are in contact with the floor over a surface area of 0.012 m². Calculate the pressure exerted by the student on the floor.

2. A box that weighs 250 N is at rest on the floor. If the pressure exerted by the box on the floor is 25,000 Pa, over what area is the box in contact with the floor?

 For more practice, visit ips.msscience.com/math_practice

SECTION 1 Pressure

Mini LAB

Interpreting Footprints

Procedure
1. Go outside to **an area of dirt, sand, or snow** where you can make footprints. Smooth the surface.
2. Make tracks in several different ways. Possible choices include walking forward, walking backward, running, jumping a short or long distance, walking carrying a load, and tiptoeing.

Analysis
1. Measure the depth of each type of track at two points: the ball of the foot and the heel. Compare the depths of the different tracks.
2. The depth of the track corresponds to the pressure on the ground. In your **Science Journal**, explain how different means of motion put pressure on different parts of the sole.
3. Have one person make a track while the other looks away. Then have the second person determine what the motion was.

Pressure and Weight To calculate the pressure that is exerted on a surface, you need to know the force and the area over which it is applied. Sometimes the force that is exerted is the weight of an object, such as when you are standing on sand, snow, or a floor. Suppose you are holding a 2-kg book in the palm of your hand. To find out how much pressure is being exerted on your hand, you first must know the force that the book is exerting on your hand—its weight.

$$\text{Weight} = \text{mass} \times \text{acceleration due to gravity}$$
$$W = (2 \text{ kg}) \times (9.8 \text{ m/s}^2)$$
$$W = 19.6 \text{ N}$$

If the area of contact between your hand and the book is 0.003 m², the pressure that is exerted on your hand by the book is:

$$P = \frac{F}{A}$$
$$P = \frac{(19.6 \text{ N})}{(0.003 \text{ m}^2)}$$
$$P = 6{,}533 \text{ Pa} = 6.53 \text{ kPa}$$

Pressure and Area One way to change the pressure that is exerted on an object is to change the area over which the force is applied. Imagine trying to drive a nail into a piece of wood, as shown in **Figure 2**. Why is the tip of a nail pointed instead of flat? When you hit the nail with a hammer, the force you apply is transmitted through the nail from the head to the tip. The tip of the nail comes to a point and is in contact with the wood over a small area. Because the contact area is so small, the pressure that is exerted by the nail on the wood is large—large enough to push the wood fibers apart. This allows the nail to move downward into the wood.

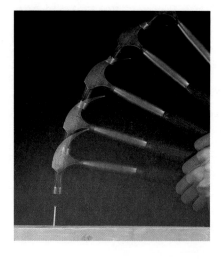
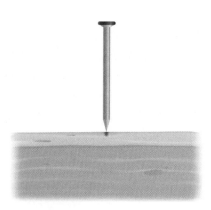

Figure 2 The force applied to the head of the nail by the hammer is the same as the force that the tip of the nail applies to the wood. However, because the area of the tip is small, the pressure applied to the wood is large.

Fluids

What do the substances in **Figure 3** have in common? Each takes the shape of its container and can flow from one place to another. A **fluid** is any substance that has no definite shape and has the ability to flow. You might think of a fluid as being a liquid, such as water or motor oil. But gases are also fluids. When you are outside on a windy day, you can feel the air flowing past you. Because air can flow and has no definite shape, air is a fluid. Gases, liquids, and the state of matter called plasma, which is found in the Sun and other stars, are fluids and can flow.

Pressure in a Fluid

Suppose you placed an empty glass on a table. The weight of the glass exerts pressure on the table. If you fill the glass with water, the weight of the water and glass together exert a force on the table. So the pressure exerted on the table increases.

Because the water has weight, the water itself also exerts pressure on the bottom of the glass. This pressure is the weight of the water divided by the area of the glass bottom. If you pour more water into the glass, the height of the water in the glass increases and the weight of the water increases. As a result, the pressure exerted by the water increases.

Figure 3 Fluids all have the ability to flow and take the shape of their containers.
Classify *What are some other examples of fluids?*

Plasma The Sun is a star with a core temperature of about 16 million°C. At this temperature, the particles in the Sun move at tremendous speeds, crashing into each other in violent collisions that tear atoms apart. As a result, the Sun is made of a type of fluid called a plasma. A plasma is a gas made of electrically charged particles.

SECTION 1 Pressure **343**

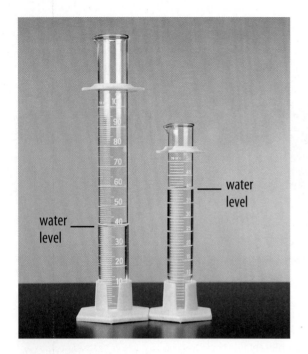

Pressure and Fluid Height Suppose you poured the same amount of water into a small and a large graduated cylinder, as shown in **Figure 4.** Notice that the height of the water in the small cylinder is greater than in the large cylinder. Is the water pressure the same at the bottom of each cylinder? The weight of the water in each cylinder is the same, but the contact area at the bottom of the small cylinder is smaller. Therefore, the pressure is greater at the bottom of the small cylinder.

The height of the water can increase if more water is added to a container or if the same amount of water is added to a narrower container. In either case, when the height of the fluid is greater, the pressure at the bottom of the container is greater. This is always true for any fluid or any container. The greater the height of a fluid above a surface, the greater the pressure exerted by the fluid on that surface. The pressure exerted at the bottom of a container doesn't depend on the shape of the container, but only on the height of the fluid above the bottom, as **Figure 5** shows.

Figure 4 Even though each graduated cylinder contains the same volume of water, the pressure exerted by the higher column of water is greater.
Infer *how the pressure exerted by a water column would change as the column becomes narrower.*

Pressure Increases with Depth If you swim underwater, you might notice that you can feel pressure in your ears. As you go deeper, you can feel this pressure increase. This pressure is exerted by the weight of the water above you. As you go deeper in a fluid, the height of the fluid above you increases. As the height of the fluid above you increases, the weight of the fluid above you also increases. As a result, the pressure exerted by the fluid increases with depth.

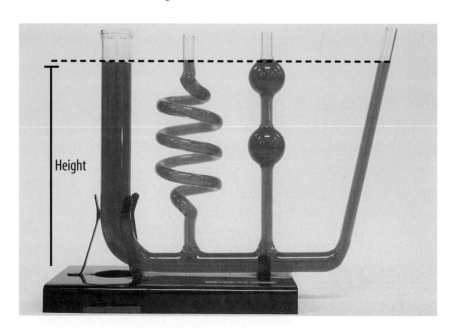

Figure 5 Pressure depends only on the height of the fluid above a surface, not on the shape of the container. The pressure at the bottom of each section of the tube is the same.

344 CHAPTER 12 Forces and Fluids

Pressure in All Directions If the pressure that is exerted by a fluid is due to the weight of the fluid, is the pressure in a fluid exerted only downward? **Figure 6** shows a small, solid cube in a fluid. The fluid exerts a pressure on each side of this cube, not just on the top. The pressure on each side is perpendicular to the surface, and the amount of pressure depends only on the depth in the fluid. As shown in **Figure 6,** this is true for any object in a fluid, no matter how complicated the shape. The pressure at any point on the object is perpendicular to the surface of the object at that point.

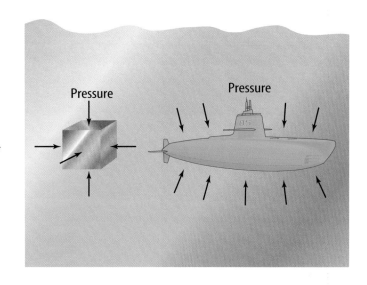

Figure 6 The pressure on all objects in a fluid is exerted on all sides, perpendicular to the surface of the object, no matter what its shape.

 *In what direction is pressure exerted by a fluid on a surface?*

Atmospheric Pressure

Even though you don't feel it, you are surrounded by a fluid that exerts pressure on you constantly. That fluid is the atmosphere. The atmosphere at Earth's surface is only about one-thousandth as dense as water. However, the thickness of the atmosphere is large enough to exert a large pressure on objects at Earth's surface. For example, look at **Figure 7.** When you are sitting down, the force pushing down on your body due to atmospheric pressure can be equal to the weight of several small cars. Atmospheric pressure is approximately 100,000 Pa at sea level. This means that the weight of Earth's atmosphere exerts about 100,000 N of force over every square meter on Earth.

Why doesn't this pressure cause you to be crushed? Your body is filled with fluids such as blood that also exert pressure. The pressure exerted outward by the fluids inside your body balances the pressure exerted by the atmosphere.

Going Higher As you go higher in the atmosphere, atmospheric pressure decreases as the amount of air above you decreases. The same is true in an ocean, lake, or pond. The water pressure is highest at the ocean floor and decreases as you go upward. The changes in pressure at varying heights in the atmosphere and depths in the ocean are illustrated in **Figure 8.**

Figure 7 Atmospheric pressure on your body is a result of the weight of the atmosphere exerting force on your body.
Infer *Why don't you feel the pressure exerted by the atmosphere?*

SECTION 1 Pressure **345**

NATIONAL GEOGRAPHIC
VISUALIZING PRESSURE AT VARYING ELEVATIONS

Figure 8

No matter where you are on Earth, you're under pressure. Air and water have weight and therefore exert pressure on your body. The amount of pressure depends on your location above or below sea level and how much air or water—or both—are exerting force on you.

▲ **HIGH ELEVATION** With increasing elevation, the amount of air above you decreases, and so does air pressure. At the 8,850-m summit of Mt. Everest, air pressure is a mere 33 kPa—about one third of the air pressure at sea level.

▲ **SEA LEVEL** Air pressure is pressure exerted by the weight of the atmosphere above you. At sea level the atmosphere exerts a pressure of about 100,000 N on every square meter of area. Called one atmosphere (atm), this pressure is also equal to 100 kPa.

▶ **REEF LEVEL** When you descend below the sea surface, pressure increases by about 1 atm every 10 meters. At 20 meters depth, you'd experience 2 atm of water pressure and 1 atm of air pressure, a total of 3 atm of pressure on your body.

▶ **VERY LOW ELEVATION** The deeper you dive, the greater the pressure. The water pressure on a submersible at a depth of 2,200 m is about 220 times greater than atmospheric pressure at sea level.

346 **CHAPTER 12** Forces and Fluids

Barometer An instrument called a barometer is used to measure atmospheric pressure. A barometer has something in common with a drinking straw. When you drink through a straw, it seems like you pull your drink up through the straw. But actually, atmospheric pressure pushes your drink up the straw. By removing air from the straw, you reduce the air pressure in the straw. Meanwhile, the atmosphere is pushing down on the surface of your drink. When you pull the air from the straw, the pressure in the straw is less than the pressure pushing down on the liquid, so atmospheric pressure pushes the drink up the straw.

One type of barometer works in a similar way, as shown in **Figure 9.** The space at the top of the tube is a vacuum. Atmospheric pressure pushes liquid up a tube. The liquid reaches a height where the pressure at the bottom of the column of liquid balances the pressure of the atmosphere. As the atmospheric pressure changes, the force pushing on the surface of the liquid changes. As a result, the height of the liquid in the tube increases as the atmospheric pressure increases.

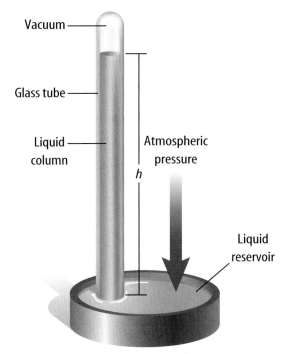

Figure 9 In this type of barometer, the height of the liquid column increases as the atmospheric pressure increases.

section 1 review

Summary

Pressure

- Pressure is the force exerted on a unit area of a surface. Pressure can be calculated from this equation:
$$P = \frac{F}{A}$$

- The SI unit for pressure is the pascal, abbreviated Pa.

Pressure in a Fluid

- The pressure exerted by a fluid depends on the depth below the fluid surface.
- The pressure exerted by a fluid on a surface is always perpendicular to the surface.

Atmospheric Pressure

- Earth's atmosphere exerts a pressure of about 100,000 Pa at sea level.
- A barometer is an instrument used to measure atmospheric pressure.

Self Check

1. **Compare** One column of water has twice the diameter as another water column. If the pressure at the bottom of each column is the same, how do the heights of the two columns compare?
2. **Explain** why the height of the liquid column in a barometer changes as atmospheric pressure changes.
3. **Classify** the following as fluids or solids: warm butter, liquid nitrogen, paper, neon gas, ice.
4. **Explain** how the pressure at the bottom of a container depends on the container shape and the fluid height.
5. **Think Critically** Explain how the diameter of a balloon changes as it rises higher in the atmosphere.

Applying Math

6. **Calculate Force** The palm of a person's hand has an area of 0.0135 m^2. If atmospheric pressure is 100,000 N/m^2, find the force exerted by the atmosphere on the person's palm.

section 2
Why do objects float?

as you read

What You'll Learn
- **Explain** how the pressure in a fluid produces a buoyant force.
- **Define** density.
- **Explain** floating and sinking using Archimedes' principle.

Why It's Important
Knowing how fluids exert forces helps you understand how boats can float.

Review Vocabulary
Newton's second law of motion: the acceleration of an object is in the direction of the total force and equals the total force divided by the object's mass

New Vocabulary
- buoyant force
- Archimedes' principle
- density

The Buoyant Force

Can you float? Think about the forces that are acting on you as you float motionless on the surface of a pool or lake. You are not moving, so according to Newton's second law of motion, the forces on you must be balanced. Earth's gravity is pulling you downward, so an upward force must be balancing your weight, as shown in **Figure 10.** This force is called the buoyant force. The **buoyant force** is an upward force that is exerted by a fluid on any object in the fluid.

What causes the buoyant force?

The buoyant force is caused by the pressure that is exerted by a fluid on an object in the fluid. **Figure 11** shows a cube-shaped object submerged in a glass of water. The water exerts pressure everywhere over the surface of the object. Recall that the pressure exerted by a fluid has two properties. One is that the direction of the pressure on a surface is always perpendicular to the surface. The other is that the pressure exerted by a fluid increases as you go deeper into the fluid.

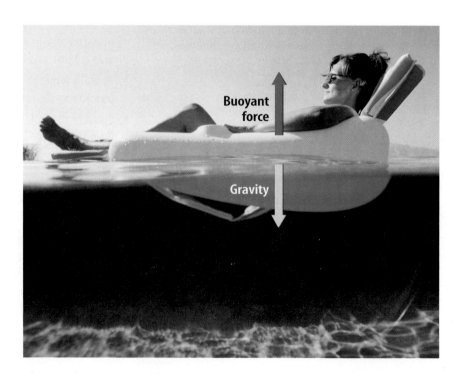

Figure 10 When you float, the forces on you are balanced. Gravity pulls you downward and is balanced by the buoyant force pushing you upward.
Infer *What is the acceleration of the person shown here?*

348 CHAPTER 12 Forces and Fluids

Buoyant Force and Unbalanced Pressure The pressure that is exerted by the water on the cube is shown in **Figure 11.** The bottom of the cube is deeper in the water. Therefore, the pressure that is exerted by the water at the bottom of the cube is greater than it is at the top of the cube. The higher pressure near the bottom means that the water exerts an upward force on the bottom of the cube that is greater than the downward force that is exerted at the top of the cube. As a result, the force that is exerted on the cube due to water pressure is not balanced, and a net upward force is acting on the cube due to the pressure of the water. This upward force is the buoyant force. A buoyant force acts on all objects that are placed in a fluid, whether they are floating or sinking.

Reading Check *When does the buoyant force act on an object?*

Sinking and Floating

If you drop a stone into a pool of water, it sinks. But if you toss a twig on the water, it floats. An upward buoyant force acts on the twig and the stone, so why does one float and one sink?

The buoyant force pushes an object in a fluid upward, but gravity pulls the object downward. If the weight of the object is greater than the buoyant force, the net force on the object is downward and it sinks. If the buoyant force is equal to the object's weight, the forces are balanced and the object floats. As shown in **Figure 12,** the fish floats because the buoyant force on it balances its weight. The rocks sink because the buoyant force acting on them is not large enough to balance their weight.

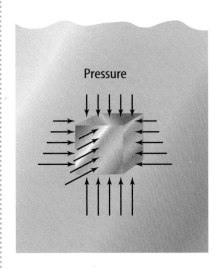

Figure 11 The pressure exerted on the bottom of the cube is greater than the pressure on the top. The fluid exerts a net upward force on the cube.

Figure 12 The weight of a rock is more than the buoyant force exerted by the water, so it sinks to the bottom.
Infer *Why do the fish float?*

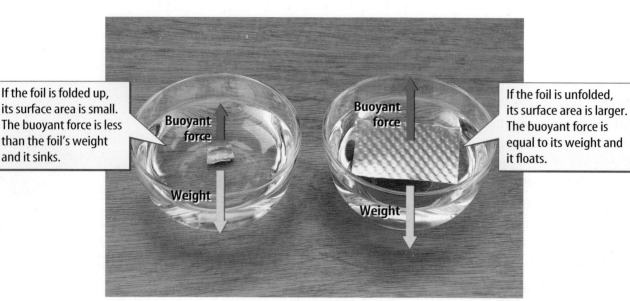

Figure 13 The buoyant force on a piece of aluminum foil increases as the surface area of the foil increases.

Figure 14 The hull of this oil tanker has a large surface area in contact with the water. As a result, the buoyant force is so large that the ship floats.

Changing the Buoyant Force

Whether an object sinks or floats depends on whether the buoyant force is smaller than its weight. The weight of an object depends only on the object's mass, which is the amount of matter the object contains. The weight does not change if the shape of the object changes. A piece of modeling clay contains the same amount of matter whether it's squeezed into a ball or pressed flat.

Buoyant Force and Shape Buoyant force does depend on the shape of the object. The fluid exerts upward pressure on the entire lower surface of the object that is in contact with the fluid. If this surface is made larger, then more upward pressure is exerted on the object and the buoyant force is greater. **Figure 13** shows how a piece of aluminum can be made to float. If the aluminum is crumpled, the buoyant force is less than the weight, so the aluminum sinks. When the aluminum is flattened into a thin, curved sheet, the buoyant force is large enough that the sheet floats. This is how large, metal ships, like the one in **Figure 14,** are able to float. The metal is formed into a curved sheet that is the hull of the ship. The contact area of the hull with the water is much greater than if the metal were a solid block. As a result, the buoyant force on the hull is greater than it would be on a metal block.

350 CHAPTER 12 Forces and Fluids

The Buoyant Force Doesn't Change with Depth

Suppose you drop a steel cube into the ocean. You might think that the cube would sink only to a depth where the buoyant force on the cube balances its weight. However, the steel sinks to the bottom, no matter how deep the ocean is.

The buoyant force on the cube is the difference between the downward force due to the water pressure on the top of the cube and the upward force due to water pressure on the bottom of the cube. **Figure 15** shows that when the cube is deeper, the pressure on the top surface increases, but the pressure on the bottom surface also increases by the same amount. As a result, the difference between the forces on the top and bottom surfaces is the same, no matter how deep the cube is submerged. The buoyant force on the submerged cube is the same at any depth.

Archimedes' Principle

A way of determining the buoyant force was given by the ancient Greek mathematician Archimedes (ar kuh MEE deez) more than 2,200 years ago. According to **Archimedes' principle,** the buoyant force on an object is equal to the weight of the fluid it displaces.

To understand Archimedes' principle, think about what happens if you drop an ice cube in a glass of water that's filled to the top. The ice cube takes the place of some of the water and causes this water to overflow, as shown in **Figure 16.** Another way to say this is that the ice cube displaced water that was in the glass.

Suppose you caught all the overflow water and weighed it. According to Archimedes' principle, the weight of the overflow, or displaced water, would be equal to the buoyant force on the ice cube. Because the ice cube is floating, the buoyant force is balanced by the weight of the ice cube. So the weight of the water that is displaced, or the buoyant force, is equal to the weight of the ice cube.

Figure 15 Because the cube on the right is deeper, the pressure on its upper surface is increased due to the weight of the water inside the dashed lines. The pressure on the bottom surface also increases by this amount.
Explain *how the buoyant force on the cube would change if it moved only to the left or right.*

A
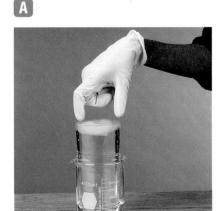

B
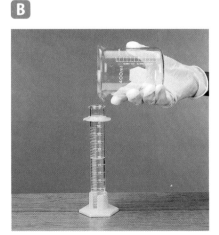

Figure 16 The buoyant force exerted on this ice cube is equal to the weight of the water displaced by the ice cube.

Naval Architect Naval architects design the ships and submarines for the U.S. Naval Fleet, Coast Guard, and Military Sealift Command. Naval architects need math, science, and English skills for designing and communicating design ideas to others.

Density Archimedes' principle leads to a way of determining whether an object placed in a fluid will float or sink. The answer depends on comparing the density of the fluid and the density of the object. The **density** of a fluid or an object is the mass of the object divided by the volume it occupies. Density can be calculated by the following formula:

Density Equation

$$\text{density (in g/cm}^3\text{)} = \frac{\text{mass (in g)}}{\text{volume (in cm}^3\text{)}}$$

$$D = \frac{m}{V}$$

For example, water has a density of 1.0 g/cm^3. The mass of any volume of a substance can be calculated by multiplying both sides of the above equation by volume. This gives the equation

$$\text{mass} = \text{density} \times \text{volume}$$

Then if the density and volume are known, the mass of the material can be calculated.

Applying Science

Layering Liquids

The density of an object or substance determines whether it will sink or float in a fluid. Just like solid objects, liquids also have different densities. If you pour vegetable oil into water, the oil doesn't mix. Instead, because the density of oil is less than the density of water, the oil floats on top of the water.

Identifying the Problem

In science class, a student is presented with five unknown liquids and their densities. He measures the volume of each and organizes his data into the table at the right. He decides to experiment with these liquids by carefully pouring them, one at a time, into a graduated cylinder.

Liquid	Color	Density (g/cm³)	Volume (cm³)
A	red	2.40	32.0
B	blue	2.90	15.0
C	green	1.20	20.0
D	yellow	0.36	40.0
E	purple	0.78	19.0

Solving the Problem

1. Assuming the liquids don't mix with each other, draw a diagram and label the colors, illustrating how these liquids would look when poured into a graduated cylinder. If 30 cm³ of water were added to the graduated cylinder, explain how your diagram would change.
2. Use the formula for density to calculate the mass of each of the unknown liquids in the chart.

Sinking and Density Suppose you place a copper block with a volume of 1,000 cm^3 into a container of water. This block weighs about 88 N. As the block sinks, it displaces water, and an upward buoyant force acts on it. If the block is completely submerged, the volume of water it has displaced is 1,000 cm^3—the same as its own volume. This is the maximum amount of water the block can displace. The weight of 1,000 cm^3 of water is about 10 N, and this is the maximum buoyant force that can act on the block. This buoyant force is less than the weight of the copper, so the copper block continues to sink.

The copper block and the displaced water had the same volume. Because the copper block had a greater density, the mass of the copper block was greater than the mass of the displaced water. As a result, the copper block weighed more than the displaced water because its density was greater. Any material with a density that is greater than the density of water will weigh more than the water that it displaces, and it will sink. This is true for any object and any fluid. Any object that has a density greater than the density of the fluid it is placed in will sink.

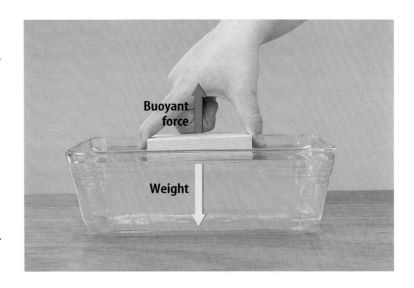

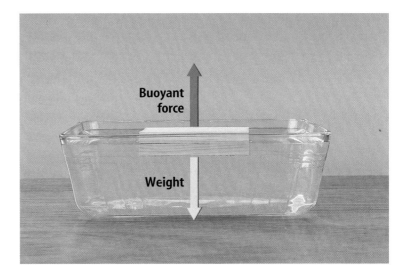

Figure 17 An object, such as this block of wood, will continue to sink in a fluid until it has displaced an amount of fluid that is equal to its mass. Then the buoyant force equals its weight.

Floating and Density Suppose you place a block of wood with a volume of 1,000 cm^3 into a container of water. This block weighs about 7 N. The block starts to sink and displaces water. However, it stops sinking and floats before it is completely submerged, as shown in **Figure 17**. The density of the wood was less than the density of the water. So the wood was able to displace an amount of water equal to its weight before it was completely submerged. It stopped sinking after it had displaced about 700 cm^3 of water. That much water has a weight of about 7 N, which is equal to the weight of the block. Any object with a density less than the fluid it is placed in will float.

Reading Check *How can you determine whether an object will float or sink?*

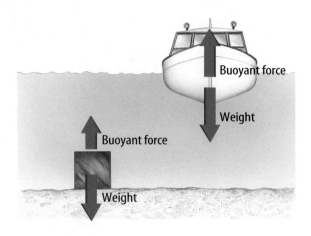

Figure 18 Even though the boat and the cube have the same mass, the boat displaces more water because of its shape. Therefore the boat floats, but the cube sinks.

Boats

Archimedes' principle provides another way to understand why boats that are made of metal can float. Look at **Figure 18.** By making a piece of steel into a boat that occupies a large volume, more water is displaced by the boat than by the piece of steel. According to Archimedes' principle, increasing the weight of the water that is displaced increases the buoyant force. By making the volume of the boat large enough, enough water can be displaced so that the buoyant force is greater than the weight of the steel.

How does the density of the boat compare to the density of the piece of steel? The steel now surrounds a volume that is filled with air that has little mass. The mass of the boat is nearly the same as the mass of the steel, but the volume of the boat is much larger. As a result, the density of the boat is much less than the density of the steel. The boat floats when its volume becomes large enough that its density is less than the density of water.

section 2 review

Summary

The Buoyant Force
- The buoyant force is an upward force that is exerted by a fluid on any object in the fluid.
- The buoyant force is caused by the increase in pressure with depth in a fluid.
- Increasing the surface area in contact with a fluid increases the buoyant force on an object.

Sinking and Floating
- An object sinks when the buoyant force on an object is less than the object's weight.
- An object floats when the buoyant force on an object equals the object's weight.

Archimedes' Principle
- Archimedes' principle states that the buoyant force on a object equals the weight of the fluid the object displaces.
- According to Archimedes' principle, an object will float in a fluid only if the density of the object is less than the density of the fluid.

Self Check

1. **Explain** whether the buoyant force on a submerged object depends on the weight of the object.
2. **Determine** whether an object will float or sink in water if it has a density of 1.5 g/cm^3. Explain.
3. **Compare** the buoyant force on an object when it is partially submerged and when it's completely submerged.
4. **Explain** how the buoyant force acting on an object placed in water can be measured.
5. **Think Critically** A submarine changes its mass by adding or removing seawater from tanks inside the sub. Explain how this can enable the sub to dive or rise to the surface.

Applying Math

6. **Calculate Buoyant Force** A ship displaces 80,000 L of water. One liter of water weighs 9.8 N. What is the buoyant force on the ship?
7. **Density** The density of 14k gold is 13.7 g/cm^3. What is the density of a ring with a mass of 7.21 g and a volume of 0.65 cm^3. Is it made from 14k gold?

ips.msscience.com/self_check_quiz

Measuring Buoyant Force

The total force on an object in a fluid is the difference between the object's weight and the buoyant force. In this lab, you will measure the buoyant force on an object and compare it to the weight of the water displaced.

Real-World Question

How is the buoyant force related to the weight of the water that an object displaces?

Goals
- **Measure** the buoyant force on an object.
- **Compare** the buoyant force to the weight of the water displaced by the object.

Materials
aluminum pan
spring scale
500-mL beaker
graduated cylinder
funnel
metal object

Safety Precautions

Procedure

1. Place the beaker in the aluminum pan and fill the beaker to the brim with water.
2. Hang the object from the spring scale and record its weight.
3. With the object hanging from the spring scale, completely submerge the object in the water. The object should not be touching the bottom or the sides of the beaker.
4. **Record** the reading on the spring scale while the object is in the water. Calculate the buoyant force by subtracting this reading from the object's weight.
5. Use the funnel to carefully pour the water from the pan into the graduated cylinder. Record the volume of this water in cm^3.
6. **Calculate** the weight of the water displaced by multiplying the volume of water by 0.0098 N.

Conclude and Apply

1. **Explain** how the total force on the object changed when it was submerged in water.
2. **Compare** the weight of the water that is displaced with the buoyant force.
3. **Explain** how the buoyant force would change if the object were submersed halfway in water.

Communicating Your Data

Make a poster of an empty ship, a heavily loaded ship, and an overloaded, sinking ship. Explain how Archimedes' principle applies in each case. **For more help, refer to the** Science Skill Handbook.

LAB **355**

section 3
Doing Work with Fluids

as you read

What You'll Learn
- **Explain** how forces are transmitted through fluids.
- **Describe** how a hydraulic system increases force.
- **Describe** Bernoulli's principle.

Why It's Important
Fluids can exert forces that lift heavy objects and enable aircraft to fly.

Review Vocabulary
work: the product of the force applied to an object and the distance the object moves in the direction of the force

New Vocabulary
- Pascal's principle
- hydraulic system
- Bernoulli's principle

Using Fluid Forces

You might have watched a hydraulic lift raise a car off the ground. It might surprise you to learn that the force pushing the car upward is being exerted by a fluid. When a huge jetliner soars through the air, a fluid exerts the force that holds it up. Fluids at rest and fluids in motion can be made to exert forces that do useful work, such as pumping water from a well, making cars stop, and carrying people long distances through the air. How are these forces produced by fluids?

Pushing on a Fluid The pressure in a fluid can be increased by pushing on the fluid. Suppose a watertight, movable cover, or piston, is sitting on top of a column of fluid in a container. If you push on the piston, the fluid can't escape past the piston, so the height of the fluid in the container doesn't change. As a result, the piston doesn't move. But now the force exerted on the bottom of the container is the weight of the fluid plus the force pushing the piston down. Because the force exerted by the fluid at the bottom of the container has increased, the pressure exerted by the fluid also has increased. **Figure 19** shows how the force exerted on a brake pedal is transmitted to a fluid.

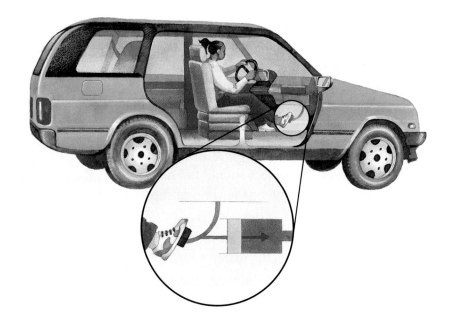

Figure 19 Because the fluid in this piston can't escape, it transmits the force you apply throughout the fluid.

Pascal's Principle

Suppose you fill a plastic bottle with water and screw the cap back on. If you poke a hole in the bottle near the top, water will leak out of the hole. However, if you squeeze the bottle near the bottom, as shown in **Figure 20,** water will shoot out of the hole. When you squeezed the bottle, you applied a force on the fluid. This increased the pressure in the fluid and pushed the water out of the hole faster.

No matter where you poke the hole in the bottle, squeezing the bottle will cause the water to flow faster out of the hole. The force you exert on the fluid by squeezing has been transmitted to every part of the bottle. This is an example of Pascal's principle. According to **Pascal's principle,** when a force is applied to a fluid in a closed container, the pressure in the fluid increases everywhere by the same amount.

Hydraulic Systems

Pascal's principle is used in building hydraulic systems like the ones used by car lifts. A **hydraulic system** uses a fluid to increase an input force. The fluid enclosed in a hydraulic system transfers pressure from one piston to another. An example is shown in **Figure 21.** An input force that is applied to the small piston increases the pressure in the fluid. This pressure increase is transmitted throughout the fluid and acts on the large piston. The force the fluid exerts on the large piston is the pressure in the fluid times the area of the piston. Because the area of the large piston is greater than the area of the small piston, the output force exerted on the large piston is greater than the input force exerted on the small piston.

Figure 20 When you squeeze the bottle, the pressure you apply is distributed throughout the fluid, forcing the water out the hole.

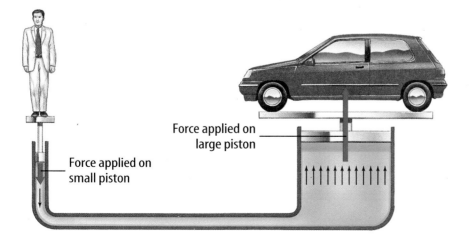

Figure 21 A hydraulic system uses Pascal's principle to make the output force applied on the large piston greater than the input force applied on the small piston.
Infer *how the force on the large piston would change if its area increased.*

Topic: Hydraulic Systems
Visit ips.msscience.com for Web links to information about hydraulic systems.

Activity Draw a diagram showing how one of the hydraulic systems that you find works. Share your diagram with the class.

Increasing Force What is the force pushing upward on the larger piston? For example, suppose that the area of the small piston is 1 m² and the area of the large piston is 2 m². If you push on the small piston with a force of 10 N, the increase in pressure at the bottom of the small piston is

$$P = F/A$$
$$= (10 \text{ N})/(1 \text{ m}^2)$$
$$= 10 \text{ Pa}$$

According to Pascal's principle, this increase in pressure is transmitted throughout the fluid. This causes the force exerted by the fluid on the larger piston to increase. The increase in the force on the larger piston can be calculated by multiplying both sides of the above formula by *A*.

$$F = P \times A$$
$$= 10 \text{ Pa} \times 2 \text{ m}^2$$
$$= 20 \text{ N}$$

The force pushing upward on the larger piston is twice as large as the force pushing downward on the smaller piston. What happens if the larger piston increases in size? Look at the calculation above. If the area of the larger piston increases to 5 m², the force pushing up on this piston increases to 50 N. So a small force pushing down on the left piston as in **Figure 21** can be made much larger by increasing the size of the piston on the right.

 *How does a hydraulic system increase force?*

Pressure in a Moving Fluid

What happens to the pressure in a fluid if the fluid is moving? Try the following experiment. Place an empty soda can on the desktop and blow to the right of the can, as shown in **Figure 22**. In which direction will the can move?

When you blow to the right of the can, the can moves to the right, toward the moving air. The air pressure exerted on the right side of the can, where the air is moving, is less than the air pressure on the left side of the can, where the air is not moving. As a result, the force exerted by the air pressure on the left side is greater than the force exerted on the right side, and the can is pushed to the right. What would happen if you blew between two empty cans?

Figure 22 By blowing on one side of the can, you decrease the air pressure on that side. Because the pressure on the opposite side is now greater, the can moves toward the side you're blowing on.

Bernoulli's Principle

The reason for the surprising behavior of the can in **Figure 22** was discovered by the Swiss scientist Daniel Bernoulli in the eighteenth century. It is an example of Bernoulli's principle. According to **Bernoulli's principle,** when the speed of a fluid increases, the pressure exerted by the fluid decreases. When you blew across the side of the can, the pressure exerted by the air on that side of the can decreased because the air was moving faster than it was on the other side. As a result, the can was pushed toward the side you blew across.

Chimneys and Bernoulli's Principle In a fireplace the hotter, less dense air above the fire is pushed upward by the cooler, denser air in the room. Wind outside of the house can increase the rate at which the smoke rises. Look at **Figure 23.** Air moving across the top of the chimney causes the air pressure above the chimney to decrease according to Bernoulli's principle. As a result, more smoke is pushed upward by the higher pressure of the air in the room.

Damage from High Winds You might have seen photographs of people preparing for a hurricane by closing shutters over windows or nailing boards across the outside of windows. In a hurricane, the high winds blowing outside the house cause the pressure outside the house to be less than the pressure inside. This difference in pressure can be large enough to cause windows to be pushed out and to shatter.

Hurricanes and other high winds sometimes can blow roofs from houses. When wind blows across the roof of a house, the pressure outside the roof decreases. If the wind outside is blowing fast enough, the outside pressure can become so low that the roof can be pushed off the house by the higher pressure of the still air inside.

Wings and Flight

You might have placed your hand outside the open window of a moving car and felt the push on it from the air streaming past. If you angled your hand so it tilted upward into the moving air, you would have felt your hand pushed upward. If you increased the tilt of your hand, you felt the upward push increase. You might not have realized it, but your hand was behaving like an airplane wing. The force that lifted your hand was provided by a fluid—the air.

Mini LAB

Observing Bernoulli's Principle

Procedure
1. Tie a piece of **string** to the handle of a **plastic spoon.**
2. Turn on a faucet to make a stream of water.
3. Holding the string, bring the spoon close to the stream of water.

Analysis
Use Bernoulli's principle to explain the motion of the spoon.

Try at Home

Figure 23 The air moving past the chimney lowers the air pressure above the chimney. As a result, smoke is forced up the chimney faster than when air above the chimney is still.

SECTION 3 Doing Work with Fluids

Producing Lift How is the upward force, or lift, on an airplane wing produced? A jet engine pushes the plane forward, or a propeller pulls the plane forward. Air flows over the wings as the plane moves. The wings are tilted upward into the airflow, just like your hand was tilted outside the car window. **Figure 24** shows how the tilt of the wing causes air that has flowed over the wing's upper and lower surfaces to be directed downward.

Lift is created by making the air flow downward. To understand this, remember that air is made of different types of molecules. The motion of these molecules is changed only when a force acts on them. When the air is directed downward, a force is being exerted on the air molecules by the wing.

However, according to Newton's third law of motion, for every action force there is an equal but opposite reaction force. The wing exerts a downward action force on the air. So the air must exert an upward reaction force on the wing. This reaction force is the lift that enables paper airplanes and jet airliners to fly.

Airplane Wings Airplanes have different wing shapes, depending on how the airplane is used. The lift on a wing depends on the amount of air that the wing deflects downward and how fast that air is moving. Lift can be increased by increasing the size or surface area of the wing. A larger wing is able to deflect more air downward.

Look at the planes in **Figure 25.** A plane designed to fly at high speeds, such as a jet fighter, can have small wings. A large cargo plane that carries heavy loads needs large wings to provide a great deal of lift. A glider flies at low speeds and uses long wings that have a large surface area to provide the lift it needs.

Reading Check *How can a wing's lift be increased?*

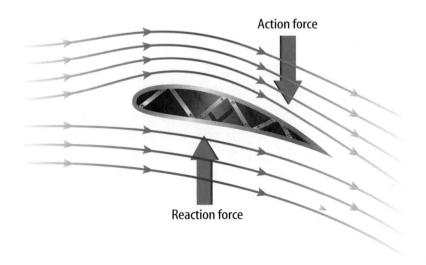

Figure 24 An airplane wing forces air to be directed downward. As a result, the air exerts an upward reaction force on the wing, producing lift.

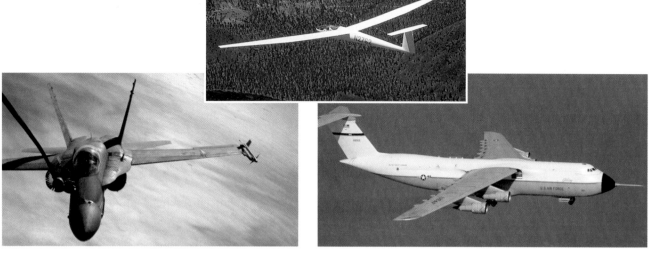

 **Birds' Wings** A bird's wing provides lift in the same way that an airplane wing does. The wings also act as propellers that pull the bird forward when it flaps its wings up and down. Bird wings also have different shapes depending on the type of flight. Seabirds have long, narrow wings, like the wings of a glider, that help them glide long distances. Forest and field birds, such as pheasants, have short, rounded wings that enable them to take off quickly and make sharp turns. Swallows, swifts, and falcons, which fly at high speeds, have small, narrow, tapered wings like those on a jet fighter.

Figure 25 Different wing shapes are used for different types of planes. Larger wings provide more lift.

section 3 review

Summary

Pascal's Principle
- Pascal's principle states that when a force is applied to a fluid in a closed container, the pressure in the fluid increases by the same amount everywhere in the fluid.
- Hydraulic systems use Pascal's principle to produce an output force that is greater than an applied input force.
- Increasing the surface area in contact with a fluid increases the buoyant force on an object.

Bernoulli's Principle
- Bernoulli's principle states that when the speed of a fluid increases, the pressure exerted by the fluid decreases.

Wings and Flight
- An airplane wing exerts a force on air and deflects it downward. By Newton's third law, the air exerts an upward reaction force.

Self Check

1. **Explain** why making an airplane wing larger enables the wing to produce more lift.
2. **Infer** If you squeeze a plastic water-filled bottle, where is the pressure change in the water the greatest?
3. **Explain** Use Bernoulli's principle to explain why a car passing a truck tends to be pushed toward the truck.
4. **Infer** why a sheet of paper rises when you blow across the top of the paper.
5. **Think Critically** Explain why the following statement is false: In a hydraulic system, because the increase in the pressure on both pistons is the same, the increase in the force on both pistons is the same.

Applying Math

6. **Calculate Force** The small piston of a hydraulic lift, has an area of 0.01 m². If a force of 250 N is applied to the small piston, find the force on the large piston if it has an area of 0.05 m².

ips.msscience.com/self_check_quiz

LAB Use the Internet

Barometric Pressure and Weather

Goals
- **Collect** barometric pressure and other weather data.
- **Compare** barometric pressure to weather conditions.
- **Predict** weather patterns based on barometric pressure, wind speed and direction, and visual conditions.

Data Source

Visit **ips.msscience.com/ internet_lab** for more information about barometric pressure, weather information, and data collected by other students.

Real-World Question

What is the current barometric pressure where you are? How would you describe the weather today where you are? What is the weather like in the region to the west of you? To the east of you? What will your weather be like tomorrow? The atmosphere is a fluid and flows from one place to another as weather patterns change. Changing weather conditions also cause the atmospheric pressure to change. By collecting barometric pressure data and observing weather conditions, you will be able to make a prediction about the next day's weather.

Make a Plan

1. Visit the Web site on the left for links to information about weather in the region where you live.
2. Find and record the current barometric pressure and whether the pressure is rising, falling, or remaining steady. Also record the wind speed and direction.
3. **Observe and record** other weather conditions, such as whether rain is falling, the Sun is shining, or the sky is cloudy.
4. Based on the data you collect and your observations, predict what you think tomorrow's weather will be. Record your prediction.
5. Repeat the data collection and observation for a total of five days.

362 CHAPTER 12 Forces and Fluids

Using Scientific Methods

Barometric Pressure Weather Data	
Location of weather station	
Barometric pressure	
Status of barometric pressure	
Wind speed	Do not write in this book.
Wind direction	
Current weather conditions	
Predictions of tomorrow's weather conditions	

▶ Follow Your Plan

1. Make sure your teacher approves your plan before you start.
2. Visit the link below to post your data.

▶ Analyze Your Data

1. **Analyze** Look at your data. What was the weather the day after the barometric pressure increased? The day after the barometric pressure decreased? The day after the barometric pressure was steady?
2. **Draw Conclusions** How accurate were your weather predictions?

▶ Conclude and Apply

1. **Infer** What is the weather to the west of you today? How will that affect the weather in your area tomorrow?
2. **Compare** What was the weather to the east of you today? How does that compare to the weather in your area yesterday?
3. **Evaluate** How does increasing, decreasing, or steady barometric pressure affect the weather?

Communicating Your Data

Find this lab using the link below. Use the data on the Web site to predict the weather where you are two days from now.

ips.msscience.com/internet_lab

Science and Language Arts

"Hurricane"
by John Balaban

Near dawn our old live oak sagged over
then crashed on the tool shed
rocketing off rakes paintcans flower pots.

All night, rain slashed the shutters until
it finally quit and day arrived in queer light,
silence, and ozoned air. Then voices calling

as neighbors crept out to see the snapped trees,
leaf mash and lawn chairs driven in heaps
with roof bits, siding, sodden birds, dead snakes.

For days, bulldozers clanked by our houses
in sickening August heat as heavy cranes
scraped the rotting tonnage from the streets.

Then our friend Elling drove in from Sarasota
in his old . . . van packed with candles, with
dog food, cat food, flashlights and batteries . . . .

Understanding Literature

Sense Impressions In this poem, John Balaban uses sense impressions to place the reader directly into the poem's environment. For example, the words *rotting tonnage* evoke the sense of smell. Give examples of other sense impressions mentioned.

Respond to the Reading

1. What kinds of damage did the hurricane cause?
2. Why do you think the poet felt relief when his friend, Elling, arrived?
3. **Linking Science and Writing** Write a poem describing a natural phenomenon involving forces and fluids. Use words that evoke at least one of the five sense impressions.

 In the poem, bits of roofs and siding from houses are part of the debris that is everywhere in heaps. According to Bernoulli's principle, the high winds in a hurricane blowing past a house causes the air pressure outside the house to be less than the air pressure inside. In some cases, the forces exerted by the air inside causes the roof to be pushed off the house, and the walls to be blown outward.

chapter 12 Study Guide

Reviewing Main Ideas

Section 1 Pressure

1. Pressure equals force divided by area.
2. Liquids and gases are fluids that flow.
3. Pressure increases with depth and decreases with elevation in a fluid.
4. The pressure exerted by a fluid on a surface is always perpendicular to the surface.

Section 2 Why do objects float?

1. A buoyant force is an upward force exerted on all objects placed in a fluid.
2. The buoyant force depends on the shape of the object.
3. According to Archimedes' principle, the buoyant force on the object is equal to the weight of the fluid displaced by the object.
4. An object floats when the buoyant force exerted by the fluid is equal to the object's weight.
5. An object will float if it is less dense than the fluid it is placed in.

Section 3 Doing Work with Fluids

1. Pascal's principle states that the pressure applied at any point to a confined fluid is transmitted unchanged throughout the fluid.
2. Bernoulli's principle states that when the velocity of a fluid increases, the pressure exerted by the fluid decreases.
3. A wing provides lift by forcing air downward.

Visualizing Main Ideas

Copy and complete the following table.

Relationships Among Forces and Fluids		
Idea	**What does it relate?**	**How?**
Density	mass and volume	
Pressure		force/area
Archimedes' principle	buoyant force and weight of fluid that is displaced	
Bernoulli's principle		velocity increases, pressure decreases
Pascal's principle	pressure applied to enclosed fluid at one point and pressure at other points in a fluid	

chapter 12 Review

Using Vocabulary

Archimedes' principle p. 351
Bernoulli's principle p. 359
buoyant force p. 348
density p. 352
fluid p. 343
hydraulic system p. 357
Pascal's principle p. 357
pressure p. 340

Answer each of the following questions using complete sentences that include vocabulary from the list above.

1. How would you describe a substance that can flow?
2. When the area over which a force is applied decreases, what increases?
3. What principle relates the weight of displaced fluid to the buoyant force?
4. How is a fluid used to lift heavy objects?
5. If you increase an object's mass but not its volume, what have you changed?
6. How is a log able to float in a river?
7. What principle explains why hurricanes can blow the roof off of a house?

Checking Concepts

Choose the word or phrase that best answers the question.

8. Which always equals the weight of the fluid displaced by an object?
   A) the weight of the object
   B) the force of gravity on the object
   C) the buoyant force on the object
   D) the net force on the object

9. What is the net force on a rock that weighs 500 N if the weight of the water it displaces is 300 N?
   A) 200 N downward   C) 800 N downward
   B) 200 N upward   D) 800 N upward

10. The pressure exerted by a fluid on a surface is always in which direction?
    A) upward
    B) downward
    C) parallel to the surface
    D) perpendicular to the surface

Use the photo below to answer question 11.

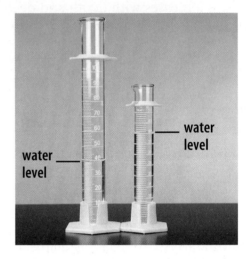

11. Each graduated cylinder contains the same amount of water. Which of the following statements is true?
    A) The pressure is greater at the bottom of the large cylinder.
    B) The pressure is greater at the bottom of the small cylinder.
    C) The pressure is equal at the bottom of both cylinders.
    D) There is zero pressure at the bottom of both cylinders.

12. Which would increase the lift provided by an airplane wing?
    A) decreasing the volume of the wing
    B) increasing the area of the wing
    C) decreasing the length of the wing
    D) increasing the mass of the wing

13. An airplane wing produces lift by forcing air in which direction?
    A) upward   C) under the wing
    B) downward   D) over the wing

Thinking Critically

14. **Explain** A sandbag is dropped from a hot-air balloon and the balloon rises. Explain why this happens.

15. **Determine** whether or not this statement is true: Heavy objects sink, and light objects float. Explain your answer.

16. **Explain** why a leaking boat sinks.

17. **Explain** why the direction of the buoyant force on a submerged cube is upward and not left or right.

18. **Recognizing Cause and Effect** A steel tank and a balloon are the same size and contain the same amount of helium. Explain why the balloon rises and the steel tank doesn't.

19. **Make and Use Graphs** Graph the pressure exerted by a 75-kg person wearing different shoes with areas of 0.01 m^2, 0.02 m^2, 0.03 m^2, 0.04 m^2, and 0.05 m^2. Plot pressure on the vertical axis and area on the horizontal axis.

20. **Explain** why it is easier to lift an object that is underwater, than it is to lift the object when it is out of the water.

21. **Infer** Two objects with identical shapes are placed in water. One object floats and the other object sinks. Infer the difference between the two objects that causes one to sink and the other to float.

22. **Compare** Two containers with different diameters are filled with water to the same height. Compare the force exerted by the fluid on the bottom of the two containers.

Performance Activities

23. **Oral Presentation** Research the different wing designs in birds or aircraft. Present your results to the class.

24. **Experiment** Partially fill a plastic dropper with water until it floats just below the surface of the water in a bowl or beaker. Place the dropper inside a plastic bottle, fill the bottle with water, and seal the top. Now squeeze the bottle. What happens to the water pressure in the bottle? How does the water level in the dropper change? How does the density of the dropper change? Use your answers to these questions to explain how the dropper moves when you squeeze the bottle.

Applying Math

25. **Buoyant Force** A rock is attached to a spring scale that reads 10 N. If the rock is submerged in water, the scale reads 6 N. What is the buoyant force on the submerged rock?

26. **Hydraulic Force** A hydraulic lift with a large piston area of 0.04 m^2 exerts a force of 5,000 N. If the smaller piston has an area of 0.01 m^2, what is the force on it?

Use the table below to answer questions 27 and 28.

Material Density	
Substance	Density (g/cm³)
Ice	0.92
Lead	11.34
Balsa wood	0.12
Sugar	1.59

27. **Density** Classify which of the above substances will and will not float in water.

28. **Volume** Find the volumes of each material, if each has a mass of 25 g.

29. **Pressure** What is the pressure due to a force of 100 N on an area 4 m^2?

30. **Pressure** A bottle of lemonade sitting on a table weighs 6 N. The bottom of the bottle has a surface area of 0.025 m^2. Calculate the pressure the bottle of lemonade exerts on the table.

chapter 12 Standardized Test Practice

Part 1 | Multiple Choice

Record your answers on the answer sheet provided by your teacher or on a sheet of paper.

1. A force of 15 N is exerted on an area of 0.1 m². What is the pressure?
   A. 150 N
   B. 150 Pa
   C. 1.5 Pa
   D. 0.007 Pa

Use the illustration below to answer questions 2 and 3.

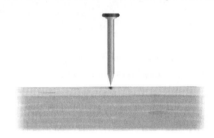

2. Which statement is TRUE?
   A. The contact area between the board and the nail tip is large.
   B. The contact area between the nail tip and the board is greater than the contact area between the nail head and the hammer head.
   C. The nail exerts no pressure on the board because its weight is so small.
   D. The pressure exerted by the nail is large because the contact area between the nail tip and the board is small.

3. Which increases the pressure exerted by the nail tip on the board?
   A. increasing the area of the nail tip
   B. decreasing the area of the nail tip
   C. increasing the length of the nail
   D. decreasing the length of the nail

Test-Taking Tip

Check the Answer Number For each question, double check that you are filling in the correct answer bubble for the question number you are working on.

4. A 15-g block of aluminum has a volume of 5.5 cm³. What is the density?
   A. 2.7 g/cm³
   B. 82.5 g/cm³
   C. 0.37 g/cm³
   D. 2.7 cm³/g

Use the illustration below to answer questions 5 and 6.

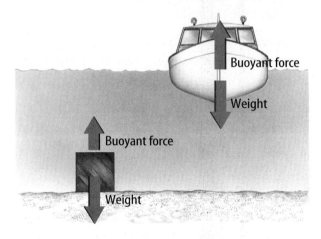

5. Assume the boat and cube have the same mass. Which of these is correct?
   A. The boat displaces less water than the cube.
   B. The densities of the boat and cube are equal.
   C. The density of the boat is less than the density of the cube.
   D. The density of the boat is greater than the density of the water.

6. Which of the following would make the cube more likely to float?
   A. increasing its volume
   B. increasing its density
   C. increasing its weight
   D. decreasing its volume

7. Which of the following instruments is used to measure atmospheric pressure?
   A. altimeter
   B. hygrometer
   C. barometer
   D. anemometer

368 STANDARDIZED TEST PRACTICE

Part 2 Short Response/Grid In

Record your answers on the answer sheet provided by your teacher or on a sheet of paper.

8. Explain why wearing snowshoes makes it easier to walk over deep, soft snow.

9. Why is a gas, such as air condidered to be a fluid?

10. Explain why the pressure exerted by the atmosphere does not crush your body.

Use the illustration below to answer questions 11 and 12.

11. How does the buoyant force on the boat change if the boat is loaded so that it floats lower in the water?

12. What changes in the properties of this boat could cause it to sink?

13. People sometimes prepare for a coming hurricane by nailing boards over the outside of windows. How can high winds damage windows?

14. What factors influence the amount of lift on an airplane wing?

15. The pressure inside the fluid of a hydraulic system is increased by 1,000 Pa. What is the increase in the force exerted on a piston that has an area of 0.05 m^2?

Part 3 Open Ended

Record your answers on a sheet of paper.

16. Explain why the interior of an airplane is pressurized when flying at high altitude. If a hole were punctured in an exterior wall, what would happen to the air pressure inside the plane?

17. In an experiment, you design small boats using aluminum foil. You add pennies to each boat until it sinks. Sketch several different shapes you might create. Which will likely hold the most pennies? Why?

18. You squeeze a round, air-filled balloon slightly, changing its shape. Describe how the pressure inside the balloon changes.

Use the illustration below to answer questions 19 and 20.

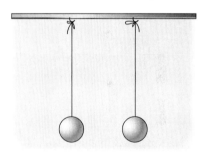

19. Describe the motion of the table tennis balls when air is blown between the two balls. Explain.

20. Describe the motion of the table tennis balls if air blown to the left of the ball on the left and to the right of the ball on the right at the same time. Explain.

21. Compare the pressure you exert on a floor when you are standing on the floor and when you are lying on the floor.

22. In order to drink a milk shake through a straw, how must the air pressure inside the straw change, compared with the air pressure outside the straw?

unit 5

Energy

How Are Train Schedules & Oil Pumps Connected?

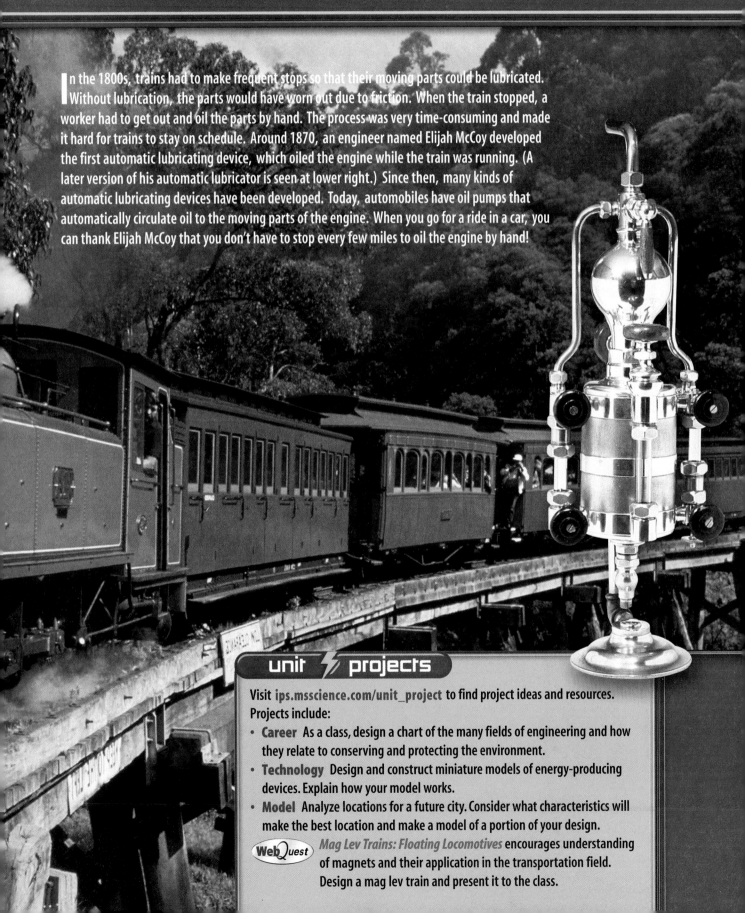

NATIONAL GEOGRAPHIC

In the 1800s, trains had to make frequent stops so that their moving parts could be lubricated. Without lubrication, the parts would have worn out due to friction. When the train stopped, a worker had to get out and oil the parts by hand. The process was very time-consuming and made it hard for trains to stay on schedule. Around 1870, an engineer named Elijah McCoy developed the first automatic lubricating device, which oiled the engine while the train was running. (A later version of his automatic lubricator is seen at lower right.) Since then, many kinds of automatic lubricating devices have been developed. Today, automobiles have oil pumps that automatically circulate oil to the moving parts of the engine. When you go for a ride in a car, you can thank Elijah McCoy that you don't have to stop every few miles to oil the engine by hand!

unit projects

Visit ips.msscience.com/unit_project to find project ideas and resources. Projects include:

- **Career** As a class, design a chart of the many fields of engineering and how they relate to conserving and protecting the environment.
- **Technology** Design and construct miniature models of energy-producing devices. Explain how your model works.
- **Model** Analyze locations for a future city. Consider what characteristics will make the best location and make a model of a portion of your design.

WebQuest *Mag Lev Trains: Floating Locomotives* encourages understanding of magnets and their application in the transportation field. Design a mag lev train and present it to the class.

chapter 13

Energy and Energy Resources

The BIG Idea
Changes occur when energy is transferred from one place to another.

SECTION 1
What is energy?
Main Idea Energy can have different forms.

SECTION 2
Energy Transformations
Main Idea Energy can change form, but can never be created or destroyed.

SECTION 3
Sources of Energy
Main Idea Energy contained in different energy sources is transformed into useful forms of energy.

Blowing Off Steam
The electrical energy you used today might have been produced by a coal-burning power plant like this one. Energy contained in coal is transformed into heat, and then into electrical energy. As boiling water heated by the burning coal is cooled, steam rises from these cone-shaped cooling towers.

Science Journal Choose three devices that use electricity, and identify the function of each device.

Start-Up Activities

Marbles and Energy

What's the difference between a moving marble and one at rest? A moving marble can hit something and cause a change to occur. How can a marble acquire energy—the ability to cause change?

1. Make a track on a table by slightly separating two metersticks placed side by side.
2. Using a book, raise one end of the track slightly and measure the height.
3. Roll a marble down the track. Measure the distance from its starting point to where it hits the floor. Repeat. Calculate the average of the two measurements.
4. Repeat steps 2 and 3 for three different heights. Predict what will happen if you use a heavier marble. Test your prediction and record your observations.
5. **Think Critically** In your Science Journal, describe how the distance traveled by the marble is related to the height of the ramp. How is the motion of the marble related to the ramp height?

Energy Make the following Foldable to help identify what you already know, what you want to know, and what you learned about energy.

STEP 1 Fold a vertical sheet of paper from side to side. Make the front edge about 1 cm shorter than the back edge.

STEP 2 Turn lengthwise and fold into thirds

STEP 3 Unfold, cut, and label each tab for only the top layer along both folds to make three tabs.

Identify Questions Before you read the chapter, write what you know and what you want to know about the types, sources, and transformation of energy under the appropriate tabs. As you read the chapter, correct what you have written and add more questions under the *Learned* tab.

Preview this chapter's content and activities at
ips.msscience.com

Get Ready to Read

Take Notes

① Learn It! The best way for you to remember information is to write it down, or take notes. Good note-taking is useful for studying and research. When you are taking notes, it is helpful to
- phrase the information in your own words;
- restate ideas in short, memorable phrases;
- stay focused on main ideas and only the most important supporting details.

② Practice It! Make note-taking easier by using a chart to help you organize information clearly. Write the main ideas in the left column. Then write at least three supporting details in the right column. Read the text from Section 2 of this chapter under the heading *Energy Changes Form*, pages 381–383. Then take notes using a chart, such as the one below.

Main Idea	Supporting Details
	1. 2. 3. 4. 5.
	1. 2. 3. 4. 5.

③ Apply It! As you read this chapter, make a chart of the main ideas. Next to each main idea, list at least two supporting details.

Target Your Reading

Reading Tip
Read one or two paragraphs first and take notes after you read. You are likely to take down too much information if you take notes as you read.

Use this to focus on the main ideas as you read the chapter.

1 Before you read the chapter, respond to the statements below on your worksheet or on a numbered sheet of paper.
- Write an **A** if you **agree** with the statement.
- Write a **D** if you **disagree** with the statement.

2 After you read the chapter, look back to this page to see if you've changed your mind about any of the statements.
- If any of your answers changed, explain why.
- Change any false statements into true statements.
- Use your revised statements as a study guide.

Science Online
Print out a worksheet of this page at ips.msscience.com

Before You Read A or D		Statement	After You Read A or D
	1	Only objects that are moving have energy.	
	2	Energy of motion depends only on the object's mass and speed.	
	3	The amount of thermal energy an object has depends on its temperature.	
	4	Energy changes from one form to another when a ball is thrown upward.	
	5	Living organisms change chemical energy to thermal energy.	
	6	The total amount of energy increases when a candle burns.	
	7	Coal is being made faster than it is being used.	
	8	Solar energy is the least expensive energy source.	
	9	A heat pump can heat and cool a building.	

374 B

section 1
What is energy?

as you read

What You'll Learn
- **Explain** what energy is.
- **Distinguish** between kinetic energy and potential energy.
- **Identify** the various forms of energy.

Why It's Important
Energy is involved whenever a change occurs.

Review Vocabulary
mass: a measure of the amount of matter in an object

New Vocabulary
- energy
- kinetic energy
- potential energy
- thermal energy
- chemical energy
- radiant energy
- electrical energy
- nuclear energy

The Nature of Energy

What comes to mind when you hear the word *energy*? Do you picture running, leaping, and spinning like a dancer or a gymnast? How would you define energy? When an object has energy, it can make things happen. In other words, **energy** is the ability to cause change. What do the items shown in **Figure 1** have in common?

Look around and notice the changes that are occurring—someone walking by or a ray of sunshine that is streaming through the window and warming your desk. Maybe you can see the wind moving the leaves on a tree. What changes are occurring?

Transferring Energy You might not realize it, but you have a large amount of energy. In fact, everything around you has energy, but you notice it only when a change takes place. Anytime a change occurs, energy is transferred from one object to another. You hear a footstep because energy is transferred from a foot hitting the ground to your ears. Leaves are put into motion when energy in the moving wind is transferred to them. The spot on the desktop becomes warmer when energy is transferred to it from the sunlight. In fact, all objects, including leaves and desktops, have energy.

Figure 1 Energy is the ability to cause change.
Explain *how these objects cause change.*

374 CHAPTER 13 Energy and Energy Resources

Energy of Motion

Things that move can cause change. A bowling ball rolls down the alley and knocks down some pins, as in **Figure 2A.** Is energy involved? A change occurs when the pins fall over. The bowling ball causes this change, so the bowling ball has energy. The energy in the motion of the bowling ball causes the pins to fall. As the ball moves, it has a form of energy called kinetic energy. **Kinetic energy** is the energy an object has due to its motion. If an object isn't moving, it doesn't have kinetic energy.

Kinetic Energy and Speed If you roll the bowling ball so it moves faster, what happens when it hits the pins? It might knock down more pins, or it might cause the pins to go flying farther. A faster ball causes more change to occur than a ball that is moving slowly. Look at **Figure 2B.** The professional bowler rolls a fast-moving bowling ball. When her ball hits the pins, pins go flying faster and farther than for a slower-moving ball. All that action signals that her ball has more energy. The faster the ball goes, the more kinetic energy it has. This is true for all moving objects. Kinetic energy increases as an object moves faster.

Reading Check *How does kinetic energy depend on speed?*

Kinetic Energy and Mass Suppose, as shown in **Figure 2C,** you roll a volleyball down the alley instead of a bowling ball. If the volleyball travels at the same speed as a bowling ball, do you think it will send pins flying as far? The answer is no. The volleyball might not knock down any pins. Does the volleyball have less energy than the bowling ball even though they are traveling at the same speed?

An important difference between the volleyball and the bowling ball is that the volleyball has less mass. Even though the volleyball is moving at the same speed as the bowling ball, the volleyball has less kinetic energy because it has less mass. Kinetic energy also depends on the mass of a moving object. Kinetic energy increases as the mass of the object increases.

Figure 2 The kinetic energy of an object depends on the mass and speed of the object.

A This ball has kinetic energy because it is rolling down the alley.

B This ball has more kinetic energy because it has more speed.

C This ball has less kinetic energy because it has less mass.

Figure 3 The potential energy of an object depends on its mass and height above the ground.
Determine which vase has more potential energy, the red one or the blue one.

Energy of Position

An object can have energy even though it is not moving. For example, a glass of water sitting on the kitchen table doesn't have any kinetic energy because it isn't moving. If you accidentally nudge the glass and it falls on the floor, changes occur. Gravity pulls the glass downward, and the glass has energy of motion as it falls. Where did this energy come from?

When the glass was sitting on the table, it had potential (puh TEN chul) energy. **Potential energy** is the energy stored in an object because of its position. In this case, the position is the height of the glass above the floor. The potential energy of the glass changes to kinetic energy as the glass falls. The potential energy of the glass is greater if it is higher above the floor. Potential energy also depends on mass. The more mass an object has, the more potential energy it has. Which object in **Figure 3** has the most potential energy?

Forms of Energy

Food, sunlight, and wind have energy, yet they seem different because they contain different forms of energy. Food and sunlight contain forms of energy different from the kinetic energy in the motion of the wind. The warmth you feel from sunlight is another type of energy that is different from the energy of motion or position.

Thermal Energy The feeling of warmth from sunlight signals that your body is acquiring more thermal energy. All objects have **thermal energy** that increases as its temperature increases. A cup of hot chocolate has more thermal energy than a cup of cold water, as shown in **Figure 4.** Similarly, the cup of water has more thermal energy than a block of ice of the same mass. Your body continually produces thermal energy. Many chemical reactions that take place inside your cells produce thermal energy. Where does this energy come from? Thermal energy released by chemical reactions comes from another form of energy called chemical energy.

Figure 4 The hotter an object is, the more thermal energy it has. A cup of hot chocolate has more thermal energy than a cup of cold water, which has more thermal energy than a block of ice with the same mass.

376 CHAPTER 13 Energy and Energy Resources

Figure 5 Complex chemicals in food store chemical energy. During activity, the chemical energy transforms into kinetic energy.

Chemical Energy When you eat a meal, you are putting a source of energy inside your body. Food contains chemical energy that your body uses to provide energy for your brain, to power your movements, and to fuel your growth. As in **Figure 5,** food contains chemicals, such as sugar, which can be broken down in your body. These chemicals are made of atoms that are bonded together, and energy is stored in the bonds between atoms. **Chemical energy** is the energy stored in chemical bonds. When chemicals are broken apart and new chemicals are formed, some of this energy is released. The flame of a candle is the result of chemical energy stored in the wax. When the wax burns, chemical energy is transformed into thermal energy and light energy.

Reading Check *When is chemical energy released?*

Light Energy Light from the candle flame travels through the air at an incredibly fast speed of 300,000 km/s. This is fast enough to circle Earth almost eight times in 1 s. When light strikes something, it can be absorbed, transmitted, or reflected. When the light is absorbed by an object, the object can become warmer. The object absorbs energy from the light and this energy is transformed into thermal energy. Then energy carried by light is called **radiant energy**. **Figure 6** shows a coil of wire that produces radiant energy when it is heated. To heat the metal, another type of energy can be used—electrical energy.

Figure 6 Electrical energy is transformed into thermal energy in the metal heating coil. As the metal becomes hotter, it emits more radiant energy.

SECTION 1 What is energy? **377**

Figure 7 Complex power plants are required to obtain useful energy from the nucleus of an atom.

Electrical Energy Electrical lighting is one of the many ways electrical energy is used. Look around at all the devices that use electricity. Electric current flows in these devices when they are connected to batteries or plugged into an electric outlet. **Electrical energy** is the energy that is carried by an electric current. An electric device uses the electrical energy provided by the current flowing in the device. Large electric power plants generate the enormous amounts of electrical energy used each day. About 20 percent of the electrical energy used in the United States is generated by nuclear power plants.

Nuclear Energy Nuclear power plants use the energy stored in the nucleus of an atom to generate electricity. Every atomic nucleus contains energy—**nuclear energy**—that can be transformed into other forms of energy. However, releasing the nuclear energy is a difficult process. It involves the construction of complex power plants, shown in **Figure 7.** In contrast, all that is needed to release chemical energy from wood is a lighted match.

section 1 review

Summary

The Nature of Energy
- Energy is the ability to cause change.
- Kinetic energy is the energy an object has due to its motion. Kinetic energy depends on an object's speed and mass.
- Potential energy is the energy an object has due to its position. Potential energy depends on an object's height and mass.

Forces of Energy
- Thermal energy increases as temperature increases.
- Chemical energy is the energy stored in chemical bonds in molecules.
- Light energy, also called radiant energy, is the energy contained in light.
- Electrical energy is the energy carried by electric current.
- Nuclear energy is the energy contained in the nucleus of an atom.

Self Check

1. **Explain** why a high-speed collision between two cars would cause more damage than a low-speed collision between the same two cars.
2. **Describe** the energy transformations that occur when a piece of wood is burned.
3. **Identify** the form of energy that is converted into thermal energy by your body.
4. **Explain** how, if two vases are side by side on a shelf, one could have more potential energy.
5. **Think Critically** A golf ball and a bowling ball are moving and both have the same kinetic energy. Which one is moving faster? If they move at the same speed, which one has more kinetic energy?

Applying Skills

6. **Communicate** In your Science Journal, record different ways the word *energy* is used. Which ways of using the word *energy* are closest to the definition of energy given in this section?

ips.msscience.com/self_check_quiz

section 2

Energy Transformations

Changing Forms of Energy

Chemical, thermal, radiant, and electrical are some of the forms that energy can have. In the world around you, energy is transforming continually between one form and another. You observe some of these transformations by noticing a change in your environment. Forest fires are a dramatic example of an environmental change that can occur naturally as a result of lightning strikes. A number of changes occur that involve energy as the mountain biker in **Figure 8** pedals up a hill. What energy transformations cause these changes to occur?

Tracking Energy Transformations As the mountain biker pedals, his leg muscles transform chemical energy into kinetic energy. The kinetic energy of his leg muscles transforms into kinetic energy of the bicycle as he pedals. Some of this energy transforms into potential energy as he moves up the hill. Also, some energy is transformed into thermal energy. His body is warmer because chemical energy is being released. Because of friction, the mechanical parts of the bicycle are warmer, too. Thermal energy is almost always produced by an energy transformation. The energy transformations that occur when people exercise, when cars run, when living things grow and even when stars explode, all produce thermal energy.

as you read

What You'll Learn
- **Apply** the law of conservation of energy to energy transformations.
- **Identify** how energy changes form.
- **Describe** how electric power plants produce energy.

Why It's Important
Changing energy from one form to another is what makes cars run, furnaces heat, telephones work, and plants grow.

Review Vocabulary
transformation: a change in composition or structure

New Vocabulary
- law of conservation of energy
- generator
- turbine

Figure 8 The ability to transform energy allows the biker to climb the hill.
Identify *all the forms of energy that are represented in the photograph.*

379

The Law of Conservation of Energy

It can be a challenge to track energy as it moves from object to object. However, one extremely important principle can serve as a guide as you trace the flow of energy. According to the **law of conservation of energy,** energy is never created or destroyed. The only thing that changes is the form in which energy appears. When the biker is resting at the summit, all his original energy is still around. Some of the energy is in the form of potential energy, which he will use as he coasts down the hill. Some of this energy was changed to thermal energy by friction in the bike. Chemical energy was also changed to thermal energy in the biker's muscles, making him feel hot. As he rests, this thermal energy moves from his body to the air around him. No energy is missing—it can all be accounted for.

Reading Check *Can energy ever be lost? Why or why not?*

Changing Kinetic and Potential Energy

The law of conservation of energy can be used to identify the energy changes in a system. For example, tossing a ball into the air and catching it is a simple system. As shown in **Figure 9,** as the ball leaves your hand, most of its energy is kinetic. As the ball rises, it slows and its kinetic energy decreases. But, the total energy of the ball hasn't changed. The decrease in kinetic energy equals the increase in potential energy as the ball flies higher in the air. The total amount of energy remains constant. Energy moves from place to place and changes form, but it never is created or destroyed.

Topic: Energy Transformations
Visit ips.msscience.com for Web links to information about energy transformations that occur during different activities and processes.

Activity Choose an activity or process and make a graph showing how the kinetic and potential energy change during it.

Figure 9 During the flight of the baseball, energy is transforming between kinetic and potential energy.
Determine *where the ball has the most kinetic energy. Where does the ball have the most total energy?*

380 CHAPTER 13

Figure 10 Hybrid cars that use an electric motor and a gasoline engine for power are now available. Hybrid cars make energy transformations more efficient.

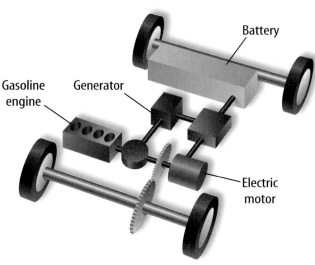

Energy Changes Form

Energy transformations occur constantly all around you. Many machines are devices that transform energy from one form to another. For example, an automobile engine transforms the chemical energy in gasoline into energy of motion. However, not all of the chemical energy is converted into kinetic energy. Instead, some of the chemical energy is converted into thermal energy, and the engine becomes hot. An engine that converts chemical energy into more kinetic energy is a more efficient engine. New types of cars, like the one shown in **Figure 10,** use an electric motor along with a gasoline engine. These engines are more efficient so the car can travel farther on a gallon of gas.

 **Transforming Chemical Energy** Inside your body, chemical energy also is transformed into kinetic energy. Look at **Figure 11.** The transformation of chemical to kinetic energy occurs in muscle cells. There, chemical reactions take place that cause certain molecules to change shape. Your muscle contracts when many of these changes occur, and a part of your body moves.

The matter contained in living organisms, also called biomass, contains chemical energy. When organisms die, chemical compounds in their biomass break down. Bacteria, fungi, and other organisms help convert these chemical compounds to simpler chemicals that can be used by other living things.

Thermal energy also is released as these changes occur. For example, a compost pile can contain plant matter, such as grass clippings and leaves. As the compost pile decomposes, chemical energy is converted into thermal energy. This can cause the temperature of a compost pile to reach 60°C.

Analyzing Energy Transformations

Procedure
1. Place soft **clay** on the floor and smooth out its surface.
2. Hold a **marble** 1.5 m above the clay and drop it. Measure the depth of the crater made by the marble.
3. Repeat this procedure using a **golf ball** and a **plastic golf ball.**

Analysis
1. Compare the depths of the craters to determine which ball had the most kinetic energy as it hit the clay.
2. Explain how potential energy was transformed into kinetic energy during your activity.

SECTION 2 Energy Transformations

NATIONAL GEOGRAPHIC VISUALIZING ENERGY TRANSFORMATIONS

Figure 11

Paddling a raft, throwing a baseball, playing the violin — your skeletal muscles make these and countless other body movements possible. Muscles work by pulling, or contracting. At the cellular level, muscle contractions are powered by reactions that transform chemical energy into kinetic energy.

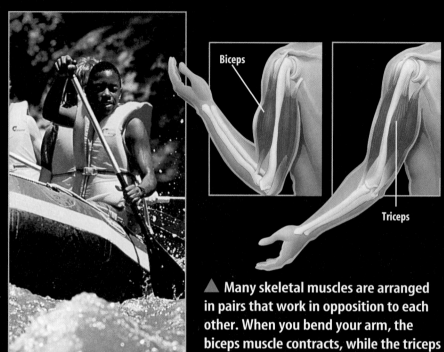

▶ Energy transformations taking place in your muscles provide the power to move.

▲ Many skeletal muscles are arranged in pairs that work in opposition to each other. When you bend your arm, the biceps muscle contracts, while the triceps relaxes. When you extend your arm the triceps contracts, and the biceps relaxes.

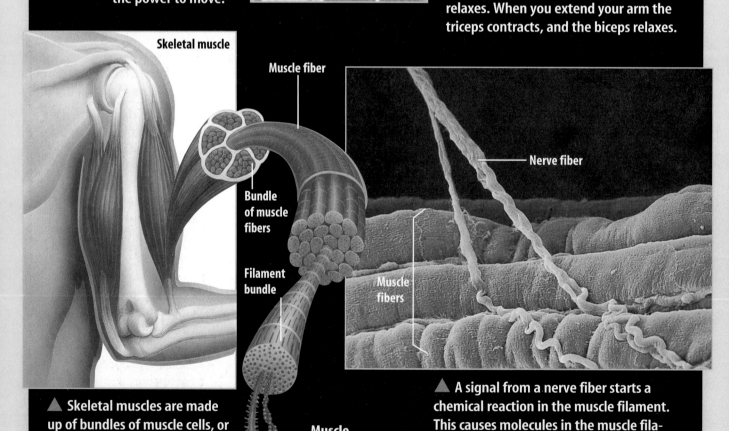

▲ Skeletal muscles are made up of bundles of muscle cells, or fibers. Each fiber is composed of many bundles of muscle filaments.

▲ A signal from a nerve fiber starts a chemical reaction in the muscle filament. This causes molecules in the muscle filament to gain energy and move. Many filaments moving together cause the muscle to contract.

Figure 12 The simple act of listening to a radio involves many energy transformations. A few are diagrammed here.

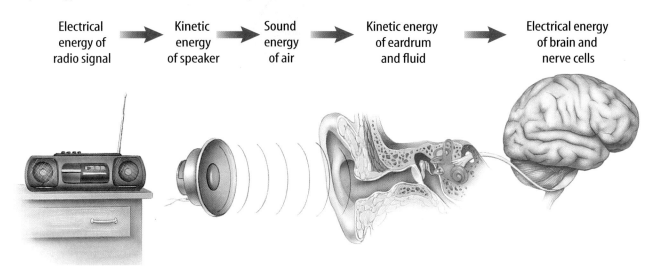

Transforming Electrical Energy Every day you use electrical energy. When you flip a light switch, or turn on a radio or television, or use a hair drier, you are transforming electrical energy to other forms of energy. Every time you plug something into a wall outlet, or use a battery, you are using electrical energy. **Figure 12** shows how electrical energy is transformed into other forms of energy when you listen to a radio. A loudspeaker in the radio converts electrical energy into sound waves that travel to your ear—energy in motion. The energy that is carried by the sound waves causes parts of the ear to move also. This energy of motion is transformed again into chemical and electrical energy in nerve cells, which send the energy to your brain. After your brain interprets this energy as a voice or music, where does the energy go? The energy finally is transformed into thermal energy.

Transforming Thermal Energy Different forms of energy can be transformed into thermal energy. For example, chemical energy changes into thermal energy when something burns. Electrical energy changes into thermal energy when a wire that is carrying an electric current gets hot. Thermal energy can be used to heat buildings and keep you warm. Thermal energy also can be used to heat water. If water is heated to its boiling point, it changes to steam. This steam can be used to produce kinetic energy by steam engines, like the steam locomotives that used to pull trains. Thermal energy also can be transformed into radiant energy. For example, when a bar of metal is heated to a high temperature, it glows and gives off light.

Controlling Body Temperature Most organisms have some adaptation for controlling the amount of thermal energy in their bodies. Some living in cooler climates have thick fur coats that help prevent thermal energy from escaping, and some living in desert regions have skin that helps keep thermal energy out. Research some of the adaptations different organisms have for controlling the thermal energy in their bodies.

SECTION 2 Energy Transformations **383**

How Thermal Energy Moves Thermal energy can move from one place to another. Look at **Figure 13.** The hot chocolate has thermal energy that moves from the cup to the cooler air around it, and to the cooler spoon. Thermal energy only moves from something at a higher temperature to something at a lower temperature.

Generating Electrical Energy

The enormous amount of electrical energy that is used every day is too large to be stored in batteries. The electrical energy that is available for use at any wall socket must be generated continually by power plants. Every power plant works on the same principle—energy is used to turn a large generator. A **generator** is a device that transforms kinetic energy into electrical energy. In fossil fuel power plants, coal, oil, or natural gas is burned to boil water. As the hot water boils, the steam rushes through a **turbine,** which contains a set of narrowly spaced fan blades. The steam pushes on the blades and turns the turbine, which in turn rotates a shaft in the generator to produce the electrical energy, as shown in **Figure 14.**

Reading Check *What does a generator do?*

Figure 13 Thermal energy moves from the hot chocolate to the cooler surroundings. **Explain** *what happens to the hot chocolate as it loses thermal energy.*

Figure 14 A coal-burning power plant transforms the chemical energy in coal into electrical energy. **List** *some of the other energy sources that power plants use.*

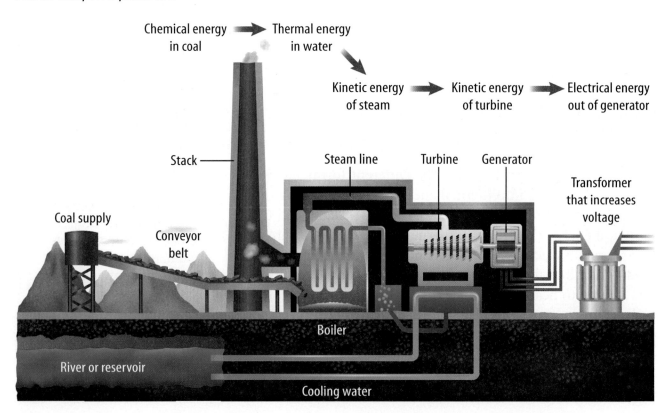

384 CHAPTER 13 Energy and Energy Resources

Power Plants Almost 90 percent of the electrical energy generated in the United States is produced by nuclear and fossil fuel power plants, as shown in **Figure 15.** Other types of power plants include hydroelectric (hi droh ih LEK trihk) and wind. Hydroelectric power plants transform the kinetic energy of moving water into electrical energy. Wind power plants transform the kinetic energy of moving air into electrical energy. In these power plants, a generator converts the kinetic energy of moving water or wind to electrical energy.

To analyze the energy transformations in a power plant, you can diagram the energy changes using arrows. A coal-burning power plant generates electrical energy through the following series of energy transformations.

chemical energy of coal → thermal energy of water → kinetic energy of steam → kinetic energy of turbine → electrical energy out of generator

Nuclear power plants use a similar series of transformations. Hydroelectric plants, however, skip the steps that change water into steam because the water strikes the turbine directly.

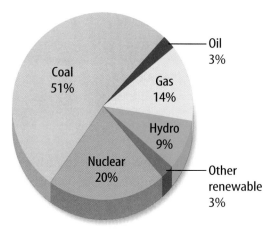

Figure 15 The graph shows sources of electrical energy in the United States.
Name the energy source that you think is being used to provide the electricity for the lights overhead.

section 2 review

Summary

Changing Forms of Energy
- Heat usually is one of the forms of energy produced in energy transformations.
- The law of conservation of energy states that energy cannot be created or destroyed; it can only change form.
- The total energy doesn't change when an energy transformation occurs.
- As an object rises and falls, kinetic and potential energy are transformed into each other, but the total energy doesn't change.

Generating Electrical Energy
- A generator converts kinetic energy into electrical energy.
- Burning fossil fuels produces thermal energy that is used to boil water and produce steam.
- In a power plant, steam is used to spin a turbine which then spins an electric generator.

Self Check

1. **Describe** the conversions between potential and kinetic energy that occur when you shoot a basketball at a basket.
2. **Explain** whether your body gains or loses thermal energy if your body temperature is 37°C and the temperature around you is 25°C.
3. **Describe** a process that converts chemical energy to thermal energy.
4. **Think Critically** A lightbulb converts 10 percent of the electrical energy it uses into radiant energy. Make a hypothesis about the other form of energy produced.

Applying Math

5. **Use a Ratio** How many times greater is the amount of electrical energy produced in the United States by coal-burning power plants than the amount produced by nuclear power plants?

Hearing with Your Jaw

You probably have listened to music using speakers or headphones. Have you ever considered how energy is transferred to get the energy from the radio or CD player to your brain? What type of energy is needed to power the radio or CD player? Where does this energy come from? How does that energy become sound? How does the sound get to you? In this activity, the sound from a radio or CD player is going to travel through a motor before entering your body through your jaw instead of your ears.

Real-World Question
How can energy be transferred from a radio or CD player to your brain?

Goals
- **Identify** energy transfers and transformations.
- **Explain** your observations using the law of conservation of energy.

Materials
radio or CD player
small electrical motor
headphone jack

Procedure

1. Go to one of the places in the room with a motor/radio assembly.
2. Turn on the radio or CD player so that you hear the music.
3. Push the headphone jack into the headphone plug on the radio or CD player.
4. Press the axle of the motor against the side of your jaw.

Conclude and Apply

1. **Describe** what you heard in your Science Journal.
2. **Identify** the form of energy produced by the radio or CD player.
3. **Draw** a diagram to show all of the energy transformations taking place.
4. **Evaluate** Did anything get hotter as a result of this activity? Explain.
5. **Explain** your observations using the law of conservation of energy.

Communicating Your Data

Compare your conclusions with those of other students in your class. **For more help, refer to the** Science Skill Handbook.

section 3
Sources of Energy

Using Energy

Every day, energy is used to provide light and to heat and cool homes, schools, and workplaces. According to the law of conservation of energy, energy can't be created or destroyed. Energy only can change form. If a car or refrigerator can't create the energy they use, then where does this energy come from?

Energy Resources

Energy cannot be made, but must come from the natural world. As you can see in **Figure 16,** the surface of Earth receives energy from two sources—the Sun and radioactive atoms in Earth's interior. The amount of energy Earth receives from the Sun is far greater than the amount generated in Earth's interior. Nearly all the energy you used today can be traced to the Sun, even the gasoline used to power the car or school bus you came to school in.

as you read

What You'll Learn
- **Explain** what renewable, non-renewable, and alternative resources are.
- **Describe** the advantages and disadvantages of using various energy sources.

Why It's Important
Energy is vital for survival and making life comfortable. Developing new energy sources will improve modern standards of living.

Review Vocabulary
resource: a natural feature or phenomenon that enhances the quality of life

New Vocabulary
- nonrenewable resource
- renewable resource
- alternative resource
- inexhaustible resource
- photovoltaic

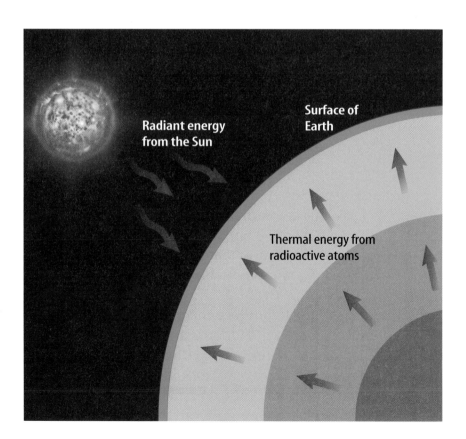

Figure 16 All the energy you use can be traced to one of two sources—the Sun or radioactive atoms in Earth's interior.

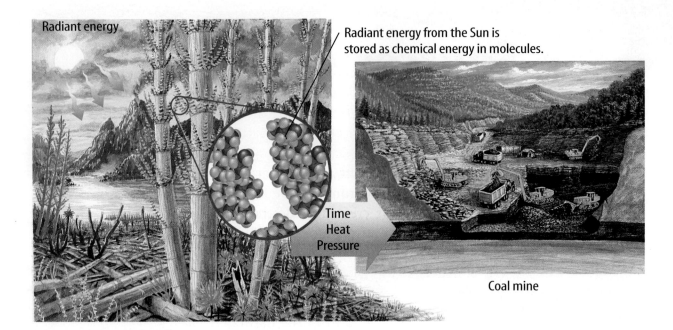

Figure 17 Coal is formed after the molecules in ancient plants are heated under pressure for millions of years. The energy stored by the molecules in coal originally came from the Sun.

Energy Source Origins The kinds of fossil fuels found in the ground depend on the kinds of organisms (animal or plant) that died and were buried in that spot. Research coal, oil, and natural gas to find out what types of organisms were primarily responsible for producing each.

Fossil Fuels

Fossil fuels are coal, oil, and natural gas. Oil and natural gas were made from the remains of microscopic organisms that lived in Earth's oceans millions of years ago. Heat and pressure gradually turned these ancient organisms into oil and natural gas. Coal was formed by a similar process from the remains of ancient plants that once lived on land, as shown in **Figure 17.**

Through the process of photosynthesis, ancient plants converted the radiant energy in sunlight to chemical energy stored in various types of molecules. Heat and pressure changed these molecules into other types of molecules as fossil fuels formed. Chemical energy stored in these molecules is released when fossil fuels are burned.

Using Fossil Fuels The energy used when you ride in a car, turn on a light, or use an electric appliance usually comes from burning fossil fuels. However, it takes millions of years to replace each drop of gasoline and each lump of coal that is burned. This means that the supply of oil on Earth will continue to decrease as oil is used. An energy source that is used up much faster than it can be replaced is a **nonrenewable resource.** Fossil fuels are nonrenewable resources.

Burning fossil fuels to produce energy also generates chemical compounds that cause pollution. Each year billions of kilograms of air pollutants are produced by burning fossil fuels. These pollutants can cause respiratory illnesses and acid rain. Also, the carbon dioxide gas formed when fossil fuels are burned might cause Earth's climate to warm.

Nuclear Energy

Can you imagine running an automobile on 1 kg of fuel that releases almost 3 million times more energy than 1 L of gas? What could supply so much energy from so little mass? The answer is the nuclei of uranium atoms. Some of these nuclei are unstable and break apart, releasing enormous amounts of energy in the process. This energy can be used to generate electricity by heating water to produce steam that spins an electric generator, as shown in **Figure 18.** Because no fossil fuels are burned, generating electricity using nuclear energy helps make the supply of fossil fuels last longer. Also, unlike fossil fuel power plants, nuclear power plants produce almost no air pollution. In one year, a typical nuclear power plant generates enough energy to supply 600,000 homes with power and produces only 1 m^3 of waste.

Nuclear Wastes Like all energy sources, nuclear energy has its advantages and disadvantages. One disadvantage is the amount of uranium in Earth's crust is nonrenewable. Another is that the waste produced by nuclear power plants is radioactive and can be dangerous to living things. Some of the materials in the nuclear waste will remain radioactive for many thousands of years. As a result the waste must be stored so no radioactivity is released into the environment for a long time. One method is to seal the waste in a ceramic material, place the ceramic in protective containers, and then bury the containers far underground. However, the burial site would have to be chosen carefully so underground water supplies aren't contaminated. Also, the site would have to be safe from earthquakes and other natural disasters that might cause radioactive material to be released.

Figure 18 To obtain electrical energy from nuclear energy, a series of energy transformations must occur.

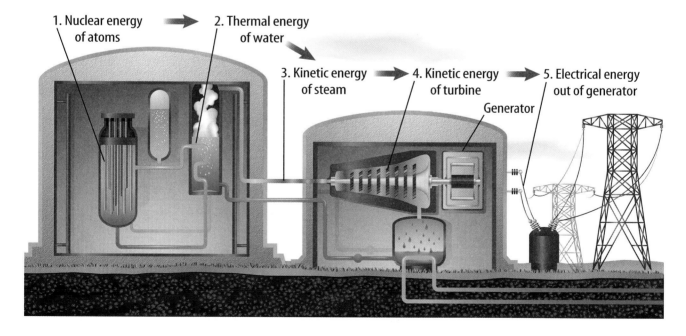

Hydroelectricity

Currently, transforming the potential energy of water that is trapped behind dams supplies the world with almost 20 percent of its electrical energy. Hydroelectricity is the largest renewable source of energy. A **renewable resource** is an energy source that is replenished continually. As long as enough rain and snow fall to keep rivers flowing, hydroelectric power plants can generate electrical energy, as shown in **Figure 19**.

Although production of hydroelectricity is largely pollution free, it has one major problem. It disrupts the life cycle of aquatic animals, especially fish. This is particularly true in the Northwest where salmon spawn and run. Because salmon return to the spot where they were hatched to lay their eggs, the development of dams has hindered a large fraction of salmon from reproducing. This has greatly reduced the salmon population. Efforts to correct the problem have resulted in plans to remove a number of dams. In an attempt to help fish bypass some dams, fish ladders are being installed. Like most energy sources, hydroelectricity has advantages and disadvantages.

Topic: Hydroelectricity
Visit ips.msscience.com for Web links to information about the use of hydroelectricity in various parts of the world.

Activity On a map of the world, show where the use of hydroelectricity is the greatest.

Applying Science

Is energy consumption outpacing production?

You use energy every day—to get to school, to watch TV, and to heat or cool your home. The amount of energy consumed by an average person has increased over time. Consequently, more energy must be produced.

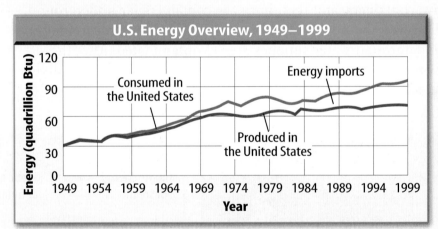

Identifying the Problem

The graph above shows the energy produced and consumed in the United States from 1949 to 1999. How does energy that is consumed by Americans compare with energy that is produced in the United States?

Solving the Problem

1. Determine the approximate amount of energy produced in 1949 and in 1999 and how much it has increased in 50 years. Has it doubled or tripled?
2. Do the same for consumption. Has it doubled or tripled?
3. Using your answers for steps 1 and 2 and the graph, where does the additional energy that is needed come from? Give some examples.

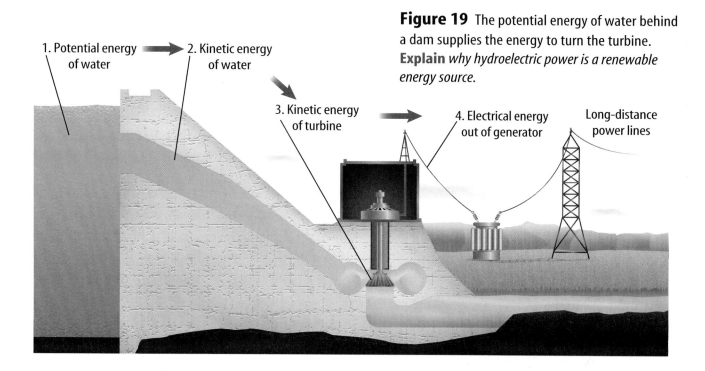

Figure 19 The potential energy of water behind a dam supplies the energy to turn the turbine. **Explain** *why hydroelectric power is a renewable energy source.*

Alternative Sources of Energy

Electrical energy can be generated in several ways. However, each has disadvantages that can affect the environment and the quality of life for humans. Research is being done to develop new sources of energy that are safer and cause less harm to the environment. These sources often are called **alternative resources.** These alternative resources include solar energy, wind, and geothermal energy.

Solar Energy

The Sun is the origin of almost all the energy that is used on Earth. Because the Sun will go on producing an enormous amount of energy for billions of years, the Sun is an inexhaustible source of energy. An **inexhaustible resource** is an energy source that can't be used up by humans.

Each day, on average, the amount of solar energy that strikes the United States is more than the total amount of energy used by the entire country in a year. However, less than 0.1 percent of the energy used in the United States comes directly from the Sun. One reason is that solar energy is more expensive to use than fossil fuels. However, as the supply of fossil fuels decreases, the cost of finding and mining these fuels might increase. Then, it may be cheaper to use solar energy or other energy sources to generate electricity and heat buildings than to use fossil fuels.

Reading Check *What is an inexhaustible energy source?*

Building a Solar Collector

Procedure
1. Line a **large pot** with **black plastic** and fill with **water.**
2. Stretch **clear-plastic wrap** over the pot and tape it taut.
3. Make a slit in the top and slide a **thermometer** or a **computer probe** into the water.
4. Place your solar collector in direct sunlight and monitor the temperature change every 3 min for 15 min.
5. Repeat your experiment without using any black plastic.

Analysis
1. Graph the temperature changes in both setups.
2. Explain how your solar collector works.

SECTION 3 Sources of Energy **391**

Collecting the Sun's Energy Two types of collectors capture the Sun's rays. If you look around your neighborhood, you might see large, rectangular panels attached to the roofs of buildings or houses. If, as in **Figure 20,** pipes come out of the panel, it is a thermal collector. Using a black surface, a thermal collector heats water by directly absorbing the Sun's radiant energy. Water circulating in this system can be heated to about 70°C. The hot water can be pumped through the house to provide heat. Also, the hot water can be used for washing and bathing. If the panel has no pipes, it is a photovoltaic (foh toh vol TAY ihk) collector, like the one pictured in **Figure 20**. A **photovoltaic** is a device that transforms radiant energy directly into electrical energy. Photovoltaics are used to power calculators and satellites, including the *International Space Station*.

Figure 20 Solar energy can be collected and utilized by individuals using thermal collectors or photovoltaic collectors.

Reading Check *What does a photovoltaic do?*

Geothermal Energy

Imagine you could take a journey to the center of Earth—down to about 6,400 km below the surface. As you went deeper and deeper, you would find the temperature increasing. In fact, after going only about 3 km, the temperature could have increased enough to boil water. At a depth of 100 km, the temperature could be over 900°C. The heat generated inside Earth is called geothermal energy. Some of this heat is produced when unstable radioactive atoms inside Earth decay, converting nuclear energy to thermal energy.

At some places deep within Earth the temperature is hot enough to melt rock. This molten rock, or magma, can rise up close to the surface through cracks in the crust. During a volcanic eruption, magma reaches the surface. In other places, magma gets close to the surface and heats the rock around it.

Geothermal Reservoirs In some regions where magma is close to the surface, rainwater and water from melted snow can seep down to the hot rock through cracks and other openings in Earth's surface. The water then becomes hot and sometimes can form steam. The hot water and steam can be trapped under high pressure in cracks and pockets called geothermal reservoirs. In some places, the hot water and steam are close enough to the surface to form hot springs and geysers.

Geothermal Power Plants In places where the geothermal reservoirs are less than several kilometers deep, wells can be drilled to reach them. The hot water and steam produced by geothermal energy then can be used by geothermal power plants, like the one in **Figure 21,** to generate electricity.

Most geothermal reservoirs contain hot water under high pressure. **Figure 22** shows how these reservoirs can be used to generate electricity. While geothermal power is an inexhaustible source of energy, geothermal power plants can be built only in regions where geothermal reservoirs are close to the surface, such as in the western United States.

Heat Pumps Geothermal heat helps keep the temperature of the ground at a depth of several meters at a nearly constant temperature of about 10° to 20°C. This constant temperature can be used to cool and heat buildings by using a heat pump.

A heat pump contains a water-filled loop of pipe that is buried to a depth where the temperature is nearly constant. In summer the air is warmer than this underground temperature. Warm water from the building is pumped through the pipe down into the ground. The water cools and then is pumped back to the house where it absorbs more heat, and the cycle is repeated. During the winter, the air is cooler than the ground below. Then, cool water absorbs heat from the ground and releases it into the house.

Figure 21 This geothermal power plant in Nevada produces enough electricity to power about 50,000 homes.

Figure 22 The hot water in a geothermal reservoir is used to generate electricity in a geothermal power plant.

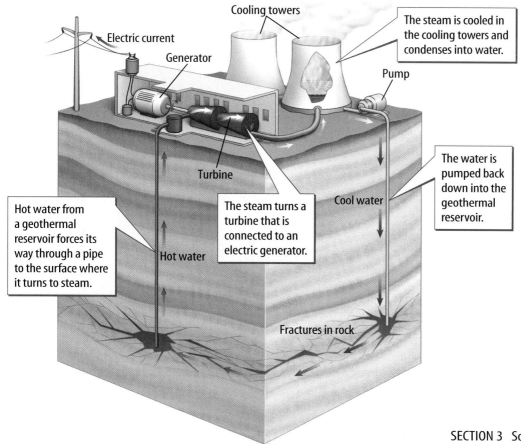

SECTION 3 Sources of Energy

Energy from the Oceans

The ocean is in constant motion. If you've been to the seashore you've seen waves roll in. You may have seen the level of the ocean rise and fall over a period of about a half day. This rise and fall in the ocean level is called a tide. The constant movement of the ocean is an inexhaustible source of mechanical energy that can be converted into electric energy. While methods are still being developed to convert the motion in ocean waves to electric energy, several electric power plants using tidal motion have been built.

Figure 23 This tidal power plant in Annapolis Royal, Nova Scotia, is the only operating tidal power plant in North America.

Using Tidal Energy A high tide and a low tide each occur about twice a day. In most places the level of the ocean changes by less than a few meters. However, in some places the change is much greater. In the Bay of Fundy in Eastern Canada, the ocean level changes by 16 m between high tide and low tide. Almost 14 trillion kg of water move into or out of the bay between high and low tide.

Figure 23 shows an electric power plant that has been built along the Bay of Fundy. This power plant generates enough electric energy to power about 12,000 homes. The power plant is constructed so that as the tide rises, water flows through a turbine that causes an electric generator to spin, as shown in **Figure 24A**. The water is then trapped behind a dam. When the tide goes out, the trapped water behind the dam is released through the turbine to generate more electricity, as shown in **Figure 24B**. Each day electric power is generated for about ten hours when the tide is rising and falling.

While tidal energy is a nonpolluting, inexhaustible energy source, its use is limited. Only in a few places is the difference between high and low tide large enough to enable a large electric power plant to be built.

Figure 24 A tidal power plant can generate electricity when the tide is coming in and going out.

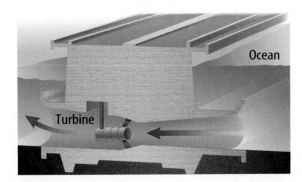

A As the tide comes in, it turns a turbine connected to a generator. When high tide occurs, gates are closed that trap water behind a dam.

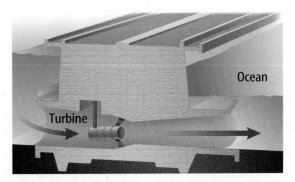

B As the tide goes out and the ocean level drops, the gates are opened and water from behind the dam flows through the turbine, causing it to spin and turn a generator.

Wind

Wind is another inexhaustible supply of energy. Modern windmills, like the ones in **Figure 25,** convert the kinetic energy of the wind to electrical energy. The propeller is connected to a generator so that electrical energy is generated when wind spins the propeller. These windmills produce almost no pollution. Some disadvantages are that windmills produce noise and that large areas of land are needed. Also, studies have shown that birds sometimes are killed by windmills.

Conserving Energy

Fossil fuels are a valuable resource. Not only are they burned to provide energy, but oil and coal also are used to make plastics and other materials. One way to make the supply of fossil fuels last longer is to use less energy. Reducing the use of energy is called conserving energy.

You can conserve energy and also save money by turning off lights and appliances such as televisions when you are not using them. Also keep doors and windows closed tightly when it's cold or hot to keep heat from leaking out of or into your house. Energy could also be conserved if buildings are properly insulated, especially around windows. The use of oil could be reduced if cars were used less and made more efficient, so they went farther on a liter of gas. Recycling materials such as aluminum cans and glass also helps conserve energy.

Figure 25 Windmills work on the same basic principles as a power plant. Instead of steam turning a turbine, wind turns the rotors. **Describe** *some of the advantages and disadvantages of using windmills.*

section 3 review

Summary

Nonrenewable Resources

- All energy resources have advantages and disadvantages.
- Nonrenewable energy resources are used faster than they are replaced.
- Fossil fuels include oil, coal, and natural gas and are nonrenewable resources. Nuclear energy is a nonrenewable resource.

Renewable and Alternative Resources

- Renewable energy resources, such as hydroelectricity, are resources that are replenished continually.
- Alternative energy sources include solar energy, wind energy, and geothermal energy.

Self Check

1. **Diagram** the energy conversions that occur when coal is formed, and then burned to produce thermal energy.
2. **Explain** why solar energy is considered an inexhaustible source of energy.
3. **Explain** how a heat pump is used to both heat and cool a building.
4. **Think Critically** Identify advantages and disadvantages of using fossil fuels, hydroelectricity, and solar energy as energy sources.

Applying Math

5. **Use a Ratio** Earth's temperature increases with depth. Suppose the temperature increase inside Earth is 500°C at a depth of 50 km. What is the temperature increase at a depth of 10 km?

LAB Use the Internet

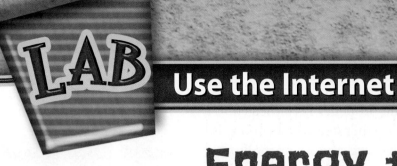

Energy to Power Your Life

Goals
- **Identify** how energy you use is produced and delivered.
- **Investigate** alternative sources for the energy you use.
- **Outline** a plan for how these alternative sources of energy could be used.

Data Source

Science Online

Visit ips.msscience.com/internet_lab for more information about sources of energy and for data collected by other students.

Real-World Question

Over the past 100 years, the amount of energy used in the United States and elsewhere has greatly increased. Today, a number of energy sources are available, such as coal, oil, natural gas, nuclear energy, hydroelectric power, wind, and solar energy. Some of these energy sources are being used up and are nonrenewable, but others are replaced as fast as they are used and, therefore, are renewable. Some energy sources are so vast that human usage has almost no effect on the amount available. These energy sources are inexhaustible.

Think about the types of energy you use at home and school every day. In this lab, you will investigate how and where energy is produced, and how it gets to you. You will also investigate alternative ways energy can be produced, and whether these sources are renewable, nonrenewable, or inexhaustible. What are the sources of the energy you use every day?

Local Energy Information	
Energy Type	
Where is that energy produced?	
How is that energy produced?	**Do not write in this book.**
How is that energy delivered to you?	
Is the energy source renewable, nonrenewable, or inexhaustible?	
What type of alternative energy source could you use instead?	

CHAPTER 13 Energy and Energy Resources

Using Scientific Methods

Make a Plan

1. Think about the activities you do every day and the things you use. When you watch television, listen to the radio, ride in a car, use a hair drier, or turn on the air conditioning, you use energy. Select one activity or appliance that uses energy.
2. **Identify** the type of energy that is used.
3. **Investigate** how that energy is produced and delivered to you.
4. **Determine** if the energy source is renewable, nonrenewable, or inexhaustible.
5. If your energy source is nonrenewable, describe how the energy you use could be produced by renewable sources.

Follow Your Plan

1. Make sure your teacher approves your plan before you start.
2. Organize your findings in a data table, similar to the one that is shown.

Analyze Your Data

1. **Describe** the process for producing and delivering the energy source you researched. How is it created, and how does it get to you?
2. How much energy is produced by the energy source you investigated?
3. Is the energy source you researched renewable, nonrenewable, or inexhaustible? Why?

Conclude and Apply

1. **Describe** If the energy source you investigated is nonrenewable, how can the use of this energy source be reduced?
2. **Organize** What alternative sources of energy could you use for everyday energy needs? On the computer, create a plan for using renewable or inexhaustible sources.

Communicating Your Data

Find this lab using the link below. Post your data in the table that is provided. **Compare** your data to those of other students. **Combine** your data with those of other students and make inferences using the combined data.

ips.msscience.com/internet_lab

SCIENCE Stats

Energy to Burn

Did you know...

... The energy released by the average hurricane is equal to about 200 times the total energy produced by all of the world's power plants. Almost all of this energy is released as heat when raindrops form.

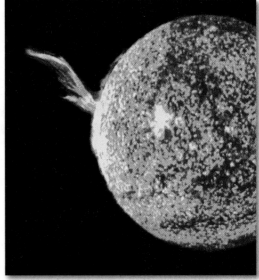

... The energy Earth gets each half hour from the Sun is enough to meet the world's demands for a year. Renewable and inexhaustible resources, including the Sun, account for only 18 percent of the energy that is used worldwide.

... The Calories in one medium apple will give you enough energy to walk for about 15 min, swim for about 10 min, or jog for about 9 min.

Applying Math If walking for 15 min requires 80 Calories of fuel (from food), how many Calories would someone need to consume to walk for 1 h?

Write About It

Where would you place solar collectors in the United States? Why? For more information on solar energy, go to ips.msscience.com/science_stats.

chapter 13 Study Guide

Reviewing Main Ideas

Section 1 What is energy?

1. Energy is the ability to cause change.
2. A moving object has kinetic energy that depends on the object's mass and speed.
3. Potential energy is energy due to position and depends on an object's mass and height.
4. Light carries radiant energy, electric current carries electrical energy, and atomic nuclei contain nuclear energy.

Section 2 Energy Transformations

1. Energy can be transformed from one form to another. Thermal energy is usually produced when energy transformations occur.
2. The law of conservation of energy states that energy cannot be created or destroyed.
3. Electric power plants convert a source of energy into electrical energy. Steam spins a turbine which spins an electric generator.

Section 3 Sources of Energy

1. The use of an energy source has advantages and disadvantages.
2. Fossil fuels and nuclear energy are nonrenewable energy sources that are consumed faster than they can be replaced.
3. Hydroelectricity is a renewable energy source that is continually being replaced.
4. Alternative energy sources include solar, wind, and geothermal energy. Solar energy is an inexhaustible energy source.

Visualizing Main Ideas

Copy and complete the concept map using the following terms: fossil fuels, hydroelectric, solar, wind, oil, coal, photovoltaic, *and* nonrenewable resources.

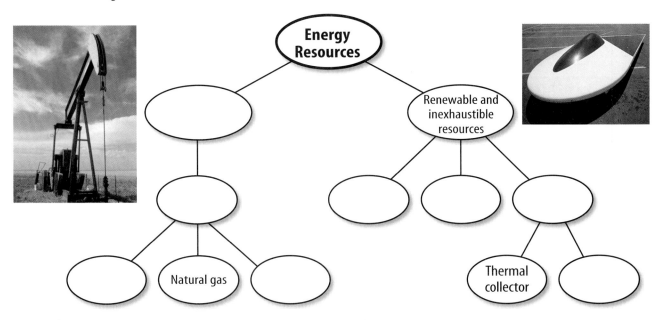

Chapter 13 Review

Using Vocabulary

alternative resource p. 391
chemical energy p. 377
electrical energy p. 378
energy p. 374
generator p. 384
inexhaustible resource p. 391
kinetic energy p. 375
law of conservation of energy p. 380
nonrenewable resource p. 388
nuclear energy p. 378
photovoltaic p. 392
potential energy p. 376
radiant energy p. 377
renewable resource p. 390
thermal energy p. 376
turbine p. 384

For each of the terms below, explain the relationship that exists.

1. electrical energy—nuclear energy
2. turbine—generator
3. photovoltaic—radiant energy—electrical energy
4. renewable resource—inexhaustible resource
5. potential energy—kinetic energy
6. kinetic energy—electrical energy—generator
7. thermal energy—radiant energy
8. law of conservation of energy—energy transformations
9. nonrenewable resource—chemical energy

Checking Concepts

Choose the word or phrase that best answers the question.

10. Objects that are able to fall have what type of energy?
    A) kinetic   C) potential
    B) radiant   D) electrical

11. Which form of energy does light have?
    A) electrical   C) kinetic
    B) nuclear   D) radiant

12. Muscles perform what type of energy transformation?
    A) kinetic to potential
    B) kinetic to electrical
    C) thermal to radiant
    D) chemical to kinetic

13. Photovoltaics perform what type of energy transformation?
    A) thermal to radiant
    B) kinetic to electrical
    C) radiant to electrical
    D) electrical to thermal

14. The form of energy that food contains is which of the following?
    A) chemical   C) radiant
    B) potential   D) electrical

15. Solar energy, wind, and geothermal are what type of energy resource?
    A) inexhaustible   C) nonrenewable
    B) inexpensive   D) chemical

16. Which of the following is a nonrenewable source of energy?
    A) hydroelectricity
    B) nuclear
    C) wind
    D) solar

17. A generator is NOT required to generate electrical energy when which of the following energy sources is used?
    A) solar   C) hydroelectric
    B) wind   D) nuclear

18. Which of the following are fossil fuels?
    A) gas   C) oil
    B) coal   D) all of these

19. Almost all of the energy that is used on Earth's surface comes from which of the following energy sources?
    A) radioactivity   C) chemicals
    B) the Sun   D) wind

400 CHAPTER REVIEW   ips.msscience.com/vocabulary_puzzlemaker

chapter 13 Review

Thinking Critically

20. **Explain** how the motion of a swing illustrates the transformation between potential and kinetic energy.

21. **Explain** what happens to the kinetic energy of a skateboard that is coasting along a flat surface, slows down, and comes to a stop.

22. **Describe** the energy transformations that occur in the process of toasting a bagel in an electric toaster.

23. **Compare and contrast** the formation of coal and the formation of oil and natural gas.

24. **Explain** the difference between the law of conservation of energy and conserving energy. How can conserving energy help prevent energy shortages?

25. **Make a Hypothesis** about how spacecraft that travel through the solar system obtain the energy they need to operate. Do research to verify your hypothesis.

26. **Concept Map** Copy and complete this concept map about energy.

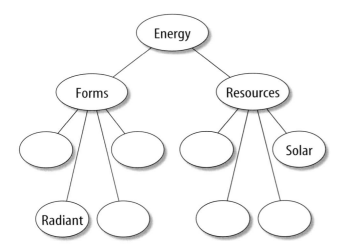

27. **Diagram** the energy transformations that occur when you rub sandpaper on a piece of wood and the wood becomes warm.

Performance Activities

28. **Multimedia Presentation** Alternative sources of energy that weren't discussed include biomass energy, wave energy, and hydrogen fuel cells. Research an alternative energy source and then prepare a digital slide show about the information you found. Use the concepts you learned from this chapter to inform your classmates about the future prospects of using such an energy source on a large scale.

Applying Math

29. **Calculate Number of Power Plants** A certain type of power plant is designed to provide energy for 10,000 homes. How many of these power plants would be needed to provide energy for 300,000 homes?

Use the table below to answer questions 30 and 31.

Energy Sources Used in the United States	
Energy Source	Percent of Energy Used
Coal	23%
Oil	39%
Natural gas	23%
Nuclear	8%
Hydroelectric	4%
Other	3%

30. **Use Percentages** According to the data in the table above, what percentage of the energy used in the United States comes from fossil fuels?

31. **Calculate a Ratio** How many times greater is the amount of energy that comes from fossil fuels than the amount of energy from all other energy sources?

Chapter 13 Standardized Test Practice

Part 1 Multiple Choice

Record your answers on the answer sheet provided by your teacher or on a sheet of paper.

1. The kinetic energy of a moving object increases if which of the following occurs?
   A. Its mass decreases.
   B. Its speed increases.
   C. Its height above the ground increases.
   D. Its temperature increases.

Use the graph below to answer questions 2–4.

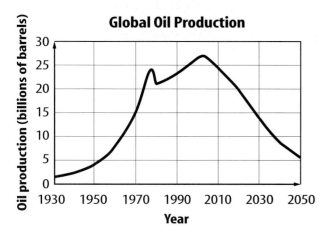

2. According to the graph above, in which year will global oil production be at a maximum?
   A. 1974
   B. 2002
   C. 2010
   D. 2050

3. Approximately how many times greater was oil production in 1970 than oil production in 1950?
   A. 2 times
   B. 10 times
   C. 6 times
   D. 3 times

4. In which year will the production of oil be equal to the oil production in 1970?
   A. 2010
   B. 2015
   C. 2022
   D. 2028

5. Which of the following energy sources is being used faster than it can be replaced?
   A. tidal
   B. wind
   C. fossil fuels
   D. hydroelectric

Use the circle graph below to answer question 6.

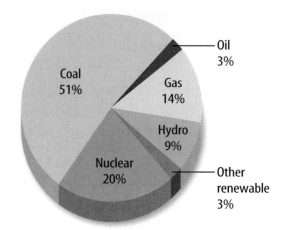

6. The circle graph shows the sources of electrical energy in the United States. In 2002, the total amount of electrical energy produced in the United States was 38.2 quads. How much electrical energy was produced by nuclear power plants?
   A. 3.0 quads
   B. 3.8 quads
   C. 7.6 quads
   D. 35.1 quads

7. When chemical energy is converted into thermal energy, which of the following must be true?
   A. The total amount of thermal energy plus chemical energy changes.
   B. Only the amount of chemical energy changes.
   C. Only the amount of thermal energy changes.
   D. The total amount of thermal energy plus chemical energy doesn't change.

8. A softball player hits a fly ball. Which of the following describes the energy conversion that occurs as it falls from its highest point?
   A. kinetic to potential
   B. potential to kinetic
   C. thermal to potential
   D. thermal to kinetic

Part 2 Short Response/Grid In

Record your answers on the answer sheet provided by your teacher or on a sheet of paper.

9. Why is it impossible to build a machine that produces more energy than it uses?

10. You toss a ball upward and then catch it on the way down. The height of the ball above the ground when it leaves your hand on the way up and when you catch it is the same. Compare the ball's kinetic energy when it leaves your hand and just before you catch it.

11. A basket ball is dropped from a height of 2 m and another identical basketball is dropped from a height of 4 m. Which ball has more kinetic energy just before it hits the ground?

Use the graph below to answer questions 12 and 13.

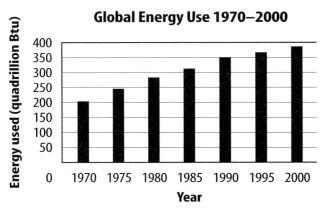

12. According to the graph above, by about how many times did the global use of energy increase from 1970 to 2000?

13. Over which five-year time period was the increase in global energy use the largest?

Test-Taking Tip

Do Your Studying Regularly Do not "cram" the night before the test. It can hamper your memory and make you tired.

Part 3 Open Ended

Record your answers on a sheet of paper.

14. When you drop a tennis ball, it hits the floor and bounces back up. But it does not reach the same height as released, and each successive upward bounce is smaller than the one previous. However, you notice the tennis ball is slightly warmer after it finishes bouncing. Explain how the law of conservation of energy is obeyed.

Use the graph below to answer questions 15–17.

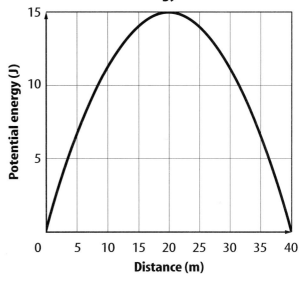

15. The graph shows how the potential energy of a batted ball depends on distance from the batter. At what distances is the kinetic energy of the ball the greatest?

16. At what distance from the batter is the height of the ball the greatest?

17. How much less is the kinetic energy of the ball at a distance of 20 m from the batter than at a distance of 0 m?

18. List advantages and disadvantages of the following energy sources: fossil fuels, nuclear energy, and geothermal energy.

chapter 14

Work and Simple Machines

The BIG Idea
A machine makes doing a job easier.

SECTION 1
Work and Power
Main Idea Work is done when a force causes an object to move in the same direction as the force.

SECTION 2
Using Machines
Main Idea A machine can change the force needed to do a job.

SECTION 3
Simple Machines
Main Idea There are six types of simple machines.

Heavy Lifting

It took the ancient Egyptians more than 100 years to build the pyramids without machines like these. But now, even tall skyscrapers can be built in a few years. Complex or simple, machines have the same purpose. They make doing a job easier.

Science Journal Describe three machines you used today, and how they made doing a task easier.

Start-Up Activities

Compare Forces

Two of the world's greatest structures were built using different tools. The Great Pyramid at Giza in Egypt was built nearly 5,000 years ago using blocks of limestone moved into place by hand with ramps and levers. In comparison, the Sears Tower in Chicago was built in 1973 using tons of steel that were hoisted into place by gasoline-powered cranes. How do machines such as ramps, levers, and cranes change the forces needed to do a job?

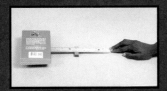

1. Place a ruler on an eraser. Place a book on one end of the ruler.
2. Using one finger, push down on the free end of the ruler to lift the book.
3. Repeat the experiment, placing the eraser in various positions beneath the ruler. Observe how much force is needed in each instance to lift the book.
4. **Think Critically** In your Science Journal, describe your observations. How did changing the distance between the book and the eraser affect the force needed to lift the book?

FOLDABLES
Study Organizer

Simple Machines Many of the devices that you use every day are simple machines. Make the following Foldable to help you understand the characteristics of simple machines.

STEP 1 Draw a mark at the midpoint of a sheet of paper along the side edge. Then **fold** the top and bottom edges in to touch the midpoint.

STEP 2 **Fold** in half from side to side.

STEP 3 **Turn** the paper vertically. **Open** and cut along the inside fold lines to form four tabs.

STEP 4 Label the tabs *Inclined Plane, Lever, Wheel and Axle,* and *Pulley.*

Read for Main Ideas As you read the chapter, list the characteristics of inclined planes, levers, wheels and axles, and pulleys under the appropriate tab.

Preview this chapter's content and activities at
ips.msscience.com

Get Ready to Read

Questions and Answers

① Learn It! Knowing how to find answers to questions will help you on reviews and tests. Some answers can be found in the textbook, while other answers require you to go beyond the textbook. These answers might be based on knowledge you already have or things you have experienced.

② Practice It! Read the excerpt below. Answer the following questions and then discuss them with a partner.

> Did you use a machine today? When you think of a machine, you might think of a device, such as a car, with many moving parts powered by an engine or an electric motor. But if you used a pair of scissors or a broom, or cut your food with a knife, you used a machine. A machine is simply a device that makes doing work easier. Even a sloping surface can be a machine.
>
> —*from page 412*

- Describe how using a broom makes cleaning a floor easier.
- How is pushing a box up a smooth ramp easier than lifting the box upward?
- Why does a screwdriver make it easier to tighten a screw?

③ Apply It! Look at some questions in the text. Which questions can be answered directly from the text? Which require you to go beyond the text?

Target Your Reading

Reading Tip

As you read, keep track of questions you answer in the chapter. This will help you remember what you read.

Use this to focus on the main ideas as you read the chapter.

1. **Before you read** the chapter, respond to the statements below on your worksheet or on a numbered sheet of paper.
   - Write an **A** if you **agree** with the statement.
   - Write a **D** if you **disagree** with the statement.

2. **After you read** the chapter, look back to this page to see if you've changed your mind about any of the statements.
   - If any of your answers changed, explain why.
   - Change any false statements into true statements.
   - Use your revised statements as a study guide.

Science Online
Print out a worksheet of this page at ips.msscience.com

Before You Read A or D	Statement	After You Read A or D
	1 Friction is caused by atoms or molecules of one object bonding to atoms or molecules in another object.	
	2 Power measures how fast work is done.	
	3 The fulcrum of a lever is always between the input force and the output force.	
	4 Efficiency is the ratio of output work to input work.	
	5 When you do work on an object, you transfer energy to the object.	
	6 A car is a combination of simple machines.	
	7 A wedge and a screw are both types of inclined planes.	
	8 Work is done anytime a force is applied.	
	9 Mechanical advantage can never be less than 1.	

section 1

Work and Power

as you read

What You'll Learn
- **Recognize** when work is done.
- **Calculate** how much work is done.
- **Explain** the relation between work and power.

Why It's Important
If you understand work, you can make your work easier.

Review Vocabulary
force: a push or a pull

New Vocabulary
- work
- power

What is work?

What does the term *work* mean to you? You might think of household chores; a job at an office, a factory, a farm; or the homework you do after school. In science, the definition of work is more specific. **Work** is done when a force causes an object to move in the same direction that the force is applied.

Can you think of a way in which you did work today? Maybe it would help to know that you do work when you lift your books, turn a doorknob, raise window blinds, or write with a pen or pencil. You also do work when you walk up a flight of stairs or open and close your school locker. In what other ways do you do work every day?

Work and Motion Your teacher has asked you to move a box of books to the back of the classroom. Try as you might, though, you just can't budge the box because it is too heavy. Although you exerted a force on the box and you feel tired from it, you have not done any work. In order for you to do work, two things must occur. First, you must apply a force to an object. Second, the object must move in the same direction as your applied force. You do work on an object only when the object moves as a result of the force you exert. The girl in **Figure 1** might think she is working by holding the bags of groceries. However, if she is not moving, she is not doing any work because she is not causing something to move.

 To do work, how must a force make an object move?

Figure 1 This girl is holding bags of groceries, yet she isn't doing any work.
Explain what must happen for work to be done.

Figure 2 To do work, an object must move in the direction a force is applied.

The boy's arms do work when they exert an upward force on the basket and the basket moves upward.

The boy's arms still exert an upward force on the basket. But when the boy walks forward, no work is done by his arms.

Applying Force and Doing Work Picture yourself lifting the basket of clothes in **Figure 2.** You can feel your arms exerting a force upward as you lift the basket, and the basket moves upward in the direction of the force your arms applied. Therefore, your arms have done work. Now, suppose you carry the basket forward. You can still feel your arms applying an upward force on the basket to keep it from falling, but now the basket is moving forward instead of upward. Because the direction of motion is not in the same direction of the force applied by your arms, no work is done by your arms.

Force in Two Directions Sometimes only part of the force you exert moves an object. Think about what happens when you push a lawn mower. You push at an angle to the ground as shown in **Figure 3.** Part of the force is to the right and part of the force is downward. Only the part of the force that is in the same direction as the motion of the mower—to the right—does work.

Figure 3 When you exert a force at an angle, only part of your force does work—the part that is in the same direction as the motion of the object.

SECTION 1 Work and Power

James Prescott Joule This English physicist experimentally verified the law of conservation of energy. He showed that various forms of energy—mechanical, electrical, and thermal—are essentially the same and can be converted one into another. The SI unit of energy and work, the joule, is named after him. Research the work of Joule and write what you learn in your Science Journal.

Calculating Work

Work is done when a force makes an object move. More work is done when the force is increased or the object is moved a greater distance. Work can be calculated using the work equation below. In SI units, the unit for work is the joule, named for the nineteenth-century scientist James Prescott Joule.

Work Equation

work (in joules) = **force** (in newtons) × **distance** (in meters)

$$W = Fd$$

Work and Distance Suppose you give a book a push and it slides across a table. To calculate the work you did, the distance in the above equation is not the distance the book moved. The distance in the work equation is the distance an object moves while the force is being applied. So the distance in the work equation is the distance the book moved while you were pushing.

Applying Math — Solve a One-Step Equation

CALCULATING WORK A painter lifts a can of paint that weighs 40 N a distance of 2 m. How much work does she do? *Hint: to lift a can weighing 40 N, the painter must exert a force of 40 N.*

Solution

1 *This is what you know:*
- force: $F = 40$ N
- distance: $d = 2$ m

2 *This is what you need to find out:* work: $W = ?$ J

3 *This is the procedure you need to use:* Substitute the known values $F = 40$ N and $d = 2$ m into the work equation:

$W = Fd = (40 \text{ N})(2 \text{ m}) = 80 \text{ N·m} = 80$ J

4 *Check your answer:* Check your answer by dividing the work you calculated by the distance given in the problem. The result should be the force given in the problem.

Practice Problems

1. As you push a lawn mower, the horizontal force is 300 N. If you push the mower a distance of 500 m, how much work do you do?

2. A librarian lifts a box of books that weighs 93 N a distance of 1.5 m. How much work does he do?

For more practice, visit ips.msscience.com/math_practice

408 CHAPTER 14 Work and Simple Machines

What is power?

What does it mean to be powerful? Imagine two weightlifters lifting the same amount of weight the same vertical distance. They both do the same amount of work. However, the amount of power they use depends on how long it took to do the work. **Power** is how quickly work is done. The weightlifter who lifted the weight in less time is more powerful.

Calculating Power Power can be calculated by dividing the amount of work done by the time needed to do the work.

Power Equation

$$\text{power (in watts)} = \frac{\text{work (in joules)}}{\text{time (in seconds)}}$$

$$P = \frac{W}{t}$$

In SI units, the unit of power is the watt, in honor of James Watt, a nineteenth-century British scientist who invented a practical version of the steam engine.

Mini LAB

Work and Power

Procedure
1. Weigh yourself on a **scale**.
2. Multiply your weight in pounds by 4.45 to convert your weight to newtons.
3. Measure the vertical height of a **stairway**. **WARNING:** *Make sure the stairway is clear of all objects.*
4. Time yourself walking slowly and quickly up the stairway.

Analysis
Calculate and compare the work and power in each case.

Try at Home

Applying Math — Solve a One-Step Equation

CALCULATING POWER You do 200 J of work in 12 s. How much power did you use?

Solution

1 *This is what you know:*
- work: $W = 200$ J
- time: $t = 12$ s

2 *This is what you need to find out:*
- power: $P = ?$ watts

3 *This is the procedure you need to use:*

Substitute the known values $W = 200$ J and $t = 12$ s into the power equation:

$$P = \frac{W}{t} = \frac{200 \text{ J}}{12 \text{ s}} = 17 \text{ watts}$$

4 *Check your answer:*

Check your answer by multiplying the power you calculated by the time given in the problem. The result should be the work given in the problem.

Practice Problems

1. In the course of a short race, a car does 50,000 J of work in 7 s. What is the power of the car during the race?
2. A teacher does 140 J of work in 20 s. How much power did he use?

 For more practice, visit ips.msscience.com/math_practice

SECTION 1 Work and Power

Topic: James Watt

Visit ips.msscience.com for Web links to information about James Watt and his steam engine.

Activity Draw a diagram showing how his steam engine worked.

Work and Energy If you push a chair and make it move, you do work on the chair and change its energy. Recall that when something is moving it has energy of motion, or kinetic energy. By making the chair move, you increase its kinetic energy.

You also change the energy of an object when you do work and lift it higher. An object has potential energy that increases when it is higher above Earth's surface. By lifting an object, you do work and increase its potential energy.

Power and Energy When you do work on an object you increase the energy of the object. Because energy can never be created or destroyed, if the object gains energy then you must lose energy. When you do work on an object you transfer energy to the object, and your energy decreases. The amount of work done is the amount of energy transferred. So power is also equal to the amount of energy transferred in a certain amount of time.

Sometimes energy can be transferred even when no work is done, such as when heat flows from a warm to a cold object. In fact, there are many ways energy can be transferred even if no work is done. Power is always the rate at which energy is transferred, or the amount of energy transferred divided by the time needed.

section 1 review

Summary

What is work?

- Work is done when a force causes an object to move in the same direction that the force is applied.
- If the movement caused by a force is at an angle to the direction the force is applied, only the part of the force in the direction of motion does work.
- Work can be calculated by multiplying the force applied by the distance:

    $W = Fd$

- The distance in the work equation is the distance an object moves while the force is being applied.

What is power?

- Power is how quickly work is done. Something is more powerful if it can do a given amount of work in less time.
- Power can be calculated by dividing the work done by the time needed to do the work:

    $P = \dfrac{W}{t}$

Self Check

1. **Describe** a situation in which work is done on an object.
2. **Evaluate** which of the following situations involves more power: 200 J of work done in 20 s or 50 J of work done in 4 s? Explain your answer.
3. **Determine** two ways power can be increased.
4. **Calculate** how much power, in watts, is needed to cut a lawn in 50 min if the work involved is 100,000 J.
5. **Think Critically** Suppose you are pulling a wagon with the handle at an angle. How can you make your task easier?

Applying Math

6. **Calculate Work** How much work was done to lift a 1,000-kg block to the top of the Great Pyramid, 146 m above ground?
7. **Calculate Work Done by an Engine** An engine is used to lift a beam weighing 9,800 N up to 145 m. How much work must the engine do to lift this beam? How much work must be done to lift it 290 m?

Building the Pyramids

Imagine moving 2.3 million blocks of limestone, each weighing more than 1,000 kg. That is exactly what the builders of the Great Pyramid at Giza did. Although no one knows for sure exactly how they did it, they probably pulled the blocks most of the way.

Work Done Using Different Ramps		
Distance (cm)	Force (N)	Work (J)
Do not write in this book.		

Real-World Question

How is the force needed to lift a block related to the distance it travels?

Goals
- **Compare** the force needed to lift a block with the force needed to pull it up a ramp.

Materials
wood block
tape
spring scale
ruler
thin notebooks
meterstick
several books

Safety Precautions

Procedure

1. Stack several books together on a tabletop to model a half-completed pyramid. Measure the height of the books in centimeters. Record the height on the first row of the data table under *Distance*.
2. Use the wood block as a model for a block of stone. Use tape to attach the block to the spring scale.
3. Place the block on the table and lift it straight up the side of the stack of books until the top of the block is even with the top of the books. Record the force shown on the scale in the data table under *Force*.
4. **Arrange** a notebook so that one end is on the stack of books and the other end is on the table. Measure the length of the notebook and record this length as distance in the second row of the data table under *Distance*.
5. **Measure** the force needed to pull the block up the ramp. Record the force in the data table.
6. Repeat steps 4 and 5 using a longer notebook to make the ramp longer.
7. **Calculate** the work done in each row of the data table.

Conclude and Apply

1. **Evaluate** how much work you did in each instance.
2. **Determine** what happened to the force needed as the length of the ramp increased.
3. **Infer** How could the builders of the pyramids have designed their task to use less force than they would lifting the blocks straight up? Draw a diagram to support your answer.

Communicating Your Data

Add your data to that found by other groups. **For more help, refer to the** Science Skill Handbook.

LAB **411**

section 2
Using Machines

as you read

What You'll Learn
- **Explain** how a machine makes work easier.
- **Calculate** the mechanical advantages and efficiency of a machine.
- **Explain** how friction reduces efficiency.

Why It's Important
Machines can't change the amount of work you need to do, but they can make doing work easier.

Review Vocabulary
friction: force that opposes motion between two touching surfaces

New Vocabulary
- input force
- output force
- mechanical advantage
- efficiency

What is a machine?

Did you use a machine today? When you think of a machine you might think of a device, such as a car, with many moving parts powered by an engine or an electric motor. But if you used a pair of scissors or a broom, or cut your food with a knife, you used a machine. A machine is simply a device that makes doing work easier. Even a sloping surface can be a machine.

Mechanical Advantage

Even though machines make work easier, they don't decrease the amount of work you need to do. Instead, a machine changes the way in which you do work. When you use a machine, you exert a force over some distance. For example, you exert a force to move a rake or lift the handles of a wheelbarrow. The force that you apply on a machine is the **input force.** The work you do on the machine is equal to the input force times the distance over which your force is applied. The work that you do on the machine is the input work.

The machine also does work by exerting a force to move an object over some distance. A rake, for example, exerts a force to move leaves. Sometimes this force is called the resistance force because the machine is trying to overcome some resistance. The force that the machine applies is the **output force.** The work that the machine does is the output work. **Figure 4** shows how a machine transforms input work to output work.

When you use a machine, the output work can never be greater than the input work. So what is the advantage of using a machine? A machine makes work easier by changing the amount of force you need to exert, the distance over which the force is exerted, or the direction in which you exert your force.

Figure 4 No matter what type of machine is used, the output work is never greater than the input work.

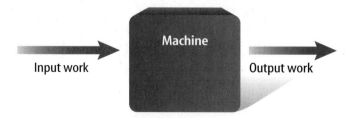

Changing Force Some machines make doing a job easier by reducing the force you have to apply to do the job. For this type of machine the output force is greater than the input force. How much larger the output force is compared to the input force is the **mechanical advantage** of the machine. The mechanical advantage of a machine is the ratio of the output force to the input force and can be calculated from this equation:

Mechanical Advantage Equation

$$\text{mechanical advantage} = \frac{\text{output force (in newtons)}}{\text{input force (in newtons)}}$$

$$MA = \frac{F_{out}}{F_{in}}$$

Mechanical advantage does not have any units, because it is the ratio of two numbers with the same units.

Topic: Historical Tools
Visit ips.msscience.com for Web links to information about early types of tools and how they took advantage of simple machines.

Activity Write a paragraph describing how simple machines were used to design early tools.

Applying Math — Solve a One-Step Equation

CALCULATING MECHANICAL ADVANTAGE To pry the lid off a paint can, you apply a force of 50 N to the handle of the screwdriver. What is the mechanical advantage of the screwdriver if it applies a force of 500 N to the lid?

Solution

1 *This is what you know:*
- input force: $F_{in} = 50$ N
- output force: $F_{out} = 500$ N

2 *This is what you need to find out:* mechanical advantage: $MA = ?$

3 *This is the procedure you need to use:* Substitute the known values $F_{in} = 50$ N and $F_{out} = 500$ N into the mechanical advantage equation:

$$MA = \frac{F_{out}}{F_{in}} = \frac{500 \text{ N}}{50 \text{ N}} = 10$$

4 *Check your answer:* Check your answer by multiplying the mechanical advantage you calculated by the input force given in the problem. The result should be the output force given in the problem.

Practice Problems

1. To open a bottle, you apply a force of 50 N to the bottle opener. The bottle opener applies a force of 775 N to the bottle cap. What is the mechanical advantage of the bottle opener?

2. To crack a pecan, you apply a force of 50 N to the nutcracker. The nutcracker applies a force of 750 N to the pecan. What is the mechanical advantage of the nutcracker?

For more practice, visit ips.msscience.com/math_practice

Figure 5 Changing the direction or the distance that a force is applied can make a task easier.

Sometimes it is easier to exert your force in a certain direction. This boy would rather pull down on the rope to lift the flag than to climb to the top of the pole and pull up.

When you rake leaves, you move your hands a short distance, but the end of the rake moves over a longer distance.

Changing Distance Some machines allow you to exert your force over a shorter distance. In these machines, the output force is less than the input force. The rake in **Figure 5** is this type of machine. You move your hands a small distance at the top of the handle, but the bottom of the rake moves a greater distance as it moves the leaves. The mechanical advantage of this type of machine is less than one because the output force is less than the input force.

Changing Direction Sometimes it is easier to apply a force in a certain direction. For example, it is easier to pull down on the rope in **Figure 5** than to pull up on it. Some machines enable you to change the direction of the input force. In these machines neither the force nor the distance is changed. The mechanical advantage of this type of machine is equal to one because the output force is equal to the input force. The three ways machines make doing work easier are summarized in **Figure 6**.

Figure 6 Machines are useful because they can increase force, increase distance, or change the direction in which a force is applied.

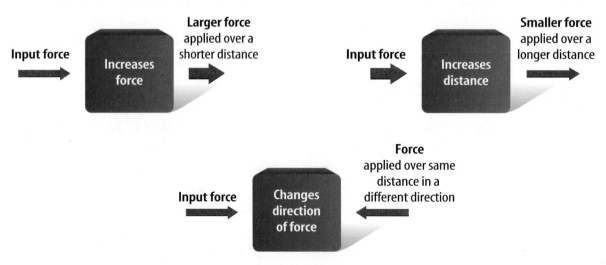

414 CHAPTER 14 Work and Simple Machines

Efficiency

A machine can't make the output work greater than the input work. In fact, for a real machine, the output work is always less than the input work. In a real machine, there is friction as parts of the machine move. Friction converts some of the input work into heat, so that the output work is reduced. The **efficiency** of a machine is the ratio of the output work to the input work, and can be calculated from this equation:

Efficiency Equation

$$\text{efficiency (in percent)} = \frac{\text{output work (in joules)}}{\text{input work (in joules)}} \times 100\%$$

$$\textit{eff} = \frac{W_{out}}{W_{in}} \times 100\%$$

If the amount of friction in the machine is reduced, the efficiency of the machine increases.

Body Temperature
Chemical reactions that enable your muscles to move also produce heat that helps maintain your body temperature. When you shiver, rapid contraction and relaxation of muscle fibers produces a large amount of heat that helps raise your body temperature. This causes the efficiency of your muscles to decrease as more energy is converted into heat.

Applying Math — Solve a One-Step Equation

CALCULATING EFFICIENCY Using a pulley system, a crew does 7,500 J of work to load a box that requires 4,500 J of work. What is the efficiency of the pulley system?

Solution

1. *This is what you know:*
   - input work: $W_{in} = 7{,}500$ J
   - output work: $W_{out} = 4{,}500$ J

2. *This is what you need to find out:* efficiency: $\textit{eff} = ?\ \%$

3. *This is the procedure you need to use:* Substitute the known values $W_{in} = 7{,}500$ J and $W_{out} = 4{,}500$ J into the efficiency equation:

   $$\textit{eff} = \frac{W_{out}}{W_{in}} = \frac{4{,}500\ \text{J}}{7{,}500\ \text{J}} \times 100\% = 60\%$$

4. *Check your answer:* Check your answer by dividing the efficiency by 100% and then multiplying your answer times the work input. The product should be the work output given in the problem.

Practice Problems

1. You do 100 J of work in pulling out a nail with a claw hammer. If the hammer does 70 J of work, what is the hammer's efficiency?

2. You do 150 J of work pushing a box up a ramp. If the ramp does 105 J of work, what is the efficiency of the ramp?

For more practice, visit
ips.msscience.com/math_practice

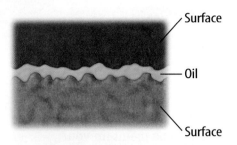

Figure 7 Lubrication can reduce the friction between two surfaces. Two surfaces in contact can stick together where the high spots on each surface come in contact. Adding oil or another lubricant separates the surface so that fewer high spots make contact.

Friction To help understand friction, imagine pushing a heavy box up a ramp. As the box begins to move, the bottom surface of the box slides across the top surface of the ramp. Neither surface is perfectly smooth—each has high spots and low spots, as shown in **Figure 7**.

As the two surfaces slide past each other, high spots on the two surfaces come in contact. At these contact points, shown in **Figure 7**, atoms and molecules can bond together. This makes the contact points stick together. The attractive forces between all the bonds in the contact points added together is the frictional force that tries to keep the two surfaces from sliding past each other.

To keep the box moving, a force must be applied to break the bonds between the contact points. Even after these bonds are broken and the box moves, new bonds form as different parts of the two surfaces come into contact.

Friction and Efficiency One way to reduce friction between two surfaces is to add oil. **Figure 7** shows how oil fills the gaps between the surfaces, and keeps many of the high spots from making contact. Because there are fewer contact points between the surfaces, the force of friction is reduced. More of the input work then is converted to output work by the machine.

section 2 review

Summary

What is a machine?

- A machine is a device that makes doing work easier.
- A machine can make doing work easier by reducing the force exerted, changing the distance over which the force is exerted, or changing the direction of the force.
- The output work done by a machine can never be greater than the input work done on the machine.

Mechanical Advantage and Efficiency

- The mechanical advantage of a machine is the number of times the machine increases the input force:
$$MA = \frac{F_{out}}{F_{in}}$$

- The efficiency of a machine is the ratio of the output work to the input work:
$$eff = \frac{W_{out}}{W_{in}} \times 100\%$$

Self Check

1. **Identify** three specific situations in which machines make work easier.
2. **Infer** why the output force exerted by a rake must be less than the input force.
3. **Explain** how the efficiency of an ideal machine compares with the efficiency of a real machine.
4. **Explain** how friction reduces the efficiency of machines.
5. **Think Critically** Can a machine be useful even if its mechanical advantage is less than one? Explain and give an example.

Applying Math

6. **Calculate Efficiency** Find the efficiency of a machine if the input work is 150 J and the output work is 90 J.
7. **Calculate Mechanical Advantage** To lift a crate, a pulley system exerts a force of 2,750 N. Find the mechanical advantage of the pulley system if the input force is 250 N.

section 3

Simple Machines

What is a simple machine?

What do you think of when you hear the word *machine?* Many people think of machines as complicated devices such as cars, elevators, or computers. However, some machines are as simple as a hammer, shovel, or ramp. A **simple machine** is a machine that does work with only one movement. The six simple machines are the inclined plane, lever, wheel and axle, screw, wedge, and pulley. A machine made up of a combination of simple machines is called a **compound machine.** A can opener is a compound machine. The bicycle in **Figure 8** is a familiar example of another compound machine.

Inclined Plane

Ramps might have enabled the ancient Egyptians to build their pyramids. To move limestone blocks weighing more than 1,000 kg each, archaeologists hypothesize that the Egyptians built enormous ramps. A ramp is a simple machine known as an inclined plane. An **inclined plane** is a flat, sloped surface. Less force is needed to move an object from one height to another using an inclined plane than is needed to lift the object. As the inclined plane becomes longer, the force needed to move the object becomes smaller.

as you read

What You'll Learn

- **Distinguish** among the different simple machines.
- **Describe** how to find the mechanical advantage of each simple machine.

Why It's Important

All machines, no matter how complicated, are made of simple machines.

Review Vocabulary
compound: made of separate pieces or parts

New Vocabulary
- simple machine
- compound machine
- inclined plane
- wedge
- screw
- lever
- wheel and axle
- pulley

Figure 8 Devices that use combinations of simple machines, such as this bicycle, are called compound machines.

SECTION 3 Simple Machines **417**

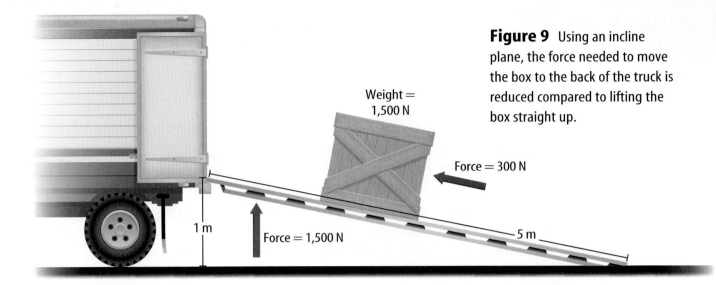

Figure 9 Using an incline plane, the force needed to move the box to the back of the truck is reduced compared to lifting the box straight up.

Using Inclined Planes Imagine having to lift a box weighing 1,500 N to the back of a truck that is 1 m off the ground. You would have to exert a force of 1,500 N, the weight of the box, over a distance of 1 m, which equals 1,500 J of work. Now suppose that instead you use a 5-m-long ramp, as shown in **Figure 9.** The amount of work you need to do does not change. You still need to do 1,500 J of work. However, the distance over which you exert your force becomes 5 m. You can calculate the force you need to exert by dividing both sides of the equation for work by distance.

$$\text{Force} = \frac{\text{work}}{\text{distance}}$$

If you do 1,500 J of work by exerting a force over 5 m, the force is only 300 N. Because you exert the input force over a distance that is five times as long, you can exert a force that is five times less.

The mechanical advantage of an inclined plane is the length of the inclined plane divided by its height. In this example, the ramp has a mechanical advantage of 5.

Wedge An inclined plane that moves is called a **wedge.** A wedge can have one or two sloping sides. The knife shown in **Figure 10** is an example of a wedge. An axe and certain types of doorstops are also wedges. Just as for an inclined plane, the mechanical advantage of a wedge increases as it becomes longer and thinner.

Figure 10 This chef's knife is a wedge that slices through food.

Figure 11 Wedge-shaped teeth help tear food.

Your front teeth help tear an apple apart.

The wedge-shaped teeth of this *Tyrannosaurus rex* show that it was a carnivore.

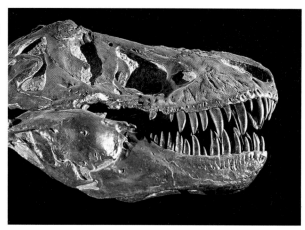

Wedges in Your Body You have wedges in your body. The bite marks on the apple in **Figure 11** show how your front teeth are wedge shaped. A wedge changes the direction of the applied effort force. As you push your front teeth into the apple, the downward effort force is changed by your teeth into a sideways force that pushes the skin of the apple apart.

The teeth of meat eaters, or carnivores, are more wedge shaped than the teeth of plant eaters, or herbivores. The teeth of carnivores are used to cut and rip meat, while herbivores' teeth are used for grinding plant material. By examining the teeth of ancient animals, such as the dinosaur in **Figure 11,** scientists can determine what the animal ate when it was living.

The Screw Another form of the inclined plane is a screw. A **screw** is an inclined plane wrapped around a cylinder or post. The inclined plane on a screw forms the screw threads. Just like a wedge changes the direction of the effort force applied to it, a screw also changes the direction of the applied force. When you turn a screw, the force applied is changed by the threads to a force that pulls the screw into the material. Friction between the threads and the material holds the screw tightly in place. The mechanical advantage of the screw is the length of the inclined plane wrapped around the screw divided by the length of the screw. The more tightly wrapped the threads are, the easier it is to turn the screw. Examples of screws are shown in **Figure 12.**

Figure 12 The thread around a screw is an inclined plane. Many familiar devices use screws to make work easier.

 *How are screws related to the inclined plane?*

SECTION 3 Simple Machines **419**

Figure 13 The mechanical advantage of a lever changes as the position of the fulcrum changes. The mechanical advantage increases as the fulcrum is moved closer to the output force.

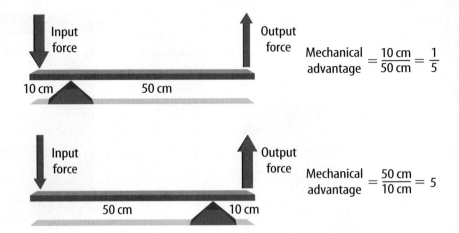

Figure 14 A faucet handle is a wheel and axle. A wheel and axle is similar to a circular lever. The center is the fulcrum, and the wheel and axle turn around it. **Explain** *how you can increase the mechanical advantage of a wheel and axle.*

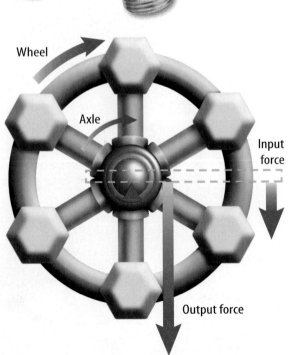

Lever

You step up to the plate. The pitcher throws the ball and you swing your lever to hit the ball? That's right! A baseball bat is a type of simple machine called a lever. A **lever** is any rigid rod or plank that pivots, or rotates, about a point. The point about which the lever pivots is called a fulcrum.

The mechanical advantage of a lever is found by dividing the distance from the fulcrum to the input force by the distance from the fulcrum to the output force, as shown in **Figure 13**. When the fulcrum is closer to the output force than the input force, the mechanical advantage is greater than one.

Levers are divided into three classes according to the position of the fulcrum with respect to the input force and output force. **Figure 15** shows examples of three classes of levers.

Wheel and Axle

Do you think you could turn a doorknob easily if it were a narrow rod the size of a pencil? It might be possible, but it would be difficult. A doorknob makes it easier for you to open a door because it is a simple machine called a wheel and axle. A **wheel and axle** consists of two circular objects of different sizes that are attached in such a way that they rotate together. As you can see in **Figure 14,** the larger object is the wheel and the smaller object is the axle.

The mechanical advantage of a wheel and axle is usually greater than one. It is found by dividing the radius of the wheel by the radius of the axle. For example, if the radius of the wheel is 12 cm and the radius of the axle is 4 cm, the mechanical advantage is 3.

420 CHAPTER 14 Work and Simple Machines

NATIONAL GEOGRAPHIC VISUALIZING LEVERS

Figure 15

Levers are among the simplest of machines, and you probably use them often in everyday life without even realizing it. A lever is a bar that pivots around a fixed point called a fulcrum. As shown here, there are three types of levers—first class, second class, and third class. They differ in where two forces—an input force and an output force—are located in relation to the fulcrum.

  Fulcrum
  Input force
  Output force

In a first-class lever, the fulcrum is between the input force and the output force. First-class levers, such as scissors and pliers, multiply force or distance depending on where the fulcrum is placed. They always change the direction of the input force, too.

First-class lever

In a second-class lever, such as a wheelbarrow, the output force is between the input force and the fulcrum. Second-class levers always multiply the input force but don't change its direction.

Second-class lever

Third-class lever

In a third-class lever, such as a baseball bat, the input force is between the output force and the fulcrum. For a third-class lever, the output force is less than the input force, but is in the same direction.

SECTION 3 Simple Machines **421**

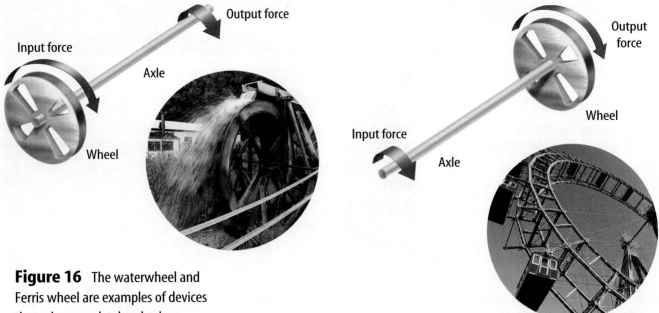

Figure 16 The waterwheel and Ferris wheel are examples of devices that rely on a wheel and axle. **Compare and contrast** waterwheels and Ferris wheels in terms of wheels and axles.

Mini LAB

Observing Pulleys

Procedure
1. Obtain two **broomsticks**. Tie a 3-m-long **rope** to the middle of one stick. Wrap the rope around both sticks four times.
2. Have two students pull the broomsticks apart while a third pulls on the rope.
3. Repeat with two wraps of rope.

Analysis
1. Compare the results.
2. Predict whether it will be easier to pull the broomsticks together with ten wraps of rope.

Using Wheels and Axles In some devices, the input force is used to turn the wheel and the output force is exerted by the axle. Because the wheel is larger than the axle, the mechanical advantage is greater than one. So the output force is greater than the input force. A doorknob, a steering wheel, and a screwdriver are examples of this type of wheel and axle.

In other devices, the input force is applied to turn the axle and the output force is exerted by the wheel. Then the mechanical advantage is less than one and the output force is less than the input force. A fan and a ferris wheel are examples of this type of wheel and axle. **Figure 16** shows an example of each type of wheel and axle.

Pulley

To raise a sail, a sailor pulls down on a rope. The rope uses a simple machine called a pulley to change the direction of the force needed. A **pulley** consists of a grooved wheel with a rope or cable wrapped over it.

Fixed Pulleys Some pulleys, such as the one on a sail, a window blind, or a flagpole, are attached to a structure above your head. When you pull down on the rope, you pull something up. This type of pulley, called a fixed pulley, does not change the force you exert or the distance over which you exert it. Instead, it changes the direction in which you exert your force, as shown in **Figure 17.** The mechanical advantage of a fixed pulley is 1.

 *How does a fixed pulley affect the input force?*

422 CHAPTER 14 Work and Simple Machines

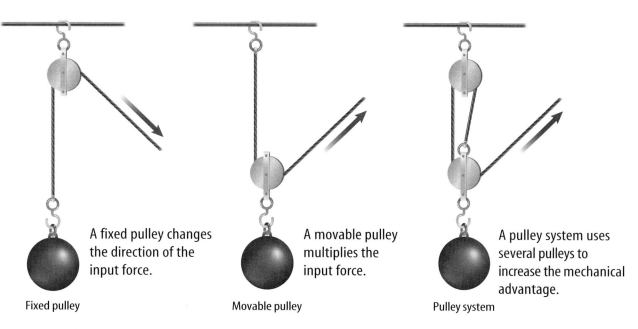

A fixed pulley changes the direction of the input force.

Fixed pulley

A movable pulley multiplies the input force.

Movable pulley

A pulley system uses several pulleys to increase the mechanical advantage.

Pulley system

Movable Pulleys Another way to use a pulley is to attach it to the object you are lifting, as shown in **Figure 17.** This type of pulley, called a movable pulley, allows you to exert a smaller force to lift the object. The mechanical advantage of a movable pulley is always 2.

More often you will see combinations of fixed and movable pulleys. Such a combination is called a pulley system. The mechanical advantage of a pulley system is equal to the number of sections of rope pulling up on the object. For the pulley system shown in **Figure 17** the mechanical advantage is 3.

Figure 17 Pulleys can change force and direction.

section 3 review

Summary

Simple and Compound Machines
- A simple machine is a machine that does work with only one movement.
- A compound machine is made from a combination of simple machines.

Types of Simple Machines
- An inclined plane is a flat, sloped surface.
- A wedge is an inclined plane that moves.
- A screw is an inclined plane that is wrapped around a cylinder or post.
- A lever is a rigid rod that pivots around a fixed point called the fulcrum.
- A wheel and axle consists of two circular objects of different sizes that rotate together.
- A pulley is a grooved wheel with a rope or cable wrapped over it.

Self Check

1. **Determine** how the mechanical advantage of a ramp changes as the ramp becomes longer.
2. **Explain** how a wedge changes an input force.
3. **Identify** the class of lever for which the fulcrum is between the input force and the output force.
4. **Explain** how the mechanical advantage of a wheel and axle change as the size of the wheel increases.
5. **Think Critically** How are a lever and a wheel and axle similar?

Applying Math

6. **Calculate Length** The Great Pyramid is 146 m high. How long is a ramp from the top of the pyramid to the ground that has a mechanical advantage of 4?
7. **Calculate Force** Find the output force exerted by a moveable pulley if the input force is 50 N.

LAB Design Your Own

Pulley Power

Goals
- **Design** a pulley system.
- **Measure** the mechanical advantage and efficiency of the pulley system.

Possible Materials
single- and multiple-pulley systems
nylon rope
steel bar to support the pulley system
meterstick
*metric tape measure
variety of weights to test pulleys
force spring scale
brick
*heavy book
balance
*scale
*Alternate materials

Safety Precautions

WARNING: *The brick could be dangerous if it falls. Keep your hands and feet clear of it.*

Real-World Question

Imagine how long it might have taken to build the Sears Tower in Chicago without the aid of a pulley system attached to a crane. Hoisting the 1-ton I beams to a maximum height of 110 stories required large lifting forces and precise control of the beam's movement.

Construction workers also use smaller pulleys that are not attached to cranes to lift supplies to where they are needed. Pulleys are not limited to construction sites. They also are used to lift automobile engines out of cars, to help load and unload heavy objects on ships, and to lift heavy appliances and furniture. How can you use a pulley system to reduce the force needed to lift a load?

Form a Hypothesis

Write a hypothesis about how pulleys can be combined to make a system of pulleys to lift a heavy load, such as a brick. Consider the efficiency of your system.

Test Your Hypothesis

Make a Plan

1. Decide how you are going to support your pulley system. What materials will you use?
2. How will you measure the effort force and the resistance force? How will you determine the mechanical advantage? How will you measure efficiency?
3. **Experiment** by lifting small weights with a single pulley, double pulley, and so on. How efficient are the pulleys? In what ways can you increase the efficiency of your setup?

424 CHAPTER 14 Work and Simple Machines

Using Scientific Methods

4. Use the results of step 3 to design a pulley system to lift the brick. Draw a diagram of your design. Label the different parts of the pulley system and use arrows to indicate the direction of movement for each section of rope.

Follow Your Plan

1. Make sure your teacher approves your plan before you start.
2. Assemble the pulley system you designed. You might want to test it with a smaller weight before attaching the brick.
3. **Measure** the force needed to lift the brick. How much rope must you pull to raise the brick 10 cm?

▶ Analyze Your Data

1. **Calculate** the ideal mechanical advantage of your design.
2. **Calculate** the actual mechanical advantage of the pulley system you built.
3. **Calculate** the efficiency of your pulley system.
4. How did the mechanical advantage of your pulley system compare with those of your classmates?

▶ Conclude and Apply

1. **Explain** how increasing the number of pulleys increases the mechanical advantage.
2. **Infer** How could you modify the pulley system to lift a weight twice as heavy with the same effort force used here?
3. **Compare** this real machine with an ideal machine.

Communicating Your Data

Show your design diagram to the class. Review the design and point out good and bad characteristics of your pulley system. **For more help, refer to the** Science Skill Handbook.

TIME SCIENCE AND Society
SCIENCE ISSUES THAT AFFECT YOU!

Bionic People

Artificial limbs can help people lead normal lives

People in need of transplants usually receive human organs. But many people's medical problems can only be solved by receiving artificial body parts. These synthetic devices, called prostheses, are used to replace anything from a heart valve to a knee joint. Bionics is the science of creating artificial body parts. A major focus of bionics is the replacement of lost limbs. Through accident, birth defect, or disease, people sometimes lack hands or feet, or even whole arms or legs.

For centuries, people have used prostheses to replace limbs. In the past, physically challenged people used devices like peg legs or artificial arms that ended in a pair of hooks. These prostheses didn't do much to replace lost functions of arms and legs.

The knowledge that muscles respond to electricity has helped create more effective prostheses. One such prostheses is the myoelectric arm. This battery-powered device connects muscle nerves in an amputated arm to a sensor.

The sensor detects when the arm tenses, then transmits the signal to an artificial hand, which opens or closes. New prosthetic hands even give a sense of touch, as well as cold and heat.

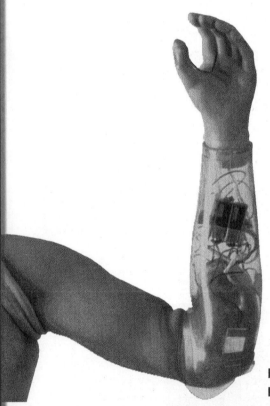

Myoelectric arms make life easier for people who have them.

Research Use your school's media center to find other aspects of robotics such as walking machines or robots that perform planetary exploration. What are they used for? How do they work? You could take it one step further and learn about cyborgs. Report to the class.

For more information, visit
ips.msscience.com/time

chapter 14 Study Guide

Reviewing Main Ideas

Section 1 Work and Power

1. Work is done when a force exerted on an object causes the object to move.
2. A force can do work only when it is exerted in the same direction as the object moves.
3. Work is equal to force times distance, and the unit of work is the joule.
4. Power is the rate at which work is done, and the unit of power is the watt.

Section 2 Using Machines

1. A machine can change the size or direction of an input force or the distance over which it is exerted.
2. The mechanical advantage of a machine is its output force divided by its input force.

Section 3 Simple Machines

1. A machine that does work with only one movement is a simple machine. A compound machine is a combination of simple machines.
2. Simple machines include the inclined plane, lever, wheel and axle, screw, wedge, and pulley.
3. Wedges and screws are inclined planes.
4. Pulleys can be used to multiply force and change direction.

Visualizing Main Ideas

Copy and complete the following concept map on simple machines.

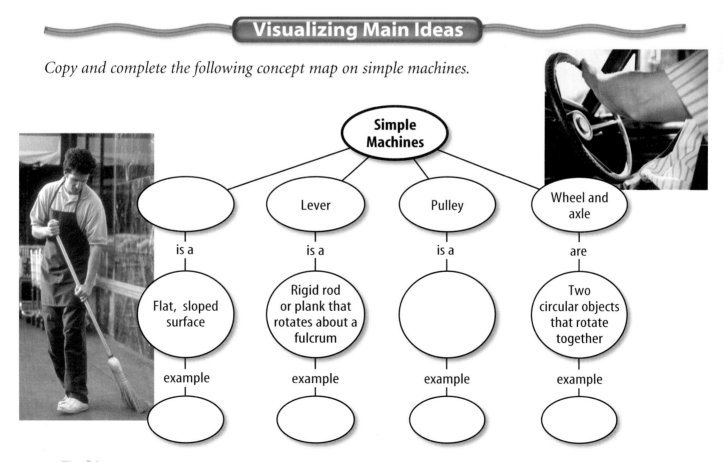

chapter 14 Review

Using Vocabulary

compound machine p. 417
efficiency p. 415
inclined plane p. 417
input force p. 412
lever p. 420
mechanical advantage p. 413
output force p. 412
power p. 409
pulley p. 422
screw p. 419
simple machine p. 417
wedge p. 418
wheel and axle p. 420
work p. 406

Each phrase below describes a vocabulary word. Write the vocabulary word that matches the phrase describing it.

1. percentage of work in to work out
2. force put into a machine
3. force exerted by a machine
4. two rigidly attached wheels
5. output force divided by input force
6. a machine with only one movement
7. an inclined plane that moves
8. a rigid rod that rotates about a fulcrum
9. a flat, sloped surface
10. amount of work divided by time

Checking Concepts

Choose the word or phrase that best answers the question.

11. Which of the following is a requirement for work to be done?
    A) Force is exerted.
    B) Object is carried.
    C) Force moves an object.
    D) Machine is used.

12. How much work is done when a force of 30 N moves an object a distance of 3 m?
    A) 3 J
    B) 10 J
    C) 30 J
    D) 90 J

13. How much power is used when 600 J of work are done in 10 s?
    A) 6 W
    B) 60 W
    C) 600 W
    D) 610 W

14. Which is a simple machine?
    A) baseball bat
    B) bicycle
    C) can opener
    D) car

15. Mechanical advantage can be calculated by which of the following expressions?
    A) input force/output force
    B) output force/input force
    C) input work/output work
    D) output work/input work

16. What is the ideal mechanical advantage of a machine that changes only the direction of the input force?
    A) less than 1
    B) zero
    C) 1
    D) greater than 1

Use the illustration below to answer question 17.

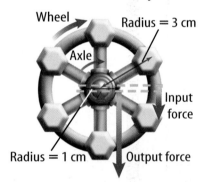

17. What is the output force if the input force on the wheel is 100 N?
    A) 5 N
    B) 200 N
    C) 500 N
    D) 2,000 N

18. Which of the following is a form of the inclined plane?
    A) pulley
    B) screw
    C) wheel and axle
    D) lever

19. For a given input force, a ramp increases which of the following?
    A) height
    B) output force
    C) output work
    D) efficiency

chapter 14 Review

Thinking Critically

Use the illustration below to answer question 20.

20. **Evaluate** Would a 9-N force applied 2 m from the fulcrum lift the weight? Explain.

21. **Explain** why the output work for any machine can't be greater than the input work.

22. **Explain** A doorknob is an example of a wheel and axle. Explain why turning the knob is easier than turning the axle.

23. **Infer** On the Moon, the force of gravity is less than on Earth. Infer how the mechanical advantage of an inclined plane would change if it were on the Moon, instead of on Earth.

24. **Make and Use Graphs** A pulley system has a mechanical advantage of 5. Make a graph with the input force on the x-axis and the output force on the y-axis. Choose five different values of the input force, and plot the resulting output force on your graph.

Use the diagram below to answer question 25.

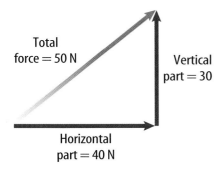

25. **Work** The diagram above shows a force exerted at an angle to pull a sled. How much work is done if the sled moves 10 m horizontally?

Performance Activities

26. **Identify** You have levers in your body. Your muscles and tendons provide the input force. Your joints act as fulcrums. The output force is used to move everything from your head to your hands. Describe and draw any human levers you can identify.

27. **Display** Make a display of everyday devices that are simple and compound machines. For devices that are simple machines, identify which simple machine it is. For compound machines, identify the simple machines that compose it.

Applying Math

28. **Mechanical Advantage** What is the mechanical advantage of a 6-m long ramp that extends from a ground-level sidewalk to a 2-m high porch?

29. **Input Force** How much input force is required to lift an 11,000-N beam using a pulley system with a mechanical advantage of 20?

30. **Efficiency** The input work done on a pulley system is 450 J. What is the efficiency of the pulley system if the output work is 375 J?

Use the table below to answer question 31.

Output Force Exerted by Machines		
Machine	Input Force (N)	Output Force (N)
A	500	750
B	300	200
C	225	225
D	800	1,100
E	75	110

31. **Mechanical Advantage** According to the table above, which of the machines listed has the largest mechanical advantage?

chapter 14 Standardized Test Practice

Part 1 Multiple Choice

Record your answers on the answer sheet provided by your teacher or on a sheet of paper.

1. The work done by a boy pulling a snow sled up a hill is 425 J. What is the power expended by the boy if he pulls on the sled for 10.5 s?
   A. 24.7 W
   B. 40.5 W
   C. 247 W
   D. 4460 W

Use the illustration below to answer questions 2 and 3.

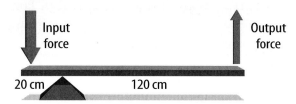

2. What is the mechanical advantage of the lever shown above?
   A. $\frac{1}{6}$
   B. $\frac{1}{2}$
   C. 2
   D. 6

3. What would the mechanical advantage of the lever be if the triangular block were moved to a position 35 cm from the edge of the output force side of the plank?
   A. $\frac{1}{4}$
   B. $\frac{1}{3}$
   C. 3
   D. 4

4. Which of the following causes the efficiency of a machine to be less than 100%?
   A. work
   B. power
   C. mechanical advantage
   D. friction

Test-Taking Tip

Simplify Diagrams Write directly on complex charts, such as a Punnett square.

Use the illustration below to answer questions 5 and 6.

5. The pulley system in the illustration above uses several pulleys to increase the mechanical advantage. What is the mechanical advantage of this system?
   A. 1
   B. 2
   C. 3
   D. 4

6. Suppose the lower pulley was removed so that the object was supported only by the upper pulley. What would the mechanical advantage be?
   A. 0
   B. 1
   C. 2
   D. 3

7. You push a shopping cart with a force of 12 N for a distance of 1.5 m. You stop pushing the cart, but it continues to roll for 1.1 m. How much work did you do?
   A. 8.0 J
   B. 13 J
   C. 18 J
   D. 31 J

8. What is the mechanical advantage of a wheel with a radius of 8.0 cm connected to an axle with a radius of 2.5 cm?
   A. 0.31
   B. 2.5
   C. 3.2
   D. 20

9. You push a 5-kg box across the floor with a force of 25 N. How far do you have to push the box to do 63 J of work?
   A. 0.40 m
   B. 1.6 m
   C. 2.5 m
   D. 13 m

Standardized Test Practice

Part 2 — Short Response/Grid In

Record your answers on the answer sheet provided by your teacher or on a sheet of paper.

10. What is the name of the point about which a lever rotates?

11. Describe how you can determine the mechanical advantage of a pulley or a pulley system.

Use the figure below to answer questions 12 and 13.

12. What type of simple machine is the tip of the dart in the photo above?

13. Would the mechanical advantage of the dart tip change if the tip were longer and thinner? Explain.

14. How much energy is used by a 75-W lightbulb in 15 s?

15. The input and output forces are applied at the ends of the lever. If the lever is 3 m long and the output force is applied 1 m from the fulcrum, what is the mechanical advantage?

16. Your body contains simple machines. Name one part that is a wedge and one part that is a lever.

17. Explain why applying a lubricant, such as oil, to the surfaces of a machine causes the efficiency of the machine to increase.

18. Apply the law of conservation of energy to explain why the output work done by a real machine is always less than the input work done on the machine.

Part 3 — Open Ended

Record your answers on a sheet of paper.

19. The output work of a machine can never be greater than the input work. However, the advantage of using a machine is that it makes work easier. Describe and give an example of the three ways a machine can make work easier.

20. A wheel and axle may have a mechanical advantage that is either greater than 1 or less than 1. Describe both types and give some examples of each.

21. Draw a sketch showing the cause of friction as two surfaces slide past each other. Explain your sketch, and describe how lubrication can reduce the friction between the two surfaces.

22. Draw the two types of simple pulleys and an example of a combination pulley. Draw arrows to show the direction of force on your sketches.

Use the figure below to answer question 23.

23. Identify two simple machines in the photo above and describe how they make riding a bicycle easier.

24. Explain why the mechanical advantage of an inclined plane can never be less than 1.

chapter 15

Thermal Energy

The BIG Idea
Thermal energy flows from areas of higher temperature to areas of lower temperature.

SECTION 1
Temperature and Thermal Energy
Main Idea Atoms and molecules in an object are moving in all directions with different speeds.

SECTION 2
Heat
Main Idea Thermal energy moves by conduction, convection, and radiation.

SECTION 3
Engines and Refrigerators
Main Idea Engines convert thermal energy to mechanical energy; refrigerators transfer thermal energy from one place to another.

Fastest to the Finish Line
In order to reach an extraordinary speed in a short distance, this dragster depends on more than an aerodynamic design. Its engine must transform the thermal energy produced by burning fuel to mechanical energy, which propels the dragster down the track.

Science Journal Describe five things that you do to make yourself feel warmer or cooler.

Start-Up Activities

Measuring Temperature

When you leave a glass of ice water on a kitchen table, the ice gradually melts and the temperature of the water increases. What is temperature, and why does the temperature of the ice water increase? In this lab you will explore one way of determining temperature.

1. Obtain three pans. Fill one pan with lukewarm water. Fill a second pan with cold water and crushed ice. Fill a third pan with very warm tap water. Label each pan.
2. Soak one of your hands in the warm water for one minute. Remove your hand from the warm water and put it in the lukewarm water. Does the lukewarm water feel cool or warm?
3. Now soak your hand in the cold water for one minute. Remove your hand from the cold water and place it in the lukewarm water. Does the lukewarm water feel cool or warm?
4. **Think Critically** Write a paragraph in your Science Journal discussing whether your sense of touch would make a useful thermometer.

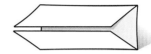

Thermal Energy Make the following Foldable to help you identify how thermal energy, heat, and temperature are related.

STEP 1 Fold a vertical piece of paper into thirds.

STEP 2 Turn the paper horizontally. Unfold and label the three columns as shown.

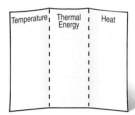

Read for Main Ideas Before you read the chapter, write down what you know about temperature, thermal energy, and heat on the appropriate tab. As you read, add to and correct what you wrote. Write what you have learned about the relationship between heat and thermal energy on the back of your Foldable.

Preview this chapter's content and activities at
ips.msscience.com

Get Ready to Read

Identify the Main Idea

① Learn It! Main ideas are the most important ideas in a paragraph, a section, or a chapter. Supporting details are facts or examples that explain the main idea. Understanding the main idea allows you to grasp the whole picture.

② Picture It! Read the following paragraph. Draw a graphic organizer like the one below to show the main idea and supporting details.

> When you heat a pot of water on a stove, thermal energy can be transferred through the water by a process other than conduction and radiation. In a gas or liquid, atoms or molecules can move much more easily than they can in a solid. As a result, these particles can travel from one place to another, carrying their energy with them. This transfer of thermal energy by the movement of atoms or molecules from one part of a material to another is called convection.
>
> —*from page 440*

Main Idea
↓ ↓ ↓
Supporting Details | Supporting Details | Supporting Details

③ Apply It! Pick a paragraph from another section of this chapter and diagram the main ideas as you did above.

Target Your Reading

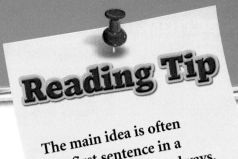

Reading Tip
The main idea is often the first sentence in a paragraph, but not always.

Use this to focus on the main ideas as you read the chapter.

1 Before you read the chapter, respond to the statements below on your worksheet or on a numbered sheet of paper.
- Write an **A** if you **agree** with the statement.
- Write a **D** if you **disagree** with the statement.

2 After you read the chapter, look back to this page to see if you've changed your mind about any of the statements.
- If any of your answers changed, explain why.
- Change any false statements into true statements.
- Use your revised statements as a study guide.

Science Online
Print out a worksheet of this page at ips.msscience.com

Before You Read A or D		Statement	After You Read A or D
	1	Temperature depends on the kinetic energy of the molecules in a material.	
	2	Heat engines can convert energy from one form to another.	
	3	Objects cannot have a temperature below zero on the Celsius scale.	
	4	In a refrigerator, the coolant gas gets cooler as it is compressed.	
	5	A conductor is any material that easily transfers thermal energy.	
	6	Energy is created by an engine.	
	7	Thermal energy from the Sun reaches Earth by conduction through space.	
	8	A car's engine converts thermal energy to mechanical energy.	
	9	Thermal energy always moves from colder objects to warmer objects.	

434 B

section 1

Temperature and Thermal Energy

as you read

What You'll Learn
- **Explain** how temperature is related to kinetic energy.
- **Describe** three scales used for measuring temperature.
- **Define** thermal energy.

Why It's Important
The movement of thermal energy toward or away from your body determines whether you feel too cold, too hot, or just right.

Review Vocabulary
kinetic energy: energy a moving object has that increases as the speed of the object increases

New Vocabulary
- temperature
- thermal energy

What is temperature?

Imagine it's a hot day and you jump into a swimming pool to cool off. When you first hit the water, you might think it feels cold. Perhaps someone else, who has been swimming for a few minutes, thinks the water feels warm. When you swim in water, touch a hot pan, or swallow a cold drink, your sense of touch tells you whether something is hot or cold. However, the words *cold*, *warm*, and *hot* can mean different things to different people.

Temperature How hot or cold something feels is related to its temperature. To understand temperature, think of a glass of water sitting on a table. The water might seem perfectly still, but water is made of molecules that are in constant, random motion. Because these molecules are always moving, they have energy of motion, or kinetic energy.

However, water molecules in random motion don't all move at the same speed. Some are moving faster and some are moving slower. **Temperature** is a measure of the average value of the kinetic energy of the molecules in random motion. The more kinetic energy the molecules have, the higher the temperature. Molecules have more kinetic energy when they are moving faster. So the higher the temperature, the faster the molecules are moving, as shown in **Figure 1.**

Figure 1 The temperature of a substance depends on how fast its molecules are moving. Water molecules are moving faster in the hot water on the left than in the cold water on the right.

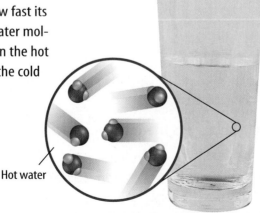

Hot water

Cold water

434 CHAPTER 15 Thermal Energy

Thermal Expansion It wasn't an earthquake that caused the sidewalk to buckle in **Figure 2**. Hot weather caused the concrete to expand so much that it cracked, and the pieces squeezed each other upward. When the temperature of an object is increased, its molecules speed up and tend to move farther apart. This causes the object to expand. When the object is cooled, its molecules slow down and move closer together. This causes the object to shrink, or contract.

Almost all substances expand when they are heated and contract when they are cooled. The amount of expansion or contraction depends on the type of material and the change in temperature. For example, liquids usually expand more than solids. Also, the greater the change in temperature, the more an object expands or contracts.

 *Why do materials expand when their temperatures increase?*

Figure 2 Most objects expand as their temperatures increase. Pieces of this concrete sidewalk forced each other upward when the concrete expanded on a hot day.

Measuring Temperature

The temperature of an object depends on the average kinetic energy of all the molecules in an object. However, molecules are so small and objects contain so many of them, that it is impossible to measure the kinetic energy of all the individual molecules.

A more practical way to measure temperature is to use a thermometer. Thermometers usually use the expansion and contraction of materials to measure temperature. One common type of thermometer uses a glass tube containing a liquid. When the temperature of the liquid increases, it expands so that the height of the liquid in the tube depends on the temperature.

Temperature Scales To be able to give a number for the temperature, a thermometer has to have a temperature scale. Two common temperature scales are the Fahrenheit and Celsius scales, shown in **Figure 3**.

On the Fahrenheit scale, the freezing point of water is given the temperature 32°F and the boiling point 212°F. The space between the boiling point and the freezing point is divided into 180 equal degrees. The Fahrenheit scale is used mainly in the United States.

On the Celsius temperature scale, the freezing point of water is given the temperature 0°C and the boiling point is given the temperature 100°C. Because there are only 100 Celsius degrees between the boiling and freezing point of water, Celsius degrees are bigger than Fahrenheit degrees.

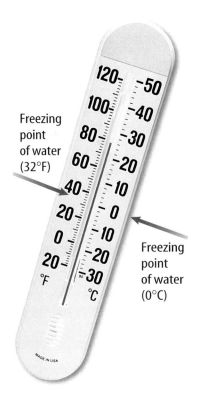

Figure 3 The Fahrenheit and Celsius scales are commonly used temperature scales.

Converting Fahrenheit and Celsius You can convert temperatures back and forth between the two temperature scales by using the following equations.

> **Temperature Conversion Equations**
>
> To convert temperature in °F to °C: $°C = \left(\frac{5}{9}\right)(°F - 32)$
>
> To convert temperature in °C to °F: $°F = \left(\frac{9}{5}\right)(°C) + 32$

For example, to convert 68°F to degrees Celsius, first subtract 32, multiply by 5, then divide by 9. The result is 20°C.

The Kelvin Scale Another temperature scale that is sometimes used is the Kelvin scale. On this scale, 0 K is the lowest temperature an object can have. This temperature is known as absolute zero. The size of a degree on the Kelvin scale is the same as on the Celsius scale. You can change from Celsius degrees to Kelvin degrees by adding 273 to the Celsius temperature.

$$K = °C + 273$$

Applying Math — Solving a Simple Equation

CONVERTING TO CELSIUS On a hot summer day, a Fahrenheit thermometer shows the temperature to be 86°F. What is this temperature on the Celsius scale?

Solution

1. *This is what you know:* Fahrenheit temperature: °F = 86

2. *This is what you need to find:* Celsius temperature: °C

3. *This is the procedure you need to use:* Substitute the Fahrenheit temperature into the equation that converts temperature in °F to °C.
   $°C = \left(\frac{5}{9}\right)(°F - 32) = \frac{5}{9}(86 - 32) = \frac{5}{9}(54) = 30°C$

4. *Check the answer:* Add 32 to your answer and multiply by 9/5. The result should be the given Fahrenheit temperature.

Practice Problems

1. A student's body temperature is 98.6°F. What is this temperature on the Celsius scale?

2. A temperature of 57°C was recorded in 1913 at Death Valley, California. What is this temperature on the Fahrenheit scale?

For more practice visit ips.msscience.com/math_practice

Thermal Energy

The temperature of an object is related to the average kinetic energy of molecules in random motion. But molecules also have potential energy. Potential energy is energy that the molecules have that can be converted into kinetic energy. The sum of the kinetic and potential energy of all the molecules in an object is the **thermal energy** of the object.

The Potential Energy of Molecules When you hold a ball above the ground, it has potential energy. When you drop the ball, its potential energy is converted into kinetic energy as the ball falls toward Earth. It is the attractive force of gravity between Earth and the ball that gives the ball potential energy.

The molecules in a material also exert attractive forces on each other. As a result, the molecules in a material have potential energy. As the molecules get closer together or farther apart, their potential energy changes.

Increasing Thermal Energy Temperature and thermal energy are different. Suppose you have two glasses filled with the same amount of milk, and at the same temperature. If you pour both glasses of milk into a pitcher, as shown in **Figure 4,** the temperature of the milk won't change. However, because there are more molecules of milk in the pitcher than in either glass, the thermal energy of the milk in the pitcher is greater than the thermal energy of the milk in either glass.

Figure 4 At the same temperature, the larger volume of milk in the pitcher has more thermal energy than the smaller volumes of milk in either glass.

section 1 review

Summary

Temperature
- Temperature is related to the average kinetic energy of the molecules an object contains.
- Most materials expand when their temperatures increase.

Measuring Temperature
- On the Celsius scale the freezing point of water is 0°C and the boiling point is 100°C.
- On the Fahrenheit scale the freezing point of water is 32°F and the boiling point is 212°F.

Thermal Energy
- The thermal energy of an object is the sum of the kinetic and potential energy of all the molecules in an object.

Self Check

1. **Explain** the difference between temperature and thermal energy. How are they related?
2. **Determine** which temperature is always larger—an object's Celsius temperature or its Kelvin temperature.
3. **Explain** how kinetic energy and thermal energy are related.
4. **Describe** how a thermometer uses the thermal expansion of a material to measure temperature.

Applying Math

5. **Convert Temperatures** A turkey cooking in an oven will be ready when the internal temperature reaches 180°F. Convert this temperature to °C and K.

section 2
Heat

as you read

What You'll Learn
- **Explain** the difference between thermal energy and heat.
- **Describe** three ways thermal energy is transferred.
- **Identify** materials that are insulators or conductors.

Why It's Important
To keep you comfortable, the flow of thermal energy into and out of your house must be controlled.

Review Vocabulary
electromagnetic wave: a wave produced by vibrating electric charges that can travel in matter and empty space

New Vocabulary
- heat
- conduction
- radiation
- convection
- conductor
- specific heat
- thermal pollution

Heat and Thermal Energy

It's the heat of the day. Heat the oven to 375°F. A heat wave has hit the Midwest. You've often heard the word *heat*, but what is it? Is it something you can see? **Heat** is the transfer of thermal energy from one object to another when the objects are at different temperatures. The amount of thermal energy that is transferred when two objects are brought into contact depends on the difference in temperature between the objects.

For example, no thermal energy is transferred when two pots of boiling water are touching, because the water in both pots is at the same temperature. However, thermal energy is transferred from the pot of hot water in **Figure 5** that is touching a pot of cold water. The hot water cools down and the cold water gets hotter. Thermal energy continues to be transferred until both objects are the same temperature.

Transfer of Thermal Energy When thermal energy is transferred, it always moves from warmer to cooler objects. Thermal energy never flows from a cooler object to a warmer object. The warmer object loses thermal energy and becomes cooler as the cooler object gains thermal energy and becomes warmer. This process of thermal energy transfer can occur in three ways—by conduction, radiation, or convection.

Figure 5 Thermal energy is transferred only when two objects are at different temperatures. Thermal energy always moves from the warmer object to the cooler object.

438 CHAPTER 15 Thermal Energy

Conduction

When you eat hot pizza, you experience conduction. As the hot pizza touches your mouth, thermal energy moves from the pizza to your mouth. This transfer of thermal energy by direct contact is called conduction. **Conduction** occurs when the particles in a material collide with neighboring particles.

Imagine holding an ice cube in your hand, as in **Figure 6.** The faster-moving molecules in your warm hand bump against the slower-moving molecules in the cold ice. In these collisions, energy is passed from molecule to molecule. Thermal energy flows from your warmer hand to the colder ice, and the slow-moving molecules in the ice move faster. As a result, the ice becomes warmer and its temperature increases. Molecules in your hand move more slowly as they lose thermal energy, and your hand becomes cooler.

Conduction usually occurs most easily in solids and liquids, where atoms and molecules are close together. Then atoms and molecules need to move only a short distance before they bump into one another. As a result, thermal energy is transferred more rapidly by conduction in solids and liquids than in gases.

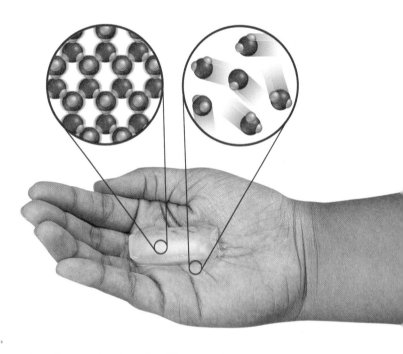

Figure 6 An ice cube in your hand melts because of conduction.

 *Why does conduction occur more easily in solids and liquids than in gases?*

Radiation

On a clear day, you walk outside and feel the warmth of the Sun. How does this transfer of thermal energy occur? Thermal energy transfer does not occur by conduction because almost no matter exists between the Sun and Earth. Instead, thermal energy is transferred from the Sun to Earth by radiation. Thermal energy transfer by **radiation** occurs when energy is transferred by electromagnetic waves. These waves carry energy through empty space, as well as through matter. The transfer of thermal energy by radiation can occur in empty space, as well as in solids, liquids, and gases.

The Sun is not the only source of radiation. All objects emit electromagnetic radiation, although warm objects emit more radiation than cool objects. The warmth you feel when you sit next to a fireplace is due to the thermal energy transferred by radiation from the fire to your skin.

Comparing Rates of Melting

Procedure
1. Prepare ice water by filling a **glass** with ice and then adding water. Let the glass sit until all the ice melts.
2. Place an ice cube in a **coffee cup.**
3. Place a similar-sized ice cube in another **coffee cup** and add ice water to a depth of about 1 cm.
4. Time how long it takes both ice cubes to melt.

Analysis
1. Which ice cube melted fastest? Why?
2. Is air or water a better insulator? Explain.

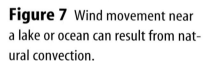

Convection

When you heat a pot of water on a stove, thermal energy can be transferred through the water by a process other than conduction and radiation. In a gas or liquid, atoms or molecules can move much more easily than they can in a solid. As a result, these particles can travel from one place to another, carrying their energy along with them. This transfer of thermal energy by the movement of atoms or molecules from one part of a material to another is called **convection.**

Transferring Thermal Energy by Convection As a pot of water is heated, thermal energy is transferred by convection. First, thermal energy is transferred to the water molecules at the bottom of the pot from the stove. These water molecules move faster as their thermal energy increases. The faster-moving molecules tend to be farther apart than the slower-moving molecules in the cooler water above. This causes the warm water to be less dense than the cooler water. As a result, the warm water rises and is replaced at the bottom of the pot by cooler water. The cooler water is heated, rises, and the cycle is repeated until all the water in the pan is at the same temperature.

Natural Convection Natural convection occurs when a warmer, less dense fluid is pushed away by a cooler, denser fluid. For example, imagine the shore of a lake. During the day, the water is cooler than the land. As shown in **Figure 7,** air above the warm land is heated by conduction. When the air gets hotter, its particles move faster and get farther from each other, making the air less dense. The cooler, denser air from over the lake flows in over the land, pushing the less dense air upward. You feel this movement of incoming cool air as wind. The cooler air then is heated by the land and also begins to rise.

Figure 7 Wind movement near a lake or ocean can result from natural convection.

440 CHAPTER 15 Thermal Energy

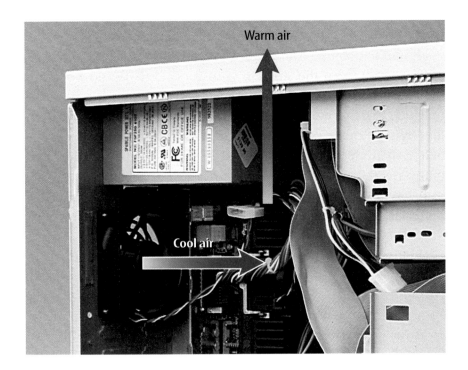

Figure 8 This computer uses forced convection to keep the electronic components surrounded by cooler air.
Identify another example of forced convection.

Forced Convection Sometimes convection can be forced. Forced convection occurs when an outside force pushes a fluid, such as air or water, to make it move and transfer thermal energy. A fan is one type of device that is used to move air. For example, computers use fans to keep their electronic components from getting too hot, which can damage them. The fan blows cool air onto the hot electronic components, as shown in **Figure 8.** Thermal energy from the electronic components is transferred to the air around them by conduction. The warm air is pushed away as cool air rushes in. The hot components then continue to lose thermal energy as the fan blows cool air over them.

Thermal Conductors

Why are cooking pans usually made of metal? Why does the handle of a metal spoon in a bowl of hot soup become warm? The answer to both questions is that metal is a good conductor. A **conductor** is any material that easily transfers thermal energy. Some materials are good conductors because of the types of atoms or chemical compounds they contain.

Reading Check *What is a conductor?*

Remember that an atom has a nucleus surrounded by one or more electrons. Certain materials, such as metals, have some electrons that are not held tightly by the nucleus and are freer to move around. These loosely held electrons can bump into other atoms and help transfer thermal energy. The best conductors of thermal energy are metals such as gold and copper.

Mini LAB

Observing Convection

Procedure

1. Fill a **250-mL beaker** with room-temperature **water** and let it stand undisturbed for at least 1 min.
2. Using a **hot plate,** heat a small amount of water in a **50-mL beaker** until it is almost boiling.
    WARNING: *Do not touch the heated hot plate.*
3. Carefully place a **penny** into the hot water and let it stand for about 1 min.
4. Take the penny out of the hot water with **metal tongs** and place it on a table. Immediately place the 250-mL beaker on the penny.
5. Using a **dropper,** gently place one drop of **food coloring** on the bottom of the 250-mL beaker of water.
6. Observe what happens in the beaker for several minutes.

Analysis
What happened when you placed the food coloring in the 250-mL beaker? Why?

SECTION 2 Heat **441**

Animal Insulation
To survive in its arctic environment, a polar bear needs good insulation against the cold. Underneath its fur, a polar bear has 10 cm of insulating blubber. Research how animals in polar regions are able to keep themselves warm. Summarize the different ways in your Science Journal.

Thermal Insulators

If you're cooking food, you want the pan to conduct thermal energy easily from the stove to your food, but you do not want the handle of the pan to become hot. An insulator is a material in which thermal energy doesn't flow easily. Most pans have handles that are made from insulators. Liquids and gases are usually better insulators than solids are. Air is a good insulator, and many insulating materials contain air spaces that reduce the transfer of thermal energy by conduction within the material. Materials that are good conductors, such as metals, are poor insulators, and poor conductors are good insulators.

Houses and buildings are made with insulating materials to reduce the flow of thermal energy between the inside and outside. Fluffy insulation like that shown in **Figure 9** is put in the walls. Some windows have double layers of glass that sandwich a layer of air or other insulating gas. This reduces the outward flow of thermal energy in the winter and the inward flow of thermal energy in the summer.

Heat Absorption

On a hot day, you can walk barefoot across the lawn, but the asphalt pavement of a street is too hot to walk on. Why is the pavement hotter than the grass? The change in temperature of an object as it is heated depends on the material it is made of.

Specific Heat The temperature change of a material as it is heated depends on the specific heat of the material. The **specific heat** of a material is the amount of thermal energy needed to raise the temperature of 1 kg of the material by 1°C.

More thermal energy is needed to change the temperature of a material with a high specific heat than one with a low specific heat. For example, sand on a beach has a lower specific heat than water. During the day, radiation from the Sun warms the sand and the water. Because of its lower specific heat, the sand heats up faster than the water. At night, however, the sand feels cool and the water feels warmer. The temperature of the water changes more slowly than the temperature of the sand as they both lose thermal energy to the cooler night air.

Figure 9 The insulation in houses and buildings helps reduce the transfer of thermal energy between the air inside and air outside.

Thermal Pollution

 Some electric power plants and factories that use water for cooling produce hot water as a by-product. If this hot water is released into an ocean, lake, or river, it will raise the temperature of the water nearby. This increase in the temperature of a body of water caused by adding warmer water is called **thermal pollution.** Rainwater that is heated after it falls on warm roads or parking lots also can cause thermal pollution if it runs off into a river or lake.

Effects of Thermal Pollution Increasing the water temperature causes fish and other aquatic organisms to use more oxygen. Because warmer water contains less dissolved oxygen than cooler water, some organisms can die due to a lack of oxygen. Also, in warmer water, many organisms become more sensitive to chemical pollutants, parasites, and diseases.

Reducing Thermal Pollution Thermal pollution can be reduced by cooling the warm water produced by factories, power plants, and runoff before it is released into a body of water. Cooling towers like the ones shown in **Figure 10** are used to cool the water used by some power plants and factories.

Figure 10 This power plant uses cooling towers to cool the warm water produced by the power plant.

section 2 review

Summary

Heat and Thermal Energy
- Heat is the transfer of thermal energy due to a temperature difference.
- Thermal energy always moves from a higher temperature to a lower temperature.

Conduction, Radiation, and Convection
- Conduction is the transfer of thermal energy when substances are in direct contact.
- Radiation is the transfer of thermal energy by electromagnetic waves.
- Convection is the transfer of thermal energy by the movement of matter.

Thermal Conductors and Specific Heat
- A thermal conductor is a material in which thermal energy moves easily.
- The specific heat of a substance is the amount of thermal energy needed to raise the temperature of 1 kg of the substance by 1°C.

Self Check

1. **Explain** why materials such as plastic foam, feathers, and fur are poor conductors of thermal energy.
2. **Explain** why the sand on a beach cools down at night more quickly than the ocean water.
3. **Infer** If a substance can contain thermal energy, can a substance also contain heat?
4. **Describe** how thermal energy is transferred from one place to another by convection.
5. **Explain** why a blanket keeps you warm.
6. **Think Critically** In order to heat a room evenly, should heating vents be placed near the floor or near the ceiling of the room? Explain.

Applying Skills

7. **Design an Experiment** to determine whether wood or iron is a better thermal conductor. Identify the dependent and independent variables in your experiment.

ips.msscience.com/self_check_quiz

Heating Up and Cooling Down

Do you remember how long it took for a cup of hot chocolate to cool before you could take a sip? The hotter the chocolate, the longer it seemed to take to cool.

▶ Real-World Question

How does the temperature of a liquid affect how quickly it warms or cools?

Goals
- Measure the temperature change of water at different temperatures.
- Infer how the rate of heating or cooling depends on the initial water temperature.

Materials
thermometers (5)
400-mL beakers (5)
stopwatch
*watch with second hand
hot plate
*Alternate materials

Safety Precautions

WARNING: *Do not use mercury thermometers. Use caution when heating with a hot plate. Hot and cold glass appears the same.*

▶ Procedure

1. Make a data table to record the temperature of water in five beakers every minute from 0 to 10 min.
2. Fill one beaker with 100 mL of water. Place the beaker on a hot plate and bring the water to a boil. Carefully remove the hot beaker from the hot plate.

3. Record the water temperature at minute 0, and then every minute for 10 min.
4. Repeat step 3 starting with hot tap water, cold tap water, refrigerated water, and ice water with the ice removed.

▶ Conclude and Apply

1. **Graph** your data. **Plot and label** lines for all five beakers on one graph.
2. **Calculate** the rate of heating or cooling for the water in each beaker by subtracting the initial temperature of the water from the final temperature and then dividing this answer by 10 min.
3. **Infer** from your results how the difference between room temperature and the initial temperature of the water affected the rate at which it heated up or cooled down.

Communicating Your Data

Share your data and graphs with other classmates and explain any differences among your data.

section 3

Engines and Refrigerators

Heat Engines

The engines used in cars, motorcycles, trucks, and other vehicles, like the one shown in **Figure 11,** are heat engines. A **heat engine** is a device that converts thermal energy into mechanical energy. Mechanical energy is the sum of the kinetic and potential energy of an object. The heat engine in a car converts thermal energy into mechanical energy when it makes the car move faster, causing the car's kinetic energy to increase.

Forms of Energy There are other forms of energy besides thermal energy and mechanical energy. For example, chemical energy is energy stored in the chemical bonds between atoms. Radiant energy is the energy carried by electromagnetic waves. Nuclear energy is energy stored in the nuclei of atoms. Electrical energy is the energy carried by electric charges as they move in a circuit. Devices such as heat engines convert one form of energy into other useful forms.

The Law of Conservation of Energy When energy is transformed from one form to another, the total amount of energy doesn't change. According to the law of conservation of energy, energy cannot be created or destroyed. Energy only can be transformed from one form to another. No device, including a heat engine, can produce energy or destroy energy.

as you read

What You'll Learn
- **Describe** what a heat engine does.
- **Explain** that energy can exist in different forms, but is never created or destroyed.
- **Describe** how an internal combustion engine works.
- **Explain** how refrigerators move thermal energy.

Why It's Important
Heat engines enable you to travel long distances.

Review Vocabulary
work: a way of transferring energy by exerting a force over a distance

New Vocabulary
- heat engine
- internal combustion engine

Figure 11 The engine in this earth mover transforms thermal energy into mechanical energy that can perform useful work.

Figure 12 Internal combustion engines are found in many tools and machines.

Topic: Automobile Engines
Visit ips.msscience.com for Web links to information on how internal combustion engines were developed for use in cars.

Activity Make a time line showing the five important events in the development of the automobile engine.

Internal Combustion Engines The heat engine you are probably most familiar with is the internal combustion engine. In **internal combustion engines,** the fuel burns in a combustion chamber inside the engine. Many machines, including cars, airplanes, buses, boats, trucks, and lawn mowers, use internal combustion engines, as shown in **Figure 12.**

Most cars have an engine with four or more combustion chambers, or cylinders. Usually the more cylinders an engine has, the more power it can produce. Each cylinder contains a piston that can move up and down. A mixture of fuel and air is injected into a combustion chamber and ignited by a spark. When the fuel mixture is ignited, it burns explosively and pushes the piston down. The up-and-down motion of the pistons turns a rod called a crankshaft, which turns the wheels of the car. **Figure 13** shows how an internal combustion engine converts thermal energy to mechanical energy in a process called the four-stroke cycle.

Several kinds of internal combustion engines have been designed. In diesel engines, the air in the cylinder is compressed to such a high pressure that the highly flammable fuel ignites without the need for a spark plug. Many lawn mowers use a two-stroke gasoline engine. The first stroke is a combination of intake and compression. The second stroke is a combination of power and exhaust.

 *How does the burning of fuel mixture cause a piston to move?*

446 CHAPTER 15 Thermal Energy

NATIONAL GEOGRAPHIC VISUALIZING THE FOUR-STROKE CYCLE

Figure 13

Most modern cars are powered by fuel-injected internal combustion engines that have a four-stroke combustion cycle. Inside the engine, thermal energy is converted into mechanical energy as gasoline is burned under pressure inside chambers known as cylinders. The steps in the four-stroke cycle are shown here.

EXHAUST STROKE

COMPRESSION STROKE

POWER STROKE

INTAKE STROKE

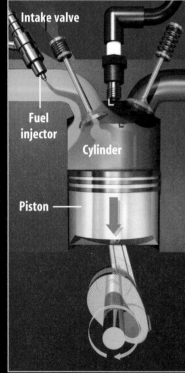

A During the intake stroke, the piston inside the cylinder moves downward. As it does, air fills the cylinder through the intake valve, and a mist of fuel is injected into the cylinder.

B The piston moves up, compressing the fuel-air mixture.

C At the top of the compression stroke, a spark ignites the fuel-air mixture. The hot gases that are produced expand, pushing the piston down and turning the crankshaft.

D The exhaust valve opens as the piston moves up, pushing the exhaust gases out of the cylinder.

SECTION 3 Engines and Refrigerators **447**

Mechanical Engineering
People who design engines and machines are mechanical engineers. Some mechanical engineers study ways to maximize the transformation of useful energy during combustion—the transformation of energy from chemical form to mechanical form.

Refrigerators

If thermal energy will only flow from something that is warm to something that is cool, how can a refrigerator be cooler inside than the air in the kitchen? A refrigerator is a heat mover. It absorbs thermal energy from the food inside the refrigerator. Then it carries the thermal energy to outside the refrigerator, where it is transferred to the surrounding air.

A refrigerator contains a material called a coolant that is pumped through pipes inside and outside the refrigerator. The coolant is the substance that carries thermal energy from the inside to the outside of the refrigerator.

Absorbing Thermal Energy Figure 14 shows how a refrigerator operates. Liquid coolant is forced up a pipe toward the freezer unit. The liquid passes through an expansion valve where it changes into a gas. When it changes into a gas, it becomes cold. The cold gas passes through pipes around the inside of the refrigerator. Because the coolant gas is so cold, it absorbs thermal energy from inside the refrigerator, and becomes warmer.

Releasing Thermal Energy However, the gas is still colder than the outside air. So, the thermal energy absorbed by the coolant cannot be transferred to the air. The coolant gas then passes through a compressor that compresses the gas. When the gas is compressed, it becomes warmer than room temperature. The gas then flows through the condenser coils, where thermal energy is transferred to the cooler air in the room. As the coolant gas cools, it changes into a liquid. The liquid is pumped through the expansion valve, changes into a gas, and the cycle is repeated.

Figure 14 A refrigerator uses a coolant to move thermal energy from inside to outside the refrigerator. The compressor supplies the energy that enables the coolant to transfer thermal energy to the room.
Diagram *how the temperature of the coolant changes as it moves in a refrigerator.*

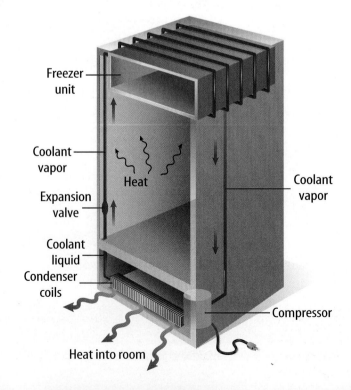

Air Conditioners Most air conditioners cool in the same way that a refrigerator does. You've probably seen air-conditioning units outside of many houses. As in a refrigerator, thermal energy from inside the house is absorbed by the coolant within pipes inside the air conditioner. The coolant then is compressed by a compressor, and becomes warmer. The warmed coolant travels through pipes that are exposed to the outside air. Here the thermal energy is transferred to the outside air.

Heat Pumps Some buildings use a heat pump for heating and cooling. Like an air conditioner or refrigerator, a heat pump moves thermal energy from one place to another. In heating mode, shown in **Figure 15,** the coolant absorbs thermal energy through the outside coils. The coolant is warmed when it is compressed and transfers thermal energy to the house through the inside coils. When a heat pump is used for cooling, it removes thermal energy from the indoor air and transfers it outdoors.

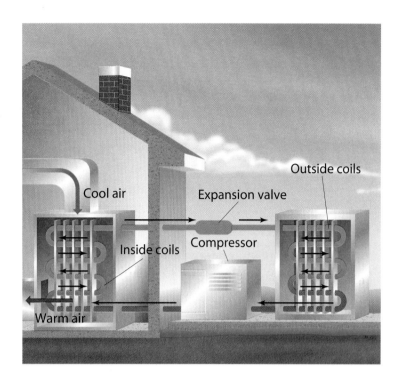

Figure 15 A heat pump heats a building by absorbing thermal energy from the outside air and transferring thermal energy to the cooler air inside.

section 3 review

Summary

Heat Engines and Energy
- A heat engine is a device that converts thermal energy into mechanical energy.
- Energy cannot be created or destroyed. It only can be transformed from one form to another.
- An internal combustion engine is a heat engine that burns fuel in a combustion chamber inside the engine.

Refrigerators and Heat Pumps
- A refrigerator uses a coolant to transfer thermal energy to outside the refrigerator.
- The coolant gas absorbs thermal energy from inside the refrigerator.
- Compressing the coolant makes it warmer than the air outside the refrigerator.
- A heat pump heats by absorbing thermal energy from the air outside, and transferring it inside a building.

Self Check

1. **Diagram** the movement of coolant and the flow of heat when a heat pump is used to cool a building.
2. **Explain** why diesel engines don't use spark plugs.
3. **Identify** the source of thermal energy in an internal combustion engine.
4. **Determine** whether you could cool a kitchen by keeping the refrigerator door open.
5. **Describe** how a refrigerator uses a coolant to keep the food compartment cool.
6. **Think Critically** Explain how an air conditioner could also be used to heat a room.

Applying Skills

7. **Make a Concept Map** Make an events-chain concept map showing the sequence of steps in a four-stroke cycle.

LAB Design Your Own

Comparing Thermal Insulators

Goals
- **Predict** the temperature change of a hot drink in various types of containers over time.
- **Design** an experiment to test the hypothesis and collect data that can be graphed.
- **Interpret** the data.

Possible Materials
hot plate
large beaker
water
100-mL graduated cylinder
alcohol thermometers
various beverage containers
material to cover the containers
stopwatch
tongs
thermal gloves or mitts

Safety Precautions

WARNING: *Use caution when heating liquids. Use tongs or thermal gloves when handling hot materials. Hot and cold glass appear the same. Treat thermometers with care and keep them away from the edges of tables.*

● Real-World Question
Insulated beverage containers are used to reduce the flow of thermal energy. What kinds of containers do you most often drink from? Aluminum soda cans? Paper, plastic, or foam cups? Glass containers? In this investigation, compare how well several different containers reduce the transfer of thermal energy. Which types of containers are most effective at keeping a hot drink hot?

● Form a Hypothesis
Predict the temperature change of a hot liquid in several containers made of different materials over a time interval.

● Test Your Hypothesis
Make a Plan
1. **Decide** what types of containers you will test. Design an experiment to test your hypothesis. This is a group activity, so make certain that everyone gets to contribute to the discussion.

450 CHAPTER 15 Thermal Energy

Using Scientific Methods

2. **List** the materials you will use in your experiment. Describe exactly how you will use these materials. Which liquid will you test? What will be its starting temperature? How will you cover the hot liquids in the container? What material will you use as a cover?
3. **Identify** the variables and controls in your experiment.
4. **Design** a data table in your Science Journal to record the observations you make.

Follow Your Plan

1. Have your teacher approve the steps of your experiment and your data table before you start.
2. To see the pattern of how well various containers retain thermal energy, you will need to graph your data. What kind of graph will you use? Make certain you take enough measurements during the experiment to make your graph.
3. The time intervals between measurements should be the same. Be sure to keep track of time as the experiment goes along. For how long will you measure the temperature?
4. Carry out your investigation and record your observations.

Analyze Your Data

1. **Graph** your data. Use one graph to show the data collected from all your containers. Label each line on your graph.
2. **Interpret Data** How can you tell by looking at your graphs which containers best reduce the flow of thermal energy?
3. **Evaluate** Did the water temperature change as you had predicted? Use your data and graph to explain your answers.

Conclude and Apply

1. **Explain** why the rate of temperature change varies among the containers. Did the size of the containers affect the rate of cooling?
2. **Conclude** which containers were the best insulators.

Compare your data and graphs with other classmates and explain any differences in your results or conclusions.

TIME SCIENCE AND Society
SCIENCE ISSUES THAT AFFECT YOU!

The Heat Is On

You may live far from water, but still live on an island—a heat island

Think about all the things that are made of asphalt and concrete in a city. As far as the eye can see, there are buildings and parking lots, sidewalks and streets. The combined effect of these paved surfaces and towering structures can make a city sizzle in the summer. There's even a name for this effect. It's called the heat island effect.

Hot Times

You can think of a city as an island surrounded by an ocean of green trees and other vegetation. In the midst of those green trees, the air can be up to 8°C cooler than it is downtown. During the day in rural areas, the Sun's energy is absorbed by plants and soil. Some of this energy causes water to evaporate, so less energy is available to heat the surroundings. This keeps the temperature lower.

Higher temperatures aren't the only problems caused by heat islands. People crank up their air conditioners for relief, so the use of energy skyrockets. Also, the added heat speeds up the rates of chemical reactions in the atmosphere. Smog is due to chemical reactions caused by the interaction of sunlight and vehicle emissions. So hotter air means more smog. And more smog means more health problems.

Cool Cures

Several U.S. cities are working with NASA scientists to come up with a cure for the summertime blues. For instance, dark materials absorb thermal energy more efficiently than light materials. So painting buildings, especially roofs, white can reduce temperature and save on cooling bills.

Dark materials, such as asphalt, absorb more thermal energy than light materials. In extreme heat, it's even possible to fry an egg on dark pavement!

Design and Research Visit the Web Site to the right to research NASA's Urban Heat Island Project. What actions are cities taking to reduce the heat-island effect? Design a city area that would help reduce this effect.

For more information, visit
ips.msscience.com/time

chapter 15 Study Guide

Reviewing Main Ideas

Section 1 — Temperature and Thermal Energy

1. Molecules of matter are moving constantly. Temperature is related to the average value of the kinetic energy of the molecules.
2. Thermometers measure temperature. Three common temperature scales are the Celsius, Fahrenheit, and Kelvin scales.
3. Thermal energy is the total kinetic and potential energy of the particles in matter.

Section 2 — Heat

1. Heat is thermal energy that is transferred from a warmer object to a colder object.
2. Thermal energy can be transferred by conduction, convection, and radiation.
3. A material that easily transfers thermal energy is a conductor. Thermal energy doesn't flow easily in an insulator.
4. The specific heat of a material is the thermal energy needed to change the temperature of 1 kg of the material 1°C.
5. Thermal pollution occurs when warm water is added to a body of water.

Section 3 — Engines and Refrigerators

1. A device that converts thermal energy into mechanical energy is an engine.
2. In an internal combustion engine, fuel is burned in combustion chambers inside the engine using a four-stroke cycle.
3. Refrigerators and air conditioners use a coolant to move thermal energy.

Visualizing Main Ideas

Copy and complete the following cycle map about the four-stroke cycle.

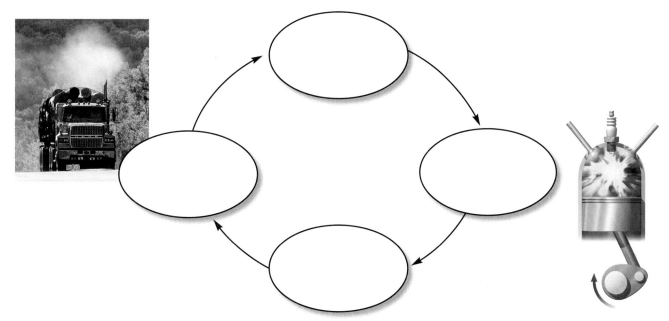

ips.msscience.com/interactive_tutor

chapter 15 Review

Using Vocabulary

conduction p. 439
conductor p. 441
convection p. 440
heat p. 438
heat engine p. 445
internal combustion engine p. 446
radiation p. 439
specific heat p. 442
temperature p. 434
thermal energy p. 437
thermal pollution p. 443

Explain the differences in the vocabulary words given below. Then explain how the words are related. Use complete sentences in your answers.

1. internal combustion engine—heat engine
2. temperature—thermal energy
3. thermal energy—thermal pollution
4. conduction—convection
5. conduction—thermal energy
6. thermal energy—specific heat
7. conduction—radiation
8. convection—radiation
9. conductor—thermal energy

Checking Concepts

Choose the word or phrase that best answers the question.

10. What source of thermal energy does an internal combustion engine use?
    A) steam
    B) hot water
    C) burning fuel
    D) refrigerant

11. What happens to most materials when they become warmer?
    A) They contract.
    B) They float.
    C) They vaporize.
    D) They expand.

12. Which occurs if two objects at different temperatures are in contact?
    A) convection
    B) radiation
    C) condensation
    D) conduction

13. Which of the following describes the thermal energy of particles in a substance?
    A) average value of all kinetic energy
    B) total value of all kinetic energy
    C) total value of all kinetic and potential energy
    D) average value of all kinetic and potential energy

14. Thermal energy moving from the Sun to Earth is an example of which process?
    A) convection
    B) expansion
    C) radiation
    D) conduction

15. Many insulating materials contain spaces filled with air because air is what type of material?
    A) conductor
    B) coolant
    C) radiator
    D) insulator

16. A recipe calls for a cake to be baked at a temperature of 350°F. What is this temperature on the Celsius scale?
    A) 162°C
    B) 177°C
    C) 194°C
    D) 212°C

17. Which of the following is true?
    A) Warm air is less dense than cool air.
    B) Warm air is as dense as cool air.
    C) Warm air has no density.
    D) Warm air is denser than cool air.

18. Which of these is the name for thermal energy that moves from a warmer object to a cooler one?
    A) kinetic energy
    B) specific heat
    C) heat
    D) temperature

19. Which of the following is an example of thermal energy transfer by conduction?
    A) water moving in a pot of boiling water
    B) warm air rising from hot pavement
    C) the warmth you feel sitting near a fire
    D) the warmth you feel holding a cup of hot cocoa

454 CHAPTER REVIEW

ips.msscience.com/vocabulary_puzzlemaker

chapter 15 Review

Thinking Critically

20. **Infer** When you heat water in a pan, the surface gets hot quickly, even though you are applying heat to the bottom of the water. Explain.

21. **Explain** why several layers of clothing often keep you warmer than a single layer.

22. **Identify** The phrase "heat rises" is sometimes used to describe the movement of thermal energy. For what type of materials is this phrase correct? Explain.

23. **Describe** When a lightbulb is turned on, the electric current in the filament causes the filament to become hot and glow. If the filament is surrounded by a gas, describe how thermal energy is transferred from the filament to the air outside the bulb.

24. **Design an Experiment** Some colors of clothing absorb radiation better than other colors. Design an experiment that will test various colors by placing them in the hot Sun for a period of time.

25. **Explain** Concrete sidewalks usually are made of slabs of concrete. Why do the concrete slabs have a space between them?

26. **Concept Map** Copy and complete the following concept map on convection in a liquid.

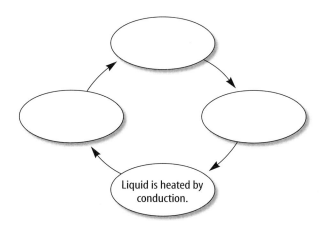

27. **Explain** A winter jacket is lined with insulating material that contains air spaces. How would the insulating properties of the jacket change if the insulating material in the jacket becomes wet? Explain.

28. **Compare** Two glasses of water are poured into a pitcher. If the temperature of the water in both glasses was the same before and after they were mixed, describe how the thermal energy of the water in the pitcher compares to the water in the glasses.

Performance Activities

29. **Poll** In the United States, the Fahrenheit temperature scale is used most often. Some people feel that Americans should switch to the Celsius scale. Take a poll of at least 20 people. Find out if they feel the switch to the Celsius scale should be made. Make a list of reasons people give for or against changing.

Applying Math

30. **Temperature Order** List the following temperatures from coldest to warmest: 80° C, 200 K, 50° F.

31. **Temperature Change** The high temperature on a summer day is 88°F and the low temperature is 61°F. What is the difference between these two temperatures in degrees Celsius?

32. **Global Temperature** The average global temperature is 286 K. Convert this temperature to degrees Celsius.

33. **Body Temperature** A doctor measures a patient's temperature at 38.4°C. Convert this temperature to degrees Fahrenheit.

ips.msscience.com/chapter_review

Part 1 Multiple Choice

Record your answers on the answer sheet provided by your teacher or on a sheet of paper.

Use the photo below to answer questions 1 and 2.

1. The temperatures of the two glasses of water shown in the photograph above are 30°C and 0°C. Which of the following is a correct statement about the two glasses of water?
   A. The cold water has a higher average kinetic energy.
   B. The warmer water has lower thermal energy.
   C. The molecules of the cold water move faster.
   D. The molecules of the warmer water have more kinetic energy.

2. The difference in temperature of the two glasses of water is 30°C. What is their difference in temperature on the Kelvin scale?
   A. 30 K
   B. 86 K
   C. 243 K
   D. 303 K

3. Which of the following describes a refrigerator?
   A. heat engine
   B. heat pump
   C. heat mover
   D. conductor

Test-Taking Tip

Avoid rushing on test day. Prepare your clothes and test supplies the night before. Wake up early and arrive at school on time on test day.

4. Which of the following is not a step in the four-stroke cycle of internal combustion engines?
   A. compression
   B. exhaust
   C. idling
   D. power

Use the table below to answer question 5.

Material	Specific Heat (J/kg °C)
aluminum	897
copper	385
lead	129
nickel	444
zinc	388

5. A sample of each of the metals in the table above is formed into a 50-g cube. If 100 J of thermal energy are applied to each of the samples, which metal would change temperature by the greatest amount?
   A. aluminum
   B. copper
   C. lead
   D. nickel

6. An internal combustion engine converts thermal energy to which of the following forms of energy?
   A. chemical
   B. mechanical
   C. radiant
   D. electrical

7. Which of the following is a statement of the law of conservation of energy?
   A. Energy never can be created or destroyed.
   B. Energy can be created, but never destroyed.
   C. Energy can be destroyed, but never created.
   D. Energy can be created and destroyed when it changes form.

Standardized Test Practice

Part 2 Short Response/Grid In

Record your answers on the answer sheet provided by your teacher or on a sheet of paper.

8. If you add ice to a glass of room-temperature ice, does the water warm the ice or does the ice cool the water? Explain.

9. Strong winds that occur during a thunderstorm are the result of temperature differences between neighboring air masses. Would you expect the warmer or the cooler air mass to rise above the other?

10. A diesel engine uses a different type of fuel than the fuel used in a gasoline engine. Explain why.

11. What are the two main events that occur while the cylinder moves downward during the intake stroke of an internal combustion engine's four-stroke cycle?

Use the photo below to answer questions 12 and 13.

12. Why are cooking pots like the one in the photograph above often made of metal? Why isn't the handle made of metal?

13. When heating water in the pot, electrical energy from the cooking unit is changed to what other type of energy?

Part 3 Open Ended

Record your answers on a sheet of paper.

Use the illustration below to answer questions 14 and 15.

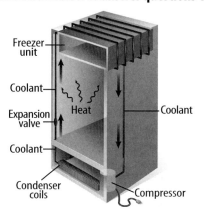

14. The illustration above shows the parts of a refrigerator and how coolant flows through the refrigerator. Explain how thermal energy is transferred to the coolant inside the refrigerator and then transferred from the coolant to the outside air.

15. What are the functions of the expansion valve, the condenser coils, and the compressor in the illustration?

16. Define convection. Explain the difference between natural and forced convection, and give an example of each.

17. Draw a sketch with arrows showing how conduction, convection, and radiation affect the movement and temperature of air near an ocean.

18. Define temperature and explain how it is related to the movement of molecules in a substance.

19. Explain what makes some materials good thermal conductors.

20. You place a cookie sheet in a hot oven. A few minutes later you hear a sound as the cookie sheet bends slightly. Explain what causes this.

unit 6
Waves, Sound, and Light

How Are
Radar & Popcorn
Connected?

NATIONAL GEOGRAPHIC

Radar systems—such as the one in this modern air traffic control room—use radio waves to detect objects. In the 1940s, the radio waves used for radar were generated by a device called a magnetron. One day, an engineer working on a radar project was standing near a magnetron when he noticed that the candy bar in his pocket had melted. Intrigued, the engineer got some unpopped popcorn and placed it next to the magnetron. Sure enough, the kernels began to pop. The engineer realized that the magnetron's short radio waves, called microwaves, caused the molecules in the food to move more quickly, increasing the food's temperature. Soon, magnetrons were being used in the first microwave ovens. Today, microwave ovens are used to pop popcorn—and heat many other kinds of food—in kitchens all over the world.

unit projects

Visit **ips.msscience.com/unit_project** to find project ideas and resources. Projects include:

- **History** Research tsunamis, their energy, and other characteristics. Graph and compare height, distance, and lives lost from past waves.
- **Technology** Discover how steel drums are made. Construct your own drum and experiment with sounds and patterns of vibrations.
- **Model** Create an original light show with colored lights and stick puppets expressing your new knowledge of light, mirrors, and lenses.

WebQuest *Laser Eye Surgery* provides an opportunity to be an informed consumer of the advantages and disadvantages of laser eye surgery.

chapter 16

Waves

The BIG Idea
Waves transfer energy from place to place without transferring matter.

SECTION 1
What are waves?
Main Idea Waves can carry energy through matter or through empty space.

SECTION 2
Wave Properties
Main Idea Waves can be described by their amplitude, wavelength, and frequency.

SECTION 3
Wave Behavior
Main Idea Waves can change direction at the boundary between different materials.

Catch A Wave
On a breezy day in Maui, Hawaii, windsurfers ride the ocean waves. Waves carry energy. You can see the ocean waves in this picture, but there are other waves you cannot see, such as microwaves, radio waves, and sound waves.

Science Journal Write a paragraph about some places where you have seen water waves.

Start-Up Activities

Waves and Energy

It's a beautiful autumn day. You are sitting by a pond in a park. Music from a school marching band is carried to your ears by waves. A fish jumps, making waves that spread past a leaf that fell from a tree, causing the leaf to move. In the following lab, you'll observe how waves carry energy that can cause objects to move.

1. Add water to a large, clear, plastic plate to a depth of about 1 cm.
2. Use a dropper to release a single drop of water onto the water's surface. Repeat.
3. Float a cork or straw on the water.
4. When the water is still, repeat step 2 from a height of 10 cm, then again from 20 cm.
5. **Think Critically** In your Science Journal, record your observations. How did the motion of the cork depend on the height of the dropper?

 Preview this chapter's content and activities at ips.msscience.com

Waves Make the following Foldable to compare and contrast the characteristics of transverse and compressional waves.

STEP 1 Fold one sheet of paper lengthwise.

STEP 2 Fold into thirds.

STEP 3 Unfold and draw overlapping ovals. Cut the top sheet along the folds.

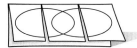

STEP 4 Label the ovals as shown.

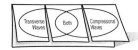

Construct a Venn Diagram As you read the chapter, list the characteristics unique to transverse waves under the left tab, those unique to compressional waves under the right tab, and those characteristics common to both under the middle tab.

Get Ready to Read

New Vocabulary

① Learn It! What should you do if you find a word you don't know or understand? Here are some suggested strategies:

1. Use context clues (from the sentence or the paragraph) to help you define it.
2. Look for prefixes, suffixes, or root words that you already know.
3. Write it down and ask for help with the meaning.
4. Guess at its meaning.
5. Look it up in the glossary or a dictionary.

② Practice It! Look at the word *rarefaction* in the following passage. See how context clues can help you understand its meaning.

Context Clue
Rarefactions are groups of molecules.

Context Clue
Rarefactions move away from a vibrating object.

Context Clue
Rarefactions are parts of a sound wave.

> When the drumhead moves downward, the molecules near it have more room and can spread farther apart. This group of molecules that are farther apart is a rarefaction. The rarefaction also moves away from the drumhead. As the drumhead vibrates up and down, it forms a series of compressions and rarefactions that move away and spread out in all directions this series of compressions and rarefactions is a sound wave.
>
> — *from page 465*

③ Apply It! Make a vocabulary bookmark with a strip of paper. As you read, keep track of words you do not know or want to learn more about.

Target Your Reading

Reading Tip
Read a paragraph containing a vocabulary word from beginning to end. Then go back to determine the meaning of the word.

Use this to focus on the main ideas as you read the chapter.

1. **Before you read** the chapter, respond to the statements below on your worksheet or on a numbered sheet of paper.
   - Write an **A** if you **agree** with the statement.
   - Write a **D** if you **disagree** with the statement.

2. **After you read** the chapter, look back to this page to see if you've changed your mind about any of the statements.
   - If any of your answers changed, explain why.
   - Change any false statements into true statements.
   - Use your revised statements as a study guide.

Science Online
Print out a worksheet of this page at ips.msscience.com

Before You Read A or D		Statement	After You Read A or D
	1	There can be no sound in outer space.	
	2	Light waves are mechanical waves that can travel only through matter.	
	3	Waves are produced by something that vibrates.	
	4	In air, sound waves travel faster than light waves.	
	5	The pitch of a sound wave depends on the frequency of the wave.	
	6	The only difference between microwaves and light waves is that microwaves have longer wavelengths.	
	7	A wave moves in a straight line, even if the speed of the wave changes.	
	8	Higher-pitched sounds bend more as they pass through an open door than lower-pitched sounds do.	
	9	When two waves overlap, they always cancel each other.	

462 B

section 1
What are waves?

as you read

What You'll Learn
- **Explain** the relationship among waves, energy, and matter.
- **Describe** the difference between transverse waves and compressional waves.

Why It's Important
Waves enable you to see and hear the world around you.

Review Vocabulary
energy: the ability to cause change

New Vocabulary
- wave
- mechanical wave
- transverse wave
- compressional wave
- electromagnetic wave

What is a wave?

When you are relaxing on an air mattress in a pool and someone does a cannonball dive off the diving board, you suddenly find yourself bobbing up and down. You can make something move by giving it a push or pull, but the person jumping didn't touch your air mattress. How did the energy from the dive travel through the water and move your air mattress? The up-and-down motion was caused by the peaks and valleys of the ripples that moved from where the splash occurred. These peaks and valleys make up water waves.

Waves Carry Energy Rhythmic disturbances that carry energy without carrying matter are called **waves**. Water waves are shown in **Figure 1**. You can see the energy of the wave from a speedboat traveling outward, but the water only moves up and down. If you've ever felt a clap of thunder, you know that sound waves can carry large amounts of energy. You also transfer energy when you throw something to a friend, as in **Figure 1**. However, there is a difference between a moving ball and a wave. A ball is made of matter, and when it is thrown, the matter moves from one place to another. So, unlike the wave, throwing a ball involves the transport of matter as well as energy.

Figure 1 Water waves and a moving ball both transfer energy from one place to another.

Water waves produced by a boat travel outward, but matter does not move along with the waves.

When the ball is thrown, energy is transferred as matter moves from place to place.

As the students pass the ball, the students' positions do not change—only the position of the ball changes.

In a water wave, water molecules bump each other and pass energy from molecule to molecule.

Figure 2 A wave transports energy without transporting matter from place to place.
Describe *other models that could be used to represent a mechanical wave.*

A Model for Waves

How does a wave carry energy without transporting matter? Imagine a line of people, as shown in **Figure 2.** The first person in line passes a ball to the second person, who passes the ball to the next person, and so on. Passing a ball down a line of people is a model for how waves can transport energy without transporting matter. Even though the ball has traveled, the people in line have not moved. In this model, you can think of the ball as representing energy. What do the people in line represent?

Think about the ripples on the surface of a pond. The energy carried by the ripples travels through the water. The water is made up of water molecules. It is the individual molecules of water that pass the wave energy, just as the people. The water molecules transport the energy in a water wave by colliding with the molecules around them, as shown in **Figure 2.**

Reading Check *What is carried by waves?*

Mechanical Waves

In the wave model, the ball could not be transferred if the line of people didn't exist. The energy of a water wave could not be transferred if no water molecules existed. These types of waves, which use matter to transfer energy, are called **mechanical waves.** The matter through which a mechanical wave travels is called a medium. For ripples on a pond, the medium is the water.

A mechanical wave travels as energy is transferred from particle to particle in the medium. For example, a sound wave is a mechanical wave that can travel through air, as well as solids, liquids, and other gases. Without a medium such as air, there would be no sound waves. In outer space sound waves can't travel because there is no air.

SECTION 1 What are waves? **463**

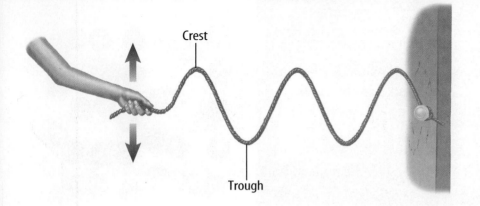

Transverse Waves In a mechanical **transverse wave**, the wave energy causes the matter in the medium to move up and down or back and forth at right angles to the direction the wave travels. You can make a model of a transverse wave. Stretch a long rope out on the ground. Hold one end in your hand. Now shake the end in your hand back and forth. As you shake the rope, you create a wave that seems to slide along the rope.

When you first started shaking the rope, it might have appeared that the rope itself was moving away from you. But it was only the wave that was moving away from your hand. The wave energy moves through the rope, but the matter in the rope doesn't travel. You can see that the wave has peaks and valleys at regular intervals. As shown in **Figure 3,** the high points of transverse waves are called crests. The low points are called troughs.

Figure 3 The high points on the wave are called crests and the low points are called troughs.

Reading Check *What are the highest points of transverse waves called?*

Figure 4 A compressional wave can travel through a coiled spring toy.

As the wave motion begins, the coils on the left are close together and the other coils are far apart.

The wave, seen in the squeezed and stretched coils, travels along the spring.

The string and coils did not travel with the wave. Each coil moved forward and then back to its original position.

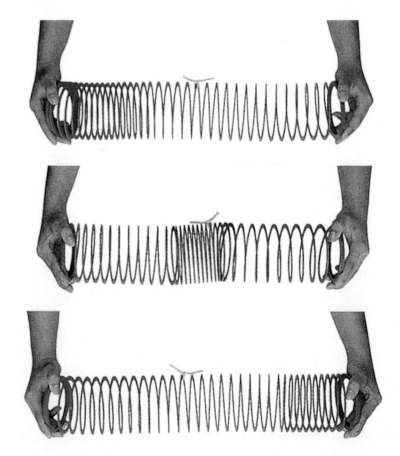

Compressional Waves Mechanical waves can be either transverse or compressional. In a **compressional wave,** matter in the medium moves forward and backward along the same direction that the wave travels. You can make a compressional wave by squeezing together and releasing several coils of a coiled spring toy, as shown in **Figure 4.**

The coils move only as the wave passes and then return to their original positions. So, like transverse waves, compressional waves carry only energy forward along the spring. In this example, the spring is the medium the wave moves through, but the spring does not move along with the wave.

Sound Waves Sound waves are compressional waves. How do you make sound waves when you talk or sing? If you hold your fingers against your throat while you hum, you can feel vibrations. These vibrations are the movements of your vocal cords. If you touch a stereo speaker while it's playing, you can feel it vibrating, too. All waves are produced by something that is vibrating.

Making Sound Waves

How do vibrating objects make sound waves? Look at the drum shown in **Figure 5.** When you hit the drumhead it starts vibrating up and down. As the drumhead moves upward, the molecules next to it are pushed closer together. This group of molecules that are closer together is a compression. As the compression is formed, it moves away from the drumhead, just as the squeezed coils move along the coiled spring toy in **Figure 4.**

When the drumhead moves downward, the molecules near it have more room and can spread farther apart. This group of molecules that are farther apart is a rarefaction. The rarefaction also moves away from the drumhead. As the drumhead vibrates up and down, it forms a series of compressions and rarefactions that move away and spread out in all directions. This series of compressions and rarefactions is a sound wave.

Mini LAB

Comparing Sounds

Procedure

1. Hold a **wooden ruler** firmly on the edge of your **desk** so that most of it extends off the edge of the desk.
2. Pluck the free end of the ruler so that it vibrates up and down. Use gentle motion at first, then pluck with more energy.
3. Repeat step 2, moving the ruler about 1 cm further onto the desk each time until only about 5 cm extend off the edge.

Analysis

1. Compare the loudness of the sounds that are made by plucking the ruler in different ways.
2. Describe the differences in the sound as the end of the ruler extended farther from the desk.

Figure 5 A vibrating drumhead makes compressions and rarefactions in the air.
Describe *how compressions and rarefactions are different.*

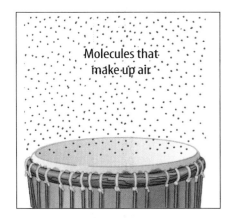
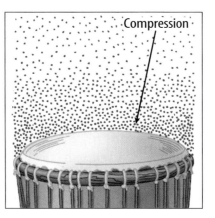
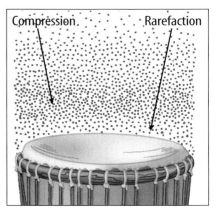

SECTION 1 What are waves? **465**

Global Positioning Systems
Maybe you've used a global positioning system (GPS) receiver to determine your location while driving, boating, or hiking. Earth-orbiting satellites send electromagnetic radio waves that transmit their exact locations and times of transmission. The GPS receiver uses information from four of these satellites to determine your location to within about 16 m.

Electromagnetic Waves

Waves that can travel through space where there is no matter are **electromagnetic waves.** There are different types of electromagnetic waves, including radio waves, infrared waves, visible light waves, ultraviolet waves, X rays, and gamma rays. These waves can travel in matter or in space. Radio waves from TV and radio stations travel through air, and may be reflected from a satellite in space. They then travel through air, through the walls of your house, and to your TV or radio.

Radiant Energy from the Sun The Sun emits electromagnetic waves that travel through space and reach Earth. The energy carried by electromagnetic waves is called radiant energy. Almost 92 percent of the radiant energy that reaches Earth from the Sun is carried by infrared and visible light waves. Infrared waves make you feel warm when you sit in sunlight, and visible light waves enable you to see. A small amount of the radiant energy that reaches Earth is carried by ultraviolet waves. These are the waves that can cause sunburn if you are exposed to sunlight for too long.

section 1 review

Summary

What is a wave?
- Waves transfer energy, but do not transfer matter.

Mechanical Waves
- Mechanical waves require a medium in which to travel.
- When a transverse wave travels, particles of the medium move at right angles to the direction the wave is traveling.
- When a compressional wave travels, particles of the medium move back and forth along the same direction the wave is traveling.
- Sound is a compressional wave.

Electromagnetic Waves
- Electromagnetic waves can travel through empty space.
- The Sun emits different types of electromagnetic waves, including infrared, visible light, and ultraviolet waves.

Self Check

1. **Describe** the movement of a floating object on a pond when struck by a wave.
2. **Explain** why a sound wave can't travel from a satellite to Earth.
3. **Compare and contrast** a transverse wave and a compressional wave. How are they similar and different?
4. **Compare and contrast** a mechanical wave and an electromagnetic wave.
5. **Think Critically** How is it possible for a sound wave to transmit energy but not matter?

Applying Skills

6. **Concept Map** Create a concept map that shows the relationships among the following: *waves, mechanical waves, electromagnetic waves, compressional waves,* and *transverse waves.*
7. **Use a Word Processor** Use word-processing software to write short descriptions of the waves you encounter during a typical day.

ips.msscience.com/self_check_quiz

section 2
Wave Properties

Amplitude

Can you describe a wave? For a water wave, one way might be to tell how high the wave rises above, or falls below, the normal level. This distance is called the wave's amplitude. The **amplitude** of a transverse wave is one-half the distance between a crest and a trough, as shown in **Figure 6.** In a compressional wave, the amplitude is greater when the particles of the medium are squeezed closer together in each compression and spread farther apart in each rarefaction.

Amplitude and Energy A wave's amplitude is related to the energy that the wave carries. For example, the electromagnetic waves that make up bright light have greater amplitudes than the waves that make up dim light. Waves of bright light carry more energy than the waves that make up dim light. In a similar way, loud sound waves have greater amplitudes than soft sound waves. Loud sounds carry more energy than soft sounds. If a sound is loud enough, it can carry enough energy to damage your hearing.

When a hurricane strikes a coastal area, the resulting water waves carry enough energy to damage almost anything that stands in their path. The large waves caused by a hurricane carry more energy than the small waves or ripples on a pond.

as you read

What You'll Learn
- **Describe** the relationship between the frequency and wavelength of a wave.
- **Explain** why waves travel at different speeds.

Why It's Important
The properties of a wave determine whether the wave is useful or dangerous.

Review Vocabulary
speed: the distance traveled divided by the time needed to travel the distance

New Vocabulary
- amplitude
- wavelength
- frequency

Figure 6 The energy carried by a wave increases as its amplitude increases.

The amplitude of a transverse wave is a measure of how high the crests are or how deep the troughs are.

A water wave of large amplitude carried the energy that caused this damage.

For transverse waves, wavelength is the distance from crest to crest or trough to trough.

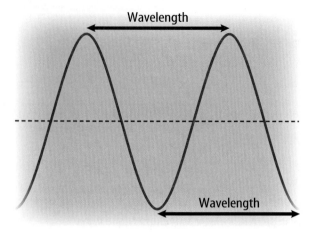

For compressional waves, wavelength is the distance from compression to compression or rarefaction to rarefaction.

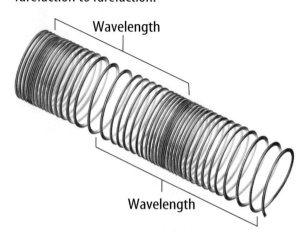

Figure 7 A transverse or a compressional wave has a wavelength.

Figure 8 The wavelengths and frequencies of electromagnetic waves vary.

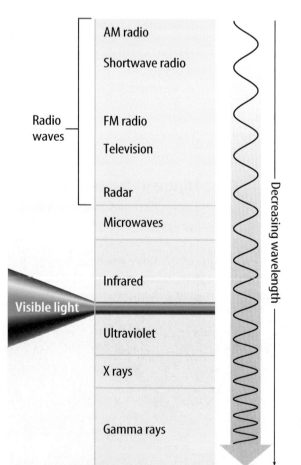

INTEGRATE Earth Science The devastating effect that a wave with large amplitude can have is seen in the aftermath of tsunamis. Tsunamis are huge sea waves that are caused by underwater earthquakes along faults on the seafloor. The movement of the seafloor along a fault produces the wave. As the wave moves toward shallow water and slows down, the amplitude of the wave grows. The tremendous amounts of energy tsunamis carry cause great damage when they move ashore.

Wavelength

Another way to describe a wave is by its wavelength. **Figure 7** shows the wavelength of a transverse wave and a compressional wave. For a transverse wave, **wavelength** is the distance from the top of one crest to the top of the next crest, or from the bottom of one trough to the bottom of the next trough. For a compressional wave, the wavelength is the distance between the center of one compression and the center of the next compression, or from the center of one rarefaction to the center of the next rarefaction.

Electromagnetic waves have wavelengths that range from kilometers, for radio waves, to less than the diameter of an atom, for X rays and gamma rays. This range is called the electromagnetic spectrum. **Figure 8** shows the names given to different parts of the electromagnetic spectrum. Visible light is only a small part of the electromagnetic spectrum. It is the wavelength of visible light waves that determines their color. For example, the wavelength of red light waves is longer than the wavelength of green light waves.

468 CHAPTER 16 Waves

Frequency

The **frequency** of a wave is the number of wavelengths that pass a given point in 1 s. The unit of frequency is the number of wavelengths per second, or hertz (Hz). Recall that waves are produced by something that vibrates. The faster the vibration is, the higher the frequency is of the wave that is produced.

Reading Check *How is the frequency of a wave measured?*

A Sidewalk Model For waves that travel with the same speed, frequency and wavelength are related. To model this relationship, imagine people on two parallel moving sidewalks in an airport, as shown in **Figure 9**. One sidewalk has four travelers spaced 4 m apart. The other sidewalk has 16 travelers spaced 1 m apart.

Now imagine that both sidewalks are moving at the same speed and approaching a pillar between them. On which sidewalk will more people go past the pillar? On the sidewalk with the shorter distance between people, four people will pass the pillar for each one person on the other sidewalk. When four people pass the pillar on the first sidewalk, 16 people pass the pillar on the second sidewalk.

Figure 9 When people are farther apart on a moving sidewalk, fewer people pass the pillar every minute.
Infer *how the number of people passing the pillar each minute would change if the sidewalk moved slower.*

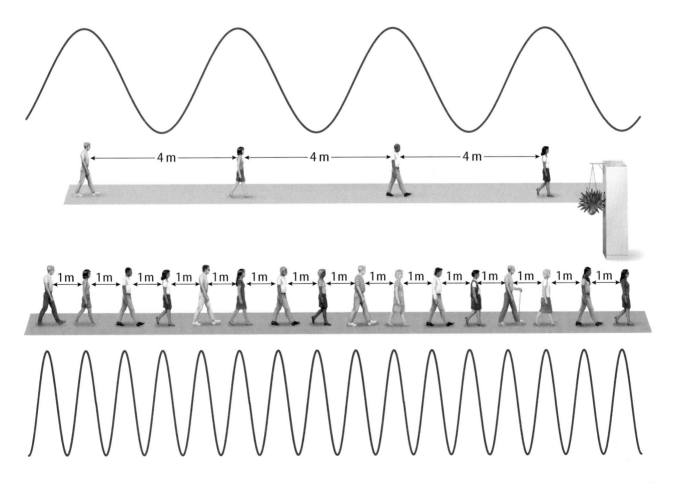

Ultrasonic Waves Sound waves with ultra-high frequencies cannot be heard by the human ear, but they are used by medical professionals in several ways. They are used to perform echocardiograms of the heart, produce ultrasound images of internal organs, break up blockages in arteries, and sterilize surgical instruments. Describe how the wavelengths of these sound waves compare to sound waves you can hear.

Figure 10 The frequency of the notes on a musical scale increases as the notes get higher in pitch, but the wavelength of the notes decreases.

Frequency and Wavelength Suppose that each person in **Figure 9** represents the crest of a wave. Then the movement of people on the first sidewalk is like a wave with a wavelength of 4 m. For the second sidewalk, the wavelength would be 1 m. On the first sidewalk, where the wavelength is longer, the people pass the pillar *less* frequently. Smaller frequencies result in longer wavelengths. On the second sidewalk, where the wavelength is shorter, the people pass the pillar *more* frequently. Higher frequencies result in shorter wavelengths. This is true for all waves that travel at the same speed. As the frequency of a wave increases, its wavelength decreases.

Reading Check How are frequency and wavelength related?

Color and Pitch Because frequency and wavelength are related, either the wavelength or frequency of a light wave determines the color of the light. For example, blue light has a larger frequency and shorter wavelength than red light.

Either the wavelength or frequency determines the pitch of a sound wave. Pitch is how high or low a sound seems to be. When you sing a musical scale, the pitch and frequency increase from note to note. Wavelength and frequency are also related for sound waves traveling in air. As the frequency of sound waves increases, their wavelength decreases. **Figure 10** shows how the frequency and wavelength change for notes on a musical scale.

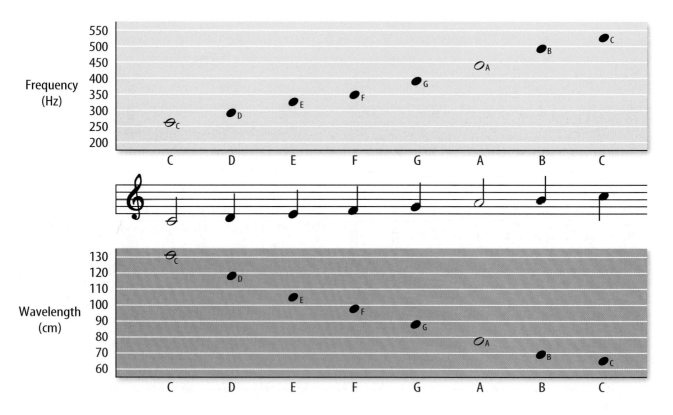

Wave Speed

You've probably watched a distant thunderstorm approach on a hot summer day. You see a bolt of lightning flash between a dark cloud and the ground. If the thunderstorm is many kilometers away, several seconds will pass between when you see the lightning and when you hear the thunder. This happens because light travels much faster in air than sound does. Light travels through air at about 300 million m/s. Sound travels through air at about 340 m/s. The speed of any wave can be calculated from this equation:

Wave Speed Equation

wave speed (in m/s) = **frequency** (in Hz) × **wavelength** (m)

$$v = f\lambda$$

In this equation, the wavelength is represented by the symbol λ, which is the Greek letter lambda.

When mechanical waves, such as sound, and electromagnetic waves, such as light, travel in different materials, they change speed. Mechanical waves usually travel fastest in solids, and slowest in gases. Electromagnetic waves travel fastest in gases and slowest in solids. For example, the speed of light is about 30 percent faster in air than in water.

Topic: Wave Speed
Visit ips.msscience.com for Web links to information about wave speed in different materials.

Activity Make a chart showing the speed of light in different materials.

section 2 review

Summary

Amplitude
- In a transverse wave, the amplitude is one-half the distance between a crest and a trough.
- The larger the amplitude, the greater the energy carried by the wave.

Wavelength
- For a transverse wave, wavelength is the distance from crest to crest, or from trough to trough.
- For a compressional wave, wavelength is the distance from compression to compression, or from rarefaction to rarefaction.

Frequency
- The frequency of a wave is the number of wavelengths that pass a given point in 1 s.
- For waves that travel at the same speed, as the frequency of the wave increases, its wavelength decreases.

Self Check

1. **Describe** how the frequency of a wave changes as its wavelength changes.
2. **Explain** why a sound wave with a large amplitude is more likely to damage your hearing than one with a small amplitude.
3. **State** what accounts for the time difference between seeing and hearing a fireworks display.
4. **Explain** why the statement "The speed of light is 300 million m/s" is not always correct.
5. **Think Critically** Explain the differences between the waves that make up bright, green light and dim, red light.

Applying Math

6. **Calculate Wave Speed** Find the speed of a wave with a wavelength of 5 m and a frequency of 68 Hz.
7. **Calculate Wavelength** Find the wavelength of a sound wave traveling in water with a speed of 1,470 m/s, and a frequency of 2,340 Hz.

Waves on a Spring

Waves are rhythmic disturbances that carry energy through matter or space. Studying waves can help you understand how the Sun's energy reaches Earth and sounds travel through the air.

▶ Real-World Question

What are some of the properties of transverse and compressional waves on a coiled spring?

Goals
- **Create** transverse and compressional waves on a coiled spring toy.
- **Investigate** wave properties such as speed and amplitude.

Materials
long, coiled spring toy
colored yarn (5 cm)
meterstick
stopwatch

Safety Precautions

WARNING: *Avoid overstretching or tangling the spring to prevent injury or damage.*

▶ Procedure

1. Prepare a data table such as the one shown.

Wave Data	
Length of stretched spring toy	Do not write in this book.
Average time for a wave to travel from end to end—step 4	
Average time for a wave to travel from end to end—step 5	

2. Work in pairs or groups and clear a place on an uncarpeted floor about 6 m × 2 m.

3. Stretch the springs between two people to the length suggested by your teacher. Measure the length.
4. Create a wave with a quick, sideways snap of the wrist. Time several waves as they travel the length of the spring. Record the average time in your data table.
5. Repeat step 4 using waves that have slightly larger amplitudes.
6. Squeeze together about 20 of the coils. Observe what happens to the unsqueezed coils. Release the coils and observe.
7. Quickly push the spring toward your partner, then pull it back.
8. Tie the yarn to a coil near the middle of the spring. Repeat step 7, observing the string.
9. **Calculate** and compare the speeds of the waves in steps 4 and 5.

▶ Conclude and Apply

1. **Classify** the wave pulses you created in each step as compressional or transverse.
2. **Classify** the unsqueezed coils in step 6 as a compression or a rarefaction.
3. **Compare and contrast** the motion of the yarn in step 8 with the motion of the wave.

𝒞ommunicating Your Data

Write a summary paragraph of how this lab demonstrated any of the vocabulary words from the first two sections of the chapter. **For more help, refer to the Science Skill Handbook.**

section 3

Wave Behavior

Reflection

What causes the echo when you yell across an empty gymnasium or down a long, empty hallway? Why can you see your face when you look in a mirror? The echo of your voice and the face you see in the mirror are caused by wave reflection.

Reflection occurs when a wave strikes an object or surface and bounces off. An echo is reflected sound. Sound reflects from all surfaces. Your echo bounces off the walls, floor, ceiling, furniture, and people. You see your face in a mirror or a still pond, as shown in **Figure 11,** because of reflection. Light waves produced by a source of light such as the Sun or a lightbulb bounce off your face, strike the mirror, and reflect back to your eyes.

When a surface is smooth and even the reflected image is clear and sharp. However, **Figure 11** shows that when light reflects from an uneven or rough surface, you can't see a sharp image because the reflected light scatters in many different directions.

Reading Check *What causes reflection?*

as you read

What You'll Learn
- **Explain** how waves can reflect from some surfaces.
- **Explain** how waves change direction when they move from one material into another.
- **Describe** how waves are able to bend around barriers.

Why It's Important
The reflection of waves enables you to see objects around you.

Review Vocabulary
echo: the repetition of a sound caused by the reflection of sound waves

New Vocabulary
- reflection
- refraction
- diffraction
- interference

The smooth surface of a still pond enables you to see a sharp, clear image of yourself.

If the surface of the pond is rough and uneven, your reflected image is no longer clear and sharp.

Figure 11 The image formed by reflection depends on the smoothness of the surface.

Mini LAB

Observing How Light Refracts

Procedure
1. Fill a **large, opaque drinking glass or cup** with water.
2. Place a **white soda straw** in the water at an angle.
3. Looking directly down into the cup from above, observe the straw where it meets the water.
4. Placing yourself so that the straw angles to your left or right, slowly back away about 1 m. Observe the straw as it appears above, at, and below the surface of the water.

Analysis
1. Describe the straw's appearance from above.
2. Compare the straw's appearance above and below the water's surface in step 4.

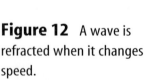

Refraction

A wave changes direction when it reflects from a surface. Waves also can change direction in another way. Perhaps you have tried to grab a sinking object when you are in a swimming pool, only to come up empty-handed. Yet you were sure you grabbed right where you saw the object. You missed grabbing the object because the light rays from the object changed direction as they passed from the water into the air. The bending of a wave as it moves from one medium into another is called **refraction.**

Refraction and Wave Speed Remember that the speed of a wave can be different in different materials. For example, light waves travel faster in air than in water. Refraction occurs when the speed of a wave changes as it passes from one substance to another, as shown in **Figure 12.** A line that is perpendicular to the water's surface is called the normal. When a light ray passes from air into water, it slows down and bends toward the normal. When the ray passes from water into air, it speeds up and bends away from the normal. The larger the change in speed of the light wave is, the larger the change in direction is.

You notice refraction when you look down into a fishbowl. Refraction makes the fish appear to be closer to the surface and farther away from you than it really is, as shown in **Figure 13.** Light rays reflected from the fish are bent away from the normal as they pass from water to air. Your brain interprets the light that enters your eyes by assuming that light rays always travel in straight lines. As a result, the light rays seem to be coming from a fish that is closer to the surface.

Figure 12 A wave is refracted when it changes speed.
Explain how the direction of the light ray changes if it doesn't change speed.

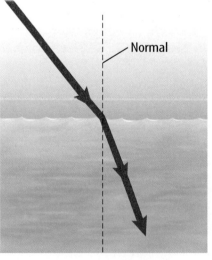

As the light ray passes from air to water, it bends toward the normal.

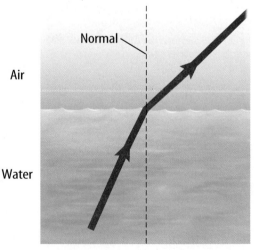

As the light ray passes from water to air, it bends away from the normal.

Color from Refraction Sunlight contains light of various wavelengths. When sunlight passes through a prism, refraction occurs twice: once when sunlight enters the prism and again when it leaves the prism and returns to the air. Violet light has the shortest wavelength and is bent the most. Red light has the longest wavelength and is bent the least. Each color has a different wavelength and is refracted a different amount. As a result, the colors of sunlight are separated when they emerge from the prism.

Figure 14 shows how refraction produces a rainbow when light waves from the Sun pass into and out of water droplets. The colors you see in a rainbow are in order of decreasing wavelength: red, orange, yellow, green, blue, indigo, and violet.

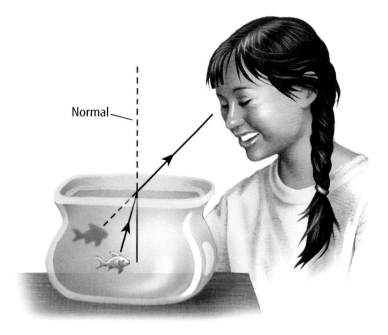

Figure 13 When you look at the goldfish in the water, the fish is in a different position than it appears.
Infer how the location of the fish would change if light traveled faster in water than in air.

Diffraction

Why can you hear music from the band room when you are down the hall? You can hear the music because the sound waves bend as they pass through an open doorway. This bending isn't caused by refraction. Instead, the bending is caused by diffraction. **Diffraction** is the bending of waves around a barrier.

Light waves can diffract, too. You can hear your friends in the band room but you can't see them until you reach the open door. Therefore, you know that light waves do not diffract as much as sound waves do. Light waves do bend around the edges of an open door. However, for an opening as wide as a door, the amount the light bends is extremely small. As a result, the diffraction of light is far too small to allow you to see around a corner.

Figure 14 Light rays refract as they enter and leave each water drop. Each color refracts at different angles because of their different wavelengths, so they separate into the colors of the visible spectrum.

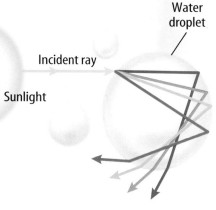

SECTION 3 Wave Behavior **475**

Diffraction and Wavelength The reason that light waves don't diffract much when they pass through an open door is that the wavelengths of visible light are much smaller than the width of the door. Light waves have wavelengths between about 400 and 700 billionths of a meter, while the width of doorway is about one meter. Sound waves that you can hear have wavelengths between a few millimeters and about 10 m. They bend more easily around the corners of an open door. A wave is diffracted more when its wavelength is similar in size to the barrier or opening.

Reading Check *Under what conditions would more diffraction of a wave occur?*

Diffraction of Water Waves Perhaps you have noticed water waves bending around barriers. For example, when water waves strike obstacles such as the islands shown in **Figure 15**, they don't stop moving. Here the size and spacing of the islands is not too different from the wavelength of the water waves. So the water waves bend around the islands, and keep on moving. They also spread out after they pass through openings between the islands. If the islands were much larger than the water wavelength, less diffraction would occur.

What happens when waves meet?

Suppose you throw two pebbles into a still pond. Ripples spread from the impact of each pebble and travel toward each other. What happens when two of these ripples meet? Do they collide like billiard balls and change direction? Waves behave differently from billiard balls when they meet. Waves pass right through each other and continue moving.

Figure 15 Water waves bend or diffract around these islands. More diffraction occurs when the object is closer in size to the wavelength.

476 CHAPTER 16 Waves

Wave Interference While two waves overlap a new wave is formed by adding the two waves together. The ability of two waves to combine and form a new wave when they overlap is called **interference.** After they overlap, the individual waves continue to travel on in their original form.

The different ways waves can interfere are shown in **Figure 16** on the next page. Sometimes when the waves meet, the crest of one wave overlaps the crest of another wave. This is called constructive interference. The amplitudes of these combining waves add together to make a larger wave while they overlap. Destructive interference occurs when the crest of one wave overlaps the trough of another wave. Then, the amplitudes of the two waves combine to make a wave with a smaller amplitude. If the two waves have equal amplitudes and meet crest to trough, they cancel each other while the waves overlap.

Waves and Particles Like waves of water, when light travels through a small opening, such as a narrow slit, the light spreads out in all directions on the other side of the slit. If small particles, instead of waves, were sent through the slit, they would continue in a straight line without spreading. The spreading, or diffraction, is only a property of waves. Interference also doesn't occur with particles. If waves meet, they reinforce or cancel each other, then travel on. If particles approach each other, they either collide and scatter or miss each other completely. Interference, like diffraction, is a property of waves.

Science online

Topic: Interference
Visit ips.msscience.com for Web links to information about wave interference.

Activity Write a paragraph about three kinds of interference you found in your research.

Applying Science

Can you create destructive interference?

Your brother is vacuuming and you can't hear the television. Is it possible to diminish the sound of the vacuum so you can hear the TV? Can you eliminate some sound waves and keep the sounds you do want to hear?

Identifying the Problem
It is possible to create a wave that will destructively interfere with one wave, but will not destructively interfere with another wave. The graph shows two waves with different wavelengths.

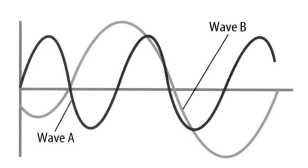

Solving the Problem
1. Create the graph of a wave that will eliminate wave **A** but not wave **B**.
2. Create the graph of a wave that would amplify wave **A**.

NATIONAL GEOGRAPHIC VISUALIZING INTERFERENCE

Figure 16

Whether they are ripples on a pond or huge ocean swells, when water waves meet they can combine to form new waves in a process called interference. As shown below, wave interference can be constructive or destructive.

Constructive Interference

In constructive interference, a wave with greater amplitude is formed.

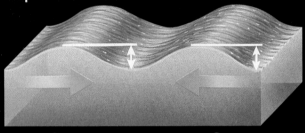

The crests of two waves—A and B—approach each other.

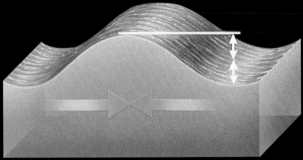

The two waves form a wave with a greater amplitude while the crests of both waves overlap.

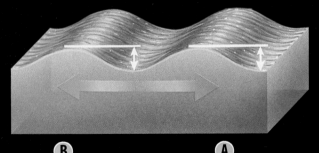

The original waves pass through each other and go on as they started.

Destructive Interference

In destructive interference, a wave with a smaller amplitude is formed.

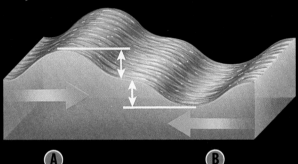

The crest of one wave approaches the trough of another.

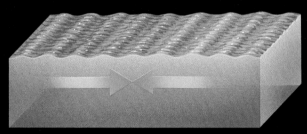

If the two waves have equal amplitude, they momentarily cancel when they meet.

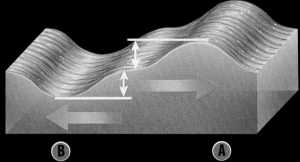

The original waves pass through each other and go on as they started.

Reducing Noise You might have seen someone use a power lawn mower or a chain saw. In the past, many people who performed these tasks damaged their hearing because of the loud noises produced by these machines.

Loud sounds have waves with larger amplitudes and carry more energy than softer sounds. The energy carried by loud sounds can damage cells in the ear that vibrate and transmit signals to the brain. Damage to the ear from loud sounds can be prevented by reducing the energy that reaches the ear. Ear protectors contain materials that absorb some of the energy carried by sound waves, so that less sound energy reaches the ear.

Pilots of small planes have a more complicated problem. If they shut out all the noise of the plane's motor, the pilots wouldn't be able to hear instructions from air-traffic controllers. To solve this problem, ear protectors have been developed, as shown in **Figure 17,** that have electronic circuits. These circuits detect noise from the aircraft and produce sound frequencies that destructively interfere with the noise. They do not interfere with human voices, so people can hear normal conversation. Destructive interference can be a benefit.

Figure 17 Some airplane pilots use special ear protectors that cancel out engine noise but don't block human voices.

section 3 review

Summary

Reflection
- Reflected sound waves can produce echoes.
- Reflected light rays produce images in a mirror.

Refraction
- The bending of waves as they pass from one medium to another is refraction.
- Refraction occurs when the wave's speed changes.
- A prism separates sunlight into the colors of the visible spectrum.

Diffraction and Interference
- The bending of waves around barriers is diffraction.
- Interference occurs when waves combine to form a new wave while they overlap.
- Destructive interference can reduce noise.

Self Check

1. **Explain** why you don't see your reflection in a building made of rough, white stone.
2. **Explain** how you are able to hear the siren of an ambulance on the other side of a building.
3. **Describe** the behavior of light that enables magnifying lenses and contact lenses to bend light rays.
4. **Define** the term *diffraction*. How does the amount of diffraction depend on wavelength?
5. **Think Critically** Why don't light rays that stream through an open window into a darkened room spread evenly through the entire room?

Applying Skills

6. **Compare and Contrast** When light rays pass from water into a certain type of glass, the rays refract toward the normal. Compare and contrast the speed of light in water and in the glass.

Design Your Own

WAVE ≈ SPEED

Goals
- **Measure** the speed of a wave within a coiled spring toy.
- **Predict** whether the speed you measured will be different in other types of coiled spring toys.

Possible Materials
long, coiled spring toy
meterstick
stopwatch
tape
*clock with a second hand
*Alternate materials

Safety Precautions

● Real-World Question

When an earthquake occurs, it produces waves that are recorded at points all over the world by instruments called seismographs. By comparing the data that they collected from these seismographs, scientists discovered that the interior of Earth must be made of layers of different materials. These data showed that the waves traveled at different speeds as they passed through different parts of Earth's interior. How can the speed of a wave be measured?

● Form a Hypothesis

In some materials, waves travel too fast for their speeds to be measured directly. Think about what you know about the relationships among the frequency, wavelength, and speed of a wave in a medium. Make a hypothesis about how you can use this relationship to measure the speed of a wave within a medium. Explain why you think the experiment will support your hypothesis.

● Test Your Hypothesis

Make a Plan

1. Make a data table in your Science Journal like the one shown.
2. In your Science Journal, write a detailed description of the coiled spring toy you are going to use. Be sure to include its mass and diameter, the width of a coil, and what it is made of.
3. **Decide** as a group how you will measure the frequency and length of waves in the spring toy. What are your variables? Which variables must be controlled? What variable do you want to measure?

480 CHAPTER 16 Waves

Using Scientific Methods

4. Repeat your experiment three times.

Follow Your Plan

1. Make sure your teacher approves your plan before you start.
2. Carry out the experiment.
3. While you are doing the experiment, record your observations and measurements in your data table.

Wave Data

	Trial 1	Trial 2	Trial 3
Length spring was stretched (m)			
Number of crests			
Wavelength (m)		Do not write in this book.	
# of vibrations timed			
# of seconds vibrations were timed			
Wave speed (m/s)			

⊙ Analyze Your Data

1. **Calculate** the frequency of the waves by dividing the number of vibrations you timed by the number of seconds you timed them. Record your results in your data table.

2. Use the following formula to calculate the speed of a wave in each trial.

    wavelength $\times$ frequency $=$ wave speed

3. Average the wave speeds from your trials to determine the average speed of a wave in your coiled spring toy.

⊙ Conclude and Apply

1. **Infer** which variables affected the wave speed in spring toys the most. Which variables affected the speed the least? Was your hypothesis supported?

2. **Analyze** what factors caused the wave speed measured in each trial to be different.

𝒞ommunicating
Your Data

Post a description of your coiled spring toy and the results of your experiment on a bulletin board in your classroom. **Compare and contrast** your results with other students in your class.

LAB **481**

SCIENCE Stats

Waves, Waves, and More Waves

Did you know...

...Radio waves from space were discovered in 1932 by Karl G. Jansky, an American engineer. His discovery led to the creation of radio astronomy, a field that explores parts of the universe that can't be seen with telescopes.

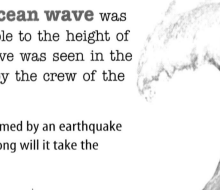

...The highest recorded ocean wave was 34 meters high, which is comparable to the height of a ten-story building. This super wave was seen in the North Pacific Ocean and recorded by the crew of the naval ship *USS Ramapo* in 1933.

Applying Math A tsunami formed by an earthquake on the ocean floor travels at 900 km/h. How long will it take the tsunami to travel 4,500 km?

...Waves let dolphins see with their ears! A dolphin sends out ultrasonic pulses, or clicks, at rates of 800 pulses per second. These sound waves are reflected back to the dolphin after they hit an obstacle or a meal. This process is called echolocation.

Graph It

Go to **ips.msscience.com/science_stats** to learn about discoveries by radio astronomers. **Make a time line showing some of these discoveries.**

Chapter 16 Study Guide

Reviewing Main Ideas

Section 1 What are waves?

1. Waves are rhythmic disturbances that carry energy but not matter.
2. Mechanical waves can travel only through matter. Electromagnetic waves can travel through matter and space.
3. In a mechanical transverse wave, matter in the medium moves back and forth at right angles to the direction the wave travels.
4. In a compressional wave, matter in the medium moves forward and backward in the same direction as the wave.

Section 2 Wave Properties

1. The amplitude of a transverse wave is the distance between the rest position and a crest or a trough.
2. The energy carried by a wave increases as the amplitude increases.
3. Wavelength is the distance between neighboring crests or neighboring troughs.
4. The frequency of a wave is the number of wavelengths that pass a given point in 1 s.
5. Waves travel through different materials at different speeds.

Section 3 Wave Behavior

1. Reflection occurs when a wave strikes an object or surface and bounces off.
2. The bending of a wave as it moves from one medium into another is called refraction. A wave changes direction, or refracts, when the speed of the wave changes.
3. The bending of waves around a barrier is called diffraction.
4. Interference occurs when two or more waves combine and form a new wave while they overlap.

Visualizing Main Ideas

Copy and complete the following spider map about waves.

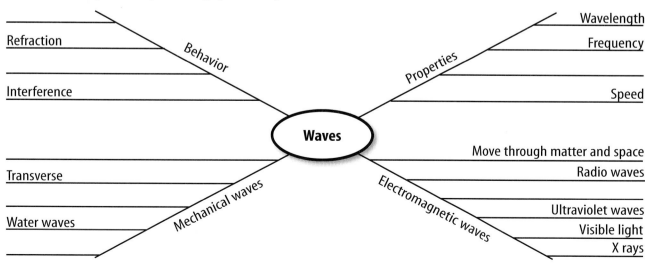

chapter 16 Review

Using Vocabulary

amplitude p. 467
compressional wave p. 465
diffraction p. 475
electromagnetic wave p. 466
frequency p. 469
interference p. 477
mechanical wave p. 463
reflection p. 473
refraction p. 474
transverse wave p. 464
wave p. 462
wavelength p. 468

Fill in the blanks with the correct word or words.

1. _____ is the change in direction of a wave going from one medium to another.
2. The type of wave that has rarefactions is a _____.
3. The distance between two adjacent crests of a transverse wave is the _____.
4. The more energy a wave carries, the greater its _____ is.
5. A(n) _____ can travel through space without a medium.

Checking Concepts

Choose the word or phrase that best answers the question.

6. What is the material through which mechanical waves travel?
   A) charged particles
   B) space
   C) a vacuum
   D) a medium

7. What is carried from particle to particle in a water wave?
   A) speed   C) energy
   B) amplitude   D) matter

8. What are the lowest points on a transverse wave called?
   A) crests   C) compressions
   B) troughs   D) rarefactions

9. What determines the pitch of a sound wave?
   A) amplitude   C) speed
   B) frequency   D) refraction

10. What is the distance between adjacent wave compressions?
    A) one wavelength
    B) 1 km
    C) 1 m/s
    D) 1 Hz

11. What occurs when a wave strikes an object or surface and bounces off?
    A) diffraction
    B) refraction
    C) a transverse wave
    D) reflection

12. What is the name for a change in the direction of a wave when it passes from one medium into another?
    A) refraction   C) reflection
    B) interference   D) diffraction

Use the figure below to answer question 13.

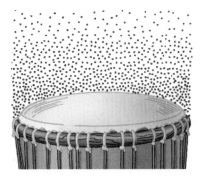

13. What type of wave is a sound wave?
    A) transverse
    B) electromagnetic
    C) compressional
    D) refracted

14. What color light has the shortest wavelength and the highest frequency?
    A) red   C) Orange
    B) green   D) Blue

chapter 16 Review

Thinking Critically

15. **Explain** what kind of wave—transverse or compressional—is produced when an engine bumps into a string of coupled railroad cars on a track.

16. **Infer** Is it possible for an electromagnetic wave to travel through a vacuum? Through matter? Explain your answers.

17. **Draw a Conclusion** Why does the frequency of a wave decrease as the wavelength increases?

18. **Explain** why you don't see your reflected image when you look at a white, rough surface?

19. **Infer** If a cannon fires at a great distance from you, why do you see the flash before you hear the sound?

20. **Form a Hypothesis** Form a hypothesis that can explain this observation. Waves A and B travel away from Earth through Earth's atmosphere. Wave A continues on into space, but wave B does not.

Use the figure below to answer question 21.

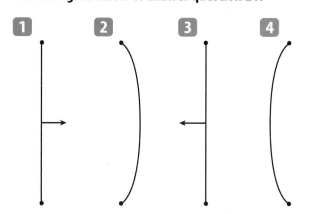

21. **Explain** how the object shown above causes compressions and rarefactions as it vibrates in air.

22. **Explain** why you can hear a person talking even if you can't see them.

23. **Compare and Contrast** AM radio waves have wavelengths between about 200 m and 600 m, and FM radio waves have wavelengths of about 3 m. Why can AM radio signals often be heard behind buildings and mountains but FM radio signals cannot?

24. **Infer** how the wavelength of a wave would change if the speed of the wave increased, but the frequency remained the same.

25. **Explain** You are motionless on a rubber raft in the middle of a pool. A friend sitting on the edge of the pool tries to make the float move to the other edge of the pool by slapping the water every second to form a wave. Explain whether the wave produced will cause you to move to the edge of the pool.

Performance Activities

26. **Make Flashcards** Work with a partner to make flashcards for the bold-faced terms in the chapter. Illustrate each term on the front of the cards. Write the term and its definition on the back of the card. Use the cards to review the terms with another team.

Applying Math

Use the following equation to answer questions 27–29.

wave speed = wavelength × frequency

27. **Wave Speed** If a wave pool generates waves with a wavelength of 3.2 m and a frequency of 0.60 Hz, how fast are the waves moving?

28. **Frequency** An earthquake wave travels at 5000 m/s and has a wavelength of 417 m. What is its frequency?

29. **Wavelength** A wave travels at a velocity of 4 m/s. It has a frequency of 3.5 Hz. What is the wavelength of the wave?

Chapter 16 Standardized Test Practice

Part 1 Multiple Choice

Record your answers on the answer sheet provided by your teacher or on a sheet of paper.

1. What do waves carry as they move?
   A. matter
   B. energy
   C. matter and energy
   D. particles and energy

Use the figure below to answer questions 2 and 3.

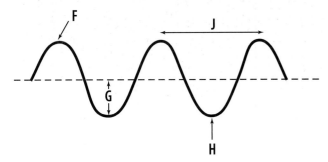

2. What property of the wave is shown at F?
   A. amplitude
   B. wavelength
   C. crest
   D. trough

3. What property of the wave is shown at J?
   A. amplitude
   B. wavelength
   C. crest
   D. trough

4. What kind of wave does NOT need a medium through which to travel?
   A. mechanical
   B. sound
   C. light
   D. refracted

5. What happens as a sound wave's energy decreases?
   A. Wave frequency decreases.
   B. Wavelength decreases.
   C. Amplitude decreases.
   D. Wave speed decreases.

6. What unit is used to measure frequency?
   A. meters
   B. meters/second
   C. decibels
   D. hertz

7. What properties of a light wave determines its color?
   A. wavelength
   B. amplitude
   C. speed
   D. interference

8. When two waves overlap and interfere constructively, what does the resulting wave have?
   A. a greater amplitude
   B. less energy
   C. a change in frequency
   D. a lower amplitude

9. What happens when light travels from air into glass?
   A. It speeds up.
   B. It slows down.
   C. It travels at 300,000 km/s.
   D. It travels at the speed of sound.

Use the figure below to answer questions 10 and 11.

10. What behavior of light waves lets you see a sharp, clear image of yourself?
    A. refraction
    B. diffraction
    C. reflection
    D. interference

11. Why can't you see a clear image of yourself if the water's surface is rough?
    A. The light bounces off the surface in only one direction.
    B. The light scatters in many different directions.
    C. There is no light shining on the water's surface.
    D. The light changes speed when it strikes the water.

Standardized Test Practice

Part 2 Short Response/Grid In

Record your answers on the answer sheet provided by your teacher or on a sheet of paper.

12. An earthquake in the middle of the Indian Ocean produces a tsunami that hits an island. Is the water that hits the island the same water that was above the place where the earthquake occurred? Explain.

Use the figure below to answer questions 13 and 14.

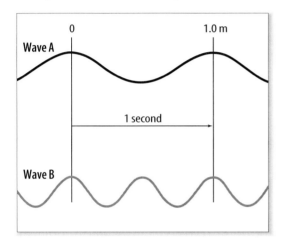

13. Compare the wavelengths and frequencies of the two waves shown.

14. If both waves are traveling through the same medium, how do their speeds compare? Explain.

15. Suppose you make waves in a pond by dipping your hand in the water with a frequency of 1 Hz. How could you make waves of a longer wavelength? How could you increase the amplitude of the waves?

16. How are all electromagnetic waves alike? How do they differ from one another?

Test-Taking Tip

Take Your Time Stay focused during the test and don't rush, even if you notice that other students are finishing the test early.

Part 3 Open Ended

Record your answers on a sheet of paper.

Use the figure below to answer questions 17 and 18.

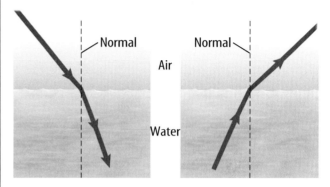

17. Why does the light ray bend toward the normal when is passes from air into water, but bend away from the normal as it passes from water into air?

18. A boy has caught a fish on his fishing line. He reels the fish in near the boat. How could the refraction of light waves affect him as he tries to net the fish while it is still in the water?

19. In a science fiction movie, a spaceship explodes. The people in a nearby spaceship see and hear the explosion. Is this realistic? Explain.

20. The speed of light in warm air is greater than its speed in cold air. The air just above a highway is warmer than the air a little higher. Will the light moving parallel to the highway be bent up or down? Explain.

21. You are standing outside a classroom with an open door. You know your friends are in the room because you can hear them talking. Explain why you can hear them talking but cannot see them.

22. How does the size of an obstacle affect the diffraction of a wave?

chapter 17

Sound

The BIG Idea

Sound waves are compressional waves produced by something that vibrates.

SECTION 1
What is sound?
Main Idea A vibrating object produces compressions and rarefactions that travel away from the object.

SECTION 2
Music
Main Idea The sound of a musical instrument depends on the natural frequencies of the materials it is made from.

An Eerie Silence

You probably have never experienced complete silence unless you've been in a room like this one. The room is lined with special materials that absorb sound waves and eliminate sound reflections.

Science Journal Write a paragraph about the quietest place you've ever been.

Start-Up Activities

Making Human Sounds

When you speak or sing, you push air from your lungs past your vocal cords, which are two flaps of tissue inside your throat. When you tighten your vocal cords, you can make the sound have a higher pitch. Do this lab to explore how you change the shape of your throat to vary the pitch of sound.

1. Hold your fingers against the front of your throat and say *Aaaah*. Notice the vibration against your fingers.
2. Now vary the pitch of this sound from low to high and back again. How do the vibrations in your throat change? Record your observations.
3. Change the sound to an *Ooooh*. What do you notice as you listen? Record your observations.
4. **Think Critically** In your Science Journal, describe how the shape of your throat changed the pitch.

 Preview this chapter's content and activities at ips.msscience.com

 **Sound** Make the following Foldable to help you answer questions about sound.

STEP 1 Fold a vertical sheet of notebook paper from side to side.

STEP 2 Cut along every third line of only the top layer to form tabs.

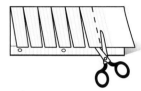

STEP 3 Write a question about sound on each tab.

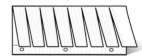

Answer Questions Before you read the chapter, write some questions you have about sound on the front of the tabs. As you read the chapter, write the answer beneath the question. You may add questions as you read.

Get Ready to Read

Monitor

① Learn It! An important strategy to help you improve your reading is monitoring, or finding your reading strengths and weaknesses. As you read, monitor yourself to make sure the text makes sense. Discover different monitoring techniques you can use at different times, depending on the type of test and situation.

② Practice It! The paragraph below appears in Section 1. Read the passage and answer the questions that follow. Discuss your answers with other students to see how they monitor their reading.

> The diffraction of lower frequencies in the human voice allows you to hear someone talking even when the person is around the corner. This is different from an echo. Echoes occur when sound waves bounce off a reflecting surface. Diffraction occurs when a wave spreads out after passing through an opening or when a wave bends around an obstacle.
>
> — *from page 498*

- What questions do you still have after reading?
- Do you understand all of the words in the passage?
- Did you have to stop reading often? Is the reading level appropriate for you?

③ Apply It! Identify one paragraph that is difficult to understand. Discuss it with a partner to improve your understanding.

Target Your Reading

Reading Tip

Monitor your reading by slowing down or speeding up, depending on your understanding of the text.

Use this to focus on the main ideas as you read the chapter.

1. **Before you read** the chapter, respond to the statements below on your worksheet or on a numbered sheet of paper.
   - Write an **A** if you **agree** with the statement.
   - Write a **D** if you **disagree** with the statement.

2. **After you read** the chapter, look back to this page to see if you've changed your mind about any of the statements.
   - If any of your answers changed, explain why.
   - Change any false statements into true statements.
   - Use your revised statements as a study guide.

Science Online
Print out a worksheet of this page at ips.msscience.com

Before You Read A or D		Statement	After You Read A or D
	1	Sound waves transfer energy only in matter.	
	2	The loudness of a sound wave increases as the frequency of a wave increases.	
	3	Sound travels faster in warm air than in cold air.	
	4	Sound usually travels faster in gases than in solids.	
	5	The pitch of a sound you hear depends on whether the source of the sound is moving relative to you.	
	6	Sound waves do not spread out when they pass through an opening.	
	7	A vibrating string whose length is fixed can produce sound waves of more than one frequency.	
	8	The body of a guitar helps make the sound of the vibrating strings louder.	
	9	Changing the length of a vibrating air column changes the pitch of the sound produced.	

section 1
What is sound?

as you read

What You'll Learn
- **Identify** the characteristics of sound waves.
- **Explain** how sound travels.
- **Describe** the Doppler effect.

Why It's Important
Sound gives important information about the world around you.

Review Vocabulary
frequency: number of wavelengths that pass a given point in one second, measured in hertz (Hz)

New Vocabulary
- loudness
- pitch
- echo
- Doppler effect

Sound and Vibration

Think of all the sounds you've heard since you awoke this morning. Did you hear your alarm clock blaring, car horns honking, or locker doors slamming? Every sound has something in common with every other sound. Each is produced by something that vibrates.

Sound Waves How does an object that is vibrating produce sound? When you speak, the vocal cords in your throat vibrate. These vibrations cause other people to hear your voice. The vibrations produce sound waves that travel to their ears. The other person's ears interpret these sound waves.

A wave carries energy from one place to another without transferring matter. An object that is vibrating in air, such as your vocal cords, produces a sound wave. The vibrating object causes air molecules to move back and forth. As these air molecules collide with those nearby, they cause other air molecules to move back and forth. In this way, energy is transferred from one place to another. A sound wave is a compressional wave, like the wave moving through the coiled spring toy in **Figure 1.** In a compressional wave, particles in the material move back and forth along the direction the wave is moving. In a sound wave, air molecules move back and forth along the direction the sound wave is moving.

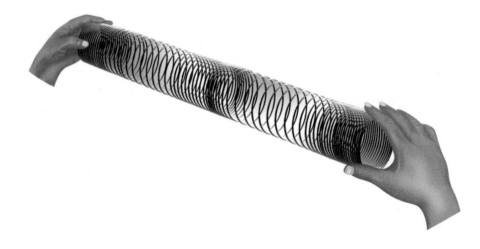

Figure 1 When the coils of a coiled spring toy are squeezed together, a compressional wave moves along the spring. The coils move back and forth as the compressional wave moves past them.

490 CHAPTER 17 Sound

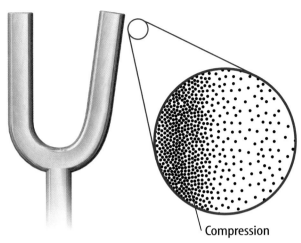

Compression

When the tuning fork vibrates outward, it forces molecules in the air next to it closer together, creating a region of compression.

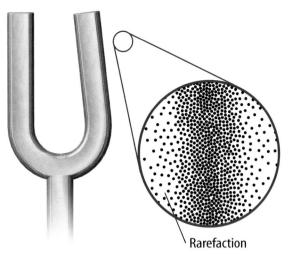

Rarefaction

When the tuning fork moves back, the molecules in the air next to it spread farther apart, creating a region of rarefaction.

Making Sound Waves When an object vibrates, it exerts a force on the surrounding air. For example, as the end of the tuning fork moves outward into the air, it pushes the molecules in the air together, as shown on the left in **Figure 2.** As a result, a region where the molecules are closer together, or more dense, is created. This region of higher density is called a compression. When the end of the tuning fork moves back, it creates a region of lower density called a rarefaction, as shown on the right in **Figure 2.** As the tuning fork continues to vibrate, a series of compressions and rarefactions is formed. The compressions and rarefactions move away from the tuning fork as molecules in these regions collide with other nearby molecules.

Like other waves, a sound wave can be described by its wavelength and frequency. The wavelength of a sound wave is shown in **Figure 3.** The frequency of a sound wave is the number of compressions or rarefactions that pass by a given point in one second. An object that vibrates faster forms a sound wave with a higher frequency.

Figure 2 A tuning fork makes a sound wave as the ends of the fork vibrate in the air.
Explain *why a sound wave cannot travel in a vacuum.*

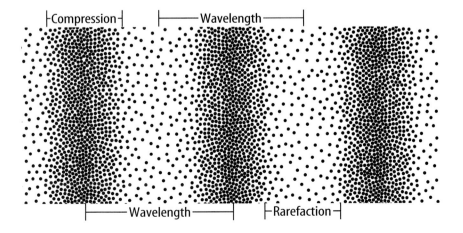

Figure 3 Wavelength is the distance from one compression to another or one rarefaction to another.

SECTION 1 What is sound? **491**

Mini LAB

Comparing and Contrasting Sounds

Procedure
1. Strike a **block of wood** with a **spoon** and listen carefully to the sound. Then press the block of wood to your ear and strike it with the spoon again. Listen carefully to the sound.
2. Tie the middle of a length of **cotton string** to a metal spoon. Strike the spoon on something to hear it ring. Now press the ends of the string against your ears and repeat the experiment. What do you hear?

Analysis
1. Did you hear sounds transmitted through wood and through string? Describe the sounds.
2. Compare and contrast the sounds in wood and in air.

Try at Home

The Speed of Sound

Sound waves can travel through other materials besides air. In fact, sound waves travel in the same way through different materials as they do in air, although they might travel at different speeds. As a sound wave travels through a material, the particles in the material collide with each other. In a solid, molecules are closer together than in liquids or gases, so collisions between molecules occur more rapidly than in liquids or gases. The speed of sound is usually fastest in solids, where molecules are closest together, and slowest in gases, where molecules are farthest apart. **Table 1** shows the speed of sound through different materials.

The Speed of Sound and Temperature The temperature of the material that sound waves are traveling through also affects the speed of sound. As a substance heats up, its molecules move faster, so they collide more frequently. The more frequent the collisions are, the faster the speed of sound is in the material. For example, the speed of sound in air at 0°C is 331 m/s; at 20°C, it is 343 m/s.

Amplitude and Loudness

What's the difference between loud sounds and quiet sounds? When you play a song at high volume and low volume, you hear the same instruments and voices, but something is different. The difference is that loud sound waves generally carry more energy than soft sound waves do.

Loudness is the human perception of how much energy a sound wave carries. Not all sound waves with the same energy are as loud. Humans hear sounds with frequencies between 3,000 Hz and 4,000 Hz as being louder than other sound waves with the same energy.

Table 1 Speed of Sound Through Different Materials	
Material	Speed (m/s)
Air	343
Water	1,483
Steel	5,940
Glass	5,640

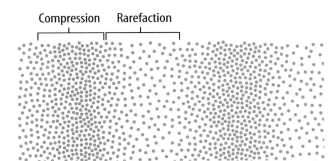
This sound wave has a lower amplitude.

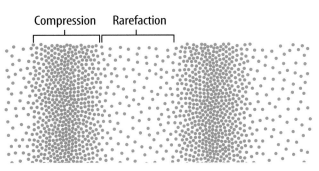
This sound wave has a higher amplitude. Particles in the material are more compressed in the compressions and more spread out in the rarefactions.

Amplitude and Energy The amount of energy a wave carries depends on its amplitude. For a compressional wave such as a sound wave, the amplitude is related to how spread out the molecules or particles are in the compressions and rarefactions, as **Figure 4** shows. The higher the amplitude of the wave is, the more compressed the particles in the compression are and the more spread out they are in the rarefactions. More energy had to be transferred by the vibrating object that created the wave to force the particles closer together or spread them farther apart. Sound waves with greater amplitude carry more energy and sound louder. Sound waves with smaller amplitude carry less energy and sound quieter.

Figure 4 The amplitude of a sound wave depends on how spread out the particles are in the compressions and rarefactions of the wave.

Reading Check What determines the loudness of different sounds?

The Decibel Scale Perhaps an adult has said to you, "Turn down your music, it's too loud! You're going to lose your hearing!" Although the perception of loudness varies from person to person, the energy carried by sound waves can be described by a scale called the decibel (dB) scale. **Figure 5** shows the decibel scale. An increase in the loudness of a sound of 10 dB means that the energy carried by the sound has increased ten times, but an increase of 20 dB means that the sound carries 100 times more energy.

Hearing damage begins to occur at sound levels of about 85 dB. The amount of damage depends on the frequencies of the sound and the length of time a person is exposed to the sound. Some music concerts produce sound levels as high as 120 dB. The energy carried by these sound waves is about 30 billion times greater than the energy carried by sound waves that are made by whispering.

Figure 5 The loudness of sound is measured on the decibel scale.

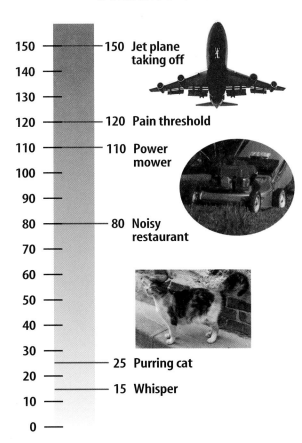

SECTION 1 What is sound? **493**

Frequency and Pitch

The **pitch** of a sound is how high or low it sounds. For example, a piccolo produces a high-pitched sound or tone, and a tuba makes a low-pitched sound. Pitch corresponds to the frequency of the sound. The higher the pitch is, the higher the frequency is. A sound wave with a frequency of 440 Hz, for example, has a higher pitch than a sound wave with a frequency of 220 Hz.

The human ear can detect sound waves with frequencies between about 20 Hz and 20,000 Hz. However, some animals can detect even higher and lower frequencies. For example, dogs can hear frequencies up to almost 50,000 Hz. Dolphins and bats can hear frequencies as high as 150,000 Hz, and whales can hear frequencies higher than those heard by humans.

Recall that frequency and wavelength are related. If two sound waves are traveling at the same speed, the wave with the shorter wavelength has a higher frequency. If the wavelength is shorter, then more compressions and rarefactions will go past a given point every second than for a wave with a longer wavelength, as shown in **Figure 6**. Sound waves with a higher pitch have shorter wavelengths than those with a lower pitch.

The Human Voice When you make a sound, you exhale past your vocal cords, causing them to vibrate. The length and thickness of your vocal cords help determine the pitch of your voice. Shorter, thinner vocal cords vibrate at higher frequencies than longer or thicker ones. This explains why children, whose vocal cords are still growing, have higher voices than adults. Muscles in the throat can stretch the vocal cords tighter, letting people vary their pitch within a limited range.

Figure 6 The upper sound wave has a shorter wavelength than the lower wave. If these two sound waves are traveling at the same speed, the upper sound wave has a higher frequency than the lower one. For this wave, more compressions and rarefactions will go past a point every second than for the lower wave.
Identify *the wave that has a higher pitch.*

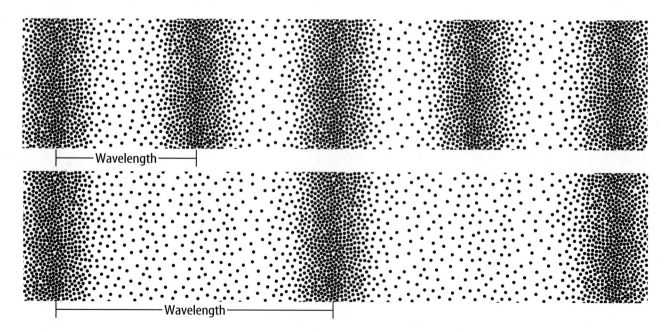

Figure 7 Sonar uses reflected sound waves to determine the location and shape of an object.

Echoes

Sound reflects off of hard surfaces, just like a water wave bounces off the side of a bath tub. A reflected sound wave is called an **echo.** If the distance between you and a reflecting surface is great enough, you might hear the echo of your voice. This is because it might take a few seconds for the sound to travel to the reflecting surface and back to your ears.

Sonar systems use sound waves to map objects underwater, as shown in **Figure 7.** The amount of time it takes an echo to return depends on how far away the reflecting surface is. By measuring the length of time between emitting a pulse of sound and hearing its echo off the ocean floor, the distance to the ocean floor can be measured. Using this method, sonar can map the ocean floor and other undersea features. Sonar also can be used to detect submarines, schools of fish, and other objects.

 *How do sonar systems measure distance?*

> **Science online**
>
> **Topic: Sonar**
> Visit ips.msscience.com for Web links to information about how sonar is used to detect objects underwater.
>
> **Activity** List and explain how several underwater discoveries were made using sonar.

 **Echolocation** Some animals use a method called echolocation to navigate and hunt. Bats, for example, emit high-pitched squeaks and listen for the echoes. The type of echo it hears helps the bat determine exactly where an insect is, as shown in **Figure 8.** Dolphins also use a form of echolocation. Their high-pitched clicks bounce off of objects in the ocean, allowing them to navigate in the same way.

People with visual impairments also use echolocation. For example, they can interpret echoes to estimate the size and shape of a room by using their ears.

Figure 8 Bats use echolocation to hunt.
Explain *why this is a good technique for hunting at night.*

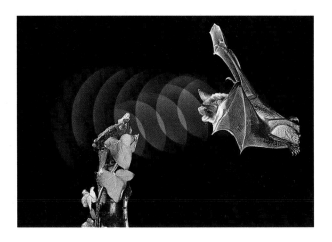

Doppler Shift of Light The frequency of light waves is also changed by the Doppler shift. If a light source is moving away from an observer, the frequencies of the emitted light waves decrease. Research how the Doppler shift is used by astronomers to determine how other objects in the universe are moving relative to Earth.

The Doppler Effect

Perhaps you've heard an ambulance siren as the ambulance speeds toward you, then goes past. You might have noticed that the pitch of the siren gets higher as the ambulance moves toward you. Then as the ambulance moves away, the pitch of the siren gets lower. The change in frequency that occurs when a source of sound is moving relative to a listener is called the **Doppler effect. Figure 9** shows why the Doppler effect occurs.

The Doppler effect occurs whether the sound source or the listener is moving. If you drive past a factory as its whistle blows, the whistle will sound higher pitched as you approach. As you move closer you encounter each sound wave a little earlier than you would if you were sitting still, so the whistle has a higher pitch. When you move away from the whistle, each sound wave takes a little longer to reach you. You hear fewer wavelengths per second, which makes the sound lower in pitch.

Radar guns that are used to measure the speed of cars and baseball pitches also use the Doppler effect. Instead of a sound wave, the radar gun sends out a radio wave. When the radio wave is reflected, its frequency changes depending on the speed of the object and whether it is moving toward the gun or away from it. The radar gun uses the change in frequency of the reflected wave to determine the object's speed.

Applying Science

How does Doppler radar work?

Doppler radar is used by the National Weather Service to detect areas of precipitation and to measure the speed at which a storm moves. Because the wind moves the rain, Doppler radar can "see" into a strong storm and expose the winds. Tornadoes that might be forming in the storm then can be identified.

Identify the Problem

An antenna sends out pulses of radio waves as it rotates. The waves bounce off raindrops and return to the antenna at a different frequency, depending on whether the rain is moving toward the antenna or away from it. The change in frequency is due to the Doppler shift.

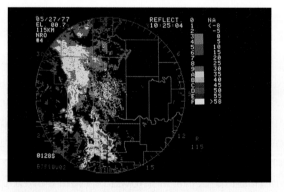

Solving the Problem
1. If the frequency of the reflected radio waves increases, how is the rain moving relative to the radar station?
2. In a tornado, winds are rotating. How would the radio waves reflected by rotating winds be Doppler-shifted?

NATIONAL GEOGRAPHIC VISUALIZING THE DOPPLER EFFECT

Figure 9

You've probably heard the siren of an ambulance as it races through the streets. The sound of the siren seems to be higher in pitch as the ambulance approaches and lower in pitch as it moves away. This is the Doppler effect, which occurs when a listener and a source of sound waves are moving relative to each other.

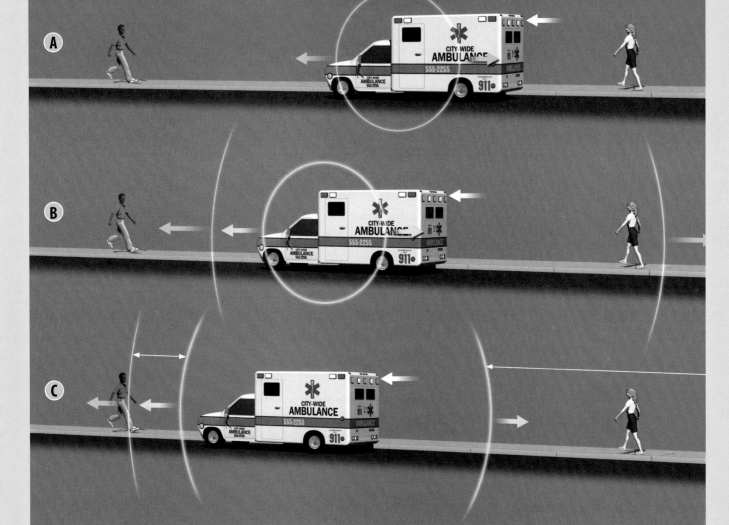

A As the ambulance speeds down the street, its siren emits sound waves. Suppose the siren emits the compression part of a sound wave as it goes past the girl.

B As the ambulance continues moving, it emits another compression. Meanwhile, the first compression spreads out from the point from which it was emitted.

C The waves traveling in the direction that the ambulance is moving have compressions closer together. As a result, the wavelength is shorter and the boy hears a higher frequency sound as the ambulance moves toward him. The waves traveling in the opposite direction have compressions that are farther apart. The wavelength is longer and the girl hears a lower frequency sound as the ambulance moves away from her.

SECTION 1 What is sound? **497**

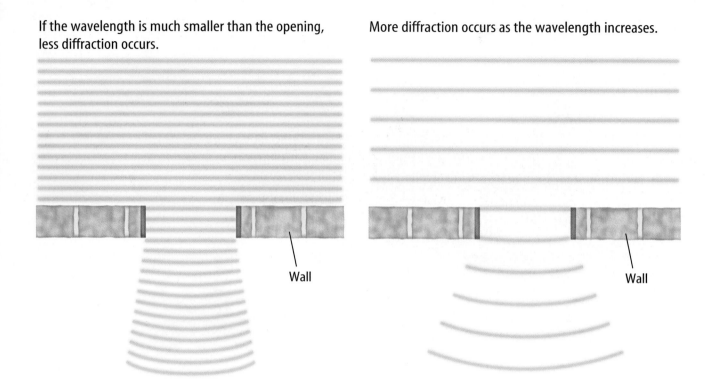

Figure 10 The spreading of a wave by diffraction depends on the wavelength and the size of the opening.

Diffraction of Sound Waves

Like other waves, sound waves diffract. This means they can bend around obstacles or spread out after passing through narrow openings. The amount of diffraction depends on the wavelength of the sound wave compared to the size of the obstacle or opening. If the wavelength is much smaller than the obstacle, almost no diffraction occurs. As the wavelength becomes closer to the size of the obstacle, the amount of diffraction increases.

You can observe diffraction of sound waves by visiting the school band room during practice. If you stand in the doorway, you will hear the band normally. However, if you stand to one side outside the door or around a corner, you will hear the lower-pitched instruments better. **Figure 10** shows why this happens. The sound waves that are produced by the lower-pitched instruments have lower frequencies and longer wavelengths. These wavelengths are closer to the size of the door opening than the higher-pitched sound waves are. As a result, the longer wavelengths diffract more, and you can hear them even when you're not standing in the doorway.

The diffraction of lower frequencies in the human voice allows you to hear someone talking even when the person is around the corner. This is different from an echo. Echoes occur when sound waves bounce off a reflecting surface. Diffraction occurs when a wave spreads out after passing through an opening, or when a wave bends around an obstacle.

Using Sound Waves

Sound waves can be used to treat certain medical problems. A process called ultrasound uses high-frequency sound waves as an alternative to some surgeries. For example, some people develop small, hard deposits in their kidneys or gallbladders. A doctor can focus ultrasound waves at the kidney or gallbladder. The ultrasound waves cause the deposits to vibrate rapidly until they break apart into small pieces. Then, the body can get rid of them.

Ultrasound can be used to make images of the inside of the body. One common use of ultrasound is to examine a developing fetus. Also, ultrasound along with the Doppler effect can be used to examine the functioning of the heart. An ultrasound image of the heart is shown in **Figure 11.** This technique can help determine if the heart valves and heart muscle are functioning properly, and how blood is flowing through the heart.

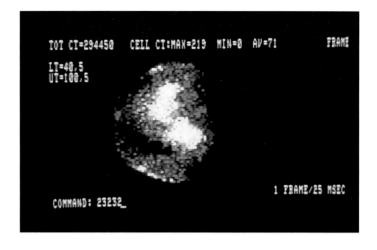

Figure 11 Ultrasound is used to make this image of the heart.
Describe *other ways ultrasound is used in medicine.*

section 1 review

Summary

Sound Waves
- Sound waves are compressional waves produced by vibrations.
- Sound travels fastest in solids and slowest in gases.
- Sound travels faster as the temperature of the medium increases.
- The energy carried by a sound wave increases as its amplitude increases.

Loudness and Pitch
- Loudness is the human perception of the energy carried by a sound wave.
- The pitch of a sound becomes higher as the frequency of the sound increases.

The Doppler Effect and Diffraction
- In the Doppler effect, the frequency of a sound wave changes if the source of the sound is moving relative to the listener.
- Diffraction occurs when sound waves bend around objects or spread out after passing through an opening.

Self Check

1. **Describe** how the loudness of a sound wave changes when the amplitude of the wave is increased.
2. **Explain** how the wavelength of a sound wave affects the diffraction of the sound wave through an open window.
3. **Describe** how echolocation could be used to measure the distance to the bottom of a lake.
4. **Discuss** how the spacing of particles in a sound wave changes as the amplitude of the wave decreases.
5. **Describe** how the wavelength of a sound wave changes if the frequency of the wave increases.
6. **Think Critically** You hear the pitch of the sound from an ambulance siren get lower, then get higher. Describe the motion of the ambulance relative to you.

Applying Math

7. **Calculate Distance** Sound travels through water at a speed of 1,483 m/s. Use the equation
$$\text{distance} = \text{speed} \times \text{time}$$
to calculate how far a sound wave in water will travel in 5 s.

Observe and Measure Reflection of Sound

Real-World Question

Like all waves, sound waves can be reflected. When sound waves strike a surface, in what direction does the reflected sound wave travel? In this activity, you'll focus sound waves using cardboard tubes to help answer this question. How are the angles made by incoming and reflected sound waves related?

Goals

- **Observe** reflection of sound waves.
- **Measure** the angles incoming and reflected sound waves make with a surface.

Materials

20-cm to 30-cm-long cardboard tubes (2)
watch that ticks audibly
protractor

Safety Precautions

Procedure

1. Work in groups of three. Each person should listen to the watch—first without a tube and then through a tube. The person who hears the watch most easily is the listener.
2. One person should hold one tube at an angle with one end above a table. Hold the watch at the other end of the tube.
3. The listener should hold the second tube at an angle, with one end near his or her ear and the other end near the end of the first tube that is just above the table. The tubes should be in the same vertical plane.
4. Move the first tube until the watch sounds loudest. The listener might need to cover the other ear to block out background noises.
5. The third person should measure the angle that each tube makes with the table.

Conclude and Apply

1. **Compare** the angles the incoming and reflected waves make with the table.
2. The normal is a line at 90 degrees to the table at the point where reflection occurs. Determine the angles the incoming and reflected waves make with the normal.
3. The law of reflection states that the angles the incoming and reflected waves make with the normal are equal. Do sound waves obey the law of reflection?

Communicating Your Data

Make a scientific illustration to show how the experiment was done. Describe your results using the illustration.

section 2
Music

What is music?

What do you like to listen to—rock 'n' roll, country, blues, jazz, rap, or classical? Music and noise are groups of sounds. Why do humans hear some sounds as music and other sounds as noise?

The answer involves patterns of sound. **Music** is a group of sounds that have been deliberately produced to make a regular pattern. Look at **Figure 12.** The sounds that make up music usually have a regular pattern of pitches, or notes. Some natural sounds such as the patter of rain on a roof, the sound of ocean waves splashing, or the songs of birds can sound musical. On the other hand, noise is usually a group of sounds with no regular pattern. Sounds you hear as noise are irregular and disorganized such as the sounds of traffic on a city street or the roar of a jet aircraft.

However, the difference between music and noise can vary from person to person. What one person considers to be music, another person might consider noise.

Natural Frequencies Music is created by vibrations. When you sing, your vocal cords vibrate. When you beat a drum, the drumhead vibrates. When you play a guitar, the strings vibrate.

If you tap on a bell with a hard object, the bell produces a sound. When you tap on a bell that is larger or smaller or has a different shape you hear a different sound. The bells sound different because each bell vibrates at different frequencies. A bell vibrates at frequencies that depend on its shape and the material it is made from. Every object will vibrate at certain frequencies called its **natural frequencies.**

as you read

What **You'll Learn**
- **Explain** the difference between music and noise.
- **Describe** how different instruments produce music.
- **Explain** how you hear.

Why **It's Important**

Music is made by people in every part of the world.

Review Vocabulary
compressional wave: a type of mechanical wave in which matter in the medium moves forward and backward in the same direction the wave travels

New Vocabulary
- music
- natural frequencies
- resonance
- fundamental frequency
- overtone
- reverberation
- eardrum

Figure 12 Music and noise have different types of sound patterns.

Noise has no specific or regular sound wave pattern.

Music is organized sound. Music has regular sound wave patterns and structures.

SECTION 2 Music **501**

Musical Instruments and Natural Frequencies Many objects vibrate at one or more natural frequencies when they are struck or disturbed. Like a bell, the natural frequencies of any object depend on the size and shape of the object and the material it is made from. Musical instruments use the natural frequencies of strings, drumheads, or columns of air contained in pipes to produce various musical notes.

Reading Check *What determines the natural frequencies?*

Resonance You may have seen the comedy routine in which a loud soprano sings high enough to shatter glass. Sometimes sound waves cause an object to vibrate. When a tuning fork is struck, it vibrates at its natural frequency and produces a sound wave with the same frequency. Suppose you have two tuning forks with the same natural frequency. You strike one tuning fork, and the sound waves it produces strike the other tuning fork. These sound waves would cause the tuning fork that wasn't struck to absorb energy and vibrate. This is an example of resonance. **Resonance** occurs when an object is made to vibrate at its natural frequencies by absorbing energy from a sound wave or another object vibrating at these frequencies.

Musical instruments use resonance to amplify their sounds. Look at **Figure 13.** The vibrating tuning fork might cause the table to vibrate at the same frequency, or resonate. The combined vibrations of the table and the tuning fork increase the loudness of the sound waves produced.

Reducing Earthquake Damage The shaking of the ground during an earthquake can cause buildings to resonate. The increased vibration of a building due to resonance could result in the collapse of the building, causing injuries and loss of life. To reduce damage during earthquakes, buildings are designed to resonate at frequencies different than those that occur during earthquakes. Research how buildings are designed to reduce damage caused by earthquakes.

Figure 13 When a vibrating tuning fork is placed against a table, resonance might cause the table to vibrate.

Overtones

Before a concert, all orchestra musicians tune their instruments by playing the same note. Even though the note has the same pitch, it sounds different for each instrument. It also sounds different from a tuning fork that vibrates at the same frequency as the note.

A tuning fork produces a single frequency, called a pure tone. However, the notes produced by musical instruments are not pure tones. Most objects have more than one natural frequency at which they can vibrate. As a result, they produce sound waves of more than one frequency.

If you play a single note on a guitar, the pitch that you hear is the lowest frequency produced by the vibrating string. The lowest frequency produced by a vibrating object is the **fundamental frequency.** The vibrating string also produces higher frequencies. These higher frequencies are **overtones.** Overtones have frequencies that are multiples of the fundamental frequency, as in **Figure 14.** The number and intensity of the overtones produced by each instrument are different and give instruments their distinctive sound quality.

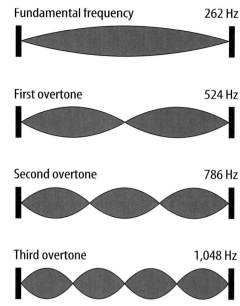

Figure 14 A string vibrates at a fundamental frequency, as well as at overtones. The overtones are multiples of that frequency.

Musical Scales

A musical instrument is a device that produces musical sounds. These sounds are usually part of a musical scale that is a sequence of notes with certain frequencies. For example, **Figure 15** shows the sequence of notes that belong to the musical scale of C. Notice that the frequency produced by the instrument doubles after eight successive notes of the scale are played. Other musical scales consist of a different sequence of frequencies.

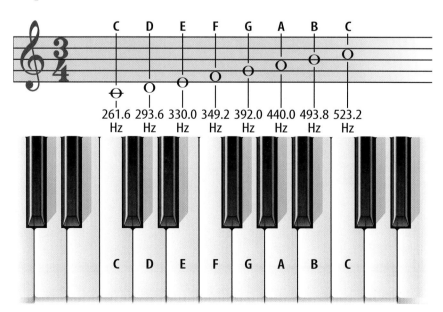

Figure 15 A piano produces a sequence of notes that are a part of a musical scale.
Describe how the frequencies of the two C notes on this scale are related.

SECTION 2 Music **503**

Mini LAB

Modeling a Stringed Instrument

Procedure
1. Stretch a **rubber band** between your fingers.
2. Pluck the rubber band. Listen to the sound and observe the shape of the vibrating band. Record what you hear and see.
3. Stretch the band farther and repeat step 2.
4. Shorten the length of the band that can vibrate by holding your finger on one point. Repeat step 2.
5. Stretch the rubber band over an open box, such as a **shoe box.** Repeat step 2.

Analysis
1. How did the sound change when you stretched the rubber band? Was this what you expected? Explain.
2. How did the sound change when you stretched the band over the box? Did you expect this? Explain.

Stringed Instruments

Stringed instruments, like the cello shown in **Figure 16,** produce music by making strings vibrate. Different methods are used to make the strings vibrate—guitar strings are plucked, piano strings are struck, and a bow is slid across cello strings. The strings often are made of wire. The pitch of the note depends on the length, diameter, and tension of the string—if the string is shorter, narrower, or tighter, the pitch increases. For example, pressing down on a vibrating guitar string shortens its length and produces a note with a higher pitch. Similarly, the thinner guitar strings produce a higher pitch than the thicker strings.

Amplifying Vibrations The sound produced by a vibrating string usually is soft. To amplify the sound, stringed instruments usually have a hollow chamber, or box, called a resonator, which contains air. The resonator absorbs energy from the vibrating string and vibrates at its natural frequencies. For example, the body of a guitar is a resonator that amplifies the sound that is produced by the vibrating strings. The vibrating strings cause the guitar's body and the air inside it to resonate. As a result, the vibrating guitar strings sound louder, just as the tuning fork that was placed against the table sounded louder.

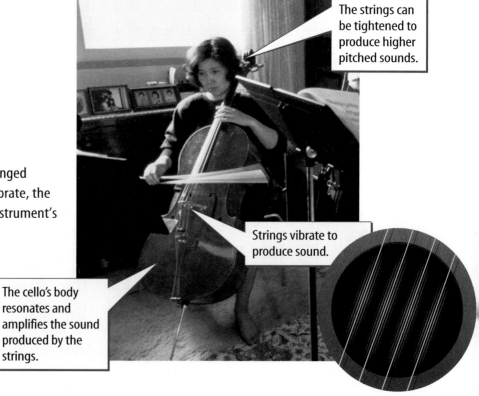

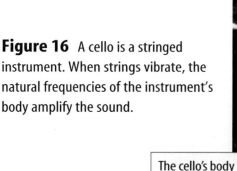

Figure 16 A cello is a stringed instrument. When strings vibrate, the natural frequencies of the instrument's body amplify the sound.

Percussion

Percussion instruments, such as the drum shown in **Figure 17,** are struck to make a sound. Striking the top surface of the drum causes it to vibrate. The vibrating drumhead is attached to a chamber that resonates and amplifies the sound.

Drums and Pitch Some drums have a fixed pitch, but some can be tuned to play different notes. For example, if the drumhead on a kettledrum is tightened, the natural frequency of the drumhead is increased. As a result, the pitches of the sounds produced by the kettledrum get higher. A steel drum, shown in **Figure 17,** plays different notes in the scale when different areas in the drum are struck. In a xylophone, wood or metal bars of different lengths are struck. The longer the bar is, the lower the note that it produces is.

Brass and Woodwinds

Just as the bars of a xylophone have different natural frequencies, so do the air columns in pipes of different lengths. Brass and woodwind instruments, such as those in **Figure 18,** are essentially pipes or tubes of different lengths that sometimes are twisted around to make them easier to hold and carry. To make music from these instruments, the air in the pipes is made to vibrate at various frequencies.

Different methods are used to make the air column vibrate. A musician playing a brass instrument, such as a trumpet, makes the air column vibrate by vibrating the lips and blowing into the mouthpiece. Woodwinds such as clarinets, saxophones, and oboes contain one or two reeds in the mouthpiece that vibrate the air column when the musician blows into the mouthpiece. Flutes also are woodwinds, but a flute player blows across a narrow opening to make the air column vibrate.

Figure 17 The sounds produced by drums depend on the material that is vibrating.

The vibrating drumhead of this drum is amplified by the resonating air in the body of the drum.

The vibrating steel surface in a steel drum produces loud sounds that don't need to be amplified by an air-filled chamber.

Figure 18 Brass and woodwind instruments produce sounds by causing a column of air to vibrate.

Figure 19 A flute changes pitch as holes are opened and closed.

By opening holes on a flute, the length of the vibrating air column is made shorter.

Changing Pitch in Woodwinds To change the note that is being played in a woodwind instrument, a musician changes the length of the resonating column of air. By making the length of the vibrating air column shorter, the pitch of the sound produced is made higher. In a woodwind such as a flute, saxophone, or clarinet, this is done by closing and opening finger holes along the length of the instrument, as shown in **Figure 19.**

Changing Pitch in Brass In brass instruments, musicians vary the pitch in other ways. One is by blowing harder to make the air resonate at a higher natural frequency. Another way is by pressing valves that change the length of the tube.

Beats

Recall that interference occurs when two waves overlap and combine to form a new wave. The new wave formed by interference can have a different frequency, wavelength, and amplitude than the two original waves.

Suppose two notes close in frequency are played at the same time. The two notes interfere to form a new sound whose loudness increases and decreases several times a second. If you were listening to the sound, you would hear a series of beats as the sound got louder and softer. The beat frequency, or the number of beats you would hear each second, is equal to the difference in the frequencies of the two notes.

For example, if the two notes have frequencies of 329 Hz and 332 Hz, the beat frequency would be 3 Hz. You would hear the sound get louder and softer—a beat—three times each second.

Figure 20 A piano can be tuned by using beats.

Beats Help Tune Instruments Beats are used to help tune instruments. For example, a piano tuner, like the one shown in **Figure 20,** might hit a tuning fork and then the corresponding key on the piano. Beats are heard when the difference in pitch is small. The piano string is tuned properly when the beats disappear. You might have heard beats while listening to an orchestra tune before a performance. You also can hear beats produced by two engines vibrating at slightly different frequencies.

Reverberation

Sound is reflected by hard surfaces. In an empty gymnasium, the sound of your voice can be reflected back and forth several times by the floor, walls, and ceiling. Repeated echoes of sound are called **reverberation.** In a gym, reverberation makes the sound of your voice linger before it dies out. Some reverberation can make voices or music sound bright and lively. Too little reverberation makes the sound flat and lifeless. However, reverberation can produce a confusing mess of noise if too many sounds linger for too long.

Concert halls and theaters, such as the one in **Figure 21,** are designed to produce the appropriate level of reverberation. Acoustical engineers use soft materials to reduce echoes. Special panels that are attached to the walls or suspended from the ceiling are designed to reflect sound toward the audience.

Topic: Controlling Reverberation
Visit ips.msscience.com for Web Links to information about how acoustical engineers control reverberation.

Activity Make a list of the materials engineers use to reduce and enhance reverberation.

Figure 21 The shape of a concert hall and the materials it contains are designed to control the reflection of sound waves.

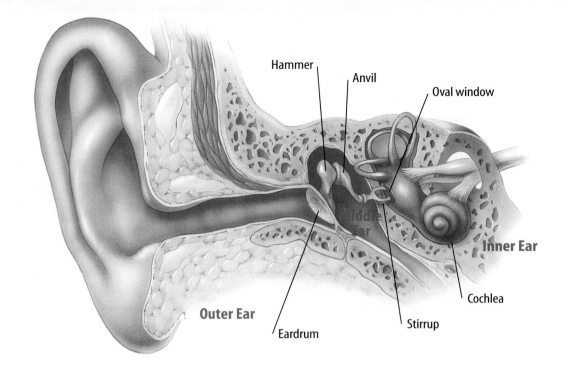

Figure 22 The human ear has three different parts—the outer ear, the middle ear, and the inner ear.

Figure 23 Animals, such as rabbits and owls, have ears that are adapted to their different needs.

The Ear

You hear sounds with your ears. The ear is a complex organ that is able to detect a wide range of sounds. The ear can detect frequencies ranging from about 20 Hz to about 20,000 Hz. The ear also can detect a wide range of sound intensities. The faintest sounds you can hear carry about one trillionth the amount of energy as the loudest sounds you can hear. The human ear is illustrated in **Figure 22.** It has three parts—the outer ear, the middle ear, and the inner ear.

The Outer Ear—Sound Collector Your outer ear collects sound waves and directs them into the ear canal. Notice that your outer ear is shaped roughly like a funnel. This shape helps collect sound waves.

Animals that rely on hearing to locate predators or prey often have larger, more adjustable ears than humans, as shown in **Figure 23.** A barn owl, which relies on its excellent hearing for hunting at night, does not have outer ears made of flesh. Instead, the arrangement of its facial feathers helps direct sound to its ears. Some sea mammals, on the other hand, have only small holes for outer ears, even though their hearing is good.

The Middle Ear—Sound Amplifier When sound waves reach the middle ear, they vibrate the **eardrum,** which is a membrane that stretches across the ear canal like a drumhead. When the eardrum vibrates, it transmits vibrations to three small connected bones—the hammer, anvil, and stirrup. The bones amplify the vibrations, just as a lever can change a small movement at one end into a larger movement at the other.

The Inner Ear—Sound Interpreter The stirrup vibrates a second membrane called the oval window. This marks the start of the inner ear, which is filled with fluid. Vibrations in the fluid are transmitted to hair-tipped cells lining the cochlea, as shown in **Figure 24**. Different sounds vibrate the cells in different ways. The cells generate signals containing information about the frequency, intensity, and duration of the sound. The nerve impulses travel along the auditory nerve and are transmitted to the part of the brain that is responsible for hearing.

 *Where are waves detected and interpreted in the ear?*

Hearing Loss

The ear can be damaged by disease, age, and exposure to loud sounds. For example, constant exposure to loud noise can damage hair cells in the cochlea. If damaged mammalian hair cells die, some loss of hearing results because mammals cannot make new hair cells. Also, some hair cells and nerve fibers in the inner ear degenerate and are lost as people age. It is estimated that about 30 percent of people over 65 have some hearing loss due to aging.

Figure 24 The inner ear contains tiny hair cells that convert vibrations into nerve impulses that travel to the brain.

section 2 review

Summary

What is music?
- Music is sound that is deliberately produced in a regular pattern.
- Objects vibrate at certain natural frequencies.
- The lowest frequency produced by a vibrating object is the object's fundamental frequency.
- The overtones produced by a vibrating object are multiples of the fundamental frequency.

Musical Instruments and Hearing
- In stringed instruments the sounds made by vibrating strings are amplified by a resonator.
- Percussion instruments produce sound by vibrating when they are struck.
- Brass and woodwind instruments produce sound by vibrating a column of air.
- The ear collects sound waves, amplifies the sound, and interprets the sound.

Self Check

1. **Describe** how music and noise are different.
2. **Infer** Two bars on a xylophone are 10 cm long and 14 cm long. Identify which bar produces a lower pitch when struck and explain why.
3. **Describe** the parts of the human ear and the function of each part in enabling you to hear sound.
4. **Predict** how the sound produced by a guitar string changes as the length of the string is made shorter.
5. **Diagram** the fundamental and the first two overtones for a vibrating string.
6. **Think Critically** How does reverberation explain why your voice sounds different in a gym than it does in your living room?

Applying Math

7. **Calculate Overtone Frequency** A guitar string has a fundamental frequency of 440 Hz. What is the frequency of the second overtone?

ips.msscience.com/self_check_quiz

LAB Design Your Own

Music

Goals
- **Design** an experiment to compare the changes that are needed in different instruments to produce a variety of different notes.
- **Observe** which changes are made when playing different notes.
- **Measure and record** these changes whenever possible.

Possible Materials
musical instruments
measuring tape
tuning forks

Safety Precautions
Properly clean the mouthpiece of any instrument before it is used by another student.

Real-World Question

The pitch of a note that is played on an instrument sometimes depends on the length of the string, the air column, or some other vibrating part. Exactly how does sound correspond to the size or length of the vibrating part? Is this true for different instruments? What causes different instruments to produce different notes?

Form a Hypothesis

Based on your reading and observations, make a hypothesis about what changes in an instrument to produce different notes.

Test Your Hypothesis

Make a Plan
1. You should do this lab as a class, using as many instruments as possible. You might want to go to the music room or invite friends and relatives who play an instrument to visit the class.

2. As a group, decide how you will measure changes in instruments. For wind instruments, can you measure the length of the vibrating air column? For stringed instruments, can you measure the length and thickness of the vibrating string?

3. Refer to the table of wavelengths and frequencies for notes in the scale. Note that no measurements are given—if you measure C to correspond to a string length of 30 cm, for example, the note G will correspond to two thirds of that length.

4. Decide which musical notes you will compare. Prepare a table to collect your data. List the notes you have selected.

Ratios of Wavelengths and Frequencies of Musical Notes		
Note	Wavelength	Frequency
C	1	1
D	8/9	9/8
E	4/5	5/4
F	3/4	4/3
G	2/3	3/2
A	3/5	5/3
B	8/15	15/8
C	1/2	2

Follow Your Plan

1. Make sure your teacher approves your plan before you start.
2. Carry out the experiment as planned.
3. While doing the experiment, record your observations and complete the data table.

Analyze Your Data

1. **Compare** the change in each instrument when the two notes are produced.
2. **Compare and contrast** the changes between instruments.
3. What were the controls in this experiment?
4. What were the variables in this experiment?
5. How did you eliminate bias?

Conclude and Apply

1. How does changing the length of the vibrating column of air in a wind instrument affect the note that is played?
2. Describe how you would modify an instrument to increase the pitch of a note that is played.

Demonstrate to another teacher or to family members how the change in the instrument produces a change in sound.

TIME SCIENCE AND Society
SCIENCE ISSUES THAT AFFECT YOU!

It's a Wrap!

No matter how quickly or slowly you open a candy wrapper, it always will make a noise

Y ou're at the movies and it's the most exciting part of the film. The audience is silent and intent on what is happening on the screen. At that moment, you decide to unwrap a piece of candy. CRACKLE! POP! SNAP! No matter how you do it, the candy wrapper makes a lot of noise.

Why can't you unwrap candy without making a racket? To test this plastics problem, researchers put some crinkly wrappers in a silent room. Then they stretched out the wrappers and recorded the sounds they made. Next, the sounds were analyzed by a computer. The research team discovered that the wrapper didn't make a continuous sound. Instead, it made many separate little popping noises, each taking only a thousandth of a second. They found that whether you open the wrapper quickly or slowly the amount of noise made by the pops will be the same. "And there's nothing you can do about it," said a member of the research team.

By understanding what makes a plastic wrapper snap when it changes shape, doctors can better understand molecules in the human body that also change shape.

The pop chart

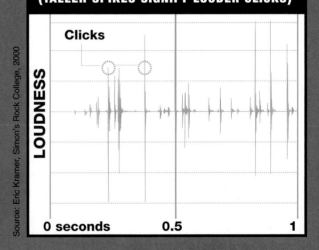

SOUND LEVEL OVER TIME
The sound that a candy wrapper makes is emitted as a series of pulses or clicks. So, opening a wrapper slowly only increases the length of time in between clicks, but the amount of noise remains the same.
(TALLER SPIKES SIGNIFY LOUDER CLICKS)

Recall and Retell Have you ever opened a candy wrapper in a quiet place? Did it bother other people? If so, did you try to open it more slowly? What happened?

For more information, visit ips.msscience.com/time

Chapter 17 Study Guide

Reviewing Main Ideas

Section 1 What is sound?

1. Sound is a compressional wave that travels through matter, such as air. Sound is produced by something that vibrates.

2. The speed of sound depends on the material in which it is traveling.

3. The larger the amplitude of a sound wave, the more energy it carries and the louder the sound.

4. The pitch of a sound wave becomes higher as its frequency increases. Sound waves can reflect and diffract.

5. The Doppler effect occurs when a source of sound and a listener are in motion relative to each other. The pitch of the sound heard by the listener changes.

Section 2 Music

1. Music is made of sounds that are used in a regular pattern. Noise is made of sounds that are irregular and disorganized.

2. Objects vibrate at their natural frequencies. These depend on the shape of the object and the material it's made of.

3. Resonance occurs when an object is made to vibrate by absorbing energy at one of its natural frequencies.

4. Musical instruments produce notes by vibrating at their natural frequencies.

5. Beats occur when two waves of nearly the same frequency interfere.

6. The ear collects sound waves and converts sound waves to nerve impulses.

Visualizing Main Ideas

Copy and complete the following concept map on sound.

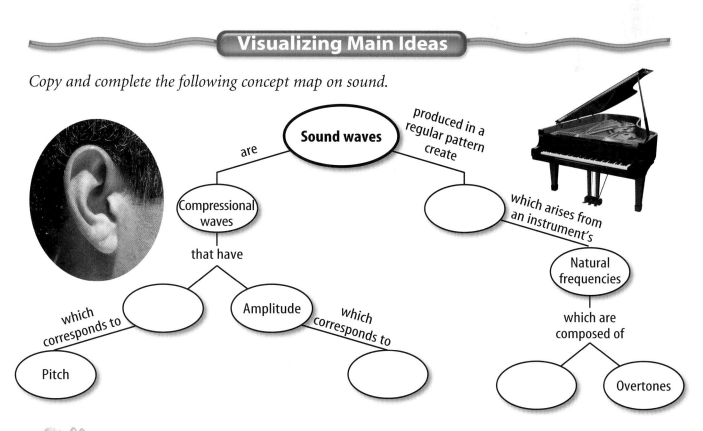

chapter 17 Review

Using Vocabulary

Doppler effect p. 496
eardrum p. 508
echo p. 495
fundamental
 frequency p. 503
loudness p. 492
music p. 501
natural frequency p. 501
overtone p. 503
pitch p. 494
resonance p. 502
reverberation p. 507

Distinguish between the terms in the following pairs.

1. overtones—fundamental frequency
2. pitch—sound wave
3. pitch—Doppler effect
4. loudness—resonance
5. fundamental frequency—natural frequency
6. loudness—amplitude
7. natural frequency—overtone
8. reverberation—resonance

Checking Concepts

Choose the word or phrase that best answers the question.

9. A tone that is lower in pitch is lower in what characteristic?
   A) frequency   C) loudness
   B) wavelength D) resonance

10. If the wave speed stays the same, which of the following decreases as the frequency increases?
    A) pitch   C) loudness
    B) wavelength D) resonance

11. What part of the ear is damaged most easily by continued exposure to loud noise?
    A) eardrum   C) oval window
    B) stirrup   D) hair cells

12. What is an echo?
    A) diffracted sound
    B) resonating sound
    C) reflected sound
    D) an overtone

13. A trumpeter depresses keys to make the column of air resonating in the trumpet shorter. What happens to the note being played?
    A) Its pitch is higher.
    B) Its pitch is lower.
    C) It is quieter.
    D) It is louder.

14. When tuning a violin, a string is tightened. What happens to a note being played on the string?
    A) Its pitch is higher.
    B) Its pitch is lower.
    C) It is quieter.
    D) It is louder.

15. As air becomes warmer, how does the speed of sound in air change?
    A) It increases.   C) It doesn't change.
    B) It decreases.   D) It oscillates.

16. Sound waves are which type of wave?
    A) slow   C) compressional
    B) transverse D) electromagnetic

17. What does the middle ear do?
    A) focuses sound
    B) interprets sound
    C) collects sound
    D) transmits and amplifies sound

18. An ambulance siren speeds away from you. What happens to the pitch of the siren?
    A) It becomes softer.
    B) It becomes louder.
    C) It decreases.
    D) It increases.

chapter 17 Review

Thinking Critically

19. **Explain** Some xylophones have open pipes of different lengths hung under each bar. The longer a bar is, the longer the pipe beneath it. Explain how these pipes help amplify the sound of the xylophone.

20. **Infer** why you don't notice the Doppler effect for a slow moving train.

21. **Predict** Suppose the movement of the bones in the middle ear were reduced. Which would be more affected—the ability to hear quiet sounds or the ability to hear high frequencies? Explain your answer.

22. **Explain** The triangle is a percussion instrument consisting of an open metal triangle hanging from a string. A chiming sound is heard when the triangle is struck by a metal rod. If the triangle is held in the hand, a quiet dull sound is heard when it is struck. Why does holding the triangle make the sound quieter?

Use the table below to answer question 23.

Speed of Sound Through Different Materials	
Material	Speed (m/s)
Air	343
Water	1,483
Steel	5,940
Glass	5,640

23. **Calculate** Using the table above, determine the total amount of time needed for a sound wave to travel 3.5 km through air and then 100.0 m through water.

24. **Predict** If the holes of a flute are all covered while playing, then all uncovered, what happens to the length of the vibrating air column? What happens to the pitch of the note?

25. **Identify Variables and Controls** Describe an experiment to demonstrate that sound is diffracted.

26. **Interpret Scientific Illustrations** The picture below shows pan pipes. How are different notes produced by blowing on pan pipes?

Performance Activities

27. **Recital** Perform a short musical piece on an instrument. Explain how your actions changed the notes that were produced.

28. **Pamphlet** Make a pamphlet describing how a hearing aid works.

29. **Interview** Interview several people over 65 with some form of hearing loss. Create a table that shows the age of each person and how their hearing has changed with age.

Applying Math

30. **Beats** Two flutes are playing at the same time. One flute plays a note with a frequency of 524 Hz. If two beats per second are heard, what are the possible frequencies the other flute is playing?

31. **Overtones** Make a table showing the first three overtones of C, which has a frequency of 262 Hz, and G, which has a frequency of 392 Hz.

Chapter 17 Standardized Test Practice

Part 1 | Multiple Choice

Record your answers on the answer sheet provided by your teacher or on a sheet of paper.

1. In which of the following materials does sound travel the fastest?
   A. empty space   C. air
   B. water   D. steel

2. How can the pitch of the sound made by a guitar string be lowered?
   A. by shortening the part of the string that vibrates
   B. by tightening the string
   C. by replacing the string with a thicker string
   D. by plucking the string harder

Use the figure below to answer questions 3 and 4.

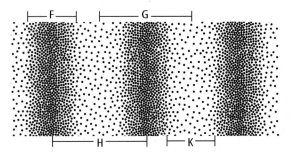

3. What part of the wave is shown at F?
   A. rarefaction   C. wavelength
   B. compression   D. amplitude

4. What part of the wave is shown at H?
   A. rarefaction   C. wavelength
   B. compression   D. amplitude

5. What happens to the particles of matter when a compressional wave moves through the matter?
   A. The particles do not move.
   B. The particles move back and forth along the wave direction.
   C. The particles move back and forth and are carried along with the wave.
   D. The particles move at right angles to the direction the wave travels.

6. If you were on a moving train, what would happen to the pitch of a bell at a crossing as you approached and then passed by the crossing?
   A. It would seem higher, then lower.
   B. It would remain the same.
   C. It would seem lower and then higher.
   D. It would keep getting lower.

Use the figure below to answer questions 7 and 8.

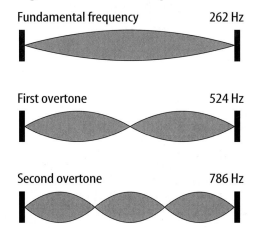

7. How are the overtone frequencies of any vibrating object related to the fundamental frequency of vibration?
   A. They are multiples of the fundamental.
   B. They are not related to the fundamental.
   C. They equal twice the fundamental.
   D. They are lower than the fundamental.

8. Which of the following is the frequency of the third overtone?
   A. 1,572 Hz   C. 1,048 Hz
   B. 1,000 Hz   D. 786 Hz

9. Which of the following is NOT related to the amplitude of a sound wave?
   A. energy carried by the wave
   B. loudness of a sound
   C. pitch of a sound
   D. how spread out the particles are in the compressions and rarefactions

Part 2 Short Response/Grid In

Record your answers on the answer sheet provided by your teacher or on a sheet of paper.

10. What is the difference between diffracted sound waves and echoes?

Use the figure below to answer questions 11–13.

Speed of Sound in Different Materials	
Material	Speed of sound (m/s)
Air	343
Water	1,483
Steel	5,940

11. A fish locator sends out a pulse of ultrasound and measures the time needed for the sound to travel to a school of fish and back to the boat. If the fish are 16 m below the boat, how long would it take sound to make the round trip in the water?

12. Suppose you are at a baseball game 150 m from home plate. How long after the batter hits the ball do you hear the sound?

13. A friend drops a stone on a steel railroad track. If the sound made by the stone hitting the track reaches you in 0.8 s, how far away is your friend?

14. Why do different objects produce different sounds when they are struck?

15. Explain how one vibrating tuning fork could make a second tuning fork also vibrate. What is this an example of?

Test-Taking Tip

Notice Units Read carefully and make note of the units used in any measurement.

Question 13 Notice the units used for time in the question and the units for speed given in the table.

Part 3 Open Ended

Record your answers on a sheet of paper.

16. Why do different musical instruments sound different even when they play a note with the same pitch?

17. Compare the way a drum and a flute produce sound waves. What acts as a resonator in each instrument?

18. Would sound waves traveling through the outer ear travel faster or slower than those traveling through the inner ear? Explain.

Use the figure below to answer questions 19.

19. Describe how the process shown in the figure can be used to map the ocean floor.

20. When a sound wave passes through an opening, what does the amount of diffraction depend on?

21. Describe how a cello produces and amplifies sounds.

22. People who work on the ground near jet runways are required to wear ear protection. Explain why this is necessary.

23. Bats use ultrasound when they echolocate prey. If ultrasound waves bounce off an insect that is flying away from the bat, how would the frequency of the wave be affected? What is this effect called?

chapter 18

Electromagnetic Waves

The BIG Idea
Electromagnetic waves are made of changing electric and magnetic fields.

SECTION 1
The Nature of Electromagnetic Waves
Main Idea Vibrating electric charges produce electromagnetic waves.

SECTION 2
The Electromagnetic Spectrum
Main Idea Different parts of the electromagnetic spectrum interact with matter in different ways.

SECTION 3
Using Electromagnetic Waves
Main Idea Electromagnetic waves are used to transmit and receive information.

Looking Through You

This color-enhanced X-ray image of a human shoulder and ribcage was made possible by electromagnetic waves. These waves are used to transmit the programs you watch on TV and they make your skin feel warm when you sit in sunlight. In fact, no matter where you go, you are always surrounded by electromagnetic waves.

Science Journal Describe how sitting in sunlight makes you feel. How can sunlight affect your skin?

Start-Up Activities

Detecting Invisible Waves

Light is a type of wave called an electromagnetic wave. You see light every day, but visible light is only one type of electromagnetic wave. Other electromagnetic waves are all around you, but you cannot see them. How can you detect electromagnetic waves that can't be seen with your eyes?

1. Cut a slit 2 cm long and 0.25 cm wide in the center of a sheet of black paper.
2. Cover a window that is in direct sunlight with the paper.
3. Position a glass prism in front of the light coming through the slit so it makes a visible spectrum on the floor or table.
4. Place one thermometer in the spectrum and a second thermometer just beyond the red light.
5. Measure the temperature in each region after 5 min.
6. **Think Critically** Write a paragraph in your Science Journal comparing the temperatures of the two regions and offer an explanation for the observed temperatures.

Electromagnetic Waves Make the following Foldable to help you understand the electromagnetic spectrum.

STEP 1 Collect 4 sheets of paper and layer them about 1 cm apart vertically. Keep the edges level.

STEP 2 Fold up the bottom edges of the paper to form 8 equal tabs.

STEP 3 Fold the papers and crease well to hold the tabs in place. Staple along the fold. Label each tab as indicated below.

Sequence Turn your Foldable so the staples are at the top. Label the tabs, in order from top to bottom, *Electromagnetic Spectrum, Radio Waves, Microwaves, Infrared Rays, Visible Light, Ultraviolet Light, X Rays,* and *Gamma Rays.* As you read, write facts you learn about each topic under the appropriate tab.

Preview this chapter's content and activities at
ips.msscience.com

Get Ready to Read

Visualize

① Learn It! Visualize by forming mental images of the text as you read. Imagine how the text descriptions look, sound, feel, smell, or taste. Look for any pictures or diagrams on the page that may help you add to your understanding.

② Practice It! Read the following paragraph. As you read, use the underlined details to form a picture in your mind.

> A radar station sends out radio waves that bounce off an object, such as an airplane. Electronic equipment measures the time it takes for the radio waves to travel to the plane, be reflected, and return. Because the speed of the radio waves is known, the distance to the airplane can be determined from the measured time.
>
> — *from page 527*

Based on the description above, try to visualiz how radar is used. Now look at the diagram on page 527.
- How closely does it match your mental picture?
- Reread the passage and look at the diagram again. Did your ideas change?
- Compare your image with what others in your class visualized.

③ Apply It! Read the chapter and list three subjects you were able to visualize. Make a rough sketch showing what you visualized.

Reading Tip

Forming your own mental images will help you remember what you read.

Target Your Reading

Use this to focus on the main ideas as you read the chapter.

1) Before you read the chapter, respond to the statements below on your worksheet or on a numbered sheet of paper.
- Write an **A** if you **agree** with the statement.
- Write a **D** if you **disagree** with the statement.

2) After you read the chapter, look back to this page to see if you've changed your mind about any of the statements.
- If any of your answers changed, explain why.
- Change any false statements into true statements.
- Use your revised statements as a study guide.

Science Online
Print out a worksheet of this page at ips.msscience.com

Before You Read A or D		Statement	After You Read A or D
	1	An electromagnetic wave is a mechanical wave.	
	2	An electromagnetic wave is produced by a moving particle.	
	3	A moving electric charge is surrounded by an electric field and a magnetic field.	
	4	All electromagnetic waves travel at the same speed in empty space.	
	5	Radio waves have the highest frequencies in the electromagnetic spectrum.	
	6	The Sun emits mostly ultraviolet waves.	
	7	Most telecommunication devices use radio waves to transmit information.	
	8	A loudspeaker converts sound waves into electromagnetic waves.	
	9	Radio waves can travel through Earth.	

SECTION 1 The Nature of Electromagnetic Waves **520 B**

section 1

The Nature of Electromagnetic Waves

as you read

What You'll Learn
- **Explain** how electromagnetic waves are produced.
- **Describe** the properties of electromagnetic waves.

Why It's Important
The energy Earth receives from the Sun is carried by electromagnetic waves.

Review Vocabulary
wave: a rhythmic disturbance that carries energy through matter or space

New Vocabulary
- electromagnetic wave
- radiant energy

Waves in Space

On a clear day you feel the warmth in the Sun's rays, and you see the brightness of its light. Energy is being transferred from the Sun to your skin and eyes. Who would guess that the way in which this energy is transferred has anything to do with radios, televisions, microwave ovens, or the X-ray pictures that are taken by a doctor or dentist? Yet the Sun and the devices shown in **Figure 1** use the same type of wave to move energy from place to place.

Transferring Energy A wave transfers energy from one place to another without transferring matter. How do waves transfer energy? Waves, such as water waves and sound waves, transfer energy by making particles of matter move. The energy is passed along from particle to particle as they collide with their neighbors. Mechanical waves are the types of waves that use matter to transfer energy.

However, mechanical waves can't travel in the almost empty space between Earth and the Sun. So how can a wave transfer energy from the Sun to Earth? A different type of wave, called an electromagnetic wave, carries energy from the Sun to Earth. An **electromagnetic wave** is a wave that can travel through empty space or through matter and is produced by charged particles that are in motion.

Figure 1 Getting a dental X ray or talking on a cell phone uses energy carried by electromagnetic waves.

520 CHAPTER 18 Electromagnetic Waves

Figure 2 A gravitational field surrounds all objects. When a ball is thrown, Earth's gravitational field exerts a downward force on the ball at every point along the ball's path.

Forces and Fields

An electromagnetic wave is made of two parts—an electric field and a magnetic field. These fields are force fields. A force field enables an object to exert forces on other objects, even though they are not touching. Earth is surrounded by a force field called the gravitational field. This field exerts the force of gravity on all objects that have mass.

Reading Check *What force field surrounds Earth?*

How does Earth's force field work? If you throw a ball in the air as high as you can, it always falls back to Earth. At every point along the ball's path, the force of gravity pulls down on the ball, as shown in **Figure 2.** In fact, at every point in space above or at Earth's surface, a ball is acted on by a downward force exerted by Earth's gravitational field. The force exerted by this field on a ball could be represented by a downward arrow at any point in space. The diagram above shows this force field that surrounds Earth and extends out into space. It is Earth's gravitational field that causes the Moon to orbit Earth.

Magnetic Fields You know that magnets repel and attract each other even when they aren't touching. Two magnets exert a force on each other when they are some distance apart because each magnet is surrounded by a force field called a magnetic field. Just as a gravitational field exerts a force on a mass, a magnetic field exerts a force on another magnet and on magnetic materials. Magnetic fields cause other magnets to line up along the direction of the magnetic field.

Earth's gravitational field extends out through space, exerting a force on all masses.
Determine *whether the forces exerted by Earth's gravitational field are attractive or repulsive.*

Topic: Force Fields
Visit to ips.msscience.com for Web links to information about Earth's gravitational and magnetic force fields.

Activity Write a paragraph comparing and contrasting the two force fields.

SECTION 1 The Nature of Electromagnetic Waves **521**

Figure 3 Force fields surround all magnets and electric charges.

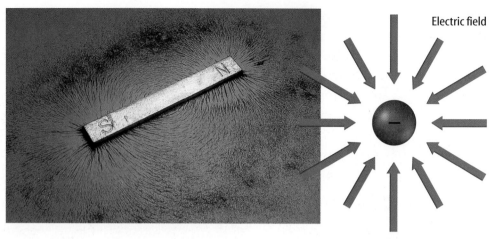

A magnetic field surrounds all magnets. The magnetic field exerts a force on iron filings, causing them to line up with the field.

The electric field around an electric charge extends out through space, exerting forces on other charged particles.

Electric Fields Recall that atoms contain protons, neutrons, and electrons. Protons and electrons have a property called electric charge. The two types of electric charge are positive and negative. Protons have positive charge and electrons have negative charge.

Just as a magnet is surrounded by a magnetic field, a particle that has electric charge, such as a proton or an electron, is surrounded by an electric field, as shown in **Figure 3.** The electric field is a force field that exerts a force on all other charged particles that are in the field.

Making Electromagnetic Waves

An electromagnetic wave is made of electric and magnetic fields. How is such a wave produced? Think about a wave on a rope. You can make a wave on a rope by shaking one end of the rope up and down. Electromagnetic waves are produced by charged particles, such as electrons, that move back and forth or vibrate.

A charged particle always is surrounded by an electric field. But a charged particle that is moving also is surrounded by a magnetic field. For example, electrons are flowing in a wire that carries an electric current. As a result, the wire is surrounded by a magnetic field, as shown in **Figure 4.** So a moving charged particle is surrounded by an electric field and a magnetic field.

Figure 4 Electrons moving in a wire produce a magnetic field in the surrounding space. This field causes iron filings to line up with the field.

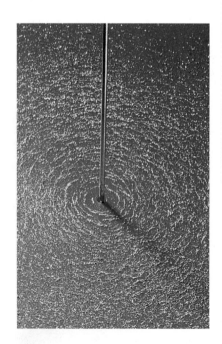

Producing Waves When you shake a rope up and down, you produce a wave that moves away from your hand. As a charged particle vibrates by moving up and down or back and forth, it produces changing electric and magnetic fields that move away from the vibrating charge in many directions. These changing fields traveling in many directions form an electromagnetic wave. **Figure 5** shows how the electric and magnetic fields change as they move along one direction.

Properties of Electromagnetic Waves

Like all waves, an electromagnetic wave has a frequency and a wavelength. You can create a wave on a rope when you move your hand up and down while holding the rope. Look at **Figure 5**. Frequency is how many times you move the rope through one complete up and down cycle in 1 s. Wavelength is the distance from one crest to the next or from one trough to the next.

Wavelength and Frequency An electromagnetic wave is produced by a vibrating charged particle. When the charge makes one complete vibration, one wavelength is created, as shown in **Figure 5**. Like a wave on a rope, the frequency of an electromagnetic wave is the number of wavelengths that pass by a point in 1 s. This is the same as the number of times in 1 s that the charged particle makes one complete vibration.

Mini LAB

Observing Electric Fields

Procedure
1. Rub a **hard, plastic comb** vigorously with a **wool sweater** or **wool flannel shirt**.
2. Turn on a **water faucet** to create the smallest possible continuous stream of water.
3. Hold the comb near the stream of water and observe.

Analysis
1. What happened to the stream of water when you held the comb near it?
2. Explain why the stream of water behaved this way.

Try at Home

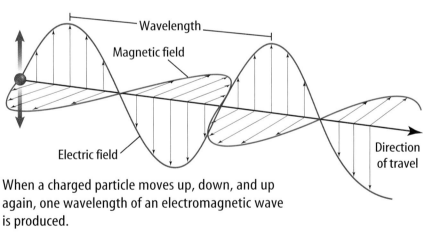

When a charged particle moves up, down, and up again, one wavelength of an electromagnetic wave is produced.

Figure 5 The vibrating motion of an electric charge produces an electromagnetic wave. One complete cycle of vibration produces one wavelength of a wave. **Determine** *the magnetic field when the electric field is zero.*

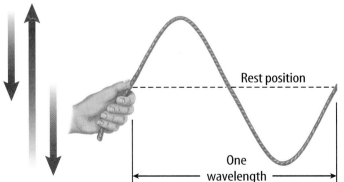

By shaking the end of a rope down, up, and down again, you make one wavelength.

SECTION 1 The Nature of Electromagnetic Waves

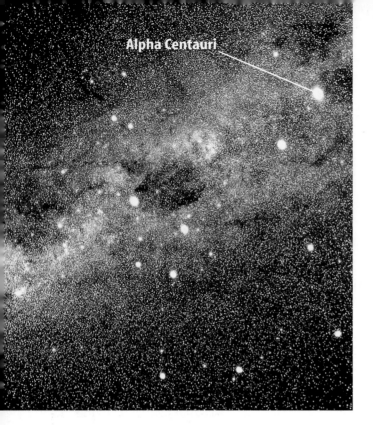

Figure 6 The light that reaches Earth today from Alpha Centauri left the star more than four years ago.

Radiant Energy The energy carried by an electromagnetic wave is called **radiant energy.** What happens if an electromagnetic wave strikes a charged particle? The electric field part of the wave exerts a force on this particle and causes it to move. Some of the radiant energy carried by the wave is transferred into the energy of motion of the particle.

Reading Check *What is radiant energy?*

The amount of energy that an electromagnetic wave carries is determined by the wave's frequency. The higher the frequency of the electromagnetic wave, the more energy it has.

The Speed of Light All electromagnetic waves travel through space at the same speed—about 300,000 km/s. This speed sometimes is called the speed of light. Even though light travels incredibly fast, stars other than the Sun are so far away that it takes years for the light they emit to reach Earth. **Figure 6** shows Alpha Centauri, one of the closest stars to our solar system. This star is more than 40 trillion km from Earth.

section 1 review

Summary

Force Fields
- A charged particle is surrounded by an electric field that exerts forces on other charged particles.
- A magnet is surrounded by a magnetic field that exerts a force on other magnets.
- A moving charged particle is surrounded by electric and magnetic fields.

Electromagnetic Waves
- The changing electric and magnetic fields made by a vibrating electric charge form an electromagnetic wave.
- Electromagnetic waves carry radiant energy.
- All electromagnetic waves travel at the speed of light, which is about 300,000 km/s in empty space.

Self Check

1. **Describe** how electromagnetic waves are produced.
2. **Compare** the energy carried by high-frequency and low-frequency electromagnetic waves.
3. **Identify** what determines the frequency of an electromagnetic wave.
4. **Compare and contrast** electromagnetic waves with mechanical waves.
5. **Think Critically** Unlike sound waves, electromagnetic waves can travel in empty space. What evidence supports this statement?

Applying Math

6. **Use Ratios** To go from Earth to Mars, light waves take four min and a spacecraft takes four months. To go to the nearest star, light takes four years. How long would it take the spacecraft to go to the nearest star?

section 2
The Electromagnetic Spectrum

Electromagnetic Waves

The room you are sitting in is bathed in a sea of electromagnetic waves. These electromagnetic waves have a wide range of wavelengths and frequencies. For example, TV and radio stations broadcast electromagnetic waves that pass through walls and windows. These waves have wavelengths from about 1 m to over 500 m. Light waves that you see are electromagnetic waves that have wavelengths more than a million times shorter than the waves broadcast by radio stations.

Classifying Electromagnetic Waves The wide range of electromagnetic waves with different frequencies and wavelengths forms the **electromagnetic spectrum.** The electromagnetic spectrum is divided into different parts. **Figure 7** shows the electromagnetic spectrum and the names given to the electromagnetic waves in different parts of the spectrum. Even though electromagnetic waves have different names, they all travel at the same speed in empty space—the speed of light. Remember that for waves that travel at the same speed, the frequency increases as the wavelength decreases. So as the frequency of electromagnetic waves increases, their wavelength decreases.

Figure 7 The electromagnetic spectrum consists of electromagnetic waves arranged in order of increasing frequency and decreasing wavelength.

as you read

What You'll Learn
- **Explain** differences among kinds of electromagnetic waves.
- **Identify** uses for different kinds of electromagnetic waves.

Why It's Important
Electromagnetic waves are used to cook food, to send and receive information, and to diagnose medical problems.

Review Vocabulary
spectrum: a continuous series of waves arranged in order of increasing or decreasing wavelength or frequency

New Vocabulary
- electromagnetic spectrum
- radio wave
- infrared wave
- visible light
- ultraviolet radiation
- X ray
- gamma ray

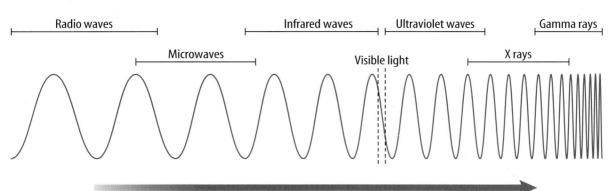

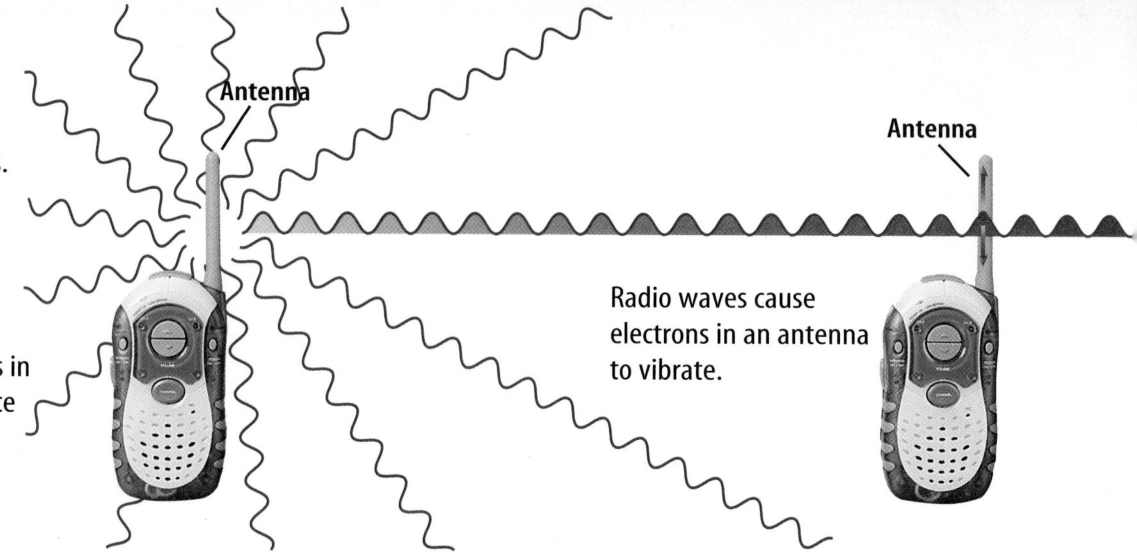

Figure 8 Antennas are used to generate and detect radio waves. **Describe** *some objects that have antennas.*

Vibrating electrons in an antenna produce radio waves.

Radio waves cause electrons in an antenna to vibrate.

Radio Waves

Electromagnetic waves with wavelengths longer than about 0.001 m are called radio waves. **Radio waves** have the lowest frequencies of all the electromagnetic waves and carry the least energy. Television signals, as well as AM and FM radio signals, are types of radio waves. Like all electromagnetic waves, radio waves are produced by moving charged particles. One way to make radio waves is to make electrons vibrate in a piece of metal, as shown in **Figure 8**. This piece of metal is called an antenna. By changing the rate at which the electrons vibrate, radio waves of different frequencies can be produced that travel outward from the antenna.

Detecting Radio Waves These radio waves can cause electrons in another piece of metal, such as another antenna, to vibrate, as shown in **Figure 8**. As the electrons in the receiving antenna vibrate, they form an alternating current. This alternating current can be used to produce a picture on a TV screen and sound from a loudspeaker. Varying the frequency of the radio waves broadcast by the transmitting antenna changes the alternating current in the receiving antenna. This produces the different pictures you see and sounds you hear on your TV.

Microwaves Radio waves with wavelengths between about 0.3 m and 0.001 m are called microwaves. They have a higher frequency and a shorter wavelength than the waves that are used in your home radio. Microwaves are used to transmit some phone calls, especially from cellular and portable phones. **Figure 9** shows a microwave tower.

Microwave ovens use microwaves to heat food. Microwaves produced inside a microwave oven cause water molecules in your food to vibrate faster, which makes the food warmer.

Figure 9 Towers such as the one shown here are used to send and receive microwaves.

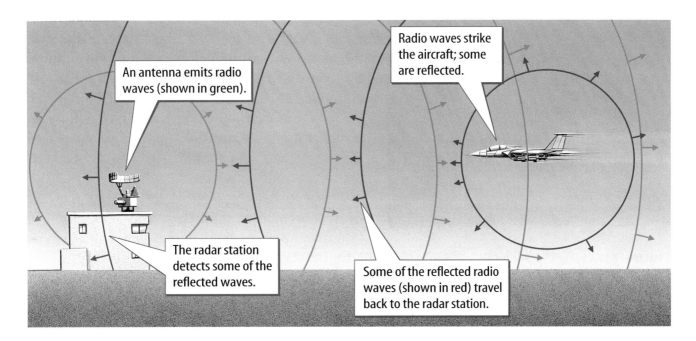

Radar You might be familiar with echolocation, in which sound waves are reflected off an object to determine its size and location. Some bats and dolphins use echolocation to navigate and hunt. Radar, an acronym for RAdio Detecting And Ranging, uses electromagnetic waves to detect objects in the same way. Radar was first used during World War II to detect and warn of incoming enemy aircraft.

Reading Check *What does radar do?*

A radar station sends out radio waves that bounce off an object such as an airplane. Electronic equipment measures the time it takes for the radio waves to travel to the plane, be reflected, and return. Because the speed of the radio waves is known, the distance to the airplane can be determined from the measured time.

An example of radar being used is shown in **Figure 10.** Because electromagnetic waves travel so quickly, the entire process takes only a fraction of a second.

Infrared Waves

You might know from experience that when you stand near the glowing coals of a barbecue or the red embers of a campfire, your skin senses the heat and becomes warm. Your skin may also feel warm near a hot object that is not glowing. The heat you are sensing with your skin is from electromagnetic waves. These electromagnetic waves are called **infrared waves** and have wavelengths between about one thousandth and 0.7 millionths of a meter.

Figure 10 Radar stations use radio waves to determine direction, distance, and speed of aircraft.

Observing the Focusing of Infrared Rays

Procedure
1. Place a **concave mirror** 2 m to 3 m away from an **electric heater.** Turn on the heater.
2. Place the palm of your hand in front of the mirror and move it back until you feel heat on your palm. Note the location of the warm area.
3. Move the heater to a new location. How does the warm area move?

Analysis
1. Did you observe the warm area? Where?
2. Compare the location of the warm area to the location of the mirror.

SECTION 2 The Electromagnetic Spectrum

Figure 11 A pit viper hunting in the dark can detect the infrared waves emitted from the warm body of its prey.

Detecting Infrared Waves Electromagnetic waves are emitted by every object. In any material, the atoms and molecules are in constant motion. Electrons in the atoms and molecules also are vibrating, and so they emit electromagnetic waves. Most of the electromagnetic waves given off by an object at room temperature are infrared waves and have a wavelength of about 0.000 01 m, or one hundred thousandth of a meter.

Infrared detectors can detect objects that are warmer or cooler than their surroundings. For example, areas covered with vegetation, such as forests, tend to be cooler than their surroundings. Using infrared detectors on satellites, the areas covered by forests and other vegetation, as well as water, rock, and soil, can be mapped. Some types of night vision devices use infrared detectors that enable objects to be seen in nearly total darkness.

Animals and Infrared Waves Some animals also can detect infrared waves. Snakes called pit vipers, such as the one shown in **Figure 11,** have a pit located between the nostril and the eye that detects infrared waves. Rattlesnakes, copperheads, and water moccasins are pit vipers. These pits help pit vipers hunt at night by detecting the infrared waves their prey emits.

Visible Light

As the temperature of an object increases, the atoms and molecules in the object move faster. The electrons also vibrate faster, and produce electromagnetic waves of higher frequency and shorter wavelength. If the temperature is high enough, the object might glow, as in **Figure 12.** Some of the electromagnetic waves that the hot object is emitting are now detectable with your eyes. Electromagnetic waves you can detect with your eyes are called **visible light.** Visible light has wavelengths between about 0.7 and 0.4 millionths of a meter. What you see as different colors are electromagnetic waves of different wavelengths. Red light has the longest wavelength (lowest frequency), and blue light has the shortest wavelength (highest frequency).

Most objects that you see do not give off visible light. They simply reflect the visible light that is emitted by a source of light, such as the Sun or a lightbulb.

Figure 12 When objects are heated, their electrons vibrate faster. When the temperature is high enough, the vibrating electrons will emit visible light.
Describe *an object that emits visible light when heated.*

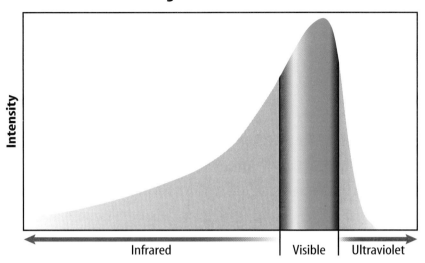

Figure 13 The Sun emits mainly infrared waves and visible light. Only about 8 percent of the electromagnetic waves emitted by the Sun are ultraviolet radiation. **Identify** *the electromagnetic waves emitted by the Sun that have the highest intensity.*

Ultraviolet Radiation

Ultraviolet radiation is higher in frequency than visible light and has even shorter wavelengths—between 0.4 millionths of a meter and about ten billionths of a meter. Ultraviolet radiation has higher frequencies than visible light and carries more energy. The radiant energy carried by an ultraviolet wave can be enough to damage the large, fragile molecules that make up living cells. Too much ultraviolet radiation can damage or kill healthy cells.

Figure 13 shows the intensity of electromagnetic waves emitted by the Sun. Too much exposure to the Sun's ultraviolet waves can cause sunburn. Exposure to these waves over a long period of time can lead to early aging of the skin and possibly skin cancer. You can reduce the amount of ultraviolet radiation you receive by wearing sunglasses and sunscreen, and staying out of the Sun when it is most intense.

Reading Check *Why can too much exposure to the Sun be harmful?*

Beneficial Uses of UV Radiation A few minutes of exposure each day to ultraviolet radiation from the Sun enables your body to produce the vitamin D it needs. Most people receive that amount during normal activity. The body's natural defense against too much ultraviolet radiation is to tan. However, a tan can be a sign that overexposure to ultraviolet radiation has occurred.

Because ultraviolet radiation can kill cells, it is used to disinfect surgical equipment in hospitals. In some chemistry labs, ultraviolet rays are used to sterilize goggles, as shown in **Figure 14.**

Figure 14 Sterilizing devices, such as this goggle sterilizer, use ultraviolet waves to kill organisms on the equipment.

SECTION 2 The Electromagnetic Spectrum

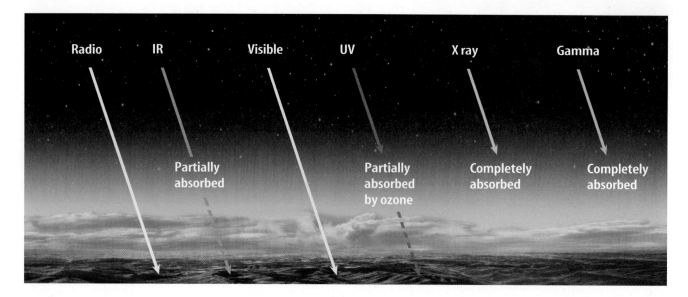

Figure 15 Earth's atmosphere serves as a shield to block some types of electromagnetic waves from reaching Earth's surface.

The Ozone Layer Much of the ultraviolet radiation arriving at Earth is absorbed in the upper atmosphere by ozone, as shown in **Figure 15.** Ozone is a molecule that has three oxygen atoms and is formed high in Earth's atmosphere.

Chemical compounds called CFCs, which are used in air conditioners and refrigerators, can react with ozone molecules and break them apart. There is evidence that these reactions play a role in forming the seasonal reduction in ozone over Antarctica, known as the ozone hole. To prevent this, the use of CFC's is being phased out.

Ultraviolet radiation is not the only type of electromagnetic wave absorbed by Earth's atmosphere. Higher energy waves of X rays and gamma rays also are absorbed. The atmosphere is transparent to radio waves and visible light and partially transparent to infrared waves.

X Rays and Gamma Rays

Ultraviolet rays can penetrate the top layer of your skin. **X rays,** with an even higher frequency than ultraviolet rays, have enough energy to go right through skin and muscle. A shield made from a dense metal, such as lead, is required to stop X rays.

Gamma rays have the highest frequency and, therefore, carry the most energy. Gamma rays are the hardest to stop. They are produced by changes in the nuclei of atoms. When protons and neutrons bond together in nuclear fusion or break apart from each other in nuclear fission, enormous quantities of energy are released. Some of this energy is released as gamma rays.

Just as too much ultraviolet radiation can hurt or kill cells, too much X-ray or gamma radiation can have the same effect. Because the energy of X rays and gamma rays is greater, the exposure that is needed to cause damage is much less.

Body Temperature
Warm-blooded animals, such as mammals, produce their own body heat. Cold-blooded animals, such as reptiles, absorb heat from the environment. Brainstorm the possible advantages of being either warm-blooded or cold-blooded. Which animals would be easier for a pit viper to detect?

Using High-Energy Electromagnetic Radiation The fact that X rays can pass through the human body makes them useful for medical diagnosis, as shown in **Figure 16.** X rays pass through the less dense tissues in skin and other organs. These X rays strike a film, creating a shadow image of the denser tissues. X-ray images help doctors detect injuries and diseases, such as broken bones and cancer. A CT scanner uses X rays to produce images of the human body as if it had been sliced like a loaf of bread.

Although the radiation received from getting one medical or dental X ray is not harmful, the cumulative effect of numerous X rays can be dangerous. The operator of the X-ray machine usually stands behind a shield to avoid being exposed to X rays. Lead shields or aprons are used to protect the parts of the patient's body that are not receiving the X rays.

Using Gamma Rays Although gamma rays are dangerous, they also have beneficial uses, just as X rays do. A beam of gamma rays focused on a cancerous tumor can kill the tumor. Gamma radiation also can kill disease-causing bacteria in food. More than 1,000 Americans die each year from *Salmonella* bacteria in poultry and *E. coli* bacteria in meat. Although gamma radiation has been used since 1963 to kill bacteria in food, this method is not widely used in the food industry.

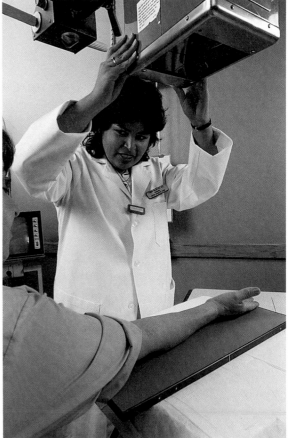

Figure 16 Dense tissues such as bone absorb more X rays than softer tissues do. Consequently, dense tissues leave a shadow on an X ray film that can be used to diagnose medical and dental conditions.

Astronomy with Different Wavelengths

Some astronomical objects produce no visible light and can be detected only through the infrared and radio waves they emit. Some galaxies emit X rays from regions that do not emit visible light. Studying stars and galaxies like these using only visible light would be like looking at only one color in a picture. **Figure 17** shows how different electromagnetic waves can be used to study the Sun.

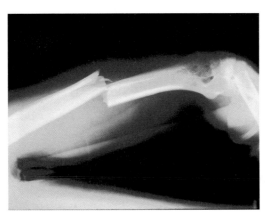

NATIONAL GEOGRAPHIC VISUALIZING THE UNIVERSE

Figure 17

For centuries, astronomers studied the universe using only the visible light coming from planets, moons, and stars. But many objects in space also emit X rays, ultraviolet and infrared radiation, and radio waves. Scientists now use telescopes that can detect these different types of electromagnetic waves. As these images of the Sun reveal, the new tools are providing more information of objects in the universe.

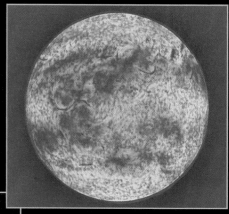

▲ **INFRARED RADIATION** An infrared telescope reveals that the Sun's surface temperature is not uniform. Some areas are hotter than others.

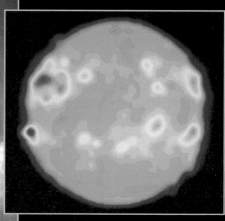

▲ **RADIO WAVES** Radio telescopes detect radio waves given off by the Sun, which have much longer wavelengths than visible light.

▲ **X RAYS** X-ray telescopes can detect the high-energy, short-wavelength X rays produced by the extreme temperatures in the Sun's outer atmosphere.

▶ **ULTRAVIOLET RADIATION** Telescopes sensitive to ultraviolet radiation—electromagnetic waves with shorter wavelengths than visible light—can "see" the Sun's outer atmosphere.

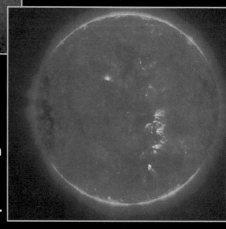

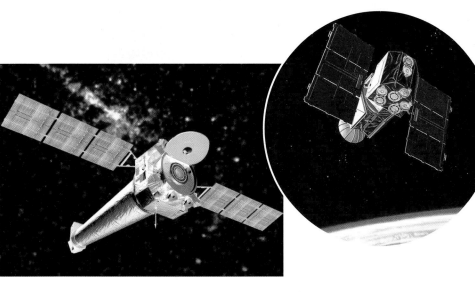

 **Satellite Observations** Recall from **Figure 15** that Earth's atmosphere blocks X rays, gamma rays, most ultraviolet rays, and some infrared rays. However, telescopes in orbit above Earth's atmosphere can detect the electromagnetic waves that can't pass through the atmosphere. **Figure 18** shows three such satellites—the Extreme Ultraviolet Explorer (EUVE), the Chandra X-Ray Observatory, and the Infrared Space Observatory (ISO).

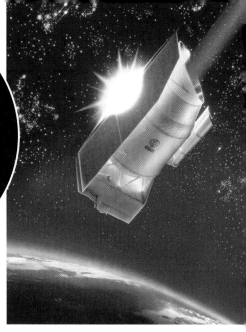

Figure 18 Launching satellite observatories above Earth's atmosphere is the only way to see the universe at electromagnetic wavelengths that are absorbed by Earth's atmosphere.

section 2 review

Summary

Radio Waves
- Radio waves have wavelengths longer than about 0.3 m.

Infrared Waves and Visible Light
- Infrared waves have wavelengths between about one thousandth and 0.7 millionths of a meter.
- The wavelengths of infrared waves emitted by an object get shorter as the object's temperature increases.
- Visible light waves have wavelengths between about 0.7 and 0.4 millionths of a meter.

Ultraviolet Waves, X Rays, and Gamma Rays
- Ultraviolet radiation has wavelengths between about 0.4 millionths of a meter and 10 billionths of a meter.
- Prolonged exposure to ultraviolet waves from the Sun can cause skin damage.
- X rays and gamma rays are the most energetic electromagnetic waves.

Self Check

1. **Explain** why ultraviolet radiation is more damaging to living cells than infrared waves.
2. **Compare and contrast** X rays and gamma rays.
3. **Describe** how infrared detectors on satellites can be used to obtain information about the location of vegetation on Earth's surface.
4. **Explain** why X rays and gamma rays coming from space do not reach Earth's surface.
5. **Explain** how the energy of electromagnetic waves change as the wavelength of the waves increase.
6. **Think Critically** Why does the Sun emit mostly infrared waves and visible light, and Earth emits infrared waves?

Applying Skills

7. **Make a table** listing five objects in your home that produce electromagnetic waves. In another column, list next to each object the type of electromagnetic wave or waves produced. In a third column describe each object's use.

SECTION 2 The Electromagnetic Spectrum

Prisms Of Light

Do you know what light is? Many would answer that light is what you turn on to see at night. However, white light is made of many different frequencies of the electromagnetic spectrum. A prism can separate white light into its different frequencies. You see different frequencies of light as different colors. What colors do you see when light passes through a prism?

Real-World Question

What happens to visible light as it passes through a prism?

Goals
- **Construct** a prism and observe the different colors that are produced.
- **Infer** how the bending of light waves depends on their wavelength.

Materials
microscope slides (3)   flashlight
transparent tape   water
clay   *prism

*Alternate materials

Safety Precautions

Procedure

1. Carefully tape the three slides together on their long sides so they form a long prism.
2. Place one end of the prism into a softened piece of clay so the prism is standing upright.
3. Fill the prism with water and put it on a table that is against a dark wall.
4. Shine a flashlight beam through the prism so the light becomes visible on the wall.

Conclude and Apply

1. **List** the order of the colors you saw on the wall.
2. **Describe** how the position of the colors on the wall changes as you change the direction of the flashlight beam.
3. **Describe** how the order of colors on the wall changes as you change the direction of the flashlight beam.
4. **Infer** which color light waves have changed direction, or have been bent, the most after passing through the prism. Which color has been bent the least?
5. **Infer** how the bending of a light wave depends on its wavelength.

Communicating Your Data

Compare your conclusions with those of other students in your class. **For more help, refer to the** Science Skill Handbook.

534   CHAPTER 18 Electromagnetic Waves

section 3
Using Electromagnetic Waves

Telecommunications

In the past week, have you spoken on the phone, watched television, done research on the Internet, or listened to the radio? Today you can talk to someone far away or transmit and receive information over long distances almost instantly. Thanks to telecommunications, the world is becoming increasingly connected through the use of electromagnetic waves.

Using Radio Waves

Radio waves usually are used to send and receive information over long distances. Using radio waves to communicate has several advantages. For example, radio waves pass through walls and windows easily. Radio waves do not interact with humans, so they are not harmful to people like ultraviolet rays or X rays are. So most telecommunication devices, such as TVs, radios, and telephones, use radio waves to transmit information such as images and sounds. **Figure 19** shows how radio waves can be used to transmit information—in this case transmitting information that enables sounds to be reproduced at a location far away.

as you read

What You'll Learn
- **Describe** different ways of using electromagnetic waves to communicate.
- **Compare and contrast** AM and FM radio signals.

Why It's Important
Using elecromagnetic waves to communicate enables you to contact others worldwide.

Review Vocabulary
satellite: a natural or artificial object that orbits a planet

New Vocabulary
- carrier wave
- Global Positioning System

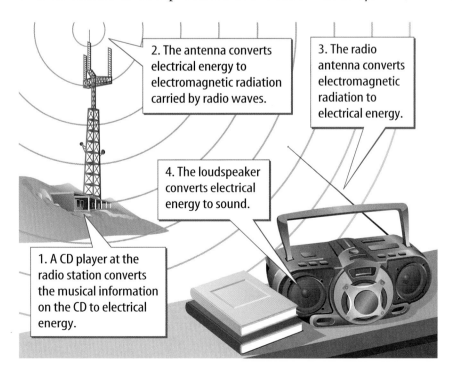

Figure 19 Radio waves are used to transmit information that can be converted to other forms of energy, such as electrical energy and sound.

535

Figure 20 A signal can be carried by a carrier wave in two ways—amplitude modulation or frequency modulation.

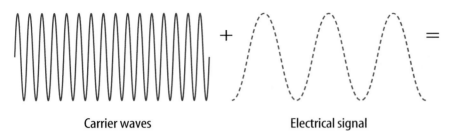
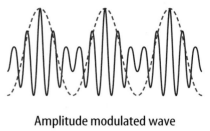
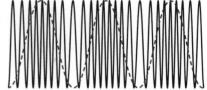

Pulsars and Little Green Men A type of collapsed star called a pulsar emits pulses of radio waves at extremely regular intervals. Pulsars were first discovered by Jocelyn Bell-Burnell and Anthony Hewish in 1967. Puzzled by a regular sequence of radio pulses they detected, they considered the possibility that the pulses might be coming from an alien civilization. They jokingly labeled the pulses LGMs, for "little green men." Soon other signals were detected that proved the pulses were coming from collapsed stars. Research the role Jocelyn Bell-Burnell played in the discovery of pulsars.

Radio Transmission How is information, such as images or sounds, broadcast by radio waves? Each radio and television station is assigned a particular frequency at which it broadcasts radio waves. The radio waves broadcast by a station at its assigned frequency are the **carrier waves** for that station. To listen to a station you tune your radio or television to the frequency of the station's carrier waves. To carry information on the carrier wave, either the amplitude or the frequency of the carrier wave is changed, or modulated.

Amplitude Modulation The letters *AM* in AM radio stand for amplitude modulation, which means that the amplitude of the carrier wave is changed to transmit information. The original sound is transformed into an electrical signal that is used to vary the amplitude of the carrier wave, as shown in **Figure 20.** Note that the frequency of the carrier wave doesn't change—only the amplitude changes. An AM receiver tunes to the frequency of the carrier wave. In the receiver, the varying amplitude of the carrier waves produces an electrical signal. The radio's loudspeaker uses this electric signal to produce the original sound.

Frequency Modulation FM radio works in much the same way as AM radio, but the frequency instead of the amplitude is modulated, as shown in **Figure 20.** An FM receiver contains electronic components that use the varying frequency of the carrier wave to produce an electric signal. As in an AM radio, this electric signal is converted into sound waves by a loudspeaker.

 *What is frequency modulation?*

Telephones

A telephone contains a microphone in the mouthpiece that converts a sound wave into an electric signal. The electric signal is carried through a wire to the telephone switching systems. There, the signal might be sent through other wires or be converted into a radio or microwave signal for transmission through the air. The electric signal also can be converted into a light wave for transmission through fiber-optic cables.

At the receiving end, the signal is converted back to an electric signal. A speaker in the earpiece of the phone changes the electric signal into a sound wave.

Reading Check *What device converts sound into an electric signal?*

Applying Math — Solve a Simple Equation

WAVELENGTH OF AN FM STATION You are listening to an FM radio station with a frequency of 94.9 MHz, which equals 94,900,000 Hz. What is the wavelength of these radio waves. Use the wave speed equation $v = \lambda f$, and assume the waves travel at the speed of light, 300,000.0 km/s.

Solution

1 *This is what you know:*
- frequency: $f = 94{,}900{,}000$ Hz
- wave speed: $v = 300{,}000.0$ km/s

2 *This is what you need to find:*
- wavelength: $\lambda = ?$ m

3 *This is the procedure you need to use:*

Solve the wave equation for wavelength, λ, by dividing each side by the frequency, f. Then substitute the known values for frequency and wave speed into the equation you derived:

$$\lambda = \frac{v}{f} = \frac{300{,}000.0 \text{ km/s}}{94{,}900{,}000 \text{ Hz}} = \frac{300{,}000.0 \text{ km } 1/s}{94{,}900{,}000 \text{ } 1/s}$$

$$= 0.00316 \text{ km} = 0.00316 \text{ km} \times (1{,}000 \text{ m/km})$$

$$= 3.16 \text{ m}$$

4 *Check your answer:*

Multiply your answer by the given frequency. The result should be the given wave speed.

Practice Problems

1. Your friend is listening to an AM station with a frequency of 1,520 kHz. What is the wavelength of these radio waves?

2. What is the frequency of the radio waves broadcast by an AM station if the wave length of the radio waves is 500.0 m?

For more practice, visit
ips.msscience.com/
math_practice

Figure 21 Cordless and cell phones use radio waves to communicate between a mobile phone and a base station.

A A cordless phone can be used more than 0.5 km from its base station.

B Cell phones communicate with a base station that can be several kilometers away, or more.

Remote Phones A telephone does not have to transmit its signal through wires. In a cordless phone, the electrical signal produced by the microphone is transmitted through an antenna in the phone to the base station. **Figure 21A** shows how incoming signals are transmitted from the base station to the phone. A cellular phone communicates with a base station that can be many kilometers away. The base station uses a large antenna, as shown in **Figure 21B,** to communicate with the cell phone and with other base stations in the cell phone network.

Pagers The base station also is used in a pager system. When you dial a pager, the signal is sent to a base station. From there, an electromagnetic signal is sent to the pager. The pager beeps or vibrates to indicate that someone has called. With a touch-tone phone, you can transmit numeric information, such as your phone number, which the pager will receive and display.

Communications Satellites

How do you send information to the other side of the world? Radio waves can't be sent directly through Earth. Instead, radio signals are sent to satellites. The satellites can communicate with other satellites or with ground stations. Some communications satellites are in geosynchronous orbit, meaning each satellite remains above the same point on the ground.

Topic: Satellite Communications
Visit ips.msscience.com for Web links to information about how satellites are used in around-the-world communications.

Activity Create a table listing satellites from several countries, their names and their communications function.

The Global Positioning System

Satellites also are used as part of the **Global Positioning System,** or GPS. GPS is used to locate objects on Earth. The system consists of satellites, ground-based stations, and portable units with receivers, as illustrated in **Figure 22.**

A GPS receiver measures the time it takes for radio waves to travel from several satellites to the receiver. This determines the distance to each satellite. The receiver then uses this information to calculate its latitude, longitude, and elevation. The accuracy of GPS receivers ranges from a few hundred meters for handheld units, to several centimeters for units that are used to measure the movements of Earth's crust.

Figure 22 The signals broadcast by GPS satellites enable portable, handheld receivers to determine the position of an object or person.

section 3 review

Summary

Using Radio Waves
- Radio waves are used for communication because they can pass through most objects.
- Amplitude modulation transmits information by modifying the amplitude of a carrier wave.
- Frequency modulation transmits information by modifying the frequency of a carrier wave.

Cordless Phones and Cell Phones
- Cordless phones use radio waves to transmit signals between the base and the handset.
- Cellular phones use radio waves to transmit signals between the phone and cell phone radio towers.

Communications Satellites
- Communications satellites in geosynchronous orbits relay radio signals from one part of the world to another.
- The Global Positioning System uses radio waves to enable a user to accurately determine their position on Earth's surface.

Self Check

1. **Describe** how a cordless phone is different from a cell phone.
2. **Explain** how a communications satellite is used.
3. **Describe** the types of information a GPS receiver provides.
4. **Describe** how an AM radio signal is used to transmit information.
5. **Think Critically** Explain why ultraviolet waves are not used to transmit signals to and from communications satellites.

Applying Skills

6. **Make an events chain** showing the sequence of energy transformations that occur when live music is broadcast by a radio station and played by a radio.
7. **Make a Diagram** showing how geosynchronous satellites and ground stations could be used to send information from you to someone on the other side of Earth.

Design Your Own

Spectrum Inspection

Goals
- **Design** an experiment that determines the relationship between brightness and the wavelengths emitted by a lightbulb.
- **Observe** the wavelengths of light emitted by a lightbulb as its brightness changes.

Possible Materials
diffraction grating
power supply with variable resistor switch
clear, tubular lightbulb and socket
red, yellow, and blue colored pencils

Safety Precautions

WARNING: *Be sure all electrical cords and connections are intact and that you have a dry working area. Do not touch the bulbs as they may be hot.*

● Real-World Question

You've heard the term "red-hot" used to describe something that is unusually hot. When a piece of metal is heated it may give off a red glow or even a yellow glow. All objects emit electromagnetic waves. How do the wavelengths of these waves depend on the temperature of the object?

● Form a Hypothesis

The brightness of a lightbulb increases as its temperature increases. Form a hypothesis describing how the wavelengths emitted by a lightbulb will change as the brightness of a lightbulb changes.

● Test Your Hypothesis

Make a Plan

1. **Decide** how you will determine the effect of lightbulb brightness on the colors of light that are emitted.
2. As shown in the photo at the right, you will look toward the light through the diffraction grating to detect the colors of light emitted by the bulb. The color spectrum will appear to the right and to the left of the bulb.
3. **List** the specific steps you will need to take to test your hypothesis. Describe precisely what you will do in each step. Will you first test the bulb at a bright or dim setting? How many settings will you test? (Try at least three.) How will you record your observations in an organized way?

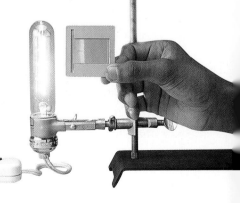

540 CHAPTER 18 Electromagnetic Waves

Using Scientific Methods

4. **List** the materials you will need for your experiment. Describe exactly how and in which order you will use these materials.
5. **Identify** any constants and variables in your experiment.

Follow Your Plan

1. Make sure your teacher approves your plan before you start.
2. **Perform** your experiment as planned.
3. While doing your experiment, write down any observations you make in your Science Journal.

Analyze Your Data

1. Use the colored pencils to draw the color spectrum emitted by the bulb at each brightness.
2. Which colors appeared as the bulb became brighter? Did any colors disappear?
3. How did the wavelengths emitted by the bulb change as the bulb became brighter?
4. **Infer** how the frequencies emitted by the lightbulb changed as it became hotter.

Conclude and Apply

1. **Infer** If an object becomes hotter, what happens to the wavelengths it emits?
2. How do the wavelengths that the bulb emits change if it is turned off?
3. **Infer** from your results whether red stars or yellow stars are hotter.

Compare your results with others in your class. How many different colors were seen?

TIME SCIENCE AND HISTORY

SCIENCE CAN CHANGE THE COURSE OF HISTORY!

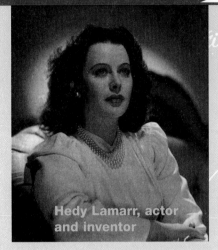
Hedy Lamarr, actor and inventor

Hopping the Frequencies

Ringggg. There it is—that familiar beep! Out come all the cellular phones. At any given moment, a million wireless signals are flying through the air—and not just cell phone signals. With radio and television signals, Internet data, and even Global Positioning System information, the air seems like a pretty crowded place. How does a cellular phone pick out its own signal from among the clutter? The answer lies in a concept developed in 1940 by Hedy Lamarr.

Lamarr was born in Vienna, Austria. In 1937, she left Austria to escape Hitler's invading Nazi army. She also left to pursue a career as an actor. And she became a famous movie star.

In 1940, Lamarr came up with an idea to keep radio signals that guided torpedoes from being jammed. Her idea, called frequency hopping, involved breaking the radio signal that was guiding the torpedo into tiny parts and rapidly changing their frequency. The enemy would not be able to keep up with the frequency changes and thus would not be able to divert the torpedo from its target.

Spread Spectrum

Lamarr's idea was ahead of its time. The digital technology that allowed efficient operation of her system wasn't invented until decades later. However, after 1962, frequency hopping was adopted and used in U.S. military communications. It was the development of wireless phones, however, that benefited the most from Lamarr's concept.

Cordless phones and other wireless technologies operate by breaking their signals into smaller parts, called packets. The frequency of the packets switches rapidly, preventing interference with other calls and enabling millions of callers to use the same narrow band of frequencies.

A torpedo is launched during World War II.

Brainstorm How are you using wireless technology in your life right now? List ways it makes your life easier. Are there drawbacks to some of the uses for wireless technology? What are they?

For more information, visit ips.msscience.com/time

chapter 18 Study Guide

Reviewing Main Ideas

Section 1 — The Nature of Electromagnetic Waves

1. Vibrating charges generate vibrating electric and magnetic fields. These vibrating fields travel through space and are called electromagnetic waves.
2. Electromagnetic waves have wavelength, frequency, amplitude, and carry energy.

Section 2 — The Electromagnetic Spectrum

1. Radio waves have the longest wavelength and lowest energy. Radar uses radio waves to locate objects.
2. All objects emit infrared waves. Most objects you see reflect the visible light emitted by a source of light.
3. Ultraviolet waves have a higher frequency and carry more energy than visible light.
4. X rays and gamma rays are highly penetrating and can be dangerous to living organisms.

Section 3 — Using Electromagnetic Waves

1. Communications systems use electromagnetic waves to transmit information.
2. Radio and TV stations use modulated carrier waves to transmit information.
3. Cordless and cell phones use radio waves to communicate between the mobile phone and a base station.
4. Radio waves are used to send information between communications satellites and ground stations on Earth.

Visualizing Main Ideas

Copy and complete the following spider map about electromagnetic waves.

Spider map — center: **Electromagnetic waves**

- properties:
  - Move at the speed of light
  - Have wavelength and frequency
- consist of two parts:
  1. _____
  2. _____
- produced by:
  3. _____
- types of:
  4. _____
  5. _____
  6. _____
  7. _____
  8. _____
  9. _____
- have energy called:
  10. _____

ips.msscience.com/interactive_tutor

Chapter 18 Review

Using Vocabulary

carrier wave p. 536
electromagnetic spectrum p. 525
electromagnetic wave p. 520
gamma ray p. 530
Global Positioning System p. 539
infrared wave p. 527
radiant energy p. 524
radio wave p. 526
ultraviolet radiation p. 529
visible light p. 528
X ray p. 530

Explain the difference between the terms in each of the following pairs.

1. infrared wave—radio wave
2. radio wave—carrier wave
3. communications satellite—Global Positioning System
4. visible light—ultraviolet radiation
5. X ray, gamma ray
6. electromagnetic wave—radiant energy
7. carrier wave—AM radio signal
8. infrared wave—ultraviolet wave

Checking Concepts

Choose the word or phrase that best answers the question.

9. Which of the following transformations can occur in a radio antenna?
   A) radio waves to sound waves
   B) radio waves to an electric signal
   C) radio waves to infrared waves
   D) sound waves to radio waves

10. Electromagnetic waves with wavelengths between about 0.7 millionths of a meter and 0.4 millionths of a meter are which of the following?
    A) gamma rays   C) radio waves
    B) microwaves   D) visible light

11. Which of the following is the speed of light in space?
    A) 186,000 km/s   C) 3,000,000 km/s
    B) 300,000 km/s   D) 30,000 km/s

12. Which of the following types of electromagnetic waves has the lowest frequency?
    A) infrared waves   C) radio waves
    B) visible light   D) gamma rays

13. Compared to an electric charge that is not moving, a moving electric charge is surrounded by which of the following additional fields?
    A) magnetic   C) electric
    B) microwave   D) gravitational

14. Most of the electromagnetic waves emitted by an object at room temperature are which of the following?
    A) visible light   C) infrared waves
    B) radio waves   D) X rays

15. Which of the following color of visible light has the highest frequency?
    A) green   C) yellow
    B) blue   D) red

16. Which type of electromagnetic waves are completely absorbed by Earth's atmosphere?
    A) radio waves   C) gamma rays
    B) infrared waves D) visible light

17. Sunburn is caused by excessive exposure to which of the following?
    A) ultraviolet waves
    B) infrared waves
    C) visible light
    D) gamma rays

18. How does the frequency of a gamma ray change as its wavelength decreases?
    A) It increases.
    B) It decreases.
    C) It doesn't change.
    D) The frequency depends on the speed.

chapter 18 Review

Thinking Critically

19. **Infer** why communications systems usually use radio waves to transmit information.

20. **Classify** List the colors of the visible light spectrum in order of increasing frequency.

21. **Compare and contrast** an electromagnetic wave with a transverse wave traveling along a rope.

22. **Explain** Some stars form black holes when they collapse. These black holes sometimes can be found by detecting X rays and gamma rays that are emitted as matter falls into the black hole. Explain why it would be difficult to detect these X rays and gamma rays using detectors at Earth's surface.

Use the table below to answer question 23.

Speed of Light in Various Materials	
Materials	Speed (km/s)
Air	300,000
Water	226,000
Polystyrene Plastic	189,000
Diamond	124,000

23. **Calculate** A radio wave has a frequency of 500,000 Hz. If the radio wave has the same frequency in air as in water, what is the ratio of the wavelength of the radio wave in air to its wavelength in water?

24. **Explain** how you could determine if there are electromagnetic waves traveling in a closed, completely dark room in a building.

25. **Infer** Light waves from a distant galaxy take 300 million years to reach Earth. How does the age of the galaxy when it emitted the light waves compare with the age of the galaxy when we see the light waves?

26. **Concept Map** Electromagnetic waves are grouped according to their frequencies. In the following concept map, write each frequency group and one way humans make use of the electromagnetic waves in that group. For example, in the second set of ovals, you might write *X rays* and *to see inside the body*. Do not write in this book.

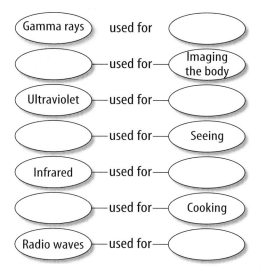

Performance Activities

27. **Oral Presentation** Explain to the class how a radio signal is generated, transmitted, and received.

28. **Poster** Make a poster showing the parts of the electromagnetic spectrum. Show how frequency, wavelength, and energy change throughout the spectrum. How is each wave generated? What are some uses of each?

Applying Math

29. **Distance** How long would it take a radio signal to travel from Earth to the Moon, a distance of 384,000 km?

30. **Wavelength** The frequency of a popular AM radio station is 720 kHz. What is the wavelength of the radio waves broadcast by this station?

Chapter 18 Standardized Test Practice

Part 1 Multiple Choice

Record your answers on the answer sheet provided by your teacher or on a sheet of paper.

1. Which of the following types of electromagnetic waves has a frequency greater than visible light?
   A. infrared waves
   B. radio waves
   C. ultraviolet waves
   D. microwaves

2. Which of the following properties of a transverse wave is the distance from one crest to the next?
   A. intensity
   B. amplitude
   C. frequency
   D. wavelength

3. Which of the following types of electromagnetic waves enables your body to produce vitamin D?
   A. gamma rays
   B. ultraviolet waves
   C. visible light
   D. infrared waves

Use the illustration below to answer question 4.

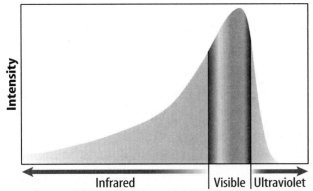

Electromagnetic Waves from the Sun

4. How does the intensity of ultraviolet waves emitted by the Sun change as the wavelength of the ultraviolet waves decreases?
   A. The intensity increases.
   B. The intensity decreases.
   C. The intensity doesn't change.
   D. The intensity increases, then decreases.

5. The color of visible light waves depends on which of the following wave properties?
   A. wavelength
   B. amplitude
   C. direction
   D. speed

6. Which of the following is NOT true about electromagnetic waves?
   A. They can travel through matter.
   B. They move by transferring matter.
   C. They are produced by vibrating charges.
   D. They can travel through empty space.

Use the illustration below to answer question 7.

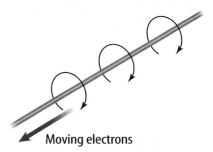

Moving electrons

7. Which of the following is represented by the circular lines around the current-carrying wire?
   A. direction of current
   B. electric and magnetic field lines
   C. magnetic field lines
   D. electric field lines

8. How are gamma rays produced?
   A. by vibrating electric fields
   B. by vibrating magnetic fields
   C. by the absorption of infrared waves
   D. by nuclear fission or fusion

9. Earth's atmosphere is transparent to which type of electromagnetic waves?
   A. gamma rays
   B. ultraviolet waves
   C. infrared waves
   D. radio waves

Standardized Test Practice

Part 2 Short Response/Grid In

Record your answers on the answer sheet provided by your teacher or on a sheet of paper.

Use the photograph below to answer question 10.

10. If the microwaves produced in a microwave oven have a frequency of 2,450 MHz, what is the wavelength of the microwaves?

11. You turn on a lamp that is plugged into an electric outlet. Does a magnetic field surround the wire that connects the lamp to the outlet? Explain.

12. A carrier wave broadcast by a radio station has a wavelength of 3.0 m. What is the frequency of the carrier wave?

13. Explain how the wavelengths of the electromagnetic waves emitted by an object change as the temperature of the object increases.

14. Explain why X rays can form images of dense tissues in the human body.

15. If the planet Mars is 80,000,000 km from Earth, how long will it take an electromagnetic wave to travel from Earth to Mars?

Test-Taking Tip

Recheck Answers Double check your answers before turning in the test.

Part 3 Open Ended

Record your answers on a sheet of paper.

16. Describe the sequence of events that occur when a radar station detects an airplane and determines the distance to the plane.

17. Explain why infrared detectors on satellites can detect regions covered by vegetation.

18. The carrier waves broadcast by a radio station are altered in order to transmit information. The two ways of altering a carrier wave are amplitude modulation (AM) and frequency modulation (FM). Draw a carrier wave, an AM wave, and an FM wave.

19. Describe the effect of the ozone layer on electromagnetic waves that strike Earth's atmosphere.

20. List the energy conversions that occur when a song recorded on a CD is broadcast as radio waves and then reproduced as sound.

Use the illustration below to answer questions 21 and 22.

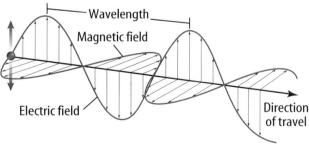

21. Explain how the vibrating electric and magnetic fields are produced.

22. Infer how the directions of the electric field and the magnetic field are related to the direction that the electromagnetic wave travels.

chapter 19

Light, Mirrors, and Lenses

The BIG Idea
Light waves can be absorbed, reflected, and transmitted by matter.

SECTION 1
Properties of Light
Main Idea A source of light gives off light rays that travel outward in all directions.

SECTION 2
Reflection and Mirrors
Main Idea When a light ray is reflected from a surface, the angle of incidence equals the angle of reflection.

SECTION 3
Refraction and Lenses
Main Idea A light ray changes direction when it moves from one material into another and changes speed.

SECTION 4
Using Mirrors and Lenses
Main Idea Lenses and mirrors are used to form images of objects that cannot be seen with the human eye.

Seeing the Light
This lighthouse at Pigeon Point, California, produces beams of light that can be seen for many miles. These intense light beams are formed in the same way as a flashlight beam. The key ingredient is a curved mirror that reflects the light from a bright source.

Science Journal Describe how you use mirrors and lenses during a typical day.

Start-Up Activities

Bending Light

Everything you see results from light waves entering your eyes. These light waves are either given off by objects, such as the Sun and lightbulbs, or reflected by objects, such as trees, books, and people. Lenses and mirrors can cause light to change direction and make objects seem larger or smaller.

1. Place two paper cups next to each other and put a penny in the bottom of each cup.
2. Fill one of the cups with water and observe how the penny looks.
3. Looking straight down at the cups, slide the cup with no water away from you just until you can no longer see the penny.
4. Pour water into this cup and observe what seems to happen to the penny.
5. **Think Critically** In your Science Journal, record your observations. Did adding water make the cup look deeper or shallower?

Preview this chapter's content and activities at ips.msscience.com

Light, Mirrors, and Lenses Make the following Foldable to help you understand the properties of and the relationship between light, mirrors, and lenses.

STEP 1 **Fold** a sheet of pape in half lengthwise. Make the back edge about 5 cm longer than the front edge.

STEP 2 **Turn** the paper so the fold is on the bottom. Then **fold** it into thirds.

STEP 3 **Unfold and cut** only the top layer along folds to make three tabs.

STEP 4 **Label** the Foldable as shown.

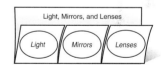

Summarize in a Table As you read the chapter, summarize the information you find about light, mirrors, lenses.

549

Get Ready to Read

Questioning

① Learn It! Asking questions helps you to understand what you read. As you read, think about the questions you'd like answered. Often you can find the answer in the next paragraph or section. Learn to ask good questions by asking who, what, when, where, why, and how.

② Practice It! Read the following passage from Section 3.

> Optical fibers are used most commonly in the communications industry. For example, television programs, computer information, and phone conversations can be coded into light signals. These signals can then be sent from one place to another using optical fibers. Because of total internal reflection, signals cannot leak from one fiber to another, causing interference. As a result, the signal is transmitted clearly.
>
> — *from page 566*

Here are some questions you might ask about this paragraph:
- How are optical fibers used by the communications industry?
- What type of signals are sent through optical fibers?
- Why are signals transmitted clearly in optical fibers?

③ Apply It! As you read the chapter, look for answers to lesson headings that are in the form of questions.

Target Your Reading

Reading Tip

Test yourself. Create questions and then read to find answers to your questions.

Use this to focus on the main ideas as you read the chapter.

1 Before you read the chapter, respond to the statements below on your worksheet or on a numbered sheet of paper.
- Write an **A** if you **agree** with the statement.
- Write a **D** if you **disagree** with the statement.

2 After you read the chapter, look back to this page to see if you've changed your mind about any of the statements.
- If any of your answers changed, explain why.
- Change any false statements into true statements.
- Use your revised statements as a study guide.

Before You Read A or D		Statement	After You Read A or D
	1	All objects give off light on their own.	
	2	You see an object when light rays travel from your eyes to the object.	
	3	The color of an object depends on the wavelengths of the light waves reflected from the object.	
	4	Light rays obey the law of reflection only if the reflecting surface is very smooth.	
	5	Light waves travel at the same speed in all materials.	
	6	A lens causes all light rays to pass through the focal point of the lens.	
	7	The image formed by a lens depends on how far the object is from the lens.	
	8	The purpose of the large concave mirror in a reflecting telescope is to magnify objects.	
	9	A laser beam contains a single wavelength of light.	

Science Online
Print out a worksheet of this page at
ips.msscience.com

550 B

section 1
Properties of Light

as you read

What You'll Learn
- **Describe** the wave nature of light.
- **Explain** how light interacts with materials.
- **Determine** why objects appear to have color.

Why It's Important
Everything you see comes from information carried by light waves.

Review Vocabulary
electromagnetic waves: waves created by vibrating electric charges that can travel through space or through matter

New Vocabulary
- light ray
- medium

What is light?

Drop a rock on the smooth surface of a pond and you'll see ripples spread outward from the spot where the rock struck. The rock produced a wave much like the one in **Figure 1.** A wave is a disturbance that carries energy through matter or space. The matter in this case is the water, and the energy originally comes from the impact of the rock. As the ripples spread out, they carry some of that energy.

Light is another type of wave that carries energy. A source of light such as the Sun or a lightbulb gives off light waves into space, just as the rock hitting the pond causes waves to form in the water. But while the water waves spread out only on the surface of the pond, light waves spread out in all directions from the light source. **Figure 1** shows how light waves travel.

Sometimes, however, it is easier to think of light in a different way. A **light ray** is a narrow beam of light that travels in a straight line. You can think of a source of light as giving off, or emitting, a countless number of light rays that are traveling away from the source in all directions.

Figure 1 Light moves away in all directions from a light source, just as ripples spread out on the surface of water.

A source of light, such as a lightbulb, gives off light rays that travel away from the light source in all directions.

Ripples on the surface of a pond are produced by an object hitting the water. The ripples spread out from the point of impact.

550 CHAPTER 19 Light, Mirrors, and Lenses

Light Travels Through Space There is, however, one important difference between light waves and the water wave ripples on a pond. If the pond dried up and had no water, ripples could not form. Waves on a pond need a material—water—in which to travel. The material through which a wave travels is called a **medium**. Light is an electromagnetic wave and doesn't need a medium in which to travel. Electromagnetic waves can travel in a vacuum, as well as through materials such as air, water, and glass.

Light and Matter

What can you see when you are in a closed room with no windows and the lights out? You can see nothing until you turn on a light or open a door to let in light from outside the room. Most objects around you do not give off light on their own. They can be seen only if light waves from another source bounce off them and into your eyes, as shown in **Figure 2**. The process of light striking an object and bouncing off is called reflection. Right now, you can see these words because light emitted by a source of light is reflecting from the page and into your eyes. Not all the light rays reflected from the page strike your eyes. Light rays striking the page are reflected in many directions, and only some of these rays enter your eyes.

Reading Check *What must happen for you to see most objects?*

Mini LAB

Observing Colors in the Dark

Procedure
1. Get six pieces of **paper** that are different colors and about 10 cm × 10 cm.
2. Darken a room and wait 10 min for your eyes to adjust to the darkness.
3. Write on each paper what color you think the paper is.
4. Turn on the lights and see if your night vision correctly detected the colors.

Analysis
1. If the room were perfectly dark, what would you see? Explain.
2. Your eyes contain rod cells and cone cells. Rod cells enable you to see in dim light, but don't detect color. Cone cells enable you to see color, but do not work in dim light. Which type of cell was working in the darkened room? Explain.

Try at Home

Figure 2 Light waves are given off by the lightbulb. Some of these light waves hit the page and are reflected. The student sees the page when some of these reflected waves enter the student's eyes.

SECTION 1 Properties of Light **551**

An opaque object allows no light to pass through it.

A translucent object allows some light to pass through it.

A transparent object allows almost all light to pass through it.

Figure 3 Materials are opaque, translucent, or transparent, depending on how much light passes through them.
Infer *which type of material reflects the least amount of light.*

Opaque, Translucent, and Transparent When light waves strike an object, some of the waves are absorbed by the object, some are reflected by it, and some might pass through it. What happens to light when it strikes the object depends on the material that the object is made of.

All objects reflect and absorb some light waves. Materials that let no light pass through them are opaque (oh PAYK). You cannot see other objects through opaque materials. On the other hand, you clearly can see other objects through materials such as glass and clear plastic that allow nearly all the light that strikes them to pass through. These materials are transparent. A third type of material allows only some light to pass through. Although objects behind these materials are visible, they are not clear. These materials, such as waxed paper and frosted glass, are translucent (trans LEW sent). Examples of opaque, translucent, and transparent objects are shown in **Figure 3**.

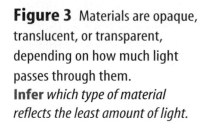

Figure 4 A beam of white light passing through a prism is separated into many colors.
Describe *the colors you see emerging from the prism.*

Color

The light from the Sun might look white, but it is a mixture of colors. Each different color of light is a light wave with a different wavelength. Red light waves have the longest wavelengths and violet light waves have the shortest wavelengths. As shown in **Figure 4,** white light is separated into different colors when it passes through a prism. The colors in white light range from red to violet. When light waves from all these colors enter the eye at the same time, the brain interprets the mixture as being white.

A pair of gym shoes and socks as seen under white light.

The same shoes and socks photographed through a red filter.

Why do objects have color? Why does grass look green or a rose look red? When a mixture of light waves strikes an object that is not transparent, the object absorbs some of the light waves. Some of the light waves that are not absorbed are reflected. If an object reflects red waves and absorbs all the other waves, it looks red. Similarly, if an object looks blue, it reflects only blue light waves and absorbs all the others. An object that reflects all the light waves that strike it looks white, while one that reflects none of the light waves that strike it looks black. **Figure 5** shows gym shoes and socks as seen under white light and as seen when viewed through a red filter that allows only red light to pass through it.

Figure 5 The color of an object depends on the light waves it reflects.
Infer *why the blue socks look black when viewed under red light.*

Primary Light Colors How many colors exist? People often say white light is made up of red, orange, yellow, green, blue, and violet light. This isn't completely true, though. Many more colors than this exist. In reality, most humans can distinguish thousands of colors, including some such as brown, pink, and purple, that are not found among the colors of the rainbow.

Light of almost any color can be made by mixing different amounts of red, green, and blue light. Red, green, and blue are known as the primary colors. Look at **Figure 6**. White light is produced where beams of red, green, and blue light overlap. Yellow light is produced where red and green light overlap. You see the color yellow because of the way your brain interprets the combination of the red and green light striking your eye. This combination of light waves looks the same as yellow light produced by a prism, even though these light waves have only a single wavelength.

Figure 6 By mixing light from the three primary colors—red, blue, and green—almost all of the visible colors can be made.

SECTION 1 Properties of Light **553**

Primary Pigment Colors Materials like paint that are used to change the color of other objects, such as the walls of a room or an artist's canvas, are called pigments. Mixing pigments together forms colors in a different way than mixing colored lights does.

Like all materials that appear to be colored, pigments absorb some light waves and reflect others. The color of the pigment you see is the color of the light waves that are reflected from it. However, the primary pigment colors are not red, blue, and green—they are yellow, magenta, and cyan. You can make almost any color by mixing different amounts of these primary pigment colors, as shown in **Figure 7**.

Although primary pigment colors are not the same as the primary light colors, they are related. Each primary pigment color results when a pigment absorbs a primary light color. For example, a yellow pigment absorbs blue light and it reflects red and green light, which you see as yellow. A magenta pigment, on the other hand, absorbs green light and reflects red and blue light, which you see as magenta. Each of the primary pigment colors is the same color as white light with one primary color removed.

Figure 7 The three primary color pigments—yellow, magenta, and cyan—can form almost all the visible colors when mixed together in various amounts.

section 1 review

Summary

Light and Matter
- Light is an electromagnetic wave that can travel in a vacuum as well as through matter.
- When light waves strike an object some light waves might be absorbed by the object, some waves might be reflected from the object, and some waves might pass through the object.
- Materials can be opaque, translucent, or transparent, depending on how much light passes through the material.

Color
- Light waves with different wavelengths have different colors.
- White light is a combination of all the colors ranging from red to violet.
- The color of an object is the color of the light waves that it reflects.
- The primary light colors are red, green, and blue. The primary pigment colors are yellow, magenta and cyan.

Self Check

1. **Diagram** the path followed by a light ray that enters one of your eyes when you are reading at night in a room.
2. **Determine** the colors that are reflected from an object that appears black.
3. **Compare and contrast** primary light colors and primary pigment colors.
4. **Describe** the difference between an opaque object and a transparent object.
5. **Think Critically** A white shirt is viewed through a filter that allows only blue light to pass through the filter. What color will the shirt appear to be?

Applying Skills

6. **Draw Conclusions** A black plastic bowl and a white plastic bowl are placed in sunlight. After 15 minutes, the temperature of the black bowl is higher than the temperature of the white bowl. Which bowl absorbs more light waves and which bowl reflects more light waves?

 ips.msscience.com/self_check_quiz

Section 2

Reflection and Mirrors

The Law of Reflection

You've probably noticed your image on the surface of a pool or lake. If the surface of the water was smooth, you could see your face clearly. If the surface of the water was wavy, however, your face might have seemed distorted. The image you saw was the result of light reflecting from the surface and traveling to your eyes. How the light was reflected determined the sharpness of the image you saw.

When a light ray strikes a surface and is reflected, as in **Figure 8,** the reflected ray obeys the law of reflection. Imagine a line that is drawn perpendicular to the surface where the light ray strikes. This line is called the normal to the surface. The incoming ray and the normal form an angle called the angle of incidence. The reflected light ray forms an angle with the normal called the angle of reflection. According to the **law of reflection,** the angle of incidence is equal to the angle of reflection. This is true for any surface, no matter what material it is made of.

Reflection from Surfaces

Why can you see your reflection in some surfaces and not others? Why does a piece of shiny metal make a good mirror, but a piece of paper does not? The answers have to do with the smoothness of each surface.

as you read

What You'll Learn
- **Explain** how light is reflected from rough and smooth surfaces.
- **Determine** how mirrors form an image.
- **Describe** how concave and convex mirrors form an image.

Why It's Important

Mirrors can change the direction of light waves and enable you to see images, such as your own face.

Review Vocabulary

normal: a line drawn perpendicular to a surface or line

New Vocabulary
- law of reflection
- focal point
- focal length

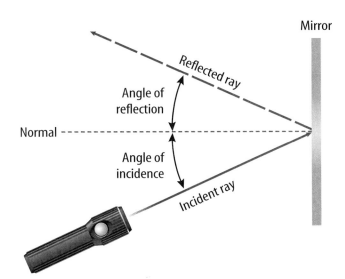

Figure 8 A light ray strikes a surface and is reflected. The angle of incidence is always equal to the angle of reflection. This is the law of reflection.

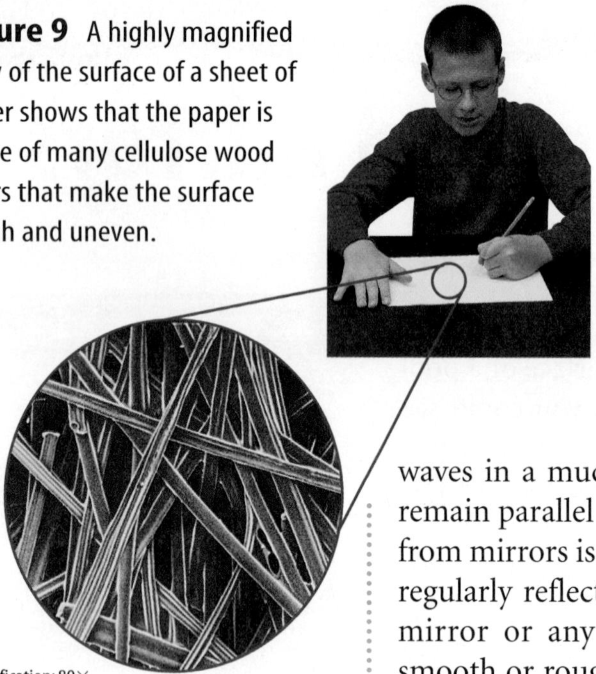

Figure 9 A highly magnified view of the surface of a sheet of paper shows that the paper is made of many cellulose wood fibers that make the surface rough and uneven.

Magnification: 80×

Regular and Diffuse Reflection Even though the surface of the paper might seem smooth, it's not as smooth as the surface of a mirror. **Figure 9** shows how rough the surface of a piece of paper looks when it is viewed under a microscope. The rough surface causes light rays to be reflected from it in many directions, as shown in **Figure 10.** This uneven reflection of light waves from a rough surface is diffuse reflection. The smoother surfaces of mirrors, as shown in **Figure 10,** reflect light waves in a much more regular way. For example, parallel rays remain parallel after they are reflected from a mirror. Reflection from mirrors is known as regular reflection. Light waves that are regularly reflected from a surface form the image you see in a mirror or any other smooth surface. Whether a surface is smooth or rough, every light ray that strikes it obeys the law of reflection.

Reading Check *Why does a rough surface cause a diffuse reflection?*

Scattering of Light When diffuse reflection occurs, light waves that were traveling in a single direction are reflected and then travel in many different directions. Scattering occurs when light waves traveling in one direction are made to travel in many different directions. Scattering also can occur when light waves strike small particles, such as dust. You may have seen dust particles floating in a beam of sunlight. When the light waves in the sunbeam strike a dust particle, they are scattered in all directions. You see the dust particles as bright specks of light when some of these scattered light waves enter your eye.

Figure 10 The roughness of a surface determines whether it looks like a mirror.

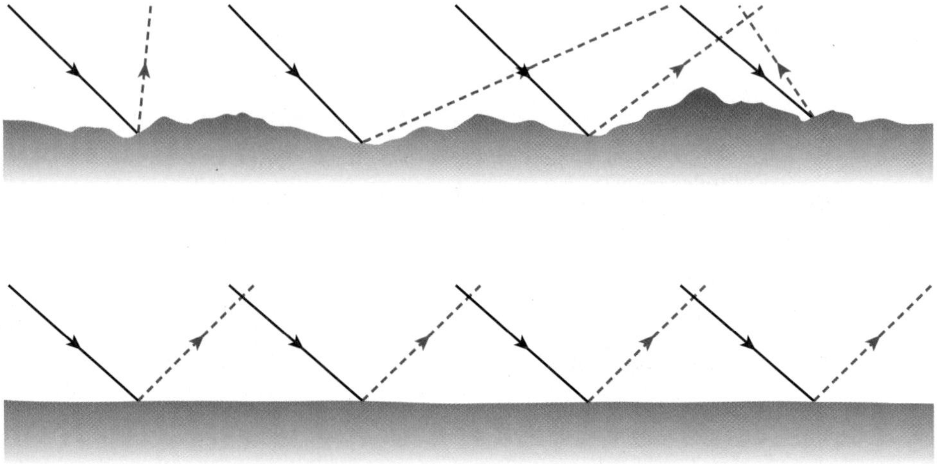

A rough surface causes parallel light rays to be reflected in many different directions.

A smooth surface causes parallel light rays to be reflected in a single direction. This type of surface looks like a mirror.

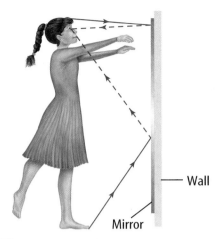
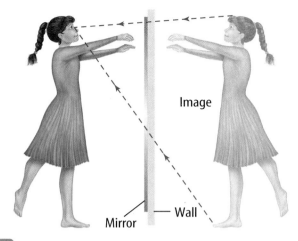

A Light rays that bounce off a person strike the mirror. Some these light rays are reflected into the person's eye.

B The light rays that are shown entering the person's eye seem to be coming from a person behind the mirror.

Reflection by Plane Mirrors Did you glance in the mirror before leaving for school this morning? If you did, you probably looked at your reflection in a plane mirror. A plane mirror is a mirror with a flat reflecting surface. In a plane mirror, your image looks much the same as it would in a photograph. However, you and your image are facing in opposite directions. This causes your left side and your right side to switch places on your mirror image. Also, your image seems to be coming from behind the mirror. How does a plane mirror form an image?

Reading Check *What is a plane mirror?*

Figure 11 shows a person looking into a plane mirror. Light waves from the Sun or another source of light strike each part of the person. These light rays bounce off the person according to the law of reflection, and some of them strike the mirror. The rays that strike the mirror also are reflected according to the law of reflection. **Figure 11A** shows the path traveled by a few of the rays that have been reflected off the person and reflected back to the person's eye by the mirror.

The Image in a Plane Mirror Why does the image you see in a plane mirror seem to be behind the mirror? This is a result of how your brain processes the light rays that enter your eyes. Although the light rays bounced off the mirror's surface, your brain interprets them as having followed the path shown by the dashed lines in **Figure 11B**. In other words, your brain always assumes that light rays travel in straight lines without changing direction. This makes the reflected light rays look as if they are coming from behind the mirror, even though no source of light is there. The image also seems to be the same distance behind the mirror as the person is in front of the mirror.

Figure 11 A plane mirror forms an image by changing the direction of light rays.
Describe *how you and your image in a plane mirror are different.*

Light Waves and Photons
When an object like a marble or a basketball bounces off a surface, it obeys the law of reflection. Because light also obeys the law of reflection, people once thought that light must be a stream of particles. Today, experiments have shown that light can behave as though it were both a wave and a stream of energy bundles called photons. Read an article about photons and write a description in your Science Journal.

SECTION 2 Reflection and Mirrors **557**

Science Online

Topic: Concave Mirrors
Visit ips.msscience.com for Web links to information about the concave mirrors used in telescopes.

Activity Make a chart showing the five largest telescope mirrors and where they are located.

Concave and Convex Mirrors

Some mirrors are not flat. A concave mirror has a surface that is curved inward, like the bowl of a spoon. Unlike plane mirrors, concave mirrors cause light rays to come together, or converge. A convex mirror, on the other hand, has a surface that curves outward, like the back of a spoon. Convex mirrors cause light waves to spread out, or diverge. These two types of mirrors form images that are different from the images that are formed by plane mirrors. Examples of a concave and a convex mirror are shown in **Figure 12.**

Reading Check *What's the difference between a concave and convex mirror?*

Concave Mirrors The way in which a concave mirror forms an image is shown in **Figure 13.** A straight line drawn perpendicular to the center of a concave or convex mirror is called the optical axis. Light rays that travel parallel to the optical axis and strike the mirror are reflected so that they pass through a single point on the optical axis called the **focal point.** The distance along the optical axis from the center of the mirror to the focal point is called the **focal length.**

The image formed by a concave mirror depends on the position of the object relative to its focal point. If the object is farther from the mirror than the focal point, the image appears to be upside down, or inverted. The size of the image decreases as the object is moved farther away from the mirror. If the object is closer to the mirror than one focal length, the image is upright and gets smaller as the object moves closer to the mirror.

A concave mirror can produce a focused beam of light if a source of light is placed at the mirror's focal point, as shown in **Figure 13.** Flashlights and automobile headlights use concave mirrors to produce directed beams of light.

Figure 12 Convex and concave mirrors have curved surfaces.

A concave mirror has a surface that's curved inward.

A convex mirror has a surface that's curved outward.

NATIONAL GEOGRAPHIC
VISUALIZING REFLECTIONS IN CONCAVE MIRRORS

Figure 13

Glance into a flat plane mirror and you'll see an upright image of yourself. But look into a concave mirror, and you might see yourself larger than life, right side up, or upside down—or not at all! This is because the way a concave mirror forms an image depends on the position of an object in front of the mirror, as shown here.

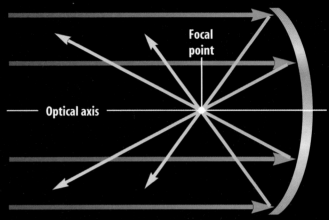

A concave mirror reflects all light rays traveling parallel to the optical axis so that they pass through the focal point.

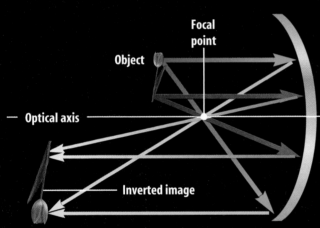

When an object, such as this flower, is placed beyond the focal point, the mirror forms an image that is inverted.

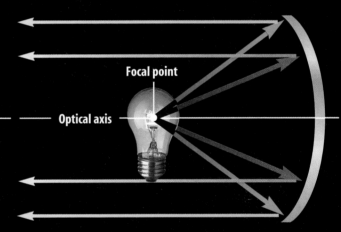

When a source of light is placed at the focal point, a beam of parallel light rays is formed. The concave mirror in a flashlight, for example, creates a beam of parallel light rays.

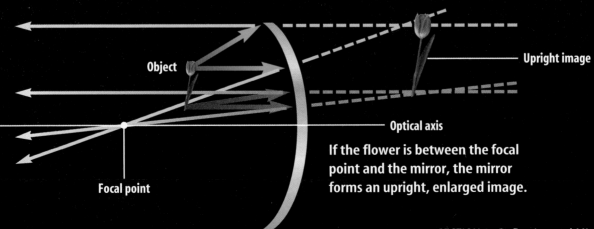

If the flower is between the focal point and the mirror, the mirror forms an upright, enlarged image.

SECTION 2 Reflection and Mirrors

Figure 14 A convex mirror is a mirror that curves outward.

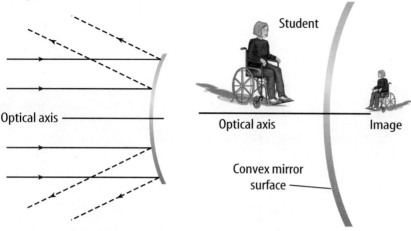

A convex mirror causes light rays that are traveling parallel to the optical axis to spread apart after they are reflected.

No matter how far an object is from a convex mirror, the image is always upright and smaller than the object.

Convex Mirrors A convex mirror has a reflecting surface that curves outward and causes light rays to spread apart, or diverge, as shown in **Figure 14.** Like the image formed by plane mirror, the image formed by a convex mirror seems to be behind the mirror. **Figure 14** shows that the image always is upright and smaller than the object.

Convex mirrors often are used as security mirrors in stores and as outside rearview mirrors on cars and other vehicles. You can see a larger area reflected in a convex mirror than in other mirrors.

section 2 review

Summary

Reflection and Plane Mirrors

- The law of reflection states that the angle of incidence equals the angle of reflection.
- A regular reflection is produced by a smooth surface, such as a mirror. A rough surface forms a diffuse reflection.
- Scattering occurs when light rays traveling in one direction are made to travel in many directions.
- A plane mirror forms a image that is reversed left to right and seems to be behind the mirror.

Concave and Convex Mirrors

- Concave mirrors curve inward and make light rays converge.
- Images formed by a concave mirror can be either upright or inverted and can vary from larger to smaller than the object.
- Convex mirrors curve outward and make light rays diverge.
- Images formed by a convex mirror are always upright and smaller than the object.

Self Check

1. **Describe** the image formed by a concave mirror when an object is less than one focal length from the mirror.
2. **Explain** why concave mirrors are used in flashlights and automobile headlights.
3. **Describe** If an object is more than one focal length from a concave mirror, how does the image formed by the mirror change as the object moves farther from the mirror?
4. **Determine** which light rays striking a concave mirror are reflected so that they pass through the focal point.
5. **Think Critically** After you wash and wax a car, you can see your reflection in the car's surface. Before you washed and waxed the car, no reflection could be seen. Explain.

Applying Skills

6. **Use a Spreadsheet** Make a table using a spreadsheet comparing the images formed by plane, concave, and convex mirrors. Include in your table how the images depend on the distance of the object from the mirror.

ips.msscience.com/self_check_quiz

Reflection from a Plane Mirror

A light ray strikes the surface of a plane mirror and is reflected. Does a relationship exist between the direction of the incoming light ray and the direction of the reflected light ray?

Real-World Question

How does the angle of incidence compare with the angle of reflection for a plane mirror?

Goals
- **Measure** the angle of incidence and the angle of reflection for a light ray reflected from a plane mirror.

Materials
flashlight
protractor
metric ruler
scissors
tape
small plane mirror, at least 10 cm on a side
black construction paper
modeling clay
white unlined paper

Safety Precautions

Procedure

1. With the scissors, cut a slit in the construction paper and tape it over the flashlight lens.
2. Place the mirror at one end of the unlined paper. Push the mirror into lumps of clay so it stands vertically, and tilt the mirror so it leans slightly toward the table.
3. **Measure** with the ruler to find the center of the bottom edge of the mirror, and mark it. Then use the protractor and the ruler to draw a line on the paper perpendicular to the mirror from the mark. Label this line *P*.
4. Draw lines on the paper from the center mark at angles of 30°, 45°, and 60° to line *P*.
5. Turn on the flashlight and place it so the beam is along the 60° line. This is the angle of incidence. Measure and record the angle that the reflected beam makes with line *P*. This is the angle of reflection. If you cannot see the reflected beam, slightly increase the tilt of the mirror.
6. Repeat step 5 for the 30°, 45°, and *P* lines.

Conclude and Apply

Infer from your results the relationship between the angle of incidence and the angle of reflection.

Communicating Your Data

Make a poster that shows your measured angles of reflection for angles of incidence of 30°, 45°, and 60°. Write the relationship between the angles of incidence and reflection at the bottom.

section 3
Refraction and Lenses

as you read

What You'll Learn
- **Determine** why light rays refract.
- **Explain** how convex and concave lenses form images.

Why It's Important
Many of the images you see every day in photographs, on TV, and in movies are made using lenses.

Review Vocabulary
refraction: bending of a wave as it changes speed, moving from one medium to another

New Vocabulary
- lens
- convex lens
- concave lens

Bending of Light Rays

Objects that are in water can sometimes look strange. A pencil in a glass of water sometimes looks as if it's bent, or as if the part of the pencil in air is shifted compared to the part in water. A penny that can't be seen at the bottom of a cup suddenly appears as you add water to the cup. Illusions such as these are due to the bending of light rays as they pass from one material to another. What causes light rays to change direction?

The Speeds of Light The speed of light in empty space is about 300 million m/s. Light passing through a material such as air, water, or glass, however, travels more slowly than this. This is because the atoms that make up the material interact with the light waves and slow them down. **Figure 15** compares the speed of light in some different materials.

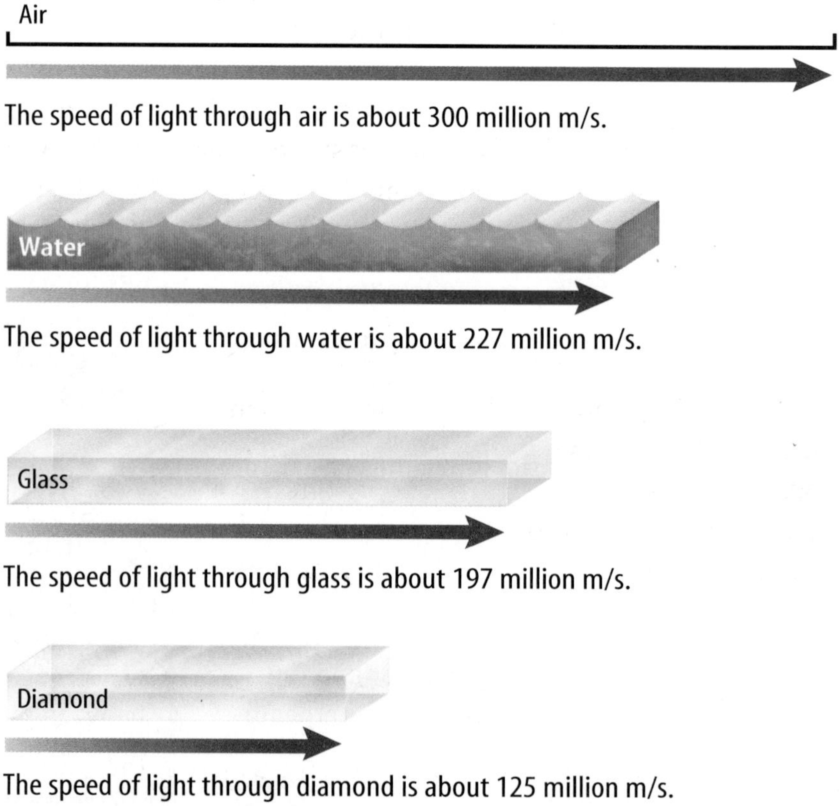

Figure 15 Light travels at different speeds in different materials.

Air
The speed of light through air is about 300 million m/s.

Water
The speed of light through water is about 227 million m/s.

Glass
The speed of light through glass is about 197 million m/s.

Diamond
The speed of light through diamond is about 125 million m/s.

The Refraction of Light Waves

Light rays from the part of a pencil that is underwater travel through water, glass, and then air before they reach your eye. The speed of light is different in each of these mediums. What happens when a light wave travels from one medium into another in which its speed is different? If the wave is traveling at an angle to the boundary between the two media, it changes direction, or bends. This bending is due to the change in speed the light wave undergoes as it moves from one medium into the other. The bending of light waves due to a change in speed is called refraction. **Figure 16** shows an example of refraction. The greater the change in speed is, the more the light wave bends, or refracts.

Figure 16 A light ray is bent as it slows down traveling from air into water.

 *What causes light to bend?*

Why does a change in speed cause the light wave to bend? Think about what happens to the wheels of a car as they move from pavement to mud at an angle, as in **Figure 17.** The wheels slip a little in the mud and don't move forward as fast as they do on the pavement. The wheel that enters the mud first gets slowed down a little, but the other wheel on that axle continues at the original speed. The difference in speed between the two wheels then causes the wheel axle to turn, so the car turns a little. Light waves behave in the same way.

Imagine again a light wave traveling at an angle from air into water. The first part of the wave to enter the water is slowed, just as the car wheel that first hit the mud was slowed. The rest of the wave keeps slowing down as it moves from the air into the water. As long as one part of the light wave is moving faster than the rest of the wave, the wave continues to bend.

Figure 17 An axle turns as the wheels cross the boundary between pavement and mud. **Predict** how the axle would turn if the wheels were going from mud to pavement.

Convex and Concave Lenses

Do you like photographing your friends and family? Have you ever watched a bird through binoculars or peered at something tiny through a magnifying glass? All of these activities involve the use of lenses. A **lens** is a transparent object with at least one curved side that causes light to bend. The amount of bending can be controlled by making the sides of the lenses more or less curved. The more curved the sides of a lens are, the more light will be bent after it enters the lens.

SECTION 3 Refraction and Lenses **563**

Figure 18 A convex lens forms an image that depends on the distance from the object to the lens.

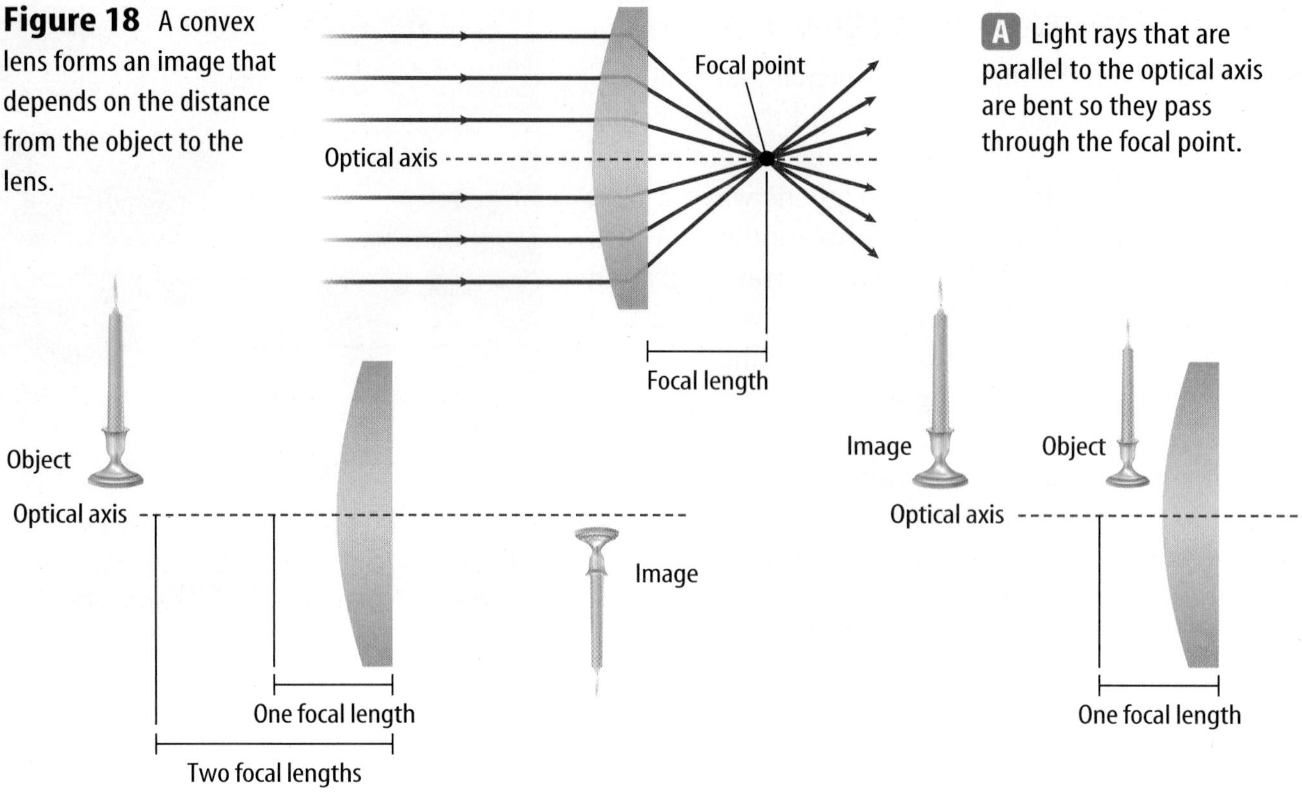

A Light rays that are parallel to the optical axis are bent so they pass through the focal point.

B If the object is more than two focal lengths from the lens, the image formed is smaller than the object and inverted.

C If the object is closer to the lens than one focal length, the image formed is enlarged and upright.

Convex Lenses A lens that is thicker in the center than at the edges is a **convex lens.** In a convex lens, light rays traveling parallel to the optical axis are bent so they pass through the focal point, as shown in **Figure 18A.** The more curved the lens is, the closer the focal point is to the lens, and so the shorter the focal length of the lens is. Because convex lenses cause light waves to meet, they also are called converging lenses.

The image formed by a convex lens is similar to the image formed by a concave mirror. For both, the type of image depends on how far the object is from the mirror or lens. Look at **Figure 18B.** If the object is farther than two focal lengths from the lens, the image seen through the lens is inverted and smaller than the object.

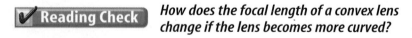

Reading Check *How does the focal length of a convex lens change if the lens becomes more curved?*

If the object is closer to the lens than one focal length, then the image formed is right-side up and larger than the object, as shown in **Figure 18C.** A magnifying glass forms an image in this way. As long as the magnifying glass is less than one focal length from the object, you can make the image appear larger by moving the magnifying glass away from the object.

Concave Lenses A lens that is thicker at the edges than in the middle is a **concave lens**. A concave lens also is called a diverging lens. **Figure 19** shows how light rays traveling parallel to the optical axis are bent after passing through a concave lens.

A concave lens causes light rays to diverge, so light rays are not brought to a focus. The type of image that is formed by a concave lens is similar to one that is formed by a convex mirror. The image is upright and smaller than the object.

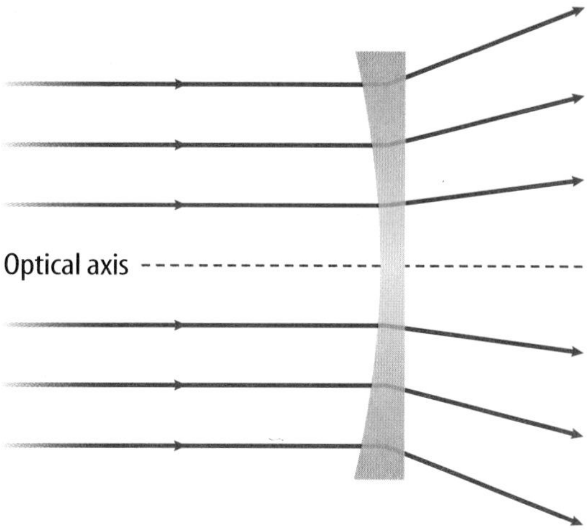

Total Internal Reflection

When you look at a glass window, you sometimes can see your reflection. You see a reflection because some of the light waves reflected from you are reflected back to your eyes when they strike the window. This is an example of a partial reflection—only some of the light waves striking the window are reflected. However, sometimes all the light waves that strike the boundary between two transparent materials can be reflected. This process is called total internal reflection.

Figure 19 A concave lens causes light rays traveling parallel to the optical axis to diverge.

The Critical Angle To see how total internal reflection occurs, look at **Figure 20**. Light travels faster in air than in water, and the refracted beam is bent away from the normal. As the angle between the incident beam and the normal increases, the refracted beam bends closer to the air-water boundary. At the same time, more of the light energy striking the boundary is reflected and less light energy passes into the air.

If a light beam in water strikes the boundary so that the angle with the normal is greater than an angle called the critical angle, total internal reflection occurs. Then all the light waves are reflected at the air-water boundary, just as if a mirror were there. The size of the critical angle depends on the two materials involved. For light passing from water to air, the critical angle is about 48 degrees.

Figure 20 When a light beam passes from one medium to another, some of its energy is reflected (red) and some is refracted (blue).

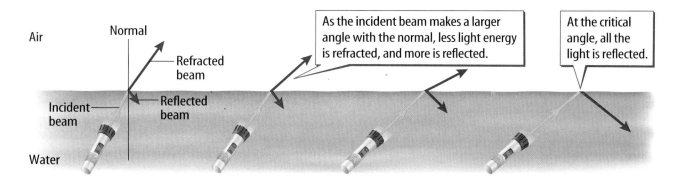

SECTION 3 Refraction and Lenses **565**

Optical Fibers Optical fibers are thin, flexible, transparent fibers. An optical fiber is like a light pipe. Even if the fiber is bent, light that enters one end of the fiber comes out the other end.

Total internal reflection makes light transmission in optical fibers possible. A thin fiber of glass or plastic is covered with another material called cladding in which light travels faster. When light strikes the boundary between the fiber and the cladding, total internal reflection can occur. In this way, the beam bounces along inside the fiber as shown in **Figure 21**.

Optical fibers are used most commonly in the communications industry. For example, television programs, computer information, and phone conversations can be coded into light signals. These signals then can be sent from one place to another using optical fibers. Because of total internal reflection, signals can't leak from one fiber to another and interfere with others. As a result, the signal is transmitted clearly. One optical fiber the thickness of a human hair can carry thousands of phone conversations.

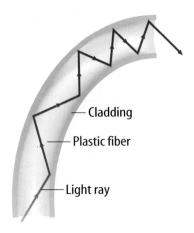

Figure 21 An optical fiber is made of materials that cause total internal reflection to occur. A light beam can travel for many kilometers through an optical fiber and lose almost no energy.

section 3 review

Summary

The Refraction of Light
- Light travels at different speeds in different materials.
- Refraction occurs when light changes speed as it travels from one material into another.

Convex and Concave Lenses
- A lens is a transparent object with at least one curved side that causes light to bend.
- A convex lens is thicker in the center than at the edges and causes light waves to converge.
- A concave lens is thinner in the center than at the edges and causes light waves to diverge.

Total Internal Reflection
- Total internal reflection occurs at the boundary between two transparent materials when light is completely reflected.
- Optical fibers use total internal reflection to transmit information over long distances with light waves.

Self Check

1. **Compare** the image formed by a concave lens and the image formed by a convex mirror.
2. **Explain** whether you would use a convex lens or a concave lens to magnify an object.
3. **Describe** the image formed by convex lens if an object is less than one focal length from the lens.
4. **Describe** how light rays traveling parallel to the optical axis are bent after they pass through a convex lens.
5. **Infer** If the speed of light were the same in all materials, would a lens cause light rays to bend?
6. **Think Critically** A light wave is bent more when it travels from air to glass than when it travels from air to water. Is the speed of light greater in water or in glass? Explain.

Applying Math

7. **Calculate Time** If light travels at 300,000 km/s and Earth is 150 million km from the Sun, how long does it take light to travel form the Sun to Earth?

section 4

Using Mirrors and Lenses

Microscopes

For almost 500 years, lenses have been used to observe objects that are too small to be seen with the unaided eye. The first microscopes were simple and magnified less than 100 times. Today, a compound microscope like the one in **Figure 22** uses a combination of lenses to magnify objects by as much as 2,500 times.

Figure 22 also shows how a microscope forms an image. An object, such as an insect or a drop of water from a pond, is placed close to a convex lens called the objective lens. This lens produces an enlarged image inside the microscope tube. The light rays from that image then pass through a second convex lens called the eyepiece lens. This lens further magnifies the image formed by the objective lens. By using two lenses, a much larger image is formed than a single lens can produce.

as you read

What You'll Learn
- **Explain** how microscopes magnify objects.
- **Explain** how telescopes make distant objects visible.
- **Describe** how a camera works.

Why It's Important
Microscopes and telescopes are used to view parts of the universe that can't be seen with the unaided eye.

Review Vocabulary
retina: region on the inner surface of the back of the eye that contains light-sensitive cells

New Vocabulary
- refracting telescope
- reflecting telescope

Figure 22 A compound microscope uses lenses to magnify objects.

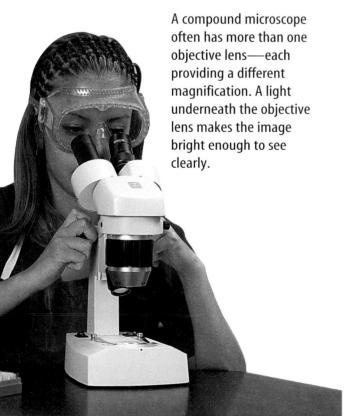

A compound microscope often has more than one objective lens—each providing a different magnification. A light underneath the objective lens makes the image bright enough to see clearly.

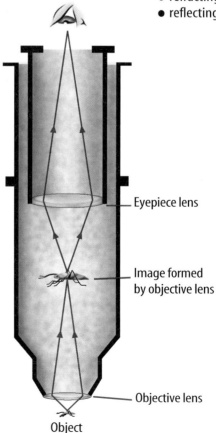

The objective lens in a compound microscope forms an enlarged image, which is then magnified by the eyepiece lens.

567

Mini LAB

Forming an Image with a Lens

Procedure
1. Fill a **glass test tube** with **water** and seal it with a **stopper**.
2. Write your name on a 10-cm × 10-cm card. Lay the test tube on the card and observe the appearance of your name.
3. Hold the test tube about 1 cm above the card and observe the appearance of your name through it again.
4. Observe what happens to your name as you slowly move the test tube away from the card.

Analysis
1. Is the water-filled test tube a concave or a convex lens?
2. Compare the images formed when the test tube was close to the card and far from the card.

Telescopes

Just as microscopes are used to magnify very small objects, telescopes are used to examine objects that are very far away. The first telescopes were made at about the same time as the first microscopes. Much of what is known about the Moon, the solar system, and the distant universe has come from images and other information gathered by telescopes.

Refracting Telescopes The simplest **refracting telescopes** use two convex lenses to form an image of a distant object. Just as in a compound microscope, light passes through an objective lens that forms an image. That image is then magnified by an eyepiece, as shown in **Figure 23.**

An important difference between a telescope and a microscope is the size of the objective lens. The main purpose of a telescope is not to magnify an image. A telescope's main purpose is to gather as much light as possible from distant objects. The larger an objective lens is, the more light can enter it. This makes images of faraway objects look brighter and more detailed when they are magnified by the eyepiece. With a large enough objective lens, it's possible to see stars and galaxies that are many trillions of kilometers away. **Figure 23** also shows the largest refracting telescope ever made.

Reading Check *How does a telescope's objective lens enable distant objects to be seen?*

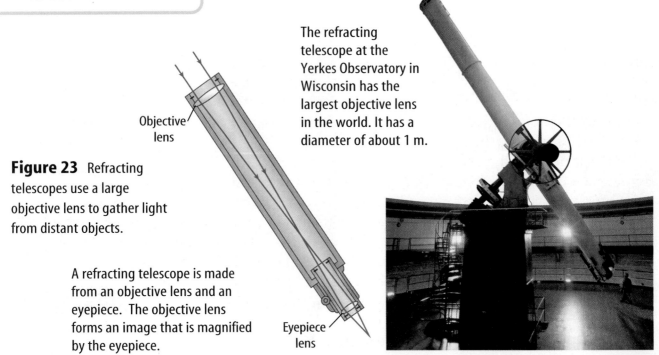

Figure 23 Refracting telescopes use a large objective lens to gather light from distant objects.

The refracting telescope at the Yerkes Observatory in Wisconsin has the largest objective lens in the world. It has a diameter of about 1 m.

A refracting telescope is made from an objective lens and an eyepiece. The objective lens forms an image that is magnified by the eyepiece.

Figure 24 Reflecting telescopes gather light by using a concave mirror.

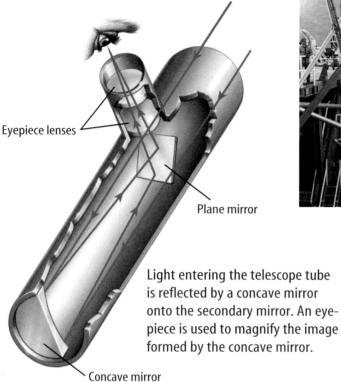

Eyepiece lenses

Plane mirror

Light entering the telescope tube is reflected by a concave mirror onto the secondary mirror. An eyepiece is used to magnify the image formed by the concave mirror.

Concave mirror

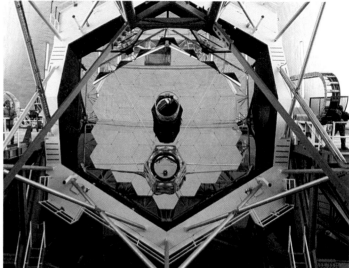

The Keck telescope in Mauna Kea, Hawaii, is the largest reflecting telescope in the world.

Reflecting Telescopes Refracting telescopes have size limitations. One problem is that the objective lens can be supported only around its edges. If the lens is extremely large, it cannot be supported enough to keep the glass from sagging slightly under its own weight. This causes the image that the lens forms to become distorted.

Reflecting telescopes can be made much larger than refracting telescopes. **Reflecting telescopes** have a concave mirror instead of a concave objective lens to gather the light from distant objects. As shown in **Figure 24**, the large concave mirror focuses light onto a secondary mirror that directs it to the eyepiece, which magnifies the image.

Because only the one reflecting surface on the mirror needs to be made carefully and kept clean, telescope mirrors are less expensive to make and maintain than lenses of a similar size. Also, mirrors can be supported not only at their edges but also on their backsides. They can be made much larger without sagging under their own weight. The Keck telescope in Hawaii, shown in **Figure 24,** is the largest reflecting telescope in the world. Its large concave mirror is 10 m in diameter, and is made of 36 six-sided segments. Each segment is 1.8 m in size and the segments are pieced together to form the mirror.

The First Telescopes
A Dutch eyeglass maker, Hans Lippershey, constructed a refracting telescope in 1608 that had a magnification of 3. In 1609 Galileo built a refracting telescope with a magnification of 20. By 1668, the first reflecting telescope was built by Isaac Newton that had a metal concave mirror with a diameter of about 5 cm. More than a century later, William Herschel built the first large reflecting telescopes with mirrors as large as 50 cm. Research the history of the telescope and make a timeline showing important events.

SECTION 4 Using Mirrors and Lenses

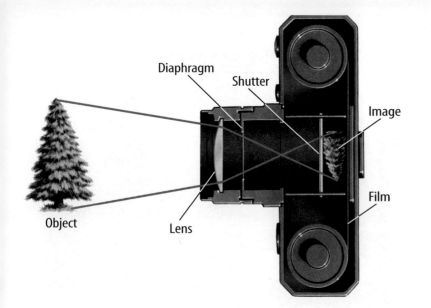

Figure 25 A camera uses a convex lens to form an image on a piece of light-sensitive film. The image formed by a camera lens is smaller than the object and is inverted.

Topic: Lasers
Visit ips.msscience.com for Web links to information about uses for lasers.

Activity Make a table listing different types of lasers and how they are used.

Cameras

You probably see photographs taken by cameras almost every day. A typical camera uses a convex lens to form an image on a section of film, just as your eye's lens focuses an image on your retina. The convex lens has a short focal length, so it forms an image that is smaller than the object and inverted on the film. Look at the camera shown in **Figure 25.** When the shutter is open, the convex lens focuses an image on a piece of film that is sensitive to light. Light-sensitive film contains chemicals that undergo chemical reactions when light hits it. The brighter parts of the image affect the film more than the darker parts do.

Reading Check *What type of lens does a camera use?*

If too much light strikes the film, the image formed on the film is overexposed and looks washed out. On the other hand, if too little light reaches the film, the photograph might be too dark. To control how much light reaches the film, many cameras have a device called a diaphragm. The diaphragm is opened to let more light onto the film and closed to reduce the amount of light that strikes the film.

Lasers

Perhaps you've seen the narrow, intense beams of laser light used in a laser light show. Intense laser beams are also used for different kinds of surgery. Why can laser beams be so intense? One reason is that a laser beam doesn't spread out as much as ordinary light as it travels.

Spreading Light Beams Suppose you shine a flashlight on a wall in a darkened room. The size of the spot of light on the wall depends on the distance between the flashlight and the wall. As the flashlight moves farther from the wall, the spot of light gets larger. This is because the beam of light produced by the flashlight spreads out as it travels. As a result, the energy carried by the light beam is spread over an increasingly larger area as the distance from the flashlight gets larger. As the energy is spread over a larger area, the energy becomes less concentrated and the intensity of the beam decreases.

Using Laser Light Laser light is different from the light produced by the flashlight in several ways, as shown in **Figure 26**. One difference is that in a beam of laser light, the crests and troughs of the light waves overlap, so the waves are in phase.

Because a laser beam doesn't spread out as much as ordinary light, a large amount of energy can be applied to a very small area. This property enables lasers to be used for cutting and welding materials and as a replacement for scalpels in surgery. Less intense laser light is used for such applications as reading and writing to CDs or in grocery store bar-code readers. Surveyors and builders use lasers to measure distances, angles, and heights. Laser beams also are used to transmit information through space or through optical fibers.

Figure 26 Laser light is different from the light produced by a lightbulb.

The light from a bulb contains waves with many different wavelengths that are out of phase and traveling in different directions.

The light from a laser contains waves with only one wavelength that are in phase and traveling in the same direction.

section 4 review

Summary

Microscopes, Telescopes, and Cameras

- A compound microscope uses an objective lens and an eyepiece lens to form an enlarged image of an object.
- A refracting telescope contains a large objective lens to gather light and a smaller eyepiece lens to magnify the image.
- A reflecting telescope uses a large concave mirror to gather light and an eyepiece lens to magnify the image.
- The image formed by a telescope becomes brighter and more detailed as the size of the objective lens or concave mirror increases.
- A camera uses a convex lens to form an image on light-sensitive film.

Laser Light

- Light from a laser contains light waves that are in phase, have only one wavelength, and travel in the same direction.
- Because laser light does not spread out much as it travels the energy it carries can be applied over a very small area.

Self Check

1. **Explain** why the concave mirror of a reflecting telescope can be made much larger than the objective lens of a refracting telescope.
2. **Describe** how a beam of laser light is different than the beam of light produced by a flashlight.
3. **Explain** why the objective lens of a refracting telescope is much larger than the objective lens of a compound microscope.
4. **Infer** how the image produced by a compound microscope would be different if the eyepiece lens were removed from the microscope.
5. **Think Critically** Explain why the intensity of the light in a flashlight beam decreases as the flashlight moves farther away.

Applying Math

6. **Calculate Image Size** The size of an image is related to the magnification of an optical instrument by the following formula:

   Image size = magnification × object size

   A blood cell has a diameter of 0.001 cm. How large is the image formed by a microscope with a magnification of 1,000?

Image Formation by a Convex Lens

Goals
- **Measure** the image distance as the object distance changes.
- **Observe** the type of image formed as the object distance changes.

Possible Materials
convex lens
modeling clay
meterstick
flashlight
masking tape
20-cm square piece of cardboard with a white surface

Safety Precautions

Real-World Question
The type of image formed by a convex lens, also called a converging lens, is related to the distance of the object from the lens. This distance is called the object distance. The location of the image also is related to the distance of the object from the lens. The distance from the lens to the image is called the image distance. How are the image distance and object distance related for a convex lens?

Procedure

1. **Design** a data table to record your data. Make three columns in your table —one column for the object distance, another for the image distance, and the third for the type of image.

Convex Lens Data

Object Distance (m)	Image Distance (m)	Image Type
Do not write in this book.		

2. Use the modeling clay to make the lens stand upright on the lab table.
3. Form the letter *F* on the glass surface of the flashlight with masking tape.
4. Turn on the flashlight and place it 1 m from the lens. Position the flashlight so the flashlight beam is shining through the lens.
5. **Record** the distance from the flashlight to the lens in the object distance column in your data table.
6. Hold the cardboard vertically upright on the other side of the lens, and move it back and forth until a sharp image of the letter *F* is obtained.

572 CHAPTER 19 Light, Mirrors, and Lenses

Using Scientific Methods

7. **Measure** the distance of the card from the lens using the meterstick, and record this distance in the Image Distance column in your data table.

8. **Record** in the third column of your data table whether the image is upright or inverted, and smaller or larger.

9. Repeat steps 4 through 8 for object distances of 0.50 m and 0.25 m and record your data in your data table.

Analyze Your Data

1. **Describe** any observed relationship between the object distance, and the image type.
2. **Identify** the variables involved in determining the image type for a convex lens.

Conclude and Apply

1. **Explain** how the image distance changed as the object distance decreased.
2. **Identify** how the image changed as the object distance decreased.
3. **Predict** what would happen to the size of the image if the flashlight were much farther away than 1 m.

Communicating Your Data

Demonstrate this lab to a third-grade class and explain how it works. **For more help, refer to the** Science Skill Handbook.

LAB 573

Oops! Accidents in SCIENCE

SOMETIMES GREAT DISCOVERIES HAPPEN BY ACCIDENT!

Eyeglasses

Inventor Unknown

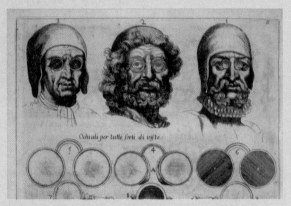
This Italian engraving from the 1600s shows some different types of glasses.

"It is not yet twenty years since the art of making spectacles, one of the most useful arts on Earth, was discovered. I, myself, have seen and conversed with the man who made them first."

This quote from an Italian monk dates back to 1306 and is one of the first historical records to refer to eyeglasses. Unfortunately, the monk, Giordano, never actually named the man he met. Thus, the inventor of eyeglasses remains unknown.

The mystery exists, in part, because different cultures in different places used some type of magnifying tool to improve their vision. For example, a rock-crystal lens, made by early Assyrians who lived 3,500 years ago in what is now Iraq, may have been used to improve vision. About 2,000 years ago, the Roman writer Seneca looked through a glass globe of water to make the letters appear bigger in the books he read. By the tenth century, glasses had been invented in China, but they were used to keep away bad luck, not to improve vision.

In the mid 1400s in Europe, eyeglasses began to appear in paintings of scholars, clergy, and the upper classes—eyeglasses were so expensive that only the rich could afford them. In the early 1700s, for example, glasses cost roughly $200, which is comparable to thousands of dollars today. By the mid-1800s, improvements in manufacturing techniques made eyeglasses much less expensive to make, and thus this important invention became widely available to people of all walks of life.

How Eyeglasses Work

Eyeglasses are used to correct farsightedness and nearsightedness, as well as other vision problems. The eye focuses light rays to form an image on a region called the retina on the back of the eye. Farsighted people have difficulty seeing things close up because light rays from nearby objects do not converge enough to form an image on the retina. This problem can be corrected by using convex lenses that cause light rays to converge before they enter the eye. Nearsighted people have problems seeing distant objects because light rays from far-away objects are focused in front of the retina. Concave lenses that cause light rays to diverge are used to correct this vision problem.

Research In many parts of the world, people have no vision care, and eye diseases and poor vision go untreated. Research the work of groups that bring eye care to people.

For more information, visit ips.msscience.com/oops

chapter 19 Study Guide

Reviewing Main Ideas

Section 1 Properties of Light

1. Light waves can be absorbed, reflected, or transmitted when they strike an object.
2. The color of an object depends on the wavelengths of light reflected by the object.

Section 2 Reflection and Mirrors

1. Light reflected from the surface of an object obeys the law of reflection—the angle of incidence equals the angle of reflection.
2. Concave mirrors cause light waves to converge, or meet. Convex mirrors cause light waves to diverge, or spread apart.

Section 3 Refraction and Lenses

1. Light waves bend, or refract, when they change speed in traveling from one medium to another.
2. A convex lens causes light waves to converge, and a concave lens causes light waves to diverge.

Section 4 Using Mirrors and Lenses

1. A compound microscope uses a convex objective lens to form an enlarged image that is further enlarged by an eyepiece.
2. A refracting telescope uses a large objective lens and an eyepiece lens to form an image of a distant object.
3. A reflecting telescope uses a large concave mirror that gathers light and an eyepiece lens to form an image of a distant object.
4. Cameras use a convex lens to form an image on light-sensitive film.

Visualizing Main Ideas

Copy and complete the following concept map.

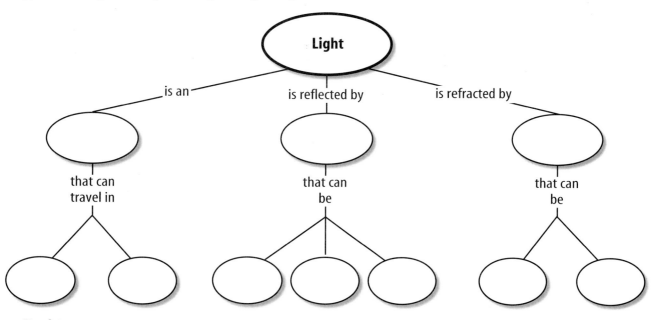

chapter 19 Review

Using Vocabulary

concave lens p. 565
convex lens p. 564
focal length p. 558
focal point p. 558
law of reflection p. 555
lens p. 563
light ray p. 550
medium p. 551
reflecting telescope p. 569
refracting telescope p. 568

Complete each statement using a word or words from the vocabulary list above.

1. A _____ is the material in which a light wave travels.
2. A narrow beam of light that travels in a straight line is a _____.
3. The _____ is the distance from a lens or a mirror to the focal point.
4. Light rays traveling parallel to the optical axis of a convex lens are bent so they pass through the _____.
5. A transparent object with at least one curved surface that causes light waves to bend is a _____.
6. A _____ is thicker in the center than it is at the edges.
7. A _____ uses a large concave mirror to gather light from distant objects.

Checking Concepts

Choose the word or phrase that best answers the question.

8. Light waves travel the fastest through which of the following?
   A) air   C) water
   B) diamond   D) a vacuum

9. Which of the following determines the color of light?
   A) a prism   C) its wavelength
   B) its refraction D) its incidence

10. If an object reflects red and green light, what color does the object appear to be?
    A) yellow   C) green
    B) red   D) purple

11. If an object absorbs all the light that hits it, what color is it?
    A) white   C) black
    B) blue   D) green

12. What type of image is formed by a plane mirror?
    A) upright   C) magnified
    B) inverted   D) all of these

13. How is the angle of incidence related to the angle of reflection?
    A) It's greater.   C) It's the same.
    B) It's smaller.   D) It's not focused.

14. Which of the following can be used to magnify objects?
    A) a concave lens C) a convex mirror
    B) a convex lens   D) all of these

15. Which of the following describes the light waves that make up laser light?
    A) same wavelength
    B) same direction
    C) in phase
    D) all of these

16. What is an object that reflects some light and transmits some light called?
    A) colored   C) opaque
    B) diffuse   D) translucent

17. What is the main purpose of the objective lens or concave mirror in a telescope?
    A) invert images   C) gather light
    B) reduce images   D) magnify images

18. Which of the following types of mirror can form an image larger than the object?
    A) convex   C) plane
    B) concave   D) all of these

576 CHAPTER REVIEW

chapter 19 Review

Thinking Critically

19. **Diagram** Suppose you can see a person's eyes in a mirror. Draw a diagram to determine whether or not that person can see you.

20. **Determine** A singer is wearing a blue outfit. What color spotlights could be used to make the outfit appear to be black?

21. **Form a hypothesis** to explain why sometimes you can see two images of yourself reflected from a window at night.

22. **Explain** why a rough surface, such as a road, becomes shiny in appearance and a better reflector when it is wet.

23. **Infer** An optical fiber is made of a material that forms the fiber and a different material that forms the outer covering. For total internal reflection to occur, how does the speed of light in the fiber compare with the speed of light in the outer covering?

Use the table below to answer question 24.

Magnification by a Convex Lens	
Object Distance (cm)	Magnification
25	4.00
30	2.00
40	1.00
60	0.50
100	0.25

24. **Use a Table** In the table above, the object distance is the distance of the object from the lens. The magnification is the image size divided by the object size. If the focal length of the lens is 20 cm, how does the size of the image change as the object gets farther from the focal point?

25. **Calculate** What is the ratio of the distance at which the magnification equals 1.00 to the focal length of the lens?

Performance Activities

26. **Oral Presentation** Investigate the types of mirrors used in fun houses. Explain how these mirrors are formed, and why they produce distorted images. Demonstrate your findings to your class.

27. **Reverse Writing** Images are reversed left to right in a plane mirror. Write a note to a friend that can be read only in a plane mirror.

28. **Design an experiment** to determine the focal length of a convex lens. Write a report describing your experiment, including a diagram.

Applying Math

Use the graph below to answer questions 29 and 30.

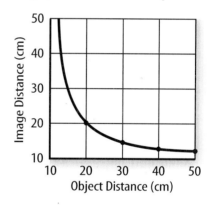

29. **Image Position** The graph shows how the distance of an image from a convex lens is related to the distance of the object from the lens. How does the position of the image change as the object gets closer to the lens?

30. **Magnification** The magnification of the image equals the image distance divided by the object distance. At what object distance does the magnification equal 2?

ips.msscience.com/chapter_review

Chapter 19 Standardized Test Practice

Part 1 Multiple Choice

Record your answers on the answer sheet provided by your teacher or on a sheet of paper.

1. Which of the following describes an object that allows no light to pass through it?
   A. transparent   C. opaque
   B. translucent   D. diffuse

2. Which statement is always true about the image formed by a concave lens?
   A. It is upside down and larger than the object.
   B. It is upside down and smaller than the object.
   C. It is upright and larger than the object.
   D. It is upright and smaller than the object.

Use the figure below to answer questions 3 and 4.

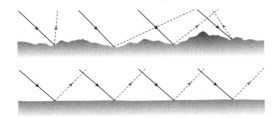

3. Which of the following describes the process occurring in the upper panel of the figure?
   A. refraction
   B. diffuse reflection
   C. regular reflection
   D. total internal reflection

4. The surface in the lower panel of the figure would be like which of the following?
   A. a mirror   C. a sheet of paper
   B. waxed paper   D. a painted wall

5. Why does a leaf look green?
   A. It reflects green light.
   B. It absorbs green light.
   C. It reflects all colors of light.
   D. It reflects all colors except green.

6. What does a refracting telescope use to form an image of a distant object?
   A. two convex lenses
   B. a concave mirror and a plane mirror
   C. two concave lenses
   D. two concave mirrors

7. Through which of the following does light travel the slowest?
   A. air   C. water
   B. diamond   D. vacuum

8. What is the bending of a light wave due to a change in speed?
   A. reflection   C. refraction
   B. diffraction   D. transmission

Use the figure below to answer questions 9 and 10.

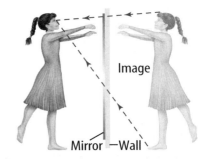

9. If the girl is standing 1 m from the mirror, where will her image seem to be located?
   A. 2 m behind the mirror
   B. 1 m behind the mirror
   C. 2 m in front of the mirror
   D. 1 m in front of the mirror

10. Which of the following describes the image of the girl formed by the plane mirror?
    A. It will be upside down.
    B. It will be in front of the mirror.
    C. It will be larger than the girl.
    D. It will be reversed left to right.

Standardized Test Practice

Part 2 Short Response/Grid In

Record your answers on the answer sheet provided by your teacher or on a sheet of paper.

11. Light travels slower in diamond than in air. Explain whether total internal reflection could occur for a light wave traveling in the diamond toward the diamond's surface.

Use the figure below to answer question 12.

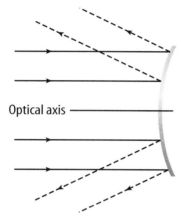

12. Identify the type of mirror shown in the figure and describe the image this mirror forms.

13. Under white light the paper of this page looks white and the print looks black. What color would the paper and the print appear to be under red light?

14. A light ray strikes a plane mirror such that the angle of incidence is 30°. What is the angle between the light ray and the surface of the mirror?

15. Contrast the light beam from a flashlight and a laser light beam.

16. An actor on stage is wearing a magenta outfit. Explain what color the outfit would appear in red light, in blue light, and in green light.

17. To use a convex lens as a magnifying lens, where must the object be located?

Part 3 Open Ended

Record your answers on a sheet of paper.

18. Explain why you can see the reflection of trees in the water of a lake on a calm day, but not a very windy day.

19. Describe how total internal reflection enables optical fibers to transmit light over long distances.

20. What happens when a source of light is placed at the focal point of a concave mirror? Give an example.

Use the illustration below to answer question 21.

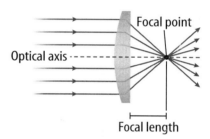

21. Describe how the position of the focal point changes as the lens becomes flatter and less curved.

22. Compare the images formed by a concave mirror when an object is between the focal point and the mirror and when an object is beyond the focal point.

23. Explain why increasing the size of the concave mirror in a reflecting telescope improves the images formed.

Test-Taking Tip

Organize Your Main Points For essay questions, spend a few minutes listing and organizing the main points that you plan to discuss. Make sure to do all of this work on your scratch paper, not your answer sheet.

Question 19 Organize your discussion points by first listing what you know about optical fibers and total internal reflection.

unit 7
Electricity & Magnetism

How Are Cone-bearing Trees & Static Electricity Connected?

NATIONAL GEOGRAPHIC

When the bark of a cone-bearing tree is broken it secretes resin, which hardens and seals the tree's wound. The resin of some ancient trees fossilized over time, forming a golden, gemlike substance called amber. The ancient Greeks prized amber highly, not only for its beauty, but also because they believed it had magical qualities. They had noticed that when amber was rubbed with wool or fur, small bits of straw or ash would stick to it. Because of amber's color and its unusual properties, some believed that amber was solidified sunshine. The Greek name for amber was *elektron* which means "substance of the Sun."

By the seventeenth century, the behavior of amber had sparked the curiosity of a number of scientists, and an explanation of amber's behavior finally emerged. When amber is rubbed by wool or fur, static electricity is produced. Today, a device called a Van de Graaff generator, like the one shown above, can produce static electricity involving millions of volts, and has been used to explore the nature of matter in atom-smashing experiments.

unit ⚡ projects

Visit ips.msscience.com/unit_project to find project ideas and resources.
Projects include:
- **History** Design a creative bookmark depicting a variety of aspects of Ben Franklin's contributions to science and his country.
- **Career** Discover how magnetic-resonance imaging is used in the medical field and how it compares to traditional X rays.
- **Model** Design an electrifying review game demonstrating your new understanding of electricity and magnetism.

 Hybrid Vehicles explores new vehicles being produced by car manufacturers. Analyze the advantages and disadvantages of hybrid vehicles.

chapter 20

Electricity

The BIG Idea

Electrical energy can be converted into other forms of energy when electric charges flow in a circuit.

SECTION 1
Electric Charge
Main Idea Electric charges are positive or negative and exert forces on each other.

SECTION 2
Electric Current
Main Idea A battery produces an electric field in a closed circuit that causes electric charges to flow.

SECTION 3
Electric Circuits
Main Idea Electrical energy can be transferred to devices connected in an electric circuit.

A Blast of Energy

This flash of lightning is an electric spark that releases an enormous amount of electrical energy in an instant. However, in homes and other buildings, electrical energy is released in a controlled way by the flow of electric currents.

Science Journal Write a paragraph describing a lightning flash you have seen. Include information about the weather conditions at the time.

Start-Up Activities

Observing Electric Forces

No computers? No CD players? No video games? Can you imagine life without electricity? Electricity also provides energy that heats and cools homes and produces light. The electrical energy that you use every day is produced by the forces that electric charges exert on each other.

1. Inflate a rubber balloon.
2. Place small bits of paper on your desktop and bring the balloon close to the bits of paper. Record your observations.
3. Charge the balloon by holding it by the knot and rubbing it on your hair or a piece of wool.
4. Bring the balloon close to the bits of paper. Record your observations.
5. Charge two balloons using the procedure in step 3. Hold each balloon by its knot and bring the balloons close to each other. Record your observations.
6. **Think Critically** Compare and contrast the force exerted on the bits of paper by the charged balloon and the force exerted by the two charged balloons on each other.

Electricity Make the following Foldable to help you understand the terms *electric charge, electric current,* and *electric circuit*.

STEP 1 **Fold** the top of a vertical piece of paper down and the bottom up to divide the paper into thirds.

STEP 2 **Turn** the paper horizontally; **unfold and label** the three columns as shown.

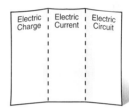

Read and Write Before you read the chapter, write a definition of electric charge, electric current, and electric circuit in the appropriate column. As you read the chapter, correct your definition and add additional information about each term.

Preview this chapter's content and activities at ips.msscience.com

Get Ready to Read

Make Predictions

1 Learn It! A prediction is an educated guess based on what you already know. One way to predict while reading is to guess what you believe the author will tell you next. As you are reading, each new topic should make sense because it is related to the previous paragraph or passage.

2 Practice It! Read the excerpt below from Section 1. Based on what you have read, make predictions about what you will read in the rest of the lesson. After you read Section 1, go back to your predictions to see if they were correct.

> Predict how the electric force depends on the amount of charge.

> Predict how the electric force would change as charged objects get farther apart.

> Can you predict how the electric force between two electrons changes as the electrons get closer together?

> The electric force between two charged objects depends on the distance between them and the **amount of charge** on each object. The electric force between two charged objects gets stronger **as the charges get closer together**. **A positive and a negative charge** are attracted to each other more strongly if they are closer together.
>
> —*from page 587*

3 Apply It! Before you read, skim the questions in the Chapter Review. Choose three questions and predict the answers.

Target Your Reading

Reading Tip

As you read, check the predictions you made to see if they were correct.

Use this to focus on the main ideas as you read the chapter.

1. **Before you read** the chapter, respond to the statements below on your worksheet or on a numbered sheet of paper.
   - Write an **A** if you **agree** with the statement.
   - Write a **D** if you **disagree** with the statement.

2. **After you read** the chapter, look back to this page to see if you've changed your mind about any of the statements.
   - If any of your answers changed, explain why.
   - Change any false statements into true statements.
   - Use your revised statements as a study guide.

Before You Read A or D		Statement	After You Read A or D
	1	Atoms become ions by gaining or losing electrons.	
	2	It is safe to take shelter under a tree during a lightning storm.	
	3	Electric current can follow only one path in a parallel circuit.	
	4	Electrons flow in straight lines through conducting wires.	
	5	Batteries produce electrical energy through nuclear reactions.	
	6	The force between electric charges always is attractive.	
	7	Electrical energy can be transformed into other forms of energy.	
	8	If the voltage in a circuit doesn't change, the current increases if the resistance decreases.	
	9	Electric charges must be touching to exert forces on each other.	

Science Online
Print out a worksheet of this page at ips.msscience.com

584 B

section 1
Electric Charge

as you read

What You'll Learn
- **Describe** how objects can become electrically charged.
- **Explain** how an electric charge affects other electric charges.
- **Distinguish** between electric conductors and insulators.
- **Describe** how electric discharges such as lightning occur.

Why It's Important
All electric phenomena result from the forces electric charges exert on each other.

Review Vocabulary
force: the push or pull one object exerts on another

New Vocabulary
- ion
- static charge
- electric force
- electric field
- insulator
- conductor
- electric discharge

Electricity

You can't see, smell, or taste electricity, so it might seem mysterious. However, electricity is not so hard to understand when you start by thinking small—very small. All solids, liquids, and gases are made of tiny particles called atoms. Atoms, as shown in **Figure 1,** are made of even smaller particles called protons, neutrons, and electrons. Protons and neutrons are held together tightly in the nucleus at the center of an atom, but electrons swarm around the nucleus in all directions. Protons and electrons have electric charge, but neutrons have no electric charge.

Positive and Negative Charge There are two types of electric charge—positive and negative. Protons have a positive charge, and electrons have a negative charge. The amount of negative charge on an electron is exactly equal to the amount of positive charge on a proton. Because atoms have equal numbers of protons and electrons, the amount of positive charge on all the protons in the nucleus of an atom is balanced by the negative charge on all the electrons moving around the nucleus. Therefore, atoms are electrically neutral, which means they have no overall electric charge.

An atom becomes negatively charged when it gains extra electrons. If an atom loses electrons it becomes positively charged. A positively or negatively charged atom is called an **ion** (I ahn).

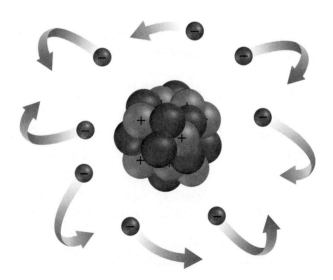

Figure 1 An atom is made of positively charged protons (orange), negatively charged electrons (red), and neutrons (blue) with no electric charge.
Identify where the protons and neutrons are located in an atom.

584 CHAPTER 20 Electricity

Figure 2 Rubbing can move electrons from one object to another. Hair holds electrons more loosely than the balloon holds them. As a result, electrons are moved from the hair to the balloon when the two make contact.
Infer *which object has become positively charged and which has become negatively charged.*

Electrons Move in Solids Electrons can move from atom to atom and from object to object. Rubbing is one way that electrons can be transferred. If you have ever taken clinging clothes from a clothes dryer, you have seen what happens when electrons are transferred from one object to another.

Suppose you rub a balloon on your hair. The atoms in your hair hold their electrons more loosely than the atoms on the balloon hold theirs. As a result, electrons are transferred from the atoms in your hair to the atoms on the surface of the balloon, as shown in **Figure 2.** Because your hair loses electrons, it becomes positively charged. The balloon gains electrons and becomes negatively charged. Your hair and the balloon become attracted to one another and make your hair stand on end. This imbalance of electric charge on an object is called a **static charge.** In solids, static charge is due to the transfer of electrons between objects. Protons cannot be removed easily from the nucleus of an atom and usually do not move from one object to another.

Reading Check *How does an object become electrically charged?*

Ions Move in Solutions Sometimes, the movement of charge can be caused by the movement of ions instead of the movement of electrons. Table salt—sodium chloride—is made of sodium ions and chloride ions that are fixed in place and cannot move through the solid. However, when salt is dissolved in water, the sodium and chloride ions break apart and spread out evenly in the water, forming a solution, as shown in **Figure 3.** Now the positive and negative ions are free to move. Solutions containing ions play an important role in enabling different parts of your body to communicate with each other. **Figure 4** shows how a nerve cell uses ions to transmit signals. These signals moving throughout your body enable you to sense, move, and even think.

Figure 3 When table salt (NaCl) dissolves in water, the sodium ions and chloride ions break apart. These ions now are able to carry electric energy.

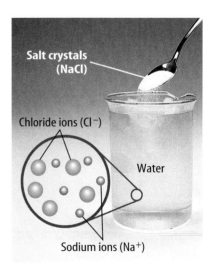

NATIONAL GEOGRAPHIC VISUALIZING NERVE IMPULSES

Figure 4

The control and coordination of all your bodily functions involves signals traveling from one part of your body to another through nerve cells. Nerve cells use ions to transmit signals from one nerve cell to another.

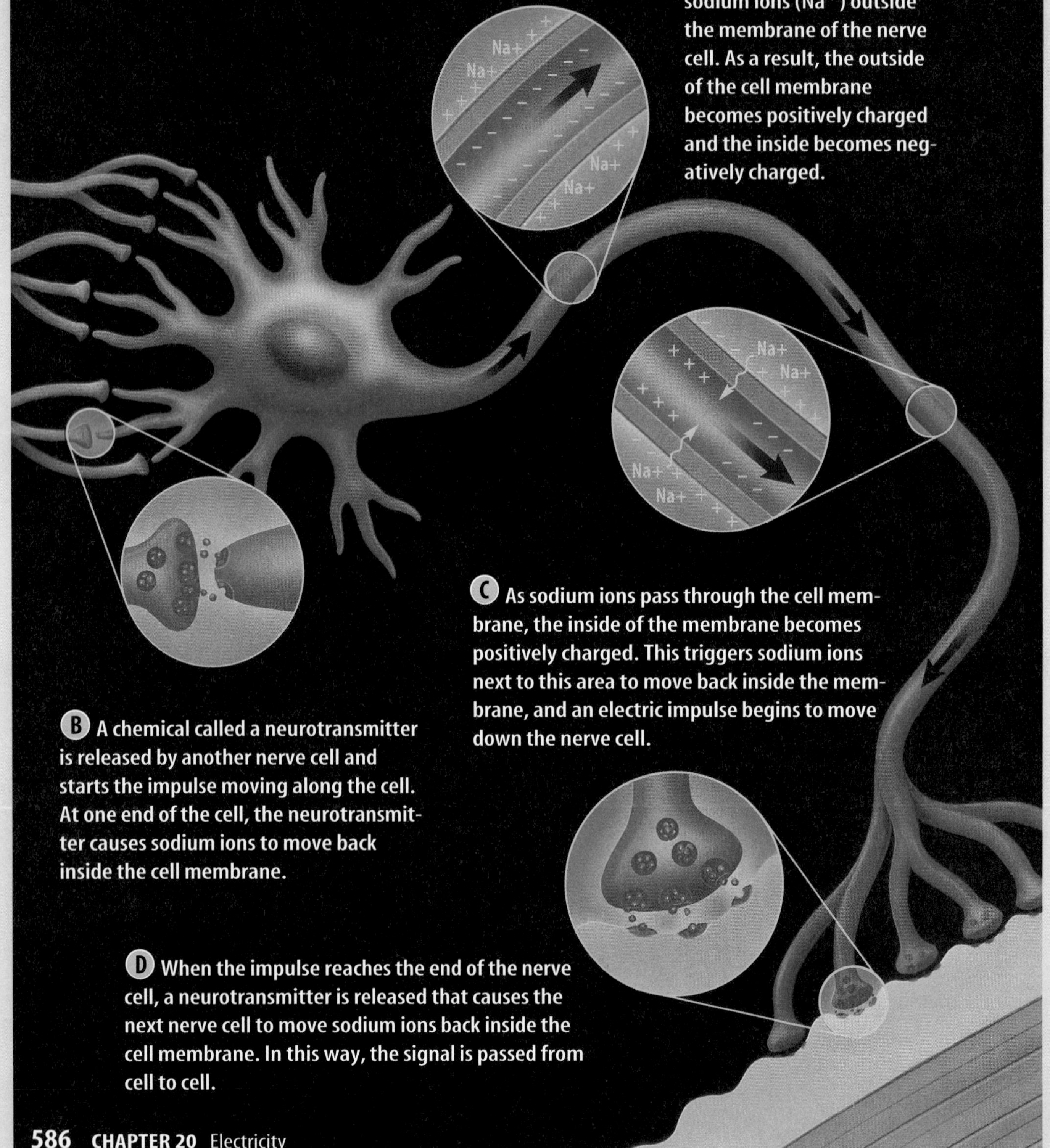

A When a nerve cell is not transmitting a signal, it moves positively charged sodium ions (Na^+) outside the membrane of the nerve cell. As a result, the outside of the cell membrane becomes positively charged and the inside becomes negatively charged.

B A chemical called a neurotransmitter is released by another nerve cell and starts the impulse moving along the cell. At one end of the cell, the neurotransmitter causes sodium ions to move back inside the cell membrane.

C As sodium ions pass through the cell membrane, the inside of the membrane becomes positively charged. This triggers sodium ions next to this area to move back inside the membrane, and an electric impulse begins to move down the nerve cell.

D When the impulse reaches the end of the nerve cell, a neurotransmitter is released that causes the next nerve cell to move sodium ions back inside the cell membrane. In this way, the signal is passed from cell to cell.

Unlike charges attract.

Like charges repel.   Like charges repel.

Figure 5 A positive charge and a negative charge attract each other. Two positive charges repel each other, as do two negative charges.

Electric Forces

The electrons in an atom swarm around the nucleus. What keeps these electrons close to the nucleus? The positively charged protons in the nucleus exert an attractive electric force on the negatively charged electrons. All charged objects exert an **electric force** on each other. The electric force between two charges can be attractive or repulsive, as shown in **Figure 5.** Objects with the same type of charge repel one another and objects with opposite charges attract one another. This rule is often stated as "like charges repel, and unlike charges attract."

The electric force between two charged objects depends on the distance between them and the amount of charge on each object. The electric force between two electric charges gets stronger as the charges get closer together. A positive and a negative charge are attracted to each other more strongly if they are closer together. Two like charges are pushed away more strongly from each other the closer they are. The electric force between two objects that are charged, such as two balloons that have been rubbed on wool, increases if the amount of charge on at least one of the objects increases.

 *How does the electric force between two charged objects depend on the distance between them?*

Electric Fields You might have noticed examples of how charged objects don't have to be touching to exert an electric force on each other. For instance, two charged balloons push each other apart even though they are not touching. How are charged objects able to exert forces on each other without touching?

Electric charges exert a force on each other at a distance through an **electric field** that exists around every electric charge. **Figure 6** shows the electric field around a positive and a negative charge. An electric field gets stronger as you get closer to a charge, just as the electric force between two charges becomes greater as the charges get closer together.

Figure 6 The lines with arrowheads represent the electric field around charges. The direction of each arrow is the direction a positive charge would move if it were placed in the field.

The electric field arrows point away from a positive charge.

The electric field arrows point toward a negative charge.
Explain *why the electric field arrows around a negative charge are in the opposite direction of the arrows around a positive charge.*

SECTION 1 Electric Charge **587**

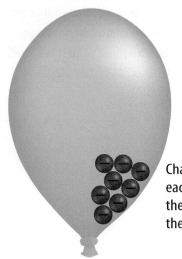

Figure 7 Electric charges move more easily through conductors than through insulators.

Charges placed on an insulator repel each other but cannot move easily on the surface of the insulator. As a result, the charges remain in one place.

The three wires in this electric cable are made of copper, which is a conductor. The wires are covered with plastic insulation that keeps the copper wires from touching each other.

Insulators and Conductors

Rubbing a balloon on your hair transfers electrons from your hair to the balloon. However, only the part of the balloon that was rubbed on your hair becomes charged, because electrons cannot move easily through rubber. As a result, the electrons that were rubbed onto the balloon tend to stay in one place, as shown in **Figure 7.** A material in which electrons cannot move easily from place to place is called an **insulator.** Examples of insulators are plastic, wood, glass, and rubber.

Materials that are **conductors** contain electrons that can move more easily in the material. The electric wire in **Figure 7** is made from a conductor coated with an insulator such as plastic. Electrons move easily in the conductor but do not move easily through the plastic insulation. This prevents electrons from moving through the insulation and causing an electric shock if someone touches the wire.

Metals as Conductors The best conductors are metals such as copper, gold, and aluminum. In a metal atom, a few electrons are not attracted as strongly to the nucleus as the other electrons, and are loosely bound to the atom. When metal atoms form a solid, the metal atoms can move only short distances. However, the electrons that are loosely bound to the atoms can move easily in the solid piece of metal. In an insulator, the electrons are bound tightly in the atoms that make up the insulator and therefore cannot move easily.

Topic: Superconductors
Visit ips.msscience.com for Web links to information about materials that are superconductors.

Activity Make a table listing five materials that can become superconductors and the critical temperature for each material.

Induced Charge

Has this ever happened to you? You walk across a carpet and as you reach for a metal doorknob, you feel an electric shock. Maybe you even see a spark jump between your fingertip and the doorknob. To find out what happened, look at **Figure 8.**

As you walk, electrons are rubbed off the rug by your shoes. The electrons then spread over the surface of your skin. As you bring your hand close to the doorknob, the electric field around the excess electrons on your hand repels the electrons in the doorknob. Because the doorknob is a good conductor, its electrons move easily away from your hand. The part of the doorknob closest to your hand then becomes positively charged. This separation of positive and negative charges due to an electric field is called an induced charge.

If the electric field between your hand and the knob is strong enough, charge can be pulled from your hand to the doorknob, as shown in **Figure 8.** This rapid movement of excess charge from one place to another is an **electric discharge.** Lightning is an example of an electric discharge. In a storm cloud, air currents sometimes cause the bottom of the cloud to become negatively charged. This negative charge induces a positive charge in the ground below the cloud. A cloud-to-ground lightning stroke occurs when electric charge moves between the cloud and the ground.

As you walk across the floor, you rub electrons from the carpet onto the bottom of your shoes. These electrons then spread out over your skin, including your hands.

As you bring your hand close to the metal doorknob, electrons on the doorknob move as far away from your hand as possible. The part of the doorknob closest to your hand is left with a positive charge.

Figure 8 A spark that jumps between your fingers and a metal doorknob starts at your feet. **Identify** *another example of an electric discharge.*

The attractive electric force between the electrons on your hand and the induced positive charge on the doorknob might be strong enough to pull electrons from your hand to the doorknob. You might see this as a spark and feel a mild electric shock.

Grounding

Lightning is an electric discharge that can cause damage and injury because a lightning bolt releases an extremely large amount of electric energy. Even electric discharges that release small amounts of energy can damage delicate circuitry in devices such as computers. One way to avoid the damage caused by electric discharges is to make the excess charges flow harmlessly into Earth's surface. Earth can be a conductor, and because it is so large, it can absorb an enormous quantity of excess charge.

The process of providing a pathway to drain excess charge into Earth is called grounding. The pathway is usually a conductor such as a wire or a pipe. You might have noticed lightning rods at the top of buildings and towers, as shown in **Figure 9**. These rods are made of metal and are connected to metal cables that conduct electric charge into the ground if the rod is struck by lightning.

Figure 9 A lightning rod can protect a building from being damaged by a lightning strike.

section 1 review

Summary

Electric Charges
- There are two types of electric charge—positive charge and negative charge.
- The amount of negative charge on an electron is equal to the amount of positive charge on a proton.
- Objects that are electrically neutral become negatively charged when they gain electrons and positively charged when they lose electrons.

Electric Forces
- Like charges repel and unlike charges attract.
- The force between two charged objects increases as they get closer together.
- A charged object is surrounded by an electric field that exerts a force on other charged objects.

Insulators and Conductors
- Electrons cannot move easily in an insulator but can move easily in a conductor.

Self Check

1. **Explain** why when objects become charged it is electrons that are transferred from one object to another rather than protons.
2. **Compare and contrast** the movement of electric charge in a solution with the transfer of electric charge between solid objects.
3. **Explain** why metals are good conductors.
4. **Compare and contrast** the electric field around a negative charge and the electric field around a positive charge.
5. **Explain** why an electric discharge occurs.
6. **Think Critically** A cat becomes negatively charged when it is brushed. How does the electric charge on the brush compare to the charge on the cat?

Applying Skills

7. **Analyze** You slide out of a car seat and as you touch the metal car door, a spark jumps between your hand and the door. Describe how the spark was formed.

590 CHAPTER 20 Electricity

ips.msscience.com/self_check_quiz

section 2

Electric Current

Flow of Charge

An electric discharge, such as a lightning bolt, can release a huge amount of energy in an instant. However, electric lights, refrigerators, TVs, and stereos need a steady source of electrical energy that can be controlled. This source of electrical energy comes from an **electric current,** which is the flow of electric charge. In solids, the flowing charges are electrons. In liquids, the flowing charges are ions, which can be positively or negatively charged. Electric current is measured in units of amperes (A). A model for electric current is flowing water. Water flows downhill because a gravitational force acts on it. Similarly, electrons flow because an electric force acts on them.

A Model for a Simple Circuit How does a flow of water provide energy? If the water is separated from Earth by using a pump, the higher water now has gravitational potential energy, as shown in **Figure 10.** As the water falls and does work on the waterwheel, the water loses potential energy and the waterwheel gains kinetic energy. For the water to flow continuously, it must flow through a closed loop. Electric charges will flow continuously only through a closed conducting loop called a **circuit.**

as you read

What You'll Learn
- **Relate** voltage to the electrical energy carried by an electric current.
- **Describe** a battery and how it produces an electric current.
- **Explain** electrical resistance.

Why It's Important
Electric current provides a steady source of electrical energy that powers the electric appliances you use every day.

Review Vocabulary
gravitational potential energy: the energy stored in an object due to its position above Earth's surface

New Vocabulary
- electric current
- voltage
- circuit
- resistance

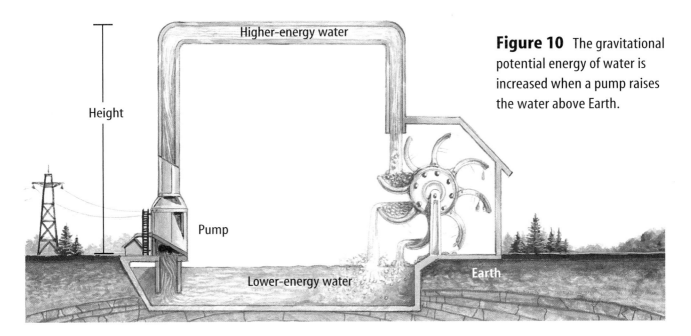

Figure 10 The gravitational potential energy of water is increased when a pump raises the water above Earth.

Figure 11 As long as there is a closed path for electrons to follow, electrons can flow in a circuit. They move away from the negative battery terminal and toward the positive terminal.

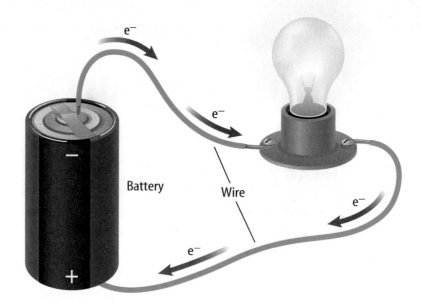

Electric Circuits The simplest electric circuit contains a source of electrical energy, such as a battery, and an electric conductor, such as a wire, connected to the battery. For the simple circuit shown in **Figure 11,** a closed path is formed by wires connected to a lightbulb and to a battery. Electric current flows in the circuit as long as none of the wires, including the glowing filament wire in the lightbulb, is disconnected or broken.

Voltage In a water circuit, a pump increases the gravitational potential energy of the water by raising the water from a lower level to a higher level. In an electric circuit, a battery increases the electrical potential energy of electrons. This electrical potential energy can be transformed into other forms of energy. The **voltage** of a battery is a measure of how much electrical potential energy each electron can gain. As voltage increases, more electrical potential energy is available to be transformed into other forms of energy. Voltage is measured in volts (V).

How a Current Flows You may think that when an electric current flows in a circuit, electrons travel completely around the circuit. Actually, individual electrons move slowly in an electric circuit. When the ends of a wire are connected to a battery, the battery produces an electric field in the wire. The electric field forces electrons to move toward the positive battery terminal. As an electron moves, it collides with other electric charges in the wire and is deflected in a different direction. After each collision, the electron again starts moving toward the positive terminal. A single electron may undergo more than ten trillion collisions each second. As a result, it may take several minutes for an electron in the wire to travel one centimeter.

Investigating the Electric Force

Procedure
1. Pour a layer of **salt** on a **plate.**
2. Sparingly sprinkle grains of **pepper** on top of the salt. Do not use too much pepper.
3. Rub a **rubber** or **plastic comb** on an article of **wool clothing.**
4. Slowly drag the comb through the salt and observe.

Analysis
1. How did the salt and pepper react to the comb?
2. Explain why the pepper reacted differently than the salt.

Batteries A battery supplies energy to an electric circuit. When the positive and negative terminals in a battery are connected in a circuit, the electric potential energy of the electrons in the circuit is increased. As these electrons move toward the positive battery terminal, this electric potential energy is transformed into other forms of energy, just as gravitational potential energy is converted into kinetic energy as water falls.

A battery supplies energy to an electric circuit by converting chemical energy to electric potential energy. For the alkaline battery shown in **Figure 12,** the two terminals are separated by a moist paste. Chemical reactions in the moist paste cause electrons to be transferred to the negative terminal from the atoms in the positive terminal. As a result, the negative terminal becomes negatively charged and the positive terminal becomes positively charged. This produces the electric field in the circuit that causes electrons to move away from the negative terminal and toward the positive terminal.

Battery Life Batteries don't supply energy forever. Maybe you know someone whose car wouldn't start after the lights had been left on overnight. Why do batteries run down? Batteries contain only a limited amount of the chemicals that react to produce chemical energy. These reactions go on as the battery is used and the chemicals are changed into other compounds. Once the original chemicals are used up, the chemical reactions stop and the battery is "dead."

Alkaline Batteries Several chemicals are used to make an alkaline battery. Zinc is a source of electrons at the negative terminal, and manganese dioxide combines with electrons at the positive terminal. The moist paste contains potassium hydroxide that helps transport electrons from the positive terminal to the negative terminal. Research dry-cell batteries and lead-acid batteries. Make a table listing the chemicals used in these batteries and their purpose.

Figure 12 When this alkaline battery is connected in an electric circuit, chemical reactions occur in the moist paste of the battery that move electrons from the positive terminal to the negative terminal.

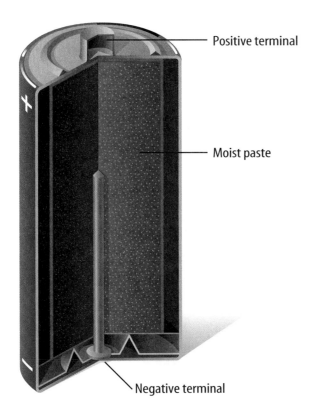

SECTION 2 Electric Current **593**

The Ohm The unit for electrical resistance was named in honor of the German physicist Georg Simon Ohm (1787–1854). Ohm is credited for discovering the relationship between current flow, voltage, and resistance. Research and find out more about Georg Ohm. Write a brief biography of him to share with the class.

Resistance

Electrons can move much more easily through conductors than through insulators, but even conductors interfere somewhat with the flow of electrons. The measure of how difficult it is for electrons to flow through a material is called **resistance**. The unit of resistance is the ohm (Ω). Insulators generally have much higher resistance than conductors.

As electrons flow through a circuit, they collide with the atoms and other electric charges in the materials that make up the circuit. Look at **Figure 13**. These collisions cause some of the electrons' electrical energy to be converted into thermal energy—heat—and sometimes into light. The amount of electrical energy that is converted into heat and light depends on the resistance of the materials in the circuit.

Buildings Use Copper Wires The amount of electrical energy that is converted into thermal energy increases as the resistance of the wire increases. Copper has low resistance and is one of the best electric conductors. Because copper is a good conductor, less heat is produced as electric current flows in copper wires, compared to wires made of other materials. As a result, copper wire is used in household wiring because the wires usually don't become hot enough to cause fires.

Resistance of Wires The electric resistance of a wire also depends on the length and thickness of the wire, as well as the material it is made from. When water flows through a hose, the water flow decreases as the hose becomes narrower or longer, as shown in **Figure 14** on the next page. The electric resistance of a wire increases as the wire becomes longer or as it becomes narrower.

Figure 13 As electrons flow through a wire, they travel in a zigzag path as they collide with atoms and other electrons. In these collisions, electrical energy is converted into other forms of energy. **Identify** *the other forms of energy that electrical energy is converted into.*

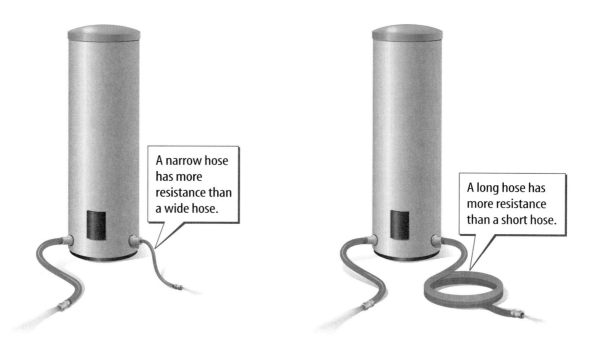

Figure 14 The resistance of a hose to the flow of water depends on the diameter and length of the hose used.
Compare and contrast water flowing in a hose and electric current flowing in a wire.

Lightbulb Filaments In a lightbulb, the filament is made of wire so narrow that it has a high resistance. When electric current flows in the filament, it becomes hot enough to emit light. The filament is made of tungsten metal, which has a much higher melting point than most other metals. This keeps the filament from melting at the high temperatures needed to produce light.

section 2 review

Flow of Charge

- Electric current is the flow of electric charges.
- Electric charges will flow continuously only through a closed conducting loop, called a circuit.
- The voltage in a circuit is a measure of the electrical potential energy of the electrons in the circuit.
- A battery supplies energy to an electric circuit by increasing the electric potential energy of electrons in the circuit.

Resistance

- Electric resistance is the measure of how difficult it is for electrons to flow through a material.
- Electric resistance is due to collisions between flowing electrons and the atoms in a material.
- Electric resistance in a circuit converts electrical energy into thermal energy and light.

Self Check

1. **Compare and contrast** an electric discharge with an electric current.
2. **Describe** how a battery causes electrons to move in a circuit.
3. **Describe** how the electric resistance of a wire changes as the wire becomes longer. How does the resistance change as the wire becomes thicker?
4. **Explain** why the electric wires in houses are usually made of copper.
5. **Think Critically** In an electric circuit, where do the electrons come from that flow in the circuit?

Applying Skills

6. **Infer** Find the voltage of various batteries such as a watch battery, a camera battery, a flashlight battery, and an automobile battery. Infer whether the voltage produced by a battery is related to its size.

section 3
Electric Circuits

as you read

What You'll Learn
- **Explain** how voltage, current, and resistance are related in an electric circuit.
- **Investigate** the difference between series and parallel circuits.
- **Determine** the electric power used in a circuit.
- **Describe** how to avoid dangerous electric shock.

Why It's Important
Electric circuits control the flow of electric current in all electrical devices.

Review Vocabulary
power: the rate at which energy is transferred; power equals the amount of energy transferred divided by the time over which the transfer occurs

New Vocabulary
- Ohm's law
- series circuit
- parallel circuit
- electric power

Controlling the Current

When you connect a conductor, such as a wire or a lightbulb, between the positive and negative terminals of a battery, electrons flow in the circuit. The amount of current is determined by the voltage supplied by the battery and the resistance of the conductor. To help understand this relationship, imagine a bucket with a hose at the bottom, as shown in **Figure 15.** If the bucket is raised, water will flow out of the hose faster than before. Increasing the height will increase the current.

Voltage and Resistance Think back to the pump and waterwheel in **Figure 10.** Recall that the raised water has energy that is lost when the water falls. Increasing the height from which the water falls increases the energy of the water. Increasing the height of the water is similar to increasing the voltage of the battery. Just as the water current increases when the height of the water increases, the electric current in a circuit increases as voltage increases.

If the diameter of the tube in **Figure 15** is decreased, resistance is greater and the flow of the water decreases. In the same way, as the resistance in an electric circuit increases, the current in the circuit decreases.

Figure 15 Raising the bucket higher increases the potential energy of the water in the bucket. This causes the water to flow out of the hose faster.

Ohm's Law A nineteenth-century German physicist, Georg Simon Ohm, carried out experiments that measured how changing the voltage in a circuit affected the current. He found a simple relationship among voltage, current, and resistance in a circuit that is now known as **Ohm's law.** In equation form, Ohm's law often is written as follows.

Ohm's Law

Voltage (in volts) = **current** (in amperes) × **resistance** (in ohms)

$$V = IR$$

According to Ohm's law, when the voltage in a circuit increases the current increases, just as water flows faster from a bucket that is raised higher. However, if the voltage in the circuit doesn't change, then the current in the circuit decreases when the resistance is increased.

Applying Math — Solving a Simple Equation

VOLTAGE FROM A WALL OUTLET A lightbulb is plugged into a wall outlet. If the lightbulb has a resistance of 220 Ω and the current in the lightbulb is 0.5 A, what is the voltage provided by the outlet?

Solution

1. *This is what you know:*
   - current: $I = 0.5$ A
   - resistance: $R = 220$ Ω

2. *This is what you need to find:* voltage: V

3. *This is the procedure you need to use:* Substitute the known values for current and resistance into Ohm's law to calculate the voltage:
   $V = IR = (0.5 \text{ A})(220 \text{ Ω}) = 110$ V

4. *Check your answer:* Divide your answer by the resistance 220 Ω. The result should be the given current 0.5 A.

Practice Problems

1. An electric iron plugged into a wall socket has a resistance of 24 Ω. If the current in the iron is 5.0 A, what is the voltage provided by the wall socket?

2. What is the current in a flashlight bulb with a resistance of 30 Ω if the voltage provided by the flashlight batteries is 3.0 V?

3. What is the resistance of a lightbulb connected to a 110-V wall outlet if the current in the lightbulb is 1.0 A?

For more practice, visit ips.msscience.com/math_practice

Identifying Simple Circuits

Procedure

1. The filament in a lightbulb is a piece of wire. For the bulb to light, an electric current must flow through the filament in a complete circuit. Examine the base of a **flashlight bulb** carefully. Where are the ends of the filament connected to the base?
2. Connect one piece of **wire,** a **battery,** and a flashlight bulb to make the bulb light. (There are four possible ways to do this.)

Analysis
Draw and label a diagram showing the path that is followed by the electrons in your circuit. Explain your diagram.

Series and Parallel Circuits

Circuits control the movement of electric current by providing paths for electrons to follow. For current to flow, the circuit must provide an unbroken path for current to follow. Have you ever been putting up holiday lights and had a string that would not light because a single bulb was missing or had burned out and you couldn't figure out which one it was? Maybe you've noticed that some strings of lights don't go out no matter how many bulbs burn out or are removed. These two strings of holiday lights are examples of the two kinds of basic circuits—series and parallel.

Wired in a Line A **series circuit** is a circuit that has only one path for the electric current to follow, as shown in **Figure 16.** If this path is broken, then the current no longer will flow and all the devices in the circuit stop working. If the entire string of lights went out when only one bulb burned out, then the lights in the string were wired as a series circuit. When the bulb burned out, the filament in the bulb broke and the current path through the entire string was broken.

Reading Check *How many different paths can electric current follow in a series circuit?*

In a series circuit, electrical devices are connected along the same current path. As a result, the current is the same through every device. However, each new device that is added to the circuit decreases the current throughout the circuit. This is because each device has electrical resistance, and in a series circuit, the total resistance to the flow of electrons increases as each additional device is added to the circuit. By Ohm's law, if the voltage doesn't change, the current decreases as the resistance increases.

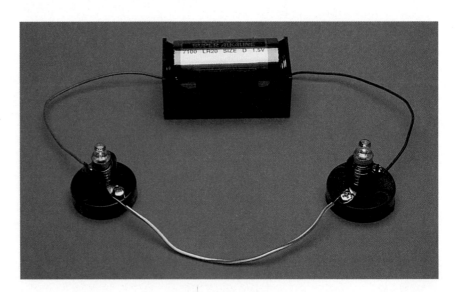

Figure 16 This circuit is an example of a series circuit. A series circuit has only one path for electric current to follow.
Predict *what will happen to the current in this circuit if any of the connecting wires are removed.*

Branched Wiring What if you wanted to watch TV and had to turn on all the lights, a hair dryer, and every other electrical appliance in the house to do so? That's what it would be like if all the electrical appliances in your house were connected in a series circuit.

Instead, houses, schools, and other buildings are wired using parallel circuits. A **parallel circuit** is a circuit that has more than one path for the electric current to follow, as shown in **Figure 17**. The current branches so that electrons flow through each of the paths. If one path is broken, electrons continue to flow through the other paths. Adding or removing additional devices in one branch does not break the current path in the other branches, so the devices on those branches continue to work normally.

In a parallel circuit, the resistance in each branch can be different, depending on the devices in the branch. The lower the resistance is in a branch, the more current flows in the branch. So the current in each branch of a parallel circuit can be different.

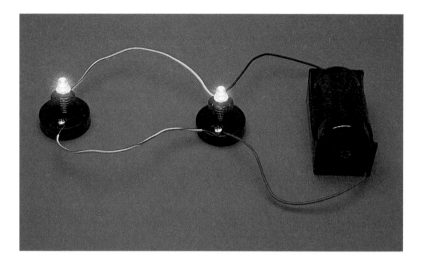

Figure 17 This circuit is an example of a parallel circuit. A parallel circuit has more than one path for electric current to follow. **Predict** what will happen to the current in the circuit if either of the wires connecting the two lightbulbs is removed.

Protecting Electric Circuits

In a parallel circuit, the current that flows out of the battery or electric outlet increases as more devices are added to the circuit. As the current through the circuit increases, the wires heat up.

To keep the wire from becoming hot enough to cause a fire, the circuits in houses and other buildings have fuses or circuit breakers like those shown in **Figure 18** that limit the amount of current in the wiring. When the current becomes larger than 15 A or 20 A, a piece of metal in the fuse melts or a switch in the circuit breaker opens, stopping the current. The cause of the overload can then be removed, and the circuit can be used again by replacing the fuse or resetting the circuit breaker.

Fuse
In some buildings, each circuit is connected to a fuse. The fuses are usually located in a fuse box.

Figure 18 You might have fuses in your home that prevent electric wires from overheating.

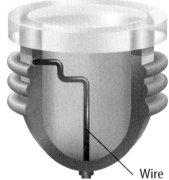

Wire
A fuse contains a piece of wire that melts and breaks when the current flowing through the fuse becomes too large.

Table 1 Power Used by Common Appliances	
Appliance	Power (W)
Computer	350
Color TV	200
Stereo	250
Refrigerator	450
Microwave	700–1,500
Hair dryer	1,000

Electric Power

When you use an appliance such as a toaster or a hair dryer, electrical energy is converted into other forms of energy. The rate at which electrical energy is converted into other forms of energy is **electric power.** In an electric appliance or in any electric circuit, the electric power that is used can be calculated from the electric power equation.

Electric Power Equation

Power (in watts) = **current** (in amperes) × **voltage** (in volts)

$$P = IV$$

The electric power is equal to the voltage provided to the appliance times the current that flows into the appliance. In the electric power equation, the SI unit of power is the watt. **Table 1** lists the electric power used by some common appliances.

Applying Math — Solving a Simple Equation

ELECTRIC POWER USED BY A LIGHTBULB A lightbulb is plugged into a 110-V wall outlet. How much electric power does the lightbulb use if the current in the bulb is 0.55 A?

Solution

1 *This is what you know:*
- voltage: $V = 110$ V
- current: $I = 0.55$ A

2 *This is what you need to find:* power: P

3 *This is the procedure you need to use:* To calculate electric power, substitute the known values for voltage and current into the equation for electric power:
$P = IV = (0.55$ A$)(110$ V$) = 60$ W

4 *Check your answer:* Divide your answer by the current 0.55 A. The result should be the given voltage 110 V.

Practice Problems

1. The batteries in a portable CD player provide 6.0 V. If the current in the CD player is 0.5 A, how much power does the CD player use?
2. What is the current in a toaster if the toaster uses 1,100 W of power when plugged into a 110-V wall outlet?
3. An electric clothes dryer uses 4,400 W of electric power. If the current in the dryer is 20.0 A, what is the voltage?

For more practice, visit ips.msscience.com/math_practice

Cost of Electric Energy Power is the rate at which energy is used, or the amount of energy that is used per second. When you use a hair dryer, the amount of electrical energy that is used depends on the power of the hair dryer and the amount of time you use it. If you used it for 5 min yesterday and 10 min today, you used twice as much energy today as yesterday.

Using electrical energy costs money. Electric companies generate electrical energy and sell it in units of kilowatt-hours to homes, schools, and businesses. One kilowatt-hour, kWh, is an amount of electrical energy equal to using 1 kW of power continuously for 1 h. This would be the amount of energy needed to light ten 100-W lightbulbs for 1 h, or one 100-W lightbulb for 10 h.

 *What does kWh stand for and what does it measure?*

An electric company usually charges its customers for the number of kilowatt-hours they use every month. The number of kilowatt-hours used in a building such as a house or a school is measured by an electric meter, which usually is attached to the outside of the building, as shown in **Figure 19**.

Figure 19 Electric meters measure the amount of electrical energy used in kilowatt-hours.
Identify *the electric meter attached to your house.*

Electrical Safety

 Have you ever had a mild electric shock? You probably felt only a mild tingling sensation, but electricity can have much more dangerous effects. In 1997, electric shocks killed an estimated 490 people in the United States. **Table 2** lists a few safety tips to help prevent electrical accidents.

Table 2 Preventing Electric Shock
Never use appliances with frayed or damaged electric cords.
Unplug appliances before working on them, such as when prying toast out of a jammed toaster.
Avoid all water when using plugged-in appliances.
Never touch power lines with anything, including kite string and ladders.
Always respect warning signs and labels.

Science Online

Topic: Cost of Electrical Energy
Visit ips.msscience.com for Web links to information about the cost of electrical energy in various parts of the world.

Activity Make a bar graph showing the cost of electrical energy for several countries on different continents.

SECTION 3 Electric Circuits **601**

Current's Effects The scale below shows how the effect of electric current on the human body depends on the amount of current that flows into the body.

0.0005 A	Tingle
0.001 A	Pain threshold
0.01 A	Inability to let go
0.025 A	
0.05 A	Difficulty breathing
0.10 A	
0.25 A	
0.50 A	Heart failure
1.00 A	

Electric Shock You experience an electric shock when an electric current enters your body. In some ways your body is like a piece of insulated wire. The fluids inside your body are good conductors of current. The electrical resistance of dry skin is much higher. Skin insulates the body like the plastic insulation around a copper wire. Your skin helps keep electric current from entering your body.

A current can enter your body when you accidentally become part of an electric circuit. Whether you receive a deadly shock depends on the amount of current that flows into your body. The current that flows through the wires connected to a 60-W lightbulb is about 0.5 A. This amount of current entering your body could be deadly. Even a current as small as 0.001 A can be painful.

Lightning Safety On average, more people are killed every year by lightning in the United States than by hurricanes or tornadoes. Most lightning deaths and injuries occur outdoors. If you are outside and can see lightning or hear thunder, take shelter indoors immediately. If you cannot go indoors, you should take these precautions: avoid high places and open fields; stay away from tall objects such as trees, flag poles, or light towers; and avoid objects that conduct current such as bodies of water, metal fences, picnic shelters, and metal bleachers.

section 3 review

Summary

Electric Circuits

- In an electric circuit, voltage, resistance, and current are related. According to Ohm's law, this relationship can be written as $V = IR$.
- A series circuit has only one path for electric current to follow.
- A parallel circuit has more than one path for current to follow.

Electric Power and Energy

- The electric power used by an appliance is the rate at which the appliance converts electrical energy to other forms of energy.
- The electric power used by an appliance can be calculated using the equation $P = IV$.
- The electrical energy used by an appliance depends on the power of the appliance and the length of time it is used. Electrical energy usually is measured in kWh.

Self Check

1. **Compare** the current in two lightbulbs wired in a series circuit.
2. **Describe** how the current in a circuit changes if the resistance increases and the voltage remains constant.
3. **Explain** why buildings are wired using parallel circuits rather than series circuits.
4. **Identify** what determines the damage caused to the human body by an electric shock.
5. **Think Critically** What determines whether a 100-W lightbulb costs more to use than a 1,200-W hair dryer costs to use?

Applying Math

6. **Calculate Energy** A typical household uses 1,000 kWh of electrical energy every month. If a power company supplies electrical energy to 1,000 households, how much electrical energy must it supply every year?

Current in a Parallel Circuit

The brightness of a lightbulb increases as the current in the bulb increases. In this lab you'll use the brightness of a lightbulb to compare the amount of current that flows in parallel circuits.

▶ Real-World Question

How does connecting devices in parallel affect the electric current in a circuit?

Goal
- **Observe** how the current in a parallel circuit changes as more devices are added.

Materials
1.5-V lightbulbs (4) battery holders (2)
1.5-V batteries (2)   minibulb sockets (4)
10-cm-long pieces of
    insulated wire (8)

Safety Precautions

▶ Procedure

1. Connect one lightbulb to the battery in a complete circuit. After you've made the bulb light, disconnect the bulb from the battery to keep the battery from running down. This circuit will be the brightness tester.

2. Make a parallel circuit by connecting two bulbs as shown in the diagram. Reconnect the bulb in the brightness tester and compare its brightness with the brightness of the two bulbs in the parallel circuit. Record your observations.

3. Add another bulb to the parallel circuit as shown in the figure. How does the brightness of the bulbs change?

4. Disconnect one bulb in the parallel circuit. Record your observations.

▶ Conclude and Apply

1. **Describe** how the brightness of each bulb depends on the number of bulbs in the circuit.
2. **Infer** how the current in each bulb depends on the number of bulbs in the circuit.

ommunicating
Your Data

Compare your conclusions with those of other students in your class. **For more help, refer to the** Science Skill Handbook.

A Model for Voltage and Current

Goal
- **Model** the flow of current in a simple circuit.

Materials
plastic funnel
rubber or plastic tubing of different diameters (1 m each)
meterstick
ring stand with ring
stopwatch
*clock displaying seconds
hose clamp
*binder clip
500-mL beakers (2)
*Alternate materials

Safety Precautions

Real-World Question
The flow of electrons in an electric circuit is something like the flow of water in a tube connected to a water tank. By raising or lowering the height of the tank, you can increase or decrease the potential energy of the water. How does the flow of water in a tube depend on the diameter of the tube and the height the water falls?

Procedure
1. **Design** a data table in which to record your data. It should be similar to the table below.
2. Connect the tubing to the bottom of the funnel and place the funnel in the ring of the ring stand.
3. **Measure** the inside diameter of the rubber tubing. Record your data.
4. Place a 500-mL beaker at the bottom of the ring stand and lower the ring so the open end of the tubing is in the beaker.
5. Use the meterstick to measure the height from the top of the funnel to the bottom of the ring stand.

Flow Rate Data				
Trial	Height (cm)	Diameter (mm)	Time (s)	Flow Rate (mL/s)
1				
2				
3		Do not write in this book.		
4				

604 CHAPTER 20 Electricity

6. Working with a classmate, pour water into the funnel fast enough to keep the funnel full but not overflowing. Measure and record the time needed for 100 mL of water to flow into the beaker. Use the hose clamp to start and stop the flow of water.
7. Connect tubing with a different diameter to the funnel and repeat steps 2 through 6.
8. Reconnect the original piece of tubing and repeat steps 4 through 6 for several lower positions of the funnel, lowering the height by 10 cm each time.

Analyze Your Data

1. **Calculate** the rate of flow for each trial by dividing 100 mL by the time measured for 100 mL of water to flow into the beaker.
2. **Make a graph** that shows how the rate of flow depends on the funnel height.

Conclude and Apply

1. **Infer** from your graph how the rate of flow depends on the height of the funnel.
2. **Explain** how the rate of flow depends on the diameter of the tubing. Is this what you expected to happen?
3. **Identify** which of the variables you changed in your trials that corresponds to the voltage in a circuit.
4. **Identify** which of the variables you changed in your trials that corresponds to the resistance in a circuit.
5. **Infer** from your results how the current in a circuit would depend on the voltage.
6. **Infer** from your results how the current in a circuit would depend on the resistance in the circuit.

Communicating Your Data

Share your graph with other students in your class. Did other students draw the same conclusions as you? **For more help, refer to the** Science Skill Handbook.

TIME SCIENCE AND Society
SCIENCE ISSUES THAT AFFECT YOU!

Fire in the Forest

Plant life returns after a forest fire in Yellowstone National Park.

Fires started by lightning may not be all bad

When lightning strikes a tree, the intense heat of the lightning bolt can set the tree on fire. The fire then can spread to other trees in the forest. Though lightning is responsible for only about ten percent of forest fires, it causes about one-half of all fire damage. For example, in 2000, fires set by lightning raged in 12 states at the same time, burning a total area roughly the size of the state of Massachusetts.

Fires sparked by lightning often start in remote, difficult-to-reach areas, such as national parks and range lands. Burning undetected for days, these fires can spread out of control. In addition to threatening lives, the fires can destroy millions of dollars worth of homes and property. Smoke from forest fires also can have harmful effects on people, especially for those with preexisting conditions, such as asthma.

People aren't the only victims of forest fires. The fires kill animals as well. Those who survive the blaze often perish because their habitats have been destroyed. Monster blazes spew carbon dioxide and other gases into the atmosphere. Some of these gases may contribute to the greenhouse effect that warms the planet. Moreover, fires cause soil erosion and loss of water reserves.

But fires caused by lightning also have some positive effects. In old, dense forests, trees often become diseased and insect-ridden. By removing these unhealthy trees, fires allow healthy trees greater access to water and nutrients. Fires also clear away a forest's dead trees, underbrush, and needles, providing space for new vegetation. Nutrients are returned to the ground as dead organic matter decays, but it can take a century for dead logs to rot completely. A fire enables nutrients to be returned to soil much more quickly. Also, the removal of these highly combustible materials prevents more widespread fires from occurring.

Research Find out more about the job of putting out forest fires. What training is needed? What gear do firefighters wear? Why would people risk their lives to save a forest? Use the media center at your school to learn more about forest firefighters and their careers.

For more information, visit ips.msscience.com/time

chapter 20 Study Guide

Reviewing Main Ideas

Section 1 Electric Charge

1. The two types of electric charge are positive and negative. Like charges repel and unlike charges attract.

2. An object becomes negatively charged if it gains electrons and positively charged if it loses electrons.

3. Electrically charged objects have an electric field surrounding them and exert electric forces on one another.

4. Electrons can move easily in conductors, but not so easily in insulators.

Section 2 Electric Current

1. Electric current is the flow of charges—usually either electrons or ions.

2. The energy carried by the current in a circuit increases as the voltage in the circuit increases.

3. In a battery, chemical reactions provide the energy that causes electrons to flow in a circuit.

4. As electrons flow in a circuit, some of their electrical energy is lost due to resistance in the circuit.

Section 3 Electric Circuits

1. In an electric circuit, the voltage, current, and resistance are related by Ohm's law.

2. The two basic kinds of electric circuits are parallel circuits and series circuits.

3. The rate at which electric devices use electrical energy is the electric power used by the device.

Visualizing Main Ideas

Copy and complete the following concept map about electricity.

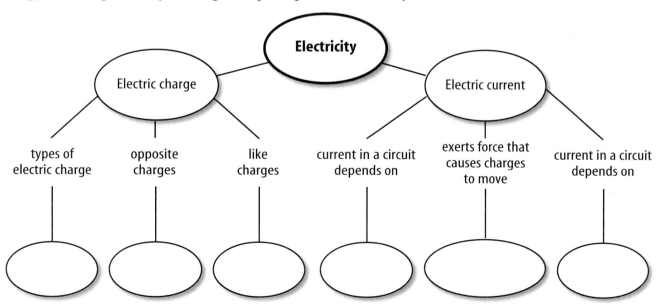

chapter 20 Review

Using Vocabulary

circuit p. 591
conductor p. 588
electric current p. 591
electric discharge p. 589
electric field p. 587
electric force p. 587
electric power p. 600
insulator p. 588
ion p. 584
Ohm's law p. 597
parallel circuit p. 599
resistance p. 594
series circuit p. 598
static charge p. 585
voltage p. 592

Answer the following questions using complete sentences.

1. What is the term for the flow of electric charge?
2. What is the relationship among voltage, current, and resistance in a circuit?
3. In what type of material do electrons move easily?
4. What is the name for the unbroken path that current follows?
5. What is an excess of electric charge on an object?
6. What is an atom that has gained or lost electrons called?
7. Which type of circuit has more than one path for electrons to follow?
8. What is the rapid movement of excess charge known as?

Checking Concepts

Choose the word or phrase that best answers the question.

9. Which of the following describes an object that is positively charged?
   A) has more neutrons than protons
   B) has more protons than electrons
   C) has more electrons than protons
   D) has more electrons than neutrons

10. Which of the following is true about the electric field around an electric charge?
    A) It exerts a force on other charges.
    B) It increases the resistance of the charge.
    C) It increases farther from the charge.
    D) It produces protons.

11. What is the force between two electrons?
    A) frictional   C) attractive
    B) neutral   D) repulsive

12. What property of a wire increases when it is made thinner?
    A) resistance
    B) voltage
    C) current
    D) static charge

13. What property does Earth have that enables grounding to drain static charges?
    A) It has a high static charge.
    B) It has a high resistance.
    C) It is a large conductor.
    D) It is like a battery.

Use the graph below to answer question 14.

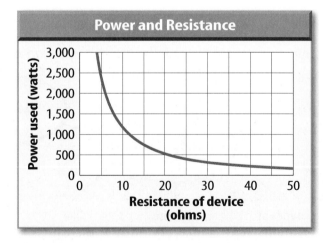

14. How does the resistance change if the power decreases from 2,500 W to 500 W?
    A) It increases four times.
    B) It decreases four times.
    C) It doubles.
    D) It doesn't change.

chapter 20 Review

Thinking Critically

15. **Determine** A metal wire is made thinner. How would you change the length of the wire to keep the electric resistance of the wire from changing?

The tables below show how the voltage and current vary in a portable radio and a portable CD player. Use these tables to answer questions 16 through 19.

Portable Radio	
Voltage (V)	Current (A)
2.0	1.0
4.0	2.0
6.0	3.0

Portable CD Player	
Voltage (V)	Current (A)
2.0	0.5
4.0	1.0
6.0	1.5

16. **Make a graph** with current plotted on the horizontal axis and voltage plotted on the vertical axis. Plot the data in the above tables for both devices on your graph.

17. **Identify** from your graph which line is more horizontal—the line for the portable radio or the line for the portable CD player.

18. **Calculate** the electric resistance using Ohm's law for each value of the current and voltage in the tables above. What is the resistance of each device?

19. **Determine** For which device is the line plotted on your graph more horizontal—the device with higher or lower resistance?

20. **Explain** why a balloon that has a static electric charge will stick to a wall.

21. **Describe** how you can tell whether the type of charge on two charged objects is the same or different.

22. **Infer** Measurements show that Earth is surrounded by a weak electric field. If the direction of this field points toward Earth, what is the type of charge on Earth's surface?

Performance Activities

23. **Design a board game** about a series or parallel circuit. The rules of the game could be based on opening or closing the circuit, adding more devices to the circuit, blowing fuses or circuit breakers, replacing fuses, or resetting circuit breakers.

Applying Math

24. **Calculate Resistance** A toaster is plugged into a 110-V outlet. What is the resistance of the toaster if the current in the toaster is 10 A?

25. **Calculate Current** A hair dryer uses 1,000 W when it is plugged into a 110-V outlet. What is the current in the hair dryer?

26. **Calculate Voltage** A lightbulb with a resistance of 30 Ω is connected to a battery. If the current in the lightbulb is 0.10 A, what is the voltage of the battery?

Use the table below to answer question 27.

Average Standby Power Used	
Appliance	Power (W)
Computer	7.0
VCR	6.0
TV	5.0

27. **Calculate Cost** The table above shows the power used by several appliances when they are turned off. Calculate the cost of the electrical energy used by each appliance in a month if the cost of electrical energy is $0.08/kWh, and each appliance is in standby mode for 600 h each month.

Chapter 20 Standardized Test Practice

Part 1 Multiple Choice

Record your answers on the answer sheet provided by your teacher or on a sheet of paper.

1. What happens when two materials are charged by rubbing against each other?
   A. both lose electrons
   B. both gain electrons
   C. one loses electrons
   D. no movement of electrons

Use the table below to answer questions 2–4.

Power Ratings of Some Appliances	
Appliance	Power (W)
Computer	350
Color TV	200
Stereo	250
Toaster	1,100
Microwave	900
Hair dryer	1,000

2. Which appliance will use the most energy if it is run for 15 minutes?
   A. microwave   C. stereo
   B. computer   D. color TV

3. What is the current in the hair dryer if it is plugged into a 110-V outlet?
   A. 110 A   C. 9 A
   B. 130,000 A   D. 1,100 A

4. Suppose using 1,000 W for 1 h costs $0.10. How much would it cost to run the color TV for 8 hours?
   A. $1.00   C. $1.60
   B. $10.00   D. $0.16

5. How does the current in a circuit change if the voltage is doubled and the resistance remains unchanged?
   A. no change   C. doubles
   B. triples   D. reduced by half

6. Which statement does NOT describe how electric changes affect each other?
   A. positive and negative charges attract
   B. positive and negative charges repel
   C. two positive charges repel
   D. two negative charges repel

Use the illustration below to answer questions 7 and 8.

7. What is the device on the chimney called?
   A. circuit breaker   C. fuse
   B. lightning rod   D. circuit

8. What is the device designed to do?
   A. stop electricity from flowing
   B. repel an electric charge
   C. turn the chimney into an insulator
   D. to provide grounding for the house

9. Which of the following is a material through which charge cannot move easily?
   A. conductor   C. wire
   B. circuit   D. insulator

10. What property of a wire increases when it is made longer?
    A. charge   C. voltage
    B. resistance   D. current

11. Which of the following materials are good insulators?
    A. copper and gold
    B. wood and glass
    C. gold and aluminum
    D. plastic and copper

Standardized Test Practice

Part 2 Short Response/Grid In

Record your answers on the answer sheet provided by your teacher or on a sheet of paper.

Use the illustration below to answer questions 12 and 13.

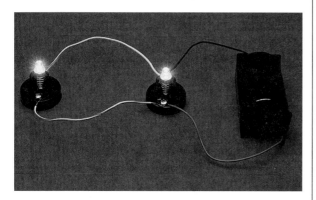

12. In this circuit, if one lightbulb is unscrewed, what happens to the current in the other lightbulb? Explain.

13. In this circuit, is the resistance and the current in each branch of the circuit always the same? Explain.

14. A 1,100-W toaster may be used for five minutes each day. A 400-W refrigerator runs all the time. Which appliance uses more electrical energy? Explain.

15. How much current does a 75-W bulb require in a 100-V circuit?

16. A series circuit containing mini-lightbulbs is opened and some of the lightbulbs are removed. What happens when the circuit is closed?

17. Suppose you plug an electric heater into the wall outlet. As soon as you turn it on, all the lights in the room go out. Explain what must have happened.

18. Explain why copper wires used in appliances or electric circuits are covered with plastic or rubber.

Part 3 Open Ended

Record your answers on a sheet of paper.

19. Why is it dangerous to use a fuse that is rated 30 A in a circuit calling for a 15-A fuse?

Use the illustration below to answer question 20.

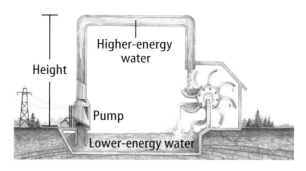

20. Compare the water pump in the water circuit above with the battery in an electric circuit.

21. Explain what causes the lightning that is associated with a thunderstorm.

22. Explain why two charged balloons push each other apart even if they are not touching.

23. Explain what can happen when you rub your feet on a carpet and then touch a metal doorknob.

24. Why does the fact that tungsten wire has a high melting point make it useful in the filaments of lightbulbs?

Test-Taking Tip

Recall Experiences Recall any hands-on experience as you read the question. Base your answer on the information given on the test.

Question 23 Recall from your personal experience the jolt you feel when you touch a doorknob after walking across a carpet.

chapter 21

Magnetism

The BIG Idea
Magnets exert forces on other magnets and on moving charges.

SECTION 1
What is magnetism?
Main Idea Moving electric charges produce magnetic fields.

SECTION 2
Electricity and Magnetism
Main Idea Magnetic fields can produce electric currents.

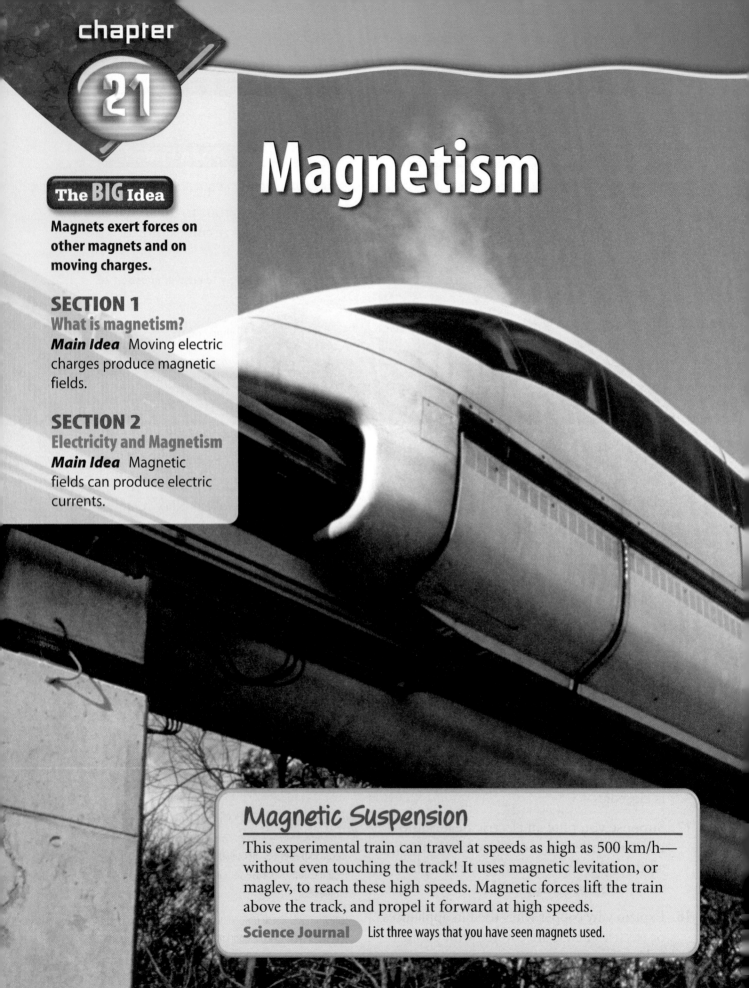

Magnetic Suspension
This experimental train can travel at speeds as high as 500 km/h—without even touching the track! It uses magnetic levitation, or maglev, to reach these high speeds. Magnetic forces lift the train above the track, and propel it forward at high speeds.

Science Journal List three ways that you have seen magnets used.

Start-Up Activities

Magnetic Forces

A maglev is moved along at high speeds by magnetic forces. How can a magnet get something to move? The following lab will demonstrate how a magnet is able to exert forces.

1. Place two bar magnets on opposite ends of a sheet of paper.
2. Slowly slide one magnet toward the other until it moves. Measure the distance between the magnets.
3. Turn one magnet around 180°. Repeat Step 2. Then turn the other magnet and repeat Step 2 again.
4. Repeat Step 2 with one magnet perpendicular to the other, in a T shape.
5. **Think Critically** In your Science Journal, record your results. In each case, how close did the magnets have to be to affect each other? Did the magnets move together or apart? How did the forces exerted by the magnets change as the magnets were moved closer together? Explain.

 **Preview this chapter's content and activities at** ips.msscience.com

 **Magnetic Forces and Fields** Make the following Foldable to help you see how magnetic forces and magnetic fields are similar and different.

STEP 1 Draw a mark at the midpoint of a vertical sheet of paper along the side edge.

STEP 2 Turn the paper horizontally and fold the outside edges in to touch at the midpoint mark.

STEP 3 Label the flaps *Magnetic Force* and *Magnetic Field*.

Compare and Contrast As you read the chapter, write information about each topic on the inside of the appropriate flap. After you read the chapter, compare and contrast the terms *magnetic force* and *magnetic field*. Write your observations under the flaps.

Get Ready to Read

Identify Cause and Effect

① Learn It! A *cause* is the reason something happens. The result of what happens is called an *effect*. Learning to identify causes and effects helps you understand why things happen. By using graphic organizers, you can sort and analyze causes and effects as you read.

② Practice It! Read the following paragraph. Then use the graphic organizer below to show what happened when the Sun ejects charged particles toward Earth.

> Sometimes the Sun ejects a large number of charged particles all at once. Most of these charged particles are deflected by Earth's magnetosphere. However, some of the ejected particles from the Sun produce other charged particles in Earth's outer atmosphere. These charged particles spiral along Earth's magnetic field lines toward Earth's magnetic poles. There they collide with atoms in the atmosphere. These collisions cause the atoms to emit light.
>
> —*from page 625*

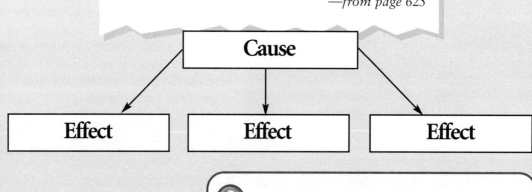

③ Apply It! As you read the chapter, be aware of causes and effects of charged particles moving in a magnetic field. Find three causes and their effects.

614 A CHAPTER 21 Magnetism

Target Your Reading

Reading Tip

Graphic organizers such as the Cause-Effect organizer help you organize what you are reading so you can remember it later.

Use this to focus on the main ideas as you read the chapter.

1. **Before you read** the chapter, respond to the statements below on your worksheet or on a numbered sheet of paper.
   - Write an **A** if you **agree** with the statement.
   - Write a **D** if you **disagree** with the statement.

2. **After you read** the chapter, look back to this page to see if you've changed your mind about any of the statements.
   - If any of your answers changed, explain why.
   - Change any false statements into true statements.
   - Use your revised statements as a study guide.

Science Online
Print out a worksheet of this page at ips.msscience.com

Before You Read A or D		Statement	After You Read A or D
	1	Opposite poles of magnets attract each other.	
	2	An electric motor converts electrical energy into kinetic energy.	
	3	Earth's magnetic field has not changed since the Earth formed.	
	4	Magnetic fields get stronger as you move away from the magnet's poles.	
	5	A wire carrying electric current is surrounded by a magnetic field.	
	6	An electromagnet is wire wrapped around a magnet.	
	7	Magnetic fields have no effect on moving electric charges.	
	8	Earth's magnetic field affects only Earth's surface.	
	9	Magnetic fields are produced by moving masses.	
	10	Transformers convert kinetic energy to electrical energy.	

614 B

section 1
What is magnetism?

as you read

What You'll Learn
- **Describe** the behavior of magnets.
- **Relate** the behavior of magnets to magnetic fields.
- **Explain** why some materials are magnetic.

Why It's Important
Magnetism is one of the basic forces of nature.

Review Vocabulary
compass: a device which uses a magnetic needle that can turn freely to determine direction

New Vocabulary
- magnetic field
- magnetic domain
- magnetosphere

Early Uses

Do you use magnets to attach papers to a metal surface such as a refrigerator? Have you ever wondered why magnets and some metals attract? Thousands of years ago, people noticed that a mineral called magnetite attracted other pieces of magnetite and bits of iron. They discovered that when they rubbed small pieces of iron with magnetite, the iron began to act like magnetite. When these pieces were free to turn, one end pointed north. These might have been the first compasses. The compass was an important development for navigation and exploration, especially at sea. Before compasses, sailors had to depend on the Sun or the stars to know in which direction they were going.

Magnets

A piece of magnetite is a magnet. Magnets attract objects made of iron or steel, such as nails and paper clips. Magnets also can attract or repel other magnets. Every magnet has two ends, or poles. One end is called the north pole and the other is the south pole. As shown in **Figure 1,** a north magnetic pole always repels other north poles and always attracts south poles. Likewise, a south pole always repels other south poles and attracts north poles.

Two north poles repel Two south poles repel

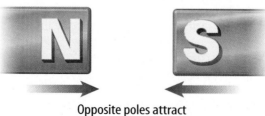

Opposite poles attract

Figure 1 Two north poles or two south poles repel each other. North and south magnetic poles are attracted to each other.

The Magnetic Field You have to handle a pair of magnets for only a short time before you can feel that magnets attract or repel without touching each other. How can a magnet cause an object to move without touching it? Recall that a force is a push or a pull that can cause an object to move. Just like gravitational and electric forces, a magnetic force can be exerted even when objects are not touching. And like these forces, the magnetic force becomes weaker as the magnets get farther apart.

This magnetic force is exerted through a **magnetic field.** Magnetic fields surround all magnets. If you sprinkle iron filings near a magnet, the iron filings will outline the magnetic field around the magnet. Take a look at **Figure 2.** The iron filings form a pattern of curved lines that start on one pole and end on the other. These curved lines are called magnetic field lines. Magnetic field lines help show the direction of the magnetic field.

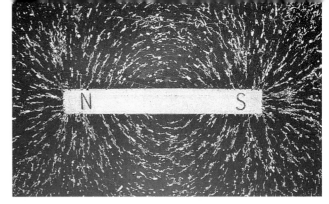

Iron filings show the magnetic field lines around a bar magnet.

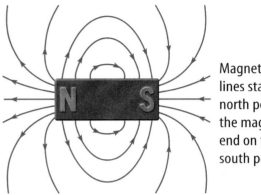

Magnetic field lines start at the north pole of the magnet and end on the south pole.

Reading Check *What is the evidence that a magnetic field exists?*

Magnetic field lines begin at a magnet's north pole and end on the south pole, as shown in **Figure 2.** The field lines are close together where the field is strong and get farther apart as the field gets weaker. As you can see in the figures, the magnetic field is strongest close to the magnetic poles and grows weaker farther from the poles.

Field lines that curve toward each other show attraction. Field lines that curve away from each other show repulsion. **Figure 3** illustrates the magnetic field lines between a north and a south pole and the field lines between two north poles.

Figure 2 A magnetic field surrounds a magnet. Where the magnetic field lines are close together, the field is strong.
Determine *for this magnet where the strongest field is.*

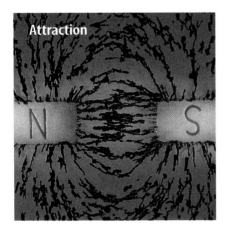

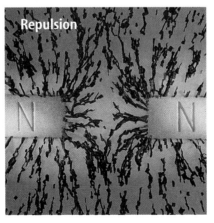

Figure 3 Magnetic field lines show attraction and repulsion.
Explain *what the field between two south poles would look like.*

SECTION 1 What is magnetism? **615**

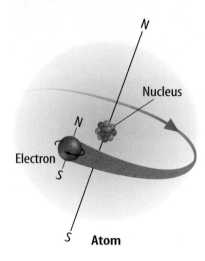

Figure 4 Movement of electrons produces magnetic fields. **Describe** *what two types of motion are shown in the illustration.*

Figure 5 Some materials can become temporary magnets.

Making Magnetic Fields Only certain materials, such as iron, can be made into magnets that are surrounded by a magnetic field. How are magnetic fields made? A moving electric charge, such as a moving electron, creates a magnetic field.

Inside every magnet are moving charges. All atoms contain negatively charged particles called electrons. Not only do these electrons swarm around the nucleus of an atom, they also spin, as shown in **Figure 4.** Because of its movement, each electron produces a magnetic field. The atoms that make up magnets have their electrons arranged so that each atom is like a small magnet. In a material such as iron, a large number of atoms will have their magnetic fields pointing in the same direction. This group of atoms, with their fields pointing in the same direction, is called a **magnetic domain.**

A material that can become magnetized, such as iron or steel, contains many magnetic domains. When the material is not magnetized, these domains are oriented in different directions, as shown in **Figure 5A.** The magnetic fields created by the domains cancel, so the material does not act like a magnet.

A magnet contains a large number of magnetic domains that are lined up and pointing in the same direction. Suppose a strong magnet is held close to a material such as iron or steel. The magnet causes the magnetic field in many magnetic domains to line up with the magnet's field, as shown in **Figure 5B.** As you can see in **Figure 5C** this process magnetizes paper clips.

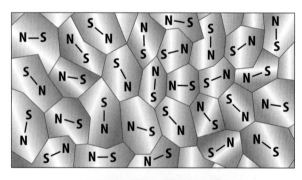

A Microscopic sections of iron and steel act as tiny magnets. Normally, these domains are oriented randomly and their magnetic fields cancel each other.

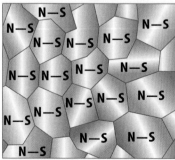

B When a strong magnet is brought near the material, the domains line up, and their magnetic fields add together.

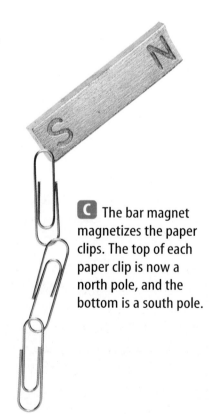

C The bar magnet magnetizes the paper clips. The top of each paper clip is now a north pole, and the bottom is a south pole.

Earth's Magnetic Field

Magnetism isn't limited to bar magnets. Earth has a magnetic field, as shown in **Figure 6.** The region of space affected by Earth's magnetic field is called the **magnetosphere** (mag NEE tuh sfihr). This deflects most of the charged particles from the Sun. The origin of Earth's magnetic field is thought to be deep within Earth in the outer core layer. One theory is that movement of molten iron in the outer core is responsible for generating Earth's magnetic field. The shape of Earth's magnetic field is similar to that of a huge bar magnet tilted about 11° from Earth's geographic north and south poles.

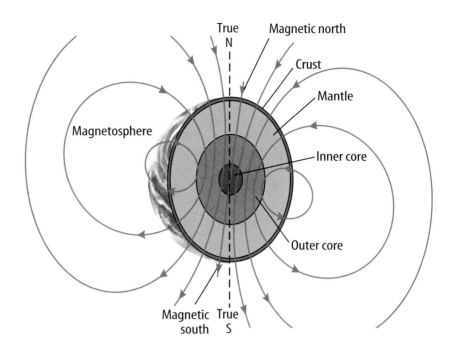

Figure 6 Earth has a magnetic field similar to the field of a bar magnet.

Applying Science

Finding the Magnetic Declination

The north pole of a compass points toward the magnetic pole, rather than true north. Imagine drawing a line between your location and the north pole, and a line between your location and the magnetic pole. The angle between these two lines is called the magnetic declination. Magnetic declination must be known if you need to know the direction to true north. However, the magnetic declination changes depending on your position.

Identifying the Problem

Suppose your location is at 50° N and 110° W. The location of the north pole is at 90° N and 110° W, and the location of the magnetic pole is at about 80° N and 105° W. What is the magnetic declination angle at your location?

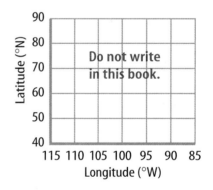

Solving the Problem

1. Draw and label a graph like the one shown above.
2. On the graph, plot your location, the location of the magnetic pole, and the location of the north pole.
3. Draw a line from your location to the north pole, and a line from your location to the magnetic pole.
4. Using a protractor, measure the angle between the two lines.

SECTION 1 What is magnetism? **617**

Figure 7 Earth's magnetic pole does not remain in one location from year to year.
Predict how you think the pole might move over the next few years.

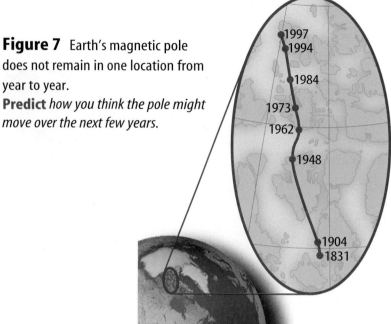

Observing Magnetic Fields

Procedure
1. Place **iron filings** in a **plastic petri dish.** Cover the dish and seal it with **clear tape.**
2. Collect **several magnets.** Place the magnets on the table and hold the dish over each one. Draw a diagram of what happens to the filings in each case.
3. Arrange two or more magnets under the dish. Observe the pattern of the filings.

Analysis
1. What happens to the filings close to the poles? Far from the poles?
2. Compare the fields of the individual magnets. How can you tell which magnet is strongest? Weakest?

Nature's Magnets Honeybees, rainbow trout, and homing pigeons have something in common with sailors and hikers. They take advantage of magnetism to find their way. Instead of using compasses, these animals and others have tiny pieces of magnetite in their bodies. These pieces are so small that they may contain a single magnetic domain. Scientists have shown that some animals use these natural magnets to detect Earth's magnetic field. They appear to use Earth's magnetic field, along with other clues like the position of the Sun or stars, to help them navigate.

Earth's Changing Magnetic Field Earth's magnetic poles do not stay in one place. The magnetic pole in the north today, as shown in **Figure 7,** is in a different place from where it was 20 years ago. In fact, not only does the position of the magnetic poles move, but Earth's magnetic field sometimes reverses direction. For example, 700 thousand years ago, a compass needle that now points north would point south. During the past 20 million years, Earth's magnetic field has reversed direction more than 70 times. The magnetism of ancient rocks contains a record of these magnetic field changes. When some types of molten rock cool, magnetic domains of iron in the rock line up with Earth's magnetic field. After the rock cools, the orientation of these domains is frozen into position. Consequently, these old rocks preserve the orientation of Earth's magnetic field as it was long ago.

Figure 8 The compass needles align with the magnetic field lines around the magnet.
Explain *what happens to the compass needles when the bar magnet is removed.*

The Compass A compass needle is a small bar magnet with a north and south magnetic pole. In a magnetic field, a compass needle rotates until it is aligned with the magnetic field line at its location. **Figure 8** shows how the orientation of a compass needle depends on its location around a bar magnet.

Earth's magnetic field also causes a compass needle to rotate. The north pole of the compass needle points toward Earth's magnetic pole that is in the north. This magnetic pole is actually a magnetic south pole. Earth's magnetic field is like that of a bar magnet with the magnet's south pole near Earth's north pole.

Topic: Compasses
Visit ips.msscience.com for Web links to information about different types of compasses.

Activity Find out how far from true north a compass points in your location.

section 1 review

Summary

Magnets
- A magnet has a north pole and a south pole.
- Like magnetic poles repel each other; unlike poles attract each other.
- A magnet is surrounded by a magnetic field that exerts forces on other magnets.
- Some materials are magnetic because their atoms behave like magnets.

Earth's Magnetic Field
- Earth is surrounded by a magnetic field similar to the field around a bar magnet.
- Earth's magnetic poles move slowly, and sometimes change places. Earth's magnetic poles now are close to Earth's geographic poles.

Self Check

1. **Explain** why atoms behave like magnets.
2. **Explain** why magnets attract iron but do not attract paper.
3. **Describe** how the behavior of electric charges is similar to that of magnetic poles.
4. **Determine** where the field around a magnet is the strongest and where it is the weakest.
5. **Think Critically** A horseshoe magnet is a bar magnet bent into the shape of the letter U. When would two horseshoe magnets attract each other? Repel? Have little effect?

Applying Skills

6. **Communicate** Ancient sailors navigated by using the Sun, stars, and following a coastline. Explain how the development of the compass would affect the ability of sailors to navigate.

SECTION 1 What is magnetism? **619**

Make a Compass

A valuable tool for hikers and campers is a compass. Almost 1,000 years ago, Chinese inventors found a way to magnetize pieces of iron. They used this method to manufacture compasses. You can use the same procedure to make a compass.

Real-World Question

How do you construct a compass?

Goals
- **Observe** induced magnetism.
- **Build** a compass.

Materials
petri dish   tape
clear bowl   marker
water   paper
sewing needle   plastic spoon
magnet   *Alternate material*

Safety Precautions

Procedure

1. Reproduce the circular protractor shown. Tape it under the bottom of your dish so it can be seen but not get wet. Add water until the dish is half full.
2. Mark one end of the needle with a marker. Magnetize a needle by placing it on the magnet aligned north and south for 1 min.
3. Float the needle in the dish using a plastic spoon to lower the needle carefully onto the water. Turn the dish so the marked part of the needle is above the 0° mark. This is your compass.
4. Bring the magnet near your compass. Observe how the needle reacts. Measure the angle the needle turns.

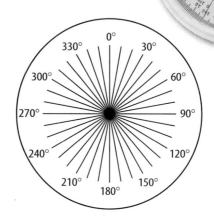

Conclude and Apply

1. **Explain** why the marked end of the needle always pointed the same way in step 3, even though you rotated the dish.
2. **Describe** the behavior of the compass when the magnet was brought close.
3. **Observe** the marked end of your needle. Does it point to the north or south pole of the bar magnet? **Infer** whether the marked end of your needle is a north or a south pole. How do you know?

Communicating Your Data

Make a half-page insert that will go into a wilderness survival guide to describe the procedure for making a compass. Share your half-page insert with your classmates. **For more help, refer to the** Science Skill Handbook.

620 CHAPTER 21 Magnetism

section 2
Electricity and Magnetism

Current Can Make a Magnet

Magnetic fields are produced by moving electric charges. Electrons moving around the nuclei of atoms produce magnetic fields. The motion of these electrons causes some materials, such as iron, to be magnetic. You cause electric charges to move when you flip a light switch or turn on a portable CD player. When electric current flows in a wire, electric charges move in the wire. As a result, a wire that contains an electric current also is surrounded by a magnetic field. **Figure 9A** shows the magnetic field produced around a wire that carries an electric current.

Electromagnets Look at the magnetic field lines around the coils of wire in **Figure 9B.** The magnetic fields around each coil of wire add together to form a stronger magnetic field inside the coil. When the coils are wrapped around an iron core, the magnetic field of the coils magnetizes the iron. The iron then becomes a magnet, which adds to the strength of the magnetic field inside the coil. A current-carrying wire wrapped around an iron core is called an **electromagnet,** as shown in **Figure 9C.**

as you read

What You'll Learn
- **Explain** how electricity can produce motion.
- **Explain** how motion can produce electricity.

Why It's Important
Electricity and magnetism enable electric motors and generators to operate.

Review Vocabulary
electric current: the flow of electric charge

New Vocabulary
- electromagnet
- motor
- aurora
- generator
- alternating current
- direct current
- transformer

Figure 9 A current-carrying wire produces a magnetic field.

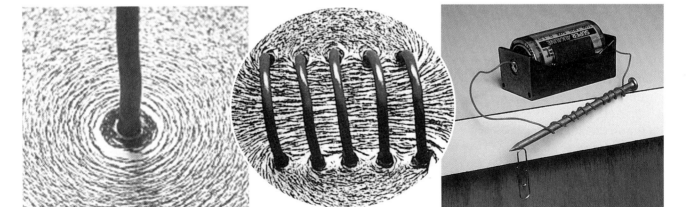

A Iron particles show the magnetic field lines around a current-carrying wire.

B When a wire is wrapped in a coil, the field inside the coil is made stronger.

C An iron core inside the coils increases the magnetic field because the core becomes magnetized.

Figure 10 An electric doorbell uses an electromagnet. Each time the electromagnet is turned on, the hammer strikes the bell. **Explain** *how the electromagnet is turned off.*

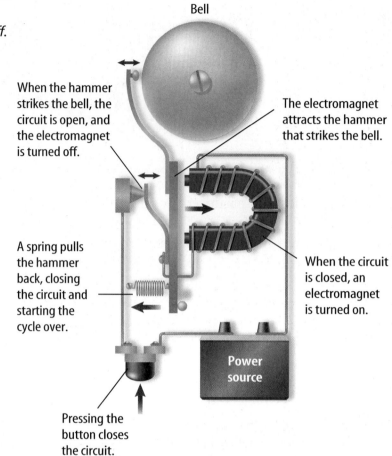

Bell

When the hammer strikes the bell, the circuit is open, and the electromagnet is turned off.

The electromagnet attracts the hammer that strikes the bell.

A spring pulls the hammer back, closing the circuit and starting the cycle over.

When the circuit is closed, an electromagnet is turned on.

Power source

Pressing the button closes the circuit.

Mini LAB

Assembling an Electromagnet

Procedure

1. Wrap a **wire** around a **16-penny steel nail** ten times. Connect one end of the wire to a **D-cell battery,** as shown in **Figure 9C.** Leave the other end loose until you use the electromagnet. **WARNING:** *When current is flowing in the wire, it can become hot over time.*
2. Connect the wire. Observe how many **paper clips** you can pick up with the magnet.
3. Disconnect the wire and rewrap the nail with 20 coils. Connect the wire and observe how many paper clips you can pick up. Disconnect the wire again.

Analysis

1. How many paper clips did you pick up each time? Did more coils make the electromagnet stronger or weaker?
2. Graph the number of coils versus number of paper clips attracted. Predict how many paper clips would be picked up with five coils of wire. Check your prediction.

Using Electromagnets The magnetic field of an electromagnet is turned on or off when the electric current is turned on or off. By changing the current, the strength and direction of the magnetic field of an electromagnet can be changed. This has led to a number of practical uses for electromagnets. A doorbell, as shown in **Figure 10,** is a familiar use of an electromagnet. When you press the button by the door, you close a switch in a circuit that includes an electromagnet. The magnet attracts an iron bar attached to a hammer. The hammer strikes the bell. When the hammer strikes the bell, the hammer has moved far enough to open the circuit again. The electromagnet loses its magnetic field, and a spring pulls the iron bar and hammer back into place. This movement closes the circuit, and the cycle is repeated as long as the button is pushed.

Some gauges, such as the gas gauge in a car, use a galvanometer to move the gauge pointer. **Figure 11** shows how a galvanometer makes a pointer move. Ammeters and voltmeters used to measure current and voltage in electric circuits also use galvanometers, as shown in **Figure 11.**

NATIONAL GEOGRAPHIC VISUALIZING VOLTMETERS AND AMMETERS

Figure 11

The gas gauge in a car uses a device called a galvanometer to make the needle of the gauge move. Galvanometers are also used in other measuring devices. A voltmeter uses a galvanometer to measure the voltage in a electric circuit. An ammeter uses a galvanometer to measure electric current. Multimeters can be used as an ammeter or voltmeter by turning a switch.

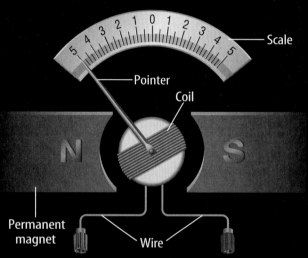

A galvanometer has a pointer attached to a coil that can rotate between the poles of a permanent magnet. When a current flows through the coil, it becomes an electromagnet. Attraction and repulsion between the magnetic poles of the electromagnet and the poles of the permanent magnet makes the coil rotate. The amount of rotation depends on the amount of current in the coil.

To measure the current in a circuit an ammeter is used. An ammeter contains a galvanometer and has low resistance. To measure current, an ammeter is connected in series in the circuit, so all the current in the circuit flows through it. The greater the current in the circuit, the more the needle moves.

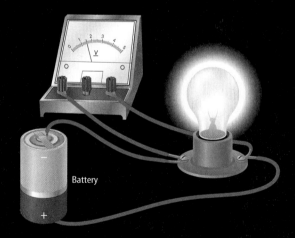

To measure the voltage in a circuit a voltmeter is used. A voltmeter also contains a galvanometer and has high resistance. To measure voltage, a voltmeter is connected in parallel in the circuit, so almost no current flows through it. The higher the voltage in the circuit, the more the needle moves.

SECTION 2 Electricity and Magnetism **623**

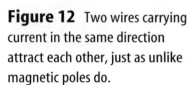
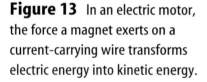

Magnets Push and Pull Currents

Look around for electric appliances that produce motion, such as a fan. How does the electric energy entering the fan become transformed into the kinetic energy of the moving fan blades? Recall that current-carrying wires produce a magnetic field. This magnetic field behaves the same way as the magnetic field that a magnet produces. Two current-carrying wires can attract each other as if they were two magnets, as shown in **Figure 12.**

Figure 12 Two wires carrying current in the same direction attract each other, just as unlike magnetic poles do.

Electric Motor Just as two magnets exert a force on each other, a magnet and a current-carrying wire exert forces on each other. The magnetic field around a current-carrying wire will cause it to be pushed or pulled by a magnet, depending on the direction the current is flowing in the wire. As a result, some of the electric energy carried by the current is converted into kinetic energy of the moving wire, as shown on the left in **Figure 13.** Any device that converts electric energy into kinetic energy is a **motor.** To keep a motor running, the current-carrying wire is formed into a loop so the magnetic field can force the wire to spin continually, as shown on the right in **Figure 13.**

Figure 13 In an electric motor, the force a magnet exerts on a current-carrying wire transforms electric energy into kinetic energy.

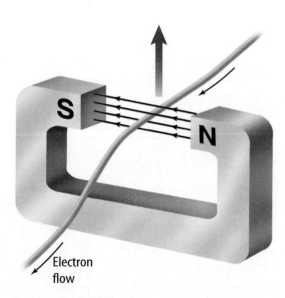

A magnetic field like the one shown will push a current-carrying wire upward.

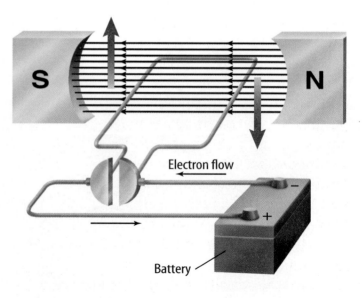

The magnetic field exerts a force on the wire loop, causing it to spin as long as current flows in the loop.

624 CHAPTER 21 Magnetism

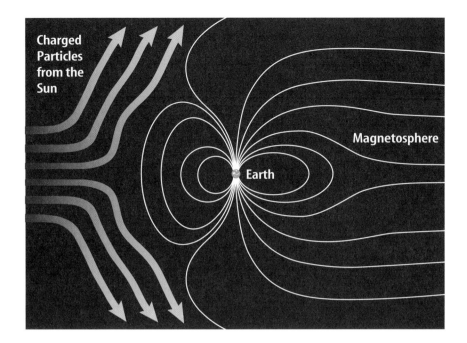

Figure 14 Earth's magnetosphere deflects most of the charged particles streaming from the Sun. **Explain** why the magnetosphere is stretched away from the Sun.

Earth's Magnetosphere The Sun emits charged particles that stream through the solar system like an enormous electric current. Just like a current-carrying wire is pushed or pulled by a magnetic field, Earth's magnetic field pushes and pulls on the electric current generated by the Sun. This causes most of the charged particles in this current to be deflected so they never strike Earth, as shown in **Figure 14.** As a result, living things on Earth are protected from damage that might be caused by these charged particles. At the same time, the solar current pushes on Earth's magnetosphere so it is stretched away from the Sun.

The Aurora Sometimes the Sun ejects a large number of charged particles all at once. Most of these charged particles are deflected by Earth's magnetosphere. However, some of the ejected particles from the Sun produce other charged particles in Earth's outer atmosphere. These charged particles spiral along Earth's magnetic field lines toward Earth's magnetic poles. There they collide with atoms in the atmosphere. These collisions cause the atoms to emit light. The light emitted causes a display known as the **aurora** (uh ROR uh), as shown in **Figure 15.** In northern latitudes, the aurora sometimes is called the northern lights.

Figure 15 An aurora is a natural light show that occurs far north and far south.

SECTION 2 Electricity and Magnetism **625**

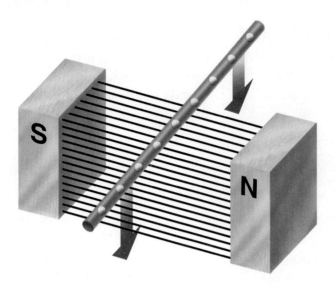

If a wire is pulled through a magnetic field, the electrons in the wire also move downward.

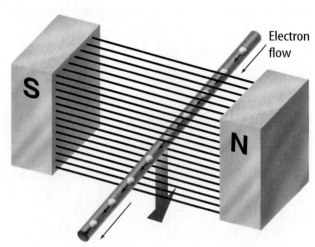

The magnetic field then exerts a force on the moving electrons, causing them to move along the wire.

Figure 16 When a wire is made to move through a magnetic field, an electric current can be produced in the wire.

Using Magnets to Create Current

In an electric motor, a magnetic field turns electricity into motion. A device called a **generator** uses a magnetic field to turn motion into electricity. Electric motors and electric generators both involve conversions between electric energy and kinetic energy. In a motor, electric energy is changed into kinetic energy. In a generator, kinetic energy is changed into electric energy. **Figure 16** shows how a current can be produced in a wire that moves in a magnetic field. As the wire moves, the electrons in the wire also move in the same direction, as shown on the left. The magnetic field exerts a force on the moving electrons that pushes them along the wire on the right, creating an electric current.

Figure 17 In a generator, a power source spins a wire loop in a magnetic field. Every half turn, the current will reverse direction. This type of generator supplies alternating current to the lightbulb.

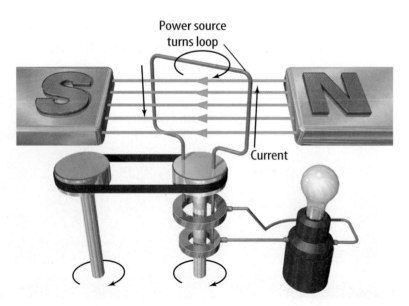

Electric Generators To produce electric current, the wire is fashioned into a loop, as in **Figure 17.** A power source provides the kinetic energy to spin the wire loop. With each half turn, the current in the loop changes direction. This causes the current to alternate from positive to negative. Such a current is called an **alternating current** (AC). In the United States, electric currents change from positive to negative to positive 60 times each second.

626 CHAPTER 21 Magnetism

Types of Current A battery produces direct current instead of alternating current. In a **direct current** (DC) electrons flow in one direction. In an alternating current, electrons change their direction of movement many times each second. Some generators are built to produce direct current instead of alternating current.

Reading Check *What type of currents can be produced by a generator?*

Power Plants Electric generators produce almost all of the electric energy used all over the world. Small generators can produce energy for one household, and large generators in electric power plants can provide electric energy for thousands of homes. Different energy sources such as gas, coal, and water are used to provide the kinetic energy to rotate coils of wire in a magnetic field. Coal-burning power plants, like the one pictured in **Figure 18,** are the most common. More than half of the electric energy generated by power plants in the United States comes from burning coal.

Topic: Power Plants
Visit ips.msscience.com for Web links to more information about the different types of power plants used in your region of the country.

Activity Describe the different types of power plants.

Voltage The electric energy produced at a power plant is carried to your home in wires. Recall that voltage is a measure of how much energy the electric charges in a current are carrying. The electric transmission lines from electric power plants transmit electric energy at a high voltage of about 700,000 V. Transmitting electric energy at a low voltage is less efficient because more electric energy is converted into heat in the wires. However, high voltage is not safe for use in homes and businesses. A device is needed to reduce the voltage.

Figure 18 Coal-burning power plants supply much of the electric energy for the world.

Figure 19 Electricity travels from a generator to your home.

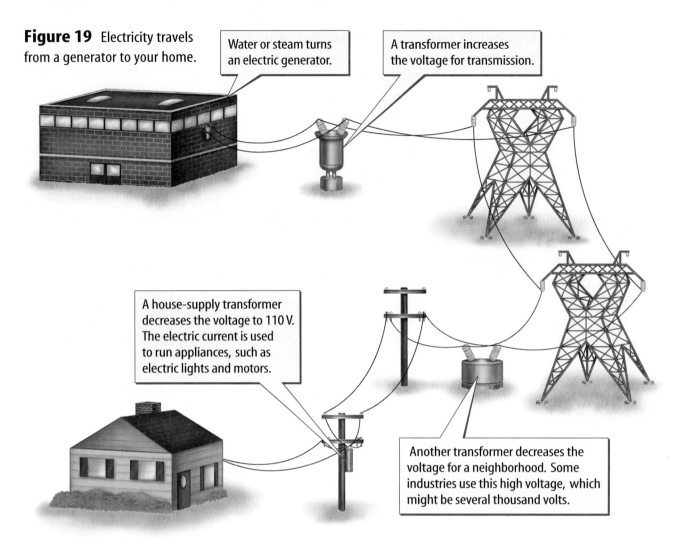

Water or steam turns an electric generator.

A transformer increases the voltage for transmission.

A house-supply transformer decreases the voltage to 110 V. The electric current is used to run appliances, such as electric lights and motors.

Another transformer decreases the voltage for a neighborhood. Some industries use this high voltage, which might be several thousand volts.

Changing Voltage

A **transformer** is a device that changes the voltage of an alternating current with little loss of energy. Transformers are used to increase the voltage before transmitting an electric current through the power lines. Other transformers are used to decrease the voltage to the level needed for home or industrial use. Such a power system is shown in **Figure 19.** Transformers also are used in power adaptors. For battery-operated devices, a power adaptor must change the 120 V from the wall outlet to the same voltage produced by the device's batteries.

Reading Check *What does a transformer do?*

A transformer usually has two coils of wire wrapped around an iron core, as shown in **Figure 20.** One coil is connected to an alternating current source. The current creates a magnetic field in the iron core, just like in an electromagnet. Because the current is alternating, the magnetic field it produces also switches direction. This alternating magnetic field in the core then causes an alternating current in the other wire coil.

Figure 20 A transformer can increase or decrease voltage. The ratio of input coils to output coils equals the ratio of input voltage to output voltage.
Determine *the output voltage if the input voltage is 60 V.*

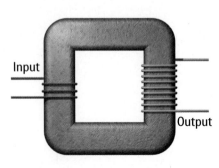

The Transformer Ratio Whether a transformer increases or decreases the input voltage depends on the number of coils on each side of the transformer. The ratio of the number of coils on the input side to the number of coils on the output side is the same as the ratio of the input voltage to the output voltage. For the transformer in **Figure 20,** the ratio of the number of coils on the input side to the number of coils on the output side is three to nine, or one to three. If the input voltage is 60 V, the output voltage will be 180 V.

In a transformer the voltage is greater on the side with more coils. If the number of coils on the input side is greater than the number of coils on the output side, the voltage is decreased. If the number of coils on the input side is less than the number on the output side, the voltage is increased.

Superconductors

Electric current can flow easily through materials, such as metals, that are electrical conductors. However, even in conductors, there is some resistance to this flow and heat is produced as electrons collide with atoms in the material.

Unlike an electrical conductor, a material known as a superconductor has no resistance to the flow of electrons. Superconductors are formed when certain materials are cooled to low temperatures. For example, aluminum becomes a superconductor at about −272°C. When an electric current flows through a superconductor, no heat is produced and no electric energy is converted into heat.

The Currents War In the late 1800s, electric power was being transmitted using a direct-current transmission system developed by Thomas Edison. To preserve his monopoly, Edison launched a public-relations war against the use of alternating-current power transmission, developed by George Westinghouse and Nikola Tesla. However, by 1893, alternating current transmission had been shown to be more efficient and economical, and quickly became the standard.

Figure 21 A small magnet floats above a superconductor. The magnet causes the superconductor to produce a magnetic field that repels the magnet.

Superconductors and Magnets Superconductors also have other unusual properties. For example, a magnet is repelled by a superconductor. As the magnet gets close to the superconductor, the superconductor creates a magnetic field that is opposite to the field of the magnet. The field created by the superconductor can cause the magnet to float above it, as shown in **Figure 21.**

Figure 22 The particle accelerator at Fermi National Accelerator Laboratory near Batavia, Illinois, accelerates atomic particles to nearly the speed of light. The particles travel in a beam only a few millimeters in diameter. Magnets made of superconductors keep the beam moving in a circular path about 2 km in diameter.

Figure 23 A patient is being placed inside an MRI machine. The strong magnetic field inside the machine enables images of tissues inside the patient's body to be made.

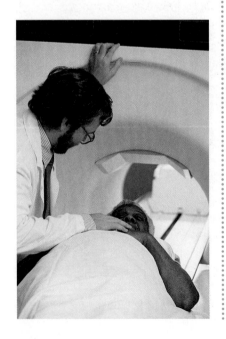

Using Superconductors Large electric currents can flow through electromagnets made from superconducting wire and can produce extremely strong magnetic fields. The particle accelerator shown in **Figure 22** uses more than 1,000 superconducting electromagnets to help accelerate subatomic particles to nearly the speed of light.

Other uses for superconductors are being developed. Transmission lines made from a superconductor could transmit electric power over long distances without having any electric energy converted to heat. It also may be possible to construct extremely fast computers using microchips made from superconductor materials.

Magnetic Resonance Imaging

A method called magnetic resonance imaging, or MRI, uses magnetic fields to create images of the inside of a human body. MRI images can show if tissue is damaged or diseased, and can detect the presence of tumors.

Unlike X-ray imaging, which uses X-ray radiation that can damage tissue, MRI uses a strong magnetic field and radio waves. The patient is placed inside a machine like the one shown in **Figure 23**. Inside the machine an electromagnet made from superconductor materials produces a magnetic field more than 20,000 times stronger than Earth's magnetic field.

Producing MRI Images About 63 percent of all the atoms in your body are hydrogen atoms. The nucleus of a hydrogen atom is a proton, which behaves like a tiny magnet. The strong magnetic field inside the MRI tube causes these protons to line up along the direction of the field. Radio waves are then applied to the part of the body being examined. The protons absorb some of the energy in the radio waves, and change the direction of their alignment.

When the radio waves are turned off, the protons realign themselves with the magnetic field and emit the energy they absorbed. The amount of energy emitted depends on the type of tissue in the body. This energy emitted is detected and a computer uses this information to form an image, like the one shown in **Figure 24.**

Connecting Electricity and Magnetism Electric charges and magnets are related to each other. Moving electric charges produce magnetic fields, and magnetic fields exert forces on moving electric charges. It is this connection that enables electric motors and generators to operate.

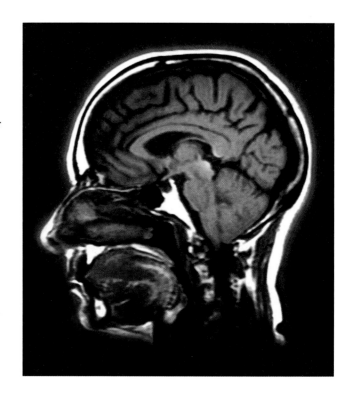

Figure 24 This MRI image shows a side view of the brain.

section 2 review

Summary

Electromagnets
- A current-carrying wire is surrounded by a magnetic field.
- An electromagnet is made by wrapping a current-carrying wire around an iron core.

Motors, Generators, and Transformers
- An electric motor transforms electrical energy into kinetic energy. An electric motor rotates when current flows in a wire loop that is surrounded by a magnetic field.
- An electric generator transforms kinetic energy into electrical energy. A generator produces current when a wire loop is rotated in a magnetic field.
- A transformer changes the voltage of an alternating current.

Self-Check

1. **Describe** how the magnetic field of an electromagnet depends on the current and the number of coils.
2. **Explain** how a transformer works.
3. **Describe** how a magnetic field affects a current-carrying wire.
4. **Describe** how alternating current is produced.
5. **Think Critically** What are some advantages and disadvantages to using superconductors as electric transmission lines?

Applying Math

6. **Calculate Ratios** A transformer has ten turns of wire on the input side and 50 turns of wire on the output side. If the input voltage is 120 V, what will the output voltage be?

How does an electric motor work?

Real-World Question

Electric motors are used in many appliances. For example, a computer contains a cooling fan and motors to spin the hard drive. A CD player contains electric motors to spin the CD. Some cars contain electric motors that move windows up and down, change the position of the seats, and blow warm or cold air into the car's interior. All these electric motors consist of an electromagnet and a permanent magnet. In this activity you will build a simple electric motor that will work for you. How can you change electric energy into motion?

Goals
- **Assemble** a small electric motor.
- **Observe** how the motor works.

Materials
22-gauge enameled wire (4 m)
steel knitting needle
*steel rod
nails (4)
hammer
ceramic magnets (2)
18-gauge insulated wire (60 cm)
masking tape
fine sandpaper
approximately 15-cm square wooden board
wooden blocks (2)
6-V battery
*1.5-V batteries connected in a series (4)
wire cutters
*scissors
*Alternate materials

Safety Precautions

WARNING: *Hold only the insulated part of a wire when it is attached to the battery. Use care when hammering nails. After cutting the wire, the ends will be sharp.*

632 **CHAPTER 21** Magnetism

Using Scientific Methods

▶ Procedure

1. Use sandpaper to strip the enamel from about 4 cm of each end of the 22-gauge wire.

2. Leaving the stripped ends free, make this wire into a tight coil of at least 30 turns. A D-cell battery or a film canister will help in forming the coil. Tape the coil so it doesn't unravel.

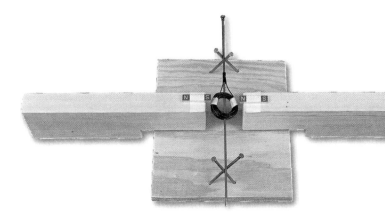

3. Insert the knitting needle through the coil. Center the coil on the needle. Pull the wire's two ends to one end of the needle.

4. Near the ends of the wire, wrap masking tape around the needle to act as insulation. Then tape one bare wire to each side of the needle at the spot where the masking tape is.

5. Tape a ceramic magnet to each block so that a north pole extends from one and a south pole from the other.

6. Make the motor. Tap the nails into the wood block as shown in the figure. Try to cross the nails at the same height as the magnets so the coil will be suspended between them.

7. Place the needle on the nails. Use bits of wood or folded paper to adjust the positions of the magnets until the coil is directly between the magnets. The magnets should be as close to the coil as possible without touching it.

8. Cut two 30-cm lengths of 18-gauge wire. Use sandpaper to strip the ends of both wires. Attach one wire to each terminal of the battery. Holding only the insulated part of each wire, place one wire against each of the bare wires taped to the needle to close the circuit. Observe what happens.

▶ Conclude and Apply

1. **Describe** what happens when you close the circuit by connecting the wires. Were the results expected?

2. **Describe** what happens when you open the circuit.

3. **Predict** what would happen if you used twice as many coils of wire.

Compare your conclusions with other students in your class. **For more help, refer to the** Science Skill Handbook.

Science and Language Arts

"Aagjuuk[1] and Sivulliit[2]"
from Intellectual Culture of the Copper Eskimos
by Knud Rasmussen, told by Tatilgak

The following are "magic words" that are spoken before the Inuit (IH noo wut) people go seal hunting. Inuit are native people that live in the arctic region. Because the Inuit live in relative darkness for much of the winter, they have learned to find their way by looking at the stars to guide them. The poem is about two constellations that are important to the Inuit people because their appearance marks the end of winter when the Sun begins to appear in the sky again.

By which way, I wonder the mornings—
You dear morning, get up!
See I am up!
By which way, I wonder,
the constellation *Aagjuuk* rises up in the sky?
By this way—perhaps—by the morning
It rises up!

Morning, you dear morning, get up!
See I am up!
By which way, I wonder,
the constellation *Sivulliit*
Has risen to the sky?
By this way—perhaps—by the morning.
It rises up!

[1] Inuit name for the constellation of stars called Aquila (A kwuh luh)
[2] Inuit name for the constellation of stars called Bootes (boh OH teez)

Understanding Literature

Ethnography Ethnography is a description of a culture. To write an ethnography, an ethnographer collects cultural stories, poems, or other oral tales from the culture that he or she is studying. Why must the Inuit be skilled in navigation?

Respond to the Reading

1. How can you tell the importance of constellations to the Inuit for telling direction?
2. How is it possible that the Inuit could see the constellations in the morning sky?
3. **Linking Science and Writing** Research the constellations in the summer sky in North America and write a paragraph describing the constellations that would help you navigate from south to north.

 Earth's magnetic field causes the north pole of a compass needle to point in a northerly direction. Using a compass helps a person to navigate and find his or her way. However, at the far northern latitudes where the Inuit live, a compass becomes more difficult to use. Some Inuit live north of Earth's northern magnetic pole. In these locations a compass needle points in a southerly direction. As a result, the Inuit developed other ways to navigate.

chapter 21 Study Guide

Reviewing Main Ideas

Section 1 — What is magnetism?

1. All magnets have two poles—north and south. Like poles repel each other and unlike poles attract.

2. A magnet is surrounded by a magnetic field that exerts forces on other magnets.

3. Atoms in magnetic materials are magnets. These materials contain magnetic domains which are groups of atoms whose magnetic poles are aligned.

4. Earth is surrounded by a magnetic field similar to the field around a bar magnet.

Section 2 — Electricity and Magnetism

1. Electric current creates a magnetic field. Electromagnets are made from a coil of wire that carries a current, wrapped around an iron core.

2. A magnetic field exerts a force on a moving charge or a current-carrying wire.

3. Motors transform electric energy into kinetic energy. Generators transform kinetic energy into electric energy.

4. Transformers are used to increase and decrease voltage in AC circuits.

Visualizing Main Ideas

Copy and complete the following concept map on magnets.

- **Magnets**
  - are made from → **Magnetic materials**
    - in which → **Moving electrons in atoms**
      - produce → ◯
        - that line up to make → ◯
  - are used by → **Electric motors**
    - in which → ◯
      - generates → ◯
        - that produces → **Kinetic energy**
  - are used by → **Generators**
    - in which → ◯
      - causes → **A wire loop to rotate**
        - that generates → ◯

ips.msscience.com/interactive_tutor

CHAPTER STUDY GUIDE 635

Chapter 21 Review

Using Vocabulary

alternating current p. 626	magnetic domain p. 616
aurora p. 625	magnetic field p. 615
direct current p. 627	magnetosphere p. 617
electromagnet p. 621	motor p. 624
generator p. 626	transformer p. 628

Explain the relationship that exists between each set of vocabulary words below.

1. generator—transformer
2. magnetic force—magnetic field
3. alternating current—direct current
4. current—electromagnet
5. motor—generator
6. electron—magnetism
7. magnetosphere—aurora
8. magnet—magnetic domain

Checking Concepts

Choose the word or phrase that best answers the question.

9. What can iron filings be used to show?
   A) magnetic field   C) gravitational field
   B) electric field   D) none of these

10. Why does the needle of a compass point to magnetic north?
    A) Earth's north pole is strongest.
    B) Earth's north pole is closest.
    C) Only the north pole attracts compasses.
    D) The compass needle aligns itself with Earth's magnetic field.

11. What will the north poles of two bar magnets do when brought together?
    A) attract
    B) create an electric current
    C) repel
    D) not interact

12. How many poles do all magnets have?
    A) one   C) three
    B) two   D) one or two

13. When a current-carrying wire is wrapped around an iron core, what can it create?
    A) an aurora   C) a generator
    B) a magnet   D) a motor

14. What does a transformer between utility wires and your house do?
    A) increases voltage
    B) decreases voltage
    C) leaves voltage the same
    D) changes DC to AC

Use the figure below to answer question 15.

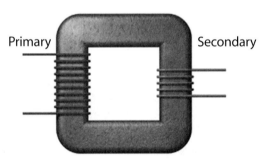

15. For this transformer which of the following describes how the output voltage compares with the input voltage?
    A) larger   C) smaller
    B) the same   D) zero voltage

16. Which energy transformation occurs in an electric motor?
    A) electrical to kinetic
    B) electrical to thermal
    C) potential to kinetic
    D) kinetic to electrical

17. What prevents most charged particles from the Sun from hitting Earth?
    A) the aurora
    B) Earth's magnetic field
    C) high-altitude electric fields
    D) Earth's atmosphere

636 CHAPTER REVIEW   ips.msscience.com/vocabulary_puzzlemaker

chapter 21 Review

Thinking Critically

18. **Concept Map** Explain how a doorbell uses an electromagnet by placing the following phrases in the cycle concept map: *circuit open, circuit closed, electromagnet turned on, electromagnet turned off, hammer attracted to magnet and strikes bell,* and *hammer pulled back by a spring.*

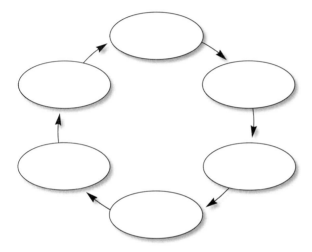

19. **Infer** A nail is magnetized by holding the south pole of a magnet against the head of the nail. Does the point of the nail become a north pole or a south pole? Include a diagram with your explanation.

20. **Explain** why an ordinary bar magnet doesn't rotate and align itself with Earth's magnetic field when you place it on a table.

21. **Determine** Suppose you were given two bar magnets. One magnet has the north and south poles labeled, and on the other magnet the magnetic poles are not labeled. Describe how you could use the labeled magnet to identify the poles of the unlabeled magnet.

22. **Explain** A bar magnet touches a paper clip that contains iron. Explain why the paper clip becomes a magnet that can attract other paper clips.

23. **Explain** why the magnetic field produced by an electromagnet becomes stronger when the wire coils are wrapped around an iron core.

24. **Predict** Magnet A has a magnetic field that is three times as strong as the field around magnet B. If magnet A repels magnet B with a force of 10 N, what is the force that magnet B exerts on magnet A?

25. **Predict** Two wires carrying electric current in the same direction are side by side and are attracted to each other. Predict how the force between the wires changes if the current in both wires changes direction.

Performance Activities

26. **Multimedia Presentation** Prepare a multimedia presentation to inform your classmates on the possible uses of superconductors.

Applying Math

Use the table below to answer questions 27 and 28.

Transformer Properties

Transformer	Number of Input Coils	Number of Output Coils
R	4	12
S	10	2
T	3	6
U	5	10

27. **Input and Output Coils** According to this table, what is the ratio of the number of input coils to the number of output coils on transformer T?

28. **Input and Output Voltage** If the input voltage is 60 V, which transformer gives an output voltage of 12 V?

Chapter 21 Standardized Test Practice

Part 1 | Multiple Choice

Record your answers on the answer sheet provided by your teacher or on a sheet of paper.

Use the figure below to answer questions 1 and 2.

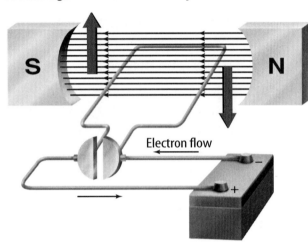

1. What is the device shown?
   A. electromagnet   C. electric motor
   B. generator   D. transformer

2. Which of the following best describes the function of this device?
   A. It transforms electrical energy into kinetic energy.
   B. It transforms kinetic energy into electrical energy.
   C. It increases voltage.
   D. It produces an alternating current.

3. How is an electromagnet different from a permanent magnet?
   A. It has north and south poles.
   B. It attracts magnetic substances.
   C. Its magnetic field can be turned off.
   D. Its poles cannot be reversed.

Test-Taking Tip

Check the Question Number For each question, double check that you are filling in the correct answer bubble for the question number you are working on.

4. Which of the following produces alternating current?
   A. electromagnet   C. generator
   B. superconductor   D. motor

5. Which statement about the domains in a magnetized substance is true?
   A. Their poles are in random directions.
   B. Their poles cancel each other.
   C. Their poles point in one direction.
   D. Their orientation cannot change.

Use the figure below to answer questions 6–8.

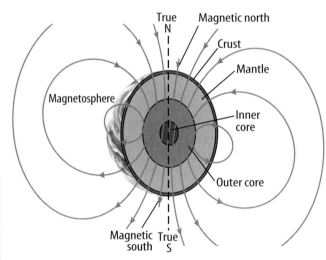

6. What is the region of space affected by Earth's magnetic field called?
   A. declination   C. aurora
   B. magnetosphere   D. outer core

7. What is the shape of Earth's magnetic field similar to?
   A. that of a horseshoe magnet
   B. that of a bar magnet
   C. that of a disk magnet
   D. that of a superconductor

8. In which of Earth's layers is Earth's magnetic field generated?
   A. crust   C. outer core
   B. mantle   D. inner core

Part 2 Short Response/Grid In

Record your answers on the answer sheet provided by your teacher or on a sheet of paper.

Use the figure below to answer questions 9 and 10.

9. Explain why the compass needles are pointed in different directions.

10. What will happen to the compass needles when the bar magnet is removed? Explain why this happens.

11. Describe the interaction between a compass needle and a wire in which an electric current is flowing.

12. What are two ways to make the magnetic field of an electromagnet stronger?

13. The input voltage in a transformer is 100 V and the output voltage is 50 V. Find the ratio of the number of wire turns on the input coil to the number of turns on the output coil.

14. Explain how you could magnetize a steel screwdriver.

15. Suppose you break a bar magnet in two. How many magnetic poles does each of the pieces have?

16. Alnico is a mixture of steel, aluminum, nickel, and cobalt. It is very hard to magnetize. However, once magnetized, it remains magnetic for a long time. Explain why it would not be a good choice for the core of an electromagnet.

Part 3 Open Ended

Record your answers on a sheet of paper.

17. Explain why the aurora occurs only near Earth's north and south poles.

18. Why does a magnet attract an iron nail to either of its poles, but attracts another magnet to only one of its poles?

19. A battery is connected to the input coil of a step-up transformer. Describe what happens when a lightbulb is connected to the output coil of the transformer.

20. Explain how electric forces and magnetic forces are similar.

Use the figure below to answer questions 21 and 22.

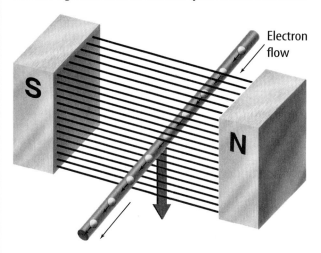

21. Describe the force that is causing the electrons to flow in the wire.

22. Infer how electrons would flow in the wire if the wire were pulled upward.

23. Explain why a nail containing iron can be magnetized, but a copper penny that contains no iron cannot be magnetized.

24. Every magnet has a north pole and a south pole. Where would the poles of a magnet that is in the shape of a disc be located?

chapter 22

Electronics and Computers

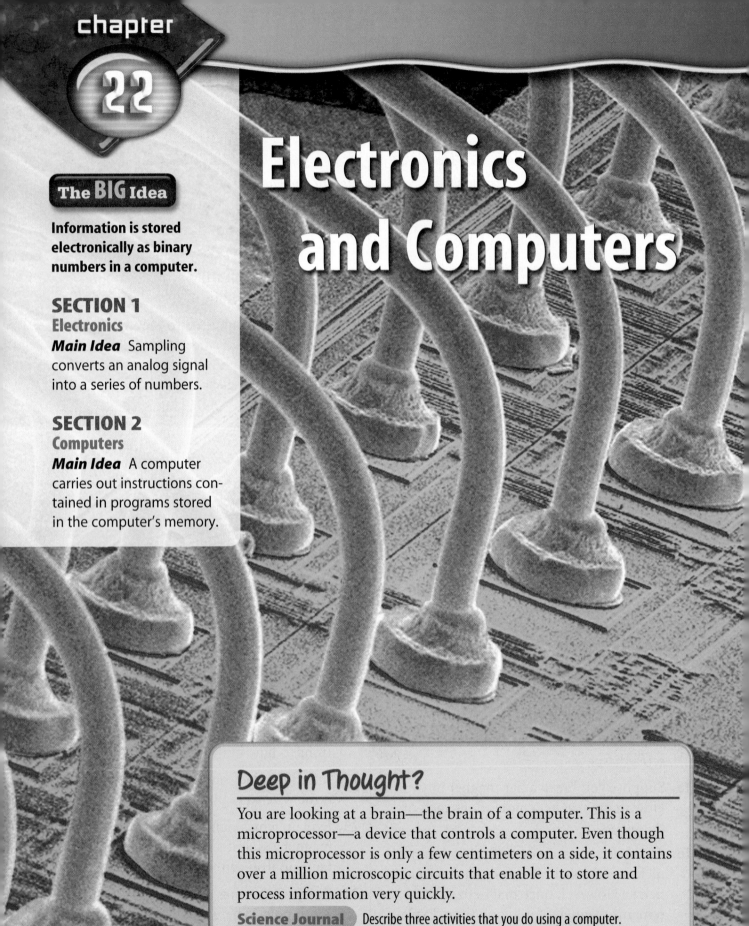

The BIG Idea

Information is stored electronically as binary numbers in a computer.

SECTION 1
Electronics
Main Idea Sampling converts an analog signal into a series of numbers.

SECTION 2
Computers
Main Idea A computer carries out instructions contained in programs stored in the computer's memory.

Deep in Thought?

You are looking at a brain—the brain of a computer. This is a microprocessor—a device that controls a computer. Even though this microprocessor is only a few centimeters on a side, it contains over a million microscopic circuits that enable it to store and process information very quickly.

Science Journal Describe three activities that you do using a computer.

Start-Up Activities

Electronic and Human Calculators

Imagine how your life would be different if you had been born before the invention of electronic devices. You could not watch television or use a computer. Besides providing entertainment, electronic devices and computers can make many tasks easier. For example, how much quicker is an electronic calculator than a human calculator?

1. Use a stopwatch to time how long it takes a volunteer to add the numbers 423, 21, 84, and 1,098.
2. Time how long it takes another volunteer to add these numbers using a calculator.
3. Repeat steps 1 and 2 this time asking the competitors to multiply 149 and 876.
4. Divide the time needed by the student calculator by the time needed by the calculator to solve each problem. How many times faster is the calculator?
5. **Think Critically** Write a paragraph describing which step in each calculation takes the most time.

Preview this chapter's content and activities at
ips.msscience.com

FOLDABLES
Study Organizer

Electronics and Computers Make the following Foldable to help you identify what you already know and what you want to learn about electronics and computers.

STEP 1 Fold a vertical sheet of paper from side to side. Make the front edge about 1 cm shorter than the back edge.

STEP 2 Turn lengthwise and fold into thirds.

STEP 3 Unfold and cut only the top layer along both folds to make three tabs.

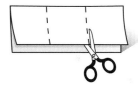

STEP 4 Label the tabs as shown.

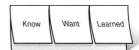

Identify Questions Before you read the chapter, write what you know under the left tab and what you want to know under the middle tab. As you read the chapter, add to and correct what you have written. After you read the chapter, write what you have learned under the right tab of your Foldable.

Get Ready to Read

Make Connections

① Learn It! Make connections between what you read and what you already know. Connections can be based on personal experiences (text-to-self), what you have read before (text-to-text), or events in other places (text-to-world).

As you read, ask connecting questions. Are you reminded of a personal experience? Have you read about the topic before? Did you think of a person, a place, or an event in another part of the world?

② Practice It! Read the excerpt below and make connections to your own knowledge and experience.

Text-to-self: What other electronic devices have you used that produce sounds or images?

Text-to-text: What have read about electric current in other chapters?

Text-to-world: Where are some places that loudspeakers are used?

> You have used another analog device if you have ever made a recording on a magnetic tape recorder. When voices or music are recorded on magnetic tape, the tape stores an analog signal of the sounds. When you play the tape, the tape recorder converts the analog signal to an electric current. This current changes smoothly with time, and causes a loudspeaker to vibrate, recreating the sounds for you to hear.
>
> — from page 643

③ Apply It! As you read this chapter, choose five words or phrases that make a connection to something you already know.

Target Your Reading

Use this to focus on the main ideas as you read the chapter.

Reading Tip

Make connections with memorable events, places, or people in your life. The better the connection, the more you will remember.

① **Before you read** the chapter, respond to the statements below on your worksheet or on a numbered sheet of paper.
- Write an **A** if you **agree** with the statement.
- Write a **D** if you **disagree** with the statement.

② **After you read** the chapter, look back to this page to see if you've changed your mind about any of the statements.
- If any of your answers changed, explain why.
- Change any false statements into true statements.
- Use your revised statements as a study guide.

Science Online
Print out a worksheet of this page at ips.msscience.com

Before You Read A or D		Statement	After You Read A or D
	1	Any electric current can carry information.	
	2	A digital signal changes smoothly with time.	
	3	A digital signal can be represented by a series of numbers.	
	4	Modern integrated circuits might contain millions of vacuum tubes.	
	5	Information is stored on a computer as numbers that contain only the digits 0 and 1.	
	6	Computer programs are tiny electronic circuits that store information.	
	7	Everything a computer does is controlled by computer programs.	
	8	A hard disk uses magnetism to store information.	
	9	A CD uses magnetism to store information.	

section 1

Electronics

as you read

What You'll Learn
- **Compare and contrast** analog and digital signals.
- **Explain** how semiconductors are used in electronic devices.

Why It's Important
You use electronic devices every day to make your life easier and more enjoyable.

Review Vocabulary
crystal: a solid substance which has a regularly repeating internal arrangement of atoms

New Vocabulary
- electronic signal
- analog signal
- digital signal
- semiconductor
- diode
- transistor
- integrated circuit

Electronic Signals

You've popped some popcorn, put a video in the VCR, and turned off the lights. Now you're ready to watch a movie. The VCR, television, and lamp shown in **Figure 1** use electricity to operate. However, unlike the lamp, the VCR and the TV are electronic devices. An electronic device uses electricity to store, process, and transfer information.

The VCR and the TV use information recorded on the videotape to produce the images and sounds you see as a movie. As the videotape moves inside the VCR, it produces a changing electric current. This changing electric current is the information the VCR uses to send signals to the TV. The TV then uses these signals to produce the images you see and the sounds you hear.

A changing electric current that carries information is an **electronic signal.** The information can be used to produce sounds, images, printed words, numbers, or other data. For example, a changing electric current causes a loudspeaker to produce sound. If the electric current didn't change, no sound would be produced by the loudspeaker. There are two types of electronic signals—analog and digital.

Analog Signals Most TVs, VCRs, radios, and telephones process and transmit information that is in the form of analog electronic signals. An **analog signal** is a signal that varies smoothly in time. In an analog electronic signal the electric current increases or decreases smoothly in time, just as your hand can move smoothly up and down.

Electronic signals are not the only types of analog signals. An analog signal can be produced by something that varies in a smooth, continuous way and contains information. For example, a person's temperature changes smoothly and contains information about a person's health.

Figure 1 A VCR sends electronic signals to the TV, which uses the information in these signals to produce images and sound.

Figure 2 Clocks can be analog or digital devices.

The information displayed on an analog device such as this clock changes continuously.

On this digital clock, the displayed time jumps from one number to another.

Analog Devices The clock with hands shown in **Figure 2** is an example of an analog device. The hands move smoothly from one number to the next to represent the time of day. Fluid-filled and dial thermometers also are analog devices. In a fluid-filled thermometer, the height of the fluid column smoothly rises or falls as the temperature changes. In a dial thermometer, a spring smoothly expands or contracts as the temperature changes.

You have used another analog device if you ever have made a recording on a magnetic tape recorder. When voices or music are recorded on magnetic tape, the tape stores an analog signal of the sounds. When you play the tape, the tape recorder converts the analog signal to an electric current. This current changes smoothly with time and causes a loudspeaker to vibrate, recreating the sounds for you to hear.

Digital Signals Some devices, such as CD players, use a different kind of electronic signal called a digital signal. Unlike an analog signal, a **digital signal** does not vary smoothly, but changes in jumps or steps. If each jump is represented by a number, a digital signal can be represented by a series of numbers.

✔ **Reading Check** *How is a digital signal different from an analog signal?*

You might have a digital clock or watch similar to the one shown on the right in **Figure 2** that displays the time as numbers. The display changes from 6:29 to 6:30 in a single jump, rather than sweeping smoothly from second to second. You might have seen digital thermometers that display temperature as a number. Some digital thermometers display temperature to the nearest whole degree, such as 23°C. The displayed temperature changes by jumps of 1°C. As a result, temperatures between two whole degrees, such as 22.7°C, are not displayed.

SECTION 1 Electronics **643**

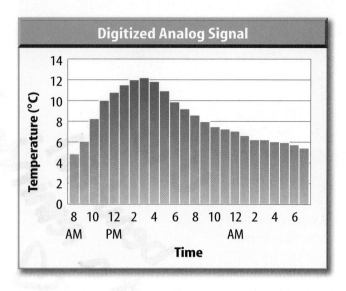

Figure 3 A temperature record made by recording the temperature every hour changes in steps and is a digital signal.

Figure 4 An analog signal can be converted to a digital signal. At a fixed time interval, the strength of the analog signal is measured and recorded. The resulting digital signal changes in steps.

Making Digital Signals A smoothly varying analog signal can be converted to a digital signal. For example, suppose you wish to create a record of how the temperature outside changed over a day. One way to do this would be read an outdoor thermometer every hour and record the temperature and time. At the end of the day your temperature record would be a series of numbers. If you used these numbers to make a graph of the temperature record, it might look like the one shown in **Figure 3**. The temperature information shown by the graph changes in steps and is a digital signal.

Sampling an Analog Signal By recording the temperature every hour, you have sampled the smoothly varying outdoor temperature. When an analog signal is sampled, a value of the signal is read and recorded at some time interval, such as every hour or every second. An example is shown in **Figure 4**. As a result, a smoothly changing analog signal is converted to a series of numbers. This series of numbers is a digital signal.

The process of converting an analog signal to a digital signal is called digitization. The analog signal on a magnetic tape can be converted to a digital signal by sampling. In this way, a song can be represented by a series of numbers.

Using Digital Signals It might seem that analog signals would be more useful than digital signals. After all, when an analog signal is converted to a digital signal, some information is lost. However, think about how analog and digital signals might be stored. Suppose a song that is stored as an analog signal on a small cassette tape were digitized and converted into a series of numbers. It might take millions of numbers to digitize a song, so how could these numbers be stored? As you will see later in this chapter, there is one electronic device that can store these numbers easily—a computer.

Once a digital signal is stored on a computer as a series of numbers, the computer can change these numbers using mathematical formulas. This process changes the signal and is called signal processing. For example, background noise can be removed from a digitized song using signal processing.

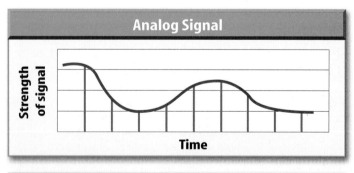

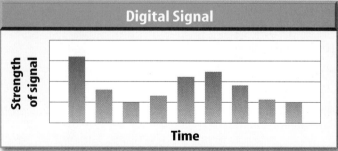

Early Television

Vacuum Tube

Modern Television

Electronic Devices

An electronic device, such as a calculator or a CD player, uses the information contained in electronic signals to do a job. For example, the job can be adding two numbers together or making sounds and images. The electronic signals are electric currents that flow through circuits in the electronic device. An electronic device, such as a calculator or a VCR, may contain hundreds or thousands of complex electric circuits.

Electronic Components The electric circuits in an electronic device usually contain electronic components. These electronic components are small devices that use the information in the electronic signals to control the flow of current in the circuits.

Early electronic devices, such as the early television shown in **Figure 5,** used electronic components called vacuum tubes, such as the one shown in the middle of **Figure 5,** to help create sounds and images. Vacuum tubes were bulky and generated a great deal of heat. As a result, early electronic devices used more electric power and were less dependable than those used today, such as the modern television shown in **Figure 5.** Today, televisions and radios no longer use vacuum tubes. Instead, they contain electronic components made from semiconductors.

Semiconductors

On the periodic table, the small number of elements found between the metals and nonmetals are called metalloids. Some metalloids, such as silicon and germanium, are semiconductors. A **semiconductor** is an element that is a poorer conductor of electricity than metals but a better conductor than nonmetals. However, semiconductors have a special property that ordinary conductors and insulators lack—their electrical conductivity can be controlled by adding impurities.

Figure 5 Because early televisions used vacuum tubes, they used more electrical power and were less reliable than their modern versions.

Topic: Semiconductor Devices
Visit ips.msscience.com for Web links to information about semiconductor devices.

Activity Choose one semiconductor device and write a paragraph explaining one way that it is used.

SECTION 1 Electronics **645**

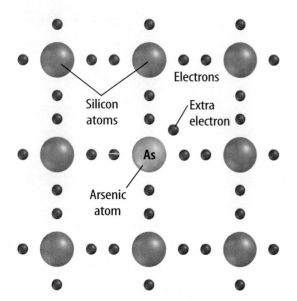

Figure 6 When arsenic atoms are added to a silicon crystal, they add extra electrons that are free to move about. This causes the electrical conductivity of the silicon crystal to increase.

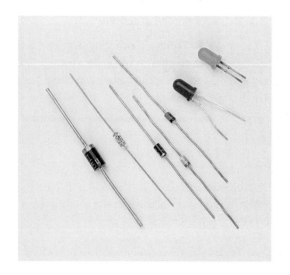

Figure 7 Diodes like these allow current to flow in only one direction.

Adding Impurities Adding even a single atom of an element such as gallium or arsenic to a million silicon atoms significantly changes the conductivity. This process of adding impurities is called doping.

Doping can produce two different kinds of semiconductors. One type of semiconductor can be created by adding atoms like arsenic to a silicon crystal, as shown in **Figure 6**. Then the silicon crystal contains extra electrons. A semiconductor with extra electrons is an n-type semiconductor.

A p-type semiconductor is produced when atoms like gallium are added to a silicon crystal. Then the silicon crystal has fewer electrons than it had before. An n-type semiconductor can give, or donate, electrons and a p-type semiconductor can take, or accept, electrons.

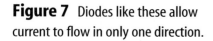 *How are n-type and p-type semiconductors different?*

Solid-State Components

The two types of semiconductors can be put together to form electronic components that can control the flow of electric current in a circuit. Combinations of n-type and p-type semiconductors can form components that behave like switches that can be turned off and on. Other combinations can form components that can increase, or amplify, the change in an electric current or voltage. Electronic components that are made from combinations of semiconductors are called solid-state components. Diodes and transistors are examples of solid-state components that often are used in electric circuits.

Diodes A **diode** is a solid-state component that, like a one-way street, allows current to flow only in one direction. In a diode, a p-type semiconductor is connected to an n-type semiconductor. Because an n-type semiconductor gives electrons and a p-type semiconductor accepts electrons, current can flow from the n-type to the p-type semiconductor, but not in the opposite direction. **Figure 7** shows common types of diodes. Diodes are useful for converting alternating current (AC) to direct current (DC). Recall that an alternating current constantly changes direction. When an alternating current reaches a diode, the diode allows the current to flow in only one direction. The result is direct current.

Transistors A **transistor** is a solid-state component that can be used to amplify signals in an electric circuit. A transistor also is used as an electronic switch. Electronic signals can cause a transistor to allow current to pass through it or to block the flow of current. **Figure 8** shows examples of transistors that are used in many electronic devices. Unlike a diode, a transistor is made from three layers of n-type and p-type semiconductor material sandwiched together.

Figure 8 Transistors such as these are used in electric circuits to amplify signals or to act as switches.

Integrated Circuits Personal computers usually contain millions of transistors, and would be many times larger if they used transistors the size of those shown in **Figure 8.** Instead, computers and other electronic devices use integrated circuits. An **integrated circuit** contains large numbers of interconnected solid-state components and is made from a single chip of semiconductor material such as silicon. An integrated circuit, like the one shown in **Figure 9,** may be smaller than 1 mm on each side and still can contain millions of transistors, diodes, and other components.

Figure 9 This tiny integrated circuit contains thousands of diodes and transistors.

section 1 review

Summary

Electronic Signals

- An electronic signal is a changing electric current that carries information.
- Analog electronic signals change continuously and digital electronic signals change in steps.
- An analog signal can be converted to a digital signal that is a series of numbers.

Solid-State Components

- Adding impurities to silicon can produce n-type semiconductors that donate electrons and p-type semiconductors that accept electrons.
- Solid-state components are electronic devices, such as diodes and transistors, made from n-type and p-type semiconductors.
- An integrated circuit contains a large number of solid-state components on a single semiconductor chip.

Self Check

1. **Explain** why the electric current that flows in a lamp is not an electronic signal.
2. **Describe** two advantages of using integrated circuits instead of vacuum tubes in electronic devices.
3. **Explain** why a digital signal can be stored on a computer.
4. **Compare and contrast** diodes and transistors.
5. **Think Critically** When an analog signal is sampled, what are the advantages and disadvantages of decreasing the time interval?

Applying Math

6. **Digital Signal** A song on a cassette tape is sampled and converted to a digital signal that is stored on a computer. The strength of the analog signal produced by the tape is sampled every 0.1 s. If the song is 3 min and 20 s long, how many numbers are in the digital signal stored on the computer?

ips.msscience.com/self_check_quiz

Investigating Diodes

Diodes are found in most electronic devices. They are used to control the flow of electrons through a circuit. Electrons will flow through a diode in only one direction, from the n-type semiconductor to the p-type semiconductor. In this lab you will use a type of diode called an LED (light-emitting diode) to observe how a diode works.

● Real-World Question

How does electric current flow through a diode?

Goals
- **Create** an electronic circuit.
- **Observe** how an LED works.

Materials
light-emitting diode   D-cell battery and holder
lightbulb and holder   wire

Safety Precautions

● Procedure

1. Set up the circuit shown below. Record your observations. Then reverse the connections so each wire is connected to the other battery terminals. Record your observations.
2. Disconnect the wires from the lightbulb and attach one wire to each end of an LED. Observe whether the LED lights up when you connect the battery.

3. Reverse the connections on the LED so the current goes into the opposite end. Observe whether the LED lights up this time. Record your observation.

● Conclude and Apply

1. **Explain** why the bulb did or did not light up each time.
2. **Explain** why the LED did or did not light up each time.
3. **Describe** how the behavior of the lightbulb is different from that of the LED.
4. **Infer** which wire on the LED is connected to the n-type semiconductor and which is connected to the p-type semiconductor based on your observations.

*C*ommunicating Your Data

Discuss your results with other students in your class. Did their LEDs behave in the same way? **For more help, refer to the Science Skill Handbook.**

1.5V Battery   Lightbulb

Section 2
Computers

What are computers?

When was the last time you used a computer? Computers are found in libraries, grocery stores, banks, and gas stations. Computers seem to be everywhere. A computer is an electronic device that can carry out a set of instructions, or a program. By changing the program, the same computer can be made to do a different job.

Compared to today's desktop and laptop computers, the first electronic computers, like the one shown in **Figure 10,** were much bigger and slower. Several of the first electronic computers were built in the United States between 1946 and 1951. Solid-state components and the integrated circuit had not been developed yet. So these early computers contained thousands of vacuum tubes that used a great deal of electric power and produced large amounts of heat.

Computers became much smaller, faster, and more efficient after integrated circuits became available in the 1960s. Today, even a game system, like the one in **Figure 10,** can carry out many more operations each second than the early computers.

as you read

What You'll Learn
- **Describe** the different parts of a computer.
- **Compare** computer hardware with computer software.
- **Discuss** the different types of memory and storage in a computer.

Why It's Important
You can do more with computers if you understand how they work.

Review Vocabulary
laser: a device that produces a concentrated beam of light

New Vocabulary
- binary system
- random-access memory
- read-only memory
- computer software
- microprocessor

Figure 10 One of the first electronic computers was ENIAC, which was built in 1946 and weighed more than 30 tons. ENIAC could do 5,000 additions per second.

This handheld game system can do millions of operations per second.

SECTION 2 Computers **649**

Using Binary Numbers

Procedure

1. Cut out **8 small paper squares.**
2. On four of the squares, draw the number zero, and on the other four, draw the number one.
3. Use the numbered squares to help determine the number of different combinations possible from four binary digits. List the combinations.

Analysis

1. From **Table 1** and your results from this MiniLAB, what happens to the number of combinations each time the number of binary digits is increased by one?
2. Infer how many combinations would be possible using five binary digits.

Try at Home

Table 1 Combinations of Binary Digits	
Number of of Binary Digits	Possible Combinations
1	0 1
2	00 01 10 11
3	000 001 010 011 100 101 110 111

Computer Information

How does a computer display images, generate sounds, and manipulate numbers and words? Every piece of information that is stored in or used by a computer must be converted to a series of numbers. The words you write with a word processor, or the numbers in a spreadsheet are stored in the computer's memory as numbers. An image or a sound file also is stored as a series of numbers. Information stored in this way is sometimes called digital information.

Binary Numbers Imagine what it would be like if you had to communicate with just two words—on and off. Could you use these words to describe your favorite music or to read a book out loud? Communication with just two words seems impossible, but that's exactly what a computer does.

All the digital information in a computer is converted to a type of number that is expressed using only two digits—0 and 1. This type of number is called a binary (BI nuh ree) number. Each 0 or 1 is called a binary digit, or bit. Because this number system uses only two digits, it is called the **binary system,** or base-2 number system.

Reading Check *Which digits are used in the binary system?*

Combining Binary Digits You might think that using only two digits would limit the amount of information you can represent. However, a small number of binary digits can be used to generate a large number of combinations, as shown in **Table 1.**

While one binary digit has only two possible combinations—0 or 1—there are four possible combinations for a group of two binary digits, as shown in **Table 1.** By using just one more binary digit the possible number of combinations is increased to eight. The number of combinations increases quickly as more binary digits are added to the group. For example, there are 65,536 combinations possible for a group of 16 binary digits.

Representing Information with Binary Digits Combinations of binary digits can be used to represent information. For example, the English alphabet has 26 letters. Suppose each letter was represented by one combination of binary digits. To represent both lowercase and uppercase letters would require a total of 52 different combinations of binary digits. Would a group of five binary digits have enough possible combinations?

Representing Letters and Numbers A common system that is used by computers represents each letter, number, or other text character by eight binary digits, or one byte. There are 256 combinations possible for a group of eight binary digits. In this system, the letter "A" is represented by the byte 01000001, while the letter "a" is represented by the byte 01100001, and a question mark is represented by 00111111.

Computer Memory

Why are digital signals stored in a computer as binary numbers? A binary number is a series of bits that can have only one of two values—0 and 1. A switch, such as a light switch on a wall, can have two positions: on or off. A switch could be used to represent the two values of a bit. A switch in the "off" position could represent a 0, and a switch in the "on" position could represent a 1. **Table 2** shows how switches could be used to represent combinations of binary digits.

Table 2 Representing Binary Digits	
Binary Number	Switches
0000	↑ ↑ ↑ ↑
0001	↑ ↑ ↑ ↓
0010	↑ ↑ ↓ ↑
0011	↑ ↑ ↓ ↓
0100	↑ ↓ ↑ ↑
1010	↓ ↑ ↓ ↑

Applying Science

How much information can be stored?

Information can be stored in a computer's memory or in storage devices such as hard disks or CDs. The amount of information that can be stored is so large that special units, shown in the table on the right, are used. Desktop computers often have hard disks that can store many gigabytes of information. How much information can be stored in one gigabyte of storage?

Size of Information Storage Units	
Information Storage Unit	Number of Bytes
kilobyte	1,024
megabyte	1,048,576
gigabyte	1,073,741,824

Identifying the Problem

When words are stored on a computer, every letter, punctuation mark, and space between words is represented by one byte. A page of text, such as this page, might contain as many as 2,900 characters. So to store a page of text on a computer might require 2,900 bytes.

If you write a page of text using a word-processing program, more bytes might be needed to store the page. This is because when the page is stored, some word-processing programs include other information along with the text.

Solving the Problem

1. If it takes 2,900 bytes to store one page of text on a computer, how many pages can be stored in 1 gigabyte of storage?
2. Suppose a book contains 400 pages of text. How many books could be stored on a 1-gigabyte hard disk?
3. A CD can hold 650 megabytes of information. How many 400-page books could be stored on a CD?

Figure 11 Computer memory is made of integrated circuits like this one. This integrated circuit can contain millions of microscopic circuits, shown here under high magnification.

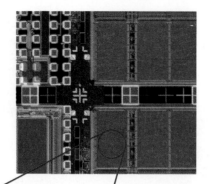
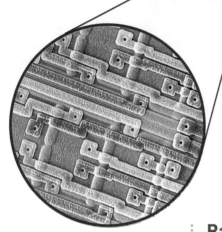

Storing Information The memory in a computer is an integrated circuit that contains millions of tiny electronic circuits, as shown in **Figure 11.** In the most commonly used type of computer memory, each circuit is able to store electric charge and can be either charged or uncharged. If the circuit is charged, it represents the bit 1 and if it is uncharged it represents the bit 0. Because computer memory contains millions of these circuits, it can store tremendous amounts of information using only the numbers 1 and 0.

What is your earliest memory? When you remember something from long ago, you use your long-term memory. On the other hand, when you work on a math problem, you may keep the numbers in your head long enough to find the answer. Like you, a computer has a long-term memory and a short-term memory that are used for different purposes.

Random-Access Memory A computer's **random-access memory,** or RAM, is short-term memory that stores documents, programs, and data while they are being used. Program instructions and data are temporarily stored in RAM while you are using a program or changing the data.

For example, a computer game is kept in RAM while you are playing it. If you are using a word-processing program to write a report, the report is temporarily held in RAM while you are working on it. Because information stored in RAM is lost when the computer is turned off, this type of memory cannot store anything that you want to use later.

The amount of RAM depends on the number of binary digits it can store. Recall that eight bits is called a byte. A megabyte is more than one million bytes. A computer that has 128 megabytes of memory can store more than 128 million bytes of information in its RAM, or nearly one billion bits.

Topic: Computer Software
Visit ips.msscience.com for Web links to information about types of computer software.

Activity Choose one type of software application and write a paragraph explaining why it is useful. Create a chart that summarizes what the software does.

 *What happens to information in RAM when the computer is turned off?*

Read-Only Memory Some information that is needed to enable the computer to operate is stored in its permanent memory. The computer can read this memory, but it cannot be changed. Memory that can't be changed and is permanently stored inside the computer is called **read-only memory,** or ROM. ROM is not lost when the computer is turned off.

Computer Programs

It's your mother's birthday and you decide to surprise her by baking a chocolate cake. You find a recipe for chocolate cake in a cookbook and follow the directions in the order the recipe tells you to. However, if the person who wrote the recipe left out any steps or put them in the wrong order, the cake probably will not turn out the way you expected. A computer program is like a recipe. A program is a series of instructions that tell the computer how to do a job. Unlike the recipe for a cake, some computer programs contain millions of instructions that tell the computer how to do many different jobs.

All the functions of a computer, such as displaying an image on the computer monitor or doing a math calculation, are controlled by programs. These instructions tell the computer how to add two numbers, how to display a word, or how to change an image on the monitor when you move a joystick. Many different programs can be stored in a computer's memory.

Computer Software When you type a report, play a video game, draw a picture, or look through an encyclopedia on a computer, you are using computer software. **Computer software** is any list of instructions for the computer. The instructions that are part of the software tell the computer what to display on the monitor. If you respond to what you see, for example by moving the mouse, the software instructions tell the computer how to respond to your action.

Computer Programming

The process of writing computer software is called computer programming. To write a computer program, you must decide what you want the computer to do, plan the best way to organize the instructions, write the instructions, and test the program to be sure it works. A person who writes computer programs is called a computer programmer. Computer programmers write software in computer languages such as Basic, C++, and Java.

Figure 12 shows part of a computer program. After the program is written, it is converted into binary digits to enable it to be stored in the computer's memory. Then the computer can carry out the program's instructions.

Mini LAB

Observing Memory

Procedure
1. Write a different five-digit number on six 3 × 5 cards.
2. Show a card to a partner for 3 s. Turn the card over and ask your partner to repeat the number. Repeat with two other cards.
3. Repeat this procedure with the last three cards, but wait 20 s before asking your partner to repeat each number.

Analysis
Is your partner's memory of the five-digit numbers more like computer RAM or ROM? Explain.

Try at Home

Figure 12 The text below is part of a computer program that directs the operation of a computer.

```
int request_dma(unsigned int dmanr, const char * device_id)
{
    if (dmanr > = MAX_DMA_CHANNELS)
        return -EINVAL;

    if (xchg(&dma_chan_busy[dmanr].lock, 1) != 0)
        return -EBUSY;

    dma_chan_busy[dmanr].device_id = device_id;

    /* old flag was 0, now contains 1 to indicate busy */
    return 0;
} /* request_dma */

void free_dma(unsigned int dmanr)
{
    if (dmanr > = MAX_DMA_CHANNELS) {
        printk("Trying to free DMA%d\n", dmanr);
        return;
    }
```

SECTION 2 Computers **653**

Computer Programmers
Computer software is written by computer programmers. It may take from a few hours to more than a year to write a program, and involve a single programmer or a team of programmers. A programmer usually must know several computer languages, such as COBOL, Java, and C++. Training in computer languages is required, and most jobs also require a college degree. Research to find the schools in your area that offer training as a computer programmer.

Computer Hardware

When you press a key on a computer's keyboard, a letter appears on the screen. This seems to occur all at once, but actually three steps are involved. In the first step, the computer receives information from an input device, such as a keyboard or mouse. For example, when you press a key on the keyboard, the computer receives and stores an electronic signal from the keyboard.

The next step is to process the input signal from the keyboard. This means to change the input signal into an electronic signal that can be understood by the computer monitor. The computer does this by following instructions contained in the programs stored in the computer's memory. The third step is to send the processed signal to the monitor.

All three steps can be carried out with a combination of hardware and software components. Computer hardware consists of input devices, output devices, storage devices, and integrated circuits for storing information. A keyboard and a mouse are examples of input devices, while a monitor, a printer, and loudspeakers are examples of output devices. Storage devices, such as floppy disks, hard disks, and CDs, are used to store information outside of the computer memory. A computer also contains a microprocessor that controls the computer hardware. Examples of computer hardware are shown in **Figure 13.**

Figure 13 Computer hardware includes input devices, output devices, and storage devices.

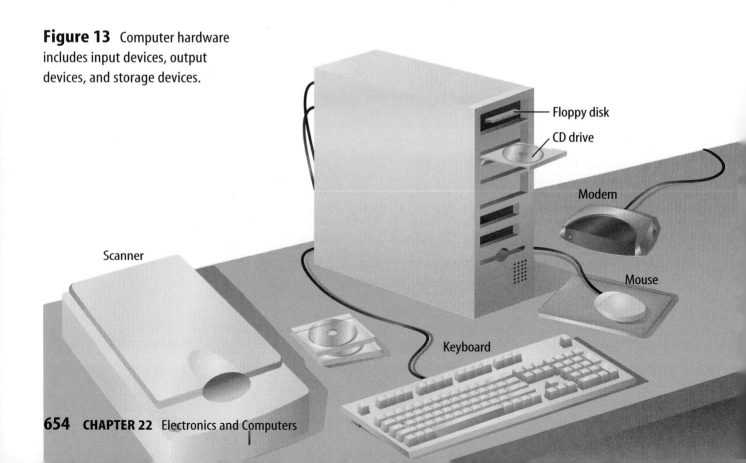

The Microprocessor Modern computers contain a microprocessor, like the one shown in **Figure 14,** that serves as the brain of the computer. A **microprocessor,** which is also called the central processing unit, or CPU, is an integrated circuit that controls the flow of information between different parts of the computer. A microprocessor can contain millions of interconnected transistors and other components. The microprocessor receives electronic signals from various parts of the computer, processes these signals, and sends electronic signals to other parts of the computer. For example, the microprocessor might tell the hard-disk drive to write data to the hard disk or the monitor to change the image on the screen. The microprocessor does this by carrying out instructions that are contained in computer programs stored in the computer's memory.

The microprocessor was developed in the late 1970s as the result of a process that made it possible to fit thousands of electronic components on a silicon chip. In the 1980s, the number of components on a silicon chip increased to hundreds of thousands. In the 1990s, microprocessors were developed that contained several million components on a single chip.

Figure 14 The pencil points to the microprocessor in the photo above. This microprocessor has dimensions of about one centimeter on a side, but contains millions of transistors and other solid-state components.

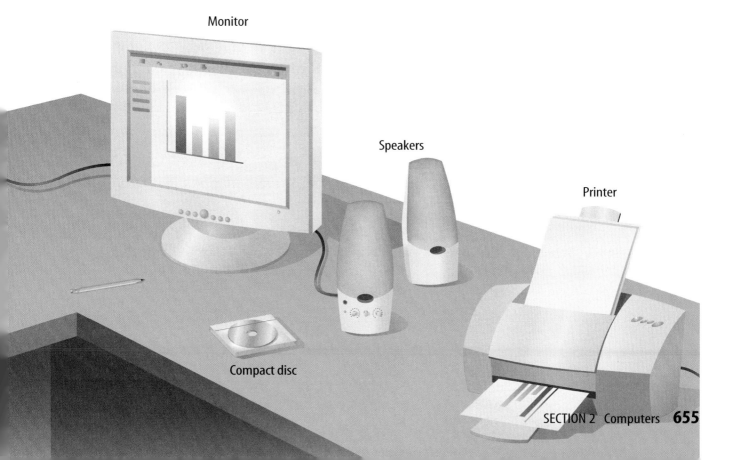

Recycling Computers Changes in computer technology occur so rapidly that computers are often replaced after being used for only a few years. What happens to old computers? Some computer parts contain lead, mercury, and other toxic substances. Research how toxic materials can be recovered from old computers, and disposed of safely. Summarize your findings in your Science Journal.

Topic: Magnetic Disks
Visit ips.msscience.com for Web links to information about storing data on magnetic disks.

Activity Write a paragraph explaining why hard disks can store more information than floppy disks.

Storing Information

You have decided to type your homework assignment on a computer. The resulting paper is quite long and you make many changes to it each time you read it. How does the computer make it possible for you to store your information and make changes to it?

Both RAM and ROM are integrated circuits inside the computer. You might wonder, then, why other types of information storage are needed. Information stored in RAM is lost when the computer is turned off, and information stored in ROM can only be read—it can't be changed. If you want to store information that can be changed but isn't lost when the computer is off, you must store that information on a storage device, such as a disk. Several different types of disks are available.

Hard Disks A hard disk is a device that stores computer information magnetically. A hard disk is usually located inside a computer. **Figure 15** shows the inside of a hard disk, and **Figure 16** shows how a hard disk stores data. The hard disk contains one or more metal disks that have magnetic particles on one surface. When you save information on a hard disk, a device called a read/write head inside the disk drive changes the orientation of the magnetic particles on the disk's surface. Orientation in one direction represents 0 and orientation in the opposite direction represents 1. When a magnetized disk is read, the read/write head converts the digital information on the disk to pulses of electric current.

Information stored magnetically cannot be read by the computer as quickly as information stored on RAM and ROM. However, because the information on a hard disk is stored magnetically rather than with electronic switches like RAM, the information isn't lost when the computer is turned off.

Figure 15 A hard disk contains a disk or platter that is coated with magnetic particles. A read/write head moves over the surface of the disk.

NATIONAL GEOGRAPHIC VISUALIZING A HARD DISK

Figure 16

Computers are useful because they can process large amounts of information quickly. Almost all desktop computers use a hard disk to store information. A hard disk is an electronic filing cabinet that can store enormous amounts of information and retrieve them quickly.

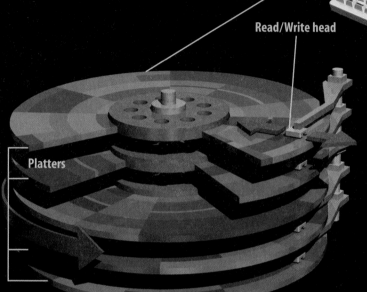

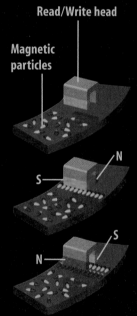

A A hard-disk drive is made of a stack of aluminum disks, called platters, that are coated with a thin layer that contains magnetic particles. Like tiny compasses, these particles will line up along magnetic field lines. The hard disk also contains read/write heads that contain electromagnets. When the hard disk is turned on, the platters spin under the heads.

B To write information on the disk, a magnetic field is created around the head by an electric current. As the platter rotates past the head, this magnetic field causes the magnetic particles on the platter to line up in bands. One direction of the bands corresponds to the digital bit 0, the other to the digital bit 1.

C To read information on the disk, no current is sent to the heads. Instead, the magnetized bands create a changing current in the head as it passes over the platter. This current is the electronic signal that represents the needed information.

SECTION 2 Computers

Floppy Disks Storing information on a hard disk is convenient, but sometimes you might want to store information that you can carry with you. The original storage device of this type was the floppy disk. A floppy disk is a thin, flexible, plastic disk. You might be confused by the term *floppy* if you have heard it used to describe disks that seem quite rigid. That is because you don't actually hold the floppy disk. Instead, you hold the harder plastic case in which the floppy disk is encased. Just as for a hard disk, the floppy disk is coated with a magnetic material that is magnetized and read by a read/write head. Floppy disks have lower storage capacity than hard disks. Also, compared to hard disks, information is read from and written to floppy disks much more slowly.

Optical Disks An optical storage disk, such as a CD, is a thin, plastic disk that has information digitally stored on it. The disk contains a series of microscopic pits and flat spots as shown on the left in **Figure 17.** A tiny laser beam shines on the surface of the disk. The information on the disk is read by measuring the intensity of the laser light reflected from the surface of the disk. This intensity will depend on whether the laser beam strikes a pit or a flat spot. The original optical storage disks, laser discs, CD-ROMs, and DVD-ROMs, were read-only. Several of these are shown on the right in **Figure 17.** However, CD-RW disks can be erased and rewritten many times. Information is written by a CD burner that causes a metal alloy in the disk to change form when heated by a laser. When the disk is read, the intensity of reflected laser light depends on which form of the alloy the beam strikes.

Science Online

Topic: Optical Disks
Visit ips.msscience.com for Web links to information about storing data on optical disks.

Activity Make a table that shows the similarities and differences between CDs and DVDs.

Figure 17 An optical storage disk stores information that is read by a laser.
Explain *the difference between a read-only disk and a reusable disk.*

Information is stored on an optical disk by a series of pits and flat spots, representing a binary 1 or 0.

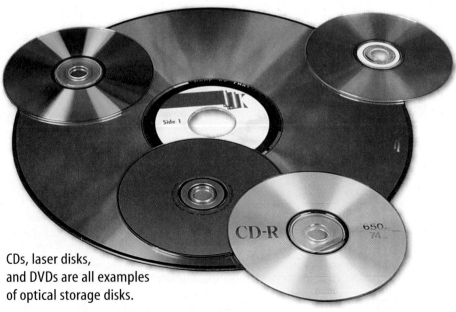

CDs, laser disks, and DVDs are all examples of optical storage disks.

658 CHAPTER 22 Electronics and Computers

Computer Networks

People can communicate using a computer if it is part of a computer network. A computer network is two or more computers that are connected to share files or other information. The computers might be linked by cables, telephone lines, or radio signals.

The Internet is a collection of computer networks from all over the world. The Internet is linked together by cable or satellite. The Internet itself has no information. No documents or files exist on the Internet, but you can use the Internet to access a tremendous amount of information by linking to other computers.

The World Wide Web is part of the Internet. The World Wide Web is the ever-changing collection of information (text, graphics, audio, and video) on computers all over the world. The computers that store these documents are called servers. When you connect with a server through the Internet, you can view any of the Web documents that are stored there, like the Web page shown in **Figure 18.** A particular collection of information that is stored in one place is known as a Web site.

Figure 18 When you connect to the Internet, you can be linked with other computers that are part of the World Wide Web. Then you can have access to the information stored at millions of Web sites.

section 2 review

Summary

Computer Information
- A binary digit can be a 0 or a 1.
- Computers store information as groups of binary digits.
- Computers use tiny electronic circuits to represent binary digits and store information.

Computer Software and Hardware
- Computer software and computer programs are lists of instructions for a computer.
- Computer programs are written in special computer languages.
- Computer hardware, such as keyboards and hard disks, is controlled by a microprocessor.

Storing Information
- Hard disks and floppy disks store information on disks coated with magnetic particles.
- Optical disks store information as a series of pits and flat spots that is read by a laser.

Self Check

1. **Explain** why the binary number system is used for storing information in computers.
2. **Compare and contrast** the Internet and the World Wide Web.
3. **Describe** what a microprocessor does with the signals it receives from various parts of a computer.
4. **Compare and contrast** three different computer information storage devices.
5. **Think Critically** Why can't computer information be stored only in RAM and ROM, making storage devices such as hard disks and optical disks unnecessary?

Applying Skills

6. **Make a Concept Map** Develop a spider map about computers. Include the following terms in your spider map: *keyboard, monitor, microprocessor, software, printer, RAM, ROM, floppy disk, hard disk, CD, Internet,* and *World Wide Web.*

ips.msscience.com/self_check_quiz

LAB
Use the Internet

Does your computer have a virus?

Goals
- **Understand** what a computer virus is.
- **Identify** different types of computer viruses.
- **Describe** how a computer virus is spread.
- **Create** a plan for protecting electronic files and computers from computer viruses.

Data Source

Science Online
Visit ips.msscience.com/internet_lab to get more information about computer viruses and for data collected by other students.

Real-World Question

The Internet has provided many ways to share information and become connected with people near and far. People can communicate ideas and information quickly and easily. Unfortunately, some people use computers and the Internet as an opportunity to create and spread computer viruses. Many new viruses are created each year that can damage information and programs on a computer. Viruses create problems for computers in homes and schools. Computer problems caused by viruses can be costly for business and government computers, as well. How can acquiring and transmitting computer viruses be prevented?

Make a Plan

People share information and ideas by exchanging electronic files with one another. Perhaps you send email to your friends and family. Many people send word processing or spreadsheet files to friends and associates. What happens if a computer file is infected with a virus? How is that virus spread among different users? How can you protect your computer and your information from being attacked by a virus?

660 CHAPTER 22 Electronics and Computers

Using Scientific Methods

◉ Follow Your Plan

1. Do research to find out what a computer virus is and the difference between various types of viruses. Also research the ways that a computer virus can damage computer files and programs.

2. After you know what a computer virus is, make a list of different types of viruses and how they are passed from computer to computer. For example, some viruses can be passed through attachments to email. Others can be passed by sharing spreadsheet files. Be specific about how a virus is passed.

3. Discover how you can protect yourself from viruses that attack your computer. Make a list of steps to follow to avoid infection.

4. Make sure your teacher approves your plan before you start.

5. Visit the link below to post your data.

◉ Analyze Your Data

1. **Explain** how computer viruses are transferred from one computer to another.

2. **Explain** how you can prevent your computer from becoming infected by a virus.

3. **Explain** how you can prevent other people from getting computer viruses.

4. **Describe** the different ways computer viruses can damage computer files and programs.

◉ Conclude and Apply

1. **List** five to eight steps a computer user should follow to prevent getting a computer virus or passing a computer virus to another computer.

2. **Discuss** how antivirus software can keep viruses from spreading. Could antivirus software always prevent you from getting a computer virus? Why or why not?

Communicating Your Data

Find this lab using the link below. Post your data in the table that is provided. **Compare** your data on types of viruses and how they infect computers with that of other students.

ips.msscience.com/internet_lab

LAB **661**

TIME SCIENCE AND Society
SCIENCE ISSUES THAT AFFECT YOU!

E-Lectrifying E-Books

Here's a look at how computers and the Internet are changing what—and how—you read

In recent years, people have been using their computers to order books from online bookstores. That's no big deal. What might become a big deal is the ordering of electronic books—books that you download to your own computer and read on the screen or print out to read later. Some famous authors are writing books just for that purpose. Some of the books are published only online—you can't find them anywhere else.

Many other Web sites, however, are selling any book anybody wishes to write—including students like you. In fact, you could start your own online bookstore with your own stories and reports. It will be up to readers to pick and choose what's good from the huge number of e-books that will be on the Web.

Curling Up with a Good Disk

Downloading books to your home computer is just one way to get an e-book. You can also buy versions of books to read on hand-held devices that are about the size of a paperback book. With one device, the books come on CD-ROM disks. With another, the books download to the device over a modem.

Current e-book devices are expensive, heavy, and awkward, and the number of books you can get for them is small. But if improvements come quickly, it might not be long before you check out of the library with a pocketful of disks instead of a heavy armload of books!

All of these stacked books can fit into one e-book.

Will Traditional Books Disappear?

Most people think that the traditional printed book will never disappear. Publishers will still be printing books on paper with soft and hard covers. But publishers also predict there will be more and more kinds of formats for books. E-books, for example, might be best for interactive works that blend video, sound, and words the way many Web sites already do. For example, an e-book biography might allow the reader to click on photos and videos of the subject, and even provide links to other sources of information.

Interview Talk to a bookstore employee to find out how book publishing and selling has changed in the last five years. Can he or she predict how people will read books in the future? Report to the class.

For more information, visit ips.msscience.com/time

chapter 22 Study Guide

Reviewing Main Ideas

Section 1 Electronics

1. A changing electric current used to carry information is an electronic signal. Electronic signals can be either analog or digital.

2. Semiconductor elements, such as silicon and germanium, conduct electricity better than nonmetals but not as well as metals. If a small amount of some impurities is added to a semiconductor, its conductivity can be controlled.

3. Diodes and transistors are solid-state components. Diodes allow current to flow in one direction only. Transistors are used as switches or amplifiers.

Section 2 Computers

1. The binary system consists of two digits, 0 and 1. Switches within electronic devices such as computers can store information by turning on (1) and off (0).

2. Electronic memory within a computer can be random-access (RAM) or read-only (ROM).

3. Computer hardware consists of the physical parts of a computer. Computer software is a list of instructions for a computer.

4. A microprocessor is a complex integrated circuit that receives signals from various parts of the computer, processes these signals, and then sends instructions to various parts of the computer.

5. Floppy disks, hard disks, and optical disks are types of computer information storage devices.

6. The Internet is a collection of linked computer networks from all over the world. The World Wide Web is part of the Internet.

Visualizing Main Ideas

Copy and complete the following concept map on computers.

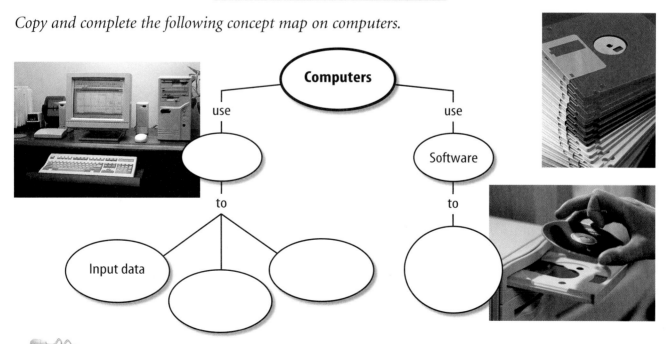

chapter 22 Review

Using Vocabulary

analog signal p. 642
binary system p. 650
computer software p. 653
digital signal p. 643
diode p. 646
electronic signal p. 642
integrated circuit p. 647
microprocessor p. 655
random access memory p. 652
read-only memory p. 652
semiconductor p. 645
transistor p. 647

Fill in the blanks with the correct vocabulary word or words.

1. _____ is a base-2 number system.
2. A(n) _____ can change AC current to DC current.
3. A(n) _____ is made from a single piece of semiconductor material and can contain thousands of solid-state components.
4. The information in a computer's _____ changes each time the computer is used.
5. An electronic device that can be used as a switch or to amplify electronic signals is a(n) _____.
6. A(n) _____ is also called a CPU.
7. An electronic signal that varies smoothly with time is a(n) _____.

Checking Concepts

Choose the word or phrase that best answers the question.

8. Which of the following best describes integrated circuits?
   A) They can be read with a laser.
   B) They use vacuum tubes as transistors and diodes.
   C) They contain pits and flat areas.
   D) They can be small and contain a large number of solid-state components.

9. Which type of elements are semiconductors?
   A) metals   C) metalloids
   B) nonmetals   D) gases

10. How is a digital signal different from an analog signal?
    A) It uses electric current.
    B) It varies continuously.
    C) It changes in steps.
    D) It is used as a switch.

11. Which of the following uses magnetic materials to store digital information.
    A) DVD   C) RAM
    B) hard disk   D) compact disk

12. Which part of a computer carries out the instructions contained in computer programs and software?
    A) RAM   C) hard disk
    B) ROM   D) microprocessor

13. Which type of computer memory is used when a computer is first turned on?
    A) ROM   C) DVD
    B) RAM   D) floppy disk

14. The instructions contained in a computer program are stored in which type of computer memory while the program is being used?
    A) ROM   C) CD
    B) RAM   D) floppy disk

Use the figure below to answer question 15.

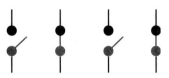

15. What binary number is represented by the positions of the switches?
    A) 1110   C) 0101
    B) 0010   D) 0001

664 CHAPTER REVIEW

chapter 22 Review

Thinking Critically

16. **Compare and contrast** an analog device and a digital device.

17. **Make and Use Tables** Copy and complete the following table that describes solid-state components.

Solid-State Components		
Component	Description	Use
Diode		
Transistor		
Integrated circuit		

18. **Explain** why the binary number system is used to store digital information in computers, instead of the decimal number system you use every day.

19. **Concept Map** Copy and complete the following events-chain map showing the sequence of events that occurs when a computer mouse is moved.

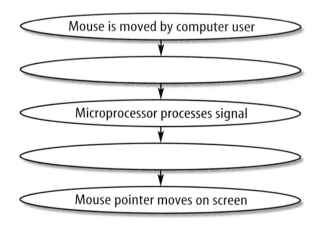

- Mouse is moved by computer user
-
- Microprocessor processes signal
- Mouse pointer moves on screen

20. **Discuss** how the development of solid-state components and integrated circuits affected devices such as TVs and computers.

21. **Make a table** to classify the different types of internal and external computer memory and storage.

Performance Activities

22. **Make a Poster** Microprocessors continue to be developed that are more complex and contain an ever-increasing number of solid state components. Visit **msscience.com** for links to information about different microprocessors and how they have changed. Make a poster that summarizes what you have learned.

Applying Math

Students at a middle school researched the storage capacity of different computer storage devices. The information is summarized in the table below. The storage capacity is listed in units of gigabytes. A gigabyte is 1,074,000,000 bytes.

Use the table below to answer questions 23–25.

Computer Storage Devices	
Device	Capacity (Gb)
Floppy disk	0.00144
Compact disc	0.650
DVD	4.7
Hard Disk A	8.60
Hard Disk B	120.2

23. **Music Files Storage** In a certain format, to store 1 min of music as a digital signal requires 10,584,000 bytes. How many minutes of music in this format can be stored on the compact disc?

24. **Digital Pictures Storage** A certain digital camera produces digital images that require 921,600 bytes to store. How many of these images could be stored on hard disk A?

25. **Documents Storage** Seven documents produced by word processing software are stored on a floppy disk. If there are 40,000 bytes of storage still available on the disk, what is the average amount of storage used by each of the documents?

ips.msscience.com/chapter_review

chapter 22 Standardized Test Practice

Part 1 Multiple Choice

Record your answers on the answer sheet provided by your teacher or on a sheet of paper.

1. What kinds of materials are used to make solid-state components?
   A. semiconductors
   B. superconductors
   C. conductors
   D. insulators

2. Which of the following are not contained in integrated circuits?
   A. semiconductors C. diodes
   B. vacuum tubes   D. transistors

Use the table below to answer questions 3 and 4.

Number of Binary Digit Combinations	
Number of Binary Digits	Total Number of Combinations
1	2
2	4
3	8
4	?
5	32

3. Which of the following is the total number of combinations of four binary digits?
   A. 64   C. 32
   B. 16   D. 8

4. Based on the data table, which of the following is the total number of combinations of six binary digits?
   A. 64   C. 32
   B. 16   D. 8

5. Which of the following best describes computer software?
   A. It is a type of temporary storage.
   B. It is a list of instructions.
   C. It contains analog information.
   D. It cannot be stored magnetically.

6. Which of the following is a computer input device?
   A. printer   C. monitor
   B. loudspeakers   D. keyboard

7. Which of the following is an optical storage device?
   A. hard disk   C. floppy disk
   B. RAM   D. CD

8. Where are elements that are semiconductors located on the periodic table?
   A. between metals and nonmetals
   B. on the right column
   C. on the left column
   D. at the bottom

9. Which of the following is not an electronic device?
   A. calculator   C. CD player
   B. television   D. light bulb

Use the figure below to answer question 10.

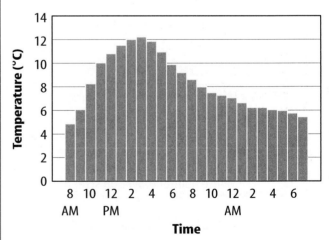

Digitized Analog Signal

10. Which of the following is a process that produces this digital signal from an analog signal?
    A. doping   C. switching
    B. programming   D. sampling

666 STANDARDIZED TEST PRACTICE

Standardized Test Practice

Part 2 Short Response/Grid In

Record your answers on the answer sheet provided by your teacher or on a sheet of paper.

11. What is the difference between a binary 1 or 0 that is stored on a platter of a hard disk?

Use the figure below to answer questions 12 and 13.

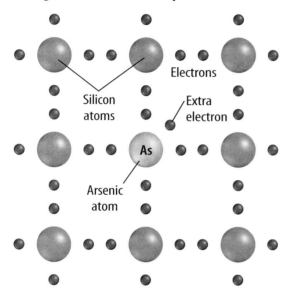

12. Describe the type of semiconductor material that is shown in the figure above.

13. Explain why atoms of other elements are added to a semiconductor material.

14. In a diode, which way does current flow between the n-type semiconductor and the p-type semiconductor?

15. Identify the process that enables background noise to be removed from a digitized music file that is stored on a hard disk.

Test-Taking Tip

Some Questions Have Qualifiers Look for qualifiers in a question. Such questions are not looking for absolute answers. Qualifiers could be words such as *most likely, most common,* or *least common.*

Part 3 Open Ended

Record your answers on a sheet of paper.

16. A barograph is a device that measures and records air pressure. A barograph contains a pen that moves up and down as the pressure changes. The pen continuously draws a line on paper attached to a drum that slowly rotates. Infer whether the barograph is an analog or digital device. Explain.

17. Classify each of the following as an input, output, or storages device: monitor, keyboard, printer, hard disk. Explain the function of each device if you are using a word processing program to write a report.

18. Compare and contrast a floppy disk with an optical disk, such as a CD.

19. Explain why electric circuits that can be charged or uncharged are used as the components of computer memory.

Use the figure below to answer questions 20 and 21.

20. Describe how the read/write heads write information on a platter.

21. Describe how the read/write heads read information that is stored on the platters.

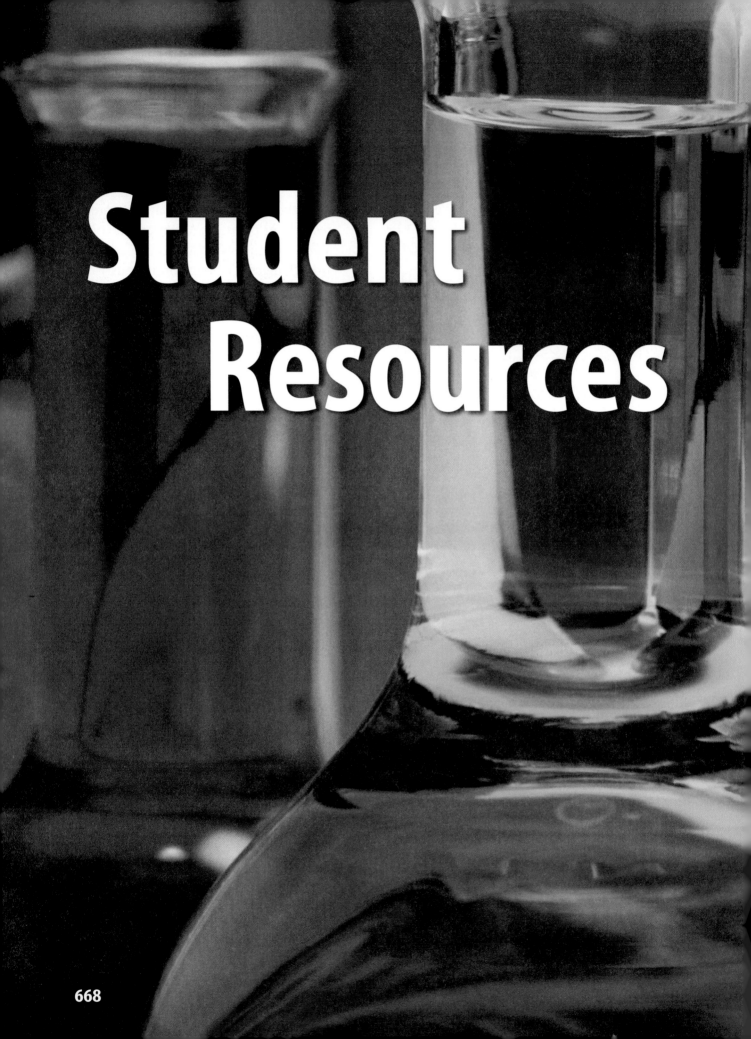
Student Resources

Student Resources

CONTENTS

Science Skill Handbook 670
Scientific Methods 670
   Identify a Question 670
   Gather and Organize
     Information 670
   Form a Hypothesis 673
   Test the Hypothesis 674
   Collect Data 674
   Analyze the Data 677
   Draw Conclusions 678
   Communicate 678
Safety Symbols 679
Safety in the Science Laboratory 680
   General Safety Rules 680
   Prevent Accidents 680
   Laboratory Work 680
   Laboratory Cleanup 681
   Emergencies 681

Extra Try at Home Labs 682
   Testing Horoscopes 682
   Disappearing Water? 682
   Comparing Atom Sizes 683
   Microscopic Crystals 683
   Colorful Liquids 684
   Human Bonding 684
   Mini Fireworks 685
   A Good Mix? 685
   Liquid Lab 686
   Measuring Momentum 686
   Friction in Traffic 687
   Submersible Egg 687
   The Heat is On 688
   Simple Machines 688
   Estimate Temperature 689
   Exploding Bag 689
   Seeing Sound 690
   Black Light 690
   Light in Liquids 691
   Bending Water 691
   Testing Magnets 692
   Pattern Counting 692

Technology Skill Handbook ...693
Computer Skills 693
   Use a Word Processing Program ...693
   Use a Database 694
   Use the Internet 694
   Use a Spreadsheet 695
   Use Graphics Software 695
Presentation Skills 696
   Develop Multimedia
     Presentations 696
   Computer Presentations 696

Math Skill Handbook 697
Math Review 697
   Use Fractions 697
   Use Ratios 700
   Use Decimals 700
   Use Proportions 701
   Use Percentages 702
   Solve One-Step Equations 702
   Use Statistics 703
   Use Geometry 704
Science Applications 707
   Measure in SI 707
   Dimensional Analysis 707
   Precision and Significant Digits ...709
   Scientific Notation 709
   Make and Use Graphs 710

Reference Handbook 712
Periodic Table of the Elements 712

English/Spanish Glossary 714

Index 731

Credits 749

Science Skill Handbook

Scientific Methods

Scientists use an orderly approach called scientific methods to solve problems. These include organizing and recording data so others can understand them. Scientists use many variations in these methods when they solve problems.

Identify a Question

The first step in a scientific investigation or experiment is to identify a question to be answered or a problem to be solved. For example, you might ask which gasoline is the most efficient.

Gather and Organize Information

After you have identified your question, begin gathering and organizing information. There are many ways to gather information, such as researching in a library, interviewing those knowledgeable about the subject, testing and working in the laboratory and field. Fieldwork is investigations and observations done outside of a laboratory.

Researching Information Before moving in a new direction, it is important to gather the information that already is known about the subject. Start by asking yourself questions to determine exactly what you need to know. Then you will look for the information in various reference sources, like the student is doing in **Figure 1.** Some sources may include textbooks, encyclopedias, government documents, professional journals, science magazines, and the Internet. Always list the sources of your information.

Figure 1 The Internet can be a valuable research tool.

Evaluate Sources of Information Not all sources of information are reliable. You should evaluate all of your sources of information, and use only those you know to be dependable. For example, if you are researching ways to make homes more energy efficient, a site written by the U.S. Department of Energy would be more reliable than a site written by a company that is trying to sell a new type of weatherproofing material. Also, remember that research always is changing. Consult the most current resources available to you. For example, a 1985 resource about saving energy would not reflect the most recent findings.

Sometimes scientists use data that they did not collect themselves, or conclusions drawn by other researchers. This data must be evaluated carefully. Ask questions about how the data were obtained, if the investigation was carried out properly, and if it has been duplicated exactly with the same results. Would you reach the same conclusion from the data? Only when you have confidence in the data can you believe it is true and feel comfortable using it.

670 STUDENT RESOURCES

Science Skill Handbook

Interpret Scientific Illustrations As you research a topic in science, you will see drawings, diagrams, and photographs to help you understand what you read. Some illustrations are included to help you understand an idea that you can't see easily by yourself, like the tiny particles in an atom in **Figure 2**. A drawing helps many people to remember details more easily and provides examples that clarify difficult concepts or give additional information about the topic you are studying. Most illustrations have labels or a caption to identify or to provide more information.

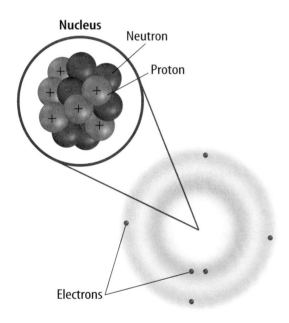

Figure 2 This drawing shows an atom of carbon with its six protons, six neutrons, and six electrons.

Concept Maps One way to organize data is to draw a diagram that shows relationships among ideas (or concepts). A concept map can help make the meanings of ideas and terms more clear, and help you understand and remember what you are studying. Concept maps are useful for breaking large concepts down into smaller parts, making learning easier.

Network Tree A type of concept map that not only shows a relationship, but how the concepts are related is a network tree, shown in **Figure 3**. In a network tree, the words are written in the ovals, while the description of the type of relationship is written across the connecting lines.

When constructing a network tree, write down the topic and all major topics on separate pieces of paper or notecards. Then arrange them in order from general to specific. Branch the related concepts from the major concept and describe the relationship on the connecting line. Continue to more specific concepts until finished.

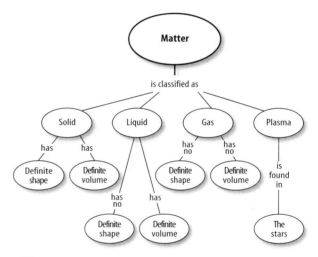

Figure 3 A network tree shows how concepts or objects are related.

Events Chain Another type of concept map is an events chain. Sometimes called a flow chart, it models the order or sequence of items. An events chain can be used to describe a sequence of events, the steps in a procedure, or the stages of a process.

When making an events chain, first find the one event that starts the chain. This event is called the initiating event. Then, find the next event and continue until the outcome is reached, as shown in **Figure 4**.

SCIENCE SKILL HANDBOOK 671

Science Skill Handbook

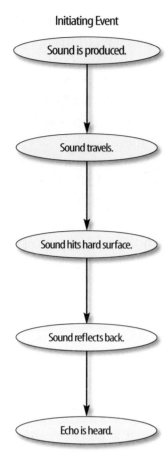

Figure 4 Events-chain concept maps show the order of steps in a process or event. This concept map shows how a sound makes an echo.

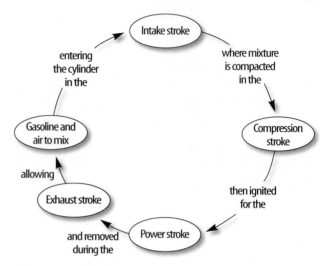

Figure 5 A cycle map shows events that occur in a cycle.

Cycle Map A specific type of events chain is a cycle map. It is used when the series of events do not produce a final outcome, but instead relate back to the beginning event, such as in **Figure 5.** Therefore, the cycle repeats itself.

To make a cycle map, first decide what event is the beginning event. This is also called the initiating event. Then list the next events in the order that they occur, with the last event relating back to the initiating event. Words can be written between the events that describe what happens from one event to the next. The number of events in a cycle map can vary, but usually contain three or more events.

Spider Map A type of concept map that you can use for brainstorming is the spider map. When you have a central idea, you might find that you have a jumble of ideas that relate to it but are not necessarily clearly related to each other. The spider map on sound in **Figure 6** shows that if you write these ideas outside the main concept, then you can begin to separate and group unrelated terms so they become more useful.

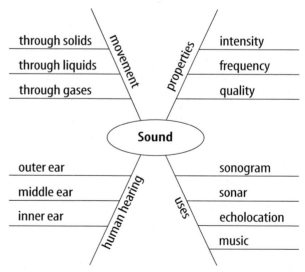

Figure 6 A spider map allows you to list ideas that relate to a central topic but not necessarily to one another.

672 STUDENT RESOURCES

Science Skill Handbook

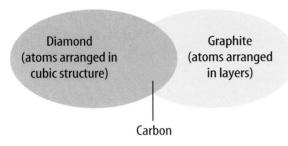

Figure 7 This Venn diagram compares and contrasts two substances made from carbon.

Venn Diagram To illustrate how two subjects compare and contrast you can use a Venn diagram. You can see the characteristics that the subjects have in common and those that they do not, shown in **Figure 7.**

To create a Venn diagram, draw two overlapping ovals that that are big enough to write in. List the characteristics unique to one subject in one oval, and the characteristics of the other subject in the other oval. The characteristics in common are listed in the overlapping section.

Make and Use Tables One way to organize information so it is easier to understand is to use a table. Tables can contain numbers, words, or both.

To make a table, list the items to be compared in the first column and the characteristics to be compared in the first row. The title should clearly indicate the content of the table, and the column or row heads should be clear. Notice that in **Table 1** the units are included.

Table 1 Recyclables Collected During Week			
Day of Week	Paper (kg)	Aluminum (kg)	Glass (kg)
Monday	5.0	4.0	12.0
Wednesday	4.0	1.0	10.0
Friday	2.5	2.0	10.0

Make a Model One way to help you better understand the parts of a structure, the way a process works, or to show things too large or small for viewing is to make a model. For example, an atomic model made of a plastic-ball nucleus and pipe-cleaner electron shells can help you visualize how the parts of an atom relate to each other. Other types of models can by devised on a computer or represented by equations.

Form a Hypothesis

A possible explanation based on previous knowledge and observations is called a hypothesis. After researching gasoline types and recalling previous experiences in your family's car you form a hypothesis—our car runs more efficiently because we use premium gasoline. To be valid, a hypothesis has to be something you can test by using an investigation.

Predict When you apply a hypothesis to a specific situation, you predict something about that situation. A prediction makes a statement in advance, based on prior observation, experience, or scientific reasoning. People use predictions to make everyday decisions. Scientists test predictions by performing investigations. Based on previous observations and experiences, you might form a prediction that cars are more efficient with premium gasoline. The prediction can be tested in an investigation.

Design an Experiment A scientist needs to make many decisions before beginning an investigation. Some of these include: how to carry out the investigation, what steps to follow, how to record the data, and how the investigation will answer the question. It also is important to address any safety concerns.

SCIENCE SKILL HANDBOOK 673

Science Skill Handbook

Test the Hypothesis

Now that you have formed your hypothesis, you need to test it. Using an investigation, you will make observations and collect data, or information. This data might either support or not support your hypothesis. Scientists collect and organize data as numbers and descriptions.

Follow a Procedure In order to know what materials to use, as well as how and in what order to use them, you must follow a procedure. **Figure 8** shows a procedure you might follow to test your hypothesis.

Procedure
1. Use regular gasoline for two weeks.
2. Record the number of kilometers between fill-ups and the amount of gasoline used.
3. Switch to premium gasoline for two weeks.
4. Record the number of kilometers between fill-ups and the amount of gasoline used.

Figure 8 A procedure tells you what to do step by step.

Identify and Manipulate Variables and Controls In any experiment, it is important to keep everything the same except for the item you are testing. The one factor you change is called the independent variable. The change that results is the dependent variable. Make sure you have only one independent variable, to assure yourself of the cause of the changes you observe in the dependent variable. For example, in your gasoline experiment the type of fuel is the independent variable. The dependent variable is the efficiency.

Many experiments also have a control—an individual instance or experimental subject for which the independent variable is not changed. You can then compare the test results to the control results. To design a control you can have two cars of the same type. The control car uses regular gasoline for four weeks. After you are done with the test, you can compare the experimental results to the control results.

Collect Data

Whether you are carrying out an investigation or a short observational experiment, you will collect data, as shown in **Figure 9**. Scientists collect data as numbers and descriptions and organize it in specific ways.

Observe Scientists observe items and events, then record what they see. When they use only words to describe an observation, it is called qualitative data. Scientists' observations also can describe how much there is of something. These observations use numbers, as well as words, in the description and are called quantitative data. For example, if a sample of the element gold is described as being "shiny and very dense" the data are qualitative. Quantitative data on this sample of gold might include "a mass of 30 g and a density of 19.3 g/cm^3."

Figure 9 Collecting data is one way to gather information directly.

674 STUDENT RESOURCES

Science Skill Handbook

Figure 10 Record data neatly and clearly so it is easy to understand.

When you make observations you should examine the entire object or situation first, and then look carefully for details. It is important to record observations accurately and completely. Always record your notes immediately as you make them, so you do not miss details or make a mistake when recording results from memory. Never put unidentified observations on scraps of paper. Instead they should be recorded in a notebook, like the one in **Figure 10.** Write your data neatly so you can easily read it later. At each point in the experiment, record your observations and label them. That way, you will not have to determine what the figures mean when you look at your notes later. Set up any tables that you will need to use ahead of time, so you can record any observations right away. Remember to avoid bias when collecting data by not including personal thoughts when you record observations. Record only what you observe.

Estimate Scientific work also involves estimating. To estimate is to make a judgment about the size or the number of something without measuring or counting. This is important when the number or size of an object or population is too large or too difficult to accurately count or measure.

Sample Scientists may use a sample or a portion of the total number as a type of estimation. To sample is to take a small, representative portion of the objects or organisms of a population for research. By making careful observations or manipulating variables within that portion of the group, information is discovered and conclusions are drawn that might apply to the whole population. A poorly chosen sample can be unrepresentative of the whole. If you were trying to determine the rainfall in an area, it would not be best to take a rainfall sample from under a tree.

Measure You use measurements everyday. Scientists also take measurements when collecting data. When taking measurements, it is important to know how to use measuring tools properly. Accuracy also is important.

Length To measure length, the distance between two points, scientists use meters. Smaller measurements might be measured in centimeters or millimeters.

Length is measured using a metric ruler or meter stick. When using a metric ruler, line up the 0-cm mark with the end of the object being measured and read the number of the unit where the object ends. Look at the metric ruler shown in **Figure 11.** The centimeter lines are the long, numbered lines, and the shorter lines are millimeter lines. In this instance, the length would be 4.50 cm.

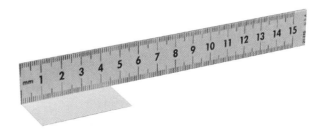

Figure 11 This metric ruler has centimeter and millimeter divisions.

SCIENCE SKILL HANDBOOK **675**

Science Skill Handbook

Mass The SI unit for mass is the kilogram (kg). Scientists can measure mass using units formed by adding metric prefixes to the unit gram (g), such as milligram (mg). To measure mass, you might use a triple-beam balance similar to the one shown in **Figure 12.** The balance has a pan on one side and a set of beams on the other side. Each beam has a rider that slides on the beam.

When using a triple-beam balance, place an object on the pan. Slide the largest rider along its beam until the pointer drops below zero. Then move it back one notch. Repeat the process for each rider proceeding from the larger to smaller until the pointer swings an equal distance above and below the zero point. Sum the masses on each beam to find the mass of the object. Move all riders back to zero when finished.

Instead of putting materials directly on the balance, scientists often take a tare of a container. A tare is the mass of a container into which objects or substances are placed for measuring their masses. To mass objects or substances, find the mass of a clean container. Remove the container from the pan, and place the object or substances in the container. Find the mass of the container with the materials in it. Subtract the mass of the empty container from the mass of the filled container to find the mass of the materials you are using.

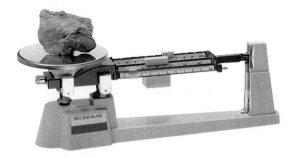

Figure 12 A triple-beam balance is used to determine the mass of an object.

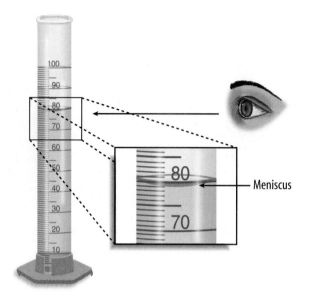

Figure 13 Graduated cylinders measure liquid volume.

Liquid Volume To measure liquids, the unit used is the liter. When a smaller unit is needed, scientists might use a milliliter. Because a milliliter takes up the volume of a cube measuring 1 cm on each side it also can be called a cubic centimeter ($cm^3 = cm \times cm \times cm$).

You can use beakers and graduated cylinders to measure liquid volume. A graduated cylinder, shown in **Figure 13,** is marked from bottom to top in milliliters. In lab, you might use a 10-mL graduated cylinder or a 100-mL graduated cylinder. When measuring liquids, notice that the liquid has a curved surface. Look at the surface at eye level, and measure the bottom of the curve. This is called the meniscus. The graduated cylinder in **Figure 13** contains 79.0 mL, or 79.0 cm^3, of a liquid.

Temperature Scientists often measure temperature using the Celsius scale. Pure water has a freezing point of 0°C and boiling point of 100°C. The unit of measurement is degrees Celsius. Two other scales often used are the Fahrenheit and Kelvin scales.

Science Skill Handbook

Analyze the Data

To determine the meaning of your observations and investigation results, you will need to look for patterns in the data. Then you must think critically to determine what the data mean. Scientists use several approaches when they analyze the data they have collected and recorded. Each approach is useful for identifying specific patterns.

Interpret Data The word *interpret* means "to explain the meaning of something." When analyzing data from an experiement, try to find out what the data show. Identify the control group and the test group to see whether or not changes in the independent variable have had an effect. Look for differences in the dependent variable between the control and test groups.

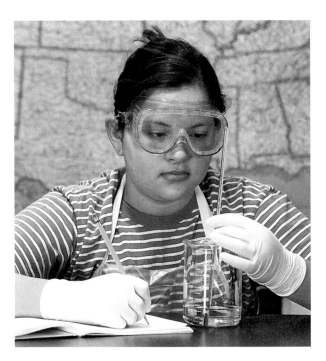

Figure 14 A thermometer measures the temperature of an object.

Scientists use a thermometer to measure temperature. Most thermometers in a laboratory are glass tubes with a bulb at the bottom end containing a liquid such as colored alcohol. The liquid rises or falls with a change in temperature. To read a glass thermometer like the thermometer in **Figure 14,** rotate it slowly until a red line appears. Read the temperature where the red line ends.

Form Operational Definitions An operational definition defines an object by how it functions, works, or behaves. For example, when you are playing hide and seek and a tree is home base, you have created an operational definition for a tree.

Objects can have more than one operational definition. For example, a ruler can be defined as a tool that measures the length of an object (how it is used). It can also be a tool with a series of marks used as a standard when measuring (how it works).

Classify Sorting objects or events into groups based on common features is called classifying. When classifying, first observe the objects or events to be classified. Then select one feature that is shared by some members in the group, but not by all. Place those members that share that feature in a subgroup. You can classify members into smaller and smaller subgroups based on characteristics. Remember that when you classify, you are grouping objects or events for a purpose. Keep your purpose in mind as you select the features to form groups and subgroups.

Compare and Contrast Observations can be analyzed by noting the similarities and differences between two more objects or events that you observe. When you look at objects or events to see how they are similar, you are comparing them. Contrasting is looking for differences in objects or events.

SCIENCE SKILL HANDBOOK **677**

Science Skill Handbook

Recognize Cause and Effect A cause is a reason for an action or condition. The effect is that action or condition. When two events happen together, it is not necessarily true that one event caused the other. Scientists must design a controlled investigation to recognize the exact cause and effect.

Draw Conclusions

When scientists have analyzed the data they collected, they proceed to draw conclusions about the data. These conclusions are sometimes stated in words similar to the hypothesis that you formed earlier. They may confirm a hypothesis, or lead you to a new hypothesis.

Infer Scientists often make inferences based on their observations. An inference is an attempt to explain observations or to indicate a cause. An inference is not a fact, but a logical conclusion that needs further investigation. For example, you may infer that a fire has caused smoke. Until you investigate, however, you do not know for sure.

Apply When you draw a conclusion, you must apply those conclusions to determine whether the data supports the hypothesis. If your data do not support your hypothesis, it does not mean that the hypothesis is wrong. It means only that the result of the investigation did not support the hypothesis. Maybe the experiment needs to be redesigned, or some of the initial observations on which the hypothesis was based were incomplete or biased. Perhaps more observation or research is needed to refine your hypothesis. A successful investigation does not always come out the way you originally predicted.

Avoid Bias Sometimes a scientific investigation involves making judgments. When you make a judgment, you form an opinion. It is important to be honest and not to allow any expectations of results to bias your judgments. This is important throughout the entire investigation, from researching to collecting data to drawing conclusions.

Communicate

The communication of ideas is an important part of the work of scientists. A discovery that is not reported will not advance the scientific community's understanding or knowledge. Communication among scientists also is important as a way of improving their investigations.

Scientists communicate in many ways, from writing articles in journals and magazines that explain their investigations and experiments, to announcing important discoveries on television and radio. Scientists also share ideas with colleagues on the Internet or present them as lectures, like the student is doing in **Figure 15**.

Figure 15 A student communicates to his peers about his investigation.

Science Skill Handbook

SAFETY SYMBOLS

	HAZARD	EXAMPLES	PRECAUTION	REMEDY
DISPOSAL	Special disposal procedures need to be followed.	certain chemicals, living organisms	Do not dispose of these materials in the sink or trash can.	Dispose of wastes as directed by your teacher.
BIOLOGICAL	Organisms or other biological materials that might be harmful to humans	bacteria, fungi, blood, unpreserved tissues, plant materials	Avoid skin contact with these materials. Wear mask or gloves.	Notify your teacher if you suspect contact with material. Wash hands thoroughly.
EXTREME TEMPERATURE	Objects that can burn skin by being too cold or too hot	boiling liquids, hot plates, dry ice, liquid nitrogen	Use proper protection when handling.	Go to your teacher for first aid.
SHARP OBJECT	Use of tools or glassware that can easily puncture or slice skin	razor blades, pins, scalpels, pointed tools, dissecting probes, broken glass	Practice common-sense behavior and follow guidelines for use of the tool.	Go to your teacher for first aid.
FUME	Possible danger to respiratory tract from fumes	ammonia, acetone, nail polish remover, heated sulfur, moth balls	Make sure there is good ventilation. Never smell fumes directly. Wear a mask.	Leave foul area and notify your teacher immediately.
ELECTRICAL	Possible danger from electrical shock or burn	improper grounding, liquid spills, short circuits, exposed wires	Double-check setup with teacher. Check condition of wires and apparatus.	Do not attempt to fix electrical problems. Notify your teacher immediately.
IRRITANT	Substances that can irritate the skin or mucous membranes of the respiratory tract	pollen, moth balls, steel wool, fiberglass, potassium permanganate	Wear dust mask and gloves. Practice extra care when handling these materials.	Go to your teacher for first aid.
CHEMICAL	Chemicals can react with and destroy tissue and other materials	bleaches such as hydrogen peroxide; acids such as sulfuric acid, hydrochloric acid; bases such as ammonia, sodium hydroxide	Wear goggles, gloves, and an apron.	Immediately flush the affected area with water and notify your teacher.
TOXIC	Substance may be poisonous if touched, inhaled, or swallowed.	mercury, many metal compounds, iodine, poinsettia plant parts	Follow your teacher's instructions.	Always wash hands thoroughly after use. Go to your teacher for first aid.
FLAMMABLE	Flammable chemicals may be ignited by open flame, spark, or exposed heat.	alcohol, kerosene, potassium permanganate	Avoid open flames and heat when using flammable chemicals.	Notify your teacher immediately. Use fire safety equipment if applicable.
OPEN FLAME	Open flame in use, may cause fire.	hair, clothing, paper, synthetic materials	Tie back hair and loose clothing. Follow teacher's instruction on lighting and extinguishing flames.	Notify your teacher immediately. Use fire safety equipment if applicable.

 Eye Safety Proper eye protection should be worn at all times by anyone performing or observing science activities.

 Clothing Protection This symbol appears when substances could stain or burn clothing.

 Animal Safety This symbol appears when safety of animals and students must be ensured.

 Handwashing After the lab, wash hands with soap and water before removing goggles.

Science Skill Handbook

Safety in the Science Laboratory

The science laboratory is a safe place to work if you follow standard safety procedures. Being responsible for your own safety helps to make the entire laboratory a safer place for everyone. When performing any lab, read and apply the caution statements and safety symbol listed at the beginning of the lab.

General Safety Rules

1. Obtain your teacher's permission to begin all investigations and use laboratory equipment.

2. Study the procedure. Ask your teacher any questions. Be sure you understand safety symbols shown on the page.

3. Notify your teacher about allergies or other health conditions which can affect your participation in a lab.

4. Learn and follow use and safety procedures for your equipment. If unsure, ask your teacher.

5. Never eat, drink, chew gum, apply cosmetics, or do any personal grooming in the lab. Never use lab glassware as food or drink containers. Keep your hands away from your face and mouth.

6. Know the location and proper use of the safety shower, eye wash, fire blanket, and fire alarm.

Prevent Accidents

1. Use the safety equipment provided to you. Goggles and a safety apron should be worn during investigations.

2. Do NOT use hair spray, mousse, or other flammable hair products. Tie back long hair and tie down loose clothing.

3. Do NOT wear sandals or other open-toed shoes in the lab.

4. Remove jewelry on hands and wrists. Loose jewelry, such as chains and long necklaces, should be removed to prevent them from getting caught in equipment.

5. Do not taste any substances or draw any material into a tube with your mouth.

6. Proper behavior is expected in the lab. Practical jokes and fooling around can lead to accidents and injury.

7. Keep your work area uncluttered.

Laboratory Work

1. Collect and carry all equipment and materials to your work area before beginning a lab.

2. Remain in your own work area unless given permission by your teacher to leave it.

680 STUDENT RESOURCES

Science Skill Handbook

3. Always slant test tubes away from yourself and others when heating them, adding substances to them, or rinsing them.

4. If instructed to smell a substance in a container, hold the container a short distance away and fan vapors towards your nose.

5. Do NOT substitute other chemicals/substances for those in the materials list unless instructed to do so by your teacher.

6. Do NOT take any materials or chemicals outside of the laboratory.

7. Stay out of storage areas unless instructed to be there and supervised by your teacher.

Laboratory Cleanup

1. Turn off all burners, water, and gas, and disconnect all electrical devices.

2. Clean all pieces of equipment and return all materials to their proper places.

3. Dispose of chemicals and other materials as directed by your teacher. Place broken glass and solid substances in the proper containers. Never discard materials in the sink.

4. Clean your work area.

5. Wash your hands with soap and water thoroughly BEFORE removing your goggles.

Emergencies

1. Report any fire, electrical shock, glassware breakage, spill, or injury, no matter how small, to your teacher immediately. Follow his or her instructions.

2. If your clothing should catch fire, STOP, DROP, and ROLL. If possible, smother it with the fire blanket or get under a safety shower. NEVER RUN.

3. If a fire should occur, turn off all gas and leave the room according to established procedures.

4. In most instances, your teacher will clean up spills. Do NOT attempt to clean up spills unless you are given permission and instructions to do so.

5. If chemicals come into contact with your eyes or skin, notify your teacher immediately. Use the eyewash or flush your skin or eyes with large quantities of water.

6. The fire extinguisher and first-aid kit should only be used by your teacher unless it is an extreme emergency and you have been given permission.

7. If someone is injured or becomes ill, only a professional medical provider or someone certified in first aid should perform first-aid procedures.

SCIENCE SKILL HANDBOOK 681

Extra Try at Home Labs

From Your Kitchen, Junk Drawer, or Yard

1 Testing Horoscopes

▶ **Real-World Question**
How can horoscopes be tested scientifically?

Possible Materials
- horoscope from previous week
- scissors
- transparent tape
- white paper
- correction fluid

▶ **Procedure**
1. Obtain a horoscope from last week and cut out the predictions for each sign. Do not cut out the zodiac signs or birth dates accompanying each prediction.
2. As you cut out a horoscope prediction, write the correct zodiac sign on the back of each prediction.
3. Develop a code for the predictions to allow you to identify them. Keep your code list in your Science Journal.
4. Scramble your predictions and tape them to a sheet of white paper. Write each prediction's code above it.
5. Ask your friends and family members to read all the predictions and choose the one that best matched their life events from the previous week. Interview at least 20 people.

▶ **Conclude and Apply**
1. Calculate the percentage of people who chose the correct sign.
2. Calculate the chances of a person choosing their correct sign randomly.

2 Disappearing Water?

▶ **Real-World Question**
How much difference does the type of measuring equipment make?

Possible Materials
- scale
- water
- measuring cups of different sizes
- measuring spoons (1 tsp = 5 mL, 1 tbsp = 15 mL)

▶ **Procedure**
1. Measure out 83 mL of water using one of the measuring devices. Transfer this amount of water to the other measuring devices.
2. Record the readings for each measuring device for the same amount of water. Do they all give the same reading, or does it seem like the amount of water changed?
3. Remember, 1 mL of water weighs 1 g. Use the scale to find out what the true amount of water is in the container.
4. Repeat steps 1–3 for different amounts of water. Try 50 mL, 128 mL, and 12 mL.

▶ **Conclude and Apply**
1. Which measuring device was the most accurate? The least?
2. Which measuring device was the most precise? The least?
3. What problem came up when you had to use the small devices several times to get up to a larger amount of water?

Extra Try at Home Labs

3 Comparing Atom Sizes

Real-World Question
How do the sizes of different types of atoms compare?

Possible Materials
- metric ruler or meterstick
- 1-m length of white paper
- transparent or masking tape
- colored pencils

Procedure
1. Tape a 1-m sheet of paper on the floor.
2. Use a scale of 1 mm: 1 picometer for measuring and drawing the relative diameters of all the atoms.
3. Study the chart of atomic sizes.
4. Use your scale to measure the relative size of a hydrogen atom on the sheet of paper. Use a red pencil to draw the relative diameter of a hydrogen atom on your paper.
5. Use your scale to measure the relative sizes of an oxygen atom, iron atom, gold atom, and francium atom. Use four other colored pencils to draw the relative diameters of these atoms on the paper.
6. Compare the relative sizes of these different atoms.

Conclude and Apply
1. Research the length of a picometer.
2. Using your scale, list the diameters of the atoms that you drew on your paper.

| Atomic Sizes (picometers) ||
Element	Diameter
Hydrogen	50
Oxygen	146
Iron	248
Gold	288
Francium	540

4 Microscopic Crystals

Real-World Question
What do crystalline and non-crystalline solids look like under a magnifying lens?

Possible Materials
- salt or sugar
- pepper
- magnifying lens
- paper
- bowl
- spoon
- measuring cup

Procedure
1. Pour 10 mL of salt into a bowl and grind the salt into small, powdery pieces with the back of the spoon.
2. Sprinkle a few grains of salt from the bowl onto a piece of paper and view the salt grains with the magnifying lens.
3. Clean out the bowl.
4. Pour 10 mL of pepper into the bowl and grind it into powder with the spoon.
5. Sprinkle a few grains of pepper from the bowl onto the paper and view the grains with the magnifying lens.

Conclude and Apply
1. Compare the difference between the salt and pepper grains under the magnifying lens.
2. Describe what a crystal is.

Adult supervision required for all labs.

Extra Try at Home Labs

5 Colorful Liquids

Real-World Question
How can the property of density be used to make a rainbow of liquids?

Possible Materials
- measuring cups (2)
- maple syrup or corn syrup
- water
- cooking oil
- food coloring
- rubbing alcohol

Procedure
1. Pour 25 mL of syrup into one measuring cup.
2. Slowly pour 25 mL of water down the sides of the same measuring cup so that the water sits on top of the syrup.
3. Slowly pour 25 mL of cooking oil down the sides of the same measuring cup so that the oil sits on top of the water.
4. Put several drops of blue food coloring into the oil and observe them for 5 minutes.
5. Slowly pour 25 mL of rubbing alcohol down the sides of the same measuring cup so that the alcohol sits on top of the oil.
6. Put several drops of red food coloring into the alcohol.

Conclude and Apply
1. Describe what happened to the drops of blue food coloring.
2. Infer what would happen if you poured the liquids into the cylinder in the reverse order.

6 Human Bonding

Real-World Question
How can humans model atoms bonding together?

Possible Materials
- family members or friends
- sheets of blank paper
- markers
- large safety pins
- large colored rubber bands

Procedure
1. Draw a large electron dot diagram of an element you choose. Have other activity participants do that too.
2. Pin the diagram to your shirt.
3. How many electrons does your element have? Gather that many rubber bands.
4. Place about half of the rubber bands on one wrist and half on the other.
5. Form bonds by finding someone who has the number of rubber bands you need to total eight. Try to form as many different compounds with different elements as you can. (You may need two or three of another element's atoms to make a compound.) Record the compounds you make in your Science Journal. Label each compound as ionic or covalent.

Conclude and Apply
1. Which elements don't form any bonds?
2. Which elements form four bonds?

Adult supervision required for all labs.

Extra Try at Home Labs

7 Mini Fireworks

Real-World Question
Where do the colors in fireworks come from?

Possible Materials
- candle
- lighter
- wooden chopsticks (or a fork or tongs)
- penny
- water in an old cup
- steel wool

Procedure
1. Light the candle.
2. Use the chopsticks to get a firm grip on the penny.
3. Hold the penny in the flame until you observe a change. *(Hint: this experiment is more fun in the bathroom with the lights off!)*
4. Drop the penny in the water when you are finished and plunge the burning end of the chopsticks or hot part of the fork into the water as well.
5. Repeat the procedure using steel wool.

Conclude and Apply
1. What color did you see?
2. Infer why copper and iron are used in fireworks.
3. Research what other elements are used in fireworks.

8 A Good Mix?

Real-World Question
What liquids will dissolve in water?

Possible Materials
- cooking oil
- water
- apple or grape juice
- rubbing alcohol
- spoon
- glass
- measuring cup

Procedure
1. Pour 100 mL of water into a large glass.
2. Pour 100 mL of apple juice into the glass and stir the water and juice together. Observe your mixture to determine whether juice is soluble in water.
3. Empty and rinse out your glass.
4. Pour 100 mL of water and 100 mL of cooking oil into the glass and stir them together. Observe your mixture to determine whether oil is soluble in water.
5. Empty and rinse out your glass.
6. Pour 100 mL of water and 100 mL of rubbing alcohol into the glass and stir them together. Observe your mixture to determine whether alcohol is soluble in water.

Conclude and Apply
1. List the liquid(s) that are soluble in water.
2. List the liquid(s) that are not soluble in water.
3. Infer why some liquids are soluble in water and others are not.

Adult supervision required for all labs.

Extra Try at Home Labs

9 Liquid Lab

Real-World Question
How do the properties of water and rubbing alcohol compare?

Possible Materials
- water
- rubbing alcohol
- vegetable oil
- glasses (2)
- ice cubes (2)
- measuring cup
- spoon

Physical Properties	Water	Rubbing Alcohol
Color		
Odor		
Viscosity		
Density		
Solubility with oil		

Procedure
1. Copy the Physical Properties chart into your Science Journal.
2. Slowly pour 200 mL of water into one glass and observe the viscosity of water. Slowly pour 200 mL of rubbing alcohol into a second glass and observe the viscosity of rubbing alcohol. Record your observations in your chart.
3. Observe the color and odor of both liquids and record your observations in your chart.
4. Drop an ice cube into each glass. Comment on the density of each liquid in your chart.
5. Pour 50 mL of vegetable oil into each glass and stir. Record your observations in your chart.

Conclude and Apply
1. Compare the properties of water and isopropyl alcohol.
2. Infer what would happen if water and isopropyl alcohol were mixed together.

10 Measuring Momentum

Real-World Question
How much momentum do rolling balls have?

Possible Materials
- meterstick
- orange cones or tape
- scale
- stopwatch
- bucket
- bowling ball
- plastic baseball
- golf ball
- tennis ball
- calculator

Procedure
1. Use a balance to measure the masses of the tennis ball, golf ball, and plastic baseball. Convert their masses from grams to kilograms.
2. Find the weight of the bowling ball in pounds. The weight should be written on the ball. Divide the ball's weight by 2.2 to calculate its mass in kilograms.
3. Go outside and measure a 10-m distance on a blacktop or concrete surface. Mark the distance with orange cones or tape.
4. Have a partner roll each ball the 10-m distance. Measure the time it takes each ball to roll 10 m.
5. Use the formula: $\text{velocity} = \frac{\text{distance}}{\text{time}}$ to calculate each ball's velocity.

Conclude and Apply
1. Calculate the momentum of each ball.
2. Infer why the momentums of the balls differed so greatly.

Extra Try at Home Labs

11 Friction in Traffic

Real-World Question
How do the various kinds of friction affect the operation of vehicles?

Possible Materials
- erasers taken from the ends of pencils (4)
- needles (2)
- small match box
- toy car

Procedure
1. Build a match box car with the materials listed, or use a toy car.
2. Invent ways to demonstrate the effects of static friction, sliding friction, and rolling friction on the car. Think of hills, ice or rain conditions, graveled roads and paved roads, etc.
3. Make drawings of how friction is acting on the car, or how the car uses friction to work.

Conclude and Apply
1. In what ways are static, sliding, and rolling friction helpful to drivers?
2. In what ways are static, sliding, and rolling friction unfavorable to car safety and operation?
3. Explain what your experiment taught you about driving in icy conditions.

12 Submersible Egg

Real-World Question
How can you make an egg float and sink again?

Possible Materials
- egg
- 10 mL (2 teaspoons) of salt
- measuring spoons
- water
- glass (250–300 mL)
- spoon
- marking pen

Procedure
1. Put 150 mL of water in the glass.
2. Gently use a marking pen to write an X on one side of the egg.
3. Put the egg in the glass. Record your observations.
4. Use the spoon to remove the egg. Add 1 mL (1/4 teaspoon) of salt and swirl or stir to dissolve. Put the egg back in. Record your observations.
5. Repeat step 4 until you have used all the salt. Remember to make observations at each step.

Conclude and Apply
1. How could you resink the egg? Try your idea. Did it work?
2. The egg always floats with the same point or side down. Why do you think this is?
3. How does your answer to question 2 relate to real-world applications?

Adult supervision required for all labs.

Extra Try at Home Labs

13 The Heat is On

▶ **Real-World Question**
How can different types of energy be transformed into thermal energy?

Possible Materials
- lamp
- incandescent light bulb
- black construction paper or cloth

▶ **Procedure**
1. Feel the temperature of a black sheet of paper. Lay the paper in direct sunlight, wait 10 min, and observe how it feels.
2. Rub the palms of your hands together quickly for 10 s and observe how they feel.
3. Switch on a lamp that has a bare light bulb. *Without touching the lightbulb,* cup your hand 2 cm above the bulb for 30 s and observe what you feel.

▶ **Conclude and Apply**
1. Infer the type of energy transformation that happened on the paper.
2. Infer the type of energy transformation that happened between the palms of your hands.
3. Infer the type of energy transformation that happened to the lightbulb.

14 Simple Machines

▶ **Real-World Question**
What types of simple machines are found in a toolbox?

Possible Materials
- box of tools

▶ **Procedure**
1. Obtain a box of tools and lay all the tools and other hardware from the box on a table.
2. Carefully examine all the tools and hardware, and separate all the items that are a type of inclined plane.
3. Carefully examine all the tools and hardware, and separate all the items that are a type of lever.
4. Identify and separate all the items that are a wheel and axle.
5. Identify any pulleys in the toolbox.
6. Identify any tools that are a combination of two or more simple machines.

▶ **Conclude and Apply**
1. List all the tools you found that were a type of inclined plane, lever, wheel and axle, or pulley.
2. List all the tools that were a combination of two or more simple machines.
3. Infer how a hammer could be used as both a first class lever and a third class lever.

Adult supervision required for all labs.

Extra Try at Home Labs

15 Estimate Temperature

Real-World Question
How can we learn to estimate temperatures?

Possible Materials
- thermometer
- bowl
- water
- ice

Procedure
1. If you have a dual-scale weather thermometer, you can learn twice as much by trying to do your estimation in degrees Fahrenheit and Celsius each time.
2. Fill a bowl with ice water. Submerge your fingers in the water and estimate the water temperature.
3. Place the thermometer in the bowl and observe the temperature.
4. Place a bowl of warm water in direct sunlight for 20 min. Submerge your fingers in the water and estimate the water temperature.
5. Place the thermometer in the bowl and observe the temperature.
6. Place the thermometer outside in a location where you can see it each day.
7. Each day for a month, step outside and estimate the temperature. Check the accuracy of your estimates with the thermometer. Record the weather conditions as well.

Conclude and Apply
1. Describe how well you can estimate air temperatures after estimating the temperature each day for a month. Did the cloudiness of the day affect your estimation skills?
2. Infer why understanding the Celsius scale might be helpful to you in the future.

16 Exploding Bag

Real-World Question
What happens when a bag pops?

Possible Materials
- paper bag or plastic produce bag

Procedure
1. Obtain a paper lunch bag. Smooth out the bag on a flat surface if it has any wrinkles.
2. Hold the neck of the bag and blow air into it until it is completely filled. The sides of the bag should be stretched out completely.
3. Twist the neck of the bag tightly to prevent air from escaping.
4. Pop the paper bag between your palms and observe what happens.
5. Examine the bag after you pop it. Observe any changes in the bag.

Conclude and Apply
1. Describe what happened when you popped the bag.
2. Infer why this happened to the bag.

Adult supervision required for all labs.

Extra Try at Home Labs

17 Seeing Sound

▶ **Real-World Question**
Is it possible to see sound waves?

Possible Materials
- scissors
- rubber band
- twigs, curved and stiff
- sewing thread
- puffed-rice cereal
- clothing hanger

▶ **Procedure**
1. Make a rubber-band bow as follows: Cut the rubber band at one point. Tie the rubber band around opposite ends of the twig so that it looks like an archery bow. Make sure the rubber band is tight like a guitar string.
2. Cut about 10 pieces of thread to equal lengths, 10 to 15 cm long.
3. With each piece of thread, tie one end around a kernel of puffed rice and the other end around the bottom of a clothing hanger. Space the hanging kernels about 1 cm apart.
4. Hook the clothing hanger over something so that the threads and cereal hang freely.
5. Hold the rubber-band bow so that the rubber band is just underneath the central hanging kernels. Pluck the rubber band, being careful not to touch the kernels. Write down your observations.

▶ **Conclude and Apply**
1. Explain how this experiment relates to the fact that sound travels through air using compression waves.

18 Black Light

▶ **Real-World Question**
How do you know that ultraviolet waves exist?

Possible Materials
- normal lamp with a white lightbulb
- ultraviolet light source (a black light)
- white paper
- laundry detergent
- glow-in-the-dark plastic toy
- a variety of rocks and minerals
- a variety of flowers and plants
- a variety of household cleaners
- different colors and materials of clothing

▶ **Procedure**
1. Dab a small amount of laundry detergent on some white paper and place it somewhere to dry.
2. Place the black light in a dark room and turn it on.
3. Write down what the detergent looks like under a normal lamp. Then, place the paper under the black light. Write down what you see.
4. Place other items under the different lights and write down what you see.

▶ **Conclude and Apply**
1. Describe the difference between the way things look under normal light and the way they look under ultraviolet light.
2. Explain how you know from this experiment that ultraviolet waves exist.

Extra Try at Home Labs

19 Light in Liquids

Real-World Question
What happens to light when it passes through different liquids found in your kitchen?

Possible Materials
- flashlight
- glass
- orange juice
- water
- milk
- maple syrup
- white vinegar
- red vinegar
- honey
- molasses
- milk
- fruit juice
- powdered drink mix
- salad dressing
- salsa

Procedure
1. Fill a glass with water and darken the room. Shine the beam of a flashlight through the glass and observe how much of the light passes through the water.
2. Identify water as an opaque, translucent, or transparent substance.
3. Repeat steps 1–2 to test a wide variety of other liquids found in your kitchen. Use the original containers when you can, but remove any labels from containers that block the light beam.

Conclude and Apply
1. Identify all the opaque liquids you tested.
2. Identify all the translucent liquids you tested.
3. Identify all the transparent liquids you tested.

20 Bending Water

Real-World Question
How can a plastic rod bend water without touching it?

Possible Materials
- plastic rod
- plastic clothes hanger
- 100% wool clothing
- water faucet

Procedure
1. Turn on a faucet until a narrow, smooth stream of water is flowing out of it. The stream of water cannot be too wide, and it cannot flow in a broken pattern.
2. Vigorously rub a plastic rod on a piece of 100% wool clothing for about 15 s.
3. Immediately hold the rod near the center of the stream of water. Move the rod close to the stream. Do not touch the water.
4. Observe what happens to the water.

Conclude and Apply
1. Describe how the plastic rod affected the stream of water.
2. Explain why the plastic rod affected the water.

Adult supervision required for all labs.

Extra Try at Home Labs

21 Testing Magnets

Real-World Question
How do the strengths of kitchen magnets compare?

Possible Materials
- several kitchen magnets
- metric ruler
- small pin or paper clip

Procedure
1. Place a small pin or paper clip on a flat, nonmetallic surface such as a wooden table.
2. Holding your metric ruler vertically, place it next to the pin with the 0 cm mark on the tabletop.
3. Hold a kitchen magnet at the 10 cm mark on the ruler.
4. Slowly lower the magnet toward the pin. At the point where the pin is attracted to the magnet, measure the height of the magnet from the table. Record the height in your Science Journal.
5. Repeat steps 2–4 to test your other kitchen magnets.

Conclude and Apply
1. Describe the results of your experiment.
2. Infer how the kitchen magnets should be used based on the results of your experiment.

22 Pattern Counting

Real-World Question
What pattern is used to count in binary?

Procedure
1. Study the pattern used for counting in 4-bit binary.
2. Describe the pattern in your own words.
3. To test your understanding of the pattern, close the book and write the counting pattern from 0–15 by using your notes.

Conclude and Apply
1. Can this pattern continue past 15? Explain.
2. Develop a pattern to count to 32.

Decimal Number	Binary Number	Decimal Number	Binary Number
0	0000	8	1000
1	0001	9	1001
2	0010	10	1010
3	0011	11	1011
4	0100	12	1100
5	0101	13	1101
6	0110	14	1110
7	0111	15	1111

Technology Skill Handbook

Computer Skills

People who study science rely on computers, like the one in **Figure 16**, to record and store data and to analyze results from investigations. Whether you work in a laboratory or just need to write a lab report with tables, good computer skills are a necessity.

Using the computer comes with responsibility. Issues of ownership, security, and privacy can arise. Remember, if you did not author the information you are using, you must provide a source for your information. Also, anything on a computer can be accessed by others. Do not put anything on the computer that you would not want everyone to know. To add more security to your work, use a password.

Use a Word Processing Program

A computer program that allows you to type your information, change it as many times as you need to, and then print it out is called a word processing program. Word processing programs also can be used to make tables.

Figure 16 A computer will make reports neater and more professional looking.

Learn the Skill To start your word processing program, a blank document, sometimes called "Document 1," appears on the screen. To begin, start typing. To create a new document, click the *New* button on the standard tool bar. These tips will help you format the document.

- The program will automatically move to the next line; press *Enter* if you wish to start a new paragraph.
- Symbols, called non-printing characters, can be hidden by clicking the *Show/Hide* button on your toolbar.
- To insert text, move the cursor to the point where you want the insertion to go, click on the mouse once, and type the text.
- To move several lines of text, select the text and click the *Cut* button on your toolbar. Then position your cursor in the location that you want to move the cut text and click *Paste*. If you move to the wrong place, click *Undo*.
- The spell check feature does not catch words that are misspelled to look like other words, like "cold" instead of "gold." Always reread your document to catch all spelling mistakes.
- To learn about other word processing methods, read the user's manual or click on the *Help* button.
- You can integrate databases, graphics, and spreadsheets into documents by copying from another program and pasting it into your document, or by using desktop publishing (DTP). DTP software allows you to put text and graphics together to finish your document with a professional look. This software varies in how it is used and its capabilities.

TECHNOLOGY SKILL HANDBOOK 693

Technology Skill Handbook

Use a Database

A collection of facts stored in a computer and sorted into different fields is called a database. A database can be reorganized in any way that suits your needs.

Learn the Skill A computer program that allows you to create your own database is a database management system (DBMS). It allows you to add, delete, or change information. Take time to get to know the features of your database software.

- Determine what facts you would like to include and research to collect your information.
- Determine how you want to organize the information.
- Follow the instructions for your particular DBMS to set up fields. Then enter each item of data in the appropriate field.
- Follow the instructions to sort the information in order of importance.
- Evaluate the information in your database, and add, delete, or change as necessary.

Use the Internet

The Internet is a global network of computers where information is stored and shared. To use the Internet, like the students in **Figure 17,** you need a modem to connect your computer to a phone line and an Internet Service Provider account.

Learn the Skill To access internet sites and information, use a "Web browser," which lets you view and explore pages on the World Wide Web. Each page is its own site, and each site has its own address, called a URL. Once you have found a Web browser, follow these steps for a search (this also is how you search a database).

Figure 17 The Internet allows you to search a global network for a variety of information.

- Be as specific as possible. If you know you want to research "gold," don't type in "elements." Keep narrowing your search until you find what you want.
- Web sites that end in *.com* are commercial Web sites; *.org, .edu,* and *.gov* are non-profit, educational, or government Web sites.
- Electronic encyclopedias, almanacs, indexes, and catalogs will help locate and select relevant information.
- Develop a "home page" with relative ease. When developing a Web site, NEVER post pictures or disclose personal information such as location, names, or phone numbers. Your school or community usually can host your Web site. A basic understanding of HTML (hypertext mark-up language), the language of Web sites, is necessary. Software that creates HTML code is called authoring software, and can be downloaded free from many Web sites. This software allows text and pictures to be arranged as the software is writing the HTML code.

Technology Skill Handbook

Use a Spreadsheet

A spreadsheet, shown in **Figure 18,** can perform mathematical functions with any data arranged in columns and rows. By entering a simple equation into a cell, the program can perform operations in specific cells, rows, or columns.

Learn the Skill Each column (vertical) is assigned a letter, and each row (horizontal) is assigned a number. Each point where a row and column intersect is called a cell, and is labeled according to where it is located—Column A, Row 1 (A1).

- Decide how to organize the data, and enter it in the correct row or column.
- Spreadsheets can use standard formulas or formulas can be customized to calculate cells.
- To make a change, click on a cell to make it activate, and enter the edited data or formula.
- Spreadsheets also can display your results in graphs. Choose the style of graph that best represents the data.

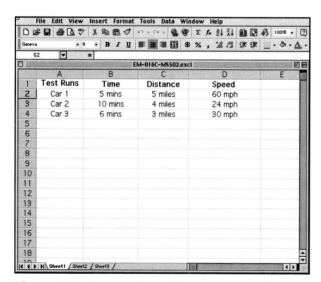

Figure 18 A spreadsheet allows you to perform mathematical operations on your data.

Use Graphics Software

Adding pictures, called graphics, to your documents is one way to make your documents more meaningful and exciting. This software adds, edits, and even constructs graphics. There is a variety of graphics software programs. The tools used for drawing can be a mouse, keyboard, or other specialized devices. Some graphics programs are simple. Others are complicated, called computer-aided design (CAD) software.

Learn the Skill It is important to have an understanding of the graphics software being used before starting. The better the software is understood, the better the results. The graphics can be placed in a word-processing document.

- Clip art can be found on a variety of internet sites, and on CDs. These images can be copied and pasted into your document.
- When beginning, try editing existing drawings, then work up to creating drawings.
- The images are made of tiny rectangles of color called pixels. Each pixel can be altered.
- Digital photography is another way to add images. The photographs in the memory of a digital camera can be downloaded into a computer, then edited and added to the document.
- Graphics software also can allow animation. The software allows drawings to have the appearance of movement by connecting basic drawings automatically. This is called in-betweening, or tweening.
- Remember to save often.

Technology Skill Handbook

Presentation Skills

Develop Multimedia Presentations

Most presentations are more dynamic if they include diagrams, photographs, videos, or sound recordings, like the one shown in **Figure 19**. A multimedia presentation involves using stereos, overhead projectors, televisions, computers, and more.

Learn the Skill Decide the main points of your presentation, and what types of media would best illustrate those points.

- Make sure you know how to use the equipment you are working with.
- Practice the presentation using the equipment several times.
- Enlist the help of a classmate to push play or turn lights out for you. Be sure to practice your presentation with him or her.
- If possible, set up all of the equipment ahead of time, and make sure everything is working properly.

Figure 19 These students are engaging the audience using a variety of tools.

Computer Presentations

There are many different interactive computer programs that you can use to enhance your presentation. Most computers have a compact disc (CD) drive that can play both CDs and digital video discs (DVDs). Also, there is hardware to connect a regular CD, DVD, or VCR. These tools will enhance your presentation.

Another method of using the computer to aid in your presentation is to develop a slide show using a computer program. This can allow movement of visuals at the presenter's pace, and can allow for visuals to build on one another.

Learn the Skill In order to create multimedia presentations on a computer, you need to have certain tools. These may include traditional graphic tools and drawing programs, animation programs, and authoring systems that tie everything together. Your computer will tell you which tools it supports. The most important step is to learn about the tools that you will be using.

- Often, color and strong images will convey a point better than words alone. Use the best methods available to convey your point.
- As with other presentations, practice many times.
- Practice your presentation with the tools you and any assistants will be using.
- Maintain eye contact with the audience. The purpose of using the computer is not to prompt the presenter, but to help the audience understand the points of the presentation.

Math Skill Handbook

Math Review

Use Fractions

A fraction compares a part to a whole. In the fraction $\frac{2}{3}$, the 2 represents the part and is the numerator. The 3 represents the whole and is the denominator.

Reduce Fractions To reduce a fraction, you must find the largest factor that is common to both the numerator and the denominator, the greatest common factor (GCF). Divide both numbers by the GCF. The fraction has then been reduced, or it is in its simplest form.

Example Twelve of the 20 chemicals in the science lab are in powder form. What fraction of the chemicals used in the lab are in powder form?

Step 1 Write the fraction.
$\frac{part}{whole} = \frac{12}{20}$

Step 2 To find the GCF of the numerator and denominator, list all of the factors of each number.
Factors of 12: 1, 2, 3, 4, 6, 12 (the numbers that divide evenly into 12)
Factors of 20: 1, 2, 4, 5, 10, 20 (the numbers that divide evenly into 20)

Step 3 List the common factors.
1, 2, 4.

Step 4 Choose the greatest factor in the list.
The GCF of 12 and 20 is 4.

Step 5 Divide the numerator and denominator by the GCF.
$\frac{12 \div 4}{20 \div 4} = \frac{3}{5}$

In the lab, $\frac{3}{5}$ of the chemicals are in powder form.

Practice Problem At an amusement park, 66 of 90 rides have a height restriction. What fraction of the rides, in its simplest form, has a height restriction?

Add and Subtract Fractions To add or subtract fractions with the same denominator, add or subtract the numerators and write the sum or difference over the denominator. After finding the sum or difference, find the simplest form for your fraction.

Example 1 In the forest outside your house, $\frac{1}{8}$ of the animals are rabbits, $\frac{3}{8}$ are squirrels, and the remainder are birds and insects. How many are mammals?

Step 1 Add the numerators.
$\frac{1}{8} + \frac{3}{8} = \frac{(1+3)}{8} = \frac{4}{8}$

Step 2 Find the GCF.
$\frac{4}{8}$ (GCF, 4)

Step 3 Divide the numerator and denominator by the GCF.
$\frac{4}{4} = 1, \frac{8}{4} = 2$

$\frac{1}{2}$ of the animals are mammals.

Example 2 If $\frac{7}{16}$ of the Earth is covered by freshwater, and $\frac{1}{16}$ of that is in glaciers, how much freshwater is not frozen?

Step 1 Subtract the numerators.
$\frac{7}{16} - \frac{1}{16} = \frac{(7-1)}{16} = \frac{6}{16}$

Step 2 Find the GCF.
$\frac{6}{16}$ (GCF, 2)

Step 3 Divide the numerator and denominator by the GCF.
$\frac{6}{2} = 3, \frac{16}{2} = 8$

$\frac{3}{8}$ of the freshwater is not frozen.

Practice Problem A bicycle rider is going 15 km/h for $\frac{4}{9}$ of his ride, 10 km/h for $\frac{2}{9}$ of his ride, and 8 km/h for the remainder of the ride. How much of his ride is he going over 8 km/h?

Math Skill Handbook

Unlike Denominators To add or subtract fractions with unlike denominators, first find the least common denominator (LCD). This is the smallest number that is a common multiple of both denominators. Rename each fraction with the LCD, and then add or subtract. Find the simplest form if necessary.

Example 1 A chemist makes a paste that is $\frac{1}{2}$ table salt (NaCl), $\frac{1}{3}$ sugar ($C_6H_{12}O_6$), and the rest water (H_2O). How much of the paste is a solid?

Step 1 Find the LCD of the fractions.
$\frac{1}{2} + \frac{1}{3}$ (LCD, 6)

Step 2 Rename each numerator and each denominator with the LCD.
$1 \times 3 = 3, \ 2 \times 3 = 6$
$1 \times 2 = 2, \ 3 \times 2 = 6$

Step 3 Add the numerators.
$\frac{3}{6} + \frac{2}{6} = \frac{(3 + 2)}{6} = \frac{5}{6}$

$\frac{5}{6}$ of the paste is a solid.

Example 2 The average precipitation in Grand Junction, CO, is $\frac{7}{10}$ inch in November, and $\frac{3}{5}$ inch in December. What is the total average precipitation?

Step 1 Find the LCD of the fractions.
$\frac{7}{10} + \frac{3}{5}$ (LCD, 10)

Step 2 Rename each numerator and each denominator with the LCD.
$7 \times 1 = 7, \ 10 \times 1 = 10$
$3 \times 2 = 6, \ 5 \times 2 = 10$

Step 3 Add the numerators.
$\frac{7}{10} + \frac{6}{10} = \frac{(7 + 6)}{10} = \frac{13}{10}$

$\frac{13}{10}$ inches total precipitation, or $1\frac{3}{10}$ inches.

Practice Problem On an electric bill, about $\frac{1}{8}$ of the energy is from solar energy and about $\frac{1}{10}$ is from wind power. How much of the total bill is from solar energy and wind power combined?

Example 3 In your body, $\frac{7}{10}$ of your muscle contractions are involuntary (cardiac and smooth muscle tissue). Smooth muscle makes $\frac{3}{15}$ of your muscle contractions. How many of your muscle contractions are made by cardiac muscle?

Step 1 Find the LCD of the fractions.
$\frac{7}{10} - \frac{3}{15}$ (LCD, 30)

Step 2 Rename each numerator and each denominator with the LCD.
$7 \times 3 = 21, \ 10 \times 3 = 30$
$3 \times 2 = 6, \ 15 \times 2 = 30$

Step 3 Subtract the numerators.
$\frac{21}{30} - \frac{6}{30} = \frac{(21 - 6)}{30} = \frac{15}{30}$

Step 4 Find the GCF.
$\frac{15}{30}$ (GCF, 15)
$\frac{1}{2}$

$\frac{1}{2}$ of all muscle contractions are cardiac muscle.

Example 4 Tony wants to make cookies that call for $\frac{3}{4}$ of a cup of flour, but he only has $\frac{1}{3}$ of a cup. How much more flour does he need?

Step 1 Find the LCD of the fractions.
$\frac{3}{4} - \frac{1}{3}$ (LCD, 12)

Step 2 Rename each numerator and each denominator with the LCD.
$3 \times 3 = 9, \ 4 \times 3 = 12$
$1 \times 4 = 4, \ 3 \times 4 = 12$

Step 3 Subtract the numerators.
$\frac{9}{12} - \frac{4}{12} = \frac{(9 - 4)}{12} = \frac{5}{12}$

$\frac{5}{12}$ of a cup of flour.

Practice Problem Using the information provided to you in Example 3 above, determine how many muscle contractions are voluntary (skeletal muscle).

Math Skill Handbook

Multiply Fractions To multiply with fractions, multiply the numerators and multiply the denominators. Find the simplest form if necessary.

Example Multiply $\frac{3}{5}$ by $\frac{1}{3}$.

Step 1 Multiply the numerators and denominators.
$$\frac{3}{5} \times \frac{1}{3} = \frac{(3 \times 1)}{(5 \times 3)} = \frac{3}{15}$$

Step 2 Find the GCF.
$$\frac{3}{15} \text{ (GCF, 3)}$$

Step 3 Divide the numerator and denominator by the GCF.
$$\frac{3}{3} = 1, \quad \frac{15}{3} = 5$$
$$\frac{1}{5}$$

$\frac{3}{5}$ multiplied by $\frac{1}{3}$ is $\frac{1}{5}$.

Practice Problem Multiply $\frac{3}{14}$ by $\frac{5}{16}$.

Find a Reciprocal Two numbers whose product is 1 are called multiplicative inverses, or reciprocals.

Example Find the reciprocal of $\frac{3}{8}$.

Step 1 Inverse the fraction by putting the denominator on top and the numerator on the bottom.
$$\frac{8}{3}$$

The reciprocal of $\frac{3}{8}$ is $\frac{8}{3}$.

Practice Problem Find the reciprocal of $\frac{4}{9}$.

Divide Fractions To divide one fraction by another fraction, multiply the dividend by the reciprocal of the divisor. Find the simplest form if necessary.

Example 1 Divide $\frac{1}{9}$ by $\frac{1}{3}$.

Step 1 Find the reciprocal of the divisor.
The reciprocal of $\frac{1}{3}$ is $\frac{3}{1}$.

Step 2 Multiply the dividend by the reciprocal of the divisor.
$$\frac{\frac{1}{9}}{\frac{1}{3}} = \frac{1}{9} \times \frac{3}{1} = \frac{(1 \times 3)}{(9 \times 1)} = \frac{3}{9}$$

Step 3 Find the GCF.
$$\frac{3}{9} \text{ (GCF, 3)}$$

Step 4 Divide the numerator and denominator by the GCF.
$$\frac{3}{3} = 1, \quad \frac{9}{3} = 3$$
$$\frac{1}{3}$$

$\frac{1}{9}$ divided by $\frac{1}{3}$ is $\frac{1}{3}$.

Example 2 Divide $\frac{3}{5}$ by $\frac{1}{4}$.

Step 1 Find the reciprocal of the divisor.
The reciprocal of $\frac{1}{4}$ is $\frac{4}{1}$.

Step 2 Multiply the dividend by the reciprocal of the divisor.
$$\frac{\frac{3}{5}}{\frac{1}{4}} = \frac{3}{5} \times \frac{4}{1} = \frac{(3 \times 4)}{(5 \times 1)} = \frac{12}{5}$$

$\frac{3}{5}$ divided by $\frac{1}{4}$ is $\frac{12}{5}$ or $2\frac{2}{5}$.

Practice Problem Divide $\frac{3}{11}$ by $\frac{7}{10}$.

Math Skill Handbook

Use Ratios

When you compare two numbers by division, you are using a ratio. Ratios can be written 3 to 5, 3:5, or $\frac{3}{5}$. Ratios, like fractions, also can be written in simplest form.

Ratios can represent probabilities, also called odds. This is a ratio that compares the number of ways a certain outcome occurs to the number of outcomes. For example, if you flip a coin 100 times, what are the odds that it will come up heads? There are two possible outcomes, heads or tails, so the odds of coming up heads are 50:100. Another way to say this is that 50 out of 100 times the coin will come up heads. In its simplest form, the ratio is 1:2.

Example 1 A chemical solution contains 40 g of salt and 64 g of baking soda. What is the ratio of salt to baking soda as a fraction in simplest form?

Step 1 Write the ratio as a fraction.
$$\frac{\text{salt}}{\text{baking soda}} = \frac{40}{64}$$

Step 2 Express the fraction in simplest form.
The GCF of 40 and 64 is 8.
$$\frac{40}{64} = \frac{40 \div 8}{64 \div 8} = \frac{5}{8}$$

The ratio of salt to baking soda in the sample is 5:8.

Example 2 Sean rolls a 6-sided die 6 times. What are the odds that the side with a 3 will show?

Step 1 Write the ratio as a fraction.
$$\frac{\text{number of sides with a 3}}{\text{number of sides}} = \frac{1}{6}$$

Step 2 Multiply by the number of attempts.
$$\frac{1}{6} \times 6 \text{ attempts} = \frac{6}{6} \text{ attempts} = 1 \text{ attempt}$$

1 attempt out of 6 will show a 3.

Practice Problem Two metal rods measure 100 cm and 144 cm in length. What is the ratio of their lengths in simplest form?

Use Decimals

A fraction with a denominator that is a power of ten can be written as a decimal. For example, 0.27 means $\frac{27}{100}$. The decimal point separates the ones place from the tenths place.

Any fraction can be written as a decimal using division. For example, the fraction $\frac{5}{8}$ can be written as a decimal by dividing 5 by 8. Written as a decimal, it is 0.625.

Add or Subtract Decimals When adding and subtracting decimals, line up the decimal points before carrying out the operation.

Example 1 Find the sum of 47.68 and 7.80.

Step 1 Line up the decimal places when you write the numbers.
```
  47.68
+  7.80
```

Step 2 Add the decimals.
```
  47.68
+  7.80
  55.48
```

The sum of 47.68 and 7.80 is 55.48.

Example 2 Find the difference of 42.17 and 15.85.

Step 1 Line up the decimal places when you write the number.
```
  42.17
- 15.85
```

Step 2 Subtract the decimals.
```
  42.17
- 15.85
  26.32
```

The difference of 42.17 and 15.85 is 26.32.

Practice Problem Find the sum of 1.245 and 3.842.

700 STUDENT RESOURCES

Math Skill Handbook

Multiply Decimals To multiply decimals, multiply the numbers like any other number, ignoring the decimal point. Count the decimal places in each factor. The product will have the same number of decimal places as the sum of the decimal places in the factors.

Example Multiply 2.4 by 5.9.

Step 1 Multiply the factors like two whole numbers.
$24 \times 59 = 1416$

Step 2 Find the sum of the number of decimal places in the factors. Each factor has one decimal place, for a sum of two decimal places.

Step 3 The product will have two decimal places.
14.16

The product of 2.4 and 5.9 is 14.16.

Practice Problem Multiply 4.6 by 2.2.

Divide Decimals When dividing decimals, change the divisor to a whole number. To do this, multiply both the divisor and the dividend by the same power of ten. Then place the decimal point in the quotient directly above the decimal point in the dividend. Then divide as you do with whole numbers.

Example Divide 8.84 by 3.4.

Step 1 Multiply both factors by 10.
$3.4 \times 10 = 34$, $8.84 \times 10 = 88.4$

Step 2 Divide 88.4 by 34.

$$\begin{array}{r} 2.6 \\ 34\overline{)88.4} \\ -68 \\ \hline 204 \\ -204 \\ \hline 0 \end{array}$$

8.84 divided by 3.4 is 2.6.

Practice Problem Divide 75.6 by 3.6.

Use Proportions

An equation that shows that two ratios are equivalent is a proportion. The ratios $\frac{2}{4}$ and $\frac{5}{10}$ are equivalent, so they can be written as $\frac{2}{4} = \frac{5}{10}$. This equation is a proportion.

When two ratios form a proportion, the cross products are equal. To find the cross products in the proportion $\frac{2}{4} = \frac{5}{10}$, multiply the 2 and the 10, and the 4 and the 5. Therefore $2 \times 10 = 4 \times 5$, or $20 = 20$.

Because you know that both proportions are equal, you can use cross products to find a missing term in a proportion. This is known as solving the proportion.

Example The heights of a tree and a pole are proportional to the lengths of their shadows. The tree casts a shadow of 24 m when a 6-m pole casts a shadow of 4 m. What is the height of the tree?

Step 1 Write a proportion.
$$\frac{\text{height of tree}}{\text{height of pole}} = \frac{\text{length of tree's shadow}}{\text{length of pole's shadow}}$$

Step 2 Substitute the known values into the proportion. Let h represent the unknown value, the height of the tree.
$$\frac{h}{6} = \frac{24}{4}$$

Step 3 Find the cross products.
$h \times 4 = 6 \times 24$

Step 4 Simplify the equation.
$4h = 144$

Step 5 Divide each side by 4.
$$\frac{4h}{4} = \frac{144}{4}$$
$h = 36$

The height of the tree is 36 m.

Practice Problem The ratios of the weights of two objects on the Moon and on Earth are in proportion. A rock weighing 3 N on the Moon weighs 18 N on Earth. How much would a rock that weighs 5 N on the Moon weigh on Earth?

Math Skill Handbook

Use Percentages

The word *percent* means "out of one hundred." It is a ratio that compares a number to 100. Suppose you read that 77 percent of the Earth's surface is covered by water. That is the same as reading that the fraction of the Earth's surface covered by water is $\frac{77}{100}$. To express a fraction as a percent, first find the equivalent decimal for the fraction. Then, multiply the decimal by 100 and add the percent symbol.

Example Express $\frac{13}{20}$ as a percent.

Step 1 Find the equivalent decimal for the fraction.

$$
\begin{array}{r}
0.65 \\
20{\overline{\smash{\big)}\,13.00}} \\
\underline{12\ 0} \\
1\ 00 \\
\underline{1\ 00} \\
0
\end{array}
$$

Step 2 Rewrite the fraction $\frac{13}{20}$ as 0.65.

Step 3 Multiply 0.65 by 100 and add the % sign.
$0.65 \times 100 = 65 = 65\%$

So, $\frac{13}{20} = 65\%$.

This also can be solved as a proportion.

Example Express $\frac{13}{20}$ as a percent.

Step 1 Write a proportion.
$\frac{13}{20} = \frac{x}{100}$

Step 2 Find the cross products.
$1300 = 20x$

Step 3 Divide each side by 20.
$\frac{1300}{20} = \frac{20x}{20}$
$65\% = x$

Practice Problem In one year, 73 of 365 days were rainy in one city. What percent of the days in that city were rainy?

Solve One-Step Equations

A statement that two things are equal is an equation. For example, $A = B$ is an equation that states that A is equal to B.

An equation is solved when a variable is replaced with a value that makes both sides of the equation equal. To make both sides equal the inverse operation is used. Addition and subtraction are inverses, and multiplication and division are inverses.

Example 1 Solve the equation $x - 10 = 35$.

Step 1 Find the solution by adding 10 to each side of the equation.
$x - 10 = 35$
$x - 10 + 10 = 35 + 10$
$x = 45$

Step 2 Check the solution.
$x - 10 = 35$
$45 - 10 = 35$
$35 = 35$

Both sides of the equation are equal, so $x = 45$.

Example 2 In the formula $a = bc$, find the value of c if $a = 20$ and $b = 2$.

Step 1 Rearrange the formula so the unknown value is by itself on one side of the equation by dividing both sides by b.

$a = bc$
$\frac{a}{b} = \frac{bc}{b}$
$\frac{a}{b} = c$

Step 2 Replace the variables a and b with the values that are given.

$\frac{a}{b} = c$
$\frac{20}{2} = c$
$10 = c$

Step 3 Check the solution.

$a = bc$
$20 = 2 \times 10$
$20 = 20$

Both sides of the equation are equal, so $c = 10$ is the solution when $a = 20$ and $b = 2$.

Practice Problem In the formula $h = gd$, find the value of d if $g = 12.3$ and $h = 17.4$.

Math Skill Handbook

Use Statistics

The branch of mathematics that deals with collecting, analyzing, and presenting data is statistics. In statistics, there are three common ways to summarize data with a single number—the mean, the median, and the mode.

The **mean** of a set of data is the arithmetic average. It is found by adding the numbers in the data set and dividing by the number of items in the set.

The **median** is the middle number in a set of data when the data are arranged in numerical order. If there were an even number of data points, the median would be the mean of the two middle numbers.

The **mode** of a set of data is the number or item that appears most often.

Another number that often is used to describe a set of data is the range. The **range** is the difference between the largest number and the smallest number in a set of data.

A **frequency table** shows how many times each piece of data occurs, usually in a survey. **Table 2** below shows the results of a student survey on favorite color.

Table 2 Student Color Choice		
Color	Tally	Frequency
red	||||	4
blue	|||||	5
black	||	2
green	|||	3
purple	||||| ||	7
yellow	||||| |	6

Based on the frequency table data, which color is the favorite?

Example The speeds (in m/s) for a race car during five different time trials are 39, 37, 44, 36, and 44.

To find the mean:

Step 1 Find the sum of the numbers.
$39 + 37 + 44 + 36 + 44 = 200$

Step 2 Divide the sum by the number of items, which is 5.
$200 \div 5 = 40$

The mean is 40 m/s.

To find the median:

Step 1 Arrange the measures from least to greatest.
36, 37, 39, 44, 44

Step 2 Determine the middle measure.
36, 37, <u>39</u>, 44, 44

The median is 39 m/s.

To find the mode:

Step 1 Group the numbers that are the same together.
44, 44, 36, 37, 39

Step 2 Determine the number that occurs most in the set.
<u>44, 44</u>, 36, 37, 39

The mode is 44 m/s.

To find the range:

Step 1 Arrange the measures from largest to smallest.
44, 44, 39, 37, 36

Step 2 Determine the largest and smallest measures in the set.
<u>44</u>, 44, 39, 37, <u>36</u>

Step 3 Find the difference between the largest and smallest measures.
$44 - 36 = 8$

The range is 8 m/s.

Practice Problem Find the mean, median, mode, and range for the data set 8, 4, 12, 8, 11, 14, 16.

Math Skill Handbook

Use Geometry

The branch of mathematics that deals with the measurement, properties, and relationships of points, lines, angles, surfaces, and solids is called geometry.

Perimeter The **perimeter** (P) is the distance around a geometric figure. To find the perimeter of a rectangle, add the length and width and multiply that sum by two, or $2(l + w)$. To find perimeters of irregular figures, add the length of the sides.

Example 1 Find the perimeter of a rectangle that is 3 m long and 5 m wide.

Step 1 You know that the perimeter is 2 times the sum of the width and length.
$P = 2(3 \text{ m} + 5 \text{ m})$

Step 2 Find the sum of the width and length.
$P = 2(8 \text{ m})$

Step 3 Multiply by 2.
$P = 16 \text{ m}$

The perimeter is 16 m.

Example 2 Find the perimeter of a shape with sides measuring 2 cm, 5 cm, 6 cm, 3 cm.

Step 1 You know that the perimeter is the sum of all the sides.
$P = 2 + 5 + 6 + 3$

Step 2 Find the sum of the sides.
$P = 2 + 5 + 6 + 3$
$P = 16$

The perimeter is 16 cm.

Practice Problem Find the perimeter of a rectangle with a length of 18 m and a width of 7 m.

Practice Problem Find the perimeter of a triangle measuring 1.6 cm by 2.4 cm by 2.4 cm.

Area of a Rectangle The **area** (A) is the number of square units needed to cover a surface. To find the area of a rectangle, multiply the length times the width, or $l \times w$. When finding area, the units also are multiplied. Area is given in square units.

Example Find the area of a rectangle with a length of 1 cm and a width of 10 cm.

Step 1 You know that the area is the length multiplied by the width.
$A = (1 \text{ cm} \times 10 \text{ cm})$

Step 2 Multiply the length by the width. Also multiply the units.
$A = 10 \text{ cm}^2$

The area is 10 cm^2.

Practice Problem Find the area of a square whose sides measure 4 m.

Area of a Triangle To find the area of a triangle, use the formula:

$A = \frac{1}{2}(\text{base} \times \text{height})$

The base of a triangle can be any of its sides. The height is the perpendicular distance from a base to the opposite endpoint, or vertex.

Example Find the area of a triangle with a base of 18 m and a height of 7 m.

Step 1 You know that the area is $\frac{1}{2}$ the base times the height.
$A = \frac{1}{2}(18 \text{ m} \times 7 \text{ m})$

Step 2 Multiply $\frac{1}{2}$ by the product of 18×7. Multiply the units.
$A = \frac{1}{2}(126 \text{ m}^2)$
$A = 63 \text{ m}^2$

The area is 63 m^2.

Practice Problem Find the area of a triangle with a base of 27 cm and a height of 17 cm.

Math Skill Handbook

Circumference of a Circle The **diameter** (d) of a circle is the distance across the circle through its center, and the **radius** (r) is the distance from the center to any point on the circle. The radius is half of the diameter. The distance around the circle is called the **circumference** (C). The formula for finding the circumference is:

$$C = 2\pi r \quad or \quad C = \pi d$$

The circumference divided by the diameter is always equal to 3.1415926... This nonterminating and nonrepeating number is represented by the Greek letter π (pi). An approximation often used for π is 3.14.

Example 1 Find the circumference of a circle with a radius of 3 m.

Step 1 You know the formula for the circumference is 2 times the radius times π.
$C = 2\pi(3)$

Step 2 Multiply 2 times the radius.
$C = 6\pi$

Step 3 Multiply by π.
$C = 19$ m

The circumference is 19 m.

Example 2 Find the circumference of a circle with a diameter of 24.0 cm.

Step 1 You know the formula for the circumference is the diameter times π.
$C = \pi(24.0)$

Step 2 Multiply the diameter by π.
$C = 75.4$ cm

The circumference is 75.4 cm.

Practice Problem Find the circumference of a circle with a radius of 19 cm.

Area of a Circle The formula for the area of a circle is:
$A = \pi r^2$

Example 1 Find the area of a circle with a radius of 4.0 cm.

Step 1 $A = \pi(4.0)^2$

Step 2 Find the square of the radius.
$A = 16\pi$

Step 3 Multiply the square of the radius by π.
$A = 50$ cm^2

The area of the circle is 50 cm^2.

Example 2 Find the area of a circle with a radius of 225 m.

Step 1 $A = \pi(225)^2$

Step 2 Find the square of the radius.
$A = 50625\pi$

Step 3 Multiply the square of the radius by π.
$A = 158962.5$

The area of the circle is 158,962 m^2.

Example 3 Find the area of a circle whose diameter is 20.0 mm.

Step 1 You know the formula for the area of a circle is the square of the radius times π, and that the radius is half of the diameter.
$A = \pi\left(\frac{20.0}{2}\right)^2$

Step 2 Find the radius.
$A = \pi(10.0)^2$

Step 3 Find the square of the radius.
$A = 100\pi$

Step 4 Multiply the square of the radius by π.
$A = 314$ mm^2

The area is 314 mm^2.

Practice Problem Find the area of a circle with a radius of 16 m.

Volume The measure of space occupied by a solid is the **volume** (V). To find the volume of a rectangular solid multiply the length times width times height, or $V = l \times w \times h$. It is measured in cubic units, such as cubic centimeters (cm^3).

Example Find the volume of a rectangular solid with a length of 2.0 m, a width of 4.0 m, and a height of 3.0 m.

Step 1 You know the formula for volume is the length times the width times the height.
$V = 2.0 \text{ m} \times 4.0 \text{ m} \times 3.0 \text{ m}$

Step 2 Multiply the length times the width times the height.
$V = 24 \text{ m}^3$

The volume is 24 m^3.

Practice Problem Find the volume of a rectangular solid that is 8 m long, 4 m wide, and 4 m high.

To find the volume of other solids, multiply the area of the base times the height.

Example 1 Find the volume of a solid that has a triangular base with a length of 8.0 m and a height of 7.0 m. The height of the entire solid is 15.0 m.

Step 1 You know that the base is a triangle, and the area of a triangle is $\frac{1}{2}$ the base times the height, and the volume is the area of the base times the height.
$V = \left[\frac{1}{2}(b \times h)\right] \times 15$

Step 2 Find the area of the base.
$V = \left[\frac{1}{2}(8 \times 7)\right] \times 15$
$V = \left(\frac{1}{2} \times 56\right) \times 15$

Step 3 Multiply the area of the base by the height of the solid.
$V = 28 \times 15$
$V = 420 \text{ m}^3$

The volume is 420 m^3.

Example 2 Find the volume of a cylinder that has a base with a radius of 12.0 cm, and a height of 21.0 cm.

Step 1 You know that the base is a circle, and the area of a circle is the square of the radius times π, and the volume is the area of the base times the height.
$V = (\pi r^2) \times 21$
$V = (\pi 12^2) \times 21$

Step 2 Find the area of the base.
$V = 144\pi \times 21$
$V = 452 \times 21$

Step 3 Multiply the area of the base by the height of the solid.
$V = 9490 \text{ cm}^3$

The volume is 9490 cm^3.

Example 3 Find the volume of a cylinder that has a diameter of 15 mm and a height of 4.8 mm.

Step 1 You know that the base is a circle with an area equal to the square of the radius times π. The radius is one-half the diameter. The volume is the area of the base times the height.
$V = (\pi r^2) \times 4.8$
$V = \left[\pi\left(\frac{1}{2} \times 15\right)^2\right] \times 4.8$
$V = (\pi 7.5^2) \times 4.8$

Step 2 Find the area of the base.
$V = 56.25\pi \times 4.8$
$V = 176.63 \times 4.8$

Step 3 Multiply the area of the base by the height of the solid.
$V = 847.8$

The volume is 847.8 mm^3.

Practice Problem Find the volume of a cylinder with a diameter of 7 cm in the base and a height of 16 cm.

Math Skill Handbook

Science Applications

Measure in SI

The metric system of measurement was developed in 1795. A modern form of the metric system, called the International System (SI), was adopted in 1960 and provides the standard measurements that all scientists around the world can understand.

The SI system is convenient because unit sizes vary by powers of 10. Prefixes are used to name units. Look at **Table 3** for some common SI prefixes and their meanings.

Table 3 Common SI Prefixes			
Prefix	Symbol	Meaning	
kilo-	k	1,000	thousand
hecto-	h	100	hundred
deka-	da	10	ten
deci-	d	0.1	tenth
centi-	c	0.01	hundredth
milli-	m	0.001	thousandth

Example How many grams equal one kilogram?

Step 1 Find the prefix *kilo* in **Table 3.**

Step 2 Using **Table 3,** determine the meaning of *kilo.* According to the table, it means 1,000. When the prefix *kilo* is added to a unit, it means that there are 1,000 of the units in a "*kilo*unit."

Step 3 Apply the prefix to the units in the question. The units in the question are grams. There are 1,000 grams in a kilogram.

Practice Problem Is a milligram larger or smaller than a gram? How many of the smaller units equal one larger unit? What fraction of the larger unit does one smaller unit represent?

Dimensional Analysis

Convert SI Units In science, quantities such as length, mass, and time sometimes are measured using different units. A process called dimensional analysis can be used to change one unit of measure to another. This process involves multiplying your starting quantity and units by one or more conversion factors. A conversion factor is a ratio equal to one and can be made from any two equal quantities with different units. If 1,000 mL equal 1 L then two ratios can be made.

$$\frac{1{,}000 \text{ mL}}{1 \text{ L}} = \frac{1 \text{ L}}{1{,}000 \text{ mL}} = 1$$

One can covert between units in the SI system by using the equivalents in **Table 3** to make conversion factors.

Example 1 How many cm are in 4 m?

Step 1 Write conversion factors for the units given. From **Table 3,** you know that 100 cm = 1 m. The conversion factors are

$$\frac{100 \text{ cm}}{1 \text{ m}} \text{ and } \frac{1 \text{ m}}{100 \text{ cm}}$$

Step 2 Decide which conversion factor to use. Select the factor that has the units you are converting from (m) in the denominator and the units you are converting to (cm) in the numerator.

$$\frac{100 \text{ cm}}{1 \text{ m}}$$

Step 3 Multiply the starting quantity and units by the conversion factor. Cancel the starting units with the units in the denominator. There are 400 cm in 4 m.

$$4 \text{ m} \times \frac{100 \text{ cm}}{1 \text{ m}} = 400 \text{ cm}$$

Practice Problem How many milligrams are in one kilogram? (Hint: You will need to use two conversion factors from **Table 3.**)

MATH SKILL HANDBOOK 707

Math Skill Handbook

Table 4 Unit System Equivalents

Type of Measurement	Equivalent
Length	1 in = 2.54 cm
	1 yd = 0.91 m
	1 mi = 1.61 km
Mass and Weight*	1 oz = 28.35 g
	1 lb = 0.45 kg
	1 ton (short) = 0.91 tonnes (metric tons)
	1 lb = 4.45 N
Volume	1 in^3 = 16.39 cm^3
	1 qt = 0.95 L
	1 gal = 3.78 L
Area	1 in^2 = 6.45 cm^2
	1 yd^2 = 0.83 m^2
	1 mi^2 = 2.59 km^2
	1 acre = 0.40 hectares
Temperature	°C = $\frac{(°F - 32)}{1.8}$
	K = °C + 273

*Weight is measured in standard Earth gravity.

Convert Between Unit Systems Table 4 gives a list of equivalents that can be used to convert between English and SI units.

Example If a meterstick has a length of 100 cm, how long is the meterstick in inches?

Step 1 Write the conversion factors for the units given. From **Table 4,** 1 in = 2.54 cm.

$$\frac{1 \text{ in}}{2.54 \text{ cm}} \text{ and } \frac{2.54 \text{ cm}}{1 \text{ in}}$$

Step 2 Determine which conversion factor to use. You are converting from cm to in. Use the conversion factor with cm on the bottom.

$$\frac{1 \text{ in}}{2.54 \text{ cm}}$$

Step 3 Multiply the starting quantity and units by the conversion factor. Cancel the starting units with the units in the denominator. Round your answer based on the number of significant figures in the conversion factor.

$$100 \text{ cm} \times \frac{1 \text{ in}}{2.54 \text{ cm}} = 39.37 \text{ in}$$

The meterstick is 39.4 in long.

Practice Problem A book has a mass of 5 lbs. What is the mass of the book in kg?

Practice Problem Use the equivalent for in and cm (1 in = 2.54 cm) to show how 1 in^3 = 16.39 cm^3.

Precision and Significant Digits

When you make a measurement, the value you record depends on the precision of the measuring instrument. This precision is represented by the number of significant digits recorded in the measurement. When counting the number of significant digits, all digits are counted except zeros at the end of a number with no decimal point such as 2,050, and zeros at the beginning of a decimal such as 0.03020. When adding or subtracting numbers with different precision, round the answer to the smallest number of decimal places of any number in the sum or difference. When multiplying or dividing, the answer is rounded to the smallest number of significant digits of any number being multiplied or divided.

Example The lengths 5.28 and 5.2 are measured in meters. Find the sum of these lengths and record your answer using the correct number of significant digits.

Step 1 Find the sum.

5.28 m	2 digits after the decimal
+ 5.2 m	1 digit after the decimal
10.48 m	

Step 2 Round to one digit after the decimal because the least number of digits after the decimal of the numbers being added is 1.

The sum is 10.5 m.

Practice Problem How many significant digits are in the measurement 7,071,301 m? How many significant digits are in the measurement 0.003010 g?

Practice Problem Multiply 5.28 and 5.2 using the rule for multiplying and dividing. Record the answer using the correct number of significant digits.

Scientific Notation

Many times numbers used in science are very small or very large. Because these numbers are difficult to work with scientists use scientific notation. To write numbers in scientific notation, move the decimal point until only one non-zero digit remains on the left. Then count the number of places you moved the decimal point and use that number as a power of ten. For example, the average distance from the Sun to Mars is 227,800,000,000 m. In scientific notation, this distance is 2.278×10^{11} m. Because you moved the decimal point to the left, the number is a positive power of ten.

The mass of an electron is about 0.000 000 000 000 000 000 000 000 000 000 911 kg. Expressed in scientific notation, this mass is 9.11×10^{-31} kg. Because the decimal point was moved to the right, the number is a negative power of ten.

Example Earth is 149,600,000 km from the Sun. Express this in scientific notation.

Step 1 Move the decimal point until one non-zero digit remains on the left.
1.496 000 00

Step 2 Count the number of decimal places you have moved. In this case, eight.

Step 3 Show that number as a power of ten, 10^8.

The Earth is 1.496×10^8 km from the Sun.

Practice Problem How many significant digits are in 149,600,000 km? How many significant digits are in 1.496×10^8 km?

Practice Problem Parts used in a high performance car must be measured to 7×10^{-6} m. Express this number as a decimal.

Practice Problem A CD is spinning at 539 revolutions per minute. Express this number in scientific notation.

Math Skill Handbook

Make and Use Graphs

Data in tables can be displayed in a graph—a visual representation of data. Common graph types include line graphs, bar graphs, and circle graphs.

Line Graph A line graph shows a relationship between two variables that change continuously. The independent variable is changed and is plotted on the *x*-axis. The dependent variable is observed, and is plotted on the *y*-axis.

Example Draw a line graph of the data below from a cyclist in a long-distance race.

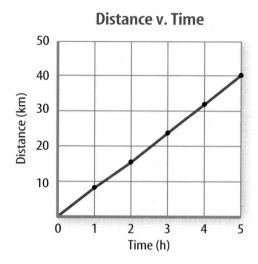

Figure 20 This line graph shows the relationship between distance and time during a bicycle ride.

Table 5 Bicycle Race Data	
Time (h)	Distance (km)
0	0
1	8
2	16
3	24
4	32
5	40

Step 1 Determine the *x*-axis and *y*-axis variables. Time varies independently of distance and is plotted on the *x*-axis. Distance is dependent on time and is plotted on the *y*-axis.

Step 2 Determine the scale of each axis. The *x*-axis data ranges from 0 to 5. The *y*-axis data ranges from 0 to 40.

Step 3 Using graph paper, draw and label the axes. Include units in the labels.

Step 4 Draw a point at the intersection of the time value on the *x*-axis and corresponding distance value on the *y*-axis. Connect the points and label the graph with a title, as shown in **Figure 20**.

Practice Problem A puppy's shoulder height is measured during the first year of her life. The following measurements were collected: (3 mo, 52 cm), (6 mo, 72 cm), (9 mo, 83 cm), (12 mo, 86 cm). Graph this data.

Find a Slope The slope of a straight line is the ratio of the vertical change, rise, to the horizontal change, run.

$$\text{Slope} = \frac{\text{vertical change (rise)}}{\text{horizontal change (run)}} = \frac{\text{change in } y}{\text{change in } x}$$

Example Find the slope of the graph in **Figure 20**.

Step 1 You know that the slope is the change in *y* divided by the change in *x*.

$$\text{Slope} = \frac{\text{change in } y}{\text{change in } x}$$

Step 2 Determine the data points you will be using. For a straight line, choose the two sets of points that are the farthest apart.

$$\text{Slope} = \frac{(40-0) \text{ km}}{(5-0) \text{ hr}}$$

Step 3 Find the change in *y* and *x*.

$$\text{Slope} = \frac{40 \text{ km}}{5 \text{ h}}$$

Step 4 Divide the change in *y* by the change in *x*.

$$\text{Slope} = \frac{8 \text{ km}}{\text{h}}$$

The slope of the graph is 8 km/h.

Math Skill Handbook

Bar Graph To compare data that does not change continuously you might choose a bar graph. A bar graph uses bars to show the relationships between variables. The *x*-axis variable is divided into parts. The parts can be numbers such as years, or a category such as a type of animal. The *y*-axis is a number and increases continuously along the axis.

Example A recycling center collects 4.0 kg of aluminum on Monday, 1.0 kg on Wednesday, and 2.0 kg on Friday. Create a bar graph of this data.

Step 1 Select the *x*-axis and *y*-axis variables. The measured numbers (the masses of aluminum) should be placed on the *y*-axis. The variable divided into parts (collection days) is placed on the *x*-axis.

Step 2 Create a graph grid like you would for a line graph. Include labels and units.

Step 3 For each measured number, draw a vertical bar above the *x*-axis value up to the *y*-axis value. For the first data point, draw a vertical bar above Monday up to 4.0 kg.

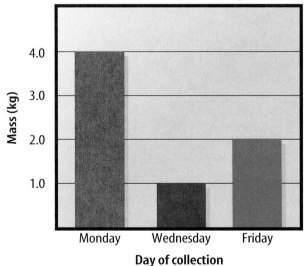

Practice Problem Draw a bar graph of the gases in air: 78% nitrogen, 21% oxygen, 1% other gases.

Circle Graph To display data as parts of a whole, you might use a circle graph. A circle graph is a circle divided into sections that represent the relative size of each piece of data. The entire circle represents 100%, half represents 50%, and so on.

Example Air is made up of 78% nitrogen, 21% oxygen, and 1% other gases. Display the composition of air in a circle graph.

Step 1 Multiply each percent by 360° and divide by 100 to find the angle of each section in the circle.

$$78\% \times \frac{360°}{100} = 280.8°$$

$$21\% \times \frac{360°}{100} = 75.6°$$

$$1\% \times \frac{360°}{100} = 3.6°$$

Step 2 Use a compass to draw a circle and to mark the center of the circle. Draw a straight line from the center to the edge of the circle.

Step 3 Use a protractor and the angles you calculated to divide the circle into parts. Place the center of the protractor over the center of the circle and line the base of the protractor over the straight line.

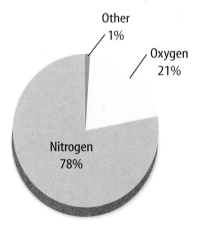

Practice Problem Draw a circle graph to represent the amount of aluminum collected during the week shown in the bar graph to the left.

MATH SKILL HANDBOOK 711

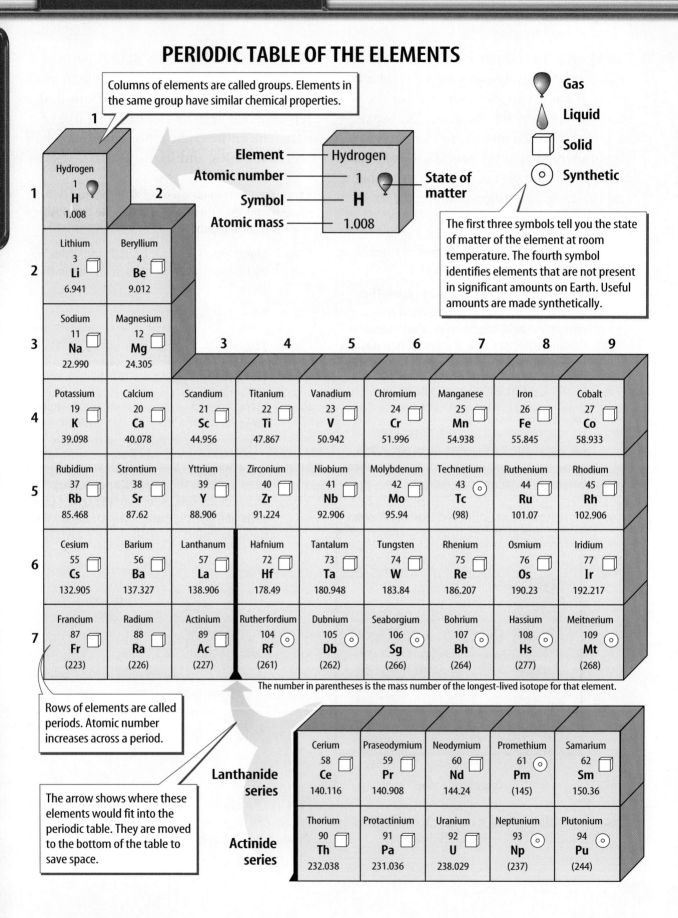

Reference Handbooks

Metal
Metalloid
Nonmetal

The color of an element's block tells you if the element is a metal, nonmetal, or metalloid.

Science Online
Visit ips.msscience.com for updates to the periodic table.

13	14	15	16	17	18
					Helium 2 **He** 4.003
Boron 5 **B** 10.811	Carbon 6 **C** 12.011	Nitrogen 7 **N** 14.007	Oxygen 8 **O** 15.999	Fluorine 9 **F** 18.998	Neon 10 **Ne** 20.180
Aluminum 13 **Al** 26.982	Silicon 14 **Si** 28.086	Phosphorus 15 **P** 30.974	Sulfur 16 **S** 32.065	Chlorine 17 **Cl** 35.453	Argon 18 **Ar** 39.948

10	11	12						
Nickel 28 **Ni** 58.693	Copper 29 **Cu** 63.546	Zinc 30 **Zn** 65.409	Gallium 31 **Ga** 69.723	Germanium 32 **Ge** 72.64	Arsenic 33 **As** 74.922	Selenium 34 **Se** 78.96	Bromine 35 **Br** 79.904	Krypton 36 **Kr** 83.798
Palladium 46 **Pd** 106.42	Silver 47 **Ag** 107.868	Cadmium 48 **Cd** 112.411	Indium 49 **In** 114.818	Tin 50 **Sn** 118.710	Antimony 51 **Sb** 121.760	Tellurium 52 **Te** 127.60	Iodine 53 **I** 126.904	Xenon 54 **Xe** 131.293
Platinum 78 **Pt** 195.078	Gold 79 **Au** 196.967	Mercury 80 **Hg** 200.59	Thallium 81 **Tl** 204.383	Lead 82 **Pb** 207.2	Bismuth 83 **Bi** 208.980	Polonium 84 **Po** (209)	Astatine 85 **At** (210)	Radon 86 **Rn** (222)
Darmstadtium 110 **Ds** (281)	Roentgenium 111 **Rg** (272)	Ununbium * 112 **Uub** (285)		Ununquadium * 114 **Uuq** (289)				

* The names and symbols for elements 112 and 114 are temporary. Final names will be selected when the elements' discoveries are verified.

Europium 63 **Eu** 151.964	Gadolinium 64 **Gd** 157.25	Terbium 65 **Tb** 158.925	Dysprosium 66 **Dy** 162.500	Holmium 67 **Ho** 164.930	Erbium 68 **Er** 167.259	Thulium 69 **Tm** 168.934	Ytterbium 70 **Yb** 173.04	Lutetium 71 **Lu** 174.967
Americium 95 **Am** (243)	Curium 96 **Cm** (247)	Berkelium 97 **Bk** (247)	Californium 98 **Cf** (251)	Einsteinium 99 **Es** (252)	Fermium 100 **Fm** (257)	Mendelevium 101 **Md** (258)	Nobelium 102 **No** (259)	Lawrencium 103 **Lr** (262)

Glossary/Glosario

Cómo usar el glosario en español:
1. Busca el término en inglés que desees encontrar.
2. El término en español, junto con la definición, se encuentran en la columna de la derecha.

Pronunciation Key

Use the following key to help you sound out words in the glossary.

a	back (BAK)		ew	food (FEWD)
ay	day (DAY)		yoo	pure (PYOOR)
ah	father (FAH thur)		yew	few (FYEW)
ow	flower (FLOW ur)		uh	comma (CAH muh)
ar	car (CAR)		u (1 con)	rub (RUB)
e	less (LES)		sh	shelf (SHELF)
ee	leaf (LEEF)		ch	nature (NAY chur)
ih	trip (TRIHP)		g	gift (GIHFT)
i (i 1 con 1 e)	idea (i DEE uh)		j	gem (JEM)
oh	go (GOH)		ing	sing (SING)
aw	soft (SAWFT)		zh	vision (VIH zhun)
or	orbit (OR buht)		k	cake (KAYK)
oy	coin (COYN)		s	seed, cent (SEED, SENT)
oo	foot (FOOT)		z	zone, raise (ZOHN, RAYZ)

English — A — Español

acceleration: equals the change in velocity divided by the time for the change to take place; occurs when an object speeds up, slows down, or turns. (p. 288)

accuracy: compares a measurement to the true value. (p. 45)

acid: substance that releases H^+ ions and produces hydronium ions when dissolved in water. (p. 232)

activation energy: minimum amount of energy needed to start a chemical reaction. (p. 201)

alternating current (AC): electric current that changes its direction repeatedly. (p. 626)

alternative resource: new renewable or inexhaustible energy source; includes solar energy, wind, and geothermal energy. (p. 391)

amino acids: building blocks of proteins; contain both an amino group and a carboxyl acid group replacing hydrogens on the same carbon atom. (p. 259)

amino (uh ME noh) group: consists of one nitrogen atom covalently bonded to two hydrogen atoms; represented by the formula $-NH_2$. (p. 259)

amplitude: for a transverse wave, one half the distance between a crest and a trough. (p. 467)

analog signal: a electronic signal that carries information and varies smoothly with time. (p. 642)

aceleración: es igual al cambio de velocidad dividido por el tiempo que toma en realizarse dicho cambio; sucede cuando un objeto aumenta su velocidad, la disminuye o gira. (p. 288)

exactitud: comparación de una medida con el valor real. (p. 45)

ácido: sustancia que libera iones H^+ y produce iones de hidronio al ser disuelta en agua. (p. 232)

energía de activación: cantidad mínima de energía necesaria para iniciar una reacción química. (p. 201)

corriente alterna (CA): corriente eléctrica que cambia de dirección repetidamente. (p. 626)

recurso alternativo: nueva fuente de energía renovable o inagotable; incluye energía solar, eólica y geotérmica. (p. 391)

aminoácidos: bloques de construcción de las proteínas que contienen un grupo amino y un grupo ácido carboxilo reemplazando hidrógenos en el mismo átomo de carbono. (p. 259)

grupo amino: consiste en un átomo de nitrógeno unido por enlaces covalentes a dos átomos de hidrógeno; se lo representa con la fórmula $-NH_2$. (p. 259)

amplitud: la mitad de la distancia entre la cresta y el valle en una onda transversal. (p. 467)

señal analógica: señal electrónica que conduce información y varía de manera uniforme con el tiempo. (p. 642)

Glossary/Glosario

aqueous/carbohydrates   **acuoso/carbohidratos**

aqueous (A kwee us): solution in which water is the solvent. (p. 224)

Archimedes' (ar kuh MEE deez) principle: states that the buoyant force on an object is equal to the weight of the fluid displaced by the object. (pp. 121, 351)

atom: a very small particle that makes up most kinds of matter and consists of smaller parts called protons, neutrons, and electrons. (p. 73)

atomic mass: average mass of an atom of an element; its unit of measure is the atomic mass unit (u), which is 1/12 the mass of a carbon-12 atom. (p. 84)

atomic number: number of protons in the nucleus of each atom of a given element; is the top number in the periodic table. (p. 83)

aurora: light display that occurs when charged particles trapped in the magnetosphere collide with Earth's atmosphere above the poles. (p. 625)

average speed: equals the total distance traveled divided by the total time taken to travel the distance. (p. 285)

acuoso: solución en la cual el agua es el solvente. (p. 224)

principio de Arquímedes: establece que la fuerza de empuje ejercida sobre un objeto es igual al peso del fluido desplazado por dicho objeto. (pp. 121, 351)

átomo: partícula muy pequeña que constituye la mayoría de los tipos de materia y que está formada por partes más pequeñas llamadas protones, neutrones y electrones. (p. 73)

masa atómica: masa promedio de un átomo de un elemento; su unidad de medida es la unidad de masa atómica (u), la cual es 1/12 de la masa de un átomo de carbono-12. (p. 84)

número atómico: número de protones en el núcleo de un átomo de determinado elemento; es el número superior en la tabla periódica. (p. 83)

aurora: despliegue de luz que se produce cuando partículas cargadas atrapadas en la magnetosfera chocan contra la atmósfera terrestre por encima de los polos. (p. 625)

velocidad promedio: es igual al total de la distancia recorrida dividida por el tiempo total necesario para recorrer dicha distancia. (p. 285)

B

balanced forces: two or more forces whose effects cancel each other out and do not change the motion of an object. (p. 311)

bar graph: a type of graph that uses bars of varying sizes to show the relationship among variables. (p. 58)

base: substance that accepts H^+ ions and produces hydroxide ions when dissolved in water. (p. 235)

Bernoulli's principle: states that when the speed of a fluid increases, the pressure exerted by the fluid decreases. (p. 359)

binary system: number system consisting of two digits, 0 and 1, that can be used by devices such as computers to store or use information. (p. 650)

buoyant force: upward force exerted on an object immersed in a fluid. (pp. 120, 348)

fuerzas balanceadas: dos o más fuerzas cuyos efectos se cancelan mutuamente sin cambiar el movimiento de un objeto. (p. 311)

gráfico de barras: tipo de gráfico que usa barras de diferentes tamaños para mostrar las diferencias entre las variables. (p. 58)

base: sustancia que acepta los iones H^+ y produce iones de hidróxido al ser disuelta en agua. (p. 235)

principio de Bernoulli: establece que cuando se incrementa la velocidad de un fluido, disminuye la presión ejercida por el mismo. (p. 359)

sistema binario: sistema numérico que consiste en dos dígitos, 0 y 1, que se puede usar con dispositivos como las computadoras para almacenar o usar información. (p. 650)

fuerza de empuje: fuerza ascendente ejercida sobre un objeto inmerso en un fluido. (pp. 120, 348)

C

carbohydrates: organic compounds containing only carbon, hydrogen, and oxygen; starches, cellulose, glycogen, sugars. (p. 264)

carbohidratos: compuestos orgánicos que sólo contienen carbono, hidrógeno y oxígeno; ejemplos son los almidones, la celulosa, el glucógeno y los azúcares. (p. 264)

Glossary/Glosario

carboxyl group/compressional wave   **grupo carboxilo/onda de compresión**

carboxyl (car BOK sul) group: consists of one carbon atom, two oxygen atoms, and one hydrogen atom; represented by the formula —COOH. (p. 259)

carrier wave: radio waves broadcast by a radio or TV station at an assigned frequency that contains information used to produce pictures and sound. (p. 536)

catalyst: substance that speeds up a chemical reaction but is not used up itself or permanently changed. (p. 205)

center of mass: point in a object that moves as if all of the object's mass were concentrated at that point. (p. 322)

chemical bond: force that holds two atoms together. (p. 169)

chemical change: any change of a material into a new material with different properties. (p. 145)

chemical energy: energy stored in chemical bonds. (p. 377)

chemical equation: shorthand form for writing what reactants are used and what products are formed in a chemical reaction; sometimes shows whether energy is produced or absorbed. (p. 192)

chemical formula: combination of chemical symbols and numbers that indicates which elements and how many atoms of each element are present in a molecule. (p. 178)

chemical property: characteristic of something that permits its change to something new. (p. 139)

chemical reaction: process that produces chemical change, resulting in new substances that have properties different from those of the original substances. (p. 190)

cholesterol: a complex lipid that is present in foods that come from animals. (p. 269)

circle graph: a type of graph that shows the parts of a whole; sometimes called a pie graph, each piece of which represents a percentage of the total. (p. 58)

circuit: closed conducting loop in which electric current can flow continually. (p. 591)

compound: a substance produced when elements combine and whose properties are different from each of the elements in it. (pp. 87, 171)

compound machine: machine made up of a combination of two or more simple machines. (p. 417)

compressional wave: a type of mechanical wave in which matter in the medium moves forward and backward along the direction the wave travels. (p. 465)

grupo carboxilo: consiste en un átomo de carbono, dos de oxígeno y uno de hidrógeno; se lo representa con la fórmula —COOH. (p. 259)

ondas conductoras: ondas de radio emitidas por una estación de radio o televisión a una frecuencia asignada, las cuales contienen información utilizada para producir imágenes y sonido. (p. 536)

catalizador: sustancia que acelera una reacción química pero que ella misma ni se agota ni sufre cambios permanentes. (p. 205)

centro de masa: punto en un objeto que se mueve como si toda la masa del objeto estuviera concentrada en ese punto. (p. 322)

enlace químico: fuerza que mantiene a dos átomos unidos. (p. 169)

cambio químico: cualquier transformación de un material en otro nuevo con propiedades diferentes. (p. 145)

energía química: energía almacenada en enlaces químicos. (p. 377)

ecuación química: forma breve para representar los reactivos utilizados y los productos que se forman en una reacción química; algunas veces muestra si se produce o absorbe energía. (p. 192)

fórmula química: combinación de símbolos y números químicos que indican cuáles elementos y cuántos átomos de cada elemento están presentes en una molécula. (p. 178)

propiedad química: característica de algo que le permite su transformación en algo nuevo. (p. 139)

reacción química: proceso que produce cambios químicos que dan como resultado nuevas sustancias cuyas propiedades son diferentes a aquellas de las sustancias originales. (p. 190)

colesterol: lípido complejo presente en alimentos de origen animal. (p. 269)

gráfico circular: tipo de gráfico que muestra las partes de un todo; algunas veces se le llama gráfico de pastel en el que cada parte representa un porcentaje del total. (p. 58)

circuito: circuito conductor cerrado en el cual la energía puede fluir continuamente. (p. 591)

compuesto: sustancia resultante de la combinación de elementos cuyas propiedades son diferentes de los elementos que la componen. (pp. 87, 171)

máquina compuesta: máquina compuesta por la combinación de dos o más máquinas. (p. 417)

onda de compresión: tipo de onda mecánica en la que la materia en el medio se mueve hacia adelante y hacia atrás en dirección de la onda. (p. 465)

Glossary/Glosario

computer software/digital signal

computer software: any list of instructions for a computer to follow that is stored in the computer's memory. (p. 653)

concave lens: lens that is thicker at its edges than in the middle. (p. 565)

concentration: describes how much solute is present in a solution compared to the amount of solvent. (pp. 203, 229)

condensation: change of matter from a gas to a liquid state. (p. 113)

conduction: transfer of thermal energy by direct contact; occurs when energy is transferred by collisions between particles. (p. 439)

conductor: material in which electrons can move or that transfers heat easily. (pp. 441, 588)

constant: variable that is not changed in an experiment. (p. 18)

controlled experiment: involves changing one factor and observing its effect on one thing while keeping all other things constant. (p. 18)

convection: transfer of thermal energy by the movement of particles from one place to another in a gas or liquid. (p. 440)

convex lens: lens that is thicker in the middle than at its edges. (p. 564)

covalent bond: chemical bond formed when atoms share electrons. (p. 173)

critical thinking: involves using knowledge and thinking skills to evaluate evidence and explanations. (p. 27)

software para computadoras/señal digital

software para computadoras: cualquier lista de instrucciones que debe realizar una computadora y que se almacena en la memoria de ésta. (p. 653)

lente cóncavo: lente que es más grueso en sus bordes que en el centro. (p. 565)

concentración: describe la cantidad de soluto presente en una solución, comparada con la cantidad de solvente. (pp. 203, 229)

condensación: cambio de estado de la materia de gas a líquido. (p. 113)

conducción: transferencia de energía térmica por contacto directo; se produce cuando la energía se transfiere mediante colisiones entre las partículas. (p. 439)

conductor: material en el cual los electrones se pueden mover o que transfiere calor fácilmente. (pp. 441, 588)

constante: variable que no cambia en un experimento. (p. 18)

experimento controlado: consiste en cambiar un factor y observar su efecto sobre algo mientras el resto de las cosas se mantiene constante. (p. 18)

convección: transferencia de energía térmica por el movimiento de partículas de un sitio a otro en un líquido o un gas. (p. 440)

lente convexo: lente que es más grueso en el centro que en sus bordes. (p. 564)

enlace covalente: enlace químico que se forma cuando los átomos comparten electrones. (p. 173)

pensamiento crítico: consiste en utilizar los conocimientos y habilidades del pensamiento para evaluar evidencias y explicaciones. (p. 27)

D

data: information gathered during an investigation; recorded in the form of descriptions, tables, graphs, or drawings. (p. 28)

density: physical property of matter that can be found by dividing the matter's mass by its volume. (pp. 121, 134, 352)

dependent variable: variable that changes as a result of a change in the independent variable. (p. 18)

diffraction: bending of waves around a barrier. (p. 475)

digital signal: electronic signal that varies information that does not vary smoothly with time, but changes in steps between certain values, and can be represented by a series of numbers. (p. 643)

datos: información recopilada durante una investigación y archivada en forma de descripciones, tablas, gráficas o planos. (p. 28)

densidad: propiedad física de la materia que se puede determinar dividendo la masa de la materia por su volumen. (pp. 121, 134, 352)

variable dependiente: variable que cambia como resultado de un cambio en la variable independiente. (p. 18)

difracción: curvatura de las ondas alrededor de una barrera. (p. 475)

señal digital: señal electrónica que varía aquella información que no varía de manera uniforme con el tiempo, pero que cambia por grados entre ciertos valores y que puede ser representada por una serie de números. (p. 643)

Glossary/Glosario

diode/electromagnetic waves   **diodo/ondas electromagnéticas**

diode: a solid-state component made from two layers of semiconductor material that allows electric current to flow in only one direction and is commonly used to change alternating current to direct current. (p. 646)

direct current (DC): electric current that flows only in one direction. (p. 627)

Doppler effect: change in the frequency of a sound wave that occurs when the sound source and the listener are in motion relative to each other. (p. 496)

diodo: componente de estado sólido conformado por dos capas de material semiconductor que permite el flujo de corriente eléctrica en una sola dirección y que comúnmente se utiliza para cambiar la corriente alterna a corriente directa. (p. 646)

corriente directa (CD): corriente eléctrica que fluye solamente en una dirección. (p. 627)

efecto Doppler: cambio en la frecuencia de una onda sonora que ocurre cuando la fuente de sonido y quien lo escucha están en movimiento relativo el uno del otro. (p. 496)

E

eardrum: membrane stretching across the ear canal that vibrates when sound waves reach the middle ear. (p. 508)

Earth science: study of Earth systems and systems in space, including weather and climate systems, and the study of nonliving things such as rocks, oceans, and planets. (p. 10)

echo: a reflected sound wave. (p. 495)

efficiency: equals the output work divided by the input work; expressed as a percentage. (p. 415)

electrical energy: energy carried by electric current. (p. 378)

electric current: the flow of electric charge, measured in amperes (A). (p. 591)

electric discharge: rapid movement of excess charge from one place to another. (p. 589)

electric field: surrounds every electric charge and exerts forces on other electric charges. (p. 587)

electric force: attractive or repulsive force exerted by all charged objects on each other. (p. 587)

electric power: rate at which electrical energy is converted into other forms of energy, measured in watts (W) or kilowatts (kW). (p. 600)

electromagnet: magnet created by wrapping a current-carrying wire around an iron core. (p. 621)

electromagnetic spectrum: range of electromagnetic waves, including radio waves, visible light, and X rays, with different frequencies and wavelengths. (p. 525)

electromagnetic waves: waves that can travel through matter or space; include radio waves, infrared waves, visible light waves, ultraviolet waves, X rays and gamma rays. (pp. 466, 520)

tímpano: membrana que se extiende a través del canal auditivo y que vibra cuando las ondas sonoras alcanzan el oído medio. (p. 508)

ciencias de la Tierra: estudio del sistema de la Tierra y de los sistemas en el espacio, incluyendo el clima y los sistemas climáticos y el estudio de los seres inanimados como las rocas, los océanos y los planetas. (p. 10)

eco: el reflejo de una onda sonora. (p. 495)

eficiencia: equivale al trabajo aplicado dividido el trabajo generado y se expresa en porcentaje. (p. 415)

energía eléctrica: energía transportada por corriente eléctrica. (p. 378)

corriente eléctrica: flujo de carga eléctrica, el cual se mide en amperios (A). (p. 591)

descarga eléctrica: movimiento rápido de carga excesiva de un lugar a otro. (p. 589)

campo eléctrico: campo que rodea a todas las cargas eléctricas y que ejerce fuerzas sobre otras cargas eléctricas. (p. 587)

fuerza eléctrica: fuerza de atracción o de repulsión que ejercen todos los objetos cargados entre ellos mismos. (p. 587)

potencia eléctrica: tasa a la cual la energía eléctrica se convierte en otras formas de energía, la cual se mide en vatios (W) o en kilovatios (kW). (p. 600)

electroimán: imán que se crea al enrollar un cable transportador de corriente alrededor de un centro de hierro. (p. 621)

espectro electromagnético: rango de ondas electromagnéticas, incluyendo las ondas de radio, luz visible, y rayos X, con diferentes frecuencias y longitudes de onda. (p. 525)

ondas electromagnéticas: ondas que pueden viajar a través de la materia o del espacio; incluyen ondas radiales, ondas infrarrojas, ondas de luz visible, ondas ultravioletas, rayos X y rayos gama. (pp. 466, 520)

Glossary/Glosario

electron/frequency

electron: invisible, negatively charged particle located in a cloudlike formation that surrounds the nucleus of an atom. (p. 76)

electron cloud: area where negatively charged electrons, arranged in energy levels, travel around an atom's nucleus. (p. 162)

electron dot diagram: chemical symbol for an element, surrounded by as many dots as there are electrons in its outer energy level. (p. 168)

electronic signal: a changing electric current that is used to carry information; can be analog or digital. (p. 642)

element: natural or synthetic material that cannot be broken down into simpler materials by ordinary means; has unique properties and is generally classified as a metal, metalloid, or nonmetal. (p. 80)

endothermic (en duh THUR mihk) reaction: chemical reaction in which heat energy is absorbed. (p. 197)

energy: the ability to cause change. (p. 374)

energy level: the different positions for an electron in an atom. (p. 163)

enzyme: catalysts that are large protein molecules which speed up reactions needed for your cells to work properly. (p. 206)

estimation: method of making an educated guess at a measurement; using the size of something familiar to guess the size of a new object. (p. 43)

exothermic (ek soh THUR mihk) reaction: chemical reaction in which heat energy is released. (p. 197)

electrón/frecuencia

electrón: partícula invisible con carga negativa, localizada en una formación parecida a una nube que rodea el núcleo de un átomo. (p. 76)

nube de electrones: área en donde los electrones cargados negativamente se distribuyen en niveles de energía y se mueven alrededor del núcleo de un átomo. (p. 162)

diagrama de punto de electrones: símbolo químico para un elemento, rodeado de tantos puntos como electrones se encuentran en su nivel exterior de energía. (p. 168)

señal electrónica: corriente eléctrica dinámica que se usa para conducir información; puede ser analógica o digital. (p. 642)

elemento: material natural o sintético que no puede ser descompuesto fácilmente en materiales más simples por medios ordinarios; tiene propiedades únicas y generalmente es clasificado como metal, metaloide o no metal. (p. 80)

reacción endotérmica: reacción química en la cual se absorbe energía calórica. (p. 197)

energía: capacidad de producir cambios. (p. 374)

nivel de energía: las diferentes posiciones de un electrón en un átomo. (p. 163)

enzimas: catalizadores que son grandes moléculas de proteínas las cuales aceleran las reacciones necesarias para que las células trabajen en forma adecuada. (p. 206)

estimación: método para hacer una suposición fundamentada en una medida, usando el tamaño de algo conocido para suponer el tamaño de un nuevo objeto. (p. 43)

reacción exotérmica: reacción química en la cual se libera energía calórica. (p. 197)

F

fluid: a substance that has no definite shape and can flow. (p. 343)

focal length: distance along the optical axis from the center of a mirror or lens to the focal point. (p. 558)

focal point: point on the optical axis of a mirror or lens where rays traveling parallel to the optical axis pass through. (p. 558)

force: a push or a pull. (p. 310)

freezing: change of matter from a liquid state to a solid state. (p. 111)

frequency: number of wavelengths that pass a given point in one second; measured in hertz (Hz). (p. 469)

fluido: sustancia que no tiene forma definida y que puede fluir. (p. 343)

distancia focal: distancia a lo largo del eje óptico desde el centro de un espejo o lente hasta el punto focal. (p. 558)

punto focal: punto en el eje óptico de un espejo o lente por el cual atraviesan los rayos que viajan en paralelo al eje óptico. (p. 558)

fuerza: presión o tracción. (p. 310)

congelación: cambio de la materia de estado líquido a sólido. (p. 111)

frecuencia: número de longitudes de onda que pasan un punto determinado en un segundo; se mide en hertz (Hz). (p. 469)

Glossary/Glosario

friction/ hydroxyl group　　　　　　　　　　　　　　　　　　　　　　　**fricción/grupo hidroxilo**

friction: force that acts to oppose sliding between two surfaces that are touching. (p. 312)

fundamental frequency: lowest natural frequency that is produced by a vibrating object, such as a string or a column of air. (p. 503)

fricción: fuerza que actúa para oponerse al deslizamiento entre dos superficies que se tocan. (p. 312)

frecuencia fundamental: frecuencia natural más baja producida por un objeto que vibra, tal como una cuerda o una columna de aire. (p. 503)

G

gamma ray: highest-energy electromagnetic waves with the shortest wavelengths and highest frequencies. (p. 530)

gas: matter that does not have a definite shape or volume; has particles that move at high speeds in all directions. (p. 106)

generator: device that uses a magnetic field to turn kinetic energy into electrical energy. (pp. 384, 626)

Global Positioning System (GPS): uses satellites, ground-based stations, and portable units with receivers to locate objects on Earth. (p. 539)

graph: used to collect, organize, and summarize data in a visual way, making it easy to use and understand. (p. 57)

rayos gama: ondas electromagnéticas que poseen la mayor cantidad de energía y las cuales presentan las longitudes de onda más cortas y las frecuencias más altas. (p. 530)

gas: materia que no tiene ni forma ni volumen definidos; tiene partículas que se mueven a altas velocidades y en todas las direcciones. (p. 106)

generador: dispositivo que utiliza un campo magnético para convertir energía cinética en energía eléctrica. (pp. 384, 626)

Sistema de Posicionamiento Global (SPG): sistema que utiliza satélites, estaciones en tierra y unidades portátiles con receptores para ubicar objetos en la Tierra. (pp. 539)

gráfico: se usa para recolectar, organizar y resumir información en forma visual, facilitando su uso y comprensión. (p. 57)

H

heat: movement of thermal energy from a substance at a higher temperature to a substance at a lower temperature. (pp. 108, 438)

heat engine: device that converts thermal energy into mechanical energy. (p. 445)

heterogeneous mixture: type of mixture where the substances are not evenly mixed. (p. 219)

homogeneous mixture: type of mixture where two or more substances are evenly mixed on a molecular level but are not bonded together. (p. 220)

hydraulic system: uses a fluid to increase an applied force. (p. 357)

hydrocarbon: organic compound that has only carbon and hydrogen atoms. (p. 251)

hydronium ion: hydrogen ion combines with a water molecule to form a hydronium ion, H_3O^+. (p. 232)

hydroxyl (hi DROK sul) group: consists of an oxygen atom and a hydrogen atom joined by a covalent bond; represented by the formula $-OH$. (p. 258)

calor: movimiento de energía térmica de una sustancia que se encuentra a una alta temperatura hacia una sustancia a una baja temperatura. (pp. 108, 438)

motor de calor: motor que transforma la energía térmica en energía mecánica. (p. 445)

mezcla heterogénea: tipo de mezcla en la cual las sustancias no están mezcladas de manera uniforme. (p. 219)

mezcla homogénea: tipo de mezcla en la cual dos o más sustancias están mezcladas en de manera uniforme a nivel molecular pero no están enlazadas. (p. 220)

sistema hidráulico: usa un fluido para incrementar una fuerza aplicada. (p. 357)

hidrocarburo: compuesto orgánico que sólo contiene átomos de carbono e hidrógeno. (p. 251)

ion de hidronio: ion de hidrógeno combinado con una molécula de agua para formar un ion de hidronio, H_3O^+. (p. 232)

grupo hidroxilo: consiste en un átomo de oxígeno y un átomo de hidrógeno unidos por un enlace covalente; se lo representa con la fórmula $-OH$. (p. 258)

hypothesis: reasonable guess that can be tested and is based on what is known and what is observed. (p. 14)

hipótesis: suposición razonable que puede ser probada y que está basada en lo que se sabe y en lo que ha sido observado. (p. 14)

inclined plane: simple machine that is a flat surface, sloped surface, or ramp. (p. 417)

independent variable: variable that is changed in an experiment. (p. 18)

indicator: compound that changes color at different pH values when it reacts with acidic or basic solutions. (p. 238)

inertia: tendency of an object to resist a change in its motion. (p. 293)

inexhaustible resource: energy source that can't be used up by humans. (p. 391)

infer: to draw a conclusion based on observation. (p. 16)

infrared wave: electromagnetic waves with wavelengths between 1 mm and 0.7 millionths of a meter. (p. 527)

inhibitor: substance that slows down a chemical reaction, making the formation of a certain amount of product take longer. (p. 204)

input force: force exerted on a machine. (p. 412)

instantaneous speed: the speed of an object at one instant of time. (p. 285)

insulator: material in which electrons cannot move easily. (p. 588)

integrated circuit: circuit that can contain millions of interconnected transistors and diodes imprinted on a single small chip of semiconductor material. (p. 647)

interference: occurs when two or more waves combine and form a new wave when they overlap. (p. 477)

internal combustion engine: heat engine in which fuel is burned in a combustion chamber inside the engine. (p. 446)

ion (I ahn): atom that is positively or negatively charged because it has gained or lost electrons. (pp. 171, 584)

ionic bond: attraction that holds oppositely charged ions close together. (p. 171)

isomers (I suh murz): compounds with the same chemical formula but different structures and different physical and chemical properties. (p. 254)

plano inclinado: máquina simple que consiste en una superficie plana, inclinada, o una rampa. (p. 417)

variable independiente: variable que cambia en un experimento. (p. 18)

indicador: compuesto que cambia de color con diferentes valores de pH al reaccionar con soluciones ácidas o básicas. (p. 238)

inercia: tendencia de un objeto a resistirse a un cambio de movimiento. (p. 293)

recurso inagotable: fuente de energía que no puede ser agotada por los seres humanos. (p. 391)

deducción: sacar una conclusión con base en una observación. (p. 16)

ondas infrarrojas: ondas electromagnéticas con longitudes de onda entre un milímetro y 0.7 millonésimas de metro. (p. 527)

inhibidor: sustancia que reduce la velocidad de una reacción química, haciendo que la formación de una determinada cantidad de producto tarde más tiempo. (p. 204)

fuerza aplicada: fuerza que se ejerce sobre una máquina. (p. 412)

velocidad instantánea: la velocidad de un objeto en un instante de tiempo. (p. 285)

aislante: material en el cual los electrones no se pueden mover fácilmente. (p. 588)

circuito integrado: circuito que puede contener millones de transistores y diodos interconectados y fijados en un solo chip de tamaño reducido y hecho de material semiconductor. (p. 647)

interferencia: ocurre cuando dos o más ondas se combinan y al sobreponerse forman una nueva onda. (p. 477)

motor de combustión interna: motor de calor en el cual el combustible es quemado en una cámara de combustión dentro del motor. (p. 446)

ion: átomo cargado positiva o negativamente a que ha ganado o perdido electrónes. (pp. 171, 584)

enlace iónico: atracción que mantiene unidos a iones con cargas opuestas. (p. 171)

isómeros: compuestos que tienen la misma fórmula química pero diferentes estructuras y propiedades físicas y químicas. (p. 254)

Glossary/Glosario

isotopes/magnetic domain

isotopes (I suh tohps): two or more atoms of the same element that have different numbers of neutrons in their nuclei. (p. 83)

isótopos/dominio magnético

isótopos: dos o más átomos del mismo elemento que tienen diferente número de neutrones en su núcleo. (p. 83)

K

Kelvin (K): SI unit for temperature. (p. 54)
kilogram (kg): SI unit for mass. (p. 53)
kinetic energy: energy an object has due to its motion. (p. 375)

Kelvin (K): unidad del SI para temperatura. (p. 54)
kilogramo (kg): unidad del SI para masa. (p. 53)
energía cinética: energía que posee un objeto debido a su movimiento. (p. 375)

L

law of conservation of energy: states that energy can change its form but is never created or destroyed. (p. 380)
law of conservation of matter: states that matter is not created or destroyed but only changes its form. (p. 74)
law of conservation of momentum: states that the total momentum of objects that collide with each other is the same before and after the collision. (p. 295)
law of reflection: states that when a wave is reflected, the angle of incidence is equal to the angle of reflection. (p. 555)
lens: transparent object that has at least one curved surface that causes light to bend. (p. 563)
lever: simple machine consisting of a rigid rod or plank that pivots or rotates about a fixed point called the fulcrum. (p. 420)
life science: study of living systems and how they interact. (p. 9)
light ray: narrow beam of light traveling in a straight line. (p. 550)
line graph: a type of graph used to show the relationship between two variables that are numbers on an *x*-axis and a *y*-axis. (p. 57)
lipids: organic compound that contains the same elements as carbohydrates but in different proportions. (p. 267)
liquid: matter with a definite volume but no definite shape that can flow from one place to another. (p. 104)
loudness: the human perception of how much energy a sound wave carries. (p. 492)

ley de la conservación de la energía: establece que la energía puede cambiar de forma pero nunca puede ser creada ni destruida. (p. 380)
ley de la conservación de la materia: establece que la materia no se crea ni se destruye, solamente cambia de forma. (p. 74)
ley de conservación de momento: establece que el momento total de los objetos que chocan entre sí es el mismo antes y después de la colisión. (p. 295)
ley de la reflexión: establece que cuando se refleja una onda, el ángulo de incidencia es igual al ángulo de reflexión. (p. 555)
lente: objeto transparente que tiene por lo menos una superficie curva que hace cambiar la dirección de la luz. (p. 563)
palanca: máquina simple que consiste en una barra rígida que puede girar sobre un punto fijo llamado punto de apoyo. (p. 420)
ciencias de la vida: estudio de los sistemas vivos y de la forma como interactúan. (p. 9)
rayo de luz: haz estrecho de luz que viaja en línea recta. (p. 550)
gráfico lineal: tipo de gráfico usado para mostrar la relación entre dos variables que son números en un eje *x* y en un eje *y*. (p. 57)
lípidos: compuestos orgánicos que contienen los mismos elementos que los carbohidratos pero en proporciones diferentes. (p. 267)
líquido: materia con volumen definido pero no con forma definida que puede fluir de un sitio a otro. (p. 104)
intensidad: percepción humana de la cantidad de energía conducida por una onda sonora. (p. 492)

M

magnetic domain: group of atoms whose fields point in the same direction. (p. 616)

dominio magnético: grupo de átomos cuyos campos apuntan en la misma dirección. (p. 616)

Glossary/Glosario

magnetic field/motor **campo magnético/motor**

magnetic field: surrounds a magnet and exerts a magnetic force on other magnets. (p. 615)

magnetosphere: region of space affected by Earth's magnetic field. (p. 617)

mass: amount of matter in an object. (pp. 53, 293)

mass number: sum of the number of protons and neutrons in the nucleus of an atom. (p. 83)

matter: anything that takes up space and has mass. (pp. 72, 102)

measurement: way to describe objects and events with numbers; for example, length, volume, mass, weight, and temperature. (p. 42)

mechanical advantage: number of times the input force is multiplied by a machine; equal to the output force divided by the input force. (p. 413)

mechanical wave: a type of wave that can travel only through matter. (p. 463)

medium: material through which a wave travels. (p. 551)

melting: change of matter from a solid state to a liquid state. (p. 109)

metal: element that is malleable, ductile, a good conductor of electricity, and generally has a shiny or metallic luster. (p. 84)

metallic bond: bond formed when metal atoms share their pooled electrons. (p. 172)

metalloid: element that has characteristics of both metals and nonmetals and is a solid at room temperature. (p. 85)

meter (m): SI unit for length. (p. 51)

microprocessor: integrated circuit that controls the flow of information between different parts of the computer; also called the central processing unit or CPU. (p. 655)

mixture: a combination of compounds and elements that has not formed a new substance and whose proportions can be changed without changing the mixture's identity. (p. 89)

model: any representation of an object or an event that is used as a tool for understanding the natural world; can communicate observations and ideas, test predictions, and save time, money, and lives. (p. 21)

molecule (MAH lih kewl): neutral particle formed when atoms share electrons. (p. 173)

momentum: a measure of how difficult it is to stop a moving object; equals the product of mass and velocity. (p. 294)

monomer: small, organic molecules that link together to form polymers. (p. 262)

motor: device that transforms electrical energy into kinetic energy. (p. 624)

campo magnético: campo que rodea a un imán y ejerce fuerza magnética sobre otros imanes. (p. 615)

magnetosfera: región del espacio afectada por el campo magnético de la Tierra. (p. 617)

masa: cantidad de materia en un objeto. (pp. 53, 293)

número de masa: suma del número de protones y neutrones en el núcleo de un átomo. (p. 83)

materia: cualquier cosa que ocupe espacio y tenga masa. (pp. 72, 102)

medida: forma para describir objetos y eventos con números; por ejemplo, longitud, volumen, masa, peso y temperatura. (p. 42)

ventaja mecánica: número de veces que la fuerza aplicada es multiplicada por una máquina; equivale a la fuerza producida dividida por la fuerza aplicada. (p. 413)

onda mecánica: tipo de onda que puede viajar únicamente a través de la materia. (p. 463)

medio: material a través del cual viaja una onda. (p. 551)

fusión: cambio de la materia de estado sólido a líquido. (p. 109)

metal: elemento maleable, dúctil y buen conductor de electricidad que generalmente tiene un lustre brillante o metálico. (p. 84)

enlace metálico: enlace que se forma cuando átomos metálicos comparten sus electrones agrupados. (p. 172)

metaloide: elemento que comparte características de los metales y de los no metales y es sólido a temperatura ambiente. (p. 85)

metro (m): unidad del SI para longitud. (p. 51)

microprocesador: circuito integrado que controla el flujo de información entre diferentes partes de una computadora; también se lo denomina la unidad central de procesamiento o CPU. (p. 655)

mezcla: combinación de compuestos y elementos que no han formado una nueva sustancia y cuyas proporciones pueden ser cambiadas sin que se pierda la identidad de la mezcla. (p. 89)

modelo: cualquier representación de un objeto o evento utilizada como herramienta para entender el mundo natural; puede comunicar observaciones e ideas, predicciones de las pruebas y ahorrar tiempo, dinero y salvar vidas. (p. 21)

molécula: partícula neutra que se forma cuando los átomos comparten electrones. (p. 173)

momento: medida de la dificultad para detener un objeto en movimiento; es igual al producto de la masa por la velocidad. (p. 294)

monómeros: moléculas orgánicas pequeñas que se unen entre sí para formar polímeros. (p. 262)

motor: dispositivo que transforma energía eléctrica en energía cinética. (p. 624)

Glossary/Glosario

natural frequencies/parallel circuit   **frecuencias naturales/circuito paralelo**

N

natural frequencies: frequencies at which an object will vibrate when it is struck or disturbed. (p. 501)

net force: combination of all forces acting on an object. (p. 311)

neutralization (new truh luh ZAY shun): reaction in which an acid reacts with a base and forms water and a salt. (p. 238)

neutron: an uncharged particle located in the nucleus of an atom. (p. 78)

Newton's first law of motion: states that if the net force acting on an object is zero, the object will remain at rest or move in a straight line with a constant speed. (p. 312)

Newton's second law of motion: states that an object acted upon by a net force will accelerate in the direction of the force, and that the acceleration equals the net force divided by the object's mass. (p. 316)

Newton's third law of motion: states that forces always act in equal but opposite pairs. (p. 323)

nonmetals: elements that are usually gases or brittle solids and poor conductors of electricity and heat; are the basis of the chemicals of life. (p. 85)

nonrenewable resource: energy resource that is used up much faster than it can be replaced. (p. 388)

nuclear energy: energy contained in atomic nuclei. (p. 378)

nucleus (NEW klee us): positively charged, central part of an atom. (p. 77)

frecuencias naturales: frecuencias a las cuales un objeto vibrará cuando es golpeado o perturbado. (p. 501)

fuerza neta: la combinación de todas las fuerzas que actúan sobre un objeto. (p. 311)

neutralización: reacción en la cual un ácido reacciona con una base para formar agua y una sal. (p. 238)

neutrón: partícula sin carga localizada en el núcleo de un átomo (p. 78)

primera ley de movimiento de Newton: establece que si la fuerza neta que actúa sobre un objeto es igual a cero, el objeto se mantendrá en reposo o se moverá en línea recta a una velocidad constante. (p. 312)

segunda ley de movimiento de Newton: establece que si una fuerza neta se ejerce sobre un objeto, éste se acelerará en la dirección de la fuerza y la aceleración es igual a la fuerza neta dividida por la masa del objeto. (p. 316)

tercera ley de movimiento de Newton: establece que las fuerzas siempre actúan en pares iguales pero opuestos. (p. 323)

no metales: elementos que por lo general son gases o sólidos frágiles y malos conductores de electricidad y calor; son la base de los compuestos químicos biológicos. (p. 85)

recurso no renovable: recurso energético que se agota mucho más rápidamente de lo que puede ser reemplazado. (p. 388)

energía nuclear: energía contenida en los núcleos de los átomos. (p. 378)

núcleo: parte central con carga positiva del átomo. (p. 77)

O

Ohm's law: states that the current in a circuit equals the voltage divided by the resistance in the circuit. (p. 597)

organic compounds: most compounds that contain carbon. (p. 250)

output force: force exerted by a machine. (p. 412)

overtones: multiples of the fundamental frequency. (p. 503)

ley de Ohm: establece que la corriente en un circuito es igual al voltaje dividido por la resistencia en el circuito. (p. 597)

compuestos orgánicos: la mayoría de compuestos que contienen carbono. (p. 250)

fuerza generada: fuerza producida por una máquina. (p. 412)

armónicos: múltiplos de la frecuencia fundamental. (p. 503)

P

parallel circuit: circuit that has more than one path for electric current to follow. (p. 599)

circuito paralelo: circuito en el cual la corriente eléctrica puede seguir más de una trayectoria. (p. 599)

Glossary/Glosario

Pascal's principle: states that when a force is applied to a confined fluid, an increase in pressure is transmitted equally to all parts of the fluid. (pp. 122, 357)

pH: measure of how acidic or basic a solution is, ranging in a scale from 0 to 14. (p. 236)

photovoltaic: device that transforms radiant energy directly into electrical energy. (p. 392)

physical change: any change in the size, shape, form, or state of matter in which the matter's identity remains the same. (p. 143)

physical property: any characteristic of matter—such as color, shape, and taste—that can be detected by the senses without changing the identity of the matter. (p. 134)

physical science: study of matter, which is anything that takes up space and has mass, and the study of energy, which is the ability to cause change. (p. 10)

pitch: how high or low a sound is. (p. 494)

polar bond: bond resulting from the unequal sharing of electrons. (p. 174)

polymer: large molecule made up of small repeating units linked by covalent bonds to form a long chain. (p. 262)

polymerization: a chemical reaction in which monomers are bonded together. (p. 262)

potential energy: energy stored in an object due to its position. (p. 376)

power: rate at which work is done; equal to the work done divided by the time it takes to do the work; measured in watts (W). (p. 409)

precipitate: solid that comes back out of its solution because of a chemical reaction or physical change. (p. 220)

precision: describes how closely measurements are to each other and how carefully measurements were made. (p. 44)

pressure: amount of force applied per unit area on an object's surface; SI unit is the Pascal (Pa). (pp. 116, 340)

product: substance that forms as a result of a chemical reaction. (p. 192)

protein: biological polymer made up of amino acids; catalyzes many cell reactions and provides structural materials for many parts of the body. (p. 263)

proton: positively charged particle located in the nucleus of an atom and that is counted to identify the atomic number. (p. 77)

principio de Pascal: establece que cuando se ejerce una fuerza sobre un fluido encerrado, se transmite un incremento de presión uniforme a todas las partes del fluido. (pp. 122, 357)

pH: medida para saber qué tan básica o ácida es una solución, en una escala de 0 a 14. (p. 236)

fotovoltaico: dispositivo que transforma la energía radiante directamente en energía eléctrica. (p. 392)

cambio físico: cualquier cambio en el tamaño, apariencia, forma o estado de la materia, en el que la identidad de la materia permanece igual. (p. 143)

propiedad física: cualquier característica de la materia, como el color, apariencia o sabor, que puede ser detectada por los sentidos sin que cambie la identidad de la materia. (p. 134)

ciencias física: estudio de la materia, lo cual es todo lo que ocupe espacio y tenga masa, y el estudio de la energía, que es la habilidad de producir cambios. (p. 10)

altura: expresa qué tan alto o bajo es un sonido. (p. 494)

enlace polar: enlace que resulta de compartir electrones en forma desigual. (p. 174)

polímero: molécula grande formada por unidades pequeñas que se repiten y están unidas por enlaces covalentes para formar una cadena larga. (p. 262)

polimerización: reacción química en la que los monómeros se unen entre sí. (p. 262)

energía potencial: energía almacenada en un objeto debido a su posición. (p. 376)

potencia: velocidad a la que se realiza un trabajo y que equivale al trabajo realizado dividido por el tiempo que toma realizar el trabajo; se mide en vatios (W). (p. 409)

precipitado: sólido que se aísla de su solución mediante una reacción química o un cambio físico. (p. 220)

precisión: describe qué tan aproximada es una medida respecto a otra y qué tan cuidadosamente fueron hechas dichas medidas. (p. 44)

presión: cantidad de fuerza aplicada por unidad de área sobre la superficie de un objeto; la unidad internacional SI es el Pascal (Pa). (pp. 116, 340)

producto: sustancia que se forma como resultado de una reacción química. (p. 192)

proteína: polímero biológico formado por aminoácidos; cataliza numerosas reacciones celulares y conforma materiales estructurales para diversas partes del cuerpo. (p. 263)

protón: partícula cargada positivamente, localizada en el núcleo de un átomo y que se cuenta para identificar el número atómico. (p. 77)

Glossary/Glosario

pulley/reverberation **polea/reverberación**

pulley: simple machine made from a grooved wheel with a rope or cable wrapped around the groove. (p. 422)

polea: máquina simple que consiste en una rueda acanalada con una cuerda o cable que corre alrededor del canal. (p. 422)

R

radiant energy: energy carried by an electromagnetic wave. (pp. 377, 524)

radiation: transfer of energy by electromagnetic waves. (p. 439)

radio waves: lowest-frequency electromagnetic waves that have wavelengths greater than about 0.3 m and are used in most forms of telecommunications technology—such as TVs, telephones, and radios. (p. 526)

random-access memory (RAM): temporary electronic memory within a computer. (p. 652)

rate: a ratio of two different kinds of measurements; the amount of change of one measurement in a given amount of time. (p. 54)

rate of reaction: measure of how fast a chemical reaction occurs. (p. 202)

reactant: substance that exists before a chemical reaction begins. (p. 192)

reactivity: describes how easily something reacts with something else. (p. 140)

read-only memory (ROM): electronic memory that is permanently stored within a computer. (p. 652)

reflecting telescope: uses a concave mirror to gather light from distant objects. (p. 569)

reflection: occurs when a wave strikes an object or surface and bounces off. (p. 473)

refracting telescope: uses two convex lenses to gather light and form an image of a distant object. (p. 568)

refraction: bending of a wave as it moves from one medium into another medium. (p. 474)

renewable resource: energy resource that is replenished continually. (p. 390)

resistance: a measure of how difficult it is for electrons to flow in a material; unit is the ohm (Ω). (p. 594)

resonance: occurs when an object is made to vibrate at its natural frequencies by absorbing energy from a sound wave or other object vibrating at this frequency. (p. 502)

reverberation: repeated echoes of sounds. (p. 507)

energía radiante: energía conducida por una onda electromagnética. (pp. 377, 524)

radiación: transferencia de energía mediante ondas electromagnéticas. (p. 439)

ondas de radio: ondas electromagnéticas con la menor frecuencia, las cuales poseen longitudes de onda mayores de unos 0.3 metros y son utilizadas en la mayoría de técnicas de telecomunicaciones, tales como televisores, teléfonos y radios. (p. 526)

memoria de acceso aleatorio (RAM): memoria electrónica temporal dentro de una computadora. (p. 652)

tasa: relación de dos diferentes tipos de medidas; los cambios en una medida en un tiempo determinado. (p. 54)

velocidad de reacción: medida de la rapidez con que se produce una reacción química. (p. 202)

reactivo: sustancia que existe antes de que comience una reacción química. (p. 192)

reactividad: describe la facilidad con la que dos cosas pueden reaccionar entre sí. (p. 140)

memoria de sólo lectura (ROM): memoria electrónica almacenada permanentemente dentro de una computadora. (p. 652)

telescopio de reflexión: utiliza un espejo cóncavo para concentrar la luz proveniente de objetos lejanos. (p. 569)

reflexión: ocurre cuando una onda choca contra un objeto o superficie y rebota. (p. 473)

telescopio de refracción: utiliza dos lentes convexos para concentrar la luz y formar una imagen de un objeto lejano. (p. 568)

refracción: curvatura de una onda a medida que se mueve de un medio a otro. (p. 474)

recurso renovable: recurso energético regenerado continuamente. (p. 390)

resistencia: medida de la dificultad que tienen los electrones para fluir en un material; se mide en ohmios (Ω). (p. 594)

resonancia: ocurre cuando se hace vibrar un objeto a sus frecuencias naturales mediante la absorción de energía de una onda sonora o de otro objeto que vibra a dicha frecuencia. (p. 502)

reverberación: ecos repetidos de los sonidos. (p. 507)

S

salts: compounds made of a metal and a nonmetal that are formed along with water when acids and bases react with each other. (p. 142)

saturated: describes a solution that holds the total amount of solute that it can hold under given conditions. (p. 228)

saturated hydrocarbon: hydrocarbon, such as methane, with only single bonds. (p. 252)

science: way of learning more about the natural world that provides possible explanations to questions and involves using a collection of skills. (p. 6)

scientific law: a rule that describes a pattern in nature but does not try to explain why something happens. (p. 7)

scientific theory: a possible explanation for repeatedly observed patterns in nature supported by observations and results from many investigations. (p. 7)

screw: simple machine that is an inclined plane wrapped around a cylinder or post. (p. 419)

semiconductor: element, such as silicon, that is a poorer electrical conductor that a metal, but a better conductor than a nonmetal, and whose electrical conductivity can be changed by adding impurities. (p. 645)

series circuit: circuit that has only one path for electric current to follow. (p. 598)

SI: International System of Units, related by multiples of ten, designed to provided a worldwide standard of physical measurement. (p. 50)

simple machine: a machine that does work with only one movement; includes the inclined plane, wedge, screw, lever, wheel and axle, and pulley. (p. 417)

size-dependent properties: physical properties—such as volume and mass—that change when the size of the object changes. (p. 136)

size-independent properties: physical properties—such as density—that do not change when the size of the object changes. (p. 136)

solid: matter with a definite shape and volume; has tightly packed particles that move mainly by vibrating. (p. 103)

solubility (sahl yuh BIH luh tee): measure of how much solute can be dissolved in a certain amount of solvent. (p. 227)

solute: substance that dissolves and seems to disappear into another substance. (p. 220)

sales: compuestos formados por un metal y un no metal que se forman junto con agua cuando reaccionan ácidos y bases entre sí. (p. 142)

saturado: describe a una solución que retiene toda la cantidad de soluto que puede retener bajo determinadas condiciones. (p. 228)

hidrocarburo saturado: hidrocarburo, como el metano, que sólo presenta enlaces sencillos. (p. 252)

ciencia: mecanismo para aprender más acerca del mundo natural, que da respuestas posibles a los interrogantes e implica hacer uso de numerosas habilidades. (p. 6)

ley científica: regla que describe un modelo en la naturaleza pero que no intenta explicar por qué suceden las cosas. (p. 7)

teoría científica: posible explicación para patrones observados repetidamente en la naturaleza y apoyada en observaciones y resultados de muchas investigaciones. (p. 7)

tornillo: máquina simple que consiste en un plano inclinado envuelto en espiral alrededor de un cilindro o poste. (p. 419)

semiconductor: elemento, como el silicio, que no es tan buen conductor de electricidad como un metal, pero que es mejor conductor que un no metal y cuya conductividad eléctrica puede ser modificada al añadirle impurezas. (p. 645)

circuito en serie: circuito en el cual la corriente eléctrica sólo puede seguir una trayectoria. (p. 598)

SI: Sistema Internacional de Unidades, se ordena en múltiplos de diez, diseñados para suministrar un estándar de medidas físicas a nivel mundial. (p. 50)

máquina simple: máquina que ejecuta el trabajo con un solo movimiento; incluye el plano inclinado, la palanca, el tornillo, la rueda y el eje y la polea. (p. 417)

propiedades dependientes del tamaño: propiedades físicas, tales como volumen y masa, que cambian cuando se modifica el tamaño del objeto. (p. 136)

propiedades independientes del tamaño: propiedades físicas, tales como la densidad, que no cambian cuando se modifica el tamaño del objeto. (p. 136)

sólido: materia con forma y volumen definidos; tiene partículas fuertemente compactadas que se mueven principalmente por vibración. (p. 103)

solubilidad: medida de la cantidad de soluto que puede disolverse en cierta cantidad de solvente. (p. 227)

soluto: sustancia que se disuelve y parece desaparecer en otra sustancia. (p. 220)

Glossary/Glosario

solution/transistor

solution: homogeneous mixture whose elements and/or compounds are evenly mixed at the molecular level but are not bonded together. (p. 220)

solvent: substance that dissolves the solute. (p. 220)

specific heat: amount of heat needed to raise the temperature of 1 kg of a substance by 1°C. (p. 442)

speed: equals the distance traveled divided by the time it takes to travel that distance. (p. 284)

starches: polymers of glucose monomers in which hundreds or thousands of glucose molecules are joined together. (p. 265)

state of matter: physical property that describes a substance as a solid, liquid, or gas. (p. 136)

static charge: imbalance of electric charge on an object. (p. 585)

substance: matter with a fixed composition whose identity can be changed by chemical processes but not by ordinary physical processes. (pp. 87, 218)

sugars: carbohydrates containing carbon atoms arranged in a ring. (p. 265)

surface tension: the uneven forces acting on the particles on the surface of a liquid. (p. 105)

system: collection of structures, cycles, and processes that relate to and interact with each other. (p. 8)

solución/transistor

solución: mezcla homogénea cuyos elementos o compuestos están mezclados de manera uniforme a nivel molecular pero no se enlazan. (p. 220)

solvente: sustancia que disuelve al soluto. (p. 220)

calor específico: cantidad de calor necesario para elevar la temperatura de 1 kilogramo de una sustancia en 1 grado centígrado. (p. 442)

rapidez: equivale a dividir la distancia recorrida por el tiempo que toma recorrer dicha distancia. (p. 284)

almidones: polímeros de monómeros de la glucosa en los que cientos o miles de moléculas de glucosa están unidas entre sí. (p. 265)

estado de la materia: propiedad física que describe a una sustancia como sólido, líquido o gas. (p. 136)

carga estática: desequilibrio de la carga eléctrica en un objeto. (p. 585)

sustancia: materia que tiene una composición fija cuya identidad puede ser cambiada mediante procesos químicos pero no mediante procesos físicos corrientes. (pp. 87, 218)

azúcares: carbohidratos que contienen átomos de carbono dispuestos en un anillo. (p. 265)

tensión superficial: fuerzas desiguales que actúan sobre las partículas que se encuentran en la superficie de un líquido. (p. 105)

sistema: colección de estructuras, ciclos y procesos relacionados que interactúan entre sí. (p. 8)

T

table: presents information in rows and columns, making it easier to read and understand. (p. 57)

technology: use of science to help people in some way. (p. 11)

temperature: measure of the average kinetic energy of the individual particles of a substance. (pp. 108, 434)

thermal energy: the sum of the kinetic and potential energy of the particles in a material. (pp. 376, 437)

thermal pollution: increase in temperature of a natural body of water; caused by adding warmer water. (p. 443)

transformer: device used to increase or decrease the voltage of an alternating current. (p. 628)

transistor: a solid-state component made from three layers of semiconductor material that can amplify the strength of an electric signal or act as an electronic switch. (p. 647)

tabla: presentación de información en filas y columnas, facilitando la lectura y comprensión. (p. 57)

tecnología: uso de la ciencia para ayudar en alguna forma a las personas. (p. 11)

temperatura: medida de la energía cinética promedio de las partículas individuales de una sustancia. (pp. 108, 434)

energía térmica: la suma de la energía cinética y potencial de las partículas en un material. (pp. 376, 437)

polución térmica: incremento de la temperatura de una masa natural de agua producido al agregarle agua a mayor temperatura. (p. 443)

transformador: dispositivo utilizado para aumentar o disminuir el voltaje de una corriente alterna. (p. 628)

transistor: componente de estado sólido formado por tres capas de material semiconductor que puede amplificar la fuerza de una señal eléctrica o actuar a manera de interruptor electrónico. (p. 647)

transverse wave: a type of mechanical wave in which the wave energy causes matter in the medium to move up and down or back and forth at right angles to the direction the wave travels. (p. 464)

turbine: set of steam-powered fan blades that spins a generator at a power plant. (p. 384)

onda transversal: tipo de onda mecánica en el cual la energía de la onda hace que la materia en el medio se mueva hacia arriba y hacia abajo o hacia adelante y hacia atrás en ángulos rectos respecto a la dirección en que viaja la onda. (p. 464)

turbina: conjunto de aspas de ventilador impulsadas por vapor que hacen girar a un generador en una planta de energía eléctrica. (p. 384)

ultraviolet radiation: electromagnetic waves with wavelengths between about 0.4 millionths of a meter and 10 billionths of a meter; has frequencies and wavelengths between visible light and X rays. (p. 529)

unbalanced forces: two or more forces acting on an object that do not cancel, and cause the object to accelerate. (p. 311)

unsaturated hydrocarbon: hydrocarbon, such as ethylene, with one or more double or triple bonds. (p. 253)

radiación ultravioleta: ondas electromagnéticas con longitudes de onda entre aproximadamente 0.4 millonésimas de metro y 10 billonésimas de metro; tienen frecuencias y longitudes de onda entre aquellas de la luz visible y los rayos X. (p. 529)

fuerzas no balanceadas: dos o más fuerzas que actúan sobre un objeto sin anularse y que hacen que el objeto se acelere. (p. 311)

hidrocarburo insaturado: hidrocarburo, como el etileno, con uno o más enlaces dobles o triples. (p. 253)

vaporization: change of matter from a liquid state to a gas. (p. 112)

variable: factor that can be changed in an experiment. (p. 18)

velocity: speed and direction of a moving object. (p. 287)

viscosity: a liquid's resistance to flow. (p. 105)

visible light: electromagnetic waves with wavelengths between 0.4 and 0.7 millionths of a meter that can be seen with your eyes. (p. 528)

voltage: a measure of the amount of electrical potential energy an electron flowing in a circuit can gain; measured in volts (V). (p. 592)

volume: the amount of space an object occupies. (p. 52)

vaporización: cambio de estado de la materia de líquido a gas. (p. 112)

variable: factor que puede cambiar en un experimento. (p. 18)

velocidad: rapidez y dirección de un objeto en movimiento. (p. 287)

viscosidad: resistencia de un líquido al flujo. (p. 105)

luz visible: ondas electromagnéticas con longitudes de onda entre 0.4 y 0.7 millonésimas de metro y que pueden ser observadas a simple vista. (p. 528)

voltaje: medida de la cantidad de energía eléctrica potencial que puede adquirir un electrón que fluye en un circuito; se mide en voltios (V). (p. 592)

volumen: la cantidad de espacio que ocupa un objeto. (p. 52)

wave: rhythmic disturbance that carries energy but not matter. (p. 462)

wavelength: for a transverse wave, the distance between the tops of two adjacent crests or the bottoms of two adjacent troughs; for a compressional wave, the distance from the centers of adjacent rarefactions or adjacent compressions. (p. 468)

onda: alteración rítmica que transporta energía pero no materia. (p. 462)

longitud de onda: en una onda transversal, es la distancia entre las puntas de dos crestas adyacentes o entre dos depresiones adyacentes; en una onda de compresión es la distancia entre los centros de dos rarefacciones adyacentes o compresiones adyacentes. (p. 468)

wedge/X ray

wedge: simple machine consisting of an inclined plane that moves; can have one or two sloping sides. (p. 418)

weight: a measurement of force that depends on gravity; measured in newtons. (pp. 53, 317)

wheel and axle: simple machine made from two circular objects of different sizes that are attached and rotate together. (p. 420)

work: is done when a force exerted on an object causes that object to move some distance; equal to force times distance; measured in joules (J). (p. 406)

cuña/rayos X

cuña: máquina simple que consiste en un plano inclinado que se mueve; puede tener uno o dos lados inclinados. (p. 418)

peso: medida de fuerza que depende de la gravedad y que se mide en Newtons. (pp. 53, 317)

rueda y eje: máquina simple compuesta por dos objetos circulares de diferentes tamaños que están interconectados y giran. (p. 420)

trabajo: se realiza cuando la fuerza ejercida sobre un objeto hace que el objeto se mueva determinada distancia; es igual a la fuerza multiplicada por la distancia y se mide en julios (J). (p. 406)

X

X ray: high-energy electromagnetic wave that is highly penetrating and can be used for medical diagnosis. (p. 530)

rayos X: ondas electromagnéticas de alta energía, las cuales son altamente penetrantes y pueden ser utilizadas para diagnósticos médicos. (p. 530)

Index

Italic numbers = illustration/photo **Bold numbers** = vocabulary term
lab = indicates a page on which the entry is used in a lab
act = indicates a page on which the entry is used in an activity

A

Acceleration, 288–292; calculating, 289–290, 290 *act*, 319, 319 *act;* equation for, 290; and force, *316,* 316–317, 320; graph of, 292, *292;* and gravity, 5 *lab;* modeling, 291, 291 *lab;* and motion, 288–289; negative, 291, *291;* positive, 291; and speed, *288,* 288–289; unit of measurement with, 317; and velocity, *288,* 288–289, *289*
Accuracy, 41 *lab,* 43, 44 *lab,* **45**–47, *46,* 51
Acetic (ethanoic) acid, 141, 192, 193, 194, 222, *222,* 237, *237,* 259, *259*
Acetylene (ethyne), 254, *254,* 255
Acetylsalicylic acid, 141
Acid(s), 232–234; chemical properties of, 141, *141;* common, 141; in environment, 233, *234;* measuring strength of, 150–151 *lab,* 236, 236–238, *237,* 240–241 *lab;* neutralizing, 238–239, *239;* physical properties of, 137, *137,* 138 *act;* properties of, 232; reaction with bases, 142, 238–239, *239;* uses of, 233, *233*
Acidophils, 236
Acid precipitation, *234*
Acid rain, *234;* effects of, 141, 148, 152, *152*
Action and reaction, 323–326, *324, 326*
Activation energy, 201, 201
Activities, Applying Math, 17, 48, 121, 135, 196, 284, 290, 294, 319, 341, 408, 409, 413, 415, 436, 537, 571, 597, 600; Applying Science, 89, 111, 167, 229, 266, 352, 390, 477, 496, 617, 651; Integrate, 9, 13, 43, 47, 51, 73, 78, 91, 104, 108, 123, 137, 140, 148, 165, 171,193, 205, 221, 225, 229, 236, 252, 264, 294, 311, 317, 324, 343, 352, 361, 381, 383, 388, 408, 415, 419, 442, 443, 448, 466, 468, 470, 495, 496, 502, 529, 530, 533, 536, 557, 593, 594, 601, 618, 630, 634, 646, 654, 656; Science Online, 18, 21, 47, 76, 81, 90, 105, 111, 113, 123, 138, 146, 164, 175, 195, 201, 219, 235, 238, 256, 267, 286, 296, 313, 324, 341, 358, 380, 390, 410, 413, 446, 471, 477, 495, 507, 521, 538, 564, 588, 619, 627, 645, 652, 656, 658; Standardized Test Practice, 38–397, 66–67, 98–99, 130–131, 156–157, 186–187, 214–215, 246–247, 276–277, 306–307, 336–337, 368–369, 402–403, 430–431, 456–457, 486–487, 516–517, 546–547, 578–579, 610–611, 638–639, 666–667
Aging, 137
Air conditioners, 449
Airplanes, flight of, 360, *360,* 361
Air pollution, and acid precipitation, *234*
Air pressure, 339 *lab.* *See also* Atmospheric pressure
Air resistance, 321
Alanine, 260, *260,* 263
Alchemist, 177
Alcohols, 258, *258,* 261 *lab*
Alkali metals, 167, *167*
Alkanes, 252
Alkenes, 253
Alkynes, 254
Alloys, 223, *223*
Alpha Centauri, *524*
Alpha-hydroxy acids, 137
Alternating current (AC), 626
Alternative resources, 391–393, 391 *lab,* *392, 393*
Aluminum, 141, *141*
Aluminum hydroxide, 141
Amines, 259, *259*
Amino acids, 259–260, *260,* 263, *263*
Amino group, 259, *259*
Ammeter, 622, *623*
Ammonia, 141, 142, 259
Ammonium chloride, 142
Amorphous solids, 104, 109, *109*
Amplitude, *467,* **467**–468, 493, *493*
Amplitude modulation (AM), 536
Analog devices, 642–643, *643*
Analog signal, 642, 644
Angle(s), critical, 565, *565*
Animal(s), effect of momentum on motion of, 294; hearing of, 508, *508;* insulation of, 442; largest, 62, *62;* speed of, 284; warm-blooded v. cold-blooded, 530
Antacid, 238
Anvil (of ear), 508, *508*
Applying Math, Acceleration of a Bus, 290; Acceleration of a Car, 319; Calculating Density, 121; Calculating Efficiency, 415; Calculating Mechanical Advantage, 413; Calculating Power, 409; Calculating Pressure, 341; Calculating Work, 408; Chapter Reviews, 33, 65, 97, 155, 185, 213, 245, 275, 305, 335, 367, 401, 429, 455, 485, 515, 545, 577, 609, 637, 665; Conserving Mass, 196; Converting to Celsius, 436; Determining Density, 135; Electric Power Used by a Lightbulb, 600; Momentum of a Bicycle, 294; Rounded Values, 48; Seasonal Temperatures, 17;

Index

Applying Science

Section Reviews, 25, 49, 54, 85, 138, 169, 199, 206, 239, 256, 287, 292, 298, 322, 328, 347, 354, 361, 395, 410, 416, 423, 437, 471, 499, 509, 524, 560, 602, 631, 647; Speed of a Swimmer, 284; Voltage from a Wall Outlet, 597; Wavelength of an FM Station, 537

Applying Science, Can you create destructive interference?, 477; Finding the Magnetic Declination, 617; How can ice save oranges?, 111; How can you compare concentrations?, 229; How does Doppler radar work?, 496; How does the periodic table help you identify properties of elements?, 167; How much information can be stored?, 651; Is energy consumption outpacing production?, 390; Layering Liquids, 352; What's the best way to desalt ocean water?, 89; Which foods are best for quick energy?, 266

Applying Skills, 11, 29, 59, 79, 91, 106, 142, 148, 178, 223, 230, 260, 269, 315, 378, 385, 443, 449, 466, 479, 533, 539, 554, 566, 590, 595, 619, 659

Aqueous solutions, 224–226, *225, 226*

Archimedes' principle, 121, *121,* 124–125 *lab,* *351,* **351**–354, *354*

Area, and pressure, 117, *117,* 342, *342*

Aristotle, 73

Arsenic, 6, *656*

Artificial body parts, 426, *426*

Ascorbic acid, 137, *137,* 141, 233

Atherosclerosis, 269, *269*

Atmospheric pressure, *117,* 117–119, *118,* 339 *lab,* 345, 345–347, *346, 347,* 362–363 *lab*

Atom(s), 73; components of, 584, *584;* electron cloud model of, 78–79, *79;* mass number of, 83; models of, 74–79, *77, 78, 79;* size of, 75, *75;* structure of, *162,* 162–163, *163,* 180–181 *lab;* symbols for, 177, *177*

Atomic mass, 84, *84*

Atomic number, 83

Atomic theory of matter, 75

Aurora, 625, *625*

Automobiles, air bags in, 332, *332;* hybrid, 381, *381,* 398, *398;* internal combustion engines in, 446, *446,* 446 *act,* 447; safety in, 300–301 *lab,* 332, *332*

Average speed, 285, *285,* 285 *lab*

Awiakta, Marilou, 182

Axle. *See* Wheel and axle

B

Bacteria, and food, 203

Baking soda, 192, 193, 194

Balaban, John, 364

Balance, 15, *15*

Balanced chemical equations, 195, 196 *act*

Balanced forces, 311, *311*

Balanced pressure, 118, *118*

Balloon races, 329 *lab*

Bar graph, 58, *58,* 59

Barium sulfate, 227

Barometer, 347, *347,* 362–363 *lab*

Barometric pressure, 362–363 *lab.* *See also* Atmospheric pressure

Base(s), 235; chemical properties of, 142, *142;* common, 141; measuring strength of, 150–151 *lab,* 236, 236–238, *237,* 240–241 *lab;* neutralizing, 238–239, *239;* physical properties of, 138, *138,* 138 *act;* properties of, 235; reaction with acids, 142, 238–239, *239;* uses of, 235, *235*

Bats, and echolocation, 495, *495*

Batteries, chemical energy in, 593, *593;* in electrical circuit, 592, *592;* life of, 593; salts in, 142, *142*

Beats, 506, *506*–507

Beeswax, 267

Benzene (cyclohexene), 256, *256,* 256 *act*

Bernoulli, Daniel, 359

Bernoulli's principle, 359, *359,* 359 *lab*

Beryllium, 165

Bicycles, 417, *417*

Binary system, 650, 650 *lab,* 651

Biological compounds, 262–269; carbohydrates, *264,* 264–266, *265;* lipids, *267,* 267 *act,* 267–269, *268, 269;* polymers, *262,* 262–263, *263;* proteins, 260, *260,* 263, *263,* 263 *act*

Biomechanics, 311

Bionics, 426

Birds, flight adaptations of, 361; how birds fly, 324 *act*

Bit, 650

Black holes, 317

Blood, as mixture, 89, *89,* 90

Blood pressure, 123 *act*

Boats, reason for floating, 350, *350,* 354, *354*

Body parts, artificial, 426, *426*

Body temperature, 383, 415, 530

Boiling point, 112, *112;* of solvent, 230

Bond(s), 170–179; with carbon, 249 *lab,* 250; chemical, **169,** *169;* covalent, **173,** 173–174, *174, 224, 224;* double, 174, *174,* 253, *253;* in hydrocarbons, 251, *251, 252, 252;* ionic, *170,* 170–172, **171,** *171, 172,* 225; metallic, **172,** *172;* polar, **174,** *174;* triple, 174, *174,* 175, 253, 254, *254*

Boomerangs, 302, *302*

Brass, 223

Brass instruments, 505, *505*–506

Bromine, 166

Bubbles, 192, *192*

Building materials, insulators, 442, *442*

Buoyant force, 120, 120–121, *121,* 124–125 *lab,* **348,** *348*–355; and Archimedes' principle, *351,* 351–354, *354;* cause of, 348–349; changing, *350,* 350–351, *351;* and depth, 351, *351;* measuring, 355 *lab;* and shape, 350, *350;* and unbalanced pressure, 349, *349*

Burning, 74, *74,* 190, *190,* 197, *197,* 201, *201*

Butadiene, 253, *253*

Butane, 252, *252,* 253, 254, *254,* 255

Index

Butyl hydroxytoluene (BHT), 204
Byte, 651

Calcium carbonate, 142
Calcium hydroxide, 141, 235, 235 *act*
Calculators, 641 *lab*
Cameras, 570, *570*
Cancer, 529
Carbohydrates, 264, **264**–266, *265*
Carbon, bonding of, 249 *lab*, 250; compounds involving. *See* Biological compounds; Hydrocarbons; Organic compounds
Carbonated beverages, 222, 228, 233 *lab*
Carbon dioxide, 88; chemical formula for, 194; covalent bond in, 174, *174*; in solution, 222, 228
Carbonic acid, *137*, 141, 233, 237
Carbon monoxide, 88
Carbon tetrachloride, 257
Carboxyl group, 259, *259*
Carboxylic acids, 259, *259*
Carlsbad Caverns, 148
Carnivores, 419
Carrier wave, 536, *536*
Car safety testing, 300–301 *lab*
Catalysts, 205, **205**–206, *206*
Catalytic converters, 205, *205*
Cathode rays, 76, *76*
Cave(s), formation of, 148, *148*; stalactites and stalagmites, 221, *221*, 233; stalactites and stalagmites in, 148, *148*
Cell(s), nerve, 585, *586*
Cello, 504, *504*
Cell phones, 542, *542*
Cellulose, 266, *266*
Celsius, 54
Celsius scale, *435*, **435**–436, 436 *act*
Central processing unit (CPU), 655, *655*
Centrifuge, 89
Chadwick, James, 78
Chalk, 142, 235

Changes, chemical, 145 *lab*, 145–149, 146 *act*, 149 *lab*; physical, *143*, **143**–144, *144*
Charge, electric, *584*, 584–590, *585*, *588*; flow of, *591*, 591–593, *592*, *593*; induced, 589, *589*; static, **585**, *585*
Chemical bonds, **169**, *169*
Chemical changes, **145**–149, 189 *lab*, 190, *190*, 192 *lab*, 207 *lab*; chemical weathering, 148, *148*; comparing, 145 *lab*; examples of, 145, *145*; formation of new materials by, 146; in nature, 147, 147–148, *148*; physical weathering, 147, *147*; signs of, 146, 146 *act*, 149 *lab*
Chemical energy, **377**, *377*, 381, *382*, 593, *593*
Chemical equations, **192**–194, 195, 195 *act*; balanced, 195, 196 *act*; energy in, 199, *199*
Chemical formulas, 88, *177*, **178**, 194
Chemical names, 192, 193
Chemical plant, 188, *188*
Chemical processes, 218
Chemical properties, **139**–142; of acids, 141, *141*; of bases, 142, *142*; choosing materials for, 140, *140*; common, 139, *139*; and pools, 140, *140*
Chemical reactions, **188**–209, *190*, *191*, 415; describing, 192–193; endothermic, **197**, 208–209 *lab*; energy in, 196–199, *197*, *198*; exothermic, **197**, *198*, 208–209 *lab*; heat absorbed in, 198, *198*; heat released in, 197–198; identifying, 189 *lab*, *192*; rates of, *202*, 202–206, *203*, *204*; slowing down, 204, *204*; speeding up, 205, 205–206, *206*; and surface area, 204, *204*
Chemical weathering, 148, *148*
Chemist, 10, 11
Chlorine, 140, *170*, 170–171, *171*, 174, *174*; isotopes of, 84, *84*; in substituted hydrocarbons, 257, *257*
Chlorofluorocarbons (CFCs), 530

Chloroform (trichloromethane), *257*
Chloromethane, 257
Chlorophyll, 193
Cholesterol, **269**, *269*
Circle graph, 58, *58*
Circuit, **591**, 596–605; electric energy in, 592, *592*; integrated, **647**, *647*, 649, 652, *652*; parallel, **599**, *599*, 603 *lab*; protecting, 599, *599*; resistance in, *594*, 594–595, *595*, *596*, *596*, 597; series, **598**, *598*; simple, *591*, 591–592, *592*, 598 *lab*
Circuit breakers, 599, *599*
Circular motion, 320–321, *321*
Cities, heat in, 452, *452*
Citric acid, 137, *137*, 233, 259
Classification, of electromagnetic waves, 525; of elements, 81, *81*, *82*; of matter, 133 *lab*; of parts of a system, 8 *lab*; of properties, 136 *lab*
Clock, analog, 643
Coal, 388, *388*
Cochlea, 509, *509*
Collisions, 281 *lab*, 295, 295–298, *296*, *297*, *298*, 299 *lab*, 300–301 *lab*
Color, 551 *lab*, 552, 552–554, *553*, *554*; and light, 470
Color pigments, 554, *554*
Combustion, 74, *74*
Communicating Your Data, 30, 32, 55, 61, 86, 93, 115, 125, 149, 151, 179, 181, 207, 209, 231, 241, 261, 271, 299, 301, 329, 331, 355, 363, 386, 397, 411, 425, 444, 451, 472, 481, 500, 511, 534, 541, 561, 573, 603, 605, 620, 633, 648, 661
Communication, of data, 55, 56–59, 61; by radio, 526, *535*, 535–536, *536*, 537 *act*; by satellite, 533, *533*, 538 *act*, 538–539; in science, 17, *17*; telecommunications, 535, 535–536, *536*; by telephone, 537–538, *538*, 542, *542*; through models, 24
Compass, 614, 619, *619*, 619 *act*, 620 *lab*

INDEX 733

Index

Compound(s), *87*, **87**–88, *88*, **171**, 218. *See* Biological compounds; Hydrocarbons; Organic compounds; comparing, 88 *lab*; ionic, 170–172, 179 *lab*; molecular, 224, 226, *226*; symbols for, 177, *177*
Compound machines, 417, *417*
Compound microscope, 567, *567*
Compression, 465, *465*, 491, *491*
Compressional waves, 464, **465**, 467, 468, *468*, 472 *lab*, 490, 490–491, *491*
Compressor, 448, *448*
Computer(s), 649–661; binary system in, 650, 650 *lab*, 651; and digital signals, 643–644, *644*; disposing of, 654; early, 649, *649*; and floppy disks, 658; hard disk of, 656, *656*, 656 *act*, 657; hardware of, 654–655, *654–655*; memory in, 651–652, *652*, 653 *lab*, 656; microprocessors in, 655, *655*; networks of, 659; software (programs) for, 652 *act*, **653**, *653*; storing information on, 651 *act*, 652, 656, 656–658, *657*, *658*; viruses affecting, 660–661 *lab*
Computer information, 650–651, 651 *act*
Computer models, 21, *21*
Computer programming, 653
Concave lens, 558 *act*, *559*, **565**, *565*
Concave mirror, 558, *558*, *559*, 569, *569*
Concentration, 203, 229 *act*, **229**–230; comparing, 229 *act*; measuring, 229–230, *230*; and rate of reaction, 203, *203*
Conclusions, 16; evaluating, 28
Condensation, *110*, **113**, *113*, 113 *act*, 144
Conduction, 439, *439*
Conductivity, of metalloids, 645–646, *646*
Conductor, 441, 588, *588*, 588 *act*
Conservation, of energy, **380**, 387, 395, 445; of mass, 194, *194*, 194 *lab*, 196; of matter, **74**; of momentum, 295, 295–298, *296*, *297*

Constant, 18
Constant speed, 285, *285*
Constructive interference, 477, *478*
Convection, 440, 440–441, *441*, 441 *lab*
Converging lens, 564
Convex lens, 564, *564*, 564 *act*, 570, *570*, 572–573 *lab*
Convex mirror, 558, 560, *560*
Coolant, 448, *448*, 449, *449*
Cooling, 444 *lab*
Copper, 200, *200*
Copper wire, 594
Cornea, 548
Covalent bond, *173*, **173**–174, *174*, 224, *224*
CPU (central processing unit), 655, *655*
Crankshaft, 446
Crash test dummies, 24, *24*
Crest, 464, *464*
Critical angle, 565, *565*
Critical thinking, 26
Crystal, 103, *103*; ionic, 175, *176*; molecular, 175, *176*; from solution, 231 *lab*; structure of, *176*
Crystalline solids, 103, *103*, 109
Crystallization, 220
Cubic meter, 52, *52*
Current(s), electric. *See* Electric current
Cycles, water, 56, *56*, 115 *lab*
Cyclohexane, 256, *256*, 256 *act*
Cyclohexene (benzene), 256, *256*, 256 *act*
Cyclopentane, 255
Cylinders, 446, *447*

Dalton, John, 75, 82
Data, communicating, 17, *17*, 30, 32; evaluating, 27–28; organizing, 15; repeatable, 28
Data communication, 55, 56–59, 61
Data Source, 362, 396, 660
Data table(s), 17 *act*, 57, 58 *act*
Decane, 255
Decibel scale, 493, *493*

Democritus, 73, 94
Density, 121, 121 *act*, **134**, 135, *135*, 135 *act*, **352**–353; calculating, 352 *act*; equation for, 352; and floating, 353, *353*; and sinking, 353, *353*
Dependent variable, 18
Depth, and buoyant force, 351, *351*; and pressure, 344
Desalination, 89 *act*, 219 *act*
Design Your Own, Car Safety Testing, 300–301; Comparing Thermal Insulators, 450–451; Design Your Own Ship, 124–125; Exothermic or Endothermic?, 208–209; Homemade pH Scale, 150–151; Modeling Motion in Two Directions, 330–331; Music, 510–511; Pace Yourself, 60–61; Pulley Power, 424–425; Spectrum Inspection, 540–541; Wave Speed, 480–481
Destructive interference, 477, *477* act, 478, 479
Diamond, 210, *210*
Dichloromethane, 257
Diffraction, 475; of light, 475–476; and wavelength, 476, *476*; of waves, 475–476, *476*, 498, *498*
Digital information, 650–651, 651 *act*
Digital signal, 643–644, *644*
Digitization, 644, *644*
Digits, number of, 48; significant, 49, *49*
Diodes, 646, *646*, 648 *lab*
Direct current (DC), 627
Direction, changing, 414, *414*; of force, 407, *407*, 414, *414*
Disk(s), floppy, 658; hard, 656, *656*, 656 *act*, 657; magnetic, 656, *656*, 656 *act*, 657; optical, 658, *658*, 658 *act*
Displacement, and distance, 283, *283*
Dissolving, 220, 225, *225*
Dissolving rates, 217 *lab*, 227. *See also* Solubility; Solution(s)
Distance, changing, 414, *414*; and displacement, 283, *283*; and work, 408, 414, *414*
Distance-time graph, 286, *286*

Dolphins

Dolphins, 482, 495
Domain, magnetic, **616,** *616*
Doping, 646, *646*
Doppler effect, 496, *497,* 499
Doppler radar, 496 *act*
Doppler shift, 496
Double bonds, 174, *174,* 253, *253*
Drawings, scale, 55 *lab;* as scientific illustrations, 56, *56*
Drum, 505, *505*
Dry ice, 114, *114*
Ductility, 84

Ear, *508,* 508–509, *509*
Eardrum, 508, *508*
Ear protectors, 479, *479*
Earth, gravitational field of, 521, *521;* magnetic field of, *617,* 617–619, *618;* magnetosphere of, 617, *617,* 625, *625*
Earthquakes, damage caused by, 502
Earth science, 10, *10*
E-books, 662, *662*
Echoes, 495, *495,* 498
Echolocation, 482, 495, *495,* 527
Efficiency, 415–416, *416;* calculating, 415 *act;* equation for, 415; and friction, 416
Einstein, Albert, 22
Electrical energy, 378, 383, *383, 384,* 384–385, *385,* 445
Electric charge, 584, 584–593, *585,* 588
Electric circuit. *See* Circuit
Electric current, 591–597, 603 *lab;* controlling, 596, 596–597; effect on body, 602; generating, *626,* 626–627, *627;* and magnetism, 621–629, 631; model for, 604–605 *lab;* in a parallel circuit, 603 *lab;* and resistance, 594, 594–595, *595,* 596, *596,* 597; types of, 626, 627
Electric discharge, 589, *589*
Electric energy, in circuit, 592, *592;* cost of, 601, 601 *act;* and resistance, 594, 594–595, *595,* 596, *596,* 597

Electric field, 522, *522,* 523 *lab,* **587,** *587*
Electric forces, 583 *lab,* **587,** *587,* 592 *lab*
Electricity, 582–605; connecting with magnetism, 631; consumption of, 390, 390 *act,* 396–397 *lab;* generating, *384,* 384–385, *385,* 390 *act,* 391, *626,* 626–627, *627;* power failure, 398; safety with, 601–602
Electric meter, 601, *601*
Electric motors, 624, *624,* 632–633 *lab*
Electric power, 600–601
Electric shock, 601–602
Electric wire, 588, 594, *594*
Electromagnet(s), *621,* **621**–622, 622, 622 *lab*
Electromagnetic rays, X rays, 520, 530–531, *531,* 532
Electromagnetic spectrum, 468, *468,* 525, **525**–533, 540–541 *lab;* gamma rays, 530, 531; infrared waves, 527 *lab,* 527–528, *528, 532;* radio waves, 526, 526–527, *527, 532,* 535, 535–536, *536,* 537 *act,* 538, *538;* ultraviolet waves, *529,* 529–530, *532;* views of universe through, 531–533, *532, 533;* visible light, 528, *528, 529;* X rays, 520, 530–531, *531, 532*
Electromagnetic waves, 466, 467, *468, 468,* 471, 518–541, **520,** *520;* classifying, 525; frequency of, 523; making, *522,* 522–523, *523;* properties of, *523,* 523–524, *524;* in telecommunications, 535, 535–536, *536;* using, 535–539; wavelength of, 523, *523,* 537 *act,* 540–541 *lab*
Electron(s), 76, 164 *act,* 584, 584–585, *585,* 592; arrangement of, *163,* 163–164, *164;* energy levels of, *163,* 163–165, *164, 165;* in magnetic fields, 616, *616;* model of energy of, 161 *lab;* movement of, 162, *162*
Electron cloud, 162, *162*
Electron cloud model of atom, 78–79, *79*

Electron dot diagrams, 168, 168 *lab,* **168**–169
Electronic books, 662, *662*
Electronic devices, 641 *lab,* 642–643, *643,* 645, *645*
Electronics, *642,* 642–648; analog devices, 642–643, *643;* calculator competition, 641 *lab;* diodes in, 646, *646,* 648 *lab;* integrated circuits in, 647, *647,* 649, 652, *652;* microprocessors, 655, *655;* semiconductors, 645 *act,* 645–646, *646,* 647; transistors in, 647, *647*
Electronic signal, 642; analog, 642, *644;* digital, **643**–644, *644*
Element(s), 80–86, 218; atomic mass of, 84, *84;* atomic number of, 83; atomic structure of, *163, 163,* 180–181 *lab;* classification of, 81, *81, 82;* halogen family of, 166, *166;* identifying characteristics of, 83–84; identifying properties of, 167 *act;* isotopes of, 83, *83,* 84, *84;* metalloids, 85, 645–646, *646;* metals, 84, *84;* noble gases, 166, *166;* nonmetals, 85, *85;* periodic table of, 81, *82, 83,* 86 *lab,* 164–165, *165,* 166, *166;* synthetic, 80, *80,* 81 *act*
Elephant(s), 134
Elevation, and pressure, 346
Elvin-Lewis, Memory, 272, *272*
Endangered species, 57, 57–59, *58, 59*
Endothermic reactions, 197, 208–209 *lab*
Energy, 372–398, 373 *lab,* **374,** *374,* 380 *act;* activation, **201,** *201;* alternative sources of, 391–393, 391 *lab, 392, 393;* and amplitude, 467–468, 493, *493;* from carbohydrates, *264,* 264–266, *265,* 266 *act;* chemical, **377,** *377,* 381, *382,* 593, *593;* in chemical reactions, 196–199, *197, 198;* conservation of, **380,** 387, 395, 445; consumption of, 390, 396–397 *lab;* electric. *See* Electric energy; electrical, **378,** 383, *383, 384,* 384–385, *385,* 445;

Index

Energy levels

in equations, 199, *199;* forms of, *376,* 376–378, *377, 378,* 445; from fossil fuels, 388, *388;* geothermal, 392–393, *393;* kinetic, **375,** *375,* 380, *380,* 381, *382;* mechanical, 445, *445, 447;* of motion, 375, *375;* nuclear, **378,** *378,* 389, *389,* 445; potential, **376,** *376,* 380, *380;* and power, 410; radiant, **377,** *377,* 445, 466, **524;** solar, 391–392, *392, 393,* 398, *398;* sources of, 264, 385, 387–395, 396–397 *lab;* storage in lipids, 267; storing, 384, *384;* thermal, 107, **107**–108, **376,** *376,* 383–385, *384.* See Thermal energy; tidal, 394, *394;* transfer of, *374;* transferring, 520; types of, 108; using, 387; and waves, 460, 461 *lab,* 462, *462;* wind, 393, *393;* and work, 410

Energy levels, of electrons, *163,* 163–165, *164, 165*

Energy transformations, *379,* 379–385, 386 *lab;* analyzing, 381 *lab;* chemical energy, 381, *382;* efficiency of, *381;* electrical energy, 383, *383;* between kinetic and potential energy, 380, *380;* thermal energy, 383–384, *384;* tracking, 379

Engines, 445, **445**–447, *446,* 446 *act, 447*

ENIAC, 649

Environment, acid in, 233, *234*

Enzymes, 206; as catalysts, 205–206, *206*

Equation(s). See Chemical equations; acceleration, 290; for density, 352; for efficiency, 415; for mechanical advantage, 413; one-step, 135 *act,* 341 *act,* 408 *act,* 409 *act,* 413 *act,* 415 *act;* for power, 409; simple, 319 *act,* 436, 537 *act,* 597 *act,* 600 *act;* for work, 408

Eruptions, volcanic, 62

Essential amino acids, 263

Estimation, 43–44, *44*

Ethane, 251, *251,* 253, *255,* 258

Ethanoic (acetic) acid, 259, *259*

Ethanol, 258

Ethene (ethylene), 253, *253, 255,* 262, *262*

Ethylene (ethene), 253, *253, 255,* 262, *262*

Ethyne (acetylene), 254, *254, 255*

Etna, Mount (Italy), *372*

Evaluation, 26–29; of conclusions, 28; of data, 27–28; of promotional materials, 29, *29;* of scientific explanation, 26–29, *28,* 30 *lab*

Evaporation, *110, 112,* 112 *lab,* 112–113, *144*

Exhaust valve, *447*

Exothermic reactions, 197, *198,* 208–209 *lab*

Expansion, thermal, 435, *435*

Experiments, 18, 18–19

Explanations, scientific, 26–29, *28,* 30 *lab*

Eye, 548

Eyeglasses, 574, *574*

Eyepiece lens, 567

Fahrenheit, 54

Fahrenheit scale, 435, 435–436, 436 *act*

Fat(s), dietary, **267,** 267 *act,* 267–269, *268*

Field(s), electric, **522,** *522,* 523 *lab,* **587,** *587;* gravitational, **521,** *521;* magnetic, 521, *522.* See Magnetic field(s)

Filaments, 166, 595

Fire. See Wildfires; chemical changes caused by, 190, *190, 191*

Fireworks, 200, *200*

First-class lever, *421*

Fixed pulleys, 422, *423*

Flammability, 139

Flight, 324 *act,* 359–361, *360, 361*

Floating, 349, *349.* See also Buoyant force; and Archimedes' principle, 354, *354;* and boats, 350, *350,* 354, *354;* and density, 353, *353*

Floppy disk, 658

Flower(s), 62, *62*

Fluid(s), 116–123, *343,* **343**–347, 356–361. See Liquid(s). See also Gas(es); and Archimedes' principle, 121, *121,* 124–125 *lab;* and buoyant force, *120,* 120–121, *121,* 124–125 *lab;* and density, 121, 121 *act;* and Pascal's principle, *122,* 122–123, *123;* and pressure, 116–120; pressure in, 343–345, *344, 345,* 356, *356,* 358, *358*

Fluid forces, 356–361; and Bernoulli's principle, 359, *359,* 359 *lab;* and hydraulic systems, *357,* 357–358, 358 *act;* increasing, 358; and Pascal's principle, 357, *357;* using, 356, *356*

Fluorine, 166, *166*

Flute, 505, 506, *506*

Focal length, 558, 571

Focal point, 558, 559

Foldables, 5, 41, 71, 101, 133, 161, 189, 217, 249, 281, 309, 339, 373, 405, 433, 461, 489, 519, 549, 583, 613, 641

Food(s), and bacteria, 203; carbohydrates in, **264,** *264;* for energy, *264,* 264–266, *265,* 266 *act;* lipids in, **267,** 267 *act;* proteins in, 263; reaction rates in, 202, *202,* 203, 204

Footprints, interpreting, 342 *lab*

Force(s), 116, **310,** 310–311. See Buoyant force; Fluid forces; Pressure; and acceleration, *316,* 316–317, 320; action and reaction, 323–326, *324, 326;* and area, 117, *117;* balanced, **311,** *311;* buoyant, *120,* **120**–121, *121,* 124–125 *lab;* changing, 413; combining, 311; comparing, 405 *lab;* direction of, 407, *407,* 414, *414;* effects of, 309 *lab;* electric, 583 *lab,* **587,** *587,* 592 *lab;* input, **412,** *412;* magnetic, 613 *lab;* measurement of, 116; net, **311,** 320; output, **412,** *412;* and pressure, 116–120; unbalanced, **311;** unit of measurement with, 317; and work, 405 *lab,* 407, *407,* 411 *lab,* 413

Force field   **Inertia**

Force field, *521, 521 act, 521–522, 522*
Force pairs, measuring, *327 lab*
Force pumps, 123, *123*
Forests, and wildfires, 606, *606*
Formic (methanoic) acid, 259, *259*
Formulas, chemical, 88, *177, 178,* 194
Fossey, Dian, 9, *9*
Fossil fuels, natural gas, 251, *251,* 252; oil (petroleum), 252; as source of energy, 388, *388*
Four-stroke cycle, 446, *447*
Free fall, 327, *327,* 328
Freezing, 101 *lab,* 110, **111**
Freezing point, 111, 111 *act*; of solvent, 230
Frequency, *469,* **469**–470, *470,* 523, 542, *542;* fundamental, **503,** *503;* natural, **501**–502; of sound waves, 491, 494, *494,* 501–502, 503, *503*
Frequency modulation (FM), 536, 537 *act*
Friction, 312, **312**–315, 314 *lab,* 416, *416;* rolling, 315, *315;* sliding, 313, 314, *314,* 315, 319, 322; static, 314
Fructose, 265, *265*
Fulcrum, 420, *421*
Fundamental frequency, 503, *503*
Fuses, 599, *599*

Galilei, Galileo, 312, 313 *act*
Gallium, 646
Galvanometer, 622, *623*
Gamma rays, 530, 531
Gas(es), 106, *106,* 136, *144. See also* Fluids; condensation of, *110,* 113, *113,* 113 *act;* natural, 251, *251,* 252; noble, 166, *166;* pressure of, *119,* 119–120, *120;* solubility of, 228, 228 *lab*
Gaseous solutions, 221, 228 *lab*
Gas-gas solution, 222
Gasoline, *139,* 387, 388
Gasoline engines, 381, *381*
Generator, 384, *384,* 626, **626**–627
Geologist, 10

Geothermal energy, 392–393, *393*
Germanium, 645
Glass, 109, *109*
Global Positioning System (GPS), 466, **539,** *539*
Glucose, 264–266, *265*
Glycerol, 267
Glycine, 260, *260,* 263
Glycogen, 266
Gold, 140, *140*
Gram, 50
Graph(s), 57, **57**–59, *58,* 58 *act,* 59; of accelerated motion, 292, *292;* distance-time, 286, *286;* of motion, 286, *286;* speed-time, 292, *292*
Gravitational field, of Earth, 521, *521*
Gravity, 317–318; and acceleration of objects, 5 *lab;* and air resistance, 321; and motion, 317, 320, *320,* 321, *321*
Great Salt Lake, 242, *242*
Grounding, 590, *590*
Guessing, 13
Guitars, 504

Halogens, 166, *166*
Hammer (of ear), 508, *508*
Hard disk, 656, *656,* 656 *act,* 657
Hardware, computer, 654–655, *654–655*
Health integration systems, 9
Hearing, *191,* 508, 508–509, *509*
Hearing loss, 509
Heart, 123, *123*
Heart disease, 268, *269*
Heat, 108, 438–444; in chemical reactions, 197–198, *198;* conduction of, 439, *439;* convection of, 440, 440–441, *441,* 441 *lab;* radiation of, 439; specific, 109, *109,* **442;** and temperature, 108–109; and thermal energy, 438–441; transfer of, 438, *438*
Heat engines, *445,* **445**–447, *446,* 446 *act, 447*
Heating, 444 *lab*

Heat island, 452, *452*
Heat pumps, 393, 449, *449*
Helium, 165, *166*
Hemoglobin, 263
Heptane, *255*
Herbivores, 419
Hertz (Hz), 469
Heterogeneous mixture, 91, **219**
Hexane, *255*
Homogeneous mixture, 91, **220,** *220*
Hurricanes, 467, *467*
Hybrid cars, 381, *381,* 398, *398*
Hydraulic systems, 122, *122, 357,* **357**–358, 358 *act*
Hydrocarbons, 251–256; bonds in, 251, *251,* 252, *252;* isomers of, 254, *254,* 254 *lab;* in rings, 256, *256;* saturated, **252,** *252;* structures of, 253; substituted, *257,* 257–260, *258, 259, 260,* 261 *lab,* 263, *263;* unsaturated, **253,** 253–254
Hydrochloric acid, 141, 142, 233, 237, *237,* 238
Hydroelectric power, 390, *391*
Hydrogen, bonding with carbon, 250. *See also* Hydrocarbons; isotopes of, 83, *83*
Hydrogen chloride, 174, *174*
Hydrogen peroxide, 88, *88*
Hydronium ions, 232, *232,* 236, 237, 239, *239*
Hydroxide ions, 236, 237, 239, *239*
Hydroxyl group, 258, *258*
Hypothesis, 14; analyzing, 14; forming, 14 *lab;* testing, 15

Ice, dry, 114, *114*
Icebergs, 132, *132*
Idea models, 22
Illustrations, scientific, *56,* 56–57
Inclined plane, 417–419, *418*
Independent variable, 18
Indicators, 238, 238 *act,* 240–241 *lab*
Indonesia, volcanoes in, 62
Induced charge, 589, *589*
Inertia, 293, *293*

INDEX 737

Index

Inexhaustible resources, 391
Inference, 16, *16, 26*
Information, digital, 650–651, 651 *act;* storing on computers, 651 *act,* 652, 656, 656–658, *657, 658*
Infrared waves, 466, 527 *lab,* 527–528, *528, 532*
Inhibitor, 201 *lab,* **204,** *204*
Inner ear, *508,* 509, *509*
Input force, 412, *412*
Instantaneous speed, 285, *285*
Insulator(s), 442, *442,* 450–451 *lab,* **588,** *588*
Integrate Astronomy, Doppler Shift of Light, 496; Measurement Accuracy, 51; Plasma, 343; Pulsars and Little Green Men, 536; Satellite Observation, 533
Integrate Career, Biologist, 13; Computer Programmer, 634; Geologist, 148; Mechanical Engineering, 448; Naval Architect, 352; Nobel Prize Winner, 165; Nutrition Science, 264; Pharmacist, 229; Physicists and Chemists, 78; Precision and Accuracy, 43
Integrate Chemistry, Adding Impurities, 646; Alkaline Batteries, 593
Integrated circuits, 647, *647,* 649, 652, *652*
Integrate Earth Science, amplitude, 468; Energy Source Origins, 388; Hydrocarbons, 252; Rocks and Minerals, 91
Integrate Environment, Earthquake Damage, 502; Recycling Computers, 656; Solutions, 225; stalactites and stalagmites, 221
Integrate Health, Aging, 137; electrical safety, 601; Health Integration Systems, 9; magnetic resonance imaging (MRI), 630; precision and accuracy in medicine, 47; Ultrasonic Waves, 470

Integrate History, Atomism, 73; Breathe Easy, 205; The Currents War, 529; Freshwater, 104; James Prescott Joule, 408; The Ohm, 594
Integrate Life Science, Animal Insulation, 442; Autumn Leaves, 193; Biomechanics, 311; bird's wings, 361; Body Temperature, 415, 530; chemical properties and pools, 140; Controlling Body Temperature, 383; Echolocation, 495; flight, 324; heart force, 123; nature's magnets, 618; Newton and Gravity, 317; pH Levels, 236; thermal pollution, 443; transforming chemical energy, 381; wedges in your body, 419
Integrate Physics, Earth's magnetic field, 634; Global Positioning System (GPS), 466; Ions, 171; Light Waves and Photons, 557; Types of Energy, 108
Integrate Social Studies, Forensics and Momentum, 294
Interference, 477 *act,* **478,** 478–479, *479*
Internal combustion engines, 446, *446,* 446 *act, 447*
International System of Units (SI), 50–54, 55 *lab*
International Union of Pure and Applied Chemistry (IUPAC), 255
Internet, 659, *659,* 660–661 *lab. See* Use the Internet
Investigations, identifying parts of, 31–32 *lab*
Iodine, *168*
Ion(s), 171, *171,* 225, **584,** 585, *585*
Ionic bond, 170, 170–172, **171,** *171, 172,* 225
Ionic compounds, 170–172, 179 *lab*
Ionic crystal, 175, *176*
Iron, 140, *140*
Isobutane, 254, *254*
Isomer(s), 254, *254*
Isopropyl alcohol, 258
Isotopes, 83, *83,* 84, *84*

Jansky, Karl G., 482
Joule, James Prescott, 415
Journal, 4, 28, 40, 70, 100, 132, 160, 188, 216, 248, 280, 308, 338, 372, 404, 432, 460, 488, 518, 548, 582, 612, 640

Kashyapa, 73, 94
Keck telescope, 569, *569*
Kelvin, 54, *54*
Kelvin scale, 436
Kilogram (kg), 50, *53,* 317
Kilometer, 51, *51*
Kilopascal (kPa), 116
Kilowatt-hour (unit of electric energy), 601
Kinetic energy, 375; and mass, 375, *375;* and speed, 375, *375;* transforming chemical energy to, 381, *382;* transforming to and from potential energy, 380, *380*
Krakatau volcano (Indonesia), 62

Lab(s), Balloon Races, 329; Building the Pyramids, 411; Collisions, 299; Conversion of Alcohols, 261; Current in a Parallel Circuit, 603; Design Your Own, 60–61, 124–125, 150–151, 208–209, 300–301, 330–331, 424–425, 450–451, 480–481, 510–511, 540–541; Elements and the Periodic Table, 86; Energy to Power Your Life, 396–397; Growing Crystals, 231; Hearing with Your Jaw, 386; Heating Up and Cooling Down, 444; How does an electric motor work?, 632–633; Identifying Parts of an Investigation, 32–33; Image Formation by a Convex

738 STUDENT RESOURCES

Index

Laboratory balance

Lens, 572–573; Investigating Diodes, 648; Ionic Compounds, 179; Launch Labs, 5, 41, 71, 101, 133, 161, 189, 217, 249, 281, 309, 339, 373, 405, 433, 461, 489, 519, 549, 583, 613, 641; Looking for Vitamin C, 270–271; Make a Compass, 620; Measuring Buoyant Force, 355; Mini Labs, 23, 44, 74, 112, 145, 168, 192, 194, 228, 233, 263, 291, 327, 342, 391, 422, 441, 465, 504, 527, 568, 598, 618, 653; Model and Invent, 180–181; Model for Voltage and Current, 604–605; Mystery Mixture, 92–93; Observe and Measure Reflection of Sound, 500; Observing Gas Solubility, 231; Physical or Chemical Change?, 207; Prisms of Light, 534; Reflection from a Plane Mirror, 561; Scale Drawing, 55; Sunset in a Bag, 149; Testing pH Using Natural Indicators, 240–241; Try at Home Mini Labs, 8, 14, 52, 88, 119, 136, 173, 204, 228, 254, 285, 314, 359, 381, 409, 440, 474, 492, 523, 551, 592, 622, 650, 653; Use the Internet, 362–363, 396–397, 660–661; The Water Cycle, 115; Waves on a Spring, 472; What is the right answer?, 31

Laboratory balance, 15, *15*
Laboratory safety, 19, *19*
Laboratory scale, 47, *47*
Lactic acid, 259
Lamarr, Hedy, 542, *542*
Land speed, 286 *act*
Lasers, 570–571, *571*
Launch Labs, Analyze a Marble Launch, 373; Bending Light, 549; Classifying Different Types of Matter, 133; Compare Forces, 405; Detecting Invisible Waves, 519; Electronic and Human Calculators, 641; Experiment with a Freezing Liquid, 101; Forces Exerted by Air, 339; Forces and Motion, 309; Identify a Chemical Change, 189; Magnetic Forces, 613; Making Human Sounds, 489; Measuring Accurately, 41; Measuring Temperature, 433; Model Carbon's Bonding, 249; Model the Energy of Electrons, 161; Motion After a Collision, 281; Observe How Gravity Accelerates Objects, 5; Observe Matter, 71; Observing Electric Forces, 583; Particle Size and Dissolving Rates, 217; Waves and Energy, 461

Lavoisier, Antoine, 74, *82*, 194
Law(s), of conservation of energy, **380,** 387, 445; of conservation of mass, 194, *194*, 194 *lab*, 196; of conservation of matter, **74;** of conservation of momentum, *295*, **295**–298, *296*, *297*; Newton's first law of motion, **312**–315, *315*; Newton's second law of motion, **316**–322, *325*; Newton's third law of motion, **323**–328, *325*, 329 *lab*; Ohm's, **597;** of reflection, **555,** *555*; scientific, **7**
Leaves, changing colors of, 147, *147*, 193
Length, measuring, 41 *lab*, 47, 50, 51, *51*
Lens, **563**–565; in camera, 570, *570*; concave, 558 *act*, *559*, **565,** *565*; converging, 564; convex, **564,** *564*, 564 *act*, 570, *570*, 572–573 *lab*; eyepiece, 567; forming an image with, 568 *lab*; in microscopes, 567, *567*; objective, 567; in reflecting telescopes, 569, *569*; in refracting telescopes, 568, *568*
Lever, 420, *420*, 421, *421*
Levi-Montalcini, Rita, 33, *33*
Life science, 9, *9*
Life scientist, 9, *9*
Light, bending of, 549 *lab*; and color, 470; diffraction of, 475–476; Doppler shift of, 496; energy of, 377, *377*; invisible, 519 *lab*; and matter, *551*, 551–552, *552*;

Magnet(s)

prisms of, 534 *lab*; properties of, 550–554; reflection of, 473, *473*, 551, *551*, 555–561, *556*, 561 *lab*; refraction of, 474, 474 *lab*, 474–475, *475*, 563, *563*; scattering of, 556; speed of, 471, 524, *524*, 562, *562*; ultraviolet, 466; visible, 466, 468, *468*, **528,** *528*, 529
Lightbulb, 166
Lightning, 471, 606
Lightning rod, 590, *590*
Light ray, 550, *550*
Light waves, 550, 550–551, *551*, 557
Lime, 235
Limestone, 148, *148*
Line graph, 57, *57*
Lipids, 267, 267 *act*, **267**–269, 268, 269
Liquid(s), 104, **104**–105, *105*, 136, *136*, 144. See also Fluids; freezing, 101 *lab*, 110, 111; layering, 352 *act*; and surface tension, 105, *105;* vaporization of, *110*, 112, 112 *lab*, 112–113; viscosity of, *105*
Liquid-gas solutions, 222, *222*
Liquid-liquid solutions, 222
Liter, 52
Lithium, 163, *163*, 165, 167
Loudness, 492–493, *493*
Luster, 84
Lye, 235

M

Machines, 412–425; compound, **417,** *417*; and efficiency, 415–416, *416;* and friction, 416, *416;* and mechanical advantage, 412, 412–414, 420, *420;* simple. *See* Simple machines
Maglev, 612, *612*, 613
Magnesium, 172, *172*
Magnesium hydroxide, 141, 238
Magnet(s), 614–616; electromagnets, 621, 621–622, *622*, 622 *lab;* poles of, 614, *614*, 615, 617 *act;*

INDEX **739**

superconductors, *629*, 629–630, *630*
Magnetic declination, 617 *act*
Magnetic disks, 656, *656*, 656 *act*, *657*
Magnetic domain, 616, *616*
Magnetic field(s), 521, *522*, 615, **615**–619; of Earth, *617*, 617–619, *618*; making, 616, *616*; observing, 618 *lab*
Magnetic field lines, 615, *615*
Magnetic force, 613 *lab*
Magnetic resonance imaging (MRI), *630*, 630–631, *631*
Magnetism, 612–631; early uses of, 614; and electric current, 621–629, *631*
Magnetite, 614
Magnetosphere, 617, *617*, 625, *625*
Malleability, 84, *84*
Map(s), 287, *287*; topographic, 21 *act*
Marble launch, analyzing, 373 *lab*
Mars Climate Orbiter, 51
Mass, 53, *53*, 134, 293, *293*; conservation of, 194, *194*, 194 *lab*, 196; and kinetic energy, 375, *375*; unit of measurement with, 317; and weight, 318
Mass number, 83
Materials, alloys, 223, *223*
Matter, 71 *lab*, **72, 102**; atomic theory of, 75; chemical changes in, *145*, 145–149, *146*, *147*, *148*, 149 *lab*; chemical properties of, *139*, 139–142, *140*, *141*, *142*; compounds, 87, 87–88, *88*; early beliefs about, 73, 94; elements in, 80–86, *82*, 86 *lab*; law of conservation of, **74**; and light, *551*, 551–552, *552*; and motion, 282; physical changes in, *143*, 143–144, *144*; physical properties of, 134–138, *135*, 135 *act*, *136*, 136 *lab*, *137*, *138*, 144, *144*; states of, **136**, *136*, **144**, 144; structure of, 72–79
Mayer, Maria Goeppert, 33, *33*
Measurement, 42–61, 47 *act*; accuracy of, 41 *lab*, 43, 44 *lab*, **45**–47, *46*, 51; of average speed, 285 *lab*; of buoyant force, 355 *lab*; of concentration, 229–230, *230*; v. estimation, 43–44, *44*; of force, 116; of force pairs, 327 *lab*; of length, 47 *act*, 50, *51*, 51; of loudness, 493, *493*; of mass, 53, *53*; precision of, 44–49, *45*, *46*, *49*; of reflection of sound, 500 *lab*; rounding, 47, 48 *act*; in SI, 50–54, 55 *lab*; of speed, 54, 60–61 *lab*; of strength of acids and bases, 150–151 *lab*, 236, 236–238, *237*, 240–241 *lab*; of temperature, 54, *54*, 435, 435–436; of time, 45, *45*, 54; units of, 50, 51, 52, 53, 54, 290, 317, 409, 594, 601; of volume, 52, *52*, 52 *lab*; of weight, 53, *53*, 327, *327*; of work, 410
Mechanical advantage, *412*, 412–414, **413**, 420, *420*
Mechanical energy, 445, *445*, 447
Mechanical waves, 463–465, *464*, 471
Medicine, magnetic resonance imaging (MRI) in, *630*, 630–631, *631*; plants as, 272, *272*; ultrasound in, 499, *499*; ultraviolet radiation in, 529, *529*; X rays in, 520, 531, *531*
Medium, 551
Melting, 109, *109*, 110, 144, *144*; comparing rates of, 440 *lab*
Melting point, 109
Memory, computer, 651–652, *652*, 653 *lab*, 656
Mendeleev, Dmitri, 82, 166
Metal(s), 84, *84*; alkali, 167, *167*; as conductors, 588
Metal alloys, 223, *223*
Metallic bond, 172, *172*
Metalloids, 85, 645; conductivity of, 645–646, *646*
Meteorologist, 10
Meter, 51; electric, 601, *601*
Methane, 173 *lab*, 251, *251*, 252, 253, *255*; substituted hydrocarbons made from, 257, *257*, 258, 259, 260
Methanoic (formic) acid, 259, *259*
Methanol, 258
Methylamine, 259, *259*

Methyl group, 251, *251*, 258, *258*
Microprocessor, 655, *655*
Microscopes, 567, *567*
Microwave(s), 526, *526*
Microwave oven, 526
Microwave tower, 526, *526*
Middle ear, 508, *508*
Mineral(s), as pure substances, 91
Mini Labs, Building a Solar Collector, 391; Comparing Chemical Changes, 145; Comparing Sounds, 465; Drawing Electron Dot Diagrams, 168; Forming an Image with a Lens, 568; Identifying Simple Circuits, 598; Interpreting Footprints, 342; Investigating the Unseen, 74; Measuring Accurately, 44; Measuring Force Pairs, 327; Modeling Acceleration, 291; Modeling a Stringed Instrument, 504; Observing Convection, 441; Observing the Law of Conservation of Mass, 194; Observing Magnetic Fields, 618; Observing a Nail in a Carbonated Drink, 233; Observing Pulleys, 422; Observing the Focusing of Infrared Rays, 527; Observing Vaporization, 112; Summing Up Proteins, 263; Thinking Like a Scientist, 23
Mining, of coal, 388
Mirror(s), concave, 558, *558*, *559*, 569, *569*; convex, 558, 560, *560*; plane, 557, *557*, 561 *lab*
Mixtures, 88, **89**, 89–93, 90 *act*, 219–220; heterogeneous, **219**; heterogeneous, 91; homogeneous, **220**, *220*; homogenous, 91; identifying, 92–93 *lab*; separating, 89, 90, 219, *219*
Model(s), 20–25; limitations of, 25, *25*; making, 22, *22*, 23; need for, 20, *20*; types of, *21*, 21–22; using, 24, *24*
Model and Invent, Atomic Structure, 180–181

Molecular compounds, 224, 226, *226*
Molecular crystal, 175, *176*
Molecules, 173, *173;* nonpolar, 175, *175,* 224, *224;* polar, *174,* 174–175, *175,* 175 *act,* 224, *224,* 226
Momentum, 294–298; calculating, 294 *act;* and collisions, 281 *lab,* 295, 295–298, *296, 297, 298,* 299 *lab,* 300–301 *lab;* conservation of, 295, 295–298, *296, 297*
Monomers, 262, *262*
Motion, 280, 282 *act,* 282–287, 308–331; and acceleration, 288–289, *316,* 316–317, 320; after a collision, 281 *lab;* and air resistance, 321; and changing position, *282,* 282–283; circular, 320–321, *321;* energy of, 375, *375;* and friction, *312,* 312–315; graphing, 286, *286,* 292, *292;* and gravity, 317, 320, *320,* 321, *321;* and matter, 282; modeling in two directions, 330–331 *lab;* and momentum, 294–298; Newton's first law of, **312**–315, *325;* Newton's second law of, **316**–322, *325;* Newton's third law of, **323**–328, *325,* 329 *lab;* on a ramp, 309 *lab;* relative, 283, *283;* and speed, 284–285, *285;* and work, 406, 406–407, *407*
Motors, electric, **624,** *624,* 632–633 *lab*
Movable pulleys, 423, *423*
MRI (magnetic resonance imaging), *630,* 630–631, *631*
Muriatic acid, 233
Muscle(s), transforming chemical energy to kinetic energy in, 381, *382*
Music, 501–507, 510–511 *lab*
Musical instruments, 503–507; brass, *505,* 505–506; modeling, 504 *lab;* and natural frequencies, 502; and overtones, 503, *503;* percussion, 505, *505;* strings, 504, *504,* 504 *lab;* tuning, *506,* 507; woodwinds, *505,* 505–506, *506*
Musical scales, 503, *503*

N

Names, chemical, 192, 193; of organic compounds, 253, 254, *255,* 256
National Geographic Unit Openers, How are Arms and Centimeters Connected?, 2; How are Charcoal and Celebrations Connected?, 158; How are City Streets and Zebra Mussels Connected?, 278; How are Cone-Bearing Trees and Static Electricity Connected?, 582; How are Radar and Popcorn Connected?, 458; How are Refrigerators and Frying Pans Connected?, 68; How are Train Schedules and Oil Pumps Connected?, 370
National Geographic Visualizing, Acid Precipitation, *234;* Chemical Reactions, *191;* Chemistry Nomenclature, 255; The Conservation of Momentum, *297;* Crystal Structure, *176;* The Doppler Effect, *497;* Energy Transformations, *382;* The Four-Stroke Cycle, *447;* The Hard Disk, *657;* Interference, *478;* Levers, *421;* The Modeling of King Tut, *24;* Nerve Impulses, *586;* Newton's Laws in Sports, *325;* The Periodic Table, *82;* Precision and Accuracy, *46;* Pressure at Varying Elevations, *346;* Reflections in Concave Mirrors, *559;* States of Matter, *110;* The Universe, *532;* Voltmeters and Ammeters, *623*
Natural frequencies, 501–502
Natural gas, 251, *251,* 252
Negative acceleration, 291, *291*
Negative charge, 584, *584*

Neon, 166, *166*
Nerve cells, 585, *586*
Net force, 311, 320
Network, computer, 659
Neurotransmitters, 586
Neutralization, 238, 238–239, *239*
Neutron(s), 78, *78,* 584, *584*
Newton (unit of force), 53, 116, 317
Newton, Isaac, 312, 313 *act*
Newton's first law of motion, 312–315, *325*
Newton's second law of motion, 316–322, *325;* and air resistance, 321; and gravity, 317–318; using, *318,* 318–320, *320*
Newton's third law of motion, 323–328, *325,* 329 *lab*
Nitric acid, 141, 233, 237
Nitrogen, 221, 222; electron dot diagram of, *168*
Noble gases, 166, *166*
Noise, 512, *512;* protection against, 479, *479*
Nonane, 255
Nonmetals, 85, *85;* noble gases, 166, *166*
Nonpolar molecules, 175, *175,* 224, *224*
Nonrenewable resources, 388, *388*
Northern lights, 625, *625*
Note-taking, 27
Novocaine, 259
N-type semiconductors, 646, 647
Nuclear energy, 378, *378,* 389, *389,* 445
Nucleus, 77, *77*
Nutrients, carbohydrates, *264,* 264–266, *265;* fats, *267,* 267 *act,* 267–269, *268;* proteins, 260, *260,* 263, *263,* 263 *lab;* vitamins, 270–271 *lab*
Nutrition, and heart disease, 268, 269, *269*

O

Objective lens, 567
Observation(s), 5 *lab,* 13, *13,* 16, 26, 27, 28; types of, 43

Index

Ocean(s), energy from, 394, *394*
Ocean water, 242, *242*; desalination of, 89 *act*; as solution, 225
Ocean waves, tsunamis, 468, 482
Octane, *255*
Ohm (unit of resistance), 594
Ohm's law, 597
Oil (petroleum), 252, 388
Oleic acid, *268*
Olympic torch, 201, *201*
One-step equations, 135 *act*, 341 *act*, 408 *act*, 409 *act*, 413 *act*, 415 *act*
Oops! Accidents in Science, Eyeglasses: Inventor Unknown, 574; The Incredible Stretching Goo, 126; What Goes Around Comes Around, 302
Opaque materials, 552, *552*
Optical disks, 658, *658*, 658 *act*
Optical fibers, 566, *566*
Orbit, of satellite, 321; weightlessness in, 328, *328*
Organic compounds, 250–261. *See* Hydrocarbons. *See also* Biological compounds; naming, 253, 254, *255*, 256
Outer ear, 508, *508*
Output force, 412, *412*
Oval window, 508, *509*
Overtones, 503, *503*
Owl, 508, *508*
Oxygen, 218
Ozone, 530
Ozone layer, 530, *530*

P

Pagers, 538
Parallel circuit, 599, *599*, 603 *lab*
Particle size, and rate of reaction, 204, *204*
Pascal (Pa), 116
Pascal's principle, 122, 122–123, *123*, **357**, *357*
Pentane, *255*
Percussion instruments, 505, *505*
Periodic table, 81, *82*, *83*, 86 *lab*; and energy levels of electrons, 164–165, *165*; halogen family on, 166, *166*; in identifying properties of elements, 167 *act*; metalloids on, 645; noble gases on, 166, *166*
Petroleum. *See* Oil (petroleum)
pH, 236–238, 240–241 *lab*
Phosphoric acid, 137
Phosphorus, 139
Photographs, scientific, 57
Photon, 557
Photovoltaic collector, **392**, *392*
pH scale, 137, 150–151 *lab*, 236, *236*
Physical changes, 143, **143**–144, *144*, 190, *190*, 207 *lab*
Physical models, 21
Physical processes, 218, *219*
Physical properties, 134–138, 136 *lab*; of acids, 137, *137*, 138 *act*; of bases, 138, *138*, 138 *act*; density, **134**, 135, *135*, 135 *act*; state of matter, 136, *136*, 144, *144*
Physical science, **10**, 10–11
Physical weathering, 147, *147*
Physicist, **10**, 11
Piano, 503, 504
Pickling, 233
Pie (circle) graph, 58, *58*
Pigments, 554, *554*
Pistons, 122, *122*, 446, *447*
Pitch, 470, *470*, **494**, *494*; varying, 489 *lab*
Plane mirrors, 557, *557*, 561 *lab*
Plant(s), chlorophyll in, 193; leaves of, 193; as medicine, 272, *272*
Plasma, 105 *act*, 343
Polar bears, 442
Polar bond, **174**, *174*
Polar molecules, 174, 174–175, *175*, 175 *act*, 224, *224*, 226
Pole(s), magnetic, 614, *614*, 615, 617 *act*
Pollution, of air, *234*; and fossil fuels, 388; thermal, **443**, *443*
Polyethylene, 262, *262*
Polymer(s), 262, **262**–263, *263*
Polymerization, **262**
Pools, and chemical properties, 140, *140*
Position, changing, 282, 282–283
Positive acceleration, 291

Positive charge, 584, *584*
Potassium, 167, *167*
Potassium chromate, 227
Potential energy, **376**, *376*, 380, *380*
Power, 396–397 *lab*, **409**–410; calculating, 409, 409 *act*; electric, **600**–601; and energy, 410; equation for, 409; geothermal, 393, *393*; hydroelectric, 390, *391*; of pulley, 424–425 *lab*; and work, 409 *lab*
Power failure, 398
Power plants, 384, 384–385, 627, *627*, 627 *act*
Precipitate, **220**, *220*, 221, 231 *lab*
Precipitation, acid, 234
Precision, 44–49, *45*, *46*, *49*
Prediction, 14, 24
Prefixes, and naming hydrocarbons, 255; in SI, 50
Pressure, 116, **116**–120, 340, **340**–347; of air, 339 *lab*; and area, 117, *117*, 342, *342*; atmospheric, 117, 117–119, *118*, 339 *lab*, 345, 345–347, *346*, *347*, 362–363 *lab*; balanced, 118, *118*; barometric, 362–363 *lab*; and buoyant force, 349, *349*; calculating, 341 *act*; and elevation, *346*; in fluid, 343–345, *344*, *345*, 356, *356*, 358, *358*; and force, 116–120; of gas, *119*, 119–120, *120*; increase with depth, 344; and solubility, 228; and temperature, 120, *120*; and volume, 119, *119*; and weight, 340, *340*, 342
Primary light colors, 553, *553*
Primary pigment colors, 554, *554*
Prism, 475, 534 *lab*, 552, *552*
Product, **192**, 193, *194*, 195
Program(s), computer, 652 *act*, 653, *653*
Programming, computer, 653
Promotional materials, evaluating, 29, *29*
Propane, 252, *252*, 253, *255*
Propene (propylene), 253, *253*
Properties, of acids, 232; of bases, 235; chemical, 139, **139**–142, *140*, *141*, *142*; classifying, 136 *lab*;

of electromagnetic waves, *523*, 523–524, *524;* of light, 550–554; physical, **134**–138, *135*, 135 *act*, *136*, 136 *lab*, *137*, *138*, 144, *144;* size-dependent, **136;** size-independent, **136;** of waves, 467–471, *468*, *469*, *470*, 472 *lab*, *477*
Propylene (propene), 253, *253*
Prostheses, 426, *426*
Proteases, 206, *206*
Proteins, 260, *260*, **263**, *263*, 263 *lab*
Proton(s), **77**, *77*, 584, *584*
P-type semiconductors, 646, *647*
Pulley, 422 *lab*, **422**–425; fixed, 422, *423;* movable, 423, *423;* power of, 424–425 *lab*
Pulley system, 423, *423*, 424–425 *lab*
Pulsars, 536
Pupil, of eye, 548
Pure tone, 503
Pyramids, building, 405 *lab*, 411 *lab*

Quarks, 79
Questioning, 13, *13*

Rabbits, hearing of, *508*
Radar, 496 *act*, 527, *527*
Radiant energy, 377, *377*, 445, 466, **524**
Radiation, 439; ultraviolet, *529*, **529**–530, *532*
Radio, 383, *383*, 386 *lab*, 526, *535*, 535–536, *536*, 537 *act*
Radioactive wastes, 389
Radio waves, 482, *526*, **526**–527, *527*, 532, *535*, 535–536, *536*, 537 *act*, 538, *538*
Rain, acid, 141, 148, 152, *152*, 234
Rainbow, *72*, 475, *475*
RAM (random-access memory), 652, 653 *lab*, 656
Rarefaction, 465, *465*, 491
Rasmussen, Knud, 634

Rate, 54
Rate of reaction, *202*, **202**–206, *203, 204*
Ratio, input coils/output coils, 629
Reactant, 192, 193, *194*, 195
Reaction(s), and action, 323–326, *324, 326*. *See also* Chemical reactions
Reaction rate, *202*, **202**–206, *203, 204*
Reactivity, 140
Reading Check, 7, 8, 10, 17, 21, 22, 43, 45, 49, 50, 53, 58, 72, 76, 83, 85, 88, 89, 102, 103, 104, 106, 108, 113, 117, 120, 137, 138, 139, 140, 143, 145, 146, 164, 165, 167, 169, 173, 174, 178, 192, 197, 201, 202, 218, 220, 222, 225, 226, 227, 228, 230, 236, 237, 238, 250, 252, 253, 256, 258, 260, 263, 265, 266, 267, 269, 283, 285, 292, 313, 314, 316, 342, 345, 349, 353, 358, 360, 375, 377, 380, 384, 391, 392, 406, 408, 413, 415, 419, 422, 435, 439, 441, 463, 464, 469, 470, 473, 475, 476, 482, 493, 495, 502, 509, 524, 527, 529, 536, 537, 551, 556, 557, 558, 563, 564, 568, 570, 585, 590, 594, 598, 601, 615, 627, 628, 643, 650, 652
Reading Strategies, 6A, 42A, 72A, 102A, 134A, 162A, 190A, 218A, 250A, 282A, 310A, 340A, 374A, 406A, 343A, 462A, 490A, 520A, 550A, 584A, 614A, 642A
Read-only memory (ROM), 652, 653 *lab*, 656
Read/write head, *657*
Real-World Questions, 30, 31, 55, 60, 86, 92, 115, 124, 149, 150, 179, 180, 207, 208, 231, 240, 261, 270, 299, 300, 329, 330, 355, 362, 386, 396, 411, 424, 444, 450, 472, 480, 500, 510, 534, 540, 561, 572, 603, 604, 620, 632, 648, 660
Recycling, of computers, 654
Reflecting telescopes, 569, *569*
Reflection, 473, 551, 555–561, *556, 561;* law of, **555,** *555;* of light,

473, *473;* and mirrors, 556, 557, *557*, 561 *lab;* of sound, 495, *495*, 500 *lab;* and surfaces, 555–556, *556;* total internal, 565, *565;* of waves, 473, *473*
Refracting telescopes, 568, *568*
Refraction, 474, 563, *563;* of light, *474*, 474 *lab*, 474–475, *475;* of waves, 474, 474–475, *475*
Refrigerators, 448, *448*
Relative motion, 283, *283*
Renewable resources, 390, *391*
Resistance, 594, **594**–595, *595*, 596, *596*, 597
Resonance, 502, *502*
Resonator, 504
Resources, alternative, 391–393, 391 *lab*, *392, 393;* inexhaustible, **391;** nonrenewable, **388,** *388;* renewable, **390,** *391*
Retina, 548
Reverberation, 507, *507*, 507 *act*
Rings, hydrocarbons in, 256, *256*
Rock(s), as mixtures, 91; physical weathering of, 147, *147*
Rocket(s), balloon, 329 *lab;* launching, 326, *326*
Rolling friction, 315, *315*
ROM (read-only memory), 652, 653 *lab*, 656
Rounding, 47, 48 *act*
Rubidium, 167
Rust(s), *145*, 198
Rutherford, Ernest, 77, 78

Safety, and air bags, 332, *332;* in automobiles, 300–301 *lab*, 332, *332;* ear protectors for, 479, *479;* with electricity, 601–602; in laboratory, 19, *19*
Safety symbols, 19
Saliva, 242
Salt(s), 142, *142;* bonding in, *170*, 170–171, *171;* crystal structure of, 103, *103;* dissolving in water, 225, *225;* movement of ions in, 585, *585*
Salt water, 219, 219 *act*, 222, 242, *242*

Index

Satellite(s), 321, 533, *533*, 538 *act*, 538–539
Saturated fats, 268, *268*
Saturated hydrocarbons, 252, *252*
Saturated solution, 228, *228*
Scale(s), 47, *47*, 53, *53*
Scale drawing, 55 *lab*
Science, 6–33; branches of, 9–11; careers in, 9–11; communication in, 17, *17*; evaluation in, 26–29, *29*, 30 *lab*; experiments in, *18*, 18–19; inference in, 16, *16, 26*; models in, *20*, 20–25, *21, 22, 23, 24, 25*; observation in, 5 *lab*, 13, *13, 16, 26, 27, 28*; safety in, *19*, 19; skills in, 12–15; systems in, *8*, 8–9; and technology, 11, *11*; women in, 33, *33*
Science and History, Ancient Views of Matter, 94; Crumbling Monuments, 152; Hopping the Frequencies, 542, *542*; Synthetic Diamonds, 210; Women in Science, 34
Science and Language Arts, "Aagjuuk and Sivulliit" (Rasmussen), 634; "Baring the Atom's Mother Heart" (Awiakta), 182; "Hurricane" (Balaban), 364
Science and Society, Air Bag Safety, 332; Bionic People, 426; E-Lectrifying E-Books, 662; The Heat Is On, 452; Fire in the Forest, 606; It's a Wrap!, 512; From Plant to Medicine, 272
Science Online, Acids and Bases, 138; Automobile Engines, 446; Blood Pressure, 123; Calcium Hydroxide, 235; Chemical Changes, 146; Chemical Equations, 195; Collisions, 296; Compasses, 619; Computer Software, 652; Condensation, 113; Controlling Reverberation, 507; Convex Lenses, 564; Cyclohexane, 256; Desalination, 219; Electrons, 164; Energy Transformations, 380; Force Fields, 521; Freezing Point Study, 111; Galileo and Newton, 313; Historical Tools, 413; How Birds Fly, 324; Hydraulic Systems, 358; Hydroelectricity, 390; Indicators, 238; Interference, 477; James Watt, 410; Land Speed Record, 286; Lipids, 267; Magnetic Disks, 656; Measurement, 47; Mixtures, 90; New Elements, 81; Olympic Torch, 201, *201*; Optical Disks, 658; Plasma, 105; Polar Molecules, 175; Power Plants, 627; Satellite Communication, 538; Scientific Data, 58; Scientific Method, 18; Semiconductor Devices, 645; Snowshoes, 341; Sonar, 495; Subatomic Particles, 76; Superconductors, 588; Topographic Maps, 21; Wave Speed, 471
Science Stats, Biggest, Tallest, Loudest, 62; Energy to Burn, 398; Salty Solutions, 242; Waves, Waves, and More waves, 482
Scientific explanations, evaluating, 26–29, *28*, 30 *lab*
Scientific illustrations, *56*, 56–57
Scientific law, 7
Scientific Methods, 12, *12*, 18 *act*, 30, 31–32, 55, 60–61, 86, 92–93, 115, 124–125, 149, 150–151, 179, 180–181, 207, 208–209, 231, 240–241, 261, 270–271, 299, 300–301, 329, 330–331, 355, 362–363, 386, *396*, 396–397, 411, 424–425, 444, 450–451, 472, 480–481, 500, 510–511, 534, 540–541, 561, 572–573, 603, 604–605, 620, 632–633, 648, 660–661; Analyze Your Data, 32, 61, 93, 125, 151, 181, 208, 271, 301, 329, 331, 363, 397, 425, 451, 472, 481, 500, 511, 541, 661; Conclude and Apply, 30, 32, 55, 61, 86, 93, 115, 125, 149, 151, 179, 181, 207, 209, 231, 241, 261, 271, 299, 301, 329, 331, 355, 363, 386, 397, 411, 425, 444, 451, 472, 481, 500, 511, 534, 541, 561, 573, 603, 605, 620, 633, 648, 661; Follow Your Plan, 61, 125, 151, 208, 331, 363, 425, 451, 481, 511, 541, 661; Form a Hypothesis, 60, 124, 150, 300, 330, 396, 424, 450, 480, 510, 540; Make a Plan, 61, 125, 151, 208, 331, 362, 425, 451, 481, 511, 541, 660; Make the Model, 181; Plan the Model, 181; Test Your Hypothesis, 61, 125, 151, 301, 330, 397, 451, 481, 511, 541
Scientific theory, 7
Screw, 419, *419*
Sea level, pressure at, 346
Second-class lever, 421
Seismic sea waves (tsunamis), 468, 482
Seismograph, 480–481 *lab*
Semiconductors, 645 *act*, **645**–646, *646*, 647
Seneca, 574
Sense(s), hearing, *191*; sight, *191*; smell, *191*; taste, *191*; touch, *191*
Series circuit, 598, *598*
Server, 659
Shape, and buoyant force, 350, *350*
Ship, designing, 124–125
Shock, electric, 601–602
Signals. See Electronic signal
Significant digits, 49, *49*
SI (International System of Units), 50–54, 55 *lab*
Silicon, 645, 646, *646*
Silver tarnish, 178, *178*, 195
Simple machines, 417–425; inclined plane, 417–419, *418;* lever, 420, *420,* 421, *421;* pulley, 422 *lab,* 422–425, *423,* 424–425 *lab;* screw, 419, *419;* wedge, *418,* 418–419, *419;* wheel and axle, 420, *420,* 422, *422*
Sinking, 349, *349,* 353, *353.* See *also* Buoyant force
Size-dependent properties, 136
Size-independent properties, 136
Skin cancer, 529

744 STUDENT RESOURCES

Sliding friction, 313, 314, *314*, *315*, 319, 322
Smell, *191*
Snakes, and electromagnetic waves, *528*
Snowshoes, 341 *act*
Soaps, 138, *138*
Soap scum, 220, *220*
Sodium, 167, *170*, 170–171, *171*
Sodium bicarbonate, 192, 193, 194
Sodium chloride, 103, *103*, 142, *142*, 225, *225*
Sodium hydroxide, 141, 142, 235
Software, computer, 652 *act*, **653**, *653*
Solar collector, 391 *lab*, 392, *392*
Solar energy, 391–392, *392*, *393*, 398, *398*
Solar system, models of, 25, *25*
Solid(s), *103*, **103**–104, 136, *136*, *144*; amorphous, 104, 109, *109*; crystalline, 103, *103*, 109; melting, 109, *109*, *110*; movement of electrons in, 585, *585*; sublimation of, 114, *114*
Solid solutions, 223, *223*
Solubility, 227–118; factors affecting, *227*, 227–228; of gas, 228, 228 *lab*; and pressure, 228; of similar and dissimilar substances, 226, *226*; and temperature, 227, *227*
Solute, 220, 222, 230
Solution(s), 220–230; aqueous, **224**–226, *225*, *226*; concentration of, 229 *act*, 229–230, *230*; crystals from, 231 *lab*; formation of, *220*, 220–221, *221*; movement of ions in, 585, *585*; saturated, **228**, *228*; supersaturated, 228, 231 *lab*; types of, 221–223, *222*, *223*; unsaturated, 228
Solvent, 220, 222; boiling point of, 230; freezing point of, 230; water as, 224–226, *225*, *226*
Sonar, 495, *495*, 495 *act*
Sound, 488–511; comparing, 465 *lab*; comparing and contrasting, 492 *lab*; and echoes, 495, *495*, 498; echolocation, 527; frequency of, 491, 494, *494*, 501–502, 503, *503*; loudness of, 492–493, *493*; and music, 501–507, 510–511 *lab*; pitch of, 470, *470*, 489 *lab*, 494, *494*; reflection of, 495, *495*, 500 *lab*; reverberation of, 507, *507*, 507 *act*; speed of, 471, 492; and vibrations, 489 *lab*, 490
Sound waves, 463, 465, *465*, 467, 470, 476, *490*, 490–491, *491*; diffraction of, 498, *498*; frequency of, 491, 494, *494*, 501–502, 503, *503*; making, 491; and resonance, 502, *502*; using, 499, *499*
Space shuttle, 328, *328*
Species, endangered, *57*, 57–59, *58*, *59*
Specific heat, 109, *109*, 442
Spectrum, electromagnetic, 468, *468*. *See* Electromagnetic spectrum
Speed, 284–285; and acceleration, *288*, 288–289; of animals, 284; average, **285**, *285*, 285 *lab*; calculating, 284 *act*; constant, 285, *285*; and distance-time graphs, 286, *286*; of heating and cooling, 444 *lab*; instantaneous, **285**, *285*; and kinetic energy, 375, *375*; land, 286 *act*; of light, 471, 524, *524*, 562, *562*; measuring, 54, 60–61 *lab*; and motion, 284–285, *285*; of sound, 471, 492; and velocity, 287; of waves, 471, 471 *act*, 474, 480–481 *lab*
Speed-time graph, 292, *292*
Sports, Newton's laws in, 325
Spring scale, 53, *53*
Stalactites, 148, *148*, 221, *221*, 233
Stalagmites, 148, 221, *221*, 233
Standardized Test Practice, 37–38, 66–67, 98–99, 130–131, 186–187, 214–215, 246–247, 276–277, 306–307, 368–369, 430–431, 456–457, 546–547, 578–579, 610–611, 638–639
Starches, 265
States of matter, 100–125, *102*, **136**, *136*; changes of, 107–115, 115 *lab*, 144, *144*; and condensation, *110*, 113, *113*, 113 *act*; and evaporation, *110*, *112*, 112 *lab*, 112–113; fluids, 116–123, *120*, *121*, 121 *act*, *122*, *123*, 124–125 *lab*; and freezing, 101 *lab*, *110*, 111; gases, 106, *106*; liquids, *104*, 104–105, *105*; and melting, 109, *109*, *110*; and pressure, 116–120; solids, *103*, 103–104; and sublimation, 114, *114*; and vaporization, *110*, *112*, 112 *lab*, 112–113
Static charge, 585, *585*
Static friction, 314
Stearic acid, *268*
Steel, 223, *223*
Stereotactic Radiotherapy (SRT), 47
Stirrup (of ear), 508, *508*
Stringed instruments, 504, *504*, 504 *lab*
Study Guide, 34, 63, 95, 127, 153, 183, 211, 243, 273, 303, 333, 365, 399, 427, 453, 483, 513, 543, 575, 607, 635, 663
Sub-atomic particles, 76 *act*
Sublimation, 114, *114*
Subscript(s), 177, 194
Substance, 87, 91, 218
Substituted hydrocarbons, *257*, 257–260; alcohols, 258, *258*, 261 *lab*; amines, 259, *259*; amino acids, 259–260, *260*, *263*, 263; carboxylic acids, 259, *259*
Sucrose, 265, *265*
Suffixes, and naming hydrocarbons, 255
Sugars, 265, *265*
Sulfuric acid, 141, 233, 234
Sun, as plasma, 343
Sundial, *45*
Sunlight, wavelengths of light in, 475
Superconductors, 588 *act*, 629, 629–630, *630*
Supersaturated solution, 228, 231 *lab*
Surface tension, 105, *105*
Switch, 651
Symbols, for atoms, 177, *177*;

Synthetic elements

for compounds, 177, *177*; for safety, 19
Synthetic elements, 80, *80*, 81 *act*
Synthetic polymers, 262
System(s), *8*, 8 *lab*, **8**–9

Table, 57, 58 *act*
Tarnish, 178, *178*, 195
Taste, *191*
Technology, 11; air conditioners, 449; airplanes, 360, *360*, *361*; alloys, 223, *223*; ammeter, 622, *623*; analog devices, 642–643, *643*; barometers, 347, *347*, 362–363 *lab*; bicycle, 417, *417*; boats, 350, *350*, 354, *354*; cameras, 570, *570*; catalytic converters, 205, *205*; cell phones, 542, *542*; centrifuge, 89; circuit breakers, 599, *599*; compass, 614, *619*, 619 *act*, 620 *lab*; computers. *See* Computer(s); concave lens, 565, *565*; convex lens, 564, *564*, 564 *act*, 570, *570*, 572–573 *lab*; diodes, 646, *646*, 648 *lab*; Doppler radar, 496 *act*; ear protectors, 479, *479*; electric meter, 601, *601*; electric motors, 624, *624*, 632–633 *lab*; electromagnets, *621*, 621–622, *622*, 622 *lab*; electronics, *642*, 642–648, *645*, 648 *lab*; eyeglasses, 574, *574*; fireworks, 200, *200*; floppy disks, 658; fuses, 599, *599*; galvanometer, 622, *623*; gasoline engines, 381, *381*; generator(s), 384, *384*, 626, 626–627; Global Positioning System (GPS), 466, 539, *539*; hard disks, 656, *656*, 656 *act*, *657*; heat pumps, 393, 449, *449*; hydraulic systems, 357, 357–358, 358 *act*; integrated circuits, 647, *647*, 649, 652, *652*; internal combustion engines, 446, *446*, 446 *act*, 447; laboratory balance, 15, *15*; lasers, 570–571, *571*; lightbulb, 166; lightning rod, 590, *590*; maglev, 612, *612*, 613; magnetic resonance imaging (MRI), 630, 630–631, *631*; *Mars Climate Orbiter,* 51; microprocessors, 655, *655*; microscopes, 567, *567*; microwave oven, 526; microwave tower, 526, *526*; optical disks, 658, *658*, 658 *act*; optical fibers, 566, *566*; pagers, 538; photovoltaic collector, 392, *392*; power plants, *384*, 384–385, 627, *627*, 627 *act*; pyramids, 405 *lab*, 411 *lab*; radar, 527, *527*; radio, 383, *383*, 386 *lab*, 526, *535*, 535–536, *536*, 537 *act*; reflecting telescopes, 569, *569*; refracting telescopes, 568, *568*; refrigerators, 448, *448*; rockets, 326, *326*, 329 *lab*; satellites, 321, 533, *533*, 538 *act*, 538–539; scales, 47, *47*, 53, *53*; and science, 11, *11*; seismographs, 480–481 *lab*; semiconductors, 645 *act*, 645–646, *646*, 647; solar collector, 391 *lab*, 392, *392*; sonar, 495, *495*, 495 *act*; space shuttle, 328, *328*; spring scale, 53, *53*; Stereotactic Radiotherapy (SRT), 47; superconductors, *629*, 629–630, *630*; telecommunications, *535*, 535–536, *536*; telephones, 537–538, *538*, 542, *542*; telescopes, 533, *533*; thermometers, 435, *435*; transformers, *628*, 628–629; transistors, 647, *647*; turbine, *384*; ultrasound, 499, *499*; vacuum tubes, 645, *645*, 649; voltmeter, 622, *623*; welding, 254, *254*; welding torch, *197*; windmill, 393, *393*; wireless, 542, *542*; World Wide Web, 659, *659*; X ray, 520, 530–531, *531*, 532

Teeth, of herbivores and carnivores, 419, *419*
Telecommunications, electromagnetic waves in, *535*, 535–536, *536*
Telephones, 537–538, *538*, 542, *542*
Telescopes, 568–569; reflecting, **569**, *569*; refracting, **568**, *568*; on satellites, 533, *533*
Television, vacuum tubes in, 645, *645*
Temperature, 108, *108,* 433 *lab*, *434*, **434**–436; of body, 383, 415, 530; and heat, 108–109; measuring, 54, *54*, 435, 435–436; and pressure, 120, *120*; and rate of reaction, *202*, 202–203, *203*; and solubility, 227, *227*; and speed of sound, 492; and thermal energy, 437, *437*
Temperature scales, 54, *54*; Celsius, *435*, 435–436, 436 *act*; converting, 436, 436 *act*; Fahrenheit, *435*, 435–436, 436 *act*; Kelvin, 436
Theory, scientific, 7
Thermal collector, 392, *392*
Thermal conductors, 441
Thermal energy, *107,* **107**–108, **376,** *376,* 383–384, *384*, 432–452, **435,** *437,* 445, *445, 447*; and heat, 438–441; and temperature, 437, *437*; transfer of, 438–440, *439, 440*
Thermal expansion, 435, *435*
Thermal insulators, 442, *442*, 450–451 *lab*
Thermal pollution, 443, *443*
Thermometer, 435, *435*
Thinking, critical, **26;** thinking like a scientist, 22 *lab*
Third-class lever, *421*
Thomson, J. J., 76, *76,* 77, *77*
Throat vibrations, 489 *lab*
Tidal energy, 394, *394*
TIME, Science and History, 33, 94, 152, 210, 542; Science and Society, 272, 332, 426, 512, 606, 662
Time, measuring, 45, *45,* 54
Tools, historical, 413 *act*
Topographic maps, 21 *act*
Total internal reflection, 565, *565*
Touch, *191*
Toxicity, *139*
Transformer, *628,* **628**–629

Index

Transistors, 647, *647*
Translucent materials, 552, *552*
Transparent materials, 552, *552*
Transverse waves, 464, *464,* 467, 468, *468,* 472 *lab*
Tree(s), changing colors of leaves, 147, *147;* tallest, 62, *62*
Trichloromethane (chloroform), *257*
Triple bonds, 174, *174,* 175, 253, 254, *254*
Trough, 464, *464*
Try at Home Mini Labs, Analyzing Energy Transformations, 381; Assembling an Electromagnet, 622; Classifying Parts of a System, 8; Classifying Properties, 136; Comparing and Contrasting Sound, 492; Comparing Compounds, 88; Comparing Rates of Melting, 440; Constructing a Model of Methane, 173; Forming a Hypothesis, 14; Identifying Inhibitors, 204; Investigating the Electric Force, 592; Measuring Average Speed, 285; Measuring Volume, 52; Modeling Isomers, 254; Observing Bernoulli's Principle, 359; Observing Chemical Processes, 228; Observing Colors in the Dark, 551; Observing Electric Fields, 523; Observing Friction, 314; Observing How Light Refracts, 474; Observing Memory, 653; Predicting a Waterfall, 119; Using Binary Numbers, 650; Work and Power, 409
Tsou Yen, 94
Tsunami, 468, 482
Tuning instruments, *506,* 507
Turbine, 384, *384*

Ultrasonic waves, 470
Ultrasound, 499, *499*
Ultraviolet light, 466

Ultraviolet radiation, *529,* **529**–530, *532*
Unbalanced forces, 311
Universe, viewing, 531–533, *532, 533*
Unknown, finding, 121 *act*
Unsaturated fats, 268, *268*
Unsaturated hydrocarbons, *253,* 253–254
Unsaturated solution, 228
Uranium, 389
Use the Internet, Barometric Pressure and Weather, 362–363; Does your computer have a virus?, 660–661; Energy To Power Your Life, 396–397

Vacuum tubes, 645, *645,* 649
Vapor, 106
Vaporization, 110, *112,* 112 *lab,* 112–113
Variables, 18
Velocity, 287; and acceleration, *288,* 288–289, *289;* and speed, 287
Vinegar (acetic acid), 192, 193, 194, 222, 259, *259*
Viruses, computer, 660–661 *lab*
Viscosity, 105
Visible light, 528, *528,* 529
Vision, *191*
Vitamin C, 270–271 *lab*
Vocal cords, 489 *lab,* 494
Voice, pitch of, 489 *lab,* 494
Volcanoes, energy from, 372, *372;* eruptions of, 62
Volcanologist, 10, *10*
Voltage, 592, *592,* 596, 597, 597 *act,* 604–605 *lab,* 627; changing, *628,* 628–629
Voltmeter, 622, *623*
Volume, 52, 134; measuring, 52, *52,* 52 *lab;* and pressure, 119, *119*

Waste(s), radioactive, 389
Water, boiling point of, 112, *112;* changes of state of, 144, *144;* as compound, 218; freshwater, 104; melting point of, 109; molecules of, 175, *175;* as solvent, 224–226, *225, 226*
Water cycle, 56, *56,* 115 *lab*
Waterfalls, 119 *act*
Water vapor, 144
Water waves, 476, *476*
Watt (W), 409
Watt, James, 409, 410 *act*
Wave(s), 460–482, **472;** amplitude of, *467,* 467–468, 493, *493;* behavior of, 473–479; carrier, **536,** *536;* compressional, *464,* **465,** 467, 468, 472 *lab,* 490, 490–491, *491;* crest of, 464, *464;* diffraction of, 475–476, *476,* 498, *498;* electromagnetic, **466,** 467, 468, *468,* 471. *See* Electromagnetic waves; and energy, 460, 461 *lab,* 462, *462;* frequency of, *469,* 469–470, *470,* 491, 494, *494,* 501–502, 503, *503,* 523, 542; infrared, 466, 527 *lab,* **527**–528, *528, 532;* and interference, 477 *act,* 477–479, *478, 479;* light, *550,* 550–551, *551,* 557; mechanical, **463**–465, *464,* 471; microwaves, 526, *526;* model for, 463, *463;* properties of, 467–471, *468, 469, 470,* 472 *lab,* 477; radio, 482, 526, **526**–527, *527, 532, 535,* 535–536, *536,* 537 *act,* 538, *538;* refraction of, *474,* 474–475, *475;* sound, 463, 465, *465,* 467, 470, 476, *490,* 490–491, *491,* 494, *494,* 498, 498–499, *499,* 501–502, *502,* 503, *503;* speed of, 471, 471 *act,* 474, 480–481 *lab;* transverse, **464,** *464,* 467, 468, 472 *lab;* trough of, 464, *464;* tsunami, 468, 482; ultrasonic, 470; ultraviolet, 466, *529,* **529**–530, *532;* visible light, 466, 468, *468;* water, 476, *476*
Wavelength, 468, *468,* 491, *491,* 494, *494,* 498, *498,* 523, *523,* 537 *act,* 540–541 *lab;* and diffraction, 476, *476;* and

Index

frequency, *469*, 470, *470*
Wave media, 551
Wax, 267
Weather, and barometric pressure, 362–363 *lab*
Weathering, chemical, 148, *148*; physical, 147, *147*
Web site, 659, *659*
Wedge, *418*, **418**–419, *419*
Weight, 53, 317–318; and mass, 318; measuring, 53, *53*, 327, *327*; and pressure, 340, *340*, 342
Weightlessness, *327*, **327**–328, *328*
Welding, 197, 254, *254*
Whales, 62, *62*
Wheel and axle, 420, *420*, 422, *422*
Wildfires, 606, *606*

Wind energy, 393, *393*
Windmill, 393, *393*
Wind tunnel, 24
Wings, 360, *360*, 361, *361*
Wire, copper, 594; electric, 588, 594, *594*
Wireless technology, 542, *542*
Women, in science, 33, *33*
Woodwind instruments, 505, 505–506, *506*
Work, 406–408; calculating, 408, 408 *act*; and distance, 408, *414*, 414; and energy, 410; equation for, 408; and force, 405 *lab*, 407, *407*, 411 *lab*, 413; measuring, 410; and mechanical advantage, *412*, 412–414; and motion, *406*, 406–407, *407*; and power, 409 *lab*

World Wide Web, 659, *659*

X rays, 520, **530**–531, *531*, 532

Yalow, Rosalyn Sussman, 33, *33*

Zewail, Ahmed H., 165, 182
Zinc, 593

Credits

Magnification Key: Magnifications listed are the magnifications at which images were originally photographed.
LM–Light Microscope
SEM–Scanning Electron Microscope
TEM–Transmission Electron Microscope

Acknowledgments: Glencoe would like to acknowledge the artists and agencies who participated in illustrating this program: Absolute Science Illustration; Andrew Evansen; Argosy; Articulate Graphics; Craig Attebery represented by Frank & Jeff Lavaty; CHK America; John Edwards and Associates; Gagliano Graphics; Pedro Julio Gonzalez represented by Melissa Turk & The Artist Network; Robert Hynes represented by Mendola Ltd.; Morgan Cain & Associates; JTH Illustration; Laurie O'Keefe; Matthew Pippin represented by Beranbaum Artist's Representative; Precision Graphics; Publisher's Art; Rolin Graphics, Inc.; Wendy Smith represented by Melissa Turk & The Artist Network; Kevin Torline represented by Berendsen and Associates, Inc.; WILDlife ART; Phil Wilson represented by Cliff Knecht Artist Representative; Zoo Botanica.

Photo Credits

Cover Mark Gamba/CORBIS; **ii** Mark Gamba/CORBIS; **vii** Aaron Haupt; **viii** John Evans; **ix** (t)PhotoDisc, (b)John Evans; **x** (l)John Evans, (r)Geoff Butler; **xi** (l)John Evans, (r)PhotoDisc; **xii** PhotoDisc; **xiii** NASA; **xiv** (t)Brenda Tharp/Photo Researchers, (b)Charles Benes/FPG International; **xv** Sovfoto/Eastfoto/PictureQuest; **xvi** PhotoDisc; **xvii** (t)Tom McHugh/Photo Researchers, (b)Duomo; **xviii** (t)Gregory G. Dimijian/Photo Researchers, (b)Douglas Peebles/CORBIS; **xix** Bjorn Backe/Papilio/CORBIS; **xx** (l)Aaron Haupt, (r)Amanita Pictures; **xxi** Amanita Pictures; **xxii** Bobby Model/National Geographic Image Collection; **xxiv** Matt Meadows; **xxvi** Morrison Photography; **1** Glencoe; **2–3** (bkgd)Wolfgang Kaehler; **2** (l)PhotoDisc; **3** (t)PhotoDisc; **4–5** David Keaton/CORBIS; **6** (l)Jack Star/Photolink/PhotoDisc, (c)Rudi Von Briel, (r)Richard T. Nowitz/CORBIS; **8** Mary Kate Denny/PhotoEdit, Inc.; **9** Peter Veit/National Geographic Image Collection; **10** (tl)G. Brad Lewis/Stone/Getty Images, (bl)Roger Ball/The Stock Market/CORBIS, (br)Will & Deni McIntyre/Photo Researchers; **11** (t)AFP/CORBIS, (b)Reuters NewMedia, Inc./CORBIS; **13 14** Richard Hutchings; **15** Matt Meadows; **16** Icon Images; **17** Richard Hutchings/PhotoEdit, Inc./PictureQuest; **18** Rudi Von Briel; **19** Bob Daemmrich; **20** Glasheen Graphics/Index Stock; **21** (cw from top)David Young-Wolff/PhotoEdit, Inc., Donald C. Johnson/The Stock Market/CORBIS, John Bavosi/Science Photo Library/Photo Researchers, A. Ramey/PhotoEdit, Inc.; **22** CORBIS/PictureQuest; **23** Todd Gipstein/CORBIS; **24** (tl cl)Betty Pat Gatliff, (tr)Richard Nowitz/Words & Pictures/PictureQuest, (bl)Michael O'Brian/Mud Island, Inc., (br)Betty Pat Gatliff; **25** (l)Carol Anne Petrachenko/CORBIS, (c)Jim Sugar Photography/CORBIS, (r)Tom Wurl/Stock Boston/PictureQuest; **26** (l)Stock Montage, (r)North Wind Picture Archives; **27** Digital Art/CORBIS; **28** SuperStock; **29** (t)Lester V. Bergman/CORBIS, (b)Bob Handelman/Stone/Getty Images; **31** Amanita Pictures; **32** (t)Aaron Haupt, (b)Matt Meadows; **35** (t)Reuters/CORBIS, (bl)UPI/Bettmann/CORBIS, (br)TIME; **37** Tim Courlas; **38** Peter Veit/National Geographic Image Collection; **39** (l)Tim Courlas/Horizons, (c r)Aaron Haupt; **40–41** Buck Miller/SuperStock; **42** Paul Almasy/CORBIS; **43** AFP/CORBIS;

44 David Young-Wolff/PhotoEdit, Inc.; **45** (tr)The Purcell Team/CORBIS, (l)Lowell D. Franga, (br)Len Delessio/Index Stock; **46** Photo by Richard T. Nowitz, imaging by Janet Dell Russell Johnson; **47** Matt Meadows; **49** Mark Burnett; **51** Tom Prettyman/PhotoEdit, Inc.; **53** (tl)Michael Dalton/Fundamental Photographs, (cl)David Young-Wolff/PhotoEdit, Inc., (cr)Dennis Potokar/Photo Researchers, (br)Matt Meadows; **55** Michael Newman/PhotoEdit, Inc.; **57** John Cancalosi/Stock Boston; **60 61** Richard Hutchings; **62** (t)Fletcher & Baylis/Photo Researchers, (b)Charles O'Rear/CORBIS; **63** Mark Burnett; **64** Chuck Liddy/AP/Wide World Photos; **66** Michael Dalton/Fundamental Photographs; **68–69** (bkgd)Stephen Frisch/Stock Boston/PictureQuest; **68** (inset)CORBIS/PictureQuest; **70–71** Russell Dohrman/Index Stock; **71** Morrison Photography; **72** (l)Gary C. Will/Visuals Unlimited, (c)Mark Burnett/Stock Boston, (r)CORBIS; **74** Mark Burnett; **75** (l)Mark Burnett, (r)NASA; **76** Van Bucher/Photo Researchers; **80** Fermi National Accelerator Laboratory/Science Photo Library/Photo Researchers; **81** Tom Stewart/The Stock Market/CORBIS; **82** (br)New York Public Library, General Research Division, Astor, Lenox, and Tilden Foundations, (others)Bettmann/CORBIS; **84** Emmanuel Scorcelletti/Liaison Agency/Getty Images; **86** Doug Martin; **87** NASA; **88** Mark Burnett; **89** Klaus Guldbrandsen/Science Photo Library/Photo Researchers; **90** (tl)Mark Thayer, (tr)CORBIS, (bl)Kenneth Mengay/Liaison Agency/Getty Images, (bc)Arthur Hill/Visuals Unlimited, (br)RMIP/Richard Haynes; **90–91** (bkgd)KS Studios; **91** (inset)Mark Burnett; **92** (t)Mark Burnett, (b)Michael Newman/PhotoEdit, Inc.; **94** (tl)Robert Essel/The Stock Market/CORBIS, (tr)John Eastcott & Yva Momatiuk/DRK Photo, (cl)Ame Hodalic/CORBIS, (cr)Diaphor Agency/Index Stock, (br)TIME; **100–101** Roger Ressmeyer/CORBIS; **102** Layne Kennedy/CORBIS; **103** (t)Telegraph Colour Library/FPG/Getty Images, (b)Paul Silverman/Fundamental Photographs; **104** Bill Aron/PhotoEdit, Inc.; **105** (l)John Serrao/Photo Researchers, (r)H. Richard Johnston; **106** Tom Tracy/Photo Network/PictureQuest; **107** Annie Griffiths Belt/CORBIS; **108** Amanita Pictures; **109** (t)David Weintraub/Stock Boston, (b)James L. Amos/Peter Arnold, Inc.; **110** Dave King/DK Images; **111** Joseph Sohm/ChromoSohm, Inc./CORBIS; **112** Michael Dalton/Fundamental Photographs; **113** Swarthout & Associates/The Stock Market/CORBIS; **114** Tony Freeman/PhotoEdit, Inc.; **116** David Young-Wolff/PhotoEdit, Inc.; **117** (b)Richard Hutchings, (t)Joshua Ets-Hokin/PhotoDisc; **118** Robbie Jack/CORBIS; **120** A. Ramey/Stock Boston; **121** Mark Burnett; **122** (t)Tony Freeman/PhotoEdit, Inc., (b)Stephen Simpson/FPG/Getty Images; **124** (t)Lester Lefkowitz/The Stock Market/CORBIS, (b)Bob Daemmrich; **125** Bob Daemmrich; **126** Daniel Belknap; **127** (l)Andrew Ward/Life File/PhotoDisc, (r)NASA/TRACE; **129** Mark Burnett; **131** Joshua Ets-Hokin/PhotoDisc; **132–133** Daryl Benson/Masterfile; **134** Steven R. Krous/Stock Boston/PictureQuest; **135** Ryan McVay/PhotoDisc; **136** David W. Hamilton/Image Bank/Getty Images; **137** (t)Morrison Photography, (b)Jose Azel/Aurora/PictureQuest; **138** Morrison Photography; **139** (l)Aaron Haupt, (r)Arthur S. Aubry/PhotoDisc; **140** (tl bl)Morrison Photography, (br)Bob Daemmrich/Stock Boston; **141** Morrison Photography; **142** Aaron Haupt; **143** AFP/CORBIS; **144** (tl)Morrison Photography, (tr)Art Montes de Oca/FPG/Getty Images, (bl)Anthony Ise/PhotoDisc, (br)Novastock/Index Stock; **145** (l)John Maher/Stock Boston/PictureQuest, (c)Matt Meadows, (r)AP/Wide World Photos/Jim McKnight; **146** Morrison Photography;

Credits

147 (t)Charles Benes/FPG/Getty Images, (b)Brenda Tharp/Photo Researchers; 148 Gerry Ellis/GLOBIO.org; 149 150 151 Morrison Photography; 152 (t)Morton Beebe, SF/CORBIS, (b)Will & Deni McIntyre/Photo Researchers; 153 (l)file photo, (r)courtesy Diamond International; 154 (tr)Michael Nelson/FPG/Getty Images, (others)Morrison Photography; 155 Amanita Pictures; 156 Rubberball Productions; 157 (l)Zefa Visual Media-Germany/Index Stock, (r)Bruce James/Getty Images; 158–159 (bkgd)PhotoDisc; 159 (inset)Stephen Frisch/Stock Boston/PictureQuest; 160–161 Christian Michel; 169 Laura Sifferlin; 170 (l)Lester V. Bergman/CORBIS, (r)Doug Martin; 175 Matt Meadows; 176 (tr cr)Kenneth Libbrecht/Caltech, (cl)Albert J. Copley/Visuals Unlimited, (bl)E.R. Degginger/Color-Pic; 178 James L. Amos/Photo Researchers; 179 180 181 Aaron Haupt; 182 Fulcrum Publishing; 187 Matt Meadows; 188–189 Simon Fraser/Science Photo Library/Photo Researchers; 190 (l)Aaron Haupt, (r)Doug Martin; 191 (tl)Patricia Lanza, (tc)Jeff J. Daly/Visuals Unlimited, (tr)Susan T. McElhinney, (bl)Craig Fuji/Seattle Times, (br)Sovfoto/Eastfoto/PictureQuest; 192 Amanita Pictures; 195 Sovfoto/Eastfoto/PictureQuest; 197 Christopher Swann/Peter Arnold, Inc.; 198 (tl)Frank Balthis, (tr)Lois Ellen Frank/CORBIS, (b)Matt Meadows; 199 David Young-Wolff/PhotoEdit/PictureQuest; 200 (l)Amanita Pictures, (r)Richard Megna/Fundamental Photographs/Photo Researchers; 201 Victoria Arocho/AP/Wide World Photos; 202 (t)Aaron Haupt, (bl)Getty Images, (br)Icon Images; 203 SuperStock; 204 (tl)Chris Arend/Alaska Stock Images/PictureQuest, (tr)Aaron Haupt, (b)Bryan F. Peterson/CORBIS; 205 courtesy General Motors; 206 207 Matt Meadows; 208 Amanita Pictures; 209 Bob Daemmrich; 210 (l)Tino Hammid Photography, (r)Joe Richard/UF News & Public Affairs; 211 David Young-Wolff/PhotoEdit, Inc.; 214 Lester V. Bergman/CORBIS; 215 Peter Walton/Index Stock; 216–217 Joseph Sohm/ChromoSohm, Inc./CORBIS; 219 (l)Stephen W. Frisch/Stock Boston, (r)Doug Martin; 220 (t)HIRB/Index Stock, (b)Doug Martin; 221 Richard Hamilton/CORBIS; 222 John Evans; 223 (l)SuperStock, (r)Annie Griffiths/CORBIS; 226 John Evans; 228 Richard Nowitz/Phototake/PictureQuest; 230 Aaron Haupt; 231 KS Studios/Mullenix; 233 John Evans; 234 (l)Joe Sohm, Chromosohm/Stock Connection/PictureQuest, (c)Andrew Popper/Phototake/PictureQuest, (r)A. Wolf/Explorer, Photo Researchers; 235 John Evans; 236 (tl tr)Elaine Shay, (tcl)Brent Turner/BLT Productions, (tcr)Matt Meadows, (bcr)Icon Images, (bl bcl)CORBIS, (br)StudiOhio; 240 241 KS Studios; 242 CORBIS; 244 Royalty-Free/CORBIS; 247 Stephen W. Frisch/Stock Boston; 248–249 Karen Su/CORBIS; 250 (l)Michael Newman/PhotoEdit, Inc., (r)Richard Price/FPG/Getty Images; 252 (l)Tony Freeman/PhotoEdit, Inc., (r)Mark Burnett; 253 (l c)Mark Burnett, (r)Will & Deni McIntyre/Photo Researchers; 254 Ted Horowitz/The Stock Market/CORBIS; 257 (l)Gary A. Conner/PhotoEdit/PictureQuest, (r)Stephen Frisch/Stock Boston/PictureQuest; 259 (t)Kim Taylor/Bruce Coleman, Inc./PictureQuest, (b)John Sims/Tony Stone Images/Getty Images; 264 (t)Elaine Shay, (b)Mitch Hrdlicka/PhotoDisc; 265 (t)KS Studios, (b)Matt Meadows; 266 267 KS Studios; 268 (l)Don Farrall/PhotoDisc, (r)KS Studios; 269 Alfred Pasieka/Peter Arnold, Inc.; 270 (t)Geoff Butler, (b)Aaron Haupt; 271 Aaron Haupt; 272 (t)Waina Cheng/Bruce Coleman, Inc., (c)David Nunuk/Science Photo Library/Photo Researchers, (b)Lee Baltermoal/FPG/Getty Images; 273 Aaron Haupt; 277 Richard Hutchings; 278–279 (bkgd)Museum of the City of New York/CORBIS; 279 (l)Lee Snider/CORBIS, (r)Scott Camazine/Photo Researchers; 280–281 William Dow/CORBIS; 282 Telegraph Colour Library/FPG/Getty Images; 283 Geoff Butler; 286 Richard Hutchings; 289 Runk/Schoenberger from Grant Heilman; 291 Mark Doolittle/Outside Images/Picturequest; 293 (l)Ed Bock/The Stock Market/CORBIS, (r)Will Hart/PhotoEdit, Inc.; 295 (t)Tom & DeeAnn McCarthy/The Stock Market/CORBIS, (bl)Jodi Jacobson/Peter Arnold, Inc., (br)Jules Frazier/PhotoDisc; 296 Mark Burnett; 298 Robert Brenner/PhotoEdit, Inc.; 299 Laura Sifferlin; 300 301 Icon Images; 302 Alexis Duclos/Liaison/Getty Images; 303 (l r)Rudi Von Briel/PhotoEdit, Inc., (c)PhotoDisc; 307 (l)Jodi Jacobson/Peter Arnold, Inc., (r)Runk/Schoenberger from Grant Heilman; 308–309 Wendell Metzen/Index Stock; 309 Richard Hutchings; 310 (l)Globus Brothers Studios, NYC, (r)Stock Boston; 311 Bob Daemmrich; 312 (t)Beth Wald/ImageState, (b)David Madison; 313 Rhoda Sidney/Stock Boston/PictureQuest; 315 (l)Myrleen Cate/PhotoEdit, Inc., (r)David Young-Wolff/PhotoEdit, Inc.; 316 Bob Daemmrich; 318 (t)Stone/Getty Images, (b)Myrleen Cate/PhotoEdit, Inc.; 320 David Madison; 322 Richard Megna/Fundamental Photographs; 323 Mary M. Steinbacher/PhotoEdit, Inc.; 324 (t)Betty Sederquist/Visuals Unlimited, (b)Jim Cummins/FPG/Getty Images; 325 (tl)Denis Boulanger/Allsport, (tr)Donald Miralle/Allsport, (b)Tony Freeman/PhotoEdit/PictureQuest; 326 (t)David Madison, (b)NASA; 328 NASA; 329 Richard Hutchings; 330 331 Mark Burnett; 332 (t)Tom Wright, (b)Didier Charre/Image Bank; 333 (tl)Philip Bailey/The Stock Market/CORBIS, (tr)Romilly Lockyer/Image Bank/Getty Images, (bl)Tony Freeman/PhotoEdit, Inc.; 337 Betty Sederquist/Visuals Unlimited; 338–339 Hughes Martin/CORBIS; 339 Matt Meadows; 340 David Young-Wolff/PhotoEdit, Inc.; 342 Runk/Schoenberger from Grant Heilman; 343 Dominic Oldershaw; 344 (t)Matt Meadows, (b)Tom Pantages; 346 (t)Bobby Model/National Geographic Image Collection, (cl)Richard Nowitz/National Geographic Image Collection, (cr)George Grall/National Geographic Image Collection, (bl)Ralph White/CORBIS, (br)CORBIS; 348 Ryan McVay/PhotoDisc; 349 CORBIS; 350 (t)Matt Meadows, (b)Vince Streano/Stone/Getty Images; 351 353 Matt Meadows; 355 John Evans; 357 KS Studios; 358 Dominic Oldershaw; 361 (t)Michael Collier/Stock Boston, (bl)George Hall/CORBIS, (br)Dean Conger/CORBIS; 362 Steve McCutcheon/Visuals Unlimited; 363 Runk/Schoenberger from Grant Heilman; 364 AP/Wide World Photos/Ray Fairall; 365 (l)D.R. & T.L. Schrichte/Stone/Getty Images, (r)CORBIS; 366 Matt Meadows; 369 Vince Streano/Stone/Getty Images; 370–371 (bkgd)Douglas Peebles/CORBIS; 371 (inset)Henry Ford Museum & Greenfield Village; 372–373 Chris Knapton/Science Photo Library/Photo Researchers; 373 Matt Meadows; 374 (l c)file photo, (r)Mark Burnett; 375 (t b)Bob Daemmrich, (c)Al Tielemans/Duomo; 376 KS Studios; 377 (l r)Bob Daemmrich, (b)Andrew McClenaghan/Science Photo Library/Photo Researchers; 378 Mark Burnett/Photo Researchers; 379 Lori Adamski Peek/Stone/Getty Images; 380 Richard Hutchings; 381 Ron Kimball/Ron Kimball Photography; 382 (t)Judy Lutz, (b)Lennart Nilsson; 384 386 KS Studios; 392 (t)Dr. Jeremy Burgess/Science Photo Library/Photo Researchers, (b)John Keating/Photo Researchers; 393 Geothermal Education Office; 394 Carsand-Mosher; 395 Billy Hustace/Stone/Getty Images; 396 SuperStock; 397 Roger Ressmeyer/CORBIS; 398 (tl)Reuters NewMedia, Inc./CORBIS, (tr)PhotoDisc, (br)Dominic Oldershaw; 399 (l)Lowell Georgia/CORBIS,

Credits

(r)Mark Richards/PhotoEdit, Inc.; **404–405** Rich Iwasaki/Getty Images; **405** Mark Burnett; **406** Mary Kate Denny/PhotoEdit, Inc.; **407** (t)Richard Hutchings, (b)Tony Freeman/PhotoEdit, Inc.; **414** (l)David Young-Wolff/PhotoEdit, Inc., (r)Frank Siteman/Stock Boston; **417** Duomo; **418** Robert Brenner/PhotoEdit, Inc.; **419** (t)Tom McHugh/Photo Researchers, (b)Amanita Pictures; **420** Amanita Pictures; **421** (t)Dorling Kindersley, (bl br)Bob Daemmrich; **422** (l)Wernher Krutein/Liaison Agency/Getty Images, (r)Siegfried Layda/Stone/Getty Images; **424** Tony Freeman/PhotoEdit, Inc.; **425** Aaron Haupt; **426** (t)Ed Kashi/CORBIS, (b)James Balog; **427** (l)Inc. Janeart/The Image Bank/Getty Images, (r)Ryan McVay/PhotoDisc; **431** (l)Comstock Images, (r)PhotoDisc; **432–433** Peter Walton/Index Stock; **434** John Evans; **435** (t)Nancy P. Alexander/Visuals Unlimited, (b)Morton & White; **437** Tom Stack & Assoc.; **438** Doug Martin; **439** Matt Meadows; **440** Jeremy Hoare/PhotoDisc; **441** Donnie Kamin/PhotoEdit, Inc.; **442** SuperStock; **443** Colin Raw/Stone/Getty Images; **444** Aaron Haupt; **445** PhotoDisc; **446** (l)Barbara Stitzer/PhotoEdit, Inc., (c)Doug Menuez/PhotoDisc, (r)Addison Geary/Stock Boston; **448** C. Squared Studios/PhotoDisc; **450 451** Morton & White; **452** (bkgd)Chip Simons/FPG/Getty Images, (inset)Joseph Sohm/CORBIS; **453** SuperStock; **456** John Evans; **457** Michael Newman/Photo Edit, Inc.; **458–459** (bkgd)Matthew Borkoski/Stock Boston/PictureQuest; **459** (inset)L. Fritz/H. Armstrong Roberts; **460–461** Douglas Peebles/CORBIS; **462** (l)file photo, (r)David Young-Wolff/PhotoEdit, Inc.; **463** David Young-Wolff/PhotoEdit, Inc.; **464** Mark Thayer; **467** Steven Starr/Stock Boston; **472** Ken Frick; **473** Mark Burnett; **475** Ernst Haas/Stone/Getty Images; **476** Peter Beattie/Liaison Agency/Getty Images; **478** D. Boone/CORBIS; **479** Seth Resnick/Stock Boston; **480–481** John Evans; **482** (t)Roger Ressmeyer/CORBIS, (b)SuperStock; **486** Mark Burnett; **488–489** Tom Wagner/CORBIS SABA; **493** (t)Joe Towers/The Stock Market/CORBIS, (c)Bob Daemmrich/Stock Boston/PictureQuest, (b)Jean-Paul Thomas/Jacana Scientific Control/Photo Researchers; **496** NOAA; **499** Spencer Grant/PhotoEdit, Inc.; **500** Timothy Fuller; **504** Dilip Mehta/Contact Press Images/PictureQuest; **505** (tl)CORBIS, (tr)Paul Seheult/Eye Ubiquitous/CORBIS, (b)Icon Images; **506** (t)William Whitehurst/The Stock Market/CORBIS, (b)G. Salter/Lebrecht Music Collection; **507** SuperStock; **508** (t)Geostock/PhotoDisc, (b)SuperStock; **509** Fred E. Hossler/Visuals Unlimited; **510** (t)Will McIntyre/Photo Researchers, (b)Oliver Benn/Stone/Getty Images; **512** Douglas Whyte/The Stock Market/CORBIS; **513** (l)The Photo Works/Photo Researchers, (r)PhotoDisc; **515** C. Squared Studios/PhotoDisc; **518–519** Maxine Hall/CORBIS; **520** (l)Bob Abraham/The Stock Market/CORBIS, (r)Jeff Greenberg/Visuals Unlimited; **521** (l)David Young-Wolff/PhotoEdit, Inc., (r)NRSC, Ltd./Science Photo Library/Photo Researchers; **522** (t)Grantpix/Photo Researchers, (b)Richard Megna/Fundamental Photographs; **524** Luke Dodd/Science Photo Library/Photo Researchers; **526** (t)Matt Meadows, (b)Jean Miele/The Stock Market/CORBIS; **528** (t)Gregory G. Dimijian/Photo Researchers, (b)Charlie Westerman/Liaison/Getty Images; **529** Aaron Haupt; **531** (l)Matt Meadows, (r)Bob Daemmrich/The Image Works; **532** (tr)Phil Degginger/Color-Pic, (l)Phil Degginger/NASA/Color-Pic, (cr)Max Planck Institute for Radio Astronomy/Science Photo Library/Photo Researchers, (br)European Space Agency/Science Photo Library/Photo Researchers; **533** (l)Harvard-Smithsonian Center for Astrophysics, (c)NASA/Science Photo Library/Photo Researchers, (r)European Space Agency; **534** Timothy Fuller; **539** Ken M. Johns/Photo Researchers; **540** (t)Michael Thomas/Stock South/PictureQuest, (b)Dominic Oldershaw; **541** Michael Thomas/Stock South/PictureQuest; **542** (bkgd)TIME, (t)Culver Pictures, (b)Hulton Archive/Getty Images; **543** (l)Macduff Everton/CORBIS, (r)NASA/Mark Marten/Photo Researchers; **547** Eric Kamp/Index Stock; **548–549** Chad Ehlers/Index Stock; **550** Dick Thomas/Visuals Unlimited; **551** John Evans; **552** (tl)Bob Woodward/The Stock Market/CORBIS, (tc)Ping Amranand/Pictor, (tr)SuperStock, (b)Runk/Schoenberger from Grant Heilman; **553** Mark Thayer; **556** (l)Susumu Nishinaga/Science Photo Library/Photo Researchers, (r)Matt Meadows; **558** (l)Matt Meadows, (r)Paul Silverman/Fundamental Photographs; **559** (l)Digital Stock, (r)Joseph Palmieri/Pictor; **561** Geoff Butler; **563** Richard Megna/Fundamental Photographs; **567** David Young-Wolff/PhotoEdit, Inc.; **568 569** Roger Ressmeyer/CORBIS; **572 573** Geoff Butler; **574** The Stapleton Collection/Bridgeman Art Library; **580–581** (bkgd)Richard Pasley/Stock Boston/PictureQuest; **580** (inset)Layne Kennedy/CORBIS; **581** (inset)Mark Burnett; **582–583** V.C.L./Getty Images; **585** (t)Richard Hutchings, (b)KS Studios; **588** Royalty Free/CORBIS; **590** J. Tinning/Photo Researchers; **593** Gary Rhijnsburger/Masterfile; **598** Doug Martin; **599** (t)Doug Martin, (b)Geoff Butler; **601** Bonnie Freer/Photo Researchers; **603** Matt Meadows; **604 605** Richard Hutchings; **606** (bkgd)Tom & Pat Leeson/Photo Researchers, (inset)William Munoz/Photo Researchers; **610** J. Tinning/Photo Researchers; **611** Doug Martin; **612–613** James Leynse/CORBIS; **615** Richard Megna/Fundamental Photographs; **616** Amanita Pictures; **619** John Evans; **620** Amanita Pictures; **621** (l)Kodansha, (c)Manfred Kage/Peter Arnold, Inc., (r)Doug Martin; **625** Bjorn Backe/Papilio/CORBIS; **627** Norbert Schafer/The Stock Market/CORBIS; **629** AT&T Bell Labs/Science Photo Library/Photo Researchers; **630** (t)Science Photo Library/Photo Researchers, (c)Fermilab/Science Photo Library/Photo Researchers, (b)SuperStock; **631** PhotoDisc; **632** (t)file photo, (b)Aaron Haupt; **633** Aaron Haupt; **634** John MacDonald; **635** (l)SIU/Peter Arnold, Inc., (r)Latent Image; **639** John Evans; **640–641** Andrew Syred/Science Photo Library/Photo Researchers; **642** Willie L. Hill, Jr./Stock Boston; **643** (l)Icon Images, (r)Doug Martin; **645** (l r)CMCD/PhotoDisc, (c)Russ Lappa; **646** Amanita Pictures; **647** (t)Amanita Pictures, (b)Charles Falco/Photo Researchers; **648** Charles Falco/Photo Researchers; **649** (l)Bettmann/CORBIS, (r)Icon Images; **652** (t)courtesy IBM/Florida State University, (b)Andrew Syred/Science Photo Library/Photo Researchers; **655** file photo; **656** Thomas Brummett/PhotoDisc; **658** (l)Dr. Dennis Kunkel/PhotoTake, NYC, (r)Aaron Haupt; **659** Timothy Fuller; **660** David Young-Wolff/PhotoEdit, Inc.; **661** Frank Cezus; **662** Tek Images/Science Photo Library/Photo Researchers; **663** (tr)Amanita Pictures, (l)Aaron Haupt, (br)Keith Brofsky/PhotoDisc; **667** Thomas Brummett/PhotoDisc; **668** PhotoDisc; **670** Tom Pantages; **674** Michelle D. Bridwell/PhotoEdit, Inc.; **675** (t)Mark Burnett, (b)Dominic Oldershaw; **676** StudiOhio; **677** Timothy Fuller; **678** Aaron Haupt; **680** KS Studios; **681** Matt Meadows; **683** (t)Dominic Oldershaw, (b)Mark Burnett; **684** John Evans; **685** (t)Amanita Pictures, (b)John Evans; **686** Mark Burnett; **688** John Evans; **690** Mark Burnett; **691** Amanita Pictures; **692** Icon Images; **693** Amanita Pictures; **694** Bob Daemmrich; **696** Davis Barber/PhotoEdit, Inc.

PERIODIC TABLE OF THE ELEMENTS

Columns of elements are called groups. Elements in the same group have similar chemical properties.

Element — Hydrogen
Atomic number — 1
Symbol — H
Atomic mass — 1.008

State of matter

- Gas
- Liquid
- Solid
- Synthetic

The first three symbols tell you the state of matter of the element at room temperature. The fourth symbol identifies elements that are not present in significant amounts on Earth. Useful amounts are made synthetically.

Group	1	2	3	4	5	6	7	8	9
1	Hydrogen 1 H 1.008								
2	Lithium 3 Li 6.941	Beryllium 4 Be 9.012							
3	Sodium 11 Na 22.990	Magnesium 12 Mg 24.305							
4	Potassium 19 K 39.098	Calcium 20 Ca 40.078	Scandium 21 Sc 44.956	Titanium 22 Ti 47.867	Vanadium 23 V 50.942	Chromium 24 Cr 51.996	Manganese 25 Mn 54.938	Iron 26 Fe 55.845	Cobalt 27 Co 58.933
5	Rubidium 37 Rb 85.468	Strontium 38 Sr 87.62	Yttrium 39 Y 88.906	Zirconium 40 Zr 91.224	Niobium 41 Nb 92.906	Molybdenum 42 Mo 95.94	Technetium 43 Tc (98)	Ruthenium 44 Ru 101.07	Rhodium 45 Rh 102.906
6	Cesium 55 Cs 132.905	Barium 56 Ba 137.327	Lanthanum 57 La 138.906	Hafnium 72 Hf 178.49	Tantalum 73 Ta 180.948	Tungsten 74 W 183.84	Rhenium 75 Re 186.207	Osmium 76 Os 190.23	Iridium 77 Ir 192.217
7	Francium 87 Fr (223)	Radium 88 Ra (226)	Actinium 89 Ac (227)	Rutherfordium 104 Rf (261)	Dubnium 105 Db (262)	Seaborgium 106 Sg (266)	Bohrium 107 Bh (264)	Hassium 108 Hs (277)	Meitnerium 109 Mt (268)

The number in parentheses is the mass number of the longest-lived isotope for that element.

Rows of elements are called periods. Atomic number increases across a period.

The arrow shows where these elements would fit into the periodic table. They are moved to the bottom of the table to save space.

Lanthanide series

Cerium 58 Ce 140.116	Praseodymium 59 Pr 140.908	Neodymium 60 Nd 144.24	Promethium 61 Pm (145)	Samarium 62 Sm 150.36

Actinide series

Thorium 90 Th 232.038	Protactinium 91 Pa 231.036	Uranium 92 U 238.029	Neptunium 93 Np (237)	Plutonium 94 Pu (244)